T0239018

Graph Neural Networks: Foundations, Frontiers, and Applications

Lingfei Wu • Peng Cui • Jian Pei • Liang Zhao
Editors

Graph Neural Networks: Foundations, Frontiers, and Applications

Springer

Editors
Lingfei Wu
JD Silicon Valley Research Center
Mountain View, CA, USA

Peng Cui
Tsinghua University
Beijing, China

Jian Pei
Simon Fraser University
Burnaby, Canada

Liang Zhao
Emory University
Atlanta, USA

ISBN 978-981-16-6056-6   ISBN 978-981-16-6054-2   (eBook)
https://doi.org/10.1007/978-981-16-6054-2

This Springer imprint is published by the registered company Springer Nature Singapore Pte Ltd.
The registered company address is: 152 Beach Road, #21-01/04 Gateway East, Singapore 189721, Singapore

Foreword

"The first comprehensive book covering the full spectrum of a young, fast-growing research field, graph neural networks (GNNs), written by authoritative authors!"

Jiawei Han (Michael Aiken Chair Professor at University of Illinois at Urbana-Champaign, ACM Fellow and IEEE Fellow)

"This book presents a comprehensive and timely survey on graph representation learning. Edited and contributed by the best group of experts in this area, this book is a must-read for students, researchers and pratictioners who want to learn anything about Graph Neural Networks."

Heung-Yeung "Harry" Shum (Former Executive Vice President for Technology and Research at Microsoft Research, ACM Fellow, IEEE Fellow, FREng)

"As the new frontier of deep learning, Graph Neural Networks offer great potential to combine probabilistic learning and symbolic reasoning, and bridge knowledge-driven and data-driven paradigms, nurturing the development of third-generation AI. This book provides a comprehensive and insightful introduction to GNN, ranging from foundations to frontiers, from algorithms to applications. It is a valuable resource for any scientist, engineer and student who wants to get into this exciting field."

Bo Zhang (Member of Chinese Academy of Science, Professor at Tsinghua University)

"Graph Neural Networks are one of the hottest areas of machine learning and this book is a wonderful in-depth resource covering a broad range of topics and applications of graph representation learning."

Jure Leskovec (Associate Professor at Stanford University, and investigator at Chan Zuckerberg Biohub).

"Graph Neural Networks are an emerging machine learning model that is already taking the scientific and industrial world by storm. The time is perfect to get in on the action – and this book is a great resource for newcomers and seasoned practitioners alike! Its chapters are very carefully written by many of the thought leaders at the forefront of the area."

Petar Veličković (Senior Research Scientist, DeepMind)

Preface

The field of graph neural networks (GNNs) has seen rapid and incredible strides over the recent years. Graph neural networks, also known as deep learning on graphs, graph representation learning, or geometric deep learning, have become one of the fastest-growing research topics in machine learning, especially deep learning. This wave of research at the intersection of graph theory and deep learning has also influenced other fields of science, including recommendation systems, computer vision, natural language processing, inductive logic programming, program synthesis, software mining, automated planning, cybersecurity, and intelligent transportation.

Although graph neural networks have achieved remarkable attention, it still faces many challenges when applying them into other domains, from the theoretical understanding of methods to the scalability and interpretability in a real system, and from the soundness of the methodology to the empirical performance in an application. However, as the field rapidly grows, it has been extremely challenging to gain a global perspective of the developments of GNNs. Therefore, we feel the urgency to bridge the above gap and have a comprehensive book on this fast-growing yet challenging topic, which can benefit a broad audience including advanced undergraduate and graduate students, postdoctoral researchers, lecturers, and industrial practitioners.

This book is intended to cover a broad range of topics in graph neural networks, from the foundations to the frontiers, and from the methodologies to the applications. Our book is dedicated to introducing the fundamental concepts and algorithms of GNNs, new research frontiers of GNNs, and broad and emerging applications with GNNs.

Book Website and Resources

The website and further resources of this book can be found at: `https://graph-neural-networks.github.io/`. The website provides online preprints and lecture slides of all the chapters. It also provides pointers to useful material and resources that are publicly available and relevant to graph neural networks.

To the Instructors

The book can be used for a one-semester graduate course for graduate students. Though it is mainly written for students with a background in computer science, students with a basic understanding of probability, statistics, graph theory, linear algebra, and machine learning techniques such as deep learning will find it easily accessible. Some chapters can be skipped or assigned as homework assignments for reviewing purposes if students have knowledge of a chapter. For example, if students have taken a deep learning course, they can skip Chapter 1. The instructors can also choose to combine Chapters 1, 2, and 3 together as a background introduction course at the very beginning.

When the course focuses more on the foundation and theories of graph neural networks, the instructor can choose to focus more on Chapters 4-8 while using Chapters 19-27 to showcase the applications, motivations, and limitations. Please refer to the Editors' Notes at the end of each chapter on how Chapters 4-8 and Chapters 19-27 are correlated. When the course focuses more on the research frontiers, Chapters 9-18 can be the pivot to organize the course. For example, an instructor can make it an advanced graduate course where the students are asked to search and present the most recent research papers in each different research frontier. They can also be asked to establish their course projects based on the applications described in Chapters 19-27 as well as the materials provided on our website.

To the Readers

This book was designed to cover a wide range of topics in the field of graph neural network field, including background, theoretical foundations, methodologies, research frontiers, and applications. Therefore, it can be treated as a comprehensive handbook for a wide variety of readers such as students, researchers, and professionals. You should have some knowledge of the concepts and terminology associated with statistics, machine learning, and graph theory. Some backgrounds of the basics have been provided and referenced in the first eight chapters. You should better also have knowledge of deep learning and some programming experience for easily accessing the most of chapters of this book. In particular, you should be able to read pseudocode and understand graph structures.

The book is well modularized and each chapter can be learned in a standalone manner based on the individual interests and needs. For those readers who want to have a solid understanding of various techniques and theories of graph neural networks, you can start from Chapters 4-9. For those who further want to perform in-depth research and advance related fields, please read those chapters of interest among Chapters 9-18, which provide comprehensive knowledge in the most recent research issues, open problems, and research frontiers. For those who want to apply graph neural networks to benefit specific domains, or aim at finding interesting applications to validate specific graph neural networks techniques, please refer to Chapters 19-27.

Acknowledgements

Graph machine learning has attracted many gifted researchers to make their seminal contributions over the last few years. We are very fortunate to discuss the challenges and opportunities, and often work with many of them on a rich variety of research topics in this exciting field. We are deeply indebted to these collaborators and colleagues from JD.COM, IBM Research, Tsinghua University, Simon Fraser University, Emory University, and elsewhere, who encouraged us to create such a book comprehensively covering various topics of Graph Neural Networks in order to educate the interested beginners and foster the advancement of the field for both academic researchers and industrial practitioners.

This book would not have been possible without the contributions of many people. We would like to give many thanks to the people who offered feedback on checking the consistency of the math notations of the entire book as well as reference editing of this book. They are people from Emory University: Ling Chen, Xiaojie Guo, and Shiyu Wang, as well as people from Tsinghua University: Yue He, Ziwei Zhang, and Haoxin Liu. We would like to give our special thanks to Dr. Xiaojie Guo, who generously offered her help in providing numerous valuable feedback on many chapters.

We also want to thank those who allowed us to reproduce images, figures, or data from their publications.

Finally, we would like to thank our families for their love, patience and support during this very unusual time when we are writing and editing this book.

Editor Biography

Dr. Lingfei Wu is a Principal Scientist at JD.COM Silicon Valley Research Center, leading a team of 30+ machine learning/natural language processing scientists and software engineers to build intelligent e-commerce personalization systems. He earned his Ph.D. degree in computer science from the College of William and Mary in 2016. Previously, he was a research staff member at IBM Thomas J. Watson Research Center and led a 10+ research scientist team for developing novel Graph Neural Networks methods and systems, which leads to the #1 AI Challenge Project in IBM Research and multiple IBM Awards including three-time Outstanding Technical Achievement Award. He has published more than 90 top-ranked conference and journal papers, and is a co-inventor of more than 40 filed US patents. Because of the high commercial value of his patents, he has received eight invention achievement awards and has been appointed as IBM Master Inventors, class of 2020. He was the recipients of the Best Paper Award and Best Student Paper Award of several conferences such as IEEE ICC'19, AAAI workshop on DLGMA'20 and KDD workshop on DLG'19. His research has been featured in numerous media outlets, including NatureNews, YahooNews, Venturebeat, TechTalks, SyncedReview, Leiphone, QbitAI, MIT News, IBM Research News, and SIAM News. He has co-organized 10+ conferences (KDD, AAAI, IEEE BigData) and is the founding co-chair for Workshops of Deep Learning on Graphs (with AAAI'21, AAAI'20, KDD'21, KDD'20, KDD'19, and IEEE BigData'19). He has currently served as Associate Editor for IEEE Transactions on Neural Networks and Learning Systems, ACM Transactions on Knowledge Discovery from Data and International Journal of Intelligent Systems, and regularly served as a SPC/PC member of the following major AI/ML/NLP conferences including KDD, IJCAI, AAAI, NIPS, ICML, ICLR, and ACL.

Dr. Peng Cui is an Associate Professor with tenure at Department of Computer Science in Tsinghua University. He obtained his PhD degree from Tsinghua University in 2010. His research interests include data mining, machine learning and multimedia analysis, with expertise on network representation learning, causal inference and stable learning, social dynamics modeling, and user behavior modeling, etc. He is keen to promote the convergence and integration of causal inference and machine learning, addressing the fundamental issues of today's AI technology, including explainability, stability and fairness issues. He is recognized as a Distinguished Scientist of ACM, Distinguished Member of CCF and Senior Member of IEEE. He has published more than 100 papers in prestigious conferences and journals in machine learning and data mining. He is one of the most cited authors in network embedding. A number of his proposed algorithms on network embedding generate substantial impact in academia and industry. His recent research won the IEEE Multimedia Best Department Paper Award, IEEE ICDM 2015 Best Student Paper Award, IEEE ICME 2014 Best Paper Award, ACM MM12 Grand Challenge Multimodal Award, MMM13 Best Paper Award, and were selected into the Best of KDD special issues in 2014 and 2016, respectively. He was PC co-chair of CIKM2019 and MMM2020, SPC or area chair of ICML, KDD, WWW, IJCAI, AAAI, etc., and Associate Editors of IEEE TKDE (2017-), IEEE TBD (2019-), ACM TIST(2018-), and ACM TOMM (2016-) etc. He received ACM China Rising Star Award in 2015, and CCF-IEEE CS Young Scientist Award in 2018.

Dr. Jian Pei is a Professor in the School of Computing Science at Simon Fraser University. He is a well-known leading researcher in the general areas of data science, big data, data mining, and database systems. His expertise is on developing effective and efficient data analysis techniques for novel data intensive applications, and transferring his research results to products and business practice. He is recognized as a Fellow of the Royal Society of Canada (Canada's national academy), the Canadian Academy of Engineering, the Association of Computing Machinery (ACM) and the Institute of Electrical and Electronics Engineers (IEEE). He is one of the most cited authors in data mining, database systems, and information retrieval. Since 2000, he has published one textbook, two monographs and over 300 research papers in refereed journals and conferences, which have been cited extensively by others. His research has generated remarkable impact substantially beyond academia. For example, his algorithms have been adopted by industry in production and popular open-source software suites. Jian Pei also demonstrated outstanding professional leadership in many academic organizations and activities. He was the editor-in-chief of the IEEE Transactions of Knowledge and Data Engineering (TKDE) in 2013-16, the chair of the Special Interest Group on Knowledge Discovery in Data (SIGKDD) of the Association for Computing Machinery (ACM) in 2017-2021, and a general co-chair or program committee co-chair of many premier conferences. He maintains a wide spectrum of industry relations with both global and local industry partners. He is an active consultant and coach for industry on enterprise data strategies, healthcare informatics, network security intelligence, computational finance, and smart retail. He received many prestigious awards, including the 2017 ACM SIGKDD Innovation Award, the 2015 ACM SIGKDD Service Award, the 2014 IEEE ICDM Research Contributions Award, the British Columbia Innovation Council 2005 Young Innovator Award, an NSERC 2008 Discovery Accelerator Supplements Award (100 awards cross the whole country), an IBM Faculty Award (2006), a KDD Best Application Paper Award (2008), an ICDE Influential Paper Award (2018), a PAKDD Best Paper Award (2014), a PAKDD Most Influential Paper Award (2009), and an IEEE Outstanding Paper Award (2007).

Dr. Liang Zhao is an assistant professor at the Department of Compute Science at Emory University. Before that, he was an assistant professor in the Department of Information Science and Technology and the Department of Computer Science at George Mason University. He obtained his PhD degree in 2016 from Computer Science Department at Virginia Tech in the United States. His research interests include data mining, artificial intelligence, and machine learning, with special interests in spatiotemporal and network data mining, deep learning on graphs, nonconvex optimization, model parallelism, event prediction, and interpretable machine learning. He received AWS Machine Learning Research Award in 2020 from Amazon Company for his research on distributed graph neural networks. He won NSF Career Award in 2020 awarded by National Science Foundation for his research on deep learning for spatial networks, and Jeffress Trust Award in 2019 for his research on deep generative models for biomolecules, awarded by Jeffress Memorial Trust Foundation and Bank of America. He won the Best Paper Award in the 19th IEEE International Conference on Data Mining (ICDM 2019) for the paper of his lab on deep graph transformation. He has also won Best Paper Award Shortlist in the 27th Web Conference (WWW 2021) for deep generative models. He was selected as "Top 20 Rising Star in Data Mining" by Microsoft Search in 2016 for his research on spatiotemporal data mining. He has also won Outstanding Doctoral Student in the Department of Computer Science at Virginia Tech in 2017. He is awarded as CIFellow Mentor 2021 by the Computing Community Consortium for his research on deep learning for spatial data. He has published numerous research papers in top-tier conferences and journals such as KDD, TKDE, ICDM, ICLR, Proceedings of the IEEE, ACM Computing Surveys, TKDD, IJCAI, AAAI, and WWW. He has been serving as organizers such as publication chair, poster chair, and session chair for many top-tier conferences such as SIGSPATIAL, KDD, ICDM, and CIKM.

List of Contributors

Miltiadis Allamanis
Microsoft Research, Cambridge, UK

Yu Chen
Facebook AI, Menlo Park, CA, USA

Yunfei Chu
Alibaba Group, Hangzhou, China

Peng Cui
Tsinghua University, Beijing, China

Tyler Derr
Vanderbilt University, Nashville, TN, USA

Keyu Duan
Texas A&M University, College Station, TX, USA

Qizhang Feng
Texas A&M University, College Station, TX, USA

Stephan Günnemann
Technical University of Munich, München, Germany

Xiaojie Guo
JD.COM Silicon Valley Research Center, Mountain View, CA, USA

Yu Hou
Weill Cornell Medicine, New York City, New York, USA

Xia Hu
Texas A&M University, College Station, TX, USA

Junzhou Huang
University of Texas at Arlington, Arlington, TA, United States

Shouling Ji

Zhejiang University, Hangzhou, China

Wei Jin
Michigan State University, East Lansing, MI, USA

Anowarul Kabir
George Mason University, Fairfax, VA, USA

Seyed Mehran Kazemi
Borealis AI, Montreal, Canada.

Jure Leskovec
Stanford University, Stanford, CA, USA

Juncheng Li
Zhejiang University, Hangzhou, China

Jiacheng Li
Zhejiang University, Hangzhou, China

Pan Li
Purdue University, Lafayette, IN, USA

Yanhua Li
Worcester Polytechnic Institute, Worcester, MA, USA

Renjie Liao
University of Toronto, Toronto, Canada

Xiang Ling
Zhejiang University, Hangzhou, China

Bang Liu
University of Montreal, Montreal, Canada

Ninghao Liu
Texas A&M University, College Station, TX, USA

Zirui Liu
Texas A&M University, College Station, TX, USA

Hehuan Ma
University of Texas at Arlington, College Station, TX, USA

Collin McMillan
University of Notre Dame, Notre Dame, IN, USA

Christopher Morris
Polytechnique Montréal, Montréal, Canada

Zongshen Mu
Zhejiang University, Hangzhou, China

Menghai Pan

Worcester Polytechnic Institute, Worcester, MA, USA

Jian Pei
Simon Fraser University, British Columbia, Canada

Yu Rong
Tencent AI Lab, Shenzhen, China

Amarda Shehu
George Mason University, Fairfax, VA, USA

Kai Shen
Zhejiang University, Hangzhou, China

Chuan Shi
Beijing University of Posts and Telecommunications, Beijing, China

Le Song
Mohamed bin Zayed University of Artificial Intelligence, Abu Dhabi, United Arab
Emirates

Chang Su
Weill Cornell Medicine, New York City, New York. USA

Jian Tang
Mila-Quebec AI Institute, HEC Montreal, Canada

Siliang Tang
Zhejiang University, Hangzhou, China

Fei Wang
Weill Cornell Medicine, New York City, New York, USA

Shen Wang
University of Illinois at Chicago, Chicago, IL, USA

Shiyu Wang
Emory University, Atlanta, GA, USA

Xiao Wang
Beijing University of Posts and Telecommunications, Beijing, China

Yu Wang
Vanderbilt University, Nashville, TN, USA

Chunming Wu
Zhejiang University, Hangzhou, China

Lingfei Wu
JD.COM Silicon Valley Research Center, Mountain View, CA, USA

Hongxia Yang
Alibaba Group, Hangzhou, China

Jiangchao Yao

Alibaba Group, Hangzhou, China

Philip S. Yu
University of Illinois at Chicago, Chicago, IL, USA

Muhan Zhang
Peking University, Beijing, China

Wenqiao Zhang
Zhejiang University, Hangzhou, China

Liang Zhao
Emory University, Atlanta, GA, USA

Chang Zhou
Alibaba Group, Hangzhou, China

Kaixiong Zhou
Texas A&M University, TX, USA

Xun Zhou
University of Iowa, Iowa City, IA, USA

Contents

Part III Frontiers of Graph Neural Networks

Part IV Broad and Emerging Applications with Graph Neural Networks

Terminologies

This chapter describes a list of definitions of terminologies related to graph neural networks used throughout this book.

1 Basic concepts of Graphs

- **Graph**: A graph is composed of a node set and an edge set, where nodes represent entities and edges represent the relationship between entities. The nodes and edges form the topology structure of the graph. Besides the graph structure, nodes, edges, and/or the whole graph can be associated with rich information represented as node/edge/graph features (also known as attributes or contents).
- **Subgraph**: A subgraph is a graph whose set of nodes and set of edges are all subsets of the original graph.
- **Centrality**: A centrality is a measurement of the importance of nodes in the graph. The basic assumption of centrality is that a node is thought to be important if many other important nodes also connect to it. Common centrality measurements include the degree centrality, the eigenvector centrality, the betweenness centrality, and the closeness centrality.
- **Neighborhood**: The neighborhood of a node generally refers to other nodes that are close to it. For example, the k-order neighborhood of a node, also called the k-step neighborhood, denotes a set of other nodes in which the shortest path distance between these nodes and the central node is no larger than k.
- **Community Structure**: A community refers to a group of nodes that are densely connected internally and less densely connected externally.
- **Graph Sampling**: Graph sampling is a technique to pick a subset of nodes and/or edges from the original graph. Graph sampling can be applied to train machine learning models on large-scale graphs while preventing severe scalability issues.

- **Heterogeneous Graphs**: Graphs are called heterogeneous if the nodes and/or edges of the graph are from different types. A typical example of heteronomous graphs is knowledge graphs where the edges are composed of different types.
- **Hypergraphs**: Hypergraphs are generalizations of graphs in which an edge can join any number of nodes.
- **Random Graph**: Random graph generally aims to model the probability distributions over graphs that the observed graphs are generated from. The most basic and well-studied random graph model, known as the Erdos–Renyi model, assumes that the node set is fixed and each edge is identically and independently generated.
- **Dynamic Graph**: Dynamic graph refers to when at least one component of the graph data changes over time, e.g., adding or deleting nodes, adding or deleting edges, changing edges weights or changing node attributes, etc. If graphs are not dynamic, we refer to them as static graphs.

2 Machine Learning on Graphs

- **Spectral Graph Theory**: Spectral graph theory analyzes matrices associated with the graph such as its adjacency matrix or Laplacian matrix using tools of linear algebra such as studying the eigenvalues and eigenvectors of the matrix.
- **Graph Signal Processing**: Graph Signal Processing (GSP) aims to develop tools for processing signals defined on graphs. A graph signal refers to a finite collection of data samples with one sample at each node in the graph.
- **Node-level Tasks**: Node-level tasks refer to machine learning tasks associated with individual nodes in the graph. Typical examples of node-level tasks include node classification and node regression.
- **Edge-level Tasks**: Edge-level tasks refer to machine learning tasks associated with a pair of nodes in the graph. A typical example of an edge-level task in link prediction.
- **Graph-level Tasks**: Graph-level tasks refer to machine learning tasks associated with the whole graph. Typical examples of graph-level tasks include graph classification and graph property prediction.
- **Transductive and Inductive Learning**: Transductive learning refers to that the targeted instances such as nodes or edges are observed at the training time (though the labels of the targeted instances remain unknown) and inductive learning aims to learn the model which is generalizable to unobserved instances.

3 Graph Neural Networks

- **Network embedding**: The goal of network embedding is to represent each node in the graph as a low-dimensional vector so that useful information such as the graph structures and some properties of the graph is preserved in the embedding vectors. Network embedding is also referred to as graph embedding and node representation learning.

- **Graph Neural Network**: Graph neural network refers to any neural network working on the graph data.
- **Graph Convolutional Network**: Graph convolutional network usually refers to a specific graph neural network proposed by Kipf and Welling Kipf and Welling (2017a). It is occasionally used as a synonym for graph neural network, i.e., referring to any neural network working on the graph data, in some literature.
- **Message-Passing**: Message-passing is a framework of graph neural networks in which the key step is to pass messages between different nodes based on graph structures in each neural network layer. The most widely adopted formulation, usually denoted as message-passing neural networks, is to only pass messages between nodes that are directly connected Gilmer et al (2017). The message passing functions are also called graph filters and graph convolutions in some literature.
- **Readout**: Readout refers to functions that summarize the information of individual nodes to form more high-level information such as forming a subgraph/supergraph or obtaining the representations of the entire graph. Readout is also called pooling and graph coarsening in some literature.
- **Graph Adversarial Attack**: Graph adversarial attacks aim to generate worst-case perturbations by manipulating the graph structure and/or node features so that the performance of some models are downgraded. Graph adversarial attacks can be categorized based on the attacker's goals, capabilities, and accessible knowledge.
- **Robustness certificates**: Methods providing formal guarantees that the prediction of a GNN is not affected even when perturbations are performed based on a certain perturbation model.

Notations

This Chapter provides a concise reference that describes the notations used throughout this book.

Numbers, Arrays, and Matrices

A scalar	x
A vector	$\mathbf{x}$
A matrix	X
An identity matrix	$\mathbf{I}$
The set of real numbers	$\mathbb{R}$
The set of complex numbers	$\mathbb{C}$
The set of integers	$\mathbb{Z}$
The set of real n-length vectors	$\mathbb{R}^n$
The set of real $m \times n$ matrices	$\mathbb{R}^{m \times n}$
The real interval including a and b	$[a,b]$
The real interval including a but excluding b	$[a,b)$
The element of the vector $\mathbf{x}$ with index i	$\mathbf{x}_i$
The element of matrix X's indexed by Row i and Column j	$X_{i,j}$

Graph Basics

A graph	$\mathcal{G}$
Edge set	$\mathcal{E}$
Vertex set	$\mathcal{V}$
Adjacent matrix of a graph	A
Laplacian matrix	L
Diagonal degree matrix	D
Isomorphism between graphs $\mathcal{G}$ and $\mathcal{H}$	$\mathcal{G} \cong \mathcal{H}$
$\mathcal{H}$ is a subgraph of graph $\mathcal{G}$	$\mathcal{H} \subseteq \mathcal{G}$
$\mathcal{H}$ is a proper subgraph of graph $\mathcal{G}$	$\mathcal{H} \subset \mathcal{G}$
Union of graphs $\mathcal{H}$ and $\mathcal{G}$	$\mathcal{G} \cup \mathcal{H}$

Intersection of graphs $\mathcal{H}$ and $\mathcal{G}$	$\mathcal{G} \cap \mathcal{H}$
Disjoint Union of graphs $\mathcal{H}$ and $\mathcal{G}$	$\mathcal{G} + \mathcal{H}$
Cartesian Product of graphs of graphs $\mathcal{H}$ and $\mathcal{G}$	$\mathcal{G} \times \mathcal{H}$
The join of graphs $\mathcal{H}$ and $\mathcal{G}$	$\mathcal{G} \vee \mathcal{H}$

Basic Operations

Transpose of matrix X	$X^{\top}$
Dot product of matrices X and Y	$X \cdot Y$ or XY
Element-wise (Hadamard) product of matrices X and Y	$X \odot Y$
Determinant of X	$\det(X)$
p-norm (also called ℓ_p norm) of $\mathbf{x}$	$\|\mathbf{x}\|_p$
Union	$\cup$
Intersection	$\cap$
Subset	$\subseteq$
Proper subset	$\subset$
Inner prodct of vector $\mathbf{x}$ and $\mathbf{y}$	$< \mathbf{x}, \mathbf{y} >$

Functions

The function f with domain $\mathbb{A}$ and range $\mathbb{B}$	$f : \mathbb{A} \rightarrow \mathbb{B}$
Derivative of y with respect to $\mathbf{x}$	$\frac{dy}{d\mathbf{x}}$
Partial derivative of y with respect to $\mathbf{x}$	$\frac{\partial y}{\partial \mathbf{x}}$
Gradient of y with respect to $\mathbf{x}$	$\nabla_{\mathbf{x}} y$
Matrix derivatives of y with respect to matrix X	$\nabla_X y$
The Hessian matrix of function f at input vector $\mathbf{x}$	$\nabla^2 f(\mathbf{x})$
Definite integral over the entire domain of $\mathbf{x}$	$\int f(\mathbf{x}) d\mathbf{x}$
Definite integral with respect to $\mathbf{x}$ over the set $\mathbb{S}$	$\int_{\mathbb{S}} f(\mathbf{x}) d\mathbf{x}$
A function of $\mathbf{x}$ parametrized by θ	$f(\mathbf{x}; \theta)$
Convolution between functions f and g	$f * g$

Probablistic Theory

A probability distribution of a	$p(\mathrm{a})$
A conditional probabilistic distribution of b given a	$p(b\|a)$
The random variables a and b are independent	$\mathrm{a} \perp \mathrm{b}$
Variables a and b are conditionally independent given c	$\mathrm{a} \perp \mathrm{b} \mid \mathrm{c}$
Random variable a has a distribution p	$\mathrm{a} \sim p$
The expectation of $f(a)$ with respect to the variable a under distribution p	$\mathbb{E}_{a \sim p}[f(a)]$
Gaussian distribution over $\mathbf{x}$ with mean μ and covariance Σ	$\mathcal{N}(\mathbf{x}; \mu, \Sigma)$

Part I
Introduction

Chapter 1
Representation Learning

Liang Zhao, Lingfei Wu, Peng Cui and Jian Pei

Abstract In this chapter, we first describe what representation learning is and why we need representation learning. Among the various ways of learning representations, this chapter focuses on deep learning methods: those that are formed by the composition of multiple non-linear transformations, with the goal of resulting in more abstract and ultimately more useful representations. We summarize the representation learning techniques in different domains, focusing on the unique challenges and models for different data types including images, natural languages, speech signals and networks. Last, we summarize this chapter.

1.1 Representation Learning: An Introduction

The effectiveness of machine learning techniques heavily relies on not only the design of the algorithms themselves, but also a good representation (feature set) of data. Ineffective data representations that lack some important information or contains incorrect or huge redundant information could lead to poor performance of the algorithm in dealing with different tasks. The goal of representation learning is to extract sufficient but minimal information from data. Traditionally, this can be achieved via human efforts based on the prior knowledge and domain expertise on the data and tasks, which is also named as feature engineering. In deploying ma-

Liang Zhao
Department of Computer Science, Emory University, e-mail: liang.zhao@emory.edu

Lingfei Wu
JD.COM Silicon Valley Research Center, e-mail: lwu@email.wm.edu

Peng Cui
Department of Computer Science, Tsinghua University, e-mail: cuip@tsinghua.edu.cn

Jian Pei
Department of Computer Science, Simon Fraser University, e-mail: jpei@cs.sfu.ca

© The Author(s), under exclusive license to Springer Nature Singapore Pte Ltd. 2022   3
L. Wu et al. (eds.), *Graph Neural Networks: Foundations, Frontiers, and Applications*,
https://doi.org/10.1007/978-981-16-6054-2_1

chine learning and many other artificial intelligence algorithms, historically a large portion of the human efforts goes into the design of prepossessing pipelines and data transformations. More specifically, feature engineering is a way to take advantage of human ingenuity and prior knowledge in the hope to extract and organize the discriminative information from the data for machine learning tasks. For example, political scientists may be asked to define a keyword list as the features of social-media text classifiers for detecting those texts on societal events. For speech transcription recognition, one may choose to extract features from raw sound waves by the operations including Fourier transformations. Although feature engineering is widely adopted over the years, its drawbacks are also salient, including: 1) Intensive labors from domain experts are usually needed. This is because feature engineering may require tight and extensive collaboration between model developers and domain experts. 2) Incomplete and biased feature extraction. Specifically, the capacity and discriminative power of the extracted features are limited by the knowledge of different domain experts. Moreover, in many domains that human beings have limited knowledge, what features to extract itself is an open questions to domain experts, such as cancer early prediction. In order to avoid these drawbacks, making learning algorithms less dependent on feature engineering has been a highly desired goal in machine learning and artificial intelligence domains, so that novel applications could be constructed faster and hopefully addressed more effectively.

The techniques of representation learning witness the development from the traditional representation learning techniques to more advanced ones. The traditional methods belong to "shallow" models and aim to learn transformations of data that make it easier to extract useful information when building classifiers or other predictors, such as Principal Component Analysis (PCA) (Wold et al, 1987), Gaussian Markov random field (GMRF) (Rue and Held, 2005), and Locality Preserving Projections (LPP) (He and Niyogi, 2004). Deep learning-based representation learning is formed by the composition of multiple non-linear transformations, with the goal of yielding more abstract and ultimately more useful representations. In the light of introducing more recent advancements and sticking to the major topic of this book, here we majorly focus on deep learning-based representation learning, which can be categorized into several types: (1) Supervised learning, where a large number of labeled data are needed for the training of the deep learning models. Given the well-trained networks, the output before the last fully-connected layers is always utilized as the final representation of the input data; (2) Unsupervised learning (including self-supervised learning), which facilitates the analysis of input data without corresponding labels and aims to learn the underlying inherent structure or distribution of data. The pre-tasks are utilized to explore the supervision information from large amounts of unlabelled data. Based on this constructed supervision information, the deep neural networks are trained to extract the meaningful representations for the future downstream tasks; (3) Transfer learning, which involves methods that utilize any knowledge resource (i.e., data, model, labels, etc.) to increase model learning and generalization for the target task. Transfer learning encompasses different scenarios including multi-task learning (MTL), model adaptation, knowledge transfer, co-variance shift, etc. There are also other important representation learning methods such as reinforcement learning, few-shot learning, and disentangled representation learning.

It is important to define what is a good representation. As the definition by Bengio (2008), representation learning is about learning the (underlying) features of the data that make it easier to extract useful information when building classifiers or other predictors. Thus, the evaluation of a learned representation is closely related to its performance on the downstream tasks. For example, in the data generation task based on a generative model, a good representation is often the one that captures the posterior distribution of the underlying explanatory factors for the observed input. While for a prediction task, a good representation is the one that captures the minimal but sufficient information of input data to correctly predict the target label. Besides the evaluation from the perspective of the downstream tasks, there are also some general properties that the good representations may hold, such as the smoothness, the linearity, capturing multiple explanatory and casual factors, holding shared factors across different tasks and simple factor dependencies.

1.2 Representation Learning in Different Areas

In this section, we summarize the development of representation learning on four different representative areas: (1) image processing; (2) speech recognition; (3) Natural language processing; and (4) network analysis. For the representation learning in each research area, we consider some of the fundamental questions that have been driving research in this area. Specifically, what makes one representation better than another, and how should we compute its representation? Why is the representation learning important in that area? Also, what are appropriate objectives for learning good representations? We also introduce the relevant typical methods and their development from the perspective of three main categories: supervised representation learning, unsupervised learning and transfer learning, respectively.

1.2.1 Representation Learning for Image Processing

Image representation learning is a fundamental problem in understanding the semantics of various visual data, such as photographs, medical images, document scans, and video streams. Normally, the goal of image representation learning for image processing is to bridge the semantic gap between the pixel data and semantics of the images. The successful achievements of image representation learning have enpowered many real-world problems, including but not limited to image search, facial recognition, medical image analysis, photo manipulation and target detection.

In recent years, we have witnessed a fast advancement of image representation learning from handcrafted feature engineering to that from scratch through deep neural network models. Traditionally, the patterns of images are extracted with the help of hand-crafted features by human beings based on prior knowledge. For example, Huang et al (2000) extracted the character's structure features from the strokes, then use them to recognize the handwritten characters. Rui (2005) adopted the morphology method to improve local feature of the characters, then use PCA to extract features of characters. However, all of these methods need to extract features from images manually and thus the prediction performances strongly rely on the prior knowledge. In the field of computer vision, manual feature extraction is very cumbersome and impractical because of the high dimensionality of feature vectors. Thus, representation learning of images which can automatically extract meaningful, hidden and complex patterns from high-dimension visual data is necessary. Deep learning-based representation learning for images is learned in an end-to-end fashion, which can perform much better than hand-crafted features in the target applications, as long as the training data is of sufficient quality and quantity.

Supervised Representation Learning for image processing. In the domain of image processing, supervised learning algorithm, such as Convolution Neural Network (CNN) and Deep Belief Network (DBN), are commonly applied in solving various tasks. One of the earliest deep-supervised-learning-based works was proposed in 2006 (Hinton et al, 2006), which is focused on the MNIST digit image classification problem, outperforming the state-of-the-art SVMs. Following this, deep convolutional neural networks (ConvNets) showed amazing performance which is greatly depends on their properties of shift in-variance, weights sharing and local pattern capturing. Different types of network architectures were developed to increase the capacity of network models, and larger and larger datasets were collected these days. Various networks including AlexNet (Krizhevsky et al, 2012), VGG (Simonyan and Zisserman, 2014b), GoogLeNet (Szegedy et al, 2015), ResNet (He et al, 2016a), and DenseNet (Huang et al, 2017a) and large scale datasets, such as ImageNet and OpenImage, have been proposed to train very deep convolutional neural networks. With the sophisticated architectures and large-scale datasets, the performance of convolutional neural networks keeps outperforming the state-of-the-arts in various computer vision tasks.

Unsupervised Representation Learning for image processing. Collection and annotation of large-scale datasets are time-consuming and expensive in both image datasets and video datasets. For example, ImageNet contains about 1.3 million labeled images covering 1,000 classes while each image is labeled by human workers with one class label. To alleviate the extensive human annotation labors, many unsupervised methods were proposed to learn visual features from large-scale unlabeled images or videos without using any human annotations. A popular solution is to propose various pretext tasks for models to solve, while the models can be trained by learning objective functions of the pretext tasks and the features are learned through this process. Various pretext tasks have been proposed for unsupervised learning, including colorizing gray-scale images (Zhang et al, 2016d) and image inpainting (Pathak et al, 2016). During the unsupervised training phase, a predefined pretext task is designed for the models to solve, and the pseudo labels for the pretext task are automatically generated based on some attributes of data. Then the models are trained according to the objective functions of the pretext tasks. When trained with pretext tasks, the shallower blocks of the deep neural network models focus on the low-level general features such as corners, edges, and textures, while the deeper blocks focus on the high-level task-specific features such as objects, scenes, and object parts. Therefore, the models trained with pretext tasks can learn kernels to capture low-level features and high-level features that are helpful for other downstream tasks. After the unsupervised training is finished, the learned visual features in this pre-trained models can be further transferred to downstream tasks (especially when only relatively small data is available) to improve performance and overcome over-fitting.

Transfer Learning for image processing. In real-world applications, due to the high cost of manual labeling, sufficient training data that belongs to the same feature space or distribution as the testing data may not always be accessible. Transfer learning mimics the human vision system by making use of sufficient amounts of prior knowledge in other related domains (i.e., source domains) when executing new tasks in the given domain (i.e., target domain). In transfer learning, both the training set and the test set can contribute to the target and source domains. In most cases, there is only one target domain for a transfer learning task, while either single or multiple source domains can exist. The techniques of transfer learning in images processing can be categorized into feature representation knowledge transfer and classifier-based knowledge transfer. Specifically, feature representation transfer methods map the target domain to the source domains by exploiting a set of extracted features, where the data divergence between the target domain and the source domains can be significantly reduced so that the performance of the task in the target domain is improved. For example, classifier-based knowledge-transfer methods usually share the common trait that the learned source domain models are utilized as prior knowledge, which are used to learn the target model together with the training samples. Instead of minimizing the cross-domain dissimilarity by updating instances' representations, classifier-based knowledge-transfer methods aim to learn a new model that minimizes the generalization error in the target domain via the provided training set from both domains and the learned model.

Other Representation Learning for Image Processing. Other types of representation learning are also commonly observed for dealing with image processing, such as reinforcement learning, and semi-supervised learning. For example, reinforcement learning are commonly explored in the task of image captioning Liu et al (2018a); Ren et al (2017) and image editing Kosugi and Yamasaki (2020), where the learning process is formalized as a sequence of actions based on a policy network.

1.2.2 Representation Learning for Speech Recognition

Nowadays, speech interfaces or systems have become widely developed and integrated into various real-life applications and devices. Services like Siri [1], Cortana [2], and Google Voice Search [3] have become a part of our daily life and are used by millions of users. The exploration in speech recognition and analysis has always been motivated by a desire to enable machines to participate in verbal human-machine interactions. The research goals of enabling machines to understand human speech, identify speakers, and detect human emotion have attracted researchers' attention for more than sixty years across several distinct research areas, including but not limited to Automatic Speech Recognition (ASR), Speaker Recognition (SR), and Speaker Emotion Recognition (SER).

Analyzing and processing speech has been a key application of machine learning (ML) algorithms. Research on speech recognition has traditionally considered the task of designing hand-crafted acoustic features as a separate distinct problem from the task of designing efficient models to accomplish prediction and classification decisions. There are two main drawbacks of this approach: First, the feature engineering is cumbersome and requires human knowledge as introduced above; and second, the designed features might not be the best for the specific speech recognition tasks at hand. This has motivated the adoption of recent trends in the speech community towards the utilization of representation learning techniques, which can learn an intermediate representation of the input signal automatically that better fits into the task at hand and hence lead to improved performance. Among all these successes, deep learning-based speech representations play an important role. One of the major reasons for the utilization of representation learning techniques in speech technology is that speech data is fundamentally different from two-dimensional image data. Images can be analyzed as a whole or in patches, but speech has to be formatted sequentially to capture temporal dependency and patterns.

Supervised representation learning for speech recognition. In the domain of speech recognition and analyzing, supervised representation learning methods are widely employed, where feature representations are learned on datasets by leveraging label information. For example, restricted Boltzmann machines (RBMs) (Jaitly and Hinton, 2011; Dahl et al, 2010) and deep belief networks (DBNs) (Cairong et al, 2016; Ali et al, 2018) are commonly utilized in learning features from speech for different tasks, including ASR, speaker recognition, and SER. For example, in 2012, Microsoft has released a new version of their MAVIS (Microsoft Audio Video Indexing Service) speech system based on context-dependent deep neural networks (Seide et al, 2011). These authors managed to reduce the word error rate on four major benchmarks by about 30% (e.g., from 27.4% to 18.5% on RT03S) com-

---

[1] Siri is an artificial intelligence assistant software that is built into Apple's iOS system.

[2] Microsoft Cortana is an intelligent personal assistant developed by Microsoft, known as "the world's first cross-platform intelligent personal assistant".

[3] Google Voice Search is a product of Google that allows you to use Google to search by speaking to a mobile phone or computer, that is, to use the legendary content on the device to be identified by the server, and then search for information based on the results of the recognition

pared to the traditional models based on Gaussian mixtures. Convolutional neural networks are another popular supervised models that are widely utilized for feature learning from speech signals in tasks such as speech and speaker recognition (Palaz et al, 2015a,b) and SER Latif et al (2019); Tzirakis et al (2018). Moreover, it has been found that LSTMs (or GRUs) can help CNNs in learning more useful features from speech by learning both the local and long-term dependency (Dahl et al, 2010).

Unsupervised Representation Learning for speech recognition. Unsupervised representation learning from large unlabelled datasets is an active area of speech recognition. In the context of speech analysis, it is able to exploit the practically available unlimited amount of unlabelled corpora to learn good intermediate feature representations, which can then be used to improve the performance of a variety of downstream supervised learning speech recognition tasks or the speech signal synthetic tasks. In the tasks of ASR and SR, most of the works are based on Variational Auto-encoder (VAEs), where a generative model and an inference model are jointly learned, which allows them to capture latent representations from observed speech data (Chorowski et al, 2019; Hsu et al, 2019, 2017). For example, Hsu et al (2017) proposed a hierarchical VAE to capture interpretable and disentangled representations from speech without any supervision. Other auto-encoding architectures like Denoised Autoencoder(DAEs) are also found very promising in finding speech representations in an unsupervised way, especially for noisy speech recognition (Feng et al, 2014; Zhao et al, 2015). Beyond the aforementioned, recently, adversarial learning (AL) is emerging as a powerful tool in learning unsupervised representation for speech, such as generative adversarial nets (GANs). It involves at least a generator and a discriminator, where the former tries to generates as realistic as possible data to obfuscate the latter which also tries its best to deobfuscate. Hence both of the generator and discriminator can be trained and improved iteratively in an adversarial way, which result in more discriminative and robust features. Among these, GANs (Chang and Scherer, 2017; Donahue et al, 2018), adversarial autoencoders (AAEs) Sahu et al (2017) are becoming mostly popular in modeling speech not only in ASR but also SR and SER.

Transfer Learning for speech recognition. Transfer learning (TL) encompasses different approaches, including MTL, model adaptation, knowledge transfer, covariance shift, etc. In the domain of speech recognition, representation learning gained much interest in these approaches of TL including but not limited to domain adaptation, multi-task learning, and self-taught learning. In terms of Domain Adaption, speech is a typical example of heterogeneous data and thus, a mismatch always exists between the probability distributions of source and target domain data. To build more robust systems for speech-related applications in real-life, domain adaptation techniques are usually applied in the training pipeline of deep neural networks to learn representations which are able to explicitly minimize the difference between the distribution of data in the source and target domains (Sun et al, 2017; Swietojanski et al, 2016). In terms of MTL, representations learned can successfully increases the performance of speech recognition without requiring contextual speech data, since speech contains multi-dimensional information (message, speaker, gender, or emotion) that can be used as auxiliary tasks. For example, In the task of ASR, by using MTL with different auxiliary tasks including gender, speaker adaptation, speech enhancement, it has been shown that the learned shared representations for different tasks can act as complementary information about the acoustic environment and give a lower word error rate (WER) (Parthasarathy and Busso, 2017; Xia and Liu, 2015).

Other Representation Learning for speech recognition. Other than the above-mentioned three categories of representation learning for speech signals, there are also some other representation learning techniques commonly explored, such as semi-supervised learning and reinforcement learning. For example, in the speech recognition for ASR, semi-supervised learning is mainly used to circumvent the lack of sufficient training data. This can be achieved either by creating features fronts ends (Thomas et al, 2013), or by using multilingual acoustic representations (Cui et al, 2015), or by extracting an intermediate representation from large unpaired datasets (Karita et al, 2018). RL is also gaining interest in the area of speech recognition, and there have been multiple approaches to model different speech problems, including dialog modeling and optimization (Levin et al, 2000), speech recognition (Shen et al, 2019), and emotion recognition (Sangeetha and Jayasankar, 2019).

1.2.3 Representation Learning for Natural Language Processing

Besides speech recognition, there are many other Natural Language Processing (NLP) applications of representation learning, such as the text representation learning. For example, Google's image search exploits huge quantities of data to map images and queries in the same space (Weston et al, 2010) based on NLP techniques. In general, there are two types of applications of representation learning in NLP. In one type, the semantic representation, such as the word embedding, is trained in a pre-training task (or directly designed by human experts) and is transferred to the model for the target task. It is trained by using language modeling objective and is taken as inputs for other down-stream NLP models. In the other type, the semantic representation lies within the hidden states of the deep learning model and directly aims for better performance of the target tasks in an end-to-end fashion. For example, many NLP tasks want to semantically compose sentence or document representation, such as tasks like sentiment classification, natural language inference, and relation extraction, which require sentence representation.

Conventional NLP tasks heavily rely on feature engineering, which requires careful design and considerable expertise. Recently, representation learning, especially deep learning-based representation learning is emerging as the most important technique for NLP. First, NLP is typically concerned with multiple levels of language entries, including but not limited to characters, words, phrases, sentences, paragraphs, and documents. Representation learning is able to represent the semantics of these multi-level language entries in a unified semantic space, and model complex semantic dependence among these language entries. Second, there are various NLP tasks that can be conducted on the same input. For example, given a sentence, we can perform multiple tasks such as word segmentation, named entity recognition, relation extraction, co-reference linking, and machine translation. In this case, it will be more efficient and robust to build a unified representation space of inputs for multiple tasks. Last, natural language texts may be collected from multiple domains, including but not limited to news articles, scientific articles, literary works, advertisement and online user-generated content such as product reviews and social media. Moreover, texts can also be collected from different languages, such as English, Chinese, Spanish, Japanese, etc. Compared to conventional NLP systems which have to design specific feature extraction algorithms for each domain according to its characteristics, representation learning enables us to build representations automatically from large-scale domain data and even add bridges among these languages from different domains. Given these advantages of representation learning for NLP in the feature engineering reduction and performance improvement, many researchers have developed efficient algorithms on representation learning, especially deep learning-based approaches, for NLP.

Supervised Representation Learning for NLP. Deep neural networks in the supervised learning setting for NLP emerge from distributed representation learning, then to CNN models, and finally to RNN models in recent years. At early stage, distributed representations are first developed in the context of statistical language modeling by Bengio (2008) in so-called neural net language models. The model is about learning a distributed representation for each word (i.e., word embedding). Following this, the need arose for an effective feature function that extracts higher-level features from constituting words or n-grams. CNNs turned out to be the natural choice given their properties of excellent performance in computer vision and speech processing tasks. CNNs have the ability to extract salient n-gram features from the input sentence to create an informative latent semantic representation of the sentence for downstream tasks. This domain was pioneered by Collobert et al (2011) and Kalchbrenner et al (2014), which led to a huge proliferation of CNN-based networks in the succeeding literature. The neural net language model was also improved by adding recurrence to the hidden layers (Mikolov et al, 2011a) (i.e., RNN), allowing it to beat the state-of-the-art (smoothed n-gram models) not only in terms of perplexity (exponential of the average negative log-likelihood of predicting the right next word) but also in terms of WER in speech recognition. RNNs use the idea of processing sequential information. The term "recurrent" applies as they perform the same computation over each token of the sequence and each step is dependent on the previous computations and results. Generally, a fixed-size vector is produced to represent a sequence by feeding tokens one by one to a recurrent unit. In a way, RNNs have "memory" over previous computations and use this information in current processing. This template is naturally suited for many NLP tasks such as language modeling (Mikolov et al, 2010, 2011b), machine translation (Liu et al, 2014; Sutskever et al, 2014), and image captioning (Karpathy and Fei-Fei, 2015).

Unsupervised Representation Learning for NLP. Unsupervised learning (including self-supervised learning) has made a great success in NLP, for the plain text itself contains abundant knowledge and patterns about languages. For example, in most deep learning based NLP models, words in sentences are first mapped to their corresponding embeddings via the techniques, such as word2vec Mikolov et al (2013b), GloVe Pennington et al (2014), and BERT Devlin et al (2019), before sending to the networks. However, there are no human-annotated "labels" for learning those word embeddings. To acquire the training objective necessary for neural networks, it is necessary to generate "labels" intrinsically from the existing data. Language modeling is a typical unsupervised learning task, which can construct the probability distribution over sequences of words and does not require human annotations. Based on the distributional hypothesis, using the language modeling objective can lead to hidden representations that encode the semantics of words. Another typical unsupervised learning model in NLP is auto-encoder (AE), which consists of a reduction (encoding) phase and a reconstruction (decoding) phase. For example, recursive auto-encoders (which generalize recurrent networks with VAE) have been used to beat the state-of-the-art at the moment of its publication in full sentence paraphrase detection (Socher et al, 2011) by almost doubling the F1 score for paraphrase detection.

Transfer Learning for NLP. Over the recent years, the field of NLP has witnessed fast growth of transfer learning methods via sequential transfer learning models and architectures, which significantly improved upon the state-of-the-arts on a wide range of NLP tasks. In terms of domain adaption, the sequential transfer learning consists of two stages: a pretraining phase in which general representations are learned on a source task or domain followed by an adaptation phase during which the learned knowledge is applied to a target task or domain. The domain adaption in NLP is categorized into model-centric, data-centric, and hybrid approaches. Model-centric methods target the approaches to augmenting the feature space, as well as altering the loss function, the architecture, or the model parameters (Blitzer et al, 2006). Data-centric methods focus on the data aspect and involve pseudo-labeling (or bootstrapping) where only small number of classes are shared between the source and target datasets (Abney, 2007). Lastly, hybrid-based methods are built by both data- and model-centric models. Similarly, great advances have also been made into the multi-task learning in NLP, where different NLP tasks can result in better representation of texts. For example, based on a convolutional architecture, Collobert et al (2011) developed the SENNA system that shares representations across the tasks of language modeling, part-of-speech tagging, chunking, named entity recognition, semantic role labeling, and syntactic parsing. SENNA approaches or sometimes even surpasses the state-of-the-art on these tasks while is simpler and much faster than traditional predictors. Moreover, learning word embeddings can be combined with learning image representations in a way that allow associating texts and images.

Other Representation Learning for NLP. In NLP tasks, when a problem gets more complicated, it requires more knowledge from domain experts to annotate training instances for fine-grained tasks and thus increases the cost of data labeling. Therefore, sometimes it requires the models or systems can be developed efficiently with (very) few labeled data. When each class has only one or a few labeled instances, the problem becomes a one/few-shot learning problem. The few-shot learning problem is derived from computer vision and has also been studied in NLP recently. For example, researchers have explored few-shot relation extractio (Han et al, 2018) where each relation has a few labeled instances, and low-resource machine translation (Zoph et al, 2016) where the size of the parallel corpus is limited.

1.2.4 Representation Learning for Networks

Beyond popular data like images, texts, and sounds, network data is another important data type that is becoming ubiquitous across a large scale of real-world applications ranging from cyber-networks (e.g., social networks, citation networks, telecommunication networks, etc.) to physical networks (e.g., transportation networks, biological networks, etc). Networks data can be formulated as graphs mathematically, where vertices and their relationships jointly characterize the network information. Networks and graphs are very powerful and flexible data formulation such that sometimes we could even consider other data types like images, and texts as special cases of it. For example, images can be considered as grids of nodes with RGB attributes which are special types of graphs, while texts can also be organized into sequential-, tree-, or graph-structured information. So in general, representation learning for networks is widely considered as a promising yet more challenging tasks that require the advancement and generalization of many techniques we developed for images, texts, and so forth. In addition to the intrinsic high complexity of network data, the efficiency of representation learning on networks is also an important issues considering the large-scale of many real-world networks, ranging from hundreds to millions or even billions of vertices. Analyzing information networks plays a crucial role in a variety of emerging applications across many disciplines. For example, in social networks, classifying users into meaningful social groups is useful for many important tasks, such as user search, targeted advertising and recommendations; in communication networks, detecting community structures can help better understand the rumor spreading process; in biological networks, inferring interactions between proteins can facilitate new treatments for diseases. Nevertheless, efficient and effective analysis of these networks heavily relies on good representations of the networks.

Traditional feature engineering on network data usually focuses on obtaining a number of predefined straightforward features in graph levels (e.g., the diameter, average path length, and clustering co-efficient), node levels (e.g., node degree and centrality), or subgraph levels (e.g., frequent subgraphs and graph motifs). Those limited number of hand-crafted, well-defined features, though describe several fundamental aspects of the graphs, discard the patterns that cannot be covered by them. Moreover, real-world network phenomena are usually highly complicated require sophisticated, unknown combinations among those predefined features or cannot be characterized by any of the existing features. In addition, traditional graph feature engineering usually involve expensive computations with super-linear or exponential complexity, which often makes many network analytic tasks computationally expensive and intractable over large-scale networks. For example, in dealing with the task of community detection, classical methods involve calculating the spectral decomposition of a matrix with at least quadratic time complexity with respect to the number of vertices. This computational overhead makes algorithms hard to scale to large-scale networks with millions of vertices.

More recently, network representation learning (NRL) has aroused a lot of research interest. NRL aims to learn latent, low-dimensional representations of network vertices, while preserving network topology structure, vertex content, and other side information. After new vertex representations are learned, network analytic tasks can be easily and efficiently carried out by applying conventional vector-based machine learning algorithms to the new representation space. Earlier work related to network representation learning dates back to the early 2000s, when researchers proposed graph embedding algorithms as part of dimensionality reduction techniques. Given a set of independent and identically distributed (i.i.d.) data points as input, graph embedding algorithms first calculate the similarity between pairwise data points to construct an affinity graph, e.g., the k-nearest neighbor graph, and then embed the affinity graph into a new space having much lower dimensionality. However, graph embedding algorithms are designed on i.i.d. data mainly for dimensionality reduction purpose, which usually have at least quadratic time complexity with respect to the number of vertices.

Since 2008, significant research efforts have shifted to the development of effective and scalable representation learning techniques that are directly designed for complex information networks. Many network representation learning algorithms (Perozzi et al, 2014; Yang et al, 2015b; Zhang et al, 2016b; Manessi et al, 2020) have been proposed to embed existing networks, showing promising performance for various applications. These methods embed a network into a latent, low-dimensional space that preserves structure proximity and attribute affinity. The resulting compact, low-dimensional vector representations can be then taken as features to any vector-based machine learning algorithms. This paves the way for a wide range of network analytic tasks to be easily and efficiently tackled in the new vector space, such as node classification (Zhu et al, 2007), link prediction (Lü and Zhou, 2011), clustering (Malliaros and Vazirgiannis, 2013), network synthesis (You et al, 2018b). The following chapters of this book will then provide a systematic and comprehensive introduction into network representation learning.

1.3 Summary

Representation learning is a very active and important field currently, which heavily influences the effectiveness of machine learning techniques. Representation learning is about learning the representations of the data that makes it easier to extract useful and discriminative information when building classifiers or other predictors. Among the various ways of learning representations, deep learning algorithms have increasingly been employed in many areas nowadays where the good representation can be learned in an efficient and automatic way based on large amount of complex and high dimensional data. The evaluation of a representation is closely related to its performance on the downstream tasks. Generally, there are also some general properties that the good representations may hold, such as the smoothness, the linearity, disentanglement, as well as capturing multiple explanatory and casual factors.

We have summarized the representation learning techniques in different domains, focusing on the unique challenges and models for different areas including the processing of images, natural language, and speech signals. For each area, there emerges many deep learning-based representation techniques from different categories, including supervised learning, unsupervised learning, transfer learning, disentangled representation learning, reinforcement learning, etc. We have also briefly mentioned about the representation learning on networks and its relations to that on images, texts, and speech, in order for the elaboration of it in the following chapters.

Chapter 2
Graph Representation Learning

Peng Cui, Lingfei Wu, Jian Pei, Liang Zhao and Xiao Wang

Abstract Graph representation learning aims at assigning nodes in a graph to low-dimensional representations and effectively preserving the graph structures. Recently, a significant amount of progress has been made toward this emerging graph analysis paradigm. In this chapter, we first summarize the motivation of graph representation learning. Afterwards and primarily, we provide a comprehensive overview of a large number of graph representation learning methods in a systematic manner, covering the traditional graph representation learning, modern graph representation learning, and graph neural networks.

2.1 Graph Representation Learning: An Introduction

Many complex systems take the form of graphs, such as social networks, biological networks, and information networks. It is well recognized that graph data is often sophisticated and thus is challenging to deal with. To process graph data effectively, the first critical challenge is to find effective graph data representation, that is, how to represent graphs concisely so that advanced analytic tasks, such as pattern discovery, analysis, and prediction, can be conducted efficiently in both time and space.

Liang Zhao
Department of Computer Science, Emory University, e-mail: liang.zhao@emory.edu

Lingfei Wu
JD.COM Silicon Valley Research Center, e-mail: lwu@email.wm.edu

Peng Cui
Department of Computer Science, Tsinghua University, e-mail: cuip@tsinghua.edu.cn

Jian Pei
Department of Computer Science, Simon Fraser University, e-mail: jpei@cs.sfu.ca

Xiao Wang
Department of Computer Science, Beijing University of Posts and Telecommunications, e-mail: xiaowang@bupt.edu.cn

L. Wu et al. (eds.), *Graph Neural Networks: Foundations, Frontiers, and Applications*,
https://doi.org/10.1007/978-981-16-6054-2_2

Traditionally, we usually represent a graph as $\mathscr{G} = (\mathscr{V}, \mathscr{E})$, where $\mathscr{V}$ is a node set and $\mathscr{E}$ is an edge set. For large graphs, such as those with billions of nodes, the traditional graph representation poses several challenges to graph processing and analysis.

(1) **High computational complexity.** These relationships encoded by the edge set E take most of the graph processing or analysis algorithms either iterative or combinatorial computation steps. For example, a popular way is to use the shortest or average path length between two nodes to represent their distance. To compute such a distance using the traditional graph representation, we have to enumerate many possible paths between two nodes, which is in nature a combinatorial problem. Such methods result in high computational complexity that prevents them from being applicable to large-scale real-world graphs.

(2) **Low parallelizability.** Parallel and distributed computing is de facto to process and analyze large-scale data. Graph data represented in the traditional way, however, casts severe difficulties to design and implementat of parallel and distributed algorithms. The bottleneck is that nodes in a graph are coupled to each other explicitly reflected by E. Thus, distributing different nodes in different shards or servers often causes demandingly high communication cost among servers, and holds back speed-up ratio.

(3) **Inapplicability of machine learning methods.** Recently, machine learning methods, especially deep learning, are very powerful in many areas. For graph data represented in the traditional way, however, most of the off-the-shelf machine learning methods may not be applicable. Those methods usually assume that data samples can be represented by independent vectors in a vector space, while the samples in graph data (i.e., the nodes) are dependant to each other to some degree determined by E. Although we can simply represent a node by its corresponding row vector in the adjacency matrix of the graph, the extremely high dimensionality of such a representation in a large graph with many nodes makes the in sequel graph processing and analysis difficult.

To tackle these challenges, substantial effort has been committed to develop novel graph representation learning, i.e., learning the dense and continuous low-dimensional vector representations for nodes, so that the noise or redundant information can be reduced and the intrinsic structure information can be preserved. In the learned representation space, the relationships among the nodes, which were originally represented by edges or other high-order topological measures in graphs, are captured by the distances between nodes in the vector space, and the structural characteristics of a node are encoded into its representation vector.

Basically, in order to make the representation space well supporting graph analysis tasks, there are two goals for graph representation learning. First, the original graph can be reconstructed from the learned representation space. It requires that, if there is an edge or relationship between two nodes, then the distance of these two nodes in the representation space should be relatively small. Second, the learned representation space can effectively support graph inference, such as predicting unseen links, identifying important nodes, and inferring node labels. It should be noted that a representation space with only the goal of graph reconstruction is not sufficient for graph inference. After the representation is obtained, downstream tasks such as node classification , node clustering , graph visualization and link prediction can be dealt with based on these representations. Overall, there are three main categories of graph representation learning methods: traditional graph embedding, modern graph embedding, and graph neural networks, which will be introduced separately in the following three sections.

2.2 Traditional Graph Embedding

Traditional graph embedding methods are originally studied as dimension reduction techniques. A graph is usually constructed from a feature represented data set, like image data set. As mentioned before, graph embedding usually has two goals, i.e. reconstructing original graph structures and support graph inference. The objective functions of traditional graph embedding methods mainly target the goal of graph reconstruction.

Specifically, Tenenbaum et al (2000) first constructs a neighborhood graph G using connectivity algorithms such as K nearest neighbors (KNN). Then based on G, the shortest path between different data can be computed. Consequently, for all the N data entries in the data set, we have the matrix of graph distances. Finally, the classical multidimensional scaling (MDS) method is applied to the matrix to obtain the coordinate vectors. The representations learned by Isomap approximately preserve the geodesic distances of the entry pairs in the low-dimensional space. The key problem of Isomap is its high complexity due to the computing of pair-wise shortest pathes. Locally linear embedding (LLE) (Roweis and Saul, 2000) is proposed to eliminate the need to estimate the pairwise distances between widely separated entries. LLE assumes that each entry and its neighbors lie on or close to a locally linear patch of a mainfold. To characterize the local geometry, each entry can be reconstructed from its neighbors. Finally, in the low-dimensional space, LLE constructs a neighborhood-preserving mapping based on locally linear reconstruction. Laplacian eigenmaps (LE) (Belkin and Niyogi, 2002) also begins with constructing a graph using ε-neighborhoods or K nearest neighbors. Then the heat kernel (Berline et al, 2003) is utilized to choose the weight of two nodes in the graph. Finally, the node representations can be obtained by based on the Laplacian matrix regularization. Furthermore, the locality preserving projection (LPP) (Berline et al, 2003), a linear approximation of the nonlinear LE, is proposed.

These methods are extended in the rich literature of graph embedding by considering different characteristics of the constructed graphs (Fu and Ma, 2012). We can find that traditional graph embedding mostly works on graphs constructed from feature represented data sets, where the proximity among nodes encoded by the edge weights is well defined in the original feature space. While, in contrast, modern graph embedding, which will be introduced in the following, mostly works on naturally formed networks, such as social networks, biology networks, and e-commerce networks. In those networks, the proximities among nodes are not explicitly or directly defined. For example, an edge between two nodes usually just implies there is a relationship between them, but cannot indicate the specific proximity. Also, even if there is no edge between two nodes, we cannot say the proximity between these two nodes is zero. The definition of node proximities depends on specific analytic tasks and application scenarios. Therefore, modern graph embedding usually incorporates rich information, such as network structures, properties, side information and advanced information, to facilitate different problems and applications. Modern graph embedding needs to target both of goals mentioned before. In view of this, traditional graph embedding can be regarded as a special case of modern graph embedding, and the recent research progress on modern graph embedding pays more attention to network inference.

2.3 Modern Graph Embedding

To well support network inference, modern graph embedding considers much richer information in a graph. According to the types of information that are preserved in graph representation learning, the existing methods can be categorized into three categories: (1) graph structures and properties preserving graph embedding, (2) graph representation learning with side information and (3) advanced information preserving graph representation learning. In technique view, different models are adopted to incorporate different types of information or address different goals. The commonly used models include matrix factorization, random walk, deep neural networks and their variations.

2.3.1 Structure-Property Preserving Graph Representation Learning

Among all the information encoded in a graph, graph structures and properties are two crucial factors that largely affect graph inference. Thus, one basic requirement of graph representation learning is to appropriately preserve graph structures and capture properties of graphs. Often, graph structures include first-order structures and higher-order structures, such as second-order structures and community structures. Graphs with different types have different properties. For example, directed graphs have the asymmetric transitivity property. The structural balance theory is widely applicable to signed graphs.

2.3.1.1 Structure Preserving Graph Representation Learning

Graph structures can be categorized into different groups that present at different granularities. The commonly exploited graph structures in graph representation learning include neighborhood structure, high-order node proximity and graph communities.

How to define the neighborhood structure in a graph is the first challenge. Based on the discovery that the distribution of nodes appearing in short random walks is similar to the distribution of words in natural language, DeepWalk (Perozzi et al, 2014) employs the random walks to capture the neighborhood structure. Then for each walk sequence generated by random walks, following Skip-Gram, DeepWalk aims to maximize the probability of the neighbors of a node in a walk sequence. Node2vec defines a flexible notion of a node's graph neighborhood and designs a second order random walks strategy to sample the neighborhood nodes, which can smoothly interpolate between breadth-first sampling (BFS) and depth-first sampling (DFS). Besides the neighborhood structure, LINE (Tang et al, 2015b) is proposed for large scale network embedding, which can preserve the first and second order proximities. The first order proximity is the observed pairwise proximity between two nodes. The second order proximity is determined by the similarity of the "contexts" (neighbors) of two nodes. Both are important in measuring the relationships between two nodes. Essentially, LINE is based on the shallow model, consequently, the representation ability is limited. SDNE (Wang et al, 2016) proposes a deep model for network embedding, which also aims at capturing the first and second order proximites. SDNE uses the deep auto-encoder architecture with multiple non-linear layers to preserve the second order proximity. To preserve the first-order proximity, the idea of Laplacian eigenmaps (Belkin and Niyogi, 2002) is adopted. Wang et al (2017g) propose a modularized nonnegative matrix factorization (M-NMF) model for graph representation learning, which aims to preserve both the microscopic structure, i.e., the first-order and second-order proximities of nodes, and the mesoscopic community structure (Girvan and Newman, 2002). They adopt the NMF model (Févotte and Idier, 2011) to preserve the microscopic structure. Meanwhile, the community structure is detected by modularity maximization (Newman, 2006a). Then, they introduce an auxiliary community representation matrix to bridge the representations of nodes with the community structure. In this way, the learned representations of nodes are constrained by both the microscopic structure and community structure.

In summary, many network embedding methods aim to preserve the local structure of a node, including neighborhood structure, high-order proximity as well as community structure, in the latent low-dimensional space. Both linear and non-linear models are attempted, demonstrating the large potential of deep models in network embedding.

2.3.1.2 Property Preserving Graph Representation Learning

Currently, most of the existing property preserving graph representation learning methods focus on graph transitivity in all types of graphs and the structural balance property in signed graphs.

We usually demonstrate that the transitivity usually exists in a graph. But meanwhile, we can find that preserving such a property is not challenging, because in a metric space, the distance between different data points naturally satisfies the triangle inequality. However, this is not always true in the real world. Ou et al (2015) aim to preserve the non-transitivity property via latent similarity components. The non-transitivity property declares that, for nodes v_1, v_2 and v_3 in a graph where $(v_1; v_2)$ and $(v_2; v_3)$ are similar pairs, $(v_1; v_3)$ may be a dissimilar pair. For example, in a social network, a student may connect with his classmates and his family, while his classmates and family are probably very different. The main idea is that they learn multiple node embeddings, and then compare different nodes based on multiple similarities, rather than one similarity. They observe that if two nodes have a large semantic similarity, at least one of the structure similarities is large, otherwise, all of the similarities are small. In a directed graph, it usually has the asymmetric transitivity property. Asymmetric transitivity indicates that, if there is a directed edge from node i to node j and a directed edge from j to v, there is likely a directed edge from i to v, but not from v to i. In order to measure this high-order proximity, HOPE (Ou et al, 2016) summarizes four measurements in a general formulation, and then utilizes a generalized SVD problem to factorize the high-order proximity (Paige and Saunders, 1981), such that the time complexity of HOPE is largely reduced, which means HOPE is scalable for large scale networks. In a signed graph with both of positive and negative edges, the social theories, such as structural balance theory (Cartwright and Harary, 1956; Cygan et al, 2012), which are very different from the unsigned graph. The structural balance theory demonstrates that users in a signed social network should be able to have their "friends" closer than their "foes". To model the structural balance phenomenon, SiNE (Wang et al, 2017f) utilizes a deep learning model consisting of two deep graphs with non-linear functions.

The importance of maintaining network properties in network embedding space, especially the properties that largely affect the evolution and formation of networks, has been well recognized. The key challenge is how to address the disparity and heterogeneity of the original network space and the embedding vector space at property level. Generally, most of the structure and property preserving methods take high order proximities of nodes into account, which demonstrate the importance of preserving high order structures in network embedding. The difference is the strategy of obtaining the high order structures. Some methods implicitly preserve highorder structure by assuming a generative mechanism from a node to its neighbors, while some other methods realize this by explicitly approximating high-order proximities in the embedding space. As topology structures are the most notable characteristic of networks, structure-preserving network methods embody a large part of the literature. Comparatively, property preserving network embedding is a relatively new research topic and is only studied lightly. As network properties usually drive the formation and evolution of networks, it shows great potential for future research and applications.

2.3.2 Graph Representation Learning with Side Information

Besides graph structures, side information is another important information source for graph representation learning. Side information in the context of graph representation learning can be divided into two categories: node content and types of nodes and edges. Their difference is the way of integrating network structures and side information.

Graph Representation Learning with Node Content. In some types of graphs, like information networks, nodes are acompanied with rich information, such as node labels, attributes or even semantic descriptions. How to combine them with the network topology in graph representation learning arouses considerable research interests. Tu et al (2016) propose a semi-supervised graph embedding algorithm, MMDW, by leveraging labeling information of nodes. MMDW is also based on the DeepWalk-derived matrix factorization. MMDW adopts support vector machines (SVM) (Hearst et al, 1998) and incorporates the label information to find an optimal classifying boundary. Yang et al (2015b) propose TADW that takes the rich information (e.g., text) associated with nodes into account when they learn the low dimensional representations of nodes. Pan et al (2016) propose a coupled deep model that incorporates graph structures, node attributes and node labels into graph embedding. Although different methods adopt different strategies to integrate node content and network topology, they all assume that node content provides additional proximity information to constrain the representations of nodes.

Heterogeneous Graph Representation Learning. Different from graphs with node content, heterogeneous graphs consist of different types of nodes and links. How to unify the heterogeneous types of nodes and links in graph embedding is also an interesting and challenging problem. Jacob et al (2014) propose a heterogeneous social graph representation learning algorithm for classifying nodes. They learn the representations of all types of nodes in a common vector space, and perform the inference in this space. Chang et al (2015) propose a deep graph representation learning algorithm for heterogeneous graphs, whose nodes have various types(e.g., images and texts). The nonlinear embeddings of images and texts are learned by a CNN model and the fully connected layers, respectively. Huang and Mamoulis (2017) propose a meta path similarity preserving heterogeneous information graph representation learning algorithm. To model a particular relationship, a meta path (Sun et al, 2011) is a sequence of object types with edge types in between.

In the methods preserving side information, side information introduces additional proximity measures so that the relationships between nodes can be learned more comprehensively. Their difference is the way of integrating network structuress and side information. Many of them are naturally extensions from structure preserving network embedding methods.

2.3.3 Advanced Information Preserving Graph Representation Learning

Different from side information, the advanced information refers to the supervised or pseudo supervised information in a specific task. The advanced information preserving network embedding usually consists of two parts. One is to preserve the network structure so as to learn the representations of nodes. The other is to establish the connection between the representations of nodes and the target task. The combination of advanced information and network embedding techniques enables representation learning for networks.

Information Diffusion. Information diffusion (Guille et al, 2013) is an ubiquitous phenomenon on the web, especially in social networks. Bourigault et al (2014) propose a graph representation learning algorithm for predicting information diffusion in social network. The goal of the proposed algorithm is to learn the representations of nodes in the latent space such that the diffusion kernel can best explain the cascades in the training set. The basic idea is to map the observed information diffusion process into a heat diffusion process modeled by a diffusion kernel in the continuous space. The kernel describes that the closer a node in the latent space is from the source node, the sooner it is infected by information from the source node. The cascade prediction problem here is defined as predicting the increment of cascade size after a given time interval (Li et al, 2017a). Li et al (2017a) argue that the previous work on cascade prediction all depends on the bag of hand-crafting features to represent the cascade and graph structures. Instead, they present an end-to-end deep learning model to solve this problem using the idea of graph embedding. The whole procedure is able to learn the representation of cascade graph in an end-to-end manner.

Anomaly Detection. Anomaly detection has been widely investigated in previous work (Akoglu et al, 2015). Anomaly detection in graphs aims to infer the structural inconsistencies, which means the anomalous nodes that connect to various diverse influential communities (Hu et al, 2016), (Burt, 2004). Hu et al (2016) propose a graph embedding based method for anomaly detection. They assume that the community memberships of two linked nodes should be similar. An anomaly node is one connecting to a set of different communities. Since the learned embedding of nodes captures the correlations between nodes and communities, based on the embedding, they propose a new measure to indicate the anomalousness level of a node. The larger the value of the measure, the higher the propensity for a node being an anomaly node.

Graph Alignment. The goal of graph alignment is to establish the correspondence between the nodes from two graphs, i.e., to predict the anchor links across two graphs. The same users who are shared by different social networks naturally form the anchor links, and these links bridge the different graphs. The anchor link prediction problem is, given a source graph, a target graph and a set of observed anchor links, to identify the hidden anchor links across the two graphs. Man et al (2016) propose a graph representation learning algorithm to solve this problem. The learned representations can preserve the graph structures and respect the observed anchor links.

Advanced information preserving graph embedding usually consists of two parts. One is to preserve the graph structures so as to learn the representations of nodes. The other is to establish the connection between the representations of nodes and the target task. The first one is similar to structure and property preserving network embedding, while the second one usually needs to consider the domain knowledge of a specific task. The domain knowledge encoded by the advanced information makes it possible to develop end-to-end solutions for network applications. Compared with the hand-crafted network features, such as numerous network centrality measures, the combination of advanced information and network embedding techniques enables representation learning for networks. Many network applications may be benefitted from this new paradigm.

2.4 Graph Neural Networks

Over the past decade, deep learning has become the "crown jewel" of artificial intelligence and machine learning, showing superior performance in acoustics, images and natural language processing, etc. Although it is well known that graphs are ubiquitous in the real world, it is very challenging to utilize deep learning methods to analyze graph data. This problem is non-trivial because of the following challenges: (1) Irregular structures of graphs. Unlike images, audio, and text, which have a clear grid structure, graphs have irregular structures, making it hard to generalize some of the basic mathematical operations to graphs. For example, defining convolution and pooling operations, which are the fundamental operations in convolutional neural networks (CNNs), for graph data is not straightforward. (2) Heterogeneity and diversity of graphs. A graph itself can be complicated, containing diverse types and properties. These diverse types, properties, and tasks require different model architectures to tackle specific problems. (3) Large-scale graphs. In the big-data era, real graphs can easily have millions or billions of nodes and edges. How to design scalable models, preferably models that have a linear time complexity with respect to the graph size, is a key problem. (4) Incorporating interdisciplinary knowledge. Graphs are often connected to other disciplines, such as biology, chemistry, and social sciences. This interdisciplinary nature provides both opportunities and challenges: domain knowledge can be leveraged to solve specific problems but integrating domain knowledge can complicate model designs.

Currently, graph neural networks have attracted considerable research attention over the past several years. The adopted architectures and training strategies vary greatly, ranging from supervised to unsupervised and from convolutional to recursive, including graph recurrent neural networks (Graph RNNs), graph convolutional networks (GCNs), graph autoencoders (GAEs), graph reinforcement learning (Graph RL), and graph adversarial methods. Specifically, Graroperty h RNNs capture recursive and sequential patterns of graphs by modeling states at either the node-level or the graph-level; GCNs define convolution and readout operations on irregular graph structures to capture common local and global structural patterns; GAEs assume low-rank graph structures and adopt unsupervised methods for node representation learning; Graph RL defines graph-based actions and rewards to obtain feedbacks on graph tasks while following constraints; Graph adversarial methods adopt adversarial training techniques to enhance the generalization ability of graphbased models and test their robustness by adversarial attacks.

There are many ongoing or future research directions which are also worthy of further study, including new models for unstudied graph structures, compositionality of existing models, dynamic graphs, interpretability and robustness, etc. On the whole, deep learning on graphs is a promising and fast-developing research field that both offers exciting opportunities and presents many challenges. Studying deep learning on graphs constitutes a critical building block in modeling relational data, and it is an important step towards a future with better machine learning and artificial intelligence techniques.

2.5 Summary

In this chapter, we introduce the motivation of graph representation learning. Then in Section 2, we discuss the traditional graph embedding methods and the modern graph embedding methods are introduced in Section 3. Basically, the structure and property preserving graph representation learning is the foundation. If one cannot preserve well the graph structures and retain the important graph properties in the representation space, serious information will be lost, which hurts the analytic tasks in sequel. Based on the structures and property preserving graph representation learning, one may apply the off-the-shelf machine learning methods. If some side information is available, it can be incorporated into graph representation learning. Furthermore, the domain knowledge of some certain applications as advanced information can be considered. As shown in Section 4, utilizing deep learning methods on graphs is a promising and fast-developing research field that both offers exciting opportunities and presents many challenges. Studying deep learning on graphs constitutes a critical building block in modeling relational data, and it is an important step towards a future with better machine learning and artificial intelligence techniques.

Chapter 3
Graph Neural Networks

Lingfei Wu, Peng Cui, Jian Pei, Liang Zhao and Le Song

Abstract Deep Learning has become one of the most dominant approaches in Artificial Intelligence research today. Although conventional deep learning techniques have achieved huge successes on Euclidean data such as images, or sequence data such as text, there are many applications that are naturally or best represented with a graph structure. This gap has driven a tide in research for deep learning on graphs, among them Graph Neural Networks (GNNs) are the most successful in coping with various learning tasks across a large number of application domains. In this chapter, we will systematically organize the existing research of GNNs along three axes: foundations, frontiers, and applications. We will introduce the fundamental aspects of GNNs ranging from the popular models and their expressive powers, to the scalability, interpretability and robustness of GNNs. Then, we will discuss various frontier research, ranging from graph classification and link prediction, to graph generation and transformation, graph matching and graph structure learning. Based on them, we further summarize the basic procedures which exploit full use of various GNNs for a large number of applications. Finally, we provide the organization of our book and summarize the roadmap of the various research topics of GNNs.

Lingfei Wu
JD.COM Silicon Valley Research Center, e-mail: lwu@email.wm.edu

Peng Cui
Department of Computer Science, Tsinghua University, e-mail: cuip@tsinghua.edu.cn

Jian Pei
Department of Computer Science, Simon Fraser University, e-mail: jpei@cs.sfu.ca

Liang Zhao
Department of Computer Science, Emory University, e-mail: liang.zhao@emory.edu

Le Song
Mohamed bin Zayed University of Artificial Intelligence, e-mail: dasongle@gmail.com

© The Author(s), under exclusive license to Springer Nature Singapore Pte Ltd. 2022   27
L. Wu et al. (eds.), *Graph Neural Networks: Foundations, Frontiers, and Applications*,
https://doi.org/10.1007/978-981-16-6054-2_3

3.1 Graph Neural Networks: An Introduction

Deep Learning has become one of the most dominant approaches in Artificial Intelligence research today. Conventional deep learning techniques, such as recurrent neural networks (Schuster and Paliwal, 1997) and convolutional neural networks (Krizhevsky et al, 2012) have achieved huge successes on Euclidean data such as images, or sequence data such as text and signals. However, in a rich variety of scientific fields, many important real-world objects and problems can be naturally or best expressed along with a complex structure, e.g., graph or manifold structure, such as social networks, recommendation systems, drug discovery and program analysis. On the one hand, these graph-structured data can encode complicated pairwise relationships for learning more informative representations; On the other hand, the structural and semantic information in original data (images or sequential texts) can be exploited to incorporate domain-specific knowledge for capturing more fine-grained relationships among the data.

In recent years, deep learning on graphs has experienced a burgeoning interest from the research community (Cui et al, 2018; Wu et al, 2019e; Zhang et al, 2020e). Among them, Graph Neural Networks (GNNs) is the most successful learning framework in coping with various tasks across a large number of application domains. Newly proposed neural network architectures on graph-structured data (Kipf and Welling, 2017a; Petar et al, 2018; Hamilton et al, 2017b) have achieved remarkable performance in some well-known domains such as social networks and bioinformatics. They have also infiltrated other fields of scientific research, including recommendation systems (Wang et al, 2019j), computer vision (Yang et al, 2019g), natural language processing (Chen et al, 2020o), program analysis (Allamanis et al, 2018b), software mining (LeClair et al, 2020), drug discovery (Ma et al, 2018), anomaly detection (Markovitz et al, 2020), and urban intelligence (Yu et al, 2018a).

Despite these successes that existing research has achieved, GNNs still face many challenges when they are used to model highly-structured data that is time-evolving, multi-relational, and multi-modal. It is also very difficult to model mapping between graphs and other highly structured data, such as sequences, trees, and graphs. One challenge with graph-structured data is that it does not show as much spatial locality and structure as image or text data does. Thus, graph-structured data is not naturally suitable for highly regularized neural structures such as convolutional and recurrent neural networks.

More importantly, new application domains for GNNs that emerge from real-world problems introduce significantly challenges for GNNs. Graphs provide a powerful abstraction that can be used to encode arbitrary data types such as multidimensional data. For example, similarity graphs, kernel matrices, and collaborative filtering matrices can also be viewed as special cases of graph structures. Therefore, a successful modeling process of graphs is likely to subsume many applications that are often used in conjunction with specialized and hand-crafted methods.

In this chapter, we will systematically organize the existing research of GNNs along three axes: foundations of GNNs, frontiers of GNNs, and GNN based applications. First of all, we will introduce the fundamental aspects of GNNs ranging from popular GNN methods and their expressive powers, to the scalability, interpretability, and robustness of GNNs. Next, we will discuss various frontier research which are built on GNNs, including graph classification, link prediction, graph generation and transformation, graph matching, graph structure learning, dynamic GNNs, heterogeneous GNNs, AutoML of GNNs and self-supervised GNNs. Based on them, we further summarize the basic procedures which exploit full use of various GNNs for a large number of applications. Finally, we provide the organization of our GNN book and summarize the roadmap of the various research topics of GNNs.

3.2 Graph Neural Networks: Overview

In this section, we summarize the development of graph neural networks along three important dimensions: (1) Foundations of GNNs; (2) Frontiers of GNNs; (3) GNN-based applications. We will first discuss the important research areas under the first two dimensions for GNNs and briefly illustrate the current progress and challenges for each research sub-domain. Then we will provide a general summarization on how to exploit the power of GNNs for a rich variety of applications.

3.2.1 Graph Neural Networks: Foundations

Conceptually, we can categorize the fundamental learning tasks of GNNs into five different directions: i) Graph Neural Networks Methods; ii) Theoretical understanding of Graph Neural Networks; iii) Scalability of Graph Neural Networks; iv) Interpretability of Graph Neural Networks; and v) Adversarial robustness of Graph Neural Networks. We will discuss these fundamental aspects of GNNs one by one in this subsection.

Graph Neural Network Methods. Graph Neural Networks are specifically designed neural architectures operated on graph-structure data. The goal of GNNs is to iteratively update the node representations by aggregating the representations of node neighbors and their own representation in the previous iteration. There are a variety of graph neural networks proposed in the literature (Kipf and Welling, 2017a; Petar et al, 2018; Hamilton et al, 2017b; Gilmer et al, 2017; Xu et al, 2019d; Velickovic et al, 2019; Kipf and Welling, 2016), which can be further categorized into supervised GNNs and unsupervised GNNs. Once the node representations are learnt, a fundamental task on graphs is node classification that tries to classify the nodes into a few predefined classes. Despite the huge successes that various GNNs have achieved, a severe issue on training deep graph neural networks has been observed to yield inferior results, namely, over-smoothing problem (Li et al, 2018b), where all the nodes have similar representations. Many recent works have been proposed with different remedies to overcome this over-smoothing issue.

Theoretical understanding of Graph Neural Networks. Rapid algorithmic developments of GNNs have aroused a significant amount of interests in theoretical analysis on the expressive power of GNNs. In particular, much efforts have been made in order to characterize the expressive power of GNNs when compared with the traditional graph algorithms (e.g. graph kernel-based methods) and how to build more powerful GNNs so as to overcome several limitations in GNNs. Specifically, Xu et al (2019d) showed that current GNN methods are able to achieve the expressive power of the 1-dimensional Weisfeiler-Lehman test (Weisfeiler and Leman., 1968), a widely used method in traditional graph kernel community (Shervashidze et al, 2011b). Much recent research has further proposed a series of design strategies in order to further reach beyond the expressive power of the Weisfeiler-Lehman test by including attaching random attributes, distance attributes, and utilizing higher-order structures.

Scalability of Graph Neural Networks. The increasing popularity of GNNs have attracted many attempts to apply various GNN methods on real-world applications, where the graph sizes are often about having one hundred million nodes and one billion edges. Unfortunately, most of the GNN methods cannot directly be applied on these large-scale graph-structured data due to large memory requirements (Hu et al, 2020b). Specifically, this is because the majority of GNNs are required to store the whole adjacent matrices and the intermediate feature matrices in the memory, rendering the significant challenges for both computer memory consumption and computational costs. In order to address these issues, many recent works have been proposed with various sampling strategies such as node-wise sampling (Hamilton et al, 2017b; Chen et al, 2018d), layer-wise sampling (Chen and Bansal, 2018; Huang et al, 2018), and graph-wise sampling (Chiang et al, 2019; Zeng et al, 2020a).

Interpretability of Graph Neural Networks. Explainable artificial intelligence are becoming increasingly popular in providing interpretable results on machine learning process, especially due to the black-box issue of deep learning techniques. As a result, there is a surge of interests in improving the interpretability of GNNs. Generally speaking, explanation results on GNNs could be important nodes, important edges, or important features of nodes or edges. Technically, white-box approximation based methods (Baldassarre and Azizpour, 2019; Sanchez-Lengeling et al, 2020) utilize the information inside the model inlucidng gradients, intermediate features, and model parameters to provide the explanation. In contrast, the black-box approximation based methods (Huang et al, 2020c; Zhang et al, 2020a; Vu and Thai, 2020) abandon the utilization of internal information of complex models but instead leverage the intrinsically interpretable simple models (e.g. linear regression and decision trees) to fit the complex models. However, most of the existing works are time-consuming, which rendering the difficulty in coping with large-scale graph. To this end, many recent efforts have been made in order to develop more efficient approaches without compromising the explanation accuracy.

Adversarial robustness of Graph Neural Networks. Trustworthy machine learning has recently attracted a significant amount of attention since the existing studies have shown that deep learning models could be deliberately fooled, evaded, misled, and stolen (Goodfellow et al, 2015). Consequently, a line of research has extensively studied the robustness of models in domains like computer vision and natural language processing, which has also influenced similar research on the robustness of GNNs. Technically, the standard approach (via adversarial examples) for studying the robustness of GNNs is to construct a small change of the input graph data and then to observe if it leads to a large change of the prediction results (i.e. node classification accuracy). There are a growing number of research works toward either adversarial attacks (Dai et al, 2018a; Wang and Gong, 2019; Wu et al, 2019b; Zügner et al, 2018; Zügner et al, 2020) or adversarial training (Xu et al, 2019c; Feng et al, 2019b; Chen et al, 2020i; Jin and Zhang, 2019). Many recent efforts have been made to provide both theoretical guarantees and new algorithmic developments in adversarial training and certified robustness.

3.2.2 Graph Neural Networks: Frontiers

Built on these aforementioned fundamental techniques of GNNs, there are various fast-growing recent research developments in coping with a variety of graph-related research problems. In this section, we will comprehensively introduce these research frontiers that are either long-standing graph learning problems with new GNN solutions or recently emerging learning problems with GNNs.

Graph Neural Networks: Graph Classification and Link Prediction. Since each layer in GNN models only produce the node-level representations, graph pooling layers are needed to further compute graph-level representation based on node-level representations. The graph-level representation, which summarizes the key characteristics of input graph-structure, is the critical component for the graph classification. Depending on the learning techniques of graph pooling layers, these methods can be generally categorized into four groups: simple flat-pooling (Duvenaud et al, 2015a; Mesquita et al, 2020), attention-based pooling (Lee et al, 2019d; Huang et al, 2019d), cluster-based pooling (Ying et al, 2018c), and other type of pooling (Zhang et al, 2018f; Bianchi et al, 2020; Morris et al, 2020b). Beside graph classification, another long-standing graph learning problem is link prediction task, which aims to predict missing or future links between any pair of nodes. Since GNNs can jointly learn from both graph structure and side information (e.g. node and edge features), it has shown great advantages over other conventional graph learning methods for link prediction. Regarding the learning types of link prediction, node-based methods (Kipf and Welling, 2016) and subgraph-based methods (Zhang and Chen, 2018a, 2020) are two popular groups of GNN based methods.

Graph Neural Networks: Graph Generation and Graph Transformation. Graph generation problem that builds probabilistic models over graphs is a classical research problem that lies at the intersection between the probability theory and the graph theory. Recent years have seen an increasing amount of interest in developing deep graph generative models that are built on modern deep learning on graphs techniques like GNNs. These deep models have proven to be a more successful approach in capturing the complex dependencies within the graph data and generating more realistic graphs. Encouraged by the great successes of Variational AutoEncoder (VAE) (Kingma and Welling, 2013) and Generative Adversarial Networks (Goodfellow et al, 2014a) (Goodfellow et al, 2014b), there are three representative GNN based learning paradigms for graph generation including GraphVAE approaches (Jin et al, 2018b; Simonovsky and Komodakis, 2018; Grover et al, 2019), GraphGAN approaches (De Cao and Kipf, 2018; You et al, 2018a) and Deep Autoregressive methods (Li et al, 2018d; You et al, 2018b; Liao et al, 2019a). Graph transformation problem can be formulated as a conditional graph generation problem, where its goal is to learn a translation mapping between the input source graph and the output target graph (Guo et al, 2018b). Such learning problem often arises in other domains such as machine translation problem in Natural Language Processing domain and image style transfer in computer Vision domain. Depending on what graph information is transformed, this problem can be generally grouped into four categories including node-level transformation (Battaglia et al, 2016; Yu et al, 2018a; Li et al, 2018e), edge-level transformation (Guo et al, 2018b; Zhu et al, 2017; Do et al, 2019), node-edge co-transformation (Maziarka et al, 2020a; Kaluza et al, 2018; Guo et al, 2019c), and graph-involved transformation (Bastings et al, 2017; Xu et al, 2018c; Li et al, 2020f).

Graph Neural Networks: Graph Matching and Graph Structure Learning. The problem of graph matching is to find the correspondence between two input graphs, which is an extensively studied problem in a variety of research fields. Conventionally, the graph matching problem is known to be NP-hard (Loiola et al, 2007), rendering this problem computationally infeasible for exact and optimum solutions for real-world large-scale problems. Due to the expressive power of GNNs, there is an increasing attention on developing various graph matching methods based on GNNs in order to improve the matching accuracy and efficiency (Zanfir and Sminchisescu, 2018; Rolínek et al, 2020; Li et al, 2019h; Ling et al, 2020). Graph matching problem aims to measure the similarity between two graph structures without changing them. In contrast, graph structure learning aims to produce an optimized graph structure by jointly learning implicit graph structure and graph node representation (Chen et al, 2020m; Franceschi et al, 2019; Velickovic et al, 2020). The learnt graph structure often can be treated as a shift compared to the intrinsic graph which is often noisy or incomplete. Graph structure learning can also be used when the initial graph is not provided while the data matrix shows correlation among data points.

Dynamic Graph Neural Networks and Heterogeneous Graph Neural Networks. In real-world applications, the graph nodes (entities) and the graph edges (relations) are often evolving over time, which naturally gives rise to dynamic graphs. Unfortunately, various GNNs cannot be directly applied to the dynamic graphs, where modeling the evolution of the graph is critical in making accurate predictions. A simple yet often effective approach is converting dynamic graphs into static graphs, leading to potential loss of information. Regarding the type of dynamic graphs, there are two major categories of GNN-based methods, including GNNs for discrete-time dynamic graphs (Seo et al, 2018; Manessi et al, 2020) and GNNs for continue-time dynamic graphs (Kazemi et al, 2019; Xu et al, 2020a). Independently, another popular graph type in real applications is heterogeneous graphs that consist of different types of graph nodes and edges. To fully exploit this information in heterogeneous graphs, different GNNs for homogeneous graphs are not applicable. As a result, a new line of research has been devoted to developing various heterogeneous graph neural networks including message passing based methods (Wang et al, 2019l; Fu et al, 2020; Hong et al, 2020b), encoder-decoder based methods (Tu et al, 2018; Zhang et al, 2019b), and adversarial based methods (Wang et al, 2018a; Hu et al, 2018a).

Graph Neural Networks: AutoML and Self-supervised Learning. Automated machine learning (AutoML) has recently drawn a significant amount of attention in both research and industrial communities, the goal of which is coping with the huge challenge of time-consuming manual tuning process, especially for complicated deep learning models. This wave of the research in AutoML also influences the research efforts in automatically identifying an optimized GNN model architecture and training hyperparameters. Most of the existing research focuses on either architecture search space (Gao et al, 2020b; Zhou et al, 2019a) or training hyperparameter search space (You et al, 2020a; Shi et al, 2020). Another important research direction of GNNs is to address the limitation of most of deep learning models that requires large amount of annotated data. As a result, self-supervised learning has been proposed which aims to design and leverage domain-specific pretext tasks on unlabeled data to pretrain a GNN model. In order to study the power of serf-supervised leanring in GNNs, there are quite a few works that systemmatically design and compare different self-supervised pretext tasks in GNNs (Hu et al, 2020c; Jin et al, 2020d; You et al, 2020c).

3.2.3 Graph Neural Networks: Applications

Due to the power of GNNs to model various data with complex structures, GNNs have been widely applied into many applications and domains, such as modern recommender systems, computer vision (CV), natural language processing (NLP), program analysis, software mining, bioinformatics, anomaly detection, and urban intelligence. Though GNNs are utilized to solve different tasks for different applications, they all consist of two important steps, namely graph construction and graph representation learning. Graph construction aims to first transform or represent the input data as graph-structured data. Based on the graphs, graph representation learning utilizes GNNs to learn the node or graph embeddings for the downstream tasks. In the following, we briefly introduce the techniques of these two steps regarding different applications.

3.2.3.1 Graph Construction

Graph construction is important in capturing the dependency among the objects in the input data. Given the various formats of input data, different applications have different graph construction techniques, while some tasks need to pre-define the semantic meaning of nodes and edges to fully express the structural information of the input data.

Input Data with Explicit Graph Structures. Some applications naturally have the structure inside the data without pre-defined nodes and the edges/relationships among them. For example, the user-item interactions in a recommender systems naturally form a graph where user-item preference is regarded as the edges between the nodes of user and item. In the task of drug design, a molecule is also naturally represented as a graph, where each node denotes an atom and an edge denotes a bond that connects two atoms. In the task of protein function prediction and interaction, the graph can also easily fit into a protein, where each amino-acid refers to a node and each edge refers to the interaction among amino-acids.

Some graphs are constructed with the node and edge attributes. For example, in dealing with the transportation in the urban intelligence, the traffic networks can be formalized as an undirected graph to predict the traffic state. Specifically, the nodes are the traffic sensing locations, e.g., sensor stations, road segments, and the edges are the intersections or road segments connecting those traffic sensing locations. Some urban traffic network can be modeled as a directed graph with attributes to predict the traffic speed, where the nodes are the road segments, and the edges are the intersections. Road segment width, length, and direction are the attributes of the nodes, and the type of intersection, and whether there are traffic lights, toll gates are the attributes of edges.

Input Data with Implicit Graph Structures. For many tasks that do not naturally involve a structured data, graph construction becomes very challenging. It is important to choose the best representation so that the nodes and edges can capture all the important things. For example, in computer vision (CV) tasks, there are three kinds of graph construction. The first is to split the image or the frame of the video into regular grids, and each grid serves as a vertex of the visual graph. The second way is to first get the preprocessed structures which can be directly borrowed for vertex representation, such as the formulation of scene graphs. The last one is about utilizing semantic information to represent visual vertexes, such as assigning pixels with similar features to the same vertex. The edges in the visual images can capture two kinds of information. One is spatial information. For example, for static methods, generating scene graphs (Xu et al, 2017a) and human skeletons (Jain et al, 2016a) is natural to choose edges between nodes in the visual graph to represent their location connection. Another is temporal information. For example, to represent the video, the model not only builds spatial relations in a frame but also captures temporal connections among adjacent frames.

In the natural language processing (NLP) tasks, the graph construction from the text data can be categorized into five categories: text graphs, syntactic graphs, semantic graphs, knowledge graphs, and hybrid graphs. Text graphs normally regard words, sentences, paragraphs, or documents as nodes and establish edges by word co-occurrence, location, or text similarities. Syntactic graphs (or trees) emphasize the syntactical dependencies between words in a sentence, such as dependency graph and constituency graph. Knowledge graphs (KGs) are graphs of data intended to accumulate and convey knowledge of the real world. Hybrid graphs contain multiple types of nodes and edges to integrate heterogeneous information. In the task of program analysis, the formulation over graph representations of programs includes syntax trees, control flow, data flow, program dependence, and call graphs, each providing different views of a program. At a high level, programs can be thought as a set of heterogeneous entities that are related through various kinds of relations. This view directly maps a program to a heterogeneous directed graph, with each entity being represented as a node and each relationship of type represented as an edge.

3.2.3.2 Graph Representation Learning

After getting the graph expression of the input data, the next step is applying GNNs for learning the graph representations. Some works directly utilize the typical GNNs, such as GCN (Kipf and Welling, 2017a), GAT (Petar et al, 2018), GGNN (Li et al, 2016a) and GraphSage (Hamilton et al, 2017b), which can be generalized to different application tasks. While some special tasks needs an additional design on the GNN architecture to better handle the specific problem. For example, in the task of recommender systems, PinSage (Ying et al, 2018a) is proposed which takes the top-k counted nodes of a node as its receptive field and utilizes weighted aggregation for aggregation. PinSage can be scalable to the web-scale recommender systems with millions of users and items. KGCN (Wang et al, 2019d) aims to enhance the item representation by performing aggregations among its corresponding entity neighborhood in a knowledge graph. KGAT (Wang et al, 2019j) shares a generally similar idea with KGCN except for incorporating an auxiliary loss for knowledge graph reconstruction. For instance, in the NLP task of KB-alignment, Xu et al (2019e) formulated it as a graph matching problem, and proposed a graph attention-based approach. It first matches all entities in two KGs, and then jointly models the local matching information to derive a graph-level matching vector. The detailed GNN techniques for each application can be found in the following chapters of this book.

3.2.4 Graph Neural Networks: Organization

The high-level organization of the book is demonstrated in Figure 1.3. The book is organized into four parts to best accommodate a variety of readers. Part I introduces basic concepts; Part II discusses the most established methods; Part III presents the most typical frontiers, and Part IV describes advances of methods and applications that tend to be important and promising for future research. Next, we briefly elaborate on each chapter.

- *Part I: Introduction.* These chapters provide the general introduction from the representation learning for different data types, to the graph representation learning. In addition, it introduces the basic ideas and typical variants of graph neural networks for the graph representation learning.
- *Part II: Foundations.* These chapters describe the foundations of the graph neural networks by introducing the properties of graph neural networks as well as several fundamental problems in this line. Specifically, this part introduces the fundamental problems in graphs: node classification, the expressive power of graph neural networks, the interpretability and scalability issues of graph neural network, and the adversarial robustness of the graph neural networks.
- *Part III: Frontiers.* In these chapters, some frontier or advanced problems in the domain of graph neural networks are proposed. Specifically, there are introductions about the techniques in graph classification, link prediction, graph generation, graph transformation, graph matching, graph structure learning. In addition, there are also introductions of several variants of GNNs for different types of graphs, such as GNNs for dynamic graphs, heterogeneous graphs. We also introduce the AutoML and self-supervised learning for GNNs.
- *Part IV: Broad and Emerging Applications.* These chapters introduce the broad and emerging applications with GNNs. Specifically, these GNNs-based applications covers modern recommender systems, tasks in computer vision and NLP, program analysis, software mining, biomedical knowledge graph mining for drug design, protein function prediction and interaction, anomaly detection, and urban intelligence.

3.3 Summary

Graph Neural Networks (GNNs) have been emerging rapidly to deal with the graph-structured data, which cannot be directly modeled by the conventional deep learning techniques that are designed for Euclidean data such as images and text. A wide range of applications can be naturally or best represented with graph structure and have been successfully handled by various graph neural networks.

In this chapter, we have systematically introduced the development and overview of GNNs, including the introduction of its foundations, frontiers, and applications. Specifically, we provide the fundamental aspects of GNNs ranging from the existing typical GNN methods and their expressive powers, to the scalability, interpretability and robustness of GNNs. These aspects motivate the research on better understanding and utilization of GNNs. Built on GNNs, recent research developments have seen a surge of interests in coping with graph-related research problems, which we called frontiers of GNNs. We have discussed various frontier research built on GNNs, ranging from graph classification and link prediction, to graph generation,

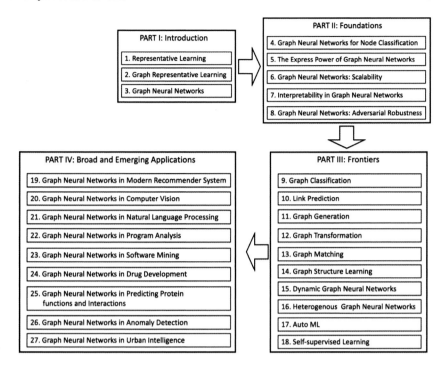

Fig. 3.1: The high-level organization of the book

transformation, matching and graph structure learning. Due to the power of GNNs to model various data with complex structures, GNNs have been widely applied into many applications and domains, such as modern recommender systems, computer vision, natural language processing, program analysis, software mining, bioinformatics, anomaly detection, and urban intelligence. Most of these tasks consist of two important steps, namely graph construction and graph representation learning. Thus, we provide the introduction of the techniques of these two steps regarding different applications. The introduction part will end here and thus a summary of the organization of this book has been provided at the end of this chapter.

Part II
Foundations of Graph Neural Networks

Chapter 4
Graph Neural Networks for Node Classification

Jian Tang and Renjie Liao

Abstract Graph Neural Networks are neural architectures specifically designed for graph-structured data, which have been receiving increasing attention recently and applied to different domains and applications. In this chapter, we focus on a fundamental task on graphs: node classification. We will give a detailed definition of node classification and also introduce some classical approaches such as label propagation. Afterwards, we will introduce a few representative architectures of graph neural networks for node classification. We will further point out the main difficulty—the oversmoothing problem—of training deep graph neural networks and present some latest advancement along this direction such as continuous graph neural networks.

4.1 Background and Problem Definition

Graph-structured data (e.g., social networks, the World Wide Web, and protein-protein interaction networks) are ubiquitous in real-world, covering a variety of applications. A fundamental task on graphs is node classification, which tries to classify the nodes into a few predefined categories. For example, in social networks, we want to predict the political bias of each user; in protein-protein interaction networks, we are interested in predicting the function role of each protein; in the World Wide Web, we may have to classify web pages into different semantic categories. To make effective prediction, a critical problem is to have very effective node representations, which largely determine the performance of node classification.

Graph neural networks are neural network architectures specifically designed for learning representations of graph-structured data including learning node represen-

Jian Tang
Mila-Quebec AI Institute, HEC Montreal, e-mail: `jian.tang@hec.ca`

Renjie Liao
University of Toronto, e-mail: `rjliao@cs.toronto.edu`

tations of big graphs (e.g., social networks and the World Wide Web) and learning representations of entire graphs (e.g., molecular graphs). In this chapter, we will focus on learning node representations for large-scale graphs and will introduce learning the whole-graph representations in other chapters. A variety of graph neural networks have been proposed (Kipf and Welling, 2017b; Veličković et al, 2018; Gilmer et al, 2017; Xhonneux et al, 2020; Liao et al, 2019b; Kipf and Welling, 2016; Veličković et al, 2019). In this chapter, we will comprehensively revisit existing graph neural networks for node classification including supervised approaches (Sec. 4.2), unsupervised approaches (Sec. 4.3), and a common problem of graph neural networks for node classification—over-smoothing (Sec. 4.4).

Problem Definition. Let us first formally define the problem of learning node representations for node classification with graph neural networks. Let $\mathscr{G} = (\mathscr{V}, \mathscr{E})$ denotes a graph, where $\mathscr{V}$ is the set of nodes and $\mathscr{E}$ is the set of edges. $A \in R^{N \times N}$ represents the adjacency matrix, where N is the total number of nodes, and $X \in R^{N \times C}$ represents the node attribute matrix, where C is the number of features for each node. The goal of graph neural networks is to learn effective node representations (denoted as $H \in R^{N \times F}$, F is the dimension of node representations) by combining the graph structure information and the node attributes, which are further used for node classification.

Table 4.1: Notations used throughout this chapter.

Concept	Notation
Graph	$\mathscr{G} = (\mathscr{V}, \mathscr{E})$
Adjacency matrix	$A \in \mathbb{R}^{N \times N}$
Node attributes	$X \in \mathbb{R}^{N \times C}$
Total number of GNN layers	K
Node representations at the k-th layer	$H^k \in \mathbb{R}^{N \times F}, k \in \{1, 2, \cdots, K\}$

4.2 Supervised Graph Neural Networks

In this section, we revisit several representative methods of graph neural networks for node classification. We will focus on the supervised methods and introduce the unsupervised methods in the next section. We will start by introducing a general framework of graph neural networks and then introduce different variants under this framework.

4.2.1 General Framework of Graph Neural Networks

The essential idea of graph neural networks is to iteratively update the node representations by combining the representations of their neighbors and their own representations. In this section, we introduce a general framework of graph neural networks in (Xu et al, 2019d). Starting from the initial node representation $H^0 = X$, in each layer we have two important functions:

- **AGGREGATE**, which tries to aggregate the information from the neighbors of each node;
- **COMBINE**, which tries to update the node representations by combining the aggregated information from neighbors with the current node representations.

Mathematically, we can define the general framework of graph neural networks as follows:

Initialization: $H^0 = X$
For $k = 1, 2, \cdots, K$,

$$a_v^k = \textbf{AGGREGATE}^k \{H_u^{k-1} : u \in N(v)\} \tag{4.1}$$
$$H_v^k = \textbf{COMBINE}^k \{H_v^{k-1}, a_v^k\}, \tag{4.2}$$

where $N(v)$ is the set of neighbors for the v-th node. The node representations H^K in the last layer can be treated as the final node representations.

Once we have the node representations, they can be used for downstream tasks. Take the node classification as an example, the label of node v (denoted as $\hat{y}_v$) can be predicted through a Softmax function, i.e.,

$$\hat{y}_v = \text{Softmax}(W H_v^\top), \tag{4.3}$$

where $W \in \mathbb{R}^{|\mathscr{L}| \times F}$, $|\mathscr{L}|$ is the number of labels in the output space.

Given a set of labeled nodes, the whole model can be trained by minimizing the following loss function:

$$O = \frac{1}{n_l} \sum_{i=1}^{n_l} \text{loss}(\hat{y}_i, y_i), \tag{4.4}$$

where y_i is the ground truth label of node i, n_l is the number of labeled nodes, $loss(\cdot, \cdot)$ is a loss function such as cross-entropy loss function. The whole neural networks can be optimized by minimizing the objective function O with backpropagation.

Above we present a general framework of graph neural networks. Next, we will introduce a few most representative instantiations or variants of graph neural networks in the literature.

4.2.2 Graph Convolutional Networks

We will start from the graph convolutional networks (GCN) (Kipf and Welling, 2017b), which is now the most popular graph neural network architecture due to its simplicity and effectiveness in a variety of tasks and applications. Specifically, the node representations in each layer is updated according to the following propagation rule:

$$H^{k+1} = \sigma(\tilde{D}^{-\frac{1}{2}}\tilde{A}\tilde{D}^{-\frac{1}{2}}H^k W^k). \tag{4.5}$$

$\tilde{A} = A + \mathbf{I}$ is the adjacency matrix of the given undirected graph $\mathscr{G}$ with self-connections, which allows to incorporate the node features itself when updating the node representations. $\mathbf{I} \in \mathbb{R}^{N \times N}$ is the identity matrix. $\tilde{D}$ is a diagonal matrix with $\tilde{D}_{ii} = \sum_j \tilde{A}_{ij}$. $\sigma(\cdot)$ is an activation function such as ReLU and Tanh. The ReLU active function is widely used, which is defined as $\text{ReLU}(x) = \max(0, x)$. $W^k \in \mathbb{R}^{F \times F'}$ (F, F' are the dimensions of node representations in the k-th, (k+1)-th layer respectively) is a laywise linear transformation matrix, which will be trained during the optimization.

We can further dissect equation equation 4.5 and understand the AGGREGATE and COMBINE function defined in GCN. For a node i, the node updating equation can be reformulated as below:

$$H_i^k = \sigma\left(\sum_{j \in \{N(i) \cup i\}} \frac{\tilde{A}_{ij}}{\sqrt{\tilde{D}_{ii}\tilde{D}_{jj}}} H_j^{k-1} W^k \right) \tag{4.6}$$

$$H_i^k = \sigma\left(\sum_{j \in N(i)} \frac{A_{ij}}{\sqrt{\tilde{D}_{ii}\tilde{D}_{jj}}} H_j^{k-1} W^k + \frac{1}{\tilde{D}_i} H_i^{k-1} W^k \right) \tag{4.7}$$

In the Equation equation 4.7, we can see that the AGGREGATE function is defined as the weighted average of the neighbor node representations. The weight of the neighbor j is determined by the weight of the edge between i and j (i.e. A_{ij} normalized by the degrees of the two nodes). The COMBINE function is defined as the summation of the aggregated messages and the node representation itself, in which the node representation is normalized by its own degree.

Connections with Spectral Graph Convolutions. Next, we discuss the connections between GCNs and traditional spectral filters defined on graphs (Defferrard et al, 2016). The spectral convolutions on graphs can be defined as a multiplication of a node-wise signal $\mathbf{x} \in \mathbb{R}^N$ with a convolutional filter $g_\theta = \text{diag}(\theta)$ ($\theta \in \mathbb{R}^N$ is the parameter of the filter) in the Fourier domain. Mathematically,

$$g_\theta \star \mathbf{x} = U g_\theta U^T \mathbf{x}. \tag{4.8}$$

U represents the matrix of the eigenvectors of the normalized graph Laplacian matrix $L = I_N - D^{-\frac{1}{2}}AD^{-\frac{1}{2}}$. $L = U\Lambda U^T$, Λ is a diagonal matrix of eigenvalues, and $U^T\mathbf{x}$ is the graph Fourier transform of the input signal $\mathbf{x}$. In practice, g_θ can be understood as a function of the eigenvalues of the normalized graph Laplacian matrix L (i.e. $g_\theta(\Lambda)$). In practice, directly calculating Eqn. equation 4.8 is very computationally expensive, which is quadratic to the number of nodes N. According to (Hammond et al, 2011), this problem can be circumvented by approximating the function $g_\theta(\Lambda)$ with a truncated expansion of Chebyshev polynomials $T_k(x)$ up to K^{th} order:

$$g_{\theta'}(\Lambda) = \sum_{k=0}^{K} \theta'_k T_k(\tilde{\Lambda}), \tag{4.9}$$

where $\tilde{\Lambda} = \frac{2}{\lambda_{max}}\Lambda - \mathbf{I}$, and λ_{max} is the largest eigenvalue of L. $\theta' \in \mathbb{R}^K$ is the vector of Chebyshev coefficients. $T_k(x)$ are Chebyshev polynomials which are recursively defined as $T_k(x) = 2xT_{k-1}(x) - T_{k-2}(x)$, with $T_0(x) = 1$ and $T_1(x) = x$. By combining Eqn. equation 4.9 and Eqn. equation 4.8, the convolution of a signal x with a filter $g_{\theta'}$ can be reformulated as below:

$$g_{\theta'} \star \mathbf{x} = \sum_{k=0}^{K} \theta'_k T_k(\tilde{L})\mathbf{x}, \tag{4.10}$$

where $\tilde{L} = \frac{2}{\lambda_{max}}L - \mathbf{I}$. From this equation, we can see that each node only depends on the information within the Kth-order neighborhood. The overall complexity of evaluating Eqn. equation 4.10 is $\mathcal{O}(|\mathscr{E}|)$ (i.e. linear to the number of edges in the original graph $\mathscr{G}$), which is very efficient.

To define a neural network based on graph convolutions, one can stack multiple convolution layers defined according to Eqn. equation 4.10 with each layer followed by a nonlinear transformation. At each layer, instead of being limited to the explicit parametrization by the Chebyshev polynomials defined in Eqn. equation 4.10, the authors of GCNs proposed to limit the number of convolutions to $K = 1$ at each layer. By doing this, at each layer, it only defines a linear function over the graph Laplacian matrix L. However, by stacking multiple such layers, we are still capable of covering a rich class of convolution filter functions on graphs. Intuitively, such a model is capable of alleviating the problem of overfitting local neighborhood structures for graphs whose node degree distribution has a high variance such as social networks, the World Wide Web, and citation networks.

At each layer, we can further approximate $\lambda_{max} \approx 2$, which could be accommodated by the neural network parameters during training. Based on al these simplifications, we have

$$g_{\theta'} \star \mathbf{x} \approx \theta'_0\mathbf{x} + \theta'_1\mathbf{x}(L - I_N)\mathbf{x} = \theta'_0\mathbf{x} - \theta'_1 D^{-\frac{1}{2}}AD^{-\frac{1}{2}}, \tag{4.11}$$

where θ'_0 and θ'_1 are too free parameters, which could be shared over the entire graph. In practice, we can further reduce the number of parameters, which allows to reduce overfitting and meanwhile minimize the number of operations per layer. As a result, the following expression can be further obtained:

$$g_\theta \star \mathbf{x} \approx \theta(\mathbf{I} + D^{-\frac{1}{2}}AD^{-\frac{1}{2}})\mathbf{x}, \tag{4.12}$$

where $\theta = \theta_0' = -\theta_1'$. One potential issue is the matrix $I_N + D^{-\frac{1}{2}}AD^{-\frac{1}{2}}$, whose eigenvalues lie in the interval of $[0, 2]$. In a deep graph convolutional neural network, repeated application of the above function will likely lead to exploding or vanishing gradients, yielding numerical instabilites. As a result, we can further renormalize this matrix by converting $\mathbf{I} + D^{-\frac{1}{2}}AD^{-\frac{1}{2}}$ to $\tilde{D}^{-\frac{1}{2}}\tilde{A}\tilde{D}^{-\frac{1}{2}}$, where $\tilde{A} = A + \mathbf{I}$, and $\tilde{D}_{ii} = \sum_j \tilde{A}_{ij}$.

In the above, we only consider the case that there is only one feature channel and one filter. This can be easily generalized to an input signal with C channels $X \in \mathbb{R}^{N \times C}$ and F filters (or number of hidden units) as follows:

$$H = \tilde{D}^{-\frac{1}{2}}\tilde{A}\tilde{D}^{-\frac{1}{2}}XW, \tag{4.13}$$

where $W \in R^{C \times F}$ is a matrix of filter parameters. H is the convolved signal matrix.

4.2.3 Graph Attention Networks

In GCNs, for a target node i, the importance of a neighbor j is determined by the weight of their edge A_{ij} (normalized by their node degrees). However, in practice, the input graph may be noisy. The edge weights may not be able to reflect the true strength between two nodes. As a result, a more principled approach would be to automatically learn the importance of each neighbor. Graph Attention Networks (a.k.a. GAT(Veličković et al, 2018)) is built on this idea and try to learn the importance of each neighbor based on the **Attention** mechanism (Bahdanau et al, 2015; Vaswani et al, 2017). Attention mechanism has been wide used in a variety of tasks in natural language understanding (e.g. machine translation and question answering) and computer vision (e.g. visual question answering and image captioning). Next, we will introduce how attention is used in graph neural networks.

Graph Attention Layer. The graph attention layer defines how to transfer the hidden node representations at layer $k - 1$ (denoted as $H^{k-1} \in \mathbb{R}^{N \times F}$) to the new node representations $H^k \in \mathbb{R}^{N \times F'}$. In order to guarantee sufficient expressive power to transform the lower-level node representations to higher-level node representations, a shared linear transformation is applied to every node, denoted as $W \in \mathbb{R}^{F \times F'}$. Afterwards, self-attention is defined on the nodes, which measures the attention coefficients for any pair of nodes through a shared attentional mechanism $a : \mathbb{R}^{F'} \times \mathbb{R}^{F'} \to R$

$$e_{ij} = a(WH_i^{k-1}, WH_j^{k-1}). \tag{4.14}$$

e_{ij} indicates the relationship strength between node i and j. Note in this subsection we use H_i^{k-1} to represent a column-wise vector instead of a row-wise vector. For each node, we can theoretically allow it to attend to every other node on the graph, which however will ignore the graph structural information. A more reasonable solution would be only to attend to the neighbors for each node. In practice, the first-order neighbors are only used (including the node itself). And to make the coefficients comparable across different nodes, the attention coefficients are usually normalized with the softmax function:

$$\alpha_{ij} = \text{Softmax}_j(\{e_{ij}\}) = \frac{\exp(e_{ij})}{\sum_{l \in N(i)} \exp(e_{il})}. \tag{4.15}$$

We can see that for a node i, α_{ij} essentially defines a multinomial distribution over the neighbors, which can also be interpreted as the transition probability from node i to each of its neighbors.

In the work by Veličković et al (2018), the attention mechanism a is defined as a single-layer feedforward neural network including a linear transformation with the weight vector $W_2 \in \mathbb{R}^{1 \times 2F'}$) and a LeakyReLU nonlinear activation function (with negative input slope $\alpha = 0.2$). More specifically, we can calculate the attention coefficients with the following architecture:

$$\alpha_{ij} = \frac{\exp(\text{LeakyReLU}(W_2[WH_i^{k-1} \| WH_j^{k-1}]))}{\sum_{l \in N(i)} \exp(\text{LeakyReLU}(W_2[WH_i^{k-1} \| WH_l^{k-1}]))}, \tag{4.16}$$

where $\|$ represents the operation of concatenating two vectors. The new node representation is a linear combination of the neighboring node representations with the weights determined by the attention coefficients (with a potential nonlinear transformation), i.e.

$$H_i^k = \sigma \left(\sum_{j \in N(i)} \alpha_{ij} W H_j^{k-1} \right). \tag{4.17}$$

Multi-head Attention.

In practice, instead of only using one single attention mechanism, *multi-head attention* can be used, each of which determines a different similarity function over the nodes. For each attention head, we can independently obtain a new node representation according to Eqn. equation 4.17. The final node representation will be a concatenation of the node representations learned by different attention heads. Mathematically, we have

$$H_i^k = \Big\|_{t=1}^T \sigma \left(\sum_{j \in N(i)} \alpha_{ij}^t W^t H_j^{k-1} \right), \tag{4.18}$$

where T is the total number of attention heads, α_{ij}^t is the attention coefficient calculated from the t-th attention head, W^t is the linear transformation matrix of the t-th attention head.

One thing that mentioned in the paper by Veličković et al (2018) is that in the final layer, when trying to combine the node representations from different attention heads, instead of using the operation concatenation, other pooling techniques could be used, e.g. simply taking the average node representations from different attention heads.

$$H_i^k = \sigma \left(\frac{1}{T} \sum_{t=1}^{T} \sum_{j \in N(i)} \alpha_{ij}^t W^t H_j^{k-1} \right). \tag{4.19}$$

4.2.4 Neural Message Passing Networks

Another very popular graph neural network architecture is the Neural Message Passing Network (MPNN) (Gilmer et al, 2017), which is originally proposed for learning molecular graph representations. However, MPNN is actually very general, provides a general framework of graph neural networks, and could be used for the task of node classification as well. The essential idea of MPNN is formulating existing graph neural networks as a general framework of neural message passing among nodes. In MPNNs, there are two important functions including **Message** and **Updating** function:

$$m_i^k = \sum_{i \in N(j)} M_k(H_i^{k-1}, H_j^{k-1}, e_{ij}), \tag{4.20}$$

$$H_i^k = U_k(H_i^{k-1}, m_i^k). \tag{4.21}$$

$M_k(\cdot, \cdot, \cdot)$ defines the message between node i and j in the k-th layer, which depends on the two node representations and the information of their edge. U_k is the node updating function in the k-th layer which combines the aggregated messages from the neighbors and the node representation itself. We can see that the MPNN framework is very similar to the general framework we introduced in Section 4.2.1. The **AGGREGATE** function defined here is simply a summation of all the messages from the neighbors. The **COMBINE** function is the same as the node **Updating** function.

4.2.5 Continuous Graph Neural Networks

The above graph neural networks iteratively update the node representations with different kinds of graph convolutional layers. Essentially, these approaches model the discrete dynamics of node representations with GNNs. Xhonneux et al (2020) proposed the continuous graph neural networks (CGNNs), which generalizes existing graph neural networks with discrete dynamics to continuous settings, i.e., trying to model the continuous dynamics of node representations. The key idea is how to characterize the continuous dynamics of node representations, i.e. the derivatives of node representation w.r.t. time. The CGNN model is inspired by the diffusion-based models on graphs such as PageRank and epidemic models on social networks. The derivatives of the node representations are defined as a combination of the node representation itself, the representations of its neighbors, and the initial status of the nodes. Specifically, two different variants of node dynamics are introduced. The first model assumes that different dimensions of node presentations (a.k.a. feature channels) are independent; the second model is more flexible, which allows different feature channels to interact with each other. Next, we give a detailed introduction to each of the two models.

Note: in this part, instead of using the original adjacency matrix A, we use the following regularized matrix for characterizing the graph structure:

$$A := \frac{\alpha}{2}\left(I + D^{-\frac{1}{2}}AD^{-\frac{1}{2}}\right), \tag{4.22}$$

where $\alpha \in (0,1)$ is a hyperparameter. D is the degree matrix of the original adjacency matrix A. With the new regularized adjacency matrix A, the eigenvalues of A will lie in the interval $[0, \alpha]$, which will make A^k converges to 0 when we increase the power of k.

Model 1: Independent Feature Channels. As different nodes in a graph are interconnected, a natural solution to model the dynamic of each feature channel should be taking the graph structure into consideration, which allows the information to propagate across different nodes. We are motivated by existing diffusion-based methods on graphs such as PageRank (Page et al, 1999) and label propagation (Zhou et al, 2004), which defines the discrete propagation of node representations (or signals on nodes) with the following step-wise propagation equations:

$$H^{k+1} = AH^k + H^0, \tag{4.23}$$

where $H^0 = X$ or the output of an encoder on the input feature X. Intuitively, at each step, the new node representation is a linear combination of its neighboring node representations as well as the initial node features. Such a mechanism allows to model the information propagation on the graph without forgetting the initial node features. We can unroll Eqn. equation 4.23 and explicitly derive the node representations at the k-th step:

$$H^k = \left(\sum_{i=0}^{k}A^i\right)H^0 = (A - \mathbf{I})^{-1}(A^{k+1} - \mathbf{I})H^0. \tag{4.24}$$

As the above equation effectively models the discrete dynamics of node representations, the CGNN model further extended it to the continuous setting, which replaces the discrete time step k to a continuous variable $t \in \mathbb{R}_0^+$. Specifically, it has been shown that Eqn. equation 4.24 is a discretization of the following ordinary differential equation (ODE):

$$\frac{dH^t}{dt} = \log AH^t + X, \tag{4.25}$$

with the initial value $H^0 = (\log A)^{-1}(A - I)X$, where X is the initial node features or the output of an encoder applied to it. We do not provide the proof here. More details can be referred to the original paper (Xhonneux et al, 2020). In Eqn. equation 4.25, as $\log A$ is intractable to compute in practice, it is approximated with the first-order of the Taylor expansion, i.e. $\log A \approx A - I$. By integrating all these information, we have the following ODE equation:

$$\frac{dH^t}{dt} = (A - I)H^t + X, \tag{4.26}$$

with the initial value $H^0 = X$, which is the first variant of the CGNN model.

The CGNN model is actually very intuitive, which has a nice connection with traditional epidemic model, which aims at studying the dynamics of infection in a population. For the epidemic model, it usually assumes that the infection of people will be affected by three different factors including the infection from neighbors, the natural recovery, and the natural characteristics of people. If we treat H^t as the number of people infected at time t, then these three factors can be naturally modeled by the three terms in Eqn. equation 4.26: AH^t for the infection from neighbors, $-H^t$ for the natural recovery, and the last one X for the natural characteristics of people.

Model 2: Modeling the Interaction of Feature Channels. The above model assumes different node feature channels are independent with each other, which is a very strong assumption and limits the capacity of the model. Inspired by the success of a linear variant of graph neural networks (i.e., Simple GCN (Wu et al, 2019a)), a more powerful discrete node dynamic model is proposed, which allows different feature channels to interact with each other as,

$$H^{k+1} = AH^k W + H^0, \tag{4.27}$$

where $W \in \mathbb{R}^{F \times F}$ is a weight matrix used to model the interactions between different feature channels. Similarly, we can also extend the above discrete dynamics into continuous case, yielding the following equation:

$$\frac{dH^t}{dt} = (A - I)H^t + H^t(W - I) + X, \tag{4.28}$$

with the initial value being $H^0 = X$. This is the second variant of CGNN with trainable weights. Similar form of ODEs defined in Eqn. equation 4.28 has been studied in the literature of control theory, which is known as Sylvester differential equation (Locatelli and Sieniutycz, 2002). The two matrices $A - I$ and $W - I$ characterize the natural solution of the system while X is the information provided to the system to drive the system into the desired state.

Discussion. The proposed continuous graph neural networks (CGNN) has multiple nice properties: (1) Recent work has shown that if we increase the number of layers K in the discrete graph neural networks, the learned node representations tend to have the problem of over-smoothing (will introduce in detail later) and hence lose the power of expressiveness. On the contrary, the continuous graph neural networks are able to train very deep graph neural networks and are experimentally robust to arbitrarily chosen integration time; (2) For some of the tasks on graphs, it is critical to model the long-range dependency between nodes, which requires training deep GNNs. Existing discrete GNNs fail to train very deep GNNs due to the over-smoothing problem. The CGNNs are able to effectively model the long-range dependency between nodes thanks to the stability w.r.t. time. (3) The hyperparameter α is very important, which controls the rate of diffusion. Specifically, it controls the rate at which high-order powers of regularized matrix A vanishes. In the work proposed by (Xhonneux et al, 2020), the authors proposed to learn a different value of α for each node, which hence allows to choose the best diffusion rates for different nodes.

4.2.6 Multi-Scale Spectral Graph Convolutional Networks

Recall the one-layer graph convolution operator used in GCNs (Kipf and Welling, 2017b) $H = LHW$, where $L = D^{-\frac{1}{2}}\tilde{A}D^{-\frac{1}{2}}$. Here we drop the superscript of the layer index to avoid the clash with the notation of the matrix power. There are two main issues with this simple graph convolution formulation. First, one such graph convolutional layer would only propagate information from any node to its nearest neighbors, *i.e.*, neighboring nodes that are one-hop away. If one would like to propagate information to M-hop away neighbors, one has to either stack M graph convolutional layers or compute the graph convolution with M-th power of the graph Laplacian, *i.e.*, $H = \sigma(L^M HW)$. When M is large, the solution of stacking layers would make the whole GCN model very deep, thus causing problems in learning like the vanishing gradient. This is similar to what people experienced in training very deep feedforward neural networks. For the matrix power solution, naively computing the M-th power of the graph Laplacian is also very costly (*e.g.*, the time complexity is $O(N^{3(M-1)})$ for graphs with N nodes). Second, there are no learnable parameters in GCNs associated with the graph Laplacian L (corresponding to the connectivities/structures). The only learnable parameter W is a linear transform applied to every node simultaneously which is not aware of the structures. Note that we typically associate learnable weights on edges while applying the convolution applied to regular graphs like grids (*e.g.*, applying 2D convolution to images). This would greatly improve the expressiveness of the model. However, it is not clear that how one can add learnable parameters to the graph Laplacian L since its size varies from graph to graph.

Algorithm 1 : Lanczos Algorithm

1: **Input:** S, x, M, ε
2: **Initialization:** $\beta_0 = 0$, $\mathbf{q}_0 = 0$, and $\mathbf{q}_1 = \mathbf{x}/\|\mathbf{x}\|$
3: **For** $j = 1, 2, \ldots, K$:
4:      $\mathbf{z} = S\mathbf{q}_j$
5:      $\gamma_j = \mathbf{q}_j^\top \mathbf{z}$
6:      $\mathbf{z} = \mathbf{z} - \gamma_j \mathbf{q}_j - \beta_{j-1} \mathbf{q}_{j-1}$
7:      $\beta_j = \|\mathbf{z}\|_2$
8:      **If** $\beta_j < \varepsilon$, quit
9:      $\mathbf{q}_{j+1} = \mathbf{z}/\beta_j$
10:
11: $Q = [\mathbf{q}_1, \mathbf{q}_2, \cdots, \mathbf{q}_M]$
12: Construct T following Eq. (4.29)
13: Eigen decomposition $T = BRB^\top$
14: Return $V = QB$ and R. $= 0$

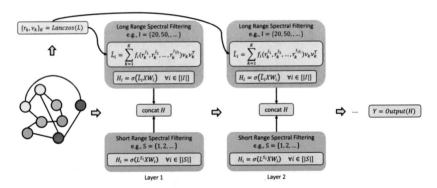

Fig. 4.1: The inference procedure of Lanczos Networks. The approximated top eigenvalues $\{r_k\}$ and eigenvectors $\{\mathbf{v}_k\}$ are computed by the Lanczos algorithm. Note that this step is only needed once per graph. The long range/scale (top blocks) graph convolutions are efficiently computed by the low-rank approximation of the graph Laplacian. One can control the ranges (*i.e.*, the exponent of eigenvalues) as hyperparameters. Learnable spectral filters are applied to the approximated top eigenvalues $\{r_k\}$. The short range/scale (bottom blocks) graph convolution is the same as GCNs. Adapted from Figure 1 of (Liao et al, 2019b).

To overcome these two problems, authors propose Lanczos Networks in (Liao et al, 2019b). Given the graph Laplacian matrix L^1 and node features X, one first

---

[1] Here we assume a symmetric graph Laplacian matrix. If it is non-symmetric (*e.g.*, for directed graphs), one can resort to the Arnoldi algorithm.

uses the M-step Lanczos algorithm (Lanczos, 1950) (listed in Alg. 1) to compute an orthogonal matrix Q and a symmetric tridiagonal matrix T, such that $Q^\top L Q = T$. We denote $Q = [\mathbf{q}_1, \cdots, \mathbf{q}_M]$ where column vector $\mathbf{q}_i$ is the i-th Lanczos vector. Note that M could be much smaller than the number of nodes N. T is illustrated as below,

$$
T = \begin{bmatrix} \gamma_1 & \beta_1 & & \\ \beta_1 & \ddots & \ddots & \\ & \ddots & \ddots & \beta_{M-1} \\ & & \beta_{M-1} & \gamma_M \end{bmatrix}. \tag{4.29}
$$

After obtaining the tridiagonal matrix T, we can compute the Ritz values and Ritz vectors which approximate the top eigenvalues and eigenvectors of L by diagonalizing the matrix T as $T = BRB^\top$, where the $K \times K$ diagonal matrix R contains the Ritz values and $B \in \mathbb{R}^{K \times K}$ is an orthogonal matrix. Here top means ranking the eigenvalues by their magnitudes in a descending order. This can be implemented via the general eigendecomposition or some fast decomposition methods specialized for tridiagonal matrices. Now we have a low rank approximation of the graph Laplacian matrix $L \approx VRV^\top$, where $V = QB$. Denoting the column vectors of V as $\{\mathbf{v}_1, \cdots, \mathbf{v}_M\}$, we can compute multi-scale graph convolution as

$$
H = \hat{L}HW
$$
$$
\hat{L} = \sum_{m=1}^{M} f_\theta(r_m^{I_1}, r_m^{I_2}, \cdots, r_m^{I_u}) v_m v_m^\top, \tag{4.30}
$$

where $\{I_1, \cdots, I_u\}$ is the set of scale/range parameters which determine how many hops (or how far) one would like to propagate the information over the graph. For example, one could easily set $\{I_1 = 50, I_2 = 100\}$ ($u = 2$ in this case) to consider the situations of propagating 50 and 100 steps respectively. Note that one only needs to compute the scalar power rather than the original matrix power. The overall complexity of the Lanczos algorithm in our context is $O(MN^2)$ which makes the whole algorithm much more efficient than naively computing the matrix power. Moreover, f_θ is a learnable spectral filter parameterized by θ and can be applied to graphs with varying sizes since we decouple the graph size and the input size of f_θ. f_θ directly acts on the graph Laplacian and greatly improves the expressiveness of the model.

Although Lanczos algorithm provides an efficient way to approximately compute arbitrary powers of the graph Laplacian, it is still a low-rank approximation which may lose certain information (*e.g.*, the high frequency one). To alleviate the problem, one can further do vanilla graph convolution with small scale parameters like $H = L^S HW$ where S could be small integers like 2 or 3. The resultant representation can be concatenated with the one obtained from the longer scale/range graph convolution in Eq. (4.30). Relying on the above design, one could add nonlinearities and stack multiple such layers to build a deep graph convolutional network (namely Lanczos Networks) just like GCNs. The overall inference procedure of Lanczos Networks is shown in Fig. 4.1. This method demonstrates strong empirical performances on a wide variety of tasks/benchmarks including molecular property prediction in quantum chemistry and document classification in citation networks. It just requires slight modifications to the implementation of the original GCNs. Nevertheless, if the input graph is extremely large (*e.g.*, some large social network), the Lanczos algorithm itself would be a computational bottleneck. How to improve this model in such a problem context would be an open question.

Here we only introduce a few representative architectures of graph neural networks for node classification. There are also many other well-known architectures including gated graph neural networks (Li et al, 2016b)—which is mainly designed for output sequences—and GraphSAGE (Hamilton et al, 2017b)—which is mainly designed for inductive setting of node classification.

4.3 Unsupervised Graph Neural Networks

In this section, we review a few representative GNN-based methods for unsupervised learning on graph-structured data, including variational graph auto-encoders (Kipf and Welling, 2016) and deep graph infomax (Veličković et al, 2019).

4.3.1 Variational Graph Auto-Encoders

Following variational auto-encoders (VAEs) (Kingma and Welling, 2014; Rezende et al, 2014) , variational graph auto-encoders (VGAEs) (Kipf and Welling, 2016) provide a framework for unsupervised learning on graph-structured data. In the following, we first review the model and then discuss its advantages and disadvantages.

4.3.1.1 Problem Setup

Suppose we are given an undirected graph $\mathscr{G} = (\mathscr{V}, \mathscr{E})$ with N nodes. Each node is associated with a node feature/attribute vector. We compactly denote all node features as a matrix $X \in \mathbb{R}^{N \times C}$. The adjacency matrix of the graph is A. We assume self-loops are added to the orignal graph $\mathscr{G}$ so that the diagonal entries of A are 1. This is a convention in graph convolutional networks (GCNs) (Kipf and Welling, 2017b) and makes the model consider a node's old representation while updating its new representation. We also assume each node is associated with a latent variable (the collection of all latent variables is again compactly denoted as a matrix $Z \in \mathbb{R}^{N \times F}$). We are interested in inferring the latent variables of nodes in the graph and decoding the edges.

4.3.1.2 Model

Similar to VAEs, the VGAE model consists of an encoder $q_\phi(Z|A,X)$, a decoder $p_\theta(A|Z)$, and a prior $p(Z)$.

Encoder The goal of the encoder is to learn a distribution of latent variables associated with each node conditioning on the node features X and the adjacency matrix A. We could instantiate $q_\phi(Z|A,X)$ as a graph neural network where the learnable parameters are ϕ. In particular, VGAE assumes an node-independent encoder as below,

$$q_\phi(Z|X,A) = \prod_{i=1}^{N} q_\phi(\mathbf{z}_i|X,A) \tag{4.31}$$

$$q_\phi(\mathbf{z}_i|X,A) = \mathcal{N}(\mathbf{z}_i|\mu_i, \mathrm{diag}(\sigma_i^2)) \tag{4.32}$$

$$\mu, \sigma = \mathrm{GCN}_\phi(X,A) \tag{4.33}$$

where $\mathbf{z}_i$, μ_i, and σ_i are the i-th rows of the matrices Z, μ, and σ respectively. Basically, we assume a multivariate Normal distribution with the diagonal covariance as the variational approximated distribution of the latent vector per node (*i.e.*, $\mathbf{z}_i$). The mean and diagonal covariance are predict by the encoder network, *i.e.*, a GCN as described in Section 4.2.2. For example, the original paper uses a two-layer GCN as follows,

$$\mu = \tilde{A}HW_\mu \tag{4.34}$$

$$\sigma = \tilde{A}HW_\sigma \tag{4.35}$$

$$H = \mathrm{ReLU}(\tilde{A}XW_0), \tag{4.36}$$

where $\tilde{A} = D^{-\frac{1}{2}}AD^{-\frac{1}{2}}$ is the symmetrically normalized adjacency matrix and D is the degree matrix. Learnable parameters are thus $\phi = [W_\mu, W_\sigma, W_0]$.

Decoder Given sampled latent variables, the decoder aims at predicting the connectivities among nodes. The original paper adopts a simple dot-product based predictor as below,

$$p(A|Z) = \prod_{i=1}^{N} \prod_{j=1}^{N} p(A_{ij}|\mathbf{z}_i, \mathbf{z}_j) \tag{4.37}$$

$$p(A_{ij}|\mathbf{z}_i, \mathbf{z}_j) = \sigma(\mathbf{z}_i^\top \mathbf{z}_j), \tag{4.38}$$

where A_{ij} denotes the (i,j)-th element and $\sigma(\cdot)$ is the logistic sigmoid function. This decoder again assumes conditional independence among all possible edges for tractability. Note that there are no learnable parameters associated with this decoder. The only way to improve the performance of the decoder is to learn good latent representations.

Prior The prior distributions over the latent variables are simply set to independent zero-mean Gaussians with unit variances,

$$p(Z) = \prod_{i=1}^{N} \mathcal{N}(\mathbf{z}_i | \mathbf{0}, \mathbf{I}). \tag{4.39}$$

This prior is fixed throughout the learning as what typical VAEs do.

Objective & Learning To learn the encoder and the decoder, one typically maximize the evidence lower bound (ELBO) as in VAEs,

$$\mathscr{L}_{\text{ELBO}} = \mathbb{E}_{q_\phi(Z|X,A)}[\log p(A|Z)] - \text{KL}(q_\phi(Z|X,A) \| p(Z)), \tag{4.40}$$

where $\text{KL}(q \| p)$ is the Kullback-Leibler divergence between distributions q and p. Note that we can not directly maximize the log likelihood since the introduction of latent variables Z induces a high-dimensional integral which is intractable. We instead maximize the ELBO in Eq. (4.40) which is a lower bound of the log likelihood. However, the first expectation term is again intractable. One often resorts to the Monte Carlo estimation by sampling a few Z from the encoder $q_\phi(Z|X,A)$ and evaluating the term using the samples. To maximize the objective, one can perform stochastic gradient descent along with the reparameterization trick (Kingma and Welling, 2014). Note that the reparameterization trick is necessary since we need to back-propagate through the sampling in the aforementioned Monte Carlo estimation term to compute the gradient w.r.t. the parameters of the encoder.

4.3.1.3 Discussion

The VGAE model is popular in the literature mainly due to its simplicity and good empirical performances. For example, since there are no learnable parameters for the prior and the decoder, the model is quite light-weight and the learning process is fast. Moreover, the VGAE model is versatile in way that once we learned a good encoder, *i.e.*, good latent representations, we can use them for predicting edges (, link prediction), node attributes, and so on. On the other side, VGAE model is still limited in the following ways. First, it can not serve as a good generative model for graphs as what VAEs do for images since the decoder is not learnable. One could simply design some learnable decoder. However, it is not clear that the goal of learning good latent representations and generating graphs with good qualities are always well-aligned. More exploration along this direction would be fruitful. Second, the independence assumption is exploited for both the encoder and the decoder which might be very limited. More structural dependence (*e.g.*, auto-regressive) would be desirable to improve the model capacity. Third, as discussed in the original paper, the prior may be potentially a poor choice. At last, for link prediction in practice, one may need to add the weighting of edges vs. non-edges in the decoder term and carefully tune it since graphs may be very sparse.

4.3.2 Deep Graph Infomax

Following Mutual Information Neural Estimation (MINE) (Belghazi et al, 2018) and Deep Infomax (Hjelm et al, 2018), Deep Graph Infomax (Veličković et al, 2019) is an unsupervised learning framework that learns graph representations via the principle of mutual information maximization.

4.3.2.1 Problem Setup

Following the original paper, we will explain the model under the single-graph setup, *i.e.*, the node feature matrix X and the graph adjacency matrix A of a single graph are provided as input. Extensions to other problem setups like transductive and inductive learning settings will be discussed in Section 4.3.2.3. The goal is to learn the node representations in an unsupervised way. After node representations are learned, one can apply some simple linear (logistic regression) classifier on top of the representations to perform supervised tasks like node classification.

4.3.2.2 Model

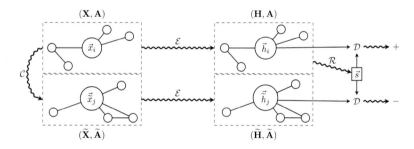

Fig. 4.2: The overall process of Deep Graph Infomax. The top path shows how the positive sample is processed, whereas the bottom shows process corresponding to the negative sample. Note that the graph representation is shared for both positive and negative samples. Subgraphs of positive and negative samples do not necessarily need to be different. Adapted from Figure 1 of (Veličković et al, 2019).

The main idea of the model is to maximize the local mutual information between a node representation (capturing local graph information) and the graph representation (capturing global graph information). By doing so, the learned node representation should capture the global graph information as much as possible. Let us denote the graph encoder as ε which could be any GNN discussed before, *e.g.*, a two-layer GCN. We can obtain all node representations as $H = \varepsilon(X, A)$ where the representation $\mathbf{h}_i$ of any node i should contain some local information near node i. Specifically, k-layer GCN should be able to leverage node information that is k-hop away. To get the global graph information, one could use a readout layer/function to process all node representations, *i.e.*, $\mathbf{s} = \mathscr{R}(H)$, where the readout function $\mathscr{R}$ could be some learnable pooling function or simply an average operator.

Objective Given the local node representation $\mathbf{h}_i$ and the global graph representation $\mathbf{s}$, the natural next-step is to compute their mutual information. Recall the definition of mutual information is as follows,

$$\text{MI}(\mathbf{h},\mathbf{s}) = \int \int p(\mathbf{h},\mathbf{s}) \log \left(\frac{p(\mathbf{h},\mathbf{s})}{p(\mathbf{h})p(\mathbf{s})} \right) d\mathbf{h}d\mathbf{s}. \tag{4.41}$$

However, maximizing the local mutual information alone is not enough to learn useful representations as shown in (Hjelm et al, 2018). To develop a more practical objective, authors in (Veličković et al, 2019) instead use a noise-contrastive type objective following Deep Infomax (Hjelm et al, 2018),

$$\mathscr{L} = \frac{1}{N+M} \left(\sum_{i=1}^{N} \mathbb{E}_{(X,A)} \left[\log \mathscr{D}(\mathbf{h}_i,\mathbf{s}) \right] + \sum_{j=1}^{M} \mathbb{E}_{(\tilde{X},\tilde{A})} \left[\log \left(1 - \mathscr{D}(\tilde{\mathbf{h}}_j,\mathbf{s}) \right) \right] \right). \tag{4.42}$$

where $\mathscr{D}$ is a binary classifier which takes both the node representation $\mathbf{h}_i$ and the graph representation $\mathbf{s}$ as input and predicts whether the pair $(\mathbf{h}_i,\mathbf{s})$ comes from the joint distribution $p(\mathbf{h},\mathbf{s})$ (positive class) or the product of marginals $p(\mathbf{h}_i)p(\mathbf{s})$ (negative class). We denote $\tilde{\mathbf{h}}_j$ as the j-th node representation from the negative sample. The numbers of positive and negative samples are N and M respectively. We will explain how to draw positive and negative samples shortly. The overall objective is thus the negative binary cross-entropy for training a probabilistic classifier. Note that this objective is the same type of distance as used in generative adversarial networks (GANs) (Goodfellow et al, 2014b) which is shown to be proportional to the Jensen-Shannon divergence (Goodfellow et al, 2014b; Nowozin et al, 2016). As verified by (Hjelm et al, 2018), maximizing the Jensen-Shannon divergence based mutual information estimator behaves similarly (*i.e.*, they have an approximately monotonic relationship) to directly maximizing the mutual information. Therefore, maximizing the objective in Eq. (4.42) is expected to maximize the mutual information. Moreover, the freedom of choosing negative samples makes the method more likely to learn useful representations than maximizing the vanilla mutual information.

Negative Sampling To generate the positive samples, one can directly sample a few nodes from the graph to construct the pairs $(\mathbf{h}_i,\mathbf{s})$. For negative samples, one can generate them via corrupting the original graph data, denoting as $(\tilde{X},\tilde{A}) = \mathscr{C}(X,A)$. In practice, one can choose various forms of this corruption function $\mathscr{C}$. For example, authors in (Veličković et al, 2019) suggest to keep the adjacency matrix to be the same and corrupt the node feature X by row-wise shuffling. Other possibilities of the corruption function include randomly sampling subgraphs and applying Dropout (Srivastava et al, 2014) to node features.

Once positive and negative samples were collected, one can learn the representations via maximizing the objective in Eq. (4.42). We summarize the training process of Deep Graph Infomax as follows:

1. Sample negative examples via the corruption function $(\tilde{X}, \tilde{A}) \sim \mathscr{C}(X, A)$.
2. Compute node representations of positive samples $H = \{\mathbf{h}_1, \cdots, \mathbf{h}_N\} = \varepsilon(X, A)$.
3. Compute node representations of negative samples $\tilde{H} = \{\tilde{\mathbf{h}}_1, \cdots, \tilde{\mathbf{h}}_M\} = \varepsilon(\tilde{X}, \tilde{A})$.
4. Compute graph representation via the readout function $\mathbf{s} = \mathscr{R}(H)$.
5. Update parameters of ε, $\mathscr{D}$, and $\mathscr{R}$ via gradient ascent to maximize Eq. (4.42).

4.3.2.3 Discussion

Deep Graph Infomax is an efficient unsupervised representation learning method for graph-structured data. The implementation of the encoder, the readout, and the binary cross-entropy type of loss are all straightforward. The mini-batch training does not necessarily need to store the whole graph since the readout can be applied to a set of subgraphs as well. Therefore, the method is memory-efficient. Also, the processing of positive and negative samples can be done in parallel. Moreover, authors prove that minimizing the cross-entropy type of classification error can be used to maximize the mutual information under certain conditions, *e.g.*, the readout function is injective and input feature comes from a finite set. However, the choice of the corruption function seems to be crucial to ensure satisfying empirical performances. There seems no such a universally good corruption function. One needs to do trial–and–error to obtain a proper one depending on the task/dataset.

4.4 Over-smoothing Problem

Training deep graph neural networks by stacking multiple layers of graph neural networks usually yields inferior results, which is a common problem observed in many different graph neural network architectures. This is mainly due to the problem of over-smoothing, which is first explicitly studied in (Li et al, 2018b). (Li et al, 2018b) showed that the graph convolutional network (Kipf and Welling, 2017b) is a special case of Laplacian smoothing:

$$Y = (1 - \gamma I)X + \gamma \tilde{A}_{rw}X, \tag{4.43}$$

where $\tilde{A}_{rw} = \tilde{D}^{-1}\tilde{A}$, which defines the transitional probabilities between nodes on graphs. The GCN corresponds to a special case of Laplacian smoothing with $\gamma = 1$ and the symmetric matrix $\tilde{A}_{sym} = \tilde{D}^{-\frac{1}{2}}\tilde{A}\tilde{D}^{-\frac{1}{2}}$ is used. The Laplacian smoothing will push nodes belonging to the same clusters to take similar representations, which are beneficial for downstream tasks such as node classification. However, when the GCNs go deep, the node representations suffer from the problem of over-smoothing, i.e., all the nodes will have similar representations. As a result, the performance on downstream tasks suffer as well. This phenomenon has later been pointed out by a few other later work as well such as (Zhao and Akoglu, 2019; Li et al, 2018b; Xu et al, 2018a; Li et al, 2019c; Rong et al, 2020b).

PairNorm (Zhao and Akoglu, 2019). Next, we will present a method called PairNorm for alleviating the problem of over-smoothing when GNNs go deep. The essential idea of PairNorm is to keep the total pairwise squared distance (TPSD) of node representations unchanged, which is the same as that of the original node feature X. Let $\tilde{H}$ be the output of the node representations by the graph convolution, which will be the input of PairNorm, and $\hat{H}$ is the output of PairNorm. The goal of PairNorm is to normalize the $\tilde{H}$ such that after normalization $\text{TPSD}(\hat{H}) = \text{TPSD}(X)$. In other words,

$$\sum_{(i,j)\in\mathscr{E}} ||\hat{H}_i - \hat{H}_j||^2 + \sum_{(i,j)\notin\mathscr{E}} ||\hat{H}_i - \hat{H}_j||^2 = \sum_{(i,j)\in\mathscr{E}} ||X_i - X_j||^2 + \sum_{(i,j)\notin\mathscr{E}} ||X_i - X_j||^2. \tag{4.44}$$

In practice, instead of measuring the TPSD of original node features X, (Zhao and Akoglu, 2019) proposed to maintain a constant TPSD value C across different graph convolutional layers. The value C will be a hyperparameter of the PairNorm layer, which can be tuned for each data set. To normalize $\tilde{H}$ into $\hat{H}$ with a constant TPSD, we must first calculate the $\text{TPSD}(\tilde{H})$. However, this is very computationally expensive, which is quadratic to the number of nodes N. We notice that the TPSD can be reformulated as:

$$\text{TPSD}(\tilde{H}) = \sum_{(i,j)\in[N]} ||\tilde{H}_i - \tilde{H}_j||^2 = 2N^2 \left(\frac{1}{N}\sum_{i=1}^{N} ||\tilde{H}_i||_2^2 - ||\frac{1}{N}\sum_{i=1}^{N}\tilde{H}_i||_2^2 \right) \tag{4.45}$$

We can further simply the above equation by substracting the row-wise mean from each $\tilde{H}_i$. In other words, $\tilde{H}_i^c = \tilde{H}_i - \frac{1}{N}\sum_{i=1}^{N}\tilde{H}_i$, which denotes the centered representation. A nice property of centering the node representation is that it will not change the TPSD and meanwhile push the second term $||\frac{1}{N}\sum_{i=1}^{N}\tilde{H}_i||_2^2$ to zero. As a result, we have

$$\text{TPSD}(\tilde{H}) = \text{TPSD}(\tilde{H}^c) = 2N||\tilde{H}^c||_F^2. \tag{4.46}$$

To summarize, the proposed PairNorm can be divded into two steps: center-and-scale,

$$\tilde{H}_i^c = \tilde{H}_i - \frac{1}{N}\sum_{i=1}^{N}\tilde{H}_i \qquad\qquad \text{(Center)} \qquad (4.47)$$

$$\hat{H}_i = s \cdot \frac{\tilde{H}_i^c}{\sqrt{\frac{1}{N}\sum_{i=1}^{N}||\tilde{H}_i^c||_2^2}} = s\sqrt{N} \cdot \frac{\tilde{H}_i^c}{\sqrt{||\tilde{H}^c||_F^2}} \qquad \text{(Scale)}, \qquad (4.48)$$

where s is a hyperparameter determining C. At the end, we have

$$\text{TPSD}(\hat{H}) = 2N||\hat{H}||_F^2 = 2N\sum_i ||s \cdot \frac{\tilde{H}_i^c}{\sqrt{\frac{1}{N}\sum_{i=1}^N ||\tilde{H}_i^c||_2^2}}||_2^2 = 2N^2 s^2 \qquad (4.49)$$

which is a constant across different graph convolutional layers.

4.5 Summary

In this chapter, we give a comprehensive introduction to different architectures of graph neural networks for node classification. These neural networks can be generally classified into two categories including supervised and unsupervised approaches. For supervised approaches, the main difference among different architectures lie in how to propagate messages between nodes, how to aggregate the messages from neighbors, and how to combine the aggregated messages from neighbors with the node representation itself. For the unsupervised approaches, the main difference comes from designing the objective function. We also discuss a common problem of training deep graph neural networks—over-smoothing, and introduce a method to tackle it. In the future, promising directions on graph neural networks include theoretical analysis for understanding the behaviors of graph neural networks, and applying them to a variety of fields and domains such as recommender systems, knowledge graphs, drug and material discovery, computer vision, and natural language understanding.

Editor's Notes: Node classification task is one of the most important tasks in Graph Neural Networks. The node representation learning techniques introduced in this chapter are the corner stone for all other tasks for the rest of the book, including graph classification task (Chapter 9), link prediction (Chapter 10), graph generation task (Chapter 11), and so on. Familiar with the learning methodologies and design principles of node representation learning is the key to deeply understanding other fundamental research directions like Theoretical analysis (Chapter 5), Scalability (Chapter 6), Explainability (Chapter 7), and Adversarial Robustness (Chapter 8).

Chapter 5
The Expressive Power of Graph Neural Networks

Pan Li and Jure Leskovec

Abstract The success of neural networks is based on their strong expressive power that allows them to approximate complex non-linear mappings from features to predictions. Since the universal approximation theorem by (Cybenko, 1989), many studies have proved that feed-forward neural networks can approximate any function of interest. However, these results have not been applied to graph neural networks (GNNs) due to the inductive bias imposed by additional constraints on the GNN parameter space. New theoretical studies are needed to better understand these constraints and characterize the expressive power of GNNs.

In this chapter, we will review the recent progress on the expressive power of GNNs in graph representation learning. We will start by introducing the most widely-used GNN framework— message passing— and analyze its power and limitations. We will next introduce some recently proposed techniques to overcome these limitations, such as injecting random attributes, injecting deterministic distance attributes, and building higher-order GNNs. We will present the key insights of these techniques and highlight their advantages and disadvantages.

5.1 Introduction

Machine learning problems can be abstracted as learning a mapping f^* from some feature space to some target space. The solution to this problem is typically given by a model f_θ that intends to approximate f^* via optimizing some parameter θ. In practice, the ground truth f^* is a priori typically unknown. Therefore, one may expect the model f_θ to approximate a rather broad range of f^*. An estimate of

Pan Li
Department of Computer Science, Purdue University, e-mail: `panli@purdue.edu`

Jure Leskovec
Department of Computer Science, Stanford University, e-mail: `jure@cs.stanford.edu`

how broad such a range could be, called the model's *expressive power*, provides an important measure of the model potential. It is desirable to have models with a more expressive power that may learn more complex mapping functions.

Neural networks (NNs) are well known for their great expressive power. Specifically, Cybenko (1989) first proved that any continuous function defined over a compact space could be uniformly approximated by neural networks with sigmoid activation functions and only one hidden layer. Later, this result got generalized to any squashing activation functions by (Hornik et al, 1989).

However, these seminal findings are insufficient to explain the current unprecedented success of NNs in practice because their strong expressive power only demonstrates that the model f_θ is able to approximate f^* but does not guarantee that the model obtained via training $\hat{f}$ indeed approximates f^*. Fig. 5.1 illustrates a well-known curve of Amount of Data vs. Performance of machine learning models (Ng, 2011). NN-based methods may only outperform traditional methods given sufficient data. One important reason is that NNs as machine learning models are still governed by the fundamental tradeoff between the data amount and model complexity (Fig. 5.2). Although NNs could be rather expressive, they are likely to overfit the

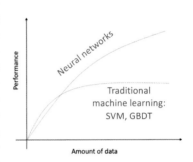

Fig. 5.1: Amount of Data *vs.* Performance of different models.

training examples when paired with more parameters. Therefore, it is necessary for practice to build NNs that can maintain strong expressive power while constraints are imposed on their parameters. At the same time, a good theoretical understanding of the expressive power of NNs with constraints on their parameters is needed.

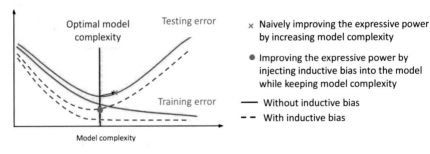

Fig. 5.2: Training and testing errors with and without inductive bias can dramatically affect the expressive power of models.

In practice, constraints on parameters are typically obtained from our prior knowledge of the data; these are referred to as inductive biases. Some significant

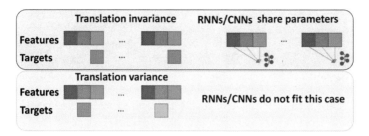

Fig. 5.3: Illustration of 1-dimensional translation invariance/variance. RNNs/CNNs use translation invariance to share parameters.

results about the expressive power of NNs with inductive bias have been shown recently. Yarotsky (2017); Liang and Srikant (2017) have proved that deep neural networks (DNNs), by stacking multiple hidden layers, can achieve good enough approximation with significantly fewer parameters than shallow NNs. The architecture of DNNs leverages the fact that data has typically a hierarchical structure. DNNs are agnostic to the type of data, while dedicated neural network architectures have been developed to support specific types of data. Recurrent neural networks (RNNs) (Hochreiter and Schmidhuber, 1997) or convolution neural networks (CNNs) (LeCun et al, 1989) were proposed to process time series and images, respectively. In these two types of data, effective patterns typically hold translation invariance in time and in space, respectively. To match this invariance, RNNs and CNNs adopt the inductive bias that their parameters have shared across time and space (Fig. 5.3). The parameter-sharing mechanism works as a constraint on the parameters and limits the expressive power of RNNs and CNNs. However, RNNs and CNNs have been shown to have sufficient expressive power to learn translation invariant functions (Siegelmann and Sontag, 1995; Cohen and Shashua, 2016; Khrulkov et al, 2018), which leads to the great practical success of RNNs and CNNs in processing time series and images.

Recently, many studies have focused on a new type of NNs, termed graph neural networks (GNNs) (Scarselli et al, 2008; Bruna et al, 2014; Kipf and Welling, 2017a; Bronstein et al, 2017; Gilmer et al, 2017; Hamilton et al, 2017b; Battaglia et al, 2018). These aim to capture the inductive bias of graphs/networks, another important type of data. Graphs are commonly used to model complex relations and interactions between multiple elements and have been widely used in machine learning applications, such as community detection, recommendation systems, molecule property prediction, and medicine design (Fortunato, 2010; Fouss et al, 2007; Pires et al, 2015). Compared to time series and images, which are well-structured and represented by tables or grids, graphs are irregular and thus introduce new challenges. A fundamental assumption behind machine learning on graphs is that the targets for prediction should be invariant to the order of nodes of the graph. To match this assumption, GNNs hold a general inductive bias termed permutation invariance. In particular, the output given by GNNs should be independent of how the node indices of a graph are assigned and thus in which order are they processed. GNNs require

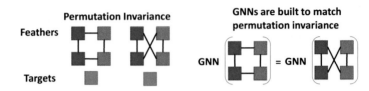

Fig. 5.4: This illustrates how GNNs are designed to maintain permutation invariance.

their parameters to be independent from the node ordering and are shared across the entire graph (Fig. 5.4). Because of this new parameter sharing mechanism in GNNs, new theoretical tools are needed to characterize their expressive power.

Analyzing the expressive power of GNNs is challenging, as this problem is closely related to some long-standing problems in graph theory. To understand this connection, consider the following example of how a GNN would predict whether a graph structure corresponds to a valid molecule. The GNN should be able to identify whether this graph structure is the same, similar, or very different from the graph structures that are known to correspond to valid molecules. Measuring whether two graphs have the same structure involves addressing the graph isomorphism problem, in which no P solutions have yet been found (Helfgott et al, 2017). In addition, measuring whether two graphs have a similar structure requires contending with the graph edit distance problem, which is even harder to address than the graph isomorphism problem (Lewis et al, 1983).

Great progress has been made recently on characterizing the expressive power of GNNs, especially on how to match their power with traditional graph algorithms and how to build more powerful GNNs that overcome the limitation of those algorithms. We will delve more into these recent efforts further along in this chapter. In particular, compared to previous introductions (Hamilton, 2020; Sato, 2020), we will focus on recent key insights and techniques that yield more powerful GNNs. Specifically, we will introduce standard message-passing GNNs that are able to achieve the limit of the 1-dimensional Weisfeiler-Lehman test (Weisfeiler and Leman, 1968), a widely-used algorithm to test for graph isomorphism. We will also discuss a number of strategies to overcome the limitations of the Weisfeiler-Lehman test — including attaching random attributes, attaching deterministic distance attributes, and leveraging higher-order structures.

In Section 5.2, we will formulate the graph representation learning problems that GNNs target. In Section 5.3, we will review the most widely used GNN framework, the message passing neural network, describing the limitations of its expressive power and discussing its efficient implementations. In Section 5.4, we will introduce a number of methods that make GNNs more powerful than the message passing neural network. In Section 5.5, we will conclude this chapter by discussing further research directions.

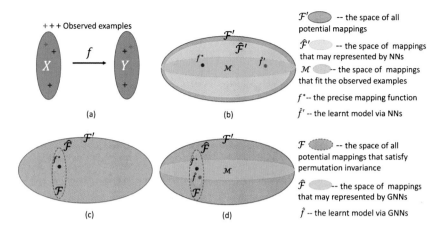

Fig. 5.5: An illustration of the expressive power of NNs and GNNs and their affects on the performance of learned models. a) Machine learning problems aim to learn the mapping from the feature space to the target space based on several observed examples. b) The expressive power of NNs refers to the gap between the two spaces $\mathscr{F}$ and $\hat{\mathscr{F}'}$. Although NNs are expressive ($\hat{\mathscr{F}'}$ is dense in $\mathscr{F}$), the learned model f' based on NNs may differ significantly from f^* due to their overfit of the limited observed data. c) Suppose f^* is known to be permutation invariant, as it works on graph-structured data. Then, the space of potential mapping functions is reduced from $\mathscr{F}'$ to a much smaller space $\mathscr{F}$ that only includes permutation invariant functions. If we adopt GNNs, the space of mapping functions that can be approximated simultaneously reduces to $\hat{\mathscr{F}}$. The gap between $\mathscr{F}$ and $\hat{\mathscr{F}}$ characterizes the expressive power of GNNs. d) Although GNNs are less expressive than general NNs ($\hat{\mathscr{F}} \subset \hat{\mathscr{F}'}$), the learned model based on GNNs f is a much better approximator of f^* as opposed to the one based on NNs $\hat{f}'$. Therefore, for graph-structured data, our understanding of the expressive power of GNNs, *i.e.*, the gap between $\mathscr{F}$ and $\hat{\mathscr{F}}$, is much more relevant than that of NNs.

5.2 Graph Representation Learning and Problem Formulation

In this section, we will set up the formal definition of graph representation learning problems, their fundamental assumption, and their inductive bias. We will also discuss relationships between different notions of graph representation learning problems frequently studied in recent literature.

First, we will start by defining graph-structured data.

Definition 5.1. (Graph-structured data) Let $\mathscr{G} = (\mathscr{V}, \mathscr{E}, X)$ denote an attributed graph, where $\mathscr{V}$ is the node set, $\mathscr{E}$ is the edge set, and $X \in \mathbb{R}^{|\mathscr{V}| \times F}$ are the node attributes. Each row of X, $X_v \in \mathbb{R}^F$ refers to the attributes on the node $v \in \mathscr{V}$. In practice, graphs are usually sparse, *i.e.*, $|\mathscr{E}| \ll |\mathscr{V}|^2$. We introduce $A \in \{0,1\}^{|\mathscr{V}| \times |\mathscr{V}|}$ to denote the adjacency matrix of G such that $A_{uv} = 1$ iff $(u,v) \in E$. Combining the

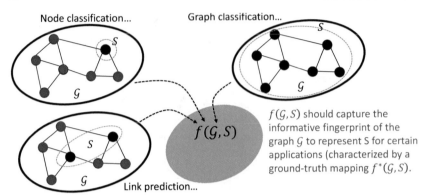

Fig. 5.6: Graph representation learning problems frequently discussed in literature.

adjacency matrix and node attributes, we may also denote $\mathscr{G} = (A, X)$. Moreover, if $\mathscr{G}$ is unattributed with no node attributes, we can assume that all elements in X are constant. Later, we also use $\mathscr{V}[\mathscr{G}]$ to denote the entire node set of a particular graph $\mathscr{G}$.

The goal of graph representation learning is to learn a model by taking graph-structured data as input and then mapping it so that certain prediction targets are matched. Different graph representation learning problems may apply to a varying number of nodes in a graph. For example, for node classification, a prediction is made for each node, for each link/relation prediction on a pair of nodes, and for each graph classification or graph property prediction on the entire node set V. We can unify all these problems as graph representation learning.

Definition 5.2. (Graph representation learning) The feature space is defined as $\mathscr{X} := \Gamma \times \mathscr{S}$, where Γ is the space of graph-structured data and $\mathscr{S}$ includes all the node subsets of interest, given a graph $\mathscr{G} \in \Gamma$. Then, a point in $\mathscr{X}$ can be denoted as $(\mathscr{G}, S)$, where S is a subset of nodes that are in $\mathscr{G}$. Later, we call $(\mathscr{G}, S)$ as a graph representation learning (GRL) example. Each GRL example $(\mathscr{G}, S) \in \mathscr{X}$ is associated with a target y in the target space $\mathscr{Y}$. Suppose the ground-truth association function between features and targets is denoted by $f^* : \mathscr{X} \to \mathscr{Y}$, $f^*(\mathscr{G}, S) = y$. Given a set of training examples $\Xi = \{(\mathscr{G}^{(i)}, S^{(i)}, y^{(i)})\}_{i=1}^k$ and a set of testing examples $\Psi = \{(\tilde{\mathscr{G}}^{(i)}, \tilde{S}^{(i)}, \tilde{y}^{(i)})\}_{i=1}^k$, a *graph representation learning* problem is to learn a function f based on Ξ such that f is close to f^* on Ψ.

The above definition is general in the sense that in a GRL example $(\mathscr{G}, S) \in \mathscr{X}$, $\mathscr{G}$ provides both raw and structural features on which some prediction for a node subset S of interest is to be made. Below, we will further list a few frequently-investigated learning problems that may be formulated as graph representation learning problems.

Remark 5.1. (Graph classification problem / Graph-level prediction) The node set S of interest is the entire node set $\mathscr{V}[\mathscr{G}]$ by default. The space of graph-structured data

Γ typically contains multiple graphs. The target space $\mathcal{Y}$ contains labels of different graphs. Later, for graph-level prediction, we will use $\mathcal{G}$ to denote a GRL example instead of $(\mathcal{G}, S)$ for notational simplicity.

Remark 5.2. (Node classification problem / Node-level prediction) In a GRL example $(\mathcal{G}, S)$, the S corresponds to one single node of interest. The corresponding $\mathcal{G}$ can be defined in different ways. On the one hand, only the nodes close to S provide effective features. In this case, $\mathcal{G}$ may be set as the induced local subgraph around S. Different $\mathcal{G}$'s for different S's may come from a single graph. On the other hand, two nodes that are far apart on one graph still hold mutual impact and can be used as a feature to make a prediction on another graph. In that case, $\mathcal{G}$ needs to include a large portion of a graph or even the entire graph.

Remark 5.3. (Link prediction problem / Node-pair-level prediction) In a GRL example $(\mathcal{G}, S)$, S corresponds to a pair of nodes of interest. Similar to the node classification problem, $\mathcal{G}$ for each example may be an induced subgraph around S or the entire graph. The target space $\mathcal{Y}$ contains 0-1 labels that indicate whether there is a probable link between two nodes. $\mathcal{Y}$ may also be generalized to include labels that reflect the types of links to be predicted.

Next, we will introduce the fundamental assumption used in most graph representation learning problems.

Definition 5.3. (Isomorphism) Consider two GRL examples $(\mathcal{G}^{(1)}, S^{(1)})$, $(\mathcal{G}^{(2)}, S^{(2)})$ $\in \mathcal{X}$. Suppose $\mathcal{G}^{(1)} = (A^{(1)}, X^{(1)})$ and $\mathcal{G}^{(2)} = (A^{(2)}, X^{(2)})$. If there exists a bijective mapping $\pi : \mathcal{V}[\mathcal{G}^{(1)}] \to \mathcal{V}[\mathcal{G}^{(2)}]$, $i \in \{1, 2\}$, such that $A_{uv}^{(1)} = A_{\pi(u)\pi(v)}^{(2)}$, $X_u^{(1)} = X_{\pi(u)}^{(2)}$ and π also gives a bijective mapping between $S^{(1)}$ and $S^{(2)}$, we call that $(\mathcal{G}^{(1)}, S^{(1)})$ and $(\mathcal{G}^{(2)}, S^{(2)})$ are isomorphic, denoted as $(\mathcal{G}^{(1)}, S^{(1)}) \cong (\mathcal{G}^{(2)}, S^{(2)})$. When the particular bijective mapping π should be highlighted, we use notation $(\mathcal{G}^{(1)}, S^{(1)}) \stackrel{\pi}{\cong} (\mathcal{G}^{(2)}, S^{(2)})$. If there is no such a π, we call that they are non-isomorphic, denoted as $(\mathcal{G}^{(1)}, S^{(1)}) \not\cong (\mathcal{G}^{(2)}, S^{(2)})$.

Assumption 1 *(Fundamental assumption in graph representation learning) Consider a graph representation learning problem with a feature space $\mathcal{X}$ and its corresponding target space $\mathcal{Y}$. Pick any two GRL examples $(\mathcal{G}^{(1)}, S^{(1)})$, $(\mathcal{G}^{(2)}, S^{(2)}) \in \mathcal{X}$. The fundamental assumption says that if $(\mathcal{G}^{(1)}, S^{(1)}) \cong (\mathcal{G}^{(2)}, S^{(2)})$, their corresponding targets in $\mathcal{Y}$ are the same.*

Due to this fundamental assumption, it is natural to introduce the corresponding *permutation invariance* as inductive bias that all models of graph representation learning should satisfy.

Definition 5.4. (Permutation invariance) A model f satisfies permutation invariance if for any $(\mathcal{G}^{(1)}, S^{(1)}) \cong (\mathcal{G}^{(2)}, S^{(2)})$, $f(\mathcal{G}^{(1)}, S^{(1)}) = f(\mathcal{G}^{(2)}, S^{(2)})$.

Now we may define the expressive power of a model for graph representation learning problems.

Definition 5.5. (Expressive power) Consider a feature space $\mathscr{X}$ of a graph representation learning problem and a model f defined on $\mathscr{X}$. Define another space $\mathscr{X}(f)$ as a subspace of the quotient space $\mathscr{X}/\cong$ such that for two GRL examples $(\mathscr{G}^{(1)}, S^{(1)})$, $(\mathscr{G}^{(2)}, S^{(2)}) \in \mathscr{X}(f)$, $f(\mathscr{G}^{(1)}, S^{(1)}) \neq f(\mathscr{G}^{(2)}, S^{(2)})$. Then, the size of $\mathscr{X}(f)$ characterizes the expressive power of f. For two models, $f^{(1)}$ and $f^{(2)}$, if $\mathscr{X}(f^{(1)}) \supset \mathscr{X}(f^{(2)})$, we say that $f^{(1)}$ is more expressive than $f^{(2)}$.

Remark 5.4. Note that the expressive power in Def. 5.5, characterized by how a model can distinguish non-isomorphic GRL examples, does not exactly match the traditional expressive power used for NNs in the sense of functional approximation. Actually, Def. 5.5 is strictly weaker because distinguishing any non-isomorphic GRL examples does not necessarily indicate that we can approximate any function f^* defined over $\mathscr{X}$. However, if a model f cannot distinguish two non-isomorphic features, f is definitely unable to approximate function f^* that maps these two examples to two different targets. Some recent studies have been able to prove some equivalence between distinguishing non-isomorphic features and permutation invariant function approximations under weak assumptions and applying involved techniques (Chen et al, 2019f; Azizian and Lelarge, 2020). Interested readers may check these references for more details.

It is trivial to provide the expressive power of a model f for graph representation learning if f does not satisfy permutation invariance. Without such a constraint, NNs can approximate all continuous functions (Cybenko, 1989), which include the continuous functions that distinguish any non-isomorphic GRL examples. Therefore, the key question we are to discuss in the chapter is: *"How to build the most expressive permutation invariant models, GNNs in particular, for graph representation learning problems?"*

5.3 The Power of Message Passing Graph Neural Networks

5.3.1 Preliminaries: Neural Networks for Sets

We will start by reviewing the NNs with sets (multisets) as their input, since a set can be viewed as a simplified-version of a graph where all edges are removed. By definition, the order of elements of a set does not impact the output; models that encode sets naturally provide an important building block for encoding the graphs. We term this approach invariant pooling.

Definition 5.6. (Multiset) A multiset is a set where its elements can be repetitive, meaning that they are present multiple times. In this chapter, we assume by default that all the sets are multisets and thus allow repetitive elements. In situations where this is not the case, we will indicate otherwise.

Definition 5.7. (Invariant pooling) Given a multiset of vectors $S = \{\mathbf{a}_1, \mathbf{a}_2, ..., \mathbf{a}_k\}$ where $\mathbf{a}_i \in \mathbb{R}^F$ and F is an arbitrary constant, an invariant pooling refers to a mapping, denoted as $q(S)$, that is invariant to the order of elements in S.

Some widely-used invariant pooling operations include: sum pooling $q(S) = \sum_{i=1}^{k} \mathbf{a}_i$, mean pooling $q(S) = \frac{1}{k} \sum_{i=1}^{k} \mathbf{a}_i$ and max pooling $[q(S)]_j = \max_{i \in [1,F]}\{a_{ij}\}$ for all $j \in [1, F]$. Zaheer et al (2017) show that any invariant poolings of a set S can be approximated by $q(S) = \phi(\sum_{i=1}^{k} \psi(\mathbf{a}_i))$, where ϕ and ψ are functions that may be approximated by fully connected NNs, provided that $\mathbf{a}_i$, $i \in [k]$ comes from a countable universe. This statement can be generalized to the case where S is a multiset (Xu et al, 2019d).

5.3.2 Message Passing Graph Neural Networks

Message passing is the most widely-used framework to build GNNs (Gilmer et al, 2017). Given a graph $\mathscr{G} = (\mathscr{V}, \mathscr{E}, X)$, the message passing framework encodes each node $v \in \mathscr{V}$ with a vector representation $\mathbf{h}_v$ and keeps updating this node representation by iteratively collecting representations of its neighbors and applying neural network layers to perform a non-linear transformation of those collections:

1. Initialize node vector representations as node attributes: $\mathbf{h}_v^{(0)} \leftarrow X_v, \forall v \in \mathscr{V}$.
2. Update each node representation based on message passing over the graph structure. In l-th layer, $l = 1, 2, ..., L$, perform the following steps:

$$\text{Message:} \quad \mathbf{m}_{vu}^{(l)} \leftarrow \text{MSG}(\mathbf{h}_v^{(l-1)}, \mathbf{h}_u^{(l-1)}), \forall (u, v) \in \mathscr{E}, \quad (5.1)$$

$$\text{Aggregation:} \quad \mathbf{a}_v^{(l)} \leftarrow \text{AGG}(\{\mathbf{m}_{vu}^{(l)} | u \in \mathscr{N}_v\}), \forall v \in \mathscr{V}, \quad (5.2)$$

$$\text{Update:} \quad \mathbf{h}_v^{(l)} \leftarrow \text{UPT}(\mathbf{h}_v^{(l-1)}, \mathbf{a}_v^{(l)}), \forall v \in \mathscr{V}. \quad (5.3)$$

where $\mathscr{N}_v$ is the set of neighbors of v.

The operations MSG, AGG, and UPT can be implemented via neural networks. Typically, MSG is implemented by a feedforward NN, e.g., $\text{MSG}(p, q) = \sigma(pW_1 + qW_2)$, where W_1 and W_2 are learnable weights, and $\sigma(\cdot)$ is an element-wise nonlinear activation. UPT can be chosen in a similar way as MSG. AGG differs as its input is a multiset of vectors and thus the order of these vectors should not affect the output. AGG is typically implemented as an invariant pooling (Def. 5.7). Each layer k can have different parameters from other layers. We will denote the GNNs that follow this message passing framework as *MP-GNN*.

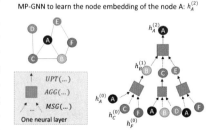

Fig. 5.7: The computing flow of MP-GNN to obtain a node representation.

MP-GNN produces representations of all the nodes, $\{\mathbf{h}_v^{(L)}|v \in V\}$. Each node representation is essentially determined by a subtree rooted at this node (Fig. 5.7). Given a specific graph representation learning problem, for example, classifying a set of nodes $S \subseteq V$, we may use the representations of relevant nodes in S to make the prediction:

$$\hat{y}_S = \text{READOUT}(\{\mathbf{h}_v^{(L)}|v \in S\}). \tag{5.4}$$

where the READOUT operation is often implemented via another invariant pooling when $|S| > 1$ plus a feed-forward NN to predict the target. Combining Eqs.equation 11.45-equation 5.4, MP-GNN builds a GNN model for graph representation learning:

$$\hat{y}_S = f_{MP-GNN}(\mathcal{G}, S). \tag{5.5}$$

We can show the permutation invariance of MP-GNN by induction over the iteration index l.

Theorem 5.1. *(Invariance of MP-GNN)* $f_{MP-GNN}(\cdot, \cdot)$ *satisfies permutation invariance (Def. 5.4) as long as the AGG and READOUT operations are invariant pooling operations (Def. 5.7).*

Proof. This can be proved trivially by induction.

MP-GNN by default leverages the inductive bias that the nodes in the graph directly affect each other only via their connected edges. The mutual effect between nodes that are not connected by an edge can be captured via paths that connect these nodes via message passing. Indeed, such inductive bias may not match the assumptions in a specific application, and MP-GNN may find it hard to capture mutual effect between two far-away nodes. However, the message-passing framework has several benefits for model implementation and practical deployment. First, it directly works on the original graph structure and no pre-processing is needed. Second, graphs in practice are typically sparse ($|\mathcal{E}| \ll |\mathcal{V}|^2$) and thus MP-GNN is able to scale to very large but sparse graphs. Third, each of the three operations MSG, AGG, and UPT can be computed in parallel across all nodes and edges, which is beneficial for parallel computing platforms such as GPUs and map-reduce systems.

Because it is natural and easy to be implemented in practice, most GNN architectures essentially follow the MP-GNN framework by adopting specific MSG, AGG, and UPT operations. Representative approaches include InteractionNet (Battaglia et al, 2016), structure2vec (Dai et al, 2016), GCN (Kipf and Welling, 2017a), GraphSAGE (Hamilton et al, 2017b), GAT (Veličković et al, 2018), GIN (Xu et al, 2019d), and many others (Kearnes et al, 2016; Zhang et al, 2018g).

5.3.3 The Expressive Power of MP-GNN

In this section, we will introduce the expressive power of MP-GNN , following the results proposed in Xu et al (2019d); Morris et al (2019).

The 1-dimensional Weisfeiler-Lehman test to distinguish $(\mathscr{G}^{(1)}, S^{(1)})$ and $(\mathscr{G}^{(2)}, S^{(2)})$:

1. Assume each node v in $\mathscr{V}[\mathscr{G}^{(i)}]$ is initialized with a color $C_v^{(i,0)} \leftarrow X_v^{(i)}$ for $i = 1, 2$. If $X_v^{(i)}$ is a vector, then an injective function is used to map it to a color.
2. For $l = 1, 2, ...$, do

$$\text{Update node colors:} \quad C_v^{(i,l)} \leftarrow \text{HASH}(C_v^{(i,l-1)}, \{C_u^{(i,l-1)} | u \in \mathscr{N}_v^{(i)}\}), \quad i \in \{1, 2\} \tag{5.6}$$

where the HASH operation can be viewed as an injective mapping where different tuples $(C_v^{(i,l-1)}, \{C_u^{(i,l-1)} | u \in \mathscr{N}_v^{(i)}\})$ are mapped to different labels.

Test: If two multisets $\{C_v^{(1,l)} | v \in S^{(1)}\}$ and $\{C_v^{(2,l)} | v \in S^{(2)}\}$ are not equal,

then return $(\mathscr{G}^{(1)}, S^{(1)}) \not\cong (\mathscr{G}^{(2)}, S^{(2)})$; else, go back to equation 5.6.

If 1-WL test returns $(\mathscr{G}^{(1)}, S^{(1)}) \not\cong (\mathscr{G}^{(2)}, S^{(2)})$, we know that $(\mathscr{G}^{(1)}, S^{(1)})$ $(\mathscr{G}^{(2)}, S^{(2)})$ are not isomorphic. However, for some non-isomorphic $(\mathscr{G}^{(1)}, S^{(1)})$ $(\mathscr{G}^{(2)}, S^{(2)})$, the 1-WL test may not return $(\mathscr{G}^{(1)}, S^{(1)}) \not\cong (\mathscr{G}^{(2)}, S^{(2)})$ (even with infinitely many iterations). In this case, the 1-WL test fails to distinguish them. Note that the 1-WL test was originally proposed to test the isomorphism of two entire graphs, i.e., $S^{(i)} = \mathscr{V}[\mathscr{G}^{(i)}]$ for $i \in \{1, 2\}$ (Weisfeiler and Leman, 1968). Here the 1-WL test is further generalized to test the case where $S^{(i)} \subset \mathscr{V}^{(i)}$, to match the general graph representation learning problems.

The expressive power we defined (Def. 5.5) is closely related to the graph isomorphism problem. This problem is challenging, as no polynomial-time algorithms have been found for it (Garey, 1979; Garey and Johnson, 2002; Babai, 2016). Despite some corner cases (Cai et al, 1992), the Weisfeiler-Lehman (WL) tests of graph isomorphism (Weisfeiler and Leman, 1968) are a family of effective and computationally efficient tests that distinguish a broad class of graphs (Babai and Kucera, 1979). Its 1-dimensional form (the 1-WL test), "naive vertex refinement", is analogous to the neighborhood aggregation in MP-GNN.

They are comparing MP-GNN with the 1-WL test, the node-representation updating procedure Eqs.equation 11.45-equation 5.3 can be viewed as an implementation of Eq.equation 5.6 and the READOUT operation in Eq.equation 5.4 can be viewed as a summary of all node representations. Although MP-GNN was not proposed to perform graph isomorphism testing, the f_{MP-GNN} can be used for this test: if $f_{MP-GNN}(\mathscr{G}^{(1)}, S^{(1)}) \neq f_{MP-GNN}(\mathscr{G}^{(2)}, S^{(2)})$, then we know that $(\mathscr{G}^{(1)}, S^{(1)}) \not\cong (\mathscr{G}^{(2)}, S^{(2)})$. Because of this analogy, the expressive power of MP-GNN can be measured by the 1-WL test. Formally, we conclude the argument in the following theorem.

Theorem 5.2. *(Lemma 2 in (Xu et al, 2019d), Theorem 1 in (Morris et al, 2019)) Consider two non-isomorphic GRL examples $(\mathscr{G}^{(1)}, S^{(1)})$ and $(\mathscr{G}^{(2)}, S^{(2)})$. If $f_{MP-GNN}(\mathscr{G}^{(1)}, S^{(1)}) \neq f_{MP-GNN}(\mathscr{G}^{(2)}, S^{(2)})$, then the 1-WL test also decides $(\mathscr{G}^{(1)}, S^{(1)})$ and $(\mathscr{G}^{(2)}, S^{(2)})$ are not isomorphic.*

Theorem 5.2 indicates that MP-GNN is at most as powerful as the 1-WL test in distinguishing different graph-structured features. Here, the 1-WL test is considered an upper bound (instead of being equal to the expressive power of MP-GNN)

Step 1: Each node is initialized with some color according to its attribute (if no attributes, use the same color).

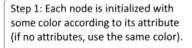

The mapping "attributes → colors" is injective.

Step 2: Each node will collect the colors from their neighbors:
   Node A: (p,{bby})
   Left node E: (b,{py});
   Right node E: (b,{pyg}) ...

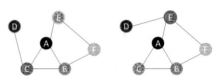

The mapping "(self-color, set of colors from neighbors) → a new color" is injective

After each iteration , check the set of node colors. Current both graphs have the same set of colors. We do step 2 again. After two iterations, we may distinguish these two graphs because left B will get a color that will not appear in the right graph, because currently left B has purple + blue in its neighborhood while no nodes in the right graph have such neighborhood.

Fig. 5.8: An illustration of steps that distinguish two graphs via the 1-dimensional Weisfeiler-Lehman test. MP-GNN follows a similar procedure and may also distinguish them.

because the updating procedure which aggregates node colors from its neighbors (Eq.equation 5.6) is injective and can distinguish between the different aggregations of node colors. This intuition is useful later to design MP-GNN that matches this upper bound.

Now that the upper bound of the representation power of MP-GNN has been established, a natural follow-up question is whether there are existing GNNs that are, in principle, as powerful as the 1-WL test. The answer is yes. As shown by Theorem 5.3: if the message passing operation (compositing Eqs.equation 11.45-equation 5.3 together) and the final READOUT (Eq.equation 5.4) are both injective, then the resulting MP-GNN is as powerful as the 1-WL test.

Theorem 5.3. *(Theorem 3 in (Xu et al, 2019d)) After sufficient iterations, MP-GNN may map any GRL examples $(\mathscr{G}^{(1)}, S^{(1)})$ and $(\mathscr{G}^{(2)}, S^{(2)})$, that the 1-WL test decides as non-isomorphic, to different representations if the following two conditions hold:*

 a) The composition of MSE, AGG and UPT (Eqs.equation 11.45-equation 5.3) constructs an injective mapping from $(\mathbf{h}_v^{(k-1)}, \{\mathbf{h}_u^{(k-1)} | u \in \mathscr{N}_v\})$ to $\mathbf{h}_v^{(k)}$.
 b) The READOUT (Eq.equation 5.4) is injective.

Although MP-GNN does not surpass the representation power of the 1-WL test, MP-GNN has important benefits over the 1-WL test from the perspective of machine learning: node colors and the final decision given by the 1-WL test are discrete (represented as node colors or a "yes/no" decision) and thus cannot capture the similarity between graph structures. In contrast, a MP-GNN satisfying the criteria in

Theorem 5.3 generalizes the 1-WL test by *learning to represent* the graph structures with vectors in a continuous space. This enables MP-GNN to not only discriminate between different structures but also to learn to map similar graph structures to similar representations, thus capturing dependencies between graph structures. Such learned representations are particularly helpful for generalizations where data contains noisy edges and the exact matching graph structures are sparse (Yanardag and Vishwanathan, 2015).

In the next subsection, we will focus on introducing the key design ideas behind MP-GNN that satisfies the conditions in Theorem 5.3.

5.3.4 MP-GNN with the Power of the 1-WL Test

Xu et al (2019d) introduced the key guidelines to satisfy the conditions in Theorem 5.3. First, to model injective multiset functions for the neighbor aggregation, the AGG operation (Eq.equation 15.16) is suggested to adopt the *sum* pooling operation, which is proved to universally represent functions defined over multisets whose elements are from a countable space (Lemma 5.1).

Lemma 5.1. *(Lemma 4 in (Xu et al, 2019d)) Suppose $\mathscr{S}$ is a countable universe of elements. Then there exists a function $q : \mathscr{S} \to \mathbb{R}^n$ such that $q(S) = \sum_{x \in S} \psi(x)$ is unique for each finite multiset $S \subset \mathscr{S}$, where ψ individually encodes each element in $\mathscr{S}$. Moreover, any multiset function g can be decomposed as $g(S) = \phi\left(\sum_{x \in S} \psi(x)\right)$ for some function ϕ.*

Remark 5.5. Note that the sum pooling operator is crucial, as some popular invariant pooling operators, such as the mean pooling operator, are not injective multiset functions. The significance of the sum pooling operation is to record the number of repetitive elements in a multiset. The mean pooling operation adopted by graph convolutional network (Kipf and Welling, 2017a) or the softmax-normalization (attention) pooling adopted by graph attention network (Veličković et al, 2018) may learn the distribution of the elements in a multiset but not the precise counts of the elements.

Thanks to the universal approximation theorem (Hornik et al, 1989), we can use multi-layer perceptrons (MLPs) to model and learn ψ and ϕ in Lemma 5.1 for universally injective AGG operation. In MP-GNN, we do not even need to explicitly model ψ and ϕ as the MSG and UPT operations — (Eqs.equation 11.45 and equation 5.3) respectfully — have already been implemented via MLPs. Therefore, using the sum pooling as the AGG operation is sufficient to achieve the most expressive MP-GNN:

$$\text{Expressive Message:} \quad \mathbf{m}_{vu}^{(k)} \leftarrow MLP_1^{(k-1)}(\mathbf{h}_v^{(k-1)} \oplus \mathbf{h}_u^{(k-1)}), \ \forall (u,v) \in \mathscr{E},$$

$$\text{Expressive Aggregation:} \quad \mathbf{a}_v^{(k)} \leftarrow \sum_{u \in \mathscr{N}_v} \mathbf{m}_{vu}^{(k)}, \ \forall v \in \mathscr{V},$$

$$\text{Expressive Update:} \quad \mathbf{h}_v^{(k)} \leftarrow MLP_2^{(k-1)}(\mathbf{h}_v^{(k-1)} \oplus \mathbf{a}_v^{(k)}), \ \forall v \in \mathscr{V}.$$

where $\oplus$ denotes concatenation. Actually, we can even simplify the procedure by using a single MLP. We can also set $\mathbf{m}_{vu}^{(k)} \rightarrow \mathbf{h}_u^{(k-1)}$, $\forall (u,v) \in E$ without decreasing the expressive power. Combining all the terms together, Xu et al (2019d) obtains the simplest update mechanism of node representations that constructs an injective mapping from $(\mathbf{h}_v^{(k-1)}, \{\mathbf{h}_u^{(k)} | u \in \mathscr{N}_v\})$ to $\mathbf{h}_v^{(k)}$:

$$\mathbf{h}_v^{(k)} \leftarrow MLP^{(k-1)}((1 + \varepsilon^{(k)})\mathbf{h}_v^{(k-1)} + \sum_{u \in \mathscr{N}_v} \mathbf{h}_u^{(k-1)}), \ \forall v \in \mathscr{V}, \tag{5.7}$$

where $\varepsilon^{(k)}$ is a learnable weight. This updating method, by using a NN-based language, is termed the graph isomorphism network (GIN) layer (Xu et al, 2019d).

Lemma 5.2 formally states that MP-GNN that adopts Eq.equation 5.7 may match the condition a) in Theorem 5.3.

Lemma 5.2. *Updating node representations by following Eq.equation 5.7 constructs an injective mapping from $(\mathbf{h}_v^{(k-1)}, \{\mathbf{h}_u^{(k)} | u \in \mathscr{N}_v\})$ to $\mathbf{h}_v^{(k)}$, if the node attributes X are from a countable space.*

Proof. Combine the proof for injectiveness of the sum aggregation with the universal approximation property of MLP (Hornik et al, 1989).

A similar idea may be adapted to the READOUT operation (Eq.5.4), which also requires an injective mapping of multisets:

$$\text{Expressive Inference:} \quad \hat{y}_S = MLP(\sum_{v \in S} \mathbf{h}_v^{(L)}). \tag{5.8}$$

Xu et al (2019d) has observed that node representations from earlier iterations may sometimes generalize better and thus also suggests using the READOUT (a counterpart to Eq.5.4) from the Jumping Knowledge Network (JK-Net) (Xu et al, 2018a), though it is not necessary from the perspective of the representation power of MP-GNN.

Overall, combining the update Eq.equation 5.7 and the READOUT Eq.equation 5.8, we may achieve an MP-GNN that is as powerful as the 1-WL test. In the next section, we introduce several techniques that allow MP-GNN to break the limitation of the 1-WL test and achieve even stronger expressive power.

5.4 Graph Neural Networks Architectures that are more Powerful than 1-WL Test

In the previous section, we characterized the representation power of MP-GNN that is bounded by the 1-WL test. In other words, if the 1-WL test cannot distinguish two GRL examples $(\mathscr{G}^{(1)}, S^{(1)})$ and $(\mathscr{G}^{(2)}, S^{(2)})$, then MP-GNN cannot distinguish them either. Although the 1-WL test cannot distinguish only a few corner graph structures, it indeed limits the applicability of GNNs in many real-world applications (You et al, 2019; Chen et al, 2020q; Ying et al, 2020b). In this section, we will introduce several approaches to overcome the above limitation of MP-GNN.

5.4.1 Limitations of MP-GNN

First, we will review several critical limitations of MP-GNN and the 1-WL test to gain the intuition for understanding the techniques that build more powerful GNNs. MP-GNN iteratively updates the representation of each node by aggregating representations of its neighbors. The obtained node representation essentially encodes the subtree rooted at Node v (Fig. 5.7). However, using this rooted subtree to represent a node may lose useful information, such as:

1. The information about the distance between multiple nodes is lost. For example, You et al (2019) noticed that MP-GNN has limited power in capturing the position/location of a given node with respect to another node in the graph. Many nodes may share similar subtrees, and thus, MP-GNN produces the same representation for them although the nodes may be located at different locations in the graph. This location information of nodes is crucial for the tasks that depend on multiple nodes, such as link prediction (Liben-Nowell and Kleinberg, 2007), as two nodes that tend to be connected with a link are typically located close to each other. An illustrative example is shown in Fig. 5.9.
2. The information about cycles is lost. Particularly, when expanding the subtree of a node, MP-GNN essentially losses track of the node identities in the subtrees. An illustrative example is shown in Fig. 5.10. The information about cycles is crucial in applications such as subgraph matching (Ying et al, 2020b) and counting (Liu et al, 2020e) because loops frequently appear in the queried subgraph patterns of a subgraph matching/counting problem. Chen et al (2020q) formally proved that MP-GNN is able to count star structures (a particular form of trees) but cannot count connected subgraphs with three or more nodes that form cycles.

Theoretically, there is a general class of graph representation learning problems that MP-GNN will fail to solve due to its limited representation. To show this, we define a class of graphs, termed attributed regular graphs.

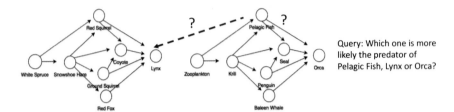

Fig. 5.9: A foodweb network example that demonstrates limitations of MP-GNN (Srinivasan and Ribeiro, 2020a). MP-GNN will associate Lynx and Orca with the same node representations, *i.e.*, $\mathbf{h}_{\text{Lynx}}^{(i)} = \mathbf{h}_{\text{Orca}}^{(i)}$, as these two nodes hold the same rooted subtree. Note that we do not consider node features. Thus, MP-GNN cannot predict whether Lynx or Orca is more likely to be the predator of Pelagic Fish (a link prediction task).

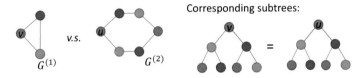

Fig. 5.10: The node representations $\mathbf{h}_v^{(L)}$ and $\mathbf{h}_u^{(L)}$ given by MP-GNN are the same, although they belong to different cycles – 3-cycle and 6-cycle, respectively.

Definition 5.8. (Attributed regular graphs) Consider an attributed graph $\mathscr{G} = (\mathscr{V}, \mathscr{E}, X)$. All nodes in $\mathscr{V}$ are partitioned according to their attributes $\mathscr{V} = \cup_{i=1}^k V_i$, such that two nodes from the same category V_i have the same attributes, while two nodes from different categories have different attributes. If for any two categories, V_i, V_j, $i, j \in [k]$, for any two nodes $u, v \in V_i$, the number of neighbors of u in V_j and the number of neighbors of v in V_j are equal, this graph can also be termed *attributed regular graph*. Denote C_i as the attribute of nodes in V_i. Also, denote the number of neighbors in V_j of a node $v \in V_i$ as r_{ij}. Then, the configuration of this attributed regular graph can be represented as a set of tuples $\text{Config}(\mathscr{G}) = \{(C_i, C_j, r_{ij})\}_{i,j \in [k]}$.

Note that the definition of attributed regular graphs is similar to k-partite regular graphs, while attributed regular graphs allow edges connecting nodes from the same partition. It can be shown that the 1-WL test will color all the nodes of one partition in the same way. Based on the bound of representation power of MP-GNN (Theorem 5.2), we can obtain the following corollary about the impossibility of MP-GNN to distinguish GRL examples defined on attributed regular graphs. Fig. 5.11 gives some examples that illustrate the impossibility. Actually, with sufficient layers (iterations), MP-GNN (the 1-WL test) will always transform any attributed graph into

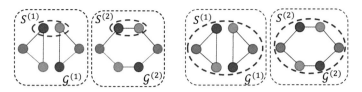

Fig. 5.11: A pair of attributed regular graphs $\mathcal{G}^{(1)}$, $\mathcal{G}^{(2)}$ with the same configuration and a proper selection of $S^{(1)}$, $S^{(2)}$: MP-GNN and the 1-WL test fail to distinguish $(\mathcal{G}^{(1)}, S^{(1)})$, $(\mathcal{G}^{(2)}, S^{(2)})$.

an attributed regular graph (Arvind et al, 2019) if we view the node representations obtained by MP-GNN as the node attributes on this transformed graph [1].

Corollary 5.1. *Consider two graph-structured features* $(\mathcal{G}^{(1)}, S^{(1)})$, $(\mathcal{G}^{(2)}, S^{(2)})$. *If two attributed regular graphs* $\mathcal{G}^{(1)}$, $\mathcal{G}^{(2)}$ *share the same configuration, i.e., Config$(\mathcal{G}^{(1)})$ = Config$(\mathcal{G}^{(2)})$, and two multisets of attributes* $\{X_v^{(1)} | v \in S^{(1)}\}$ *and* $\{X_v^{(2)} | v \in S^{(2)}\}$ *are also equal, then* $f_{MP-GNN}(\mathcal{G}^{(1)}, S^{(1)}) = f_{MP-GNN}(\mathcal{G}^{(2)}, S^{(2)})$. *Therefore, if graph representation learning problems associate* $\{X_v^{(1)} | v \in S^{(1)}\}$ *and* $\{X_v^{(2)} | v \in S^{(2)}\}$ *with different targets, MP-GNN does not hold the expressive power to distinguish them and predict their correct targets.*

Proof. The proof is obtained by tracking each iteration of the 1-WL test and performing an induction.

Next, we will introduce several approaches that address the above limitations and that further improve the expressive power of MP-GNN .

5.4.2 Injecting Random Attributes

The main reason for limitations on the expressive power of MP-GNN is that MP-GNN does not track node identities; however, different nodes with the same attributes will be initialized with the same vector representations. This condition will be maintained unless their neighbors propagate different node representations. One way to improve the expressive power of MP-GNN is to inject each node with a unique attribute. Given a GRL example $(\mathcal{G}, S)$, where $\mathcal{G} = (A, X)$,

$$g_I(\mathcal{G}, S) = (\mathcal{G}_I, S), \quad \text{where } \mathcal{G}_I = (A, X \oplus I), \tag{5.9}$$

where $\oplus$ is concatenation and I is an identity matrix, this gives each node a unique one-hot encoding and yields a new attributed graph G_I. The composite model

---

[1] Most transformed graphs have one single node per partition. In this case, two graphs that share the same configuration are isomorphic.

$f_{MP-GNN} \circ g_I$ increases expressive power as node identities are attached to the messages in the message passing framework and the distance and loop information can be learnt with sufficient iterations of message propagation.

However, the limitation of the above framework is that it is not permutation invariant (Def.5.4): given that two isomorphic GRL examples $(\mathscr{G}^{(1)}, S^{(1)}) \cong (\mathscr{G}^{(2)}, S^{(2)})$, $g_I(\mathscr{G}^{(1)}, S^{(1)})$ and $g_I(\mathscr{G}^{(2)}, S^{(2)})$ may be not isomorphic any more. Then, the composite model $f_{MP-GNN} \circ g_I(\mathscr{G}^{(1)}, S^{(1)})$ may not equal $f_{MP-GNN} \circ g_I(\mathscr{G}^{(2)}, S^{(2)})$. As the obtained model loses the fundamental inductive bias of graph representation learning, it is hard to be generalized[2].

Remark 5.6. Some other approaches may share the same limitation with g_I, e.g., using the adjacency matrix A (each row of A representing node attributes). However, Srinivasan and Ribeiro (2020a) argued that node embeddings obtained via matrix factorization, such as deepwalk (Perozzi et al, 2014) and node2vec (Grover and Leskovec, 2016), can keep the required invariance and thus are still generalizable. We will return to this concept in Sec.5.4.2.4.

To overcome the above limitation, different methods have been proposed recently. These models share the following strategy: they first design some additional random node attributes Z, use them to argue the original dataset, and then learn a GNN model over the augmented dataset (Fig. 5.13).

The obtained models will be more expressive, as the random node attributes can be viewed as unique node identities that distinguish nodes. However, if the model is only trained based on a single GRL example augmented by these random attributes, it cannot keep invariance as discussed above. Instead, the model needs to be trained over multiple GRL examples augmented by independently injected random attributes. The new augmented GRL examples have the same target as the original GRL examples from which they are generated. This training of models over augmented examples essentially regularizes the permutation variance of the models and makes them behave almost "permutation invariant."

Different methods to inject these random attributes may be adopted, but a direct way is to attach Z to X, i.e., given a graph-structured data $(\mathscr{G}, S)$, where $\mathscr{G} = (A, X)$,

$$g_Z(\mathscr{G}, S) = (\mathscr{G}_Z, S), \quad \text{where } \mathscr{G}_Z = (A, \tilde{X}_Z) \text{ and } \tilde{X}_Z \leftarrow X \oplus Z. \tag{5.10}$$

Note that for each realization Z, the composite model $f_{MP-GNN} \circ g_Z$ is not permutation invariant. Instead, all these approaches make $\mathbb{E}[f_{MP-GNN} \circ g_Z]$ permutation invariant and expect the models to keep invariant in expectation. To match such invariance in expectation, an approach must satisfy the following proposition.

Proposition 5.1. *The following two properties are needed to build a model by injecting random features Z.*

---

[2] Recent literature often states that the composite model is not inductive. Inductiveness and generalization to unobserved examples are related. In the transductive setting, $f_{MP-GNN} \circ g_I$ is less generalizable than f_{MP-GNN}, although the prediction performance of $f_{MP-GNN} \circ g_I$ may be sometimes better than f_{MP-GNN} due to the much stronger expressive power of $f_{MP-GNN} \circ g_I$.

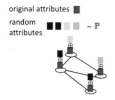

Types of random attributes	Positional information	Model & reference
Random permutations	No	RP-GNN (Murphy et al, 2019)
(Almost uniform) Discrete r.v.	No	rGIN (Sato et al, 2020)
Distances to random anchor sets	Yes	PGNN (You et al, 2019)
Graph-convoluted Gaussian r.v.	Yes	CGNN (Srinivasan & Ribeiro, 2020)
Random signed Laplacian eigenmap	Yes	LE-GNN (Dwivedi et al, 2020)

original attributes ■
random ■ ■ □ □ ~ $\mathbb{P}$
attributes

Fig. 5.12: Injecting random node attributes can improve the expressive power of GNNs. Different types of random node attributes are adopted in different works. Some random node attributes contain node positional information (the position of a node with respect to other nodes in the graph).

1. *A sufficient number of Z's should be sampled during the training stage so that the model indeed captures permutation invariance in expectation.*
2. *The randomness in Z should be agnostic to the original node identities.*

To satisfy the property 1, a method suggests that for each forward pass to compute $f_{MP-GNN} \circ g_Z$ during the training stage, one Z should be re-sampled once or multiple times to get enough data argumentation. To satisfy the property 2, four different types of random Z have been proposed as described next.

5.4.2.1 Relational Pooling - GNN (RP-GNN) (Murphy et al, 2019a)

Murphy et al (2019a) considered randomly assigning an order of nodes as their extra attributes and proposed the model relational pooling GNN (RP-GNN). We use Z_{RP} to denote additional node attributes Z used in RP-GNN. Suppose the graph $\mathscr{G}$ has n nodes, Z_{RP} is uniformly sampled from all possible permutation matrices. That is, randomly pick a bijective mapping (permutation) $\pi : V(\mathscr{G}) \to V(\mathscr{G})$, and design permutation matrix $[Z_{RP}]_{ij} = 1$ if $j = \pi(i)$ and $[Z_{RP}]_{ij} = 0$ otherwise. Then, RP-GNN adopts the composite model,

$$f_{RP-GNN} = \mathbb{E}[f_{MP-GNN} \circ g_{Z_{RP}}]. \tag{5.11}$$

Theorem 5.4. *(Theorem 2.2 (Murphy et al, 2019a)) The RP-GNN f_{RP-GNN} is strictly more powerful than the original f_{MP-GNN}.*

Computing the expectation $\mathbb{E}[f_{MP-GNN} \circ g_{Z_{RP}}]$ is intractable as one needs to compute $f_{MP-GNN} \circ g_{Z_{RP}}$ for all possible permutations $\pi : V(\mathscr{G}) \to V(\mathscr{G})$. To overcome this problem, sampling of Z_{RP} may be needed.

However, as the entire permutation space is too large, uniformly random sampling of a limited number of Z_{RP} may introduce a large variance. To reduce the potential variance, Murphy et al (2019a) also proposed to sample all π's that permute only a small subset of nodes instead of the entire set of nodes. More recently, Chen et al (2020q) further adapted RP-GNN to solve the subgraph counting problem. They suggest to use all π's that permute all the nodes of each connected local subgraph.

5.4.2.2 Random Graph Isomorphic Network (rGIN) (Sato et al, 2021)

Sato et al (2021) generalized RP-GNN by setting the additional attributes of each node sampled from an almost uniform discrete probability distribution. The key difference from RP-GNN is that the additional attributes of two nodes are set to be independent of each other (while in RP-GNN, one-time random attributes of different nodes are correlated due to the nature of permutation). We use Z_r to denote Z used in rGIN and $[Z_r]_v$ to denote the attributes of node v. For example, they set

$$f_{rGIN} = \mathbb{E}[f_{MP-GNN} \circ g_{Z_r}], \text{ where } [Z_r]_v \sim \text{Unif}(\mathscr{D}) \, i.i.d. \, \forall \, v \in \mathscr{V}[\mathscr{G}],$$

where $\mathbb{E}$ indicates expectation and $\mathscr{D}$ is a discrete space with at least $1/p$ elements for some $p > 0$. Similar to RP-GNN, f_{rGIN} can be implemented by sampling only a few Z_r's for each evaluation of $f_{MP-GNN} \circ g_Z$ (indeed, one Z_r is sampled per forwarding evaluation (Sato et al, 2021)).

Theorem 5.5. *(Theorem 4.1 (Sato et al, 2021)) Consider a GRL example $(\mathscr{G},v)$, where only a single node is contained in the node set of interest. For any graph-structured features $(\mathscr{G}',v')$, where the nodes of $\mathscr{G}'$ have a bounded maximal degree and the attributes X come from a finite space, then there exist an MP-GNN , such that:*

1. *If $(\mathscr{G}',v') \cong (\mathscr{G},v)$, $f_{MP-GNN} \circ g_{Z_r}(\mathscr{G}',v') > 0.5$ with high probability.*
2. *If $(\mathscr{G}',v') \ncong (\mathscr{G},v)$, $f_{MP-GNN} \circ g_{Z_r}(\mathscr{G}',v') < 0.5$ with high probability.*

This result can be viewed as a characterization of the expressive power of rGIN. However, this result is lessened by the fact that almost all nodes of all graphs will be associated with different representations within two iterations of the 1-WL test (so is MP-GNN) (Babai and Kucera, 1979). Moreover, the isomorphism problem of graphs with a bounded degree is known to be in P (Fortin, 1996). Instead, a very recent work was able to demonstrate the universal approximation of rGIN, which gives a stronger characterization of the expressive power of rGIN.

Theorem 5.6. *(Theorem 4.1 (Abboud et al, 2020)) Consider any invariant mapping $f^* : \mathscr{G}_n \to \mathbb{R}$, where $\mathscr{G}_n$ contains all graphs with n nodes. Then, there exists a rGIN $f_{MP-GNN} \circ g_{Z_r}$ such that*

$$p(|f_{MP-GNN} \circ g_{Z_r} - f^*| < \varepsilon) > 1 - \delta, \text{ for some given } \varepsilon > 0, \, \delta \in (0,1).$$

The above RP-GNN and rGIN adopt random attributes that are totally agnostic to the input data $(\mathscr{G},S)$. Instead, the next two methods inject random attributes that leverage the input data. Particularly, these random attributes are related to the position/location of a node in the graph, which tends to counter the loss of positional information of nodes in MP-GNN.

5.4.2.3 Position-aware GNN (PGNN) (You et al, 2019)

You et al (2019) demonstrated that MP-GNN may not capture the position/location of a node in the graph, which is critical information for applications such as link prediction. Therefore, they proposed to use node positional embeddings as extra attributes. To capture permutation invariance in the sense of expectation, node positional embeddings are generated based on randomly selected anchor node sets. We denote the random attributes adopted in PGNN as Z_P, which is constructed as follows. Considering a graph $\mathscr{G} = (\mathscr{V}, \mathscr{E}, X)$,

1. Randomly select a few anchor sets $(S_1, S_2, ..., S_K)$, where $S_k \subset \mathscr{V}$. Note that the choice of S_k is agnostic to the node identities: given a k, S_k will include each node with the same probability.
2. For some $u \in G$, set $[Z_P]_u = (d(u, S_1), ..., d(u, S_K))$ where $d(u, S_k)$, $k \in [K]$ is a distance metric between u and the anchor set S_k.

As the selection of the anchor sets is agnostic to node identities, the obtained Z_P still satisfies the property 2 in Proposition 5.1. Next, we specify the strategy to sample these anchor sets and the choice of the distance metric. The primary requirement to select those anchor sets is to keep low distortion of the two distances between nodes, where one distance is given by the original graph and the other one is given by those anchor sets. Specifically, distortion measures the faithfulness of the embeddings in preserving distances when mapping from one metric space to another metric space, which is defined as follows:

Definition 5.9. Given two metric spaces $(\mathscr{V}, d)$ and $(\mathscr{Z}, d')$ and a function $Z_P : \mathscr{V} \to \mathscr{Z}$, Z_P is said to have distortion α if $\forall u, v \in \mathscr{V}$, $\frac{1}{\alpha} d(u, v) \leq d'([Z_P]_u, [Z_P]_v) \leq d(u, v)$.

Fortunately, Bourgain (1985) showed the existence of a low distortion embedding that maps from any metric space to the l_p metric space:

Theorem 5.7. *(Bourgain's Theorem (Bourgain, 1985))* *Given any finite metric space $(\mathscr{V}, d)$ with $|\mathscr{V}| = n$, there exists an embedding of $(\mathscr{V}, d)$ into $\mathbb{R}^K$ under any l_p metric, where $K = O(\log^2 n)$, and the distortion of the embedding is $O(\log n)$.*

Based on a constructive proof of Theorem 5.7, Linial et al (1995) provide an algorithm to construct an $O(\log^2 n)$ dimensional embedding via anchor sets. This yields the selection of anchor sets and the definition of the distance metric to define Z_P, which are adopted by PGNN (You et al, 2019).

By selecting $K = c \log^2 n$, many random sets $S_{i,j} \subset \mathscr{V}, i = 1, 2, ..., \log n, j = 1, 2, ..., c \log n$, where c is a constant, $S_{i,j}$ is chosen by including each point in $\mathscr{V}$ independently with probability $\frac{1}{2^i}$. We further define

$$[Z_P]_v = \left(\frac{d(v, S_{1,1})}{k}, \frac{d(v, S_{1,2})}{k}, ..., \frac{d(v, S_{\log n, c \log n})}{k} \right) \tag{5.12}$$

where $d(v, S_{i,j}) = \min_{u \in S_{i,j}} d(v, u)$. Then, Z_P is an embedding method that satisfies Theorem 5.7.

Compared with RP-GNN and rGIN, the random attributes adopted by PGNN deal specifically with the positional information of a node in graph. Therefore, PGNN is better for the tasks that are directly related to the positions of nodes, *e.g.*, link prediction. You et al (2019) did not provide a mathematical characterization of the representation power of PGNN. However, the way to establish Z_P allows that for the two nodes u, v, the attributes $[Z_P]_u$ and $[Z_P]_v$ by definition are statistically correlated. As for the example in Fig. 5.9, such correlation gives PGNN the information that the distance between Lynx and Pelagic Fish is different from the distance between Orca and Pelagic Fish, and thus may successfully distinguish $(G, \{\text{Lynx}, \text{Pelagic Fish}\})$ and $(G, \{\text{Orca}, \text{Pelagic Fish}\})$ and making the right link prediction.

Note that the original PGNN (You et al, 2019) does not use MP-GNN as the backbone to perform message passing. Instead, PGNN allows message passing from nodes to anchor sets. As such, this approach is not directly relevant to the expressive power and is thus out of the scope of this chapter, so we will not discuss it in detail. Interested readers may refer to the original paper (You et al, 2019).

5.4.2.4 Randomized Matrix Factorization (Srinivasan and Ribeiro, 2020a)(Dwivedi et al, 2020)

Srinivasan and Ribeiro (2020a) recently made an important observation that node positional embeddings obtained via the factorization of some variants of the adjacency matrix A can be used as node attributes as long as certain random perturbation is allowed. The obtained models still keep permutation invariance in expectation. Srinivasan and Ribeiro (2020a) argue that a model that is built upon these random perturbed node positional embeddings is still inductive and holds good generalization properties. This significant observation challenges the traditional claim that models built upon these node positional embeddings are not inductive. A high-level idea of why this is true is as follows: suppose the SVD decomposition of the adjacency matrix $A = U \Sigma U^T$. When we permute the order of nodes, that is, the row and column orders of A, the row order of U will be changed simultaneously. Therefore, the models that use U as the node attributes should keep the permutation invariance. That randomly perturbed factorization is needed because such SVD decomposition is not unique.

Although Srinivasan and Ribeiro (2020a) proposed this idea, they did not explicitly compute the node positional embeddings via matrix factorization. Instead, their method samples a series of Gaussian random matrices $Z_{G,1}, Z_{G,2}, \ldots$ and let them propagate over the graph, *e.g.*, for the two hops,

$$Z_G = \psi(\hat{A} \, \psi(\hat{A} Z_{G,1}) + Z_{G,2}),$$

where ψ's are MLPs and $\hat{A}$ indicates some variant of the adjacency matrix. The rows of Z_G essentially give rough node positional embeddings. Then, these obtained node embeddings are further used as the attributes of nodes in MP-GNN.

Dwivedi et al (2020) indeed adopted matrix factorization explicitly. They proposed to use the randomly perturbed Laplacian eigenmaps as the additional attributes. Specifically, suppose the normalized Laplacian matrix is defined as

$$L = I - D^{-1/2}AD^{-1/2},$$

where D is the diagonal degree matrix. Denote the eigenvalue decomposition of L as $L = U\Sigma U^T$. The eigenvalue decomposition is not unique, so we assume that U can be arbitrarily chosen from all the potential choices. Fortunately, if there are no multiple eigenvalues, this U is unique for each column up to a $\pm$ sign. Then, we may directly set the extra node attributes as

$$Z_{LE} = U\Gamma, \text{ where } \Gamma_{ii} \sim \text{Unif}(\{-1,1\}) \text{ i.i.d. } \forall i \in [|V|], \Gamma_{ij} = 0, \forall i \neq j, \quad (5.13)$$

where Γ is a diagonal matrix where diagonal elements are uniformly independently set as 1 or -1. Here, U can be replaced with a few slices of the columns of U. Let $g_{Z_{LE}}$ denote the operation to concatenate these additional attributes Z_{LE} with the original node attributes. Then, the overall composite model becomes $f_{MP-GNN} \circ g_{Z_{LE}}$. The following lemma shows that the permutation invariance of $f_{MP-GNN} \circ g_{Z_{LE}}$ in expectation if the Laplacian matrices hold distinct eigenvalues:

Lemma 5.3. *If $(\mathscr{G}^{(1)}, S^{(1)}) \cong (\mathscr{G}^{(2)}, S^{(2)})$ and if there are no multiple eigenvalues of their corresponding normalized Laplacian matrices, then any choice of eigenvalue decomposition to obtain node embeddings will yield*

$$\mathbb{E}(f_{MP-GNN} \circ g_{Z_{LE}}(\mathscr{G}^{(1)}, S^{(1)})) = \mathbb{E}(f_{MP-GNN} \circ g_{Z_{LE}}(\mathscr{G}^{(2)}, S^{(2)})).$$

Proof. The proof can be easily seen from the above arguments.

As shown in Lemma 5.3, the composite model keeps permutation invariance in expectation for most graphs, although it may break invariance in some corner cases. Regarding the expressive power, Z_{LE} associates different nodes with distinct attributes because U is an orthogonal matrix by definition. Hence, there must exist $f_{MP-GNN} \circ g_{Z_{LE}}$ that may distinguish any node subsets from the graph:

Theorem 5.8. *For any two GRL examples $(\mathscr{G}, S^{(1)})$, $(\mathscr{G}, S^{(2)})$ over the same graph $\mathscr{G}$, even if they are isomorphic, as long as $S^{(1)} \neq S^{(2)}$, there exists an f_{MP-GNN} such that $f_{MP-GNN} \circ g_{Z_{LE}}(\mathscr{G}, S^{(1)}) \neq f_{MP-GNN} \circ g_{Z_{LE}}(\mathscr{G}, S^{(2)})$. However, if those two GRL examples are indeed isomorphic $(\mathscr{G}, S^{(1)}) \cong (\mathscr{G}, S^{(2)})$ over the same graph $\mathscr{G}$ and the normalized Laplacian matrix of $\mathscr{G}$ has no multiple same-valued eigenvalues, then $\mathbb{E}(f_{MP-GNN} \circ g_{Z_{LE}}(\mathscr{G}, S^{(1)})) = \mathbb{E}(f_{MP-GNN} \circ g_{Z_{LE}}(\mathscr{G}, S^{(2)}))$.*

Proof. The proof can be easily seen from the above arguments.

Theorem 5.8 implies the potential of $f_{MP-GNN} \circ g_{Z_{LE}}$ to distinguish different node sets from the same graph. Note that although $f_{MP-GNN} \circ g_{Z_{LE}}$ achieves great representation power, it does not always work very well for link prediction in practice (Dwivedi et al, 2020) when compared with another model SEAL (Zhang and

Chen, 2018b) (compare their performance on the COLLAB dataset in (Hu et al, 2020b)). SEAL is based on the deterministic distance attributes that are introduced in the next subsection. Whether a model is permutation invariant is a much weaker statement on characterizing the generalization of the model. Actually, when the model is paired node positional embeddings, the dimension of the parameter space increases, and thus also negatively impacts the generalization. A comprehensive investigation of this observation is left for future study.

In the next subsection, we will introduce deterministic node distance attributes, which provide a different angle to solve the above problem. Distance encoding has a solid mathematical foundation and provides the theoretical support for many empirically well-behaved models such as SEAL (Zhang and Chen, 2018b) and ID-GNN (You et al, 2021).

5.4.3 Injecting Deterministic Distance Attributes

In this subsection, we will introduce an approach that boosts the expressive power of MP-GNN by injecting deterministic distance attributes.

The key motivation behind the deterministic distance attributes is as follows. In Section 5.4.1, we have shown that MP-GNN is limited in its ability to measure the distances between different nodes, to count cycles[3], and to distinguish attributed regular graphs. All of these limitations are essentially inherited from the 1-WL test which does not capture distance information between the nodes. If MP-GNN is paired with some distance information, then the composite model must achieve more expressive power. Then, the question is how to inject the distance information properly.

There are two important pieces of intuition to design such distance attributes. First, the effective distance information is typically correlated with the tasks. For example, consider a GRL example $(\mathscr{G}, S)$. If this task is node classification ($|S| = 1$), the information of distance from this node to itself (thus the cycles containing this node) is relevant because it measures the information of the contextual structure. If the task is link prediction ($|S| = 2$), the information of distance between the two end nodes of the link is relevant as two nodes near to each other in the network tend to be connected by a link. For graph-level prediction ($S = \mathscr{V}(\mathscr{G})$), the information of distances between any pairs of nodes could be relevant as it can be viewed as a group of link predictions. Second, besides the distance between the nodes in S, the distance from S to other nodes in G may also provide useful side-information. Both two aspects inspire the design of distance attributes.

There have been a few empirically successful GNN models that leverage deterministic distance attributes, although their impact on the expressive power of GNNs

---

[3] Cycles actually carry a special type of distance information, as they describe the length of walks from one node to itself. If the distance from one node to itself is not measured by the shortest path distance but by the returning probability of random walk, this distance already contains the cycle information.

has not been characterized until very recently (Li et al, 2020e). For link prediction, Li et al (2016a) first consider annotating the two end-nodes of the link of interest. These two end-nodes are annotated with one-hot encodings and all other nodes are annotated by zeros. Such annotations can be transformed into distance information via GNN message passing. Again for link prediction, Zhang and Chen (2018b) first sample an enclosing subgraph around the queried link and then annotate each node in this subgraph with one-hot encodings of the shortest path distances (SPDs) from this node to the two end-nodes of the link. Note that deciding whether a node is in the enclosing subgraph around the queries link already gives a distance attribute. Zhang and Chen (2019) uses a similar idea to perform matrix completion which is a similar task to link prediction. For graph classification and graph-level property prediction, Chen et al (2019a) and Maziarka et al (2020a) adopt the SPDs between two nodes as edge attributes. These edge attributes can be also used as the input of MSG (Eq.equation 11.45) in MP-GNN. You et al (2021) annotates a node as 1 and other nodes as 0 to improve MP-GNN in node classification. As our focus is on the theoretical characterization of the expressive power, we will not go into detail about these empirically successful works. Interested readers are referred to the relevant papers.

Remark 5.7. (Comparison between deterministic distance attributes and random attributes) Deterministic distance attributes have some advantages. First, as there is no randomness in the input attributes, the optimization procedure of the model contains less noise. Hence, the training procedure tends to converge much faster than the model with random attributes. The model evaluation performance contains much less noise too. Some empirical evaluation of the convergence of the model training with random attributes can be found in Abboud et al (2020). Second, a model based on deterministic distance attributes typically shows better generalization in practice than the one based on random attributes. Although theoretically a model is permutation invariant when being trained based on sufficiently many examples with random attributes (as discussed in Sec.5.4.2), in practice, this could be hard to achieved due to the high complexity.

Deterministic distance attributes have some disadvantages. First, models that are paired with deterministic attributes may never achieve the universal approximation, unless the graph isomorphism problem is in P. However, random attributes may be universal in the probabilistic sense (*e.g.,* Theorem 5.6). Second, deterministic distance attributes typically depend on the information S in a GRL example $(\mathscr{G}, S)$. This introduces an issue in computation: that is, if there are two GRL examples $(\mathscr{G}^{(1)}, S^{(1)})$ and $(\mathscr{G}^{(2)}, S^{(2)})$ sharing the same graph $\mathscr{G}$ but with different node sets of interest $S^{(1)} \neq S^{(2)}$, they will be attached with different deterministic distance attributes and hence GNNs have to make inference over them separately. However, GNNs with random attributes can share intermediate node representations $\{\mathbf{h}_v^{(L)} | v \in \mathscr{V}[\mathscr{G}]\}$ in Eq.equation 5.4, between the two GRL examples, which saves intermediate computation.

5.4.3.1 Distance Encoding (Li et al, 2020e)

Suppose we aim to make prediction for a GRL example $(\mathscr{G},S)$. Li et al (2020e) defined distance encoding $\zeta(u|S)$ as an extra node attribute for node $u \in \mathscr{V}[\mathscr{G}]$.

Definition 5.10. For a GRL example $(\mathscr{G},S)$ where $\mathscr{G} = (A,X)$. Distance encoding $\zeta(u|S)$ for node u is defined as follows

$$\zeta(u|S) = \sum_{v \in S} \mathrm{MLP}(\zeta(u|v)) \tag{5.14}$$

where $\zeta(u|v)$ charaterizes a certain distance between u and v. We may choose

$$\zeta(u|v) = \mathrm{g}(\ell_{uv}), \ \ell_{uv} = (1,(W)_{uv},(W^2)_{uv},...,(W^k)_{uv},...), \tag{5.15}$$

where $W = AD^{-1}$ is the random walk matrix and $\mathrm{g}(\cdot)$ is a general function that maps ℓ_{uv} to different types of distance measures.

Note that $\zeta(u|S)$ depends on the graph structure $\mathscr{G}$, which we omit in our notation for simplicity. First, setting $\mathrm{g}(\ell_{uv})$ as the first non-zero position in ℓ_{uv} gives the *shortest-path-distance (SPD)* from v to u. Second, setting $\mathrm{g}(\ell_{uv})$ as follows gives *generalized PageRank scores* (Li et al, 2019f):

$$\zeta_{gpr}(u|v) = \sum_{k \geq 1} \gamma_k (W^k)_{uv} = \left(\sum_{k \geq 0} \gamma_k W^k\right)_{uv}, \quad \gamma_k \in \mathbb{R}, \text{ for all } k \in \mathbb{Z}_{\geq 0}. \tag{5.16}$$

Different choices of $\{\gamma_k | k \in \mathbb{Z}_{\geq 0}\}$ yield various distance measures between u and v.

$$\text{Personalized PageRank scores (Jeh and Widom, 2003):} \quad \gamma_k = \alpha^k, \alpha \in (0,1),$$
$$\text{Heat-kernel PageRank scores (Chung, 2007):} \quad \gamma_k = \beta^k e^{-\beta}/k!, \beta > 0,$$
$$\text{Inverse hitting time (Lovász et al, 1993):} \quad \gamma_k = k.$$

It is important to see that the above definition of distance encoding satisfies permutation invariance.

Lemma 5.4. *For two isomorphic GRL examples* $(\mathscr{G}^{(1)},S^{(1)}) \stackrel{\pi}{\cong} (\mathscr{G}^{(2)},S^{(2)})$, *if* $\pi(u) = \pi(v)$ *for* $u \in \mathscr{V}[\mathscr{G}^{(1)}]$ *and* $v \in \mathscr{V}[\mathscr{G}^{(2)}]$, *their distance encodings are equal* $\zeta(u|S^{(1)}) = \zeta(v|S^{(2)})$.

Proof. The proof can be easily seen by the definition of distance encoding.

Li et al (2020e) considers using distance encoding as node extra attributes. Specifically, MP-GNN can be improved by setting $\tilde{X}_v = X_v \oplus \zeta(v|S)$,where $\oplus$ is the concatenation. The obtained model is termed DE-GNN, denoted as f_{DE}.

DE-GNN has been shown to be more powerful than MP-GNN. Recall that the fundamental limit of MP-GNN is the 1-WL test for graph representation learning problems (Theorem 5.2). Corollary 5.1 further indicates that attributed regular graphs may not be distinguished by MP-GNN under certain scenarios. Li et al

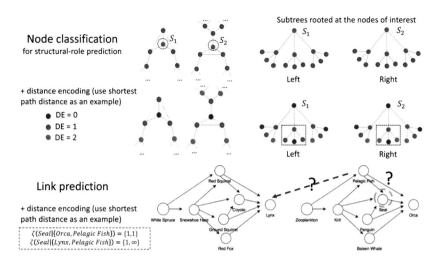

Fig. 5.13: Distance encoding can be used to distinguish non-isomorphic graph-structured examples. In the example of node classification, we consider classifying nodes based on their roles in their contextual structures, termed structural roles (Henderson et al, 2012). Nodes in S_1 and S_2 have different structure roles. However MP-GNN with two layers will confuse these two nodes; while with distance encoding, DE-GNN may distinguish them. In the example of link prediction, although two nodes {Lynx, G} and {Orca, G} are isomorphic (where we ignore the node identities), distance encoding on the node Seal allows us to distinguish node pairs {Orca, Pegagic Fish} and {Lynx, Pegagic Fish}.

(2020e) considers the scenario when the graphs are regular and do not have attributes and proved that DE-GNN can distinguish two GRL examples with high probability, which is formally stated in the following theorem.

Theorem 5.9. *(Theorem 3.3 (Li et al, 2020e)) Consider two GRL examples $(\mathscr{G}^{(1)}, S^{(1)})$ and $(\mathscr{G}^{(2)}, S^{(2)})$ where $\mathscr{G}^{(1)}$ and $\mathscr{G}^{(2)}$ are two n-sized unattributed regular graphs, and $|S^{(1)}| = |S^{(2)}|$ is a constant (independent of n). Suppose $\mathscr{G}^{(1)}$ and $\mathscr{G}^{(2)}$ are uniformly independently sampled from all n-sized r-regular graphs where $3 \leq r < (2\log n)^{1/2}$. Then, for any small constant $\varepsilon > 0$, there exists DE-GNN with certain weights within $L \leq \lceil (\frac{1}{2} + \varepsilon) \frac{\log n}{\log(r-1)} \rceil$ layers that can distinguish these two examples with high probability. Specifically, the outputs $f_{DE}((\mathscr{G}^{(1)}, S^{(1)})) \neq f_{DE}((\mathscr{G}^{(2)}, S^{(2)}))$ with probability $1 - o(n^{-1})$. The specific form of DE, i.e., g in Eq.equation 5.15, can be simply chosen as short path distance. The little-o notation here and later are w.r.t. n.*

Theorem 5.9 focuses on the node sets of unattributed regular graphs. We conjecture that the statement can be generalized to attributed regular graphs as distinct attributes can only further improve the distinguishing power of a model. Moreover, the assumption on regularity of graphs is also not crucial because the 1-WL test or MP-GNN may transform all graphs, attributed or not, into attributed regular graphs with enough iterations (Arvind et al, 2019).

Of course, DE-GNN may not distinguish any non-isomorphic GRL examples. Li et al (2020e) introduce the limitation of DE-GNN. Particularly, DE-GNN cannot distinguish nodes of distance regular graphs with the same intersection arrays, although DE-GNN may distinguish their edges (See Fig. 5.14 later). Li et al (2020e) also generalize the above results to the case that leverages distance attributes as edge attributes (to control message aggregation in MP-GNN). Interested readers can check the details in their original paper.

5.4.3.2 Identity-aware GNN (You et al, 2021)

As a concurrent work with DE-GNN, You et al (2021) studied a special type of distance encoding to improve the node representations learnt by MP-GNN. Specifically, when MP-GNN is adopted to compute the representation of node v, You et al (2021) suggests attaching each node u in the graph with an extra binary attribute $\zeta_{ID}(u|\{v\})$ to indicate the identity of node v where

$$\zeta_{ID}(u|\{v\}) = \begin{cases} 1 & \text{if } u = v, \\ 0 & \text{o.w.} \end{cases} \tag{5.17}$$

MP-GNN that leverages $\zeta_{ID}(u|\{v\})$ is termed Identity-aware GNN (ID-GNN). $\zeta_{ID}(u|\{v\})$ is a simple implementation of distance encoding (Eq. equation 5.14) when the set S contains only one node v. Although ID-GNN does not compute distance measures as DE-GNN, ID-GNN holds the same representation power as DE-GNN for node classification, as the distance information from another node u to the target node v can be learnt by ID-GNN via an extra identity attribute.

Theorem 5.10. *For two graph-structured examples* $(\mathcal{G}^{(1)}, S^{(1)})$ *and* $(\mathcal{G}^{(2)}, S^{(2)})$, *where* $|S^{(i)}| = 1$ *for* $i \in \{1,2\}$ *and* $\mathcal{G}^{(i)}$ *is unattributed, if DE-GNN can distinguish them with L layers, then ID-GNN requires at most 2L layers to distinguish them.*

Proof. ID-GNN needs the first L layers to propagate the identity attribute to capture distance information and the second L layers to let such information propagate back to finally be merged into the node representations.

Although ID-GNN adopts a specific type of DE to learn node representations, ID-GNN was also used to perform graph-level prediction (You et al, 2021). Specifically, for every node v in the graph G, ID-GNN attaches 1 to this node, 0's to other nodes and computes the node representation h_v. Iterating over all the nodes, ID-GNN collects all node representations $\{\mathbf{h}_v|v \in \mathcal{V}(\mathcal{G})\}$. Then, by following Eq.equation 5.4 (S is the entire node set $\mathcal{V}(\mathcal{G})$ here), ID-GNN can aggregate the node representations of all the nodes and further make graph-level predictions. Actually, combining the statement of Theorem 5.9 and the union bound, Li et al (2020e) indicates the

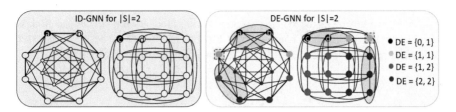

Fig. 5.14: ID-GNN *v.s.* DE-GNN makes predictions for a pair of nodes. Two graphs are the Shrikhande graph (left) and the 4×4 Rook's graph (right). ID-GNN (black nodes attached identities) cannot distinguish node pairs $\{a,b\}$ and $\{c,d\}$. DE-GNN may learn distinct representations of $\{a,b\}$ and $\{c,d\}$. In these two graphs, each node is colored with its DE that is a set of SPDs to either node in the target node sets $\{a,b\}$ or $\{c,d\}$ (Eq. equation 5.14). Note that the neighbors of nodes with DE= $\{1,1\}$ (dashed boxes) are enclosed by red ellipses which shows that the neighbors of these two nodes have different DE's. Hence, after one layer of DE-GNN, the intermediate representations of nodes with DE= $\{1,1\}$ are different between these two graphs. Using another layer, DE-GNN can distinguish the representations of $\{a,b\}$ and $\{c,d\}$.

expressive power of the above procedure for the entire graph classification problem, summarized in the following corollary.

Corollary 5.2. *(Reproduced from Corollary 3.4 (Li et al, 2020e)) Consider two GRL examples $\mathscr{G}^{(1)}$ and $\mathscr{G}^{(2)}$. Suppose $\mathscr{G}^{(1)}$ and $\mathscr{G}^{(2)}$ are uniformly independently sampled from all n-sized unattributed r-regular graphs where $3 \leq r < (2\log n)^{1/2}$. Then, ID-GNN with a sufficient number of layers can distinguish these two graphs with probability $1 - o(1)$. The little-o notation here and later are w.r.t. n.*

ID-GNN can be viewed as the simplest version of DE-GNN that achieves the same expressive power for node-level prediction. However, when the prediction tasks contain two nodes, *i.e.,* node-pair-level prediction, ID-GNN will be less powerful than DE-GNN.

To make a prediction for a GRL example $(\mathscr{G}, S)$ where $|S| = 2$, ID-GNN can adopt two different approaches. First, ID-GNN can attach the extra identity attributes to the two nodes in S separately, learn their representations separately and combine these two representations to make the final prediction. However, this approach cannot capture the distance information between the two nodes in S. Instead, ID-GNN uses an alternative approach. ID-GNN attaches the extra identity attribute to only one of nodes in S and performs message passing. Then, after a sufficient number of layers where the extra node identity is propagated from one node to another in S, the distance information between these two nodes can be captured. Finally, ID-GNN makes its prediction based on the two node representations in S. Note that although the second approach captures the distance information between the two nodes in S, it is still less powerful than DE-GNN. One example is shown in Fig. 5.14.

Up to this point, we have mostly focused on the message passing framework of GNNs, which leverages the sparsity of real-world graphs. In the next subsection, we remove the need for sparsity and discuss higher-order GNNs. These GNNs essentially mimic higher-dimensional WL tests and achieve more expressive power.

5.4.4 Higher-order Graph Neural Networks

The final collection of techniques for building GNNs, which overcome the limitation of the 1-WL test, are related to higher-dim WL test. In this subsection, for notational simplicity, we focus only on graph-level prediction learning problems, where higher-order GNNs are mostly used.

The family of WL tests forms a hierarchy for the graph isomorphism problem (Cai et al, 1992). There are different definitions of the higher-dim WL tests. We follow the terminology adopted in Maron et al (2019a) and introduce two types of WL tests: the k-forklore WL (k-FWL) test and the k-WL test.

Recall $\mathcal{G}^{(i)} = \{A^{(i)}, X^{(i)}\}$, $i \in \{1,2\}$. For both $\mathcal{G}^{(i)}$'s, $i \in \{1,2\}$, do the following steps.

1. For each k-tuple of node set $V_j = (v_{j_1}, v_{j_2}, ..., v_{j_k}) \in \mathcal{V}^k$, $j \in [n]^k$, we initialize V_j with a color denoted by $C_j^{(0)}$. These colors satisfy the condition that for two k-tuples, say $V_{j'}$, $C_j^{(0)}$ and $C_{j'}^{(0)}$ are the same if and only if: (1) $X_{v_{j_a}} = X_{v_{j'_a}}$; (2) $v_{j_a} = v_{j_b} \Leftrightarrow v_{j'_a} = v_{j'_b}$; and (3) $(v_{j_a}, v_{j_b}) \in \mathcal{E} \Leftrightarrow (v_{j'_a}, v_{j'_b}) \in \mathcal{E}$ for all $a, b \in [k]$.

2. k-**FWL:** For each k-tuple V_j and $u \in V$, define $N_{k-FWL}(V_j; u)$ as a k-tuple of k-tuples, such that $N_{k-FWL}(V_j; u) = ((u, v_{j_2}, ..., v_{j_k}), (v_{j_1}, u, ..., v_{j_k}), (v_{j_1}, v_{j_2}, ..., u))$. Then the color of V_i can be updated via the following mapping.

   Update colors: $C_j^{(l+1)} \leftarrow \text{HASH}(C_j^{(l)}, \{(C_{j'}^{(l)} | V_{j'} \in N_{k-FWL}(V_j; u))\}_{u \in V})$.   (5.18)

   k-**WL:** For each k-tuple V_j and $u \in \mathcal{V}$, define $N_{k-WL}(V_j; u)$ as a set of k-tuples such that $N_{k-WL}(V_j; u) = \{(u, v_{j_2}, ..., v_{j_k}), (v_{j_1}, u, ..., v_{j_k}), (v_{j_1}, v_{j_2}, ..., u)\}$ Then, the color of V_i can be updated via the following mapping.

   Update colors: $C_j^{(l+1)} \leftarrow \text{HASH}(C_j^{(l)}, \cup_{u \in V} \{C_{j'}^{(l)} | V_j' \in N_{k-WL}(V_j; u)\})$,   (5.19)

   where the HASH operations in both cases guarantee an injective mapping with different inputs yielding different outputs.

3. For each step l, $\{C_j^{(l)}\}_{j \in [V(G^{(i)})]^k}$ is a multi-set. If such multi-sets of the two graphs are not equal, return $\mathcal{G}^{(1)} \not\cong \mathcal{G}^{(2)}$. Otherwise, go to Eq. equation 5.19.

Similar to the 1-WL test, if the k-(F)WL test returns $\mathcal{G}^{(1)} \not\cong \mathcal{G}^{(2)}$, then it follows that $\mathcal{G}^{(1)}$, $\mathcal{G}^{(2)}$ are not isomorphic. However, the reverse is not true.

Fig. 5.15: Use k-FWL and k-WL to distinguish $\mathcal{G}^{(1)}$ and $\mathcal{G}^{(2)}$.

The key idea of these higher-dim WL tests is to color every k-tuple of nodes in the graphs and update these colors by aggregating the colors from other k-tuples that share $k-1$ nodes. The procedures of the k-FWL test and the k-WL test are shown in Fig. 5.15. Note that they perform aggregation differently, and as such, have different power to distinguish non-isomorphic graphs. These two types of tests form a nested hierarchy, as summarized in the following theorem.

Theorem 5.11. *(Cai et al, 1992; Grohe and Otto, 2015; Grohe, 2017)*

1. *The k-FWL test and the $k+1$-WL test have the same discriminatory power, for $k \geq 1$.*
2. *The 1-FWL test, the 2-WL test and the 1-WL test have the same discriminatory power.*
3. *There are some graphs that the $k+1$-WL test can distinguish while the k-WL test cannot, for $k \geq 2$.*

Because of Theorem 5.11, GNNs that are able to capture the power of these higher-dim WL tests can be strictly more powerful than the 1-WL test. Therefore, higher-order GNNs have the potential to learn even more complex functions than MP-GNN.

However, the drawback of these GNNs is their computational complexity. By the definition of higher-order WL tests, the colors of all k-tuples of nodes need to be tracked. Correspondingly, higher-order GNNs that mimic higher-order WL tests need to associate each k-tuple with a vector representation. Therefore, their memory complexity is at least $\Omega(|\mathcal{V}|^k)$, where $|\mathcal{V}|$ is the number of nodes in the graph. The computational complexity is at least $\Omega(|\mathcal{V}|^{k+1})$, which makes these higher-order GNNs prohibitively expensive for large-scaled graphs.

5.4.4.1 k-WL-induced GNNs (Morris et al, 2019)

Morris et al (2019) first proposed k-GNN by following the k-WL test. Specifically, k-GNN associates each k-tuple of nodes, denoted by $V_j, j \in \mathcal{V}^k$, with a vector representation that is initialized as $\mathbf{h}_j^{(0)}$. In order to save memory, k-GNN only considers k-tuples that contain k different nodes and ignores the order of these nodes. Therefore, each k-tuple reduces to a set of k nodes. With some modification of notation in this subsection, let $\mathcal{V}_j$ denote this set of k different nodes. The initial representation of $\mathcal{V}_j$, $\mathbf{h}_j^{(0)}$ is chosen as a one-hot encoding such that $\mathbf{h}_j^{(0)} = \mathbf{h}_{j'}^{(0)}$, if and only if the subgraphs induced by V_j and $V_{j'}$ are isomorphic.

Then, k-GNN follows the following update procedure of these representations:

$$\mathbf{h}_j^{(l+1)} = \text{MLP}(\mathbf{h}_j^{(l)} \oplus \sum_{V_{j'}:N_{k-GNN}(V_j)} \mathbf{h}_{j'}^{(l)}), \quad \forall k\text{-sized node sets } V_j, \tag{5.20}$$

where $N_{k-GNN}(V_j) = \{V_{j'}| |V_{j'} \cap V_j| = k-1\}$. Note that $N_{k-GNN}(V_j)$ defines the neighbors of V_j differently than N_{k-WL} (see Eq.equation 5.19), because V_j is now a k-sized node set instead of a k-tuple.

Eq.equation 5.20 has time complexity at least $O(|\mathcal{V}|^k)$ as the size of $N_{k-GNN}(V_j)$ is $O(|\mathcal{V}|^k)$. Recently, Morris et al (2019) also considers using a local neighborhood of V_j instead of $N_{k-GNN}(V_j)$. This local neighborhood only includes $V_{j'} \in N_{k-GNN}(V_j)$, such that the node in $V_{j'} \backslash V_j$ is connected to at least one node in V_j. Morris et al (2020b) demonstrated that a variant of this local version of k-GNN may be as powerful as the k-WL test, although a deeper architecture with more layers is needed to match the expressive power.

k-GNN is at most as powerful as the k-WL test. To be more expressive than MP-GNN, $k = 3$ is needed. Therefore, the memory complexity is at least $\Omega(|\mathcal{V}|^3)$. Subsequently, the computational complexity of k-GNN, even for their local version, is at least $\Omega(|\mathcal{V}|^3)$ per layer.

5.4.4.2 Invariant and equivariant GNNs (Maron et al, 2018, 2019b)

To build higher-order GNNs, every k-tuple needs to be associated with a vector representation. Therefore, regardless whether a local or a global neighborhood aggregation is adopted (Eq.equation 5.20), the benefit of reducing the computation by leveraging the sparse graph structure is limited, as it cannot reduce the dominant term $\Omega(|\mathcal{V}|^k)$. Moreover, to handle a sparse graph structure, these higher-order GNNs also need random memory access, which introduces additional computational overhead. Therefore, a line of research into building higher-order GNNs totally ignores graph sparsity. Graphs are viewed as tensors and NNs take these tensors as input. The NNs are designed to be invariant to the order of tensor indices.

Many approaches (Maron et al, 2018, 2019a,b; Chen et al, 2019f; Keriven and Peyré, 2019; Vignac et al, 2020a; Azizian and Lelarge, 2020) adopt this formulation to build GNNs and analyze their expressive power.

Each k-tuple $V_j \in V^k$ is associated with a vector representation $\mathbf{h}_j^{(l)}$. We assume that $\mathbf{h}_j^{(l)} \in \mathbb{R}$ for simplicity. By concatenating the k-tuple's representations together, we obtain a k-order tensor:

$$H \in \mathbb{R}^{\otimes_k |\mathcal{V}|}, \quad \text{where} \quad \mathbb{R}^{\otimes_k |\mathcal{V}|} = \mathbb{R}^{\overbrace{|\mathcal{V}| \times \cdots \times |\mathcal{V}|}^{k\,\text{times}}}.$$

Maron et al (2018) investigates linear permutation invariant and equivariant mappings defined on $\mathbb{R}^{\otimes_k |\mathcal{V}|}$.

Definition 5.11. Given a bijective mapping $\pi : \mathcal{V} \to \mathcal{V}$ and $H \in \mathbb{R}^{\otimes_k |\mathcal{V}|}$, define $\pi(H) := H'$, where $H'_{(\pi(v_1),\pi(v_2),...,\pi(v_k))} = H_{(v_1,v_2,...,v_k)}$, for all k-tuples $(v_1, v_2, ..., v_k) \in \mathcal{V}^k$.

Definition 5.12. A mapping $g : \mathbb{R}^{\otimes_k |\mathcal{V}|} \to \mathbb{R}$ is called *invariant*, if for any bijective mapping $\pi : \mathcal{V} \to \mathcal{V}$ and $H \in \mathbb{R}^{\otimes_k |\mathcal{V}|}$, $g(H) = g(\pi(H))$.

Definition 5.13. A mapping $g : \mathbb{R}^{\otimes_k |\mathcal{V}|} \to \mathbb{R}^{\otimes_k |\mathcal{V}|}$ is called *equivariant*, if for any bijective mapping $\pi : \mathcal{V} \to \mathcal{V}$ and $H \in \mathbb{R}^{\otimes_k |\mathcal{V}|}$, $\pi(g(H)) = g(\pi(H))$.

Maron et al (2018) showed that the number of the bases needed to represent all possible linear invariant mappings from $\mathbb{R}^{\otimes_k|\mathcal{V}|} \to \mathbb{R}$ is $b(k)$, where $b(k)$ is the k-th Bell number. Additionally, the number of bases, needed to represent all possible linear equivariant mappings from $\mathbb{R}^{\otimes_k|\mathcal{V}|} \to \mathbb{R}^{\otimes_{k'}|V|}$, is $b(k + k')$. To better understand this observation, consider the invariant case with $k = 1$. In this case, the linear invariant mapping $g : \mathbb{R}^{|\mathcal{V}|} \to \mathbb{R}$ is essentially an invariant pooling (Def.5.7). As $b(1) = 1$, the linear invariant mapping $g : \mathbb{R}^{|\mathcal{V}|} \to \mathbb{R}$ only holds one single base — the sum pooling, *i.e.*, g follows the form $g(a) = c\langle \mathbf{1}, a \rangle$, where c is a parameter to be learned. Consider the equivariant case, where $k = 1$ and $k' = 1$. As $b(2) = 2$, the linear equivariant mapping $g : \mathbb{R}^{|\mathcal{V}|} \to \mathbb{R}^{|\mathcal{V}|}$ holds two bases, *i.e.*, g has the form $g(a) = (c_1 I + c_2 \mathbf{1}\mathbf{1}^\top)a$, where c_1, c_2 are parameters to be learned.

Based on the above observations, GNNs can be built by compositing these linear invariant/equivariant mappings. Learning can be performed via learning the weights before the above bases. Towards this end, Maron et al (2018, 2019a) has proposed using these linear invariant/equivariant mappings to build GNNs:

$$f_{k-inv} = g_{inv} \circ g_{equ}^{(L)} \circ \sigma \circ g_{equ}^{(L-1)} \circ \sigma \cdots \circ \sigma \circ g_{equ}^{(1)}, \tag{5.21}$$

where g_{inv} is a linear invariant layer $\mathbb{R}^{\otimes_k|\mathcal{V}|} \to \mathbb{R}$, $g_{equ}^{(l)}$'s, $l \in [L]$ are linear equivariant layers from $\mathbb{R}^{\otimes_k|\mathcal{V}|} \to \mathbb{R}^{\otimes_k|\mathcal{V}|}$, and σ is an element-wise non-linear activation function. It can be shown that f_{k-inv} is an invariant mapping. Maron et al (2018); Azizian and Lelarge (2020) proved that the connection of f_{k-inv} to the k-WL test can be summarized with the following theorem.

Theorem 5.12. *(Reproduced from (Maron et al, 2018; Azizian and Lelarge, 2020)) For two non-isomorphic graphs $\mathcal{G}^{(1)} \not\cong \mathcal{G}^{(2)}$, if the k-WL test can distinguish them, then there exists f_{k-inv} that can distinguish them.*

Maron et al (2019b); Keriven and Peyré (2019) also studied whether the models f_{k-inv} may universally approximate any permutation invariant function. However, they were pessimistic in their conclusion since this would require high-order tensors, $k = \Omega(n)$, which can hardly be implemented in practice (Maron et al, 2019b).

Similar to k-GNN, f_{inv} is also at most as powerful as the k-WL test. To be more expressive than MP-GNN, f_{inv} should use at least $k = 3$. Therefore, the memory complexity is at least $\Omega(|\mathcal{V}|^3)$. Then, the number of bases of the linear equivariant layer is $b(6) = 203$. Therefore, the computation at each layer follows that: (1) a tensor in $\mathbb{R}^{\otimes_3|\mathcal{V}|}$ times $b(6)$ many tensors in $\mathbb{R}^{\otimes_6|\mathcal{V}|}$ get $b(6)$ many tensors in $\mathbb{R}^{\otimes_3|\mathcal{V}|}$; (2) these tensors get summed into a single tensor in $\mathbb{R}^{\otimes_3|\mathcal{V}|}$.

5.4.4.3 k-FWL-induced GNNs (Maron et al, 2019a; Chen et al, 2019f)

The higher-order GNNs in previous two subsections match the expressive power of the k-WL test. According to Theorem 5.11, the k-FWL test holds the same power as the $k + 1$-WL test, which is strictly more powerful than the k-WL test for $k \geq 2$, while the k-FWL test only needs to track the representations of k-tuples. Therefore, if GNNs can mimic the k-FWL test, they may hold similar memory cost as the GNNs introduced in the previous two subsections while being more expressive. Maron et al (2019a); Chen et al (2019f) proposed PPGN and Ring-GNN respectively to match the k-FWL test.

The key difference between the k-FWL test and the k-WL test is the leveraging of the neighbors of a k-tuple V_j. Note that $N_{k-FWL}(V_j; u)$ in Eq.equation 5.18 groups the neighboring tuples of V_j into a higher-level tuple, while $N_{k-WL}(V_j; u)$ skips grouping them due to the set union operation in Eq.equation 5.19. This yields the key mechanism in the GNN design to match the k-FWL test: implement the aggregating procedure in the k-FWL test of Eq.equation 5.18 via a product-sum procedure. Suppose the representation for V_j is $\mathbf{h}_j \in \mathbb{R}$. We may design the aggregation of $\{(C_{j'}^{(l)} | V_{j'} \in N_{k-FWL}(V_j; u))\}_{u \in V}$ as

$$\sum_{u \in V} \prod_{V_{j'} \in N_{k-FWL}(V_j; u)} \mathbf{h}_{j'}.$$

If we combine all these representations into a tensor $H \in \mathbb{R}^{\otimes k |V| \times F}$, the above operation can essentially be represented as tensor operation, i.e., define

$$H' := \sum_{u \in V} H_{u, \cdot, \cdots, \cdot} \odot H_{\cdot, u, \cdots, \cdot} \odot \cdots \odot H_{\cdot, \cdot, \cdots, u}, \text{ where}$$

$$[H']_{v_{j_1}, v_{j_2}, \cdots, v_{j_k}} = \sum_{u \in V} H_{u, v_{j_2}, \cdots, v_{j_k}} \cdot H_{v_{j_1}, u, \cdots, v_{j_k}} \cdots \cdot H_{v_{j_1}, v_{j_2}, \cdots, u}.$$

Based on the above observation, Maron et al (2019a) built PPGN as follows. First, for all $V_j \in \mathscr{V}^k$, initialize $\mathbf{h}_j^{(0)} \in \mathbb{R}$ such that $\mathbf{h}_j^{(0)} = \mathbf{h}_{j'}^{(0)}$, if and only if: (1) $X_{v_{j_a}} = X_{v_{j'_a}}$; (2) $v_{j_a} = v_{j_b} \Leftrightarrow v_{j'_a} = v_{j'_b}$; and (3) $(v_{j_a}, v_{j_b}) \in \mathscr{E} \Leftrightarrow (v_{j'_a}, v_{j'_b}) \in \mathscr{E}$, for all $a, b \in [k]$. Then, combine $\mathbf{h}_j^{(0)}$ into a tensor $H^{(0)} \in \mathbb{R}^{\otimes k |\mathscr{V}|}$. Perform the updating procedure for $l = 0, 1, ..., L-1$:

$$H^{(l+1)} = \tilde{H}^{(l,0)} \oplus \left[\sum_{u \in V} \tilde{H}_{u, \cdot, \cdots, \cdot}^{(l,1)} \odot \tilde{H}_{\cdot, u, \cdots, \cdot}^{(l,2)} \odot \cdots \odot \tilde{H}_{\cdot, \cdot, \cdots, u}^{(l,k)} \right],$$

$$\text{where,} \quad \tilde{H}^{(l,i)} = \text{MLP}^{(l,i)}(H^{(l)}). \tag{5.22}$$

Here, MLPs are imposed at the last dimension of these tensors. MLPs with different sup-script have different parameters. Finally, perform a READOUT $\sum_{V_j \in V^k} \mathbf{h}_j^{(L)}$ to obtain the graph representation.

Maron et al (2019a) proved that PPGN, when $k = 2$, can match the power of the 2-FWL test. Azizian and Lelarge (2020) generalized this result to an arbitrary k.

Theorem 5.13. *(Reproduced from (Azizian and Lelarge, 2020)) For two non-isomorphic graphs $G^{(1)} \not\cong G^{(2)}$, if the k-FWL test can distinguish them, then there exists a PPGN that can distinguish them.*

To be more powerful than the 1-WL test, PPGN only needs to set $k = 2$ and hence the memory complexity is just $\Omega(|\mathcal{V}|^2)$. Regarding the computation, the product-sum-type aggregation of PPGN is indeed more complex than f_{inv} in Sec.5.4.4.2. However, when $k = 2$, Eq.equation 5.22 reduces to the product of two matrices, which can be efficiently computed in parallel computing units.

5.5 Summary

Graph neural networks have recently achieved unprecedented success across many domains due to their great expressive power to learn complex functions defined over graphs and relational data. In this chapter, we provided a systematic study of the expressive power of GNNs by giving an overview of recent research results in this field.

We first established that the message passing GNN is at most as powerful as the 1-WL test to distinguish non-isomorphic graphs. The key condition that guarantees to match the limit is an injective updating function of node representations. Next, we discussed techniques that have been proposed to build more powerful GNNs. One approach to make message passing GNNs more expressive is to pair the input graphs with extra attributes. In particular, we discussed two types of extra attributes — random attributes and deterministic distance attributes. Injecting random attributes allows GNNs to distinguish any non-isomorphic graphs, though a large amount of data augmentation is required to make GNNs approximately invariant. Meanwhile, injecting deterministic distance attributes does not require the same data augmentation, but the expressive power of the resulting GNNs still holds certain limitations. Mimicking higher-dimensional WL tests is another way to build more powerful GNNs. These approaches do not track node representations. Instead, they update the representation of every k-tuple of nodes ($k \geq 2$). Overall, the message passing GNN is powerful but holds some limits in its expressive power. Different techniques make GNNs overcome these limits to a different extent while incurring different types of computational costs.

We would like to list some additional research results on the expressive power of GNNs that we were not able to cover earlier due to space limitations. Barceló et al (2019) study the expressive power of GNNs to represent Boolean classifiers, which is useful to understand how GNNs represent knowledge and logic. Vignac et al (2020a) propose a structural message passing framework for GNNs, where a matrix instead of a vector is adopted as the node representation to make GNN more expressive. Balcilar et al (2021) studied the expressivity of GNNs via the spectral analysis of GNN-based graph signal transformations. Chen et al (2020k) studies the effect of non-linearity of GNNs in the message passing procedure on their expressive power, which complements our understanding of many works that suggest a linear message passing procedure (Wu et al, 2019a; Klicpera et al, 2019a; Chien et al, 2021).

Theoretical characterization of GNNs is an important research direction, where the analysis of expressive power is only one of its aspects, perhaps the best-studied up to this point. Machine learning models hold two fundamental blocks, training and generalization. However, only a few research works have analyzed them (Garg et al, 2020; Liao et al, 2021; Xu et al, 2020c). The authors suggest that future research on building more expressive GNNs always takes these two blocks into account. A related, significant question is how to build more expressive GNNs with only a limited depth and width[4]. Note that limiting the model depth and width yields the potential of more efficient GNN training and better generalization. To conclude this chapter, let us quote Sir Winston Churchill:"Now this is not the end. It is not even the beginning of the end. But it is, perhaps, the end of the beginning." We have strong confidence that the machine learning community will put more effort on theory for GNNs in the future to match their success and break their encountered difficulties in real-world applications.

Acknowledgements The authors would like to greatly thank Jiaxuan You and Weihua Hu for sharing many materials reproduced here. The authors would like to greatly thank Rok Sosič and Natasha Sharp for commenting on and polishing the manuscript. The authors also gratefully acknowledge the support of DARPA under Nos. HR00112190039 (TAMI), N660011924033 (MCS); ARO under Nos. W911NF-16-1-0342 (MURI), W911NF-16-1-0171 (DURIP); NSF under Nos. OAC-1835598 (CINES), OAC-1934578 (HDR), CCF-1918940 (Expeditions), IIS-2030477 (RAPID), NIH under No. R56LM013365; Stanford Data Science Initiative, Wu Tsai Neurosciences Institute, Chan Zuckerberg Biohub, Amazon, JPMorgan Chase, Docomo, Hitachi, Intel, JD.com, KDDI, NVIDIA, Dell, Toshiba, Visa, and UnitedHealth Group. J. L. is a Chan Zuckerberg Biohub investigator.

Editor's Notes: The theoretical analysis of expressive power reveals how the architecture of GNNs works and gains its advantage. Hence it provides support for readers to understand the great success of GNNs in fundamental graph learning tasks, e.g. link prediction (chapter 10) and graph matching (chapter 13), various downstream tasks, e.g. recommender system (chapter 19) and natural language processing (chapter 21), as well as its relevance with other GNNs' characterizations, e.g. scalability (chapter 6) and robustness (chapter 8). Inspired by these theories, it's also probably to motivate the study of preferable GNN models that can break through unsolved challenges in existing problems, such as graph transformation (chapter 12) and drug discovery (chapter 24).

---

[4] Loukas (2020) measures the required depth and width of GNNs by viewing them as distributed algorithms, which does not assume permutation invariance. Instead, here we are talking about the expressive power that refers to the capability of learning permutation invariant functions.

Chapter 6
Graph Neural Networks: Scalability

Hehuan Ma, Yu Rong, and Junzhou Huang

Abstract Over the past decade, Graph Neural Networks have achieved remarkable success in modeling complex graph data. Nowadays, graph data is increasing exponentially in both magnitude and volume, e.g., a social network can be constituted by billions of users and relationships. Such circumstance leads to a crucial question, how to properly extend the scalability of Graph Neural Networks? There remain two major challenges while scaling the original implementation of GNN to large graphs. First, most of the GNN models usually compute the entire adjacency matrix and node embeddings of the graph, which demands a huge memory space. Second, training GNN requires recursively updating each node in the graph, which becomes infeasible and ineffective for large graphs. Current studies propose to tackle these obstacles mainly from three sampling paradigms: node-wise sampling, which is executed based on the target nodes in the graph; layer-wise sampling, which is implemented on the convolutional layers; and graph-wise sampling, which constructs sub-graphs for the model inference. In this chapter, we will introduce several representative research accordingly.

Hehuan Ma
Department of CSE, University of Texas at Arlington, e-mail: `hehuan.ma@mavs.uta.edu`

Yu Rong
Tencent AI Lab, e-mail: `yu.rong@hotmail.com`

Junzhou Huang
Department of CSE, University of Texas at Arlington, e-mail: `jzhuang@uta.edu`

6.1 Introduction

Graph Neural Network (GNN) has gained increasing popularity and obtained remarkable achievement in many fields, including social network (Freeman, 2000; Perozzi et al, 2014; Hamilton et al, 2017b; Kipf and Welling, 2017b), bioinformatics (Gilmer et al, 2017; Yang et al, 2019b; Ma et al, 2020a), knowledge graph (Liben-Nowell and Kleinberg, 2007; Hamaguchi et al, 2017; Schlichtkrull et al, 2018), etc. GNN models are powerful to capture accurate graph structure information as well as the underlying connections and interactions between nodes (Li et al, 2016b; Veličković et al, 2018; Xu et al, 2018a, 2019d). Generally, GNN models are constructed based on the features of the nodes and edges, as well as the adjacency matrix of the whole graph. However, since the graph data is growing rapidly nowadays, the graph size is increasing exponentially too. Recently published graph benchmark datasets, Open Graph Benchmark (OGB), collects several commonly used datasets for machine learning on graphs (Weihua Hu, 2020). Table 6.1 is the statistics of the datasets about node classification tasks. As observed, large-scale dataset *ogbn-papers100M* contains over one hundred million nodes and one billion edges. Even the relatively small dataset *ogbn-arxiv* still consists of fairly large nodes and edges.

Table 6.1: The statistics of node classification datasets from OGB (Weihua Hu, 2020).

Scale	Name	Number of Nodes	Number of Edges
Large	ogbn-papers100M	111,059,956	1,615,685,872
Medium	ogbn-products	2,449,029	61,859,140
Medium	ogbn-proteins	132,534	39,561,252
Medium	ogbn-mag	1,939,743	21,111,007
Small	ogbn-arxiv	169,343	1,166,243

For such large graphs, the original implementation of GNN is not suitable. There are two main obstacles, 1) large memory requirement, and 2) inefficient gradient update. First, most of the GNN models need to store the entire adjacent matrices and the feature matrices in the memory, which demand huge memory consumption. Moreover, the memory may not be adequate for handling very large graphs. Therefore, GNN cannot be applied on large graphs directly. Second, during the training phase of most GNN models, the gradient of each node is updated in every iteration, which is inefficient and infeasible for large graphs. Such scenario is similar with the gradient descent versus stochastic gradient descent, while the gradient descent may take too long to converge on large dataset, and stochastic gradient is introduced to speed up the process towards an optimum.

In order to tackle these obstacles, recent studies propose to design proper sampling algorithms on large graphs to reduce the computational cost as well as increase the scalability. In this chapter, we categorize different sampling methods according to the underlying algorithms, and introduce typical works accordingly.

6.2 Preliminary

We first briefly introduce some concepts and notations that are used in this chapter. Given a graph $\mathscr{G}(\mathscr{V}, \mathscr{E})$, $|\mathscr{V}| = n$ denotes the set of n nodes and $|\mathscr{E}| = m$ denotes a set of m edges. Node $u \in \mathscr{N}(v)$ is the neighborhood of node v, where $v \in \mathscr{V}$, and $(u, v) \in \mathscr{E}$. The vanilla GNN architecture can be summarized as:

$$\mathbf{h}^{(l+1)} = \sigma\left(A\mathbf{h}^{(l)}W^{(l)}\right),$$

where A is the normalized adjacency matrix, $\mathbf{h}^{(l)}$ represents the embedding of the node in the graph for layer/depth l, $W^{(l)}$ is the weight matrix of the neural network, and σ denotes the activation function.

For large-scaled graph learning, the problem is often referred as the node classification, where each node v is associated with a label y, and the goal is to learn from the graph and predict the labels of unseen nodes.

6.3 Sampling Paradigms

The concept of sampling aims at selecting a partition of all the samples to represent the entire sample distribution. Therefore, the sampling algorithm on large graphs refers to the approach that uses partial graph instead of the full graph to address target problems. In this chapter, we categorize different sampling algorithms into three major groups, which are node-wise sampling, layer-wise sampling and graph-wise sampling.

Node-wise sampling plays a dominant role during the early stage of implementing GCN on large graphs, such as *Graph SAmple and aggreGatE* (Graph-SAGE) (Hamilton et al, 2017b) and *Variance Reduction Graph Convolutional Networks* (VR-GCN) (Chen et al, 2018d). Later, layer-wise sampling algorithms are proposed to address the neighborhood expansion problem occurred during node-wise sampling, e.g., *Fast Learning Graph Convolutional Networks* (Fast-GCN) (Chen et al, 2018c) and *Adaptive Sampling Graph Convolutional Networks* (ASGCN) (Huang et al, 2018). Moreover, graph-wise sampling paradigms are designed to further improve the efficiency and scalability, e.g., *Cluster Graph Convolutional Networks* (Cluster-GCN) (Chiang et al, 2019) and *Graph SAmpling based INductive learning meThod* (GraphSAINT) (Zeng et al, 2020a). Fig. 6.1 illustrates a comparison between three sampling paradigms. In the node-wise sampling, the nodes are sampled based on the target node in the graph. While in the layer-wise sampling, the nodes are sampled based on the convolutional layers in the GNN

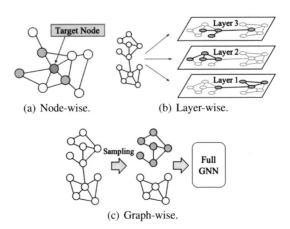

(a) Node-wise.   (b) Layer-wise.

(c) Graph-wise.

Fig. 6.1: Three sampling paradigms toward large-scale GNNs.

models. For the graph-wise sampling, the sub-graphs are sampled from the original graph, and used for the model inference.

According to these paradigms, two main issues should be addressed while constructing large-scale GNNs: 1) *how to design efficient sampling algorithms?* and 2) *how to guarantee the sampling quality?* In recent years, a lot of works have studied about how to construct large-scale GNNs and how to address the above issues properly. Fig. 6.2 displays a timeline of certain representative works in this area from the year 2017 until recent. Each work will be introduced accordingly in this chapter.

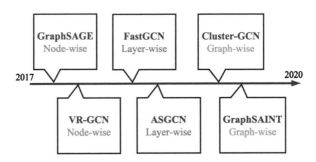

Fig. 6.2: Timeline of leading research work toward large-scale GNNs.

Other than these major sampling paradigms, more recent works have attempted to improve the scalability of large graphs from various perspectives as well. For example, heterogeneous graph has attracted more and more attention with regards to the rapid growth of data. Large graphs not only include millions of nodes but also various data types. How to train GNNs on such large graphs has become a new domain of interest. Li et al (2019a) proposes a GCN-based Anti-Spam (GAS) model to detect spams by considering both homogeneous and heterogeneous graphs. Zhang et al (2019b) designs a random walk sampling method based on all types of nodes. Hu et al (2020e) employs the transformer architecture to learn the mutual attention between nodes, and sample the nodes according to different node types.

6.3.1 Node-wise Sampling

Rather than use all the nodes in the graph, the first approach selects certain nodes through various sampling algorithms to construct large-scale GNNs. GraphSAGE (Hamilton et al, 2017b) and VR-GCN (Chen et al, 2018d) are two pivotal studies that utilize such a method.

6.3.1.1 GraphSAGE

At the early stage of GNN development, most work target at the transductive learning on a fixed-size graph (Kipf and Welling, 2017b, 2016), while the inductive setting is more practical in many cases. Yang et al (2016b) develops an inductive learning on graph embeddings, and GraphSAGE Hamilton et al (2017b) extends the study on large graphs. The overall architecture is illustrated in Fig. 6.3.

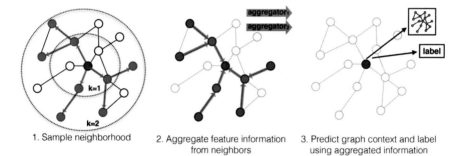

1. Sample neighborhood   2. Aggregate feature information from neighbors   3. Predict graph context and label using aggregated information

Fig. 6.3: Overview of the GraphSAGE architecture. Step 1: sample the neighborhoods of the target node; step 2: aggregate feature information from the neighbors; step 3: utilize the aggregated information to predict the graph context or label. Figure excerpted from (Hamilton et al, 2017b).

GraphSAGE can be viewed as an extension of the original Graph Convolutional Network (GCN) (Kipf and Welling, 2017b). The first extension is the generalized aggregator function. Given $\mathcal{G}(\mathcal{V}, \mathcal{E})$, $\mathcal{N}(v)$ is the neighborhood of v, $\mathbf{h}$ is the representation of the node, the embedding generation at the current $(l+1)$-th depth from the target node $v \in \mathcal{V}$ can be formulated as,

$$\mathbf{h}_{\mathcal{N}(v)}^{(l+1)} = \text{AGGREGATE}_l \left(\left\{ \mathbf{h}_u^{(l)}, \forall u \in \mathcal{N}(v) \right\} \right),$$

Different from the original mean aggregator in GCN, GraphSAGE proposes LSTM aggregator and Pooling aggregator to aggregate the information from the neighbors. The second extension is that GraphSAGE applies the concatenation function to combine the information of target node and neighborhoods instead of the summation function:

$$\mathbf{h}_v^{(l+1)} = \sigma \left(W^{(l+1)} \cdot \text{CONCAT} \left(\mathbf{h}_v^{(l)}, \mathbf{h}_{\mathcal{N}(v)}^{(l+1)} \right) \right),$$

where $W^{(l+1)}$ are the weight matrices, and σ is the activation function.

In order to make GNN suitable for the large-scale graphs, GraphSAGE introduces the mini-batch training strategy to reduce the computation cost during the training phase. Specifically, in each training iteration, only the nodes that are used by computing the representations in the batch are considered, which significantly reduces the number of sampled nodes. Take layer 2 in Fig. 6.4(a) as an example, unlike the full-batch training which takes all 11 nodes into consideration, only 6 nodes are involved for mini-batch training. However, the simple implementation of mini-batch training strategy suffers the neighborhood expansion problem. As shown in layer 1 of Fig. 6.4(a), most of the nodes are sampled since the number of sampled nodes grows exponentially if all the neighbors are sampled at each layer. Thus, all the nodes are selected eventually if the model contains many layers.

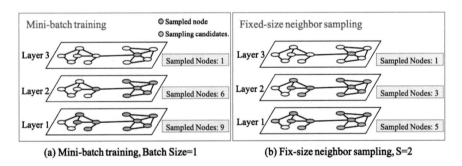

(a) Mini-batch training, Batch Size=1   (b) Fix-size neighbor sampling, S=2

Fig. 6.4: Visual comparison between mini-batch training and fixed-size neighbor sampling.

To further improve the training efficiency and eliminate the neighborhood expansion problem, GraphSAGE adopts fixed-size neighbor sampling strategy. In specific, a fixed-size set of neighbor nodes are sampled for each layer for computing, instead of using the entire neighborhood sets. For example, one can set the fixed-size set as two nodes, which is illustrated in Fig. 6.4(b), the yellow nodes represent the sampled nodes, and the blue nodes are the candidate nodes. It is observed that the number of sampled nodes is significantly reduced, especially for layer 1.

In summary, GraphSAGE is the first to consider inductive representation learning on large graphs. It introduces a generalized aggregator, the mini-batch training, and fixed-size neighbor sampling algorithm to accelerate the training process. However, fixed-size neighbor sampling strategy can not totally avoid the neighborhood expansion problem. Also, there is no theoretical guarantees for the sampling quality.

6.3.1.2 VR-GCN

In order to further reduce the size of the sampled nodes, as well as conduct a comprehensive theoretical analysis, VR-GCN (Chen et al, 2018d) proposes a Control Variate Based Estimator. It only samples an arbitrarily small size of the neighbor nodes by employing historical activations of the nodes. Fig. 6.5 compares the receptive field of one target node using different sampling strategies. For the implementation of the original GCN (Kipf and Welling, 2017b), the number of sampled nodes is increased exponentially with the number of layers. With neighbor sampling, the size of the receptive field is reduced randomly according to the preset sampling number. Compared with them, VR-GCN utilizes the historical node activations as a control variate to keep the receptive field small scaled.

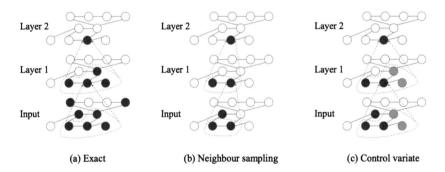

(a) Exact   (b) Neighbour sampling   (c) Control variate

Fig. 6.5: Illustration of the receptive field of a single node utilizing different sampling strategies with a two-layer graph convolutional neural network. The red circle represents the latest activation, and the blue circle indicates the historical activation. Figure excerpted from (Chen et al, 2018d).

The neighbor sampling (NS) algorithm proposed by GraphSAGE (Hamilton et al, 2017b) can be formulated as:

$$\text{NS}_v^{(l)} := R \sum_{u \in \hat{\mathcal{N}}^{(l)}(v)} A_{vu} \mathbf{h}_u^{(l)}, \quad R = \mathcal{N}(v)/d^{(l)},$$

where $\mathcal{N}(v)$ represents the neighbor set of node v, $d^{(l)}$ is the sampled size of the neighbor nodes at layer l, $\hat{\mathcal{N}}^{(l)}(v) \subset \mathcal{N}(v)$ is the sampled neighbor set of node v at layer l, and A represents the normalized adjacency matrix. Such a method has been proved to be a biased sampling, and would cause larger variance. The detailed proof can be found in (Chen et al, 2018d). Such properties result in a larger sample size $\hat{\mathcal{N}}^{(l)}(v) \subset \mathcal{N}(v)$.

To address these issues, VR-GCN proposes Control Variate Based Estimator (CV Sampler) to maintain all the historical hidden embedding $\bar{\mathbf{h}}_v^{(l)}$ of every partici- pated node. For a better estimation, since the difference between $\bar{\mathbf{h}}_v^{(l)}$ and $\mathbf{h}_v^{(l)}$ shall be small if the model weights do not change too fast. CV Sampler is capable of reducing the variance and obtaining a smaller sample size $\hat{n}^{(l)}(v)$ eventually. Thus, the feed-forward layer of VR-GCN can be defined as,

$$H^{(l+1)} = \sigma \left(A^{(l)} \left(H^{(l+1)} - \bar{H}^{(l)} \right) + A\bar{H}^{(l)} \right) W^{(l)}.$$

$A^{(l)}$ is the sampled normalized adjacency matrix at layer l, $\bar{H}^{(l)} = \{\bar{\mathbf{h}}_1^{(l)}, \cdots, \bar{\mathbf{h}}_n^{(l)}\}$ is the stack of the historical hidden embedding $\bar{\mathbf{h}}^{(l)}$, $H^{(l+1)} = \{\mathbf{h}_1^{(l+1)}, \cdots, \mathbf{h}_n^{(l+1)}\}$ is the embedding of the graph nodes in the $(l+1)$-th layer, and $W^{(l)}$ is the learnable weights matrix. In such a manner, the sampled size of $A^{(l)}$ is greatly reduced com- pared with GraphSAGE by utilizing the historical hidden embedding $\bar{\mathbf{h}}^{(l)}$, which introduces a more efficient computing method. Moreover, VR-GCN also studies how to apply the Control Variate Estimator on the dropout model. More details can be found in the paper.

In summary, VR-GCN first analyzes the variance reduction on node-wise sam- pling, and successfully reduces the size of the samples. However, the trade-off is that the additional memory consumption for storing the historical hidden embed- dings would be very large. Also, the limitation of applying GNNs on large-scale graphs is that it is not realistic to store the full adjacent matrices or the feature ma- trices. In VR-GCN, the historical hidden embeddings storage actually increases the memory cost, which is not helping from this perspective.

6.3.2 Layer-wise Sampling

Since node-wise sampling can only alleviate but not completely solve the neigh- borhood expansion problem, layer-wise sampling has been studied to address this obstacle.

6.3.2.1 FastGCN

In order to solve the neighborhood expansion problem, FastGCN (Chen et al, 2018c) first proposes to understand the GNN from the functional generalization perspective. The authors point out that training algorithms such as stochastic gradient descent are implemented according to the additivity of the loss function for independent data samples. However, GNN models generally lack sample loss independence. To solve this problem, FastGCN converts the common graph convolution view to an integral transform view by introducing a probability measure for each node. Fig. 6.6 shows the conversion between the traditional graph convolution view and the integral transform view. In the graph convolution view, a fixed number of nodes are sampled in a bootstrapping manner in each layer, and are connected if there is a connection exists. Each convolutional layer is responsible for integrating the node embeddings. The integral transform view is visualized according to the probability measure, and the integral transform (demonstrated in the yellow triangle form) is used to calculate the embedding function in the next layer. More details can be found in (Chen et al, 2018c).

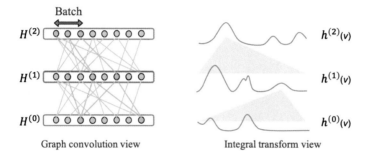

Fig. 6.6: Two views of GCN. The circles represent the nodes in the graph, while the yellow circles indicate the sampled nodes. The lines represent the connection between nodes.

Formally, given a graph $\mathscr{G}(\mathscr{V}, \mathscr{E})$, an inductive graph $\mathscr{G}'$ with respect to a possibility space $(\mathscr{V}', F, p)$ is constructed. In specific, $\mathscr{V}'$ denotes the sample space of nodes which are iid samples. The probability measure p defines a sampling distribution, and F can be any event space, e.g., $F = 2^{\mathscr{V}'}$. Take node v and u with same probability measure p, $g\left(\mathbf{h}^{(K)}(v)\right)$ as the gradient of the final embedding of node v, and E as the expectation function, the functional generalization is formulated as,

$$L = \mathrm{E}_{v \sim p}\left[g\left(\mathbf{h}^{(K)}(v)\right)\right] = \int g\left(\mathbf{h}^{(K)}(v)\right) dp(v).$$

Moreover, consider sampling t_l iid samples $u_1^{(l)}, \ldots, u_{t_l}^{(l)} \sim p$ for each layer l, $l = 0, \ldots, K - 1$, a layer-wise estimation of the loss function is admitted as,

$$L_{t_0, t_1, \ldots, t_K} := \frac{1}{t_K} \sum_{i=1}^{t_K} g\left(\mathbf{h}_{t_K}^{(K)}\left(u_i^{(K)}\right)\right),$$

which proves that FastGCN samples a fixed number of nodes at each layer.

Furthermore, in order to reduce the sampling variance, FastGCN adopts the importance sampling with respect to the weights in the normalized adjacency matrix.

$$q(u) = \|A(:,u)\|^2 / \sum_{u' \in \mathcal{V}} \|A(:,u')\|^2, \quad u \in \mathcal{V}, \tag{6.1}$$

where A is the normalized adjacency matrix of the graph. Detailed proofs can be found in (Chen et al, 2018c). According to Equation 6.1, the entire sampling process is independent for each layer, and the sampling probability keeps the same.

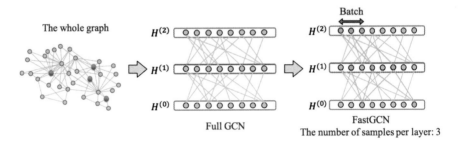

Fig. 6.7: Comparison between full GCN and FastGCN.

Compared with GraphSAGE (Hamilton et al, 2017b), FastGCN is much less computational costly. Assume t_l neighbor nodes are samples for layer l, the neighborhood expansion size is at most the sum of the t_l's for FastGCN, while could be up to the product of the t_l's for GraphSAGE. Fig. 6.7 illustrates the sampling difference between Full GCN and FastGCN. In full GCN, the connections are very sparse so that it has to compute and update all the gradients, while FastGCN only samples a fixed number of samples at each layer. Therefore, the computational cost is greatly decreased. On the other hand, FastGCN still retains most of the information according to the importance sampling method. The fixed number of nodes are randomly sampled in each training iteration, thus every node and the corresponding connections could be selected and fit into the model if the training time is long enough. Therefore, the information of the entire graph is generally retained.

In summary, FastGCN solves the neighborhood expansion problem according to the fixed-size layer sampling. Meanwhile, this sample strategy has a quality guarantee. However, since FastGCN samples each layer independently, it failed to capture the between-layer correlations, which leads to a performance compromise.

6.3.2.2 ASGCN

To better capture the between-layer correlations, ASGCN (Huang et al, 2018) proposes an adaptive layer-wise sampling strategy. In specific, the sampling probability of lower layers depends on the upper ones. As shown in Fig. 8(a), ASGCN only samples nodes from the neighbors of the sampled node (yellow node) to obtain the better between-layer correlations, while FastGCN utilizes the importance sampling among all the nodes.

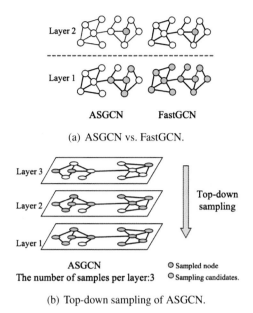

(a) ASGCN vs. FastGCN.

(b) Top-down sampling of ASGCN.

Fig. 6.8: A demonstration of the sampling strategies used in ASGCN.

Meanwhile, the sampling process of ASGCN is performed in a top-down manner. As shown in Fig. 8(b), the sampling process is first conducted in the output layer, which is the layer 3. Next, the participated nodes of the intermediate layer are sampled according to the results of the output layer. Such a sampling strategy captures dense connections between layers.

The sampling probability of lower layers depends on the upper ones. Take Fig. 6.9 as an illustration, $p(u_j \mid v_i)$ is the probability of sampling node u_j given node v_i, v_i refers to node i in the $(l+1)$-th layer while u_j denotes node j in the l-th layer, n' represents the sampled node number in every layer while n is the number of all the nodes in the graph, $q(u_j \mid v_1, \cdots, v_{n'})$ is the probability of sampling u_j given all the nodes in the current layer, and $\hat{a}(v_i, u_j)$ represents the entry of node v_i and u_j in the re-normalized adjacency matrix $\hat{A}$. The sampling probability $q(u_j)$ can be written as,

$$q(u_j) = \frac{p(u_j \mid v_i)}{q(u_j \mid v_1 \ldots v_{n'})}$$

$$p(u_j \mid v_i) = \frac{\hat{a}(v_i, u_j)}{\mathcal{N}(v_i)}, \quad \mathcal{N}(v_i) = \sum_{j=1}^{n} \hat{a}(v_i, u_j).$$

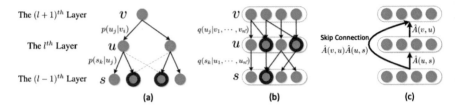

Fig. 6.9: Network construction example: (a) node-wise sampling; (b) layer-wise sampling; (c) skip connection implementation. Figure excerpted from (Huang et al, 2018).

To further reduce the sampling variance, ASGCN introduces the explicit variance reduction to optimize the sampling variance as the final objective. Consider $x(u_j)$ as the node feature of node u_j, the optimal sampling probability $q^*(u_j)$ can be formulated as,

$$q^*(u_j) = \frac{\sum_{i=1}^{n'} p(u_j \mid v_i) \left| g(x(u_j)) \right|}{\sum_{j=1}^{n} \sum_{i=1}^{n'} p(u_j \mid v_i) \left| g(x(v_j)) \right|}, \quad g(x(u_j)) = W_g x(u_j). \quad (6.2)$$

However, simply utilizing the sampler given by Equation 6.2 is not sufficient to secure a minimal variance. Thus, ASGCN designs a hybrid loss by adding the variance to the classification loss $\mathscr{L}_c$, as shown in Equation 6.3. In such a manner, the variance can be trained to achieve the minimal status.

$$\mathscr{L} = \frac{1}{n'} \sum_{i=1}^{n'} \mathscr{L}_c \left(y_i, \bar{y}(\hat{\mu}_q(v_i))\right) + \lambda \operatorname{Var}_q (\hat{\mu}_q(v_i)), \quad (6.3)$$

where y_i is the ground-truth label, $\hat{\mu}_q(v_i)$ represents the output hidden embeddings of node v_i, and $\bar{y}(\hat{\mu}_q(v_i))$ is the prediction. λ is involved as a trade-off parameter. The variance reduction term $\lambda \operatorname{Var}_q(\hat{\mu}_q(v_i))$ can also be viewed as a regularization according to the sampled instances.

ASGCN also proposes a skip connection method to obtain the information across distant nodes. As shown in Fig. 6.9 (c), the nodes in the (l-1)-th layer theoretically preserve the second-order proximity (Tang et al, 2015b), which are the 2-hop neighbors for the nodes in the (l+1)-th layer. The sampled nodes will include both 1-hop and 2-hop neighbors by adding a skip connection between the (l-1)-th layer and the (l+1)-th layer, which captures the information between distant nodes and facilitates the model training.

In summary, by introducing the adaptive sampling strategy, ASGCN has gained better performance as well as equips a better variance control. However, it also brings in the additional dependence during sampling. Take FastGCN as an example, it can perform parallel sampling to accelerate the sampling process since each layer is sampled independently. While in ASGCN, the sampling process is dependent to the upper layer, thus parallel processing is not applicable.

6.3.3 Graph-wise Sampling

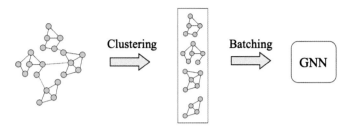

Fig. 6.10: An illustration of graph-wise sampling on large-scale graph.

Other than layer-wise sampling, the graph-wise sampling strategy is introduced recently to accomplish efficient training on large-scale graphs. As shown in Fig. 6.10, a whole graph can be sampled into several sub-graphs and fit into the GNN models, in order to reduce the computational cost.

6.3.3.1 Cluster-GCN

Cluster-GCN (Chiang et al, 2019) first proposes to extract small graph clusters based on efficient graph clustering algorithms. The intuition is that the mini-batch algorithm is correlated with the number of links between nodes in one batch. Hence, Cluster-GCN constructs mini-batch on the sub-graph level, while previous studies usually construct mini-batch based on the nodes.

Cluster-GCN extracts small clusters based on the following clustering algorithms. A graph $\mathscr{G}(\mathscr{V}, \mathscr{E})$ can be devided into c portions by grouping its nodes, where $\mathscr{V} = [\mathscr{V}_1, \cdots \mathscr{V}_c]$. The extracted sub-graphs can be defined as,

$$\bar{\mathscr{G}} = [\mathscr{G}_1, \cdots, \mathscr{G}_c] = [\{\mathscr{V}_1, \mathscr{E}_1\}, \cdots, \{\mathscr{V}_c, \mathscr{E}_c\}].$$

$(\mathscr{V}_t, \mathscr{E}_t)$ represents the nodes and the links within the t-th portion, $t \in (1, c)$. And the re-ordered adjacency matrix can be written as,

$$A = \bar{A} + \Delta = \begin{bmatrix} A_{11} & \cdots & A_{1c} \\ \vdots & \ddots & \vdots \\ A_{c1} & \cdots & A_{cc} \end{bmatrix}; \quad \bar{A} = \begin{bmatrix} A_{11} & \cdots & 0 \\ \vdots & \ddots & \vdots \\ 0 & \cdots & A_{cc} \end{bmatrix}, \Delta = \begin{bmatrix} 0 & \cdots & A_{1c} \\ \vdots & \ddots & \vdots \\ A_{c1} & \cdots & 0 \end{bmatrix}.$$

Different graph clustering algorithms can be used to partition the graph by enabling more links between nodes within the cluster. The motivation of considering sub-graph as a batch also follows the nature of graphs, which is that neighbors usually stay closely with each other.

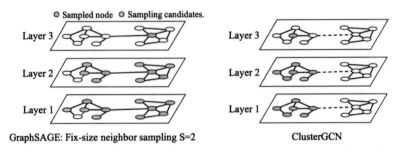

Fig. 6.11: Comparison between GraphSAGE and Cluster-GCN. In Cluster-GCN, it only samples the nodes in each sub-graph.

Obviously, this strategy can avoid the neighbor expansion problem since it only samples the nodes in the clusters, as shown in Fig. 6.11. For Cluster-GCN, since there is no connection between the sub-graphs, the nodes in other sub-graphs will not be sampled when the layer increases. In such a manner, the sampling process establishes a neighbor expansion control by sampling over the sub-graphs, while in layer-wise sampling the neighbor expansion control is implemented by fixing the neighbor sampling size.

However, there still remain two concerns with the vanilla Cluster-GCN. The first one is that the links between sub-graphs are dismissed, which may fail to capture important correlations. The second issue is that the clustering algorithm may change the original distribution of the dataset and introduce some bias. To address these concerns, the authors propose stochastic multiple partitions scheme to randomly combine clusters to a batch. In specific, the graph is first clustered into p sub-graphs; then in each epoch training, a new batch is formed by randomly combine q clusters ($q < p$), and the interactions between clusters are included too. Fig. 6.12 visualized an example when q equals to 2. As observed, the new batch is formed by 2 random clusters, along with the retained connections between the clusters.

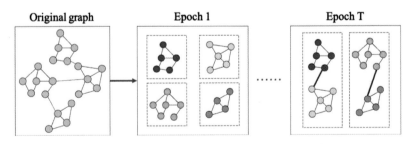

Fig. 6.12: An illustration of stochastic multiple partitions scheme.

In summary, Cluster-GCN is a practical solution based on the sub-graph batching. It has good performance and good memory usage, and can alleviate the neighborhood expansion problem in traditional mini-batch training. However, Cluster-GCN does not analyze the sampling quality, e.g., the bias and variance of this sampling strategy. In addition, the performance is highly correlated to the clustering algorithm.

6.3.3.2 GraphSAINT

Instead of using clustering algorithms to generate the sub-graphs which may bring in certain bias or noise, GraphSAINT (Zeng et al, 2020a) proposes to directly sample a sub-graph for mini-batch training according to sub-graph sampler, and employ a full GCN on the sub-graph to generate the node embeddings as well as back-propagate the loss for each node. As shown in Fig. 6.13, sub-graph $\mathcal{G}_s$ is constructed from the original graph $\mathcal{G}$ with Nodes 0, 1, 2, 3, 4, 7 included. Next, a full GCN is applied on these 6 nodes along with the corresponding connections.

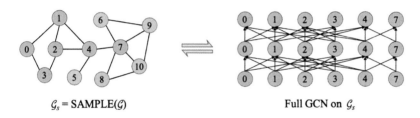

$$\mathcal{G}_s = \text{SAMPLE}(\mathcal{G}) \qquad\qquad \text{Full GCN on } \mathcal{G}_s$$

Fig. 6.13: An illustration of GraphSAINT training algorithm. The yellow circle indicates the sampled node.

GraphSAINT introduces three sub-graph sampler constructions to form the sub-graphs, which are node sampler, edge sampler and random walk sampler (Fig. 6.14). Given graph $\mathcal{G}(\mathcal{V},\mathcal{E})$, node $v \in \mathcal{V}$, edge $(u,v) \in \mathcal{E}$, the node sampler randomly samples $\mathcal{V}_s$ nodes from $\mathcal{V}$. The edge sampler selects the sub-graph based on the probability of edges in the original graph $\mathcal{G}$. The random walk sampler picks node pairs according to the probability that there exists L hops paths from node u to v.

Moreover, GraphSAINT provides comprehensive theoretical analysis on how to control the bias and variance of the sampler. First, it proposes loss normalization and aggregation normalization to eliminate the sampling bias.

$$\text{Loss normalization:} \quad \mathscr{L}_{\text{batch}} = \sum_{v \in \mathcal{G}_s} L_v/\lambda_v, \quad \lambda_v = |\mathcal{V}|p_v$$

$$\text{Aggregation normalization:} \quad a(u,v) = p_{u,v}/p_v$$

where p_v is the probability of a node $v \in \mathcal{V}$ being sampled, $p_{u,v}$ is the probability of an edge $(u,v) \in \mathcal{E}$ being sampled, L_v represents the loss of v in the output layer. Second, GraphSAINT also proposes to minimize the sampling variance by adjusting the edge sampling probability by:

$$p_{u,v} \propto 1/d_u + 1/d_v.$$

The extensive experiments demonstrate the effectiveness and efficiency of Graph-SAINT, and prove that GraphSAINT converges fast as well as achieves superior performance.

In summary, GraphSAINT proposes a highly flexible and extensible framework including the graph sampler strategies and the GNN architectures, as well as achieves good performance on both accuracy and speed.

6.3.3.3 Overall Comparison of Different Models

Table 6.2 compares and summarizes the characteristics of previously mentioned models. *Paradigm* indicates the different sampling paradigms, and *Model* defers to the proposed method in each paper. *Sampling Strategy* shows the sampling theory, and *Variance Reduction* denotes whether such analysis is conducted in the paper. *Solved Problem* represents the problem that proposed model has addressed, and *Characteristic* extracts the features of the model.

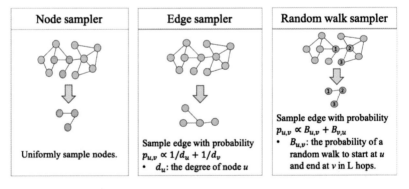

Fig. 6.14: An illustration of different samplers.

Table 6.2: The comparison between different models.

Paradigm	Model	Sampling Strategy	Variance Reduction	Solved Problem	Characteristics
Node-wise Sampling	GraphSAGE (Hamilton et al, 2017b)	Random	×	Inductive learning	Mini-batch training, reduce neighborhood expansion.
	VR-GCN (Chen et al, 2018d)	Random	✓	Neighborhood expansion	Historical activations.
Layer-wise Sampling	FastGCN (Chen et al, 2018c)	Importance	✓	Neighborhood expansion	Integral transform view.
	ASGCN (Huang et al, 2018)	Importance	✓	Between-layer correlation	Explicit variance reduction, skip connection.
Graph-wise Sampling	Cluster-GCN (Chiang et al, 2019)	Random	✓	Graph batching	Mini-batch on sub-graph.
	GraphSAINT (Zeng et al, 2020a)	Edge Probability	✓	Neighborhood expansion	Variance and bias control.

6.4 Applications of Large-scale Graph Neural Networks on Recommendation Systems

Deploying large-scale neural networks in academia has achieved remarkable success. Other than the theoretical study on *how to expand the GNNs on large graphs*, another crucial problem is *how to embed the algorithms into industrial applications*. One of the most conventional applications that demand tremendous data is the recommendation systems, which learn the user preferences and make predictions for what the users may interest in. Traditional recommendation algorithms like collaborative filtering are mainly designed according to the user-item interactions(Goldberg et al, 1992; Koren et al, 2009; Koren, 2009; He et al, 2017b). Such methods are not capable of the explosive increased web-scale data due to the extreme sparsity. Recently, graph-based deep learning algorithms have gained significant achievements on improving the prediction performance of recommendation systems by modeling the graph structures of web-scale data (Zhang et al, 2019b; Shi et al, 2018a; Wang et al, 2018b). Therefore, utilizing large-scale GNNs for recommendation has become a trend in industry (Ying et al, 2018b; Zhao et al, 2019b; Wang et al, 2020d; Jin et al, 2020b).

Recommendation systems can be typically categorized into two fields: item-item recommendation and user-item recommendation. The former one aims to find the similar items based on a user's historical interactions; while the later one directly predicts the user's preferred items by learning the user behaviors. In this chapter, we briefly introduce notable recommendation systems that are implemented on large graphs for each field.

6.4.1 Item-item Recommendation

PinSage (Ying et al, 2018b) is one of the successful applications in the early stage of utilizing large-scale GNNs on item-item recommendation systems, which is deployed on Pinterest[1]. Pinterest is a social media application that shares and discovers various content. The users mark their interested content with pins and organize them on the boards. When the users browse the website, Pinterest recommends the potentially interesting content for them. By the year 2018, the Pinterest graph contains 2 billion pins, 1 billion boards, and over 18 billion edges between pins and boards.

In order to scale the training model on such a large graph, Ying et al (2018b) proposes PinSage, a random-walk-based GCN, to implement node-wise sampling on Pinterest graph. In specific, a short random walk is used to select a fixed-number neighborhood of the target node. Fig. 6.15 demonstrates the overall architecture of PinSage. Take node A as an example, a 2-depth convolution is constructed to generate the node embedding $\mathbf{h}_A^{(2)}$. The embedding vector $\mathbf{h}_{\mathcal{N}(A)}^{(1)}$ of node A's neighbors are aggregated by node B, C, and D. Similar process is established to get the 1-hop neighbors' embedding $\mathbf{h}_B^{(1)}$, $\mathbf{h}_C^{(1)}$, and $\mathbf{h}_D^{(1)}$. An illustration of all participated nodes for each node from the input graph is shown at the bottom of Fig. 6.15. In addition, a L1-normalization is computed to sort the neighbors by their importance (Eksombatchai et al, 2018), and a curriculum training strategy is used to further improve the prediction performance by feeding harder-and-harder examples.

A series of comprehensive experiments that are conducted on Pinterest data, e.g., offline experiments, production A/B tests and user studies, have demonstrated the effectiveness of the proposed method. Moreover, with the adoption of highly efficient MapReduce inference pipeline, the entire process on the whole graph can be finished within one day.

6.4.2 User-item Recommendation

Unlike item-item recommendation, user-item recommendation systems is more complex since it aims at predicting the user's behaviors. Moreover, there remains more auxiliary information between users and users, items and items, and users and items, which leads to a heterogeneous graph problem. As shown in Fig. 6.16, there are various properties of the edges between user-user and item-item, which cannot be considered as one simple relation, e.g., user *searches* a word or *visits* a shop should be considered with different impacts.

---

[1] https://www.pinterest.com/

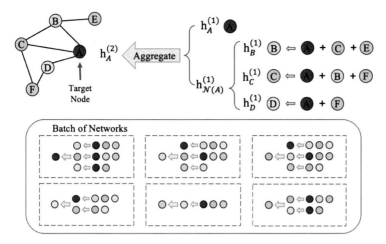

Fig. 6.15: Overview of PinSage architecture. Colored nodes are applied to illustrate the construction of graph convolutions.

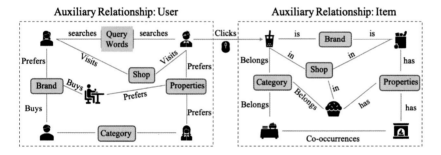

Fig. 6.16: Examples of heterogeneous auxiliary relationships on e-commerce websites.

IntentGC (Zhao et al, 2019b) proposes a GCN-based framework for large-scale user-item recommendation on e-commerce data. It explores the explicit user preferences as well as the abundant auxiliary information by graph convolutions and make predictions. E-commerce data such as Amazon contains billions of users and items, while the diverse relationships bring in more complexity. Thus, the graph structure gets larger and more complicated. Moreover, due to the sparsity of user-item graph network, sampling methods like GraphSAGE may result in a very huge sub-graph. In order to train efficient graph convolutions, IntentGC designs a faster graph convolution mechanism to boost the training, named as IntentNet.

As shown in Fig. 6.17, the bit-wise operation illustrates the traditional way of node embedding construction in GNN. In specific, consider node v as the target node, the embedding vector $\mathbf{h}_v^{(l+1)}$ is generated by concatenating the neighborhoods'

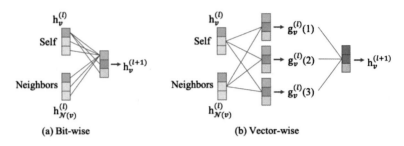

Fig. 6.17: Comparison between bit-wise and vector-wise graph convolution.

embeddings $\mathbf{h}^{(l)}_{\mathcal{N}(v)}$ and the target itself $\mathbf{h}^{(l)}_v$. Such an operation is able to capture two types of information: the interactions between target node and its neighborhoods; and the interactions between different dimensions of the embedding space. However, in user-item networks, learning the information between different feature dimensions may be less informative and unnecessary. Therefore, IntentNet designs a vector-wise convolution operation as follows:

$$\mathbf{g}^{(l)}_v(i) = \sigma\left(W^{(l)}_v(i,1) \cdot \mathbf{h}^{(l)}_v + W^{(l)}_v(i,2) \cdot \mathbf{h}^{(l)}_{\mathcal{N}(v)}\right),$$
$$\mathbf{h}^{(l+1)}_v = \sigma\left(\sum_{i=1}^L \theta^{(l)}_i \cdot \mathbf{g}^{(l)}_v(i)\right),$$

where $W^{(l)}_v(i,1)$ and $W^{(l)}_v(i,2)$ are the associated weight matrices for the *i-th* local filter. $\mathbf{g}^{(l)}_v(i)$ represents the operation that learns the interactions between the target node and its neighbor nodes in a vector-wise manner. Another vector-wise layer is applied to gather the final embedding vector of the target node for the next convolutional layer. Moreover, the output vector of the last convolutional layer is fed into a three-layer fully-connected network to further learn the node-level combinatory features. Such an operation significantly promotes the training efficiency and reduces the time complexity.

Extensive experiments are conducted on Taobao and Amazon datasets, which contain millions to billions of users and items. IntentGC outperforms other baseline methods, as well as reduces the training time for about two days compared with GraphSAGE.

6.5 Future Directions

Overall, in recent years, the scalability of GNNs has been extensively studied and has achieved fruitful results. Fig. 6.18 summarizes the development towards large-scale GNNs.

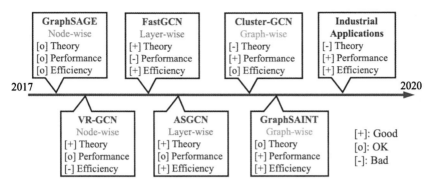

Fig. 6.18: Overall performance comparison of introduced work on large-scale GNNs.

GraphSAGE is the first to propose sampling on the graph instead of computing on the whole graph. VR-GCN designs another node sampling algorithm and provides a comprehensive theoretical analysis, but the efficiency is still limited. FastGCN and ASGCN propose to sample over layers, and both prove the efficiency with detailed analysis. Cluster-GCN first partitions the graph into sub-graphs to eliminate the neighborhood expansion problem, and boosts the performance of several benchmarks. GraphSAINT further improves the graph-wise sampling algorithm to achieve the state-of-the-art classification performance over commonly used benchmark datasets. Various industrial applications prove the effectiveness and practicability of large-scale GNNs in the real world.

However, many new open problems arise, e.g., how to balance the trade-off between variance and bias during sampling; how to deal with complex graph types such as heterogeneous/dynamic graphs; how to properly design models over complex GNN architectures. Studies toward such directions would improve the development of large-scale GNNs.

Editor's Notes: For graphs of large scale or with rapid expansibility, such as dynamic graph (chapter 15) and heterogeneous graph (chapter 16), the scalability characterization of GNNs is of vital importance to determine whether the algorithm is superior in practice. For example, graph sampling strategy is especially necessary to ensure computational efficiency in industrial scenarios, such as recommender system (chapter 19) and urban intelligence (chapter 27). With the increasing complexity and scale of the real problem, the limitation in scalability has been considered almost everywhere in the study of GNNs. Researchers devoted to graph embedding (chapter 2), graph structure learning (chapter 14) and self-supervised learning (chapter 18) put forward very remarkable works to overcome it.

Chapter 7
Interpretability in Graph Neural Networks

Ninghao Liu and Qizhang Feng and Xia Hu

Abstract Interpretable machine learning, or explainable artificial intelligence, is experiencing rapid developments to tackle the opacity issue of deep learning techniques. In graph analysis, motivated by the effectiveness of deep learning, graph neural networks (GNNs) are becoming increasingly popular in modeling graph data. Recently, an increasing number of approaches have been proposed to provide explanations for GNNs or to improve GNN interpretability. In this chapter, we offer a comprehensive survey to summarize these approaches. Specifically, in the first section, we review the fundamental concepts of interpretability in deep learning. In the second section, we introduce the post-hoc explanation methods for understanding GNN predictions. In the third section, we introduce the advances of developing more interpretable models for graph data. In the fourth section, we introduce the datasets and metrics for evaluating interpretation. Finally, we point out future directions of the topic.

7.1 Background: Interpretability in Deep Models

Deep learning has become an indispensable tool for a wide range of applications such as image processing, natural language processing, and speech recognition. Despite the success, deep models have been criticized as "black boxes" due to their complexity in processing information and making decisions. In this section, we introduce the research background of interpretability in deep models, including the

Ninghao Liu
Department of CSE, Texas A&M University, e-mail: `nhliu43@tamu.edu`

Qizhang Feng
Department of CSE, Texas A&M University, e-mail: `qf31@tamu.edu`

Xia Hu
Department of CSE, Texas A&M University, e-mail: `xiahu@tamu.edu`

© The Author(s), under exclusive license to Springer Nature Singapore Pte Ltd. 2022   121
L. Wu et al. (eds.), *Graph Neural Networks: Foundations, Frontiers, and Applications*,
https://doi.org/10.1007/978-981-16-6054-2_7

definition of interpretability/interpretation, the reasons for exploring model interpretation, the methods of obtaining interpretation in traditional deep models, the opportunities and challenges to achieve interpretability in GNN models.

7.1.1 Definition of Interpretability and Interpretation

There is no unified mathematical definition of interpretability. A commonly used (nonmathematical) definition of interpretability is given below (Miller, 2019).

Definition 7.1. *Interpretability* is the degree to which an observer can understand the cause of a decision.

There are three elements in the above definition: "understand", "cause", and "a decision". According to different scenarios, it is common that these elements are re-weighted or even some elements are replaced. First, in the context of machine learning systems where the role of humans needs to be emphasized, the definition of interpretability is usually revised to adapt to humans (Kim et al, 2016), where interpretation results that better facilitate human understanding and reasoning habits are more desirable. Second, from the term "cause" in the definition, it is natural to think that interpretation studies causality properties in models. While causality is important in developing certain types of interpretation methods, it is also common that interpretation is obtained beyond the framework of causal theories. Third, there is an increasing number of techniques that jump out of the scheme of explaining "a decision", and try to understand a broader range of entities such as model components (Olah et al, 2018) and data representations.

The interpretation is one mode in which an observer may obtain an understanding of a model or its predictions. A general and widely followed definition is as below (Montavon et al, 2018).

Definition 7.2. An *interpretation* is the mapping of an abstract concept into a domain that the human can understand.

Typical examples of human-understandable domains include arrays of pixels in images or words in texts. There are two elements that merit attention in the above definition: "concept" and "understand". First, the "concept" to be explained could refer to different aspects, such as a predicted class (i.e., the logit value of the predicted class), the perception of a model component, or the meaning of a latent dimension. Second, in specific scenarios where user experience is important, it is necessary to transfer raw interpretation to the format that facilitates user comprehension, sometimes even with the cost of sacrificing interpretation accuracy.

It is also worth noting that, in this work, we distinguish between "interpretation" and "explanation". Although their differences have not been formally defined, in literatures, explanation mainly refers to the collection of important features for a given prediction (e.g., classification or regression) (Montavon et al, 2018). Meanwhile, "explanation" is more likely to be used if we are studying post-hoc interpretation

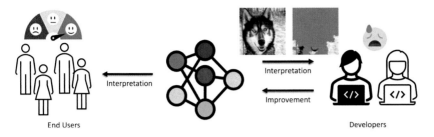

Fig. 7.1: Left: Interpretation could benefit user experiences in interaction with models. Right: Through interpretation, we could identify model behaviors that are not desirable according to humans, and work on improving the model accordingly (Ribeiro et al, 2016).

or human-understandable interpretation. "Interpretation" usually refers to a broader range of concepts, especially to emphasize that the model itself is intrinsically interpretable (i.e., the transparency of the model).

7.1.2 The Value of Interpretation

There are several pragmatic reasons that motivate people to study and improve model interpretability. Depending on who finally benefits from interpretation, we divide the reasons into model-oriented and user-oriented, as shown in Fig. 7.1.

7.1.2.1 Model-Oriented Reasons

Interpretation is an effective tool to diagnose the defects in models and provide directions on how to improve. Therefore, after several iterations of model updates, it is possible to obtain better models with particular properties coming about, and we could apply these models to our advantage. There are several properties that have been considered in literatures that are summarized as below.

1. *Credibility:* A model is regarded as credible if the rationale used behind predictions is consistent with the well-established domain knowledge. Through interpretation, we could observe whether the predictions are based on proper evidences, or they are simply from the exploitation of artifacts in data. By extracting explanations from a model and making the explanations to match human-annotated evidences in data, we are able to improve the model's credibility when making decisions (Du et al, 2019; Wang et al, 2018c).
2. *Fairness:* Machine learning systems have the risk of amplifying societal stereotypes if they rely on sensitive attributes, such as race, gender and age, in making predictions. Through interpretation, we could observe whether the predictions

are based on sensitive features that are required to be avoided in real applications.

3. *Adversarial-Attack Robustness:* Adversarial attack refers to adding carefully-crafted perturbations to input, where the perturbations are almost imperceptible to humans, but can cause the model to make wrong predictions (Goodfellow et al, 2015). Robustness against adversarial attacks is an increasingly important topic in machine learning security. Recent studies have shown how interpretation could help in discovering new attack schemes and designing defense strategies (Liu et al, 2020d).

4. *Backdoor-Attack Robustness:* Backdoor attack refers to injecting malicious functionality into a model, by either implanting additional modules or poisoning training data. The model will behave normally unless it is fed with input containing patterns that trigger the malicious functionality. Studying model robustness against backdoor attacks is attracting more interest recently. Recent research discovers that interpretation could be applied in identifying if a model has been infected by backdoors (Huang et al, 2019c; Tang et al, 2020a).

7.1.2.2 User-Oriented Reasons

The interpretation could contribute to the construction of interfaces between humans and machines.

1. *Improving User Experiences:* By providing intuitive visual information, interpretation could gain user trust, and increase a system's ease of use. For example, in healthcare-related applications, if the model could explain to patients how it makes diagnoses, the patients would be more convinced (Ahmad et al, 2018). For another example, in a recommender system, providing explanations can help users to make faster decisions and persuade users to purchase the recommended products (Li et al, 2020c).

2. *Facilitating Decision Making:* In many applications, a model plays the role as an assistant, while humans will make the final decision. In this case, interpretation helps shape human understandings towards instances, thus affecting subsequent decision-making processes. For example, in outlier detection, some outliers own malicious properties that should be handled with caution, while some are benign instances that simply happen to be "different". With interpretation, it is much easier for human decision-makers to understand whether a given outlier is malicious or benign.

7.1.3 Traditional Interpretation Methods

In general, there are two categories of techniques in improving model interpretability. Some efforts have been paid to build more transparent models, and we are able to grasp how the models (or parts of the models) work. We call this direction as

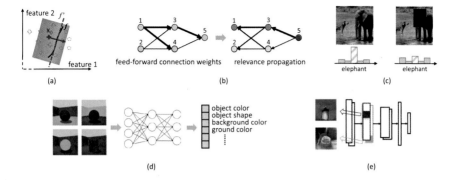

Fig. 7.2: Illustration of post-hoc interpretation methods. (a): Local approximation based interpretation. (b): Layer-wise relevance propagation. (c): Explanation based on perturbation. (d): Explaining the meaning of latent representation dimensions. (e): Explaining the meaning of neurons in a convolutional neural network via input generation.

interpretable modeling. Meanwhile, instead of elucidating the internal mechanisms by which models work, some methods investigate *post-hoc interpretation* to provide explanations to models that are already built. In this part, we introduce the techniques of the two categories. Some of the methods provide motivation for GNN interpretation which will be introduced in later sections.

7.1.3.1 Post-Hoc Interpretation

The post-hoc interpretation has received a lot of interests in both research and real applications. Flexibility is one of the advantages of post-hoc interpretation, as it put less requirement on the model types or structures. In the following paragraphs, we briefly introduce several commonly used methods. The illustration of the basic idea behind each of these methods is shown in Fig. 7.2.

The first type of methods to be introduced is *approximation-based methods*. Given a function f that is complex to understand and an input instance $\mathbf{x}^* \in \mathbb{R}^m$, we could approximate f with a simple and understandable surrogate function h (usually chosen as a linear function) locally around $\mathbf{x}^*$. Here m is the number of features in each instance. There are several ways to build h. A straightforward way is based on the first-order Taylor expansion, where:

$$f(\mathbf{x}) \approx h(\mathbf{x}) = f(\mathbf{x}^*) + \mathbf{w}^\top \cdot (\mathbf{x} - \mathbf{x}^*), \tag{7.1}$$

where $\mathbf{w} \in \mathbb{R}^m$ tells how sensitive the output is to the input features. Typically, $\mathbf{w}$ can be estimated with the gradient (Simonyan et al, 2013), so that $\mathbf{w} = \nabla_{\mathbf{x}} f(\mathbf{x}^*)$. When gradient information is not available, such as in tree-based models, we could build h through training (Ribeiro et al, 2016). The general idea is that a number of training instances $(\mathbf{x}^i, f(\mathbf{x}^i))$, $1 \leq i \leq n$ are sampled around $\mathbf{x}^*$, i.e., $\|\mathbf{x}^i - \mathbf{x}^*\| \leq \varepsilon$. The instances are then used to train h, so that h approximates f around $\mathbf{x}^*$.

Besides directly studying the sensitivity between input and output, there is another type of method called *layer-wise relevance propagation (LRP)* (Bach et al, 2015). Specifically, LRP redistributes the activation score of output neuron to its predecessor neurons, which iterates until reaching the input neurons. The redistribution of scores is based on the connection weights between neurons in adjacent layers. The share received by each input neuron is used as its contribution to the output.

Another way to understand the importance of a feature $\mathbf{x}_i$ is to answer questions like "What would have happened to f, had $\mathbf{x}_i$ not existed in input?". If $\mathbf{x}_i$ is important for predicting $f(\mathbf{x})$, then removing/weakening it will cause a significant drop in prediction confidence. This type of method is called the *perturbation method* (Fong and Vedaldi, 2017). One of the key challenges in designing perturbation methods is how to guarantee the input after perturbation is still valid. For example, it is argued that perturbation on word embedding vectors cannot explain deep language models, because texts are discrete symbols, and it is hard to identify the meaning of perturbed embeddings.

Different from the previous methods that focus on explaining prediction results, there is another type of method that tries to understand how data is represented inside a model. We call it *representation interpretation*. There is no unified definition for representation interpretation. The design of methods under this category is usually motivated by the nature of the problem or the properties of data. For example, in natural language processing, it has been shown that a word embedding could be understood as the composition of a number of basis word embeddings, where the basis words constitute a dictionary (Mathew et al, 2020).

Besides understanding predictions and data representations, another interpretation scheme is to understand the role of *model components*. A well-known example is to visualize the visual patterns that maximally activate the target neuron/layer in a CNN model (Olah et al, 2018). In this way, we understand what kind of visual signal is detected by the target component. The interpretation is usually obtained through a *generative process*, so that the result is understandable to humans.

7.1.3.2 Interpretable Modeling

Interpretable modeling is achieved via incorporating interpretability directly into the model structures or learning process. It is still an extremely challenging problem to develop models that are both transparent and could achieve state-of-the-art performances. Many efforts have been paid to improve the intrinsic interpretability of deep models. Some details are discussed as below.

A straightforward strategy is to rely on *distillation*. Specifically, we first build a complex model (e.g., a deep model) to achieve good performance. Then, we use another model, which is readily recognized as interpretable, to mimic the predictions of the complex model. The pool of interpretable models includes linear models, decision trees, rule-based models, etc. This strategy is also called *mimic learning*. The interpretable model trained in this way tends to perform better than normal training, and it is also much easier to understand than the complex model.

Attention models, originally introduced for machine translation tasks, have now become enormously popular, partially due to their interpretation properties. The intuition behind attention models can be explained using human biological systems, where we tend to selectively focus on some parts of the input, while ignoring other irrelevant parts (Xu et al, 2015). By examining attention scores, we could know which features in the input have been used for making the prediction. This is also similar to using post-hoc interpretation algorithms that find which input features are important. The major difference is that attention scores are generated during model prediction, while post-hoc interpretation is performed after prediction.

Deep models heavily rely on learning effective representations to compress information for downstream tasks. However, it is hard for humans to understand the representations as the meanings of different dimensions are unknown. To tackle this challenge, *disentangled representation learning* has been proposed. Disentangled representation learning breaks down features of different meanings and encodes them as separate dimensions in representations. As a result, we could check each dimension to understand which factors of input data are encoded. For example, after learning disentangled representations on 3D-chair images, factors such as chair leg style, width and azimuth, are separately encoded into different dimensions (Higgins et al, 2017).

7.1.4 Opportunities and Challenges

Despite the major progress made in domains such as vision, language and control, many defining characteristics of human intelligence remain out of reach for traditional deep models such as convolutional neural networks (CNNs), recurrent neural networks (RNNs) and multi-layer perceptrons (MLPs). To look for new model architectures, people believe that GNN architectures could lay the foundation for more interpretable patterns of reasoning (Battaglia et al, 2018). In this part, we discuss the advantages of GNNs and challenges to be tackled in terms of interpretability.

The GNN architecture is regarded as more interpretable because it facilitates learning about entities, relations, and rules for composing them. First, entities are discrete and usually represent high-level concepts or knowledge items, so it is regarded as easier for humans to understand than image pixels (tiny granularity) or word embeddings (latent space vectors). Second, GNN inference propagates information through links, so it is easier to find the explicit reasoning path or subgraph that contributes to the prediction result. Therefore, there is a recent trend of transforming images or text data into graphs, and then applying GNN models for predictions. For example, to build a graph from an image, we can treat objects inside the image (or different portions within an object) as nodes, and generate links based on the spatial relations between nodes. Similarly, a document can be transformed into a graph by discovering concepts (e.g., nouns, named entities) as nodes and extracting their relations as links through lexical parsing.

Although the graph data format lays a foundation for interpretable modeling, there are still several challenges that undermine GNN interpretability. First, GNN still maps nodes and links into embeddings. Therefore, similar to traditional deep models, GNN also suffers from the opacity of information processing in intermediate layers. Second, different information propagation paths or subgraphs contribute differently to the final prediction. GNN does not directly provide the most important reasoning paths for its prediction, so post-hoc interpretation methods are still needed. In the following sections, we will introduce the recent advances in tackling the above challenges to improve the explainability and interpretability of GNNs.

7.2 Explanation Methods for Graph Neural Networks

In this section, we introduce the post-hoc explanation methods for understanding GNN predictions. Similar to the categorization in Section 7.1.3, we include approximation-based methods, relevance-propagation-based methods, perturbation-based methods, and generative methods.

7.2.1 Background

Before introducing the techniques, we first provide the definition of graphs and review the fundamental formulations of a GNN model.

Graphs: In the rest of the chapter, if not specified, the graphs we discuss are limited to homogeneous graphs.

Definition 7.3. A *homogeneous graph* is defined as $\mathscr{G} = (\mathscr{V}, \mathscr{E})$, where $\mathscr{V}$ is the set of nodes and $\mathscr{E}$ is the set of edges between nodes.

Furthermore, let $A \in \mathbb{R}^{n \times n}$ be the adjacency matrix of $\mathscr{G}$, where $n = |\mathscr{V}|$. For unweighted graphs, $A_{i,j}$ is binary, where $A_{i,j} = 1$ means there exists an edge $(i, j) \in \mathscr{E}$, otherwise $A_{i,j} = 0$. For weighted graphs, each edge (i, j) is assigned a weight $w_{i,j}$, so $A_{i,j} = w_{i,j}$. In some cases, nodes are associated with features, which could be denoted as $X \in \mathbb{R}^{n \times m}$, and $X_{i,:}$ is the feature vector of node i. The number of features for each node is m. In this chapter, unless otherwise stated, we focus on GNN models on homogeneous graphs.

GNN Fundamentals: Traditional GNNs propagate information via the input graph's structure according to the propagation scheme:

$$H^{l+1} = \sigma(\tilde{D}^{-\frac{1}{2}}\tilde{A}\tilde{D}^{-\frac{1}{2}}H^l W^l), \qquad (7.2)$$

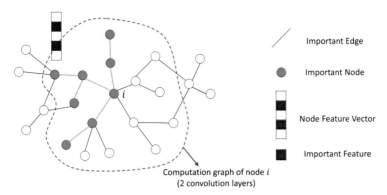

Fig. 7.3: Illustration of explanation result formats. Explanation results for graph neural networks could be the important nodes, the important edges, the important features, etc. An explanation method may return multiple types of results.

where H^l denotes the embedding matrix at layer l, and W^l denotes the trainable parameters at layer l. Also, $\tilde{A} = A + \mathbf{I}$ denotes the adjacency matrix of the graph after adding the self-loop. The matrix $\tilde{D}$ is the diagonal degree matrix of $\tilde{A}$, i.e., $\tilde{D}_{i,i} = \sum_j \tilde{A}_{i,j}$. Therefore, $\tilde{D}^{-\frac{1}{2}}\tilde{A}\tilde{D}^{-\frac{1}{2}}$ normalizes the adjacency matrix. If we only focus on the embedding update of node i, the GCN propagation scheme could be rewritten as:

$$H_{i,:}^{l+1} = \sigma\left(\sum_{j \in \mathcal{V}_i \cup \{i\}} \frac{1}{c_{i,j}} H_{j,:}^l W^l \right), \tag{7.3}$$

where $H_{j,:}$ denotes the j-th row of matrix H, and $\mathcal{V}_i$ denotes the neighbors of node i. Here $c_{i,j}$ is a normalization constant, and $\frac{1}{c_{i,j}} = (\tilde{D}^{-\frac{1}{2}}\tilde{A}\tilde{D}^{-\frac{1}{2}})_{i,j}$. Therefore, the embedding of node i at layer l can be seen as aggregating neighbor embeddings of nodes that are neighbors of node i, followed by some transformations. The embeddings in the first layer H^0 is usually set as the node features. As the layer goes deeper, the computation of each node's embedding will include further nodes. For example, in a 2-layer GNN, computing the embedding of node i will use the information of nodes within the 2-hop neighborhood of node i. The subgraph composed by these nodes is called the *computation graph* of node i, as shown in Fig. 7.3.

Target Models: There are two common tasks in graph analysis, i.e., graph-level predictions and node-level predictions. We use classification tasks as the example. In graph-level tasks, the model $f(\mathcal{G}) \in \mathbb{R}^C$ produces a single prediction for the whole graph, where C is the number of classes. The prediction score for class c could be written as $f^c(\mathcal{G})$. In node-level tasks, the model $f(\mathcal{G}) \in \mathbb{R}^{n \times C}$ returns a matrix, where each row is the prediction for a node. Some explanation methods are designed solely for graph-level tasks, some are for node-level tasks, while some could handle both scenarios. The computation graphs introduced above are commonly used in explaining node-level predictions.

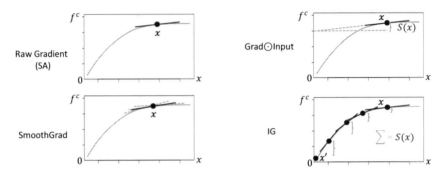

Fig. 7.4: Illustration of several gradient-based explanation methods. Methods relying on local gradients may suffer from the saturation problem or noises in input, where a feature's local sensitivity is not consistent with its overall contribution. SmoothGrad removes noises in an explanation by averaging multiple explanations on nearby points. IG is more accurate than Grad ⊙ Input in measuring feature contribution.

Interpretation Formats: According to the introduction above, there are several input modes that could be included in the explanation as shown in Fig. 7.3. Specifically, explanation methods could identify what are the *important nodes*, *important edges* and *important features* that contribute most to the prediction. Some explanation methods may identify multiple types of input modes simultaneously.

7.2.2 Approximation-Based Explanation

The approximation-based explanation has been widely used to analyze the prediction of models with complex structures. Approximation-based approaches could be further divided into white-box approximation and black-box approximation. The white-box approximation uses information inside the model, which includes but is not limited to gradients, intermediate features, model parameters, etc. The black-box approximation does not utilize information propagation inside the model. It usually uses a simple and interpretable model to fit the target model's decision on an input instance. Then, the explanation can be easily extracted from the simple model. The details of commonly used methods for both categories are introduced as below.

7.2.2.1 White-Box Approximation Method

Sensitivity Analysis (SA) Baldassarre and Azizpour (2019) study the impact of a particular change in an independent variable on a dependent variable. In the context of explanation, the dependent variable refers to the prediction, while the independent variables refer to the features. The local gradient of the model is commonly used as sensitivity scores to represent the correlation between the feature and the prediction result. The sensitivity score is defined as:

$$\mathscr{S}(\mathbf{x}) = \|\nabla_{\mathbf{x}} f(\mathscr{G})\|^2, \tag{7.4}$$

where $\mathscr{G}$ is the input instance graph to be explained, $f(\mathscr{G})$ is the model prediction function. Here $\mathbf{x}$ refers to the feature vector of a node of interest. Node features with higher sensitivity scores are more important because they can lead to drastic changes to model decisions.

Although SA is intuitive and straightforward, its effectiveness is still limited. It assumes input features are mutually independent, and does not necessarily pay attention to their correlations in the actual decision-making process. Also, sensitivity analysis only measures the impact of local changes to the decision function $f(\mathscr{G})$, rather than thoroughly explaining the decision function value itself. Explanation results provided by sensitivity analysis are usually relatively noisy and challenging to comprehend. Therefore, some follow-up techniques have been developed trying to overcome this limitation (as shown in Fig. 7.4).

GuidedBP(Baldassarre and Azizpour, 2019) is similar to SA except that it only detects the features that positively activate the neurons, with the assumption that negative gradients may confuse the contribution of important features and makes the visualizing noisy. To follow this intuition, GuideBP modifies the process of back-propagation of SA and discards all negative gradients.

Grad ⊙ Input Sanchez-Lengeling et al (2020) measures the feature contribution scores as the element-wise product of the input features and the gradients of decision function with respect to the features:

$$\mathscr{S}(\mathbf{x}) = \nabla_{\mathbf{x}}^{\top} f(\mathscr{G}) \odot \mathbf{x}. \tag{7.5}$$

Therefore, Grad ⊙ Input considers not only the feature sensitivity, but also the scale of feature values. However, the methods mentioned above all suffered from the *saturation problem*, where the scope of the local gradients is too limited to reflect the overall contribution of each feature.

Integrated Gradients (IG) Sanchez-Lengeling et al (2020) solve the saturation problem by aggregating feature contribution along a designed path in input space. This path starts from a chosen baseline point $\mathscr{G}'$ and ends at the target input $\mathscr{G}$. Specifically, the feature contribution is computed as:

$$\mathscr{S}(\mathbf{x}) = (\mathbf{x} - \mathbf{x}') \int_{\alpha=0}^{1} \nabla_{\mathbf{x}} f\left(\mathscr{G}' + \alpha\left(\mathscr{G} - \mathscr{G}'\right)\right) d\alpha \tag{7.6}$$

where $\mathbf{x}'$ denotes a feature vector in the baseline point $\mathscr{G}'$, while $\mathbf{x}$ is a feature vector in the original input $\mathscr{G}$. The choice of baseline $\mathscr{G}'$ is relatively flexible. A typical strategy is to use a null graph as the baseline, which has the same topology but its nodes use "unspecified" categorical features. This is motivated by the application of

IG in explaining image classification models (Sundararajan et al, 2017), where the baseline is usually chosen as a pure black image or an image with random noises.

The explanations obtained by the above methods usually contain a lot of noises. Therefore, Smilkov et al (2017) propose SmoothGrad to alleviate the problem. **SmoothGrad** averages attributions evaluated on a number of noise-perturbed versions of the input. This method initially aims at sharpening the saliency maps on images. Furthermore, Sanchez-Lengeling et al (2020) apply it to the Grad $\odot$ Input method by adding Gaussian noise to node and edge features, and averaging multiple explanations to a smoothed one.

Class Activation Mapping (CAM) (Pope et al, 2019) is an explanation method that is initially developed for CNNs. This method only works under a specific model architecture, where the last convolutional layer is followed by a global average pooling (GAP) layer before the final softmax layer. The feature maps (i.e., activations) in the last convolutional layer are aggregated and re-scaled to the same size as the input image, so that the activations highlight the important regions in the image. The idea of CAM can also be adapted to graph neural networks. Specifically, the GAP layer in a GNN could be defined as averaging the embeddings of all nodes in the last graph convolution layer: $\mathbf{h} = \frac{1}{n} \sum_{i=1}^{n} H_{i,:}^{L}$, where L is the last graph convolution layer. CAM treats each dimension of the final node embeddings (i.e., $H_{:,k}^{L}$) as a feature map. The logit value for class c is:

$$f^c(\mathcal{G}) = \sum_k w_k^c \mathbf{h}_k \tag{7.7}$$

where $\mathbf{h}_k$ denotes the k-th entry of $\mathbf{h}$, w_k^c is the GAP-layer weight of k-th feature map with respect to class c. Therefore, the contribution of node i to the prediction is:

$$\mathscr{S}(i) = \frac{1}{n} \sum_k w_k^c H_{i,k}^L. \tag{7.8}$$

Although CAM is simple and efficient, it only works on models with certain structures, which greatly limits its application scenarios.

Grad-CAM (Pope et al, 2019) combines gradient information with feature maps to relax the limitation of CAM. While CAM uses the GAP layer to estimate the weight of each feature map, Grad-CAM employs the gradient of output with respect to the feature maps to compute the weights, so that:

$$w_k^c = \frac{1}{n} \sum_{i=1}^{n} \frac{\partial f^c(\mathcal{G})}{\partial H_{i,k}^L}, \tag{7.9}$$

$$\mathscr{S}(i) = ReLU \left(\sum_k w_k^c H_{i,k}^L \right). \tag{7.10}$$

The *ReLU* function forces the explanation to focus on the positive influence on the class of interest. Grad-CAM is equivalent to CAM for GNNs with only one fully-connected layer before output. Compared to CAM, Grad-CAM can be applied to more GNN architectures, thus avoiding the trade-off between model explainability and capacity.

7.2.2.2 Black-Box Approximation Methods

Different from white-box approximation methods, black-box approximation methods manage to bypass the need to obtain internal information of complex models. The general idea is to use models that are intrinsically interpretable (such as linear regressions, decision trees) to fit the complex model. Then, we can explain the decision based on the simple models. The fundamental assumption behind this is that: Given an input instance, the model's decision boundary within the neighborhood of that instance can be well approximated by the interpretable model. The major challenge is how to define the neighborhood space given an input graph which is a discrete data structure.

We introduce several approaches, including GraphLime (Huang et al, 2020c), RelEx (Zhang et al, 2020a), and PGM-Explainer (Vu and Thai, 2020). These methods share a similar procedure: First, a neighborhood space is defined around the target instance. Second, data points are sampled within this space and their predictions are obtained after being fed into the target model. A training dataset is built, where each instance-label pair consists of a sampled point and its prediction. Finally, an interpretable model is trained by using the dataset. The key difference between these methods lies in two aspects, i.e., the definition of the neighborhood, and the choice of the interpretable model.

GraphLime is a local explanation method for GNN predictions on graph nodes. Given the prediction result on a target node v_t, GraphLime defines the neighborhood space as a set of nodes which are in the k-hop neighborhood of the target node in the input graph:

$$\mathcal{V}_t = \{v \mid distance(v_t, v) \leq k, v \in \mathcal{V}\}, \qquad (7.11)$$

where the k-hop neighborhood refers to the nodes which are within k hops from the target node. GraphLime collects the features of nodes in $\mathcal{V}_t$ as the corpus, and employs HSIC Lasso (Hilbert-Schmidt independence criterion Lasso) to measure the independence between features and predictions of the nodes. The top important features are selected as the explanation result, so GraphLime cannot provide explanations based on structural information of the graph.

RelEx defines the neighborhood space as a set of perturbed graphs to the computation graph of the target node. Similar to GraphLime, RelEx explains GNN predictions on nodes. The computation graph $\mathcal{G}_t$ of the target node v_t is composed of the k-hop neighbor nodes around node v_t and the edges that connect them. First, RelEx proposes a BFS sampling strategy to sample multiple perturbed graphs $\{\mathcal{G}'_{t,1}, \mathcal{G}'_{t,2}, ..., \mathcal{G}'_{t,I}\}$ from the computation graph. These perturbed graphs are fed into the original GNN f to build a training set $\{\mathcal{G}'_{t,i}, f(\mathcal{G}'_{t,i})\}_{i=1}^{I}$. Then, a new GNN f' is trained upon the training set to approximate f. After that, a mask M is trained for explanation. The mask is applied to the adjacency matrix of $\mathcal{G}_t$. The value of each mask entry is in $[0,1]$, so it is a soft mask. There are two loss terms for training the mask: (1) $f'(\mathscr{G}_t \odot M)$ is close to $f'(\mathscr{G}_t)$, (2) the mask M is sparse. The resultant mask entry values indicate the importance score of edges in $\mathscr{G}_t$, where a higher mask value means the corresponding edge is more important.

PGM-Explainer applies probabilistic graphical models to explain GNNs. To find the neighbor instances of the target, PGM-Explainer first randomly selects nodes to be perturbed from computation graphs. Then, the selected nodes' features are set to the mean value among all nodes. After that, PGM-Explainer employs a pair-wise dependence test to filter out unimportant samples, aiming at reducing the computational complexity. Finally, a Bayesian network is introduced to fit the predictions of chosen samples. Therefore, the advantage of PGM-Explainer is that it illustrates the dependency between features.

7.2.3 Relevance-Propagation Based Explanation

Relevance propagation redistributes the activation score of output neuron to its predecessor neurons, iterating until reaching the input neurons. The core of relevance propagation methods is about defining a rule for the activation redistribution between neurons. Relevance propagation has been widely used to explain models in domains such as computer vision and natural language processing. Recently, some work has been proposed to explore the possibility of revising relevance propagation method for GNNs. Some representative approaches include LRP (Layer-wise Relevance Propagation) (Baldassarre and Azizpour, 2019; Schwarzenberg et al, 2019), GNN-LRP (Schnake et al, 2020), ExcitationBP (Pope et al, 2019).

LRP is first proposed in (Bach et al, 2015) to calculate the contribution of individual pixels to the prediction result for an image classifier. The core idea of LRP is to use back propagation to recursively propagate the relevance scores of high-level neurons to low-level neurons, up to the input-level feature neurons. The relevance score of the output neuron is set as the prediction score. The relevance score that a neuron receives is proportional to its activation value, which follows the intuition that neurons with higher activation tend to contribute more to the prediction. In (Baldassarre and Azizpour, 2019; Schwarzenberg et al, 2019), the propagation rule is defined as below:

$$R_i^l = \sum_j \frac{z_{i,j}^+}{\sum_k z_{k,j}^+ + b_j^+ + \varepsilon} R_j^{l+1}$$
$$z_{i,j} = x_i^l w_{i,j}$$

(7.12)

where R_i^l, R_j^{l+1} is the relevance score of neuron i in layer l and neuron j in layer $l+1$, respectively. x_i^l is the activation of neuron i in layer l. $w_{i,j}$ is the connection weight. ε prevents the denominator from being zero. This propagation rule only allows positive activation values. Also, explanations obtained using this method are limited to nodes and node features, where graph edges are excluded. The reason is that the adjacency matrix is treated as part of the GNN model. Therefore, LRP is unable to analyze topological information which nevertheless plays an important role in graph data.

ExcitationBP is a top-down attention model originally developed for CNNs (Zhang et al, 2018d). It shares a similar idea as LRP. However, ExcitationBP defines the relevance score as a probability distribution and uses a conditional probability model to describe the relevance propagation rule.

$$P(a_j) = \sum_i P(a_j \mid a_i) P(a_i) \tag{7.13}$$

where a_j is the j-th neuron in the lower layer and a_i is the i-th parent neuron of a_j in the higher layer. When the propagation process passes through the activation function, only non-negative weights are considered and negative weights are set to zero. To extend ExcitationBP for graph data, new backward propagation schemes are designed for the softmax classifier, the GAP (global average pooling) layer and the graph convolutional operator.

GNN-LRP mitigates the weakness of traditional LRP by defining a new propagation rule. Instead of using the adjacency matrix to obtain propagation paths, GNN-LRP assigns the relevance score to a walk, which refers to a message flow path in the graph. The relevance score is defined by the T-order Taylor expansion of the model with respect to the incorporation operator (graph convolutional operator, linear message function, etc.). The intuition is that the incorporation operator with greater gradients has a greater influence on the final decision.

7.2.4 Perturbation-Based Approaches

An assumption behind prediction explanations is that important input parts significantly contribute to the output while unimportant parts have minor influences. It thus implies that masking out the unimportant parts will have a negligible impact on the output, and masking out the important parts will have a significant impact. The goal is to find a mask M to indicate graph component importance. The mask could be applied to nodes, edges or features in graphs. The mask value can either be binary $M_i \in \{0, 1\}$ or continuous $M_i \in [0, 1]$. Some recent perturbation-based approaches are introduced as below.

GNNExplainer (Ying et al, 2019) is the first perturbation-based explanation method for GNNs. Given the model's prediction on a node v, GNNExplainer tries to find a compact subgraph $\mathcal{G}_S$ from the computation graph of node v that is most crucial for the prediction. The problem is defined as maximizing the mutual information (MI) between the predictions of the original computation graph and the predictions of the subgraph:

$$\max_{\mathscr{G}_S} MI\left(Y,(\mathscr{G}_S,X_S)\right) = H(Y) - H\left(Y \mid \mathscr{G} = \mathscr{G}_S, X = X_S\right), \qquad (7.14)$$

where $\mathscr{G}_S$ and X_S is the subgraph and its nodes' features. Y is the predicted label distribution, and its entropy $H(Y)$ is a constant. To solve the optimization problem above, the authors apply a soft-mask M on adjacency matrix:

$$\min_{M} - \sum_{c=1}^{C} \mathbf{1}[y = c] \log P_{\Phi}\left(Y = y \mid G = A_c \odot \sigma(M), X = X_c\right), \qquad (7.15)$$

where A_c is the adjacency matrix of the computation graph, X_c is the corresponding feature matrix, and M denotes the trainable parameters. The sigmoid function projects the mask value in $[0,1]$. Finally, a subgragh is built by selecting the edges (and the nodes connected by these edges) corresponding to the high values in M. Besides providing explanations based on graph structures, GNNExplainer could also offer feature-wise explanations by applying a similar masking learning process on features. Moreover, regularization techniques could be applied to enforce the explanation to be sparse. As a model-agnostic approach, GNNExplainer is suitable for any graph-based machine learning tasks and GNN models.

PGExplainer (Luo et al, 2020) shares the same idea with GNNExplainer and learns a discrete mask applied on edges to explain the predictions. The main idea is to use a deep neural network to generate edge mask values:

$$M_{i,j} = \text{MLP}_{\Psi}\left([\mathbf{z}_i; \mathbf{z}_j]\right), \qquad (7.16)$$

where Ψ denotes the trainable parameters of the MLP. $\mathbf{z}^i$ and $\mathbf{z}^j$ are the embedding vector for node i and j, respectively. $[\cdot;\cdot]$ denotes concatenation. Similar to the GNNExplainer, the mask generator is trained by maximizing the mutual information between the original prediction and the new prediction.

GraphMask (Schlichtkrull et al, 2021) also produces the explanation by estimating the influences of edges. Similar to PGExplainer, GraphMask learns an erasure function that quantifies the importance of each edge. The erasure function is defined as:

$$z_{u,v}^{(k)} = g_{\pi}\left(\mathbf{h}_u^{(k)}, \mathbf{h}_v^{(k)}, \mathbf{m}_{u,v}^{(k)}\right) \qquad (7.17)$$

where $\mathbf{h}_u$, $\mathbf{h}_v$ and $\mathbf{m}_{u,v}$ refers to the hidden embedding vectors for node u, node v and the message sent through the edge in graph convolution. π denotes the parameters of function g. One difference between GraphMask and PGExplainer is that the former also takes the edge embedding as input. Another difference is that GraphMask provides the importance estimation for every graph convolution layer, and k indicates the layer that the embedding vectors belong to. Instead of directly erasing the influences of unimportant edges, the authors then propose to replace the message sent through unimportant edges as:

$$\tilde{\mathbf{m}}_{u,v}^{(k)} = z_{u,v}^{(k)} \cdot \mathbf{m}_{u,v}^{(k)} + \left(1 - z_{u,v}^{(k)}\right) \cdot \mathbf{b}^{(k)}, \qquad (7.18)$$

where $\mathbf{b}^{(k)}$ is trainable. The work shows that a large proportion of edges can be dropped without deteriorating the model performance.

Causal Screening (Wang et al, 2021) is a model-agnostic post-hoc method that identifies a subgraph of input as an explanation from the cause-effect standpoint. Causal Screening exerts causal effect of candidate subgraph as the metric:

$$S(\mathcal{G}_k) = MI(\text{do}(\mathcal{G} = \mathcal{G}_k); \hat{y}) - MI(\text{do}(\mathcal{G} = \emptyset); \hat{y}) \qquad (7.19)$$

where $\mathcal{G}_k$ is the candidate subgraph, k is the number of edges and MI is the mutual information. The intervention $\text{do}(\mathcal{G} = \mathcal{G}_k)$ and $\text{do}(\mathcal{G} = \emptyset)$ means the model input receives treatment (feeding $\mathcal{G}_k$ into the model) and control (feeding $\emptyset$ into the model), respectively. $\hat{y}$ denotes the prediction when feeding the original graph into the model. Causal Screening uses a greedy algorithm to search for the explanation. Starting from an empty set, at each step, it adds one edge with the highest causal effect into the candidate subgraph.

CF-GNNExplainer (Lucic et al, 2021) also proposes to generate counterfactual explanations for GNNs. Different from previous methods that try to find a sparse subgraph to preserve the correct prediction, CF-GNNExplainer proposes to find the minimal number edges to be removed such that the prediction changes. Similar to GNNExplainer, CF-GNNExplainer employs the soft mask as well. Therefore, it also suffers from the "introduced evidence" problem (Dabkowski and Gal, 2017), which means that non-zero or non-one values may introduce unnecessary information or noises, and thus influence the explanation result.

7.2.5 Generative Explanation

Many methods introduced in previous subsections define the explanation as selecting sub-graphs that contains part of nodes, edges or features of the original input. Recently, XGNN (Yuan et al, 2020b) proposes to obtain explanation by *generating* a graph that maximizes the prediction of the given GNN model. Some methods that share a similar idea have been proposed for computer vision tasks. For example, the role of a neuron could be understood by finding the input prototypes that maximally activates the neuron's activation (Olah et al, 2018). The problem of finding prototype samples can be defined as an optimization problem, which can be solved by gradient ascent. However, this method can not be directly used on GNNs because the gradient ascent method is not compatible with the discrete and topological nature of graph data. To tackle this problem, XGNN defines graph generation as a reinforcement learning task.

To be more specific, the generator follows the steps below. First, it randomly picks one node as the initial graph. Second, given an intermediate graph, the generator adds a new edge to the graph. This action is carried out in two steps: choosing the edge's starting point as well as the end point. XGNN employs another GNN as the policy to determine the action. The GNN learns nodes features, and two MLPs then take the learned features as input to predict the possibility of a start point and an endpoint. The endpoint and the edge between the two points are added to update the intermediate graph as an action. Finally, it calculates the reward of the action, so that we can train the generator via policy gradient algorithms. The reward consists of two terms. The first term is the score of the intermediate graph after feeding it to the target GNN model. The second one is a regularization term that guarantees the validity of the intermediate graph. The above steps are executed repeatedly until the number of action steps reaches the predefined upper limit. As a generative explanation method, XGNN provides a holistic explanation for graph classification. There could be more generative explanation methods for other graph analysis tasks to be explored in the future.

7.3 Interpretable Modeling on Graph Neural Networks

Following the introduction in Section 7.1.3.2, we introduce two categories of interpretable modeling approaches, i.e., GNN models with attention mechanism and disentangled representation learning on graphs.

7.3.1 GNN-Based Attention Models

Attention mechanisms benefit model interpretability by highlighting relevant parts of the graph for the given task through attention scores. According to the graph types, we introduce attention models built upon homogeneous graphs and heterogeneous graphs, respectively.

7.3.1.1 Attention Models for Homogeneous Graphs

Graph Attention Networks (GATs) enable assigning different weights to different node embeddings in a neighborhood when aggregating information (Veličković et al, 2018). Specifically, let $\mathbf{h}^i$ denote the column-wise embedding of node i, then the embedding update is written as:

$$\mathbf{h}^i_{l+1} = \sigma\Big(\sum_{j \in \mathscr{V}_i \cup \{i\}} \alpha_{i,j} W \mathbf{h}^j_l\Big), \tag{7.20}$$

where $\alpha_{i,j}$ is the attention score, and $\mathscr{V}_i$ denotes the set of neighbors of node i. Also, GAT uses a shared parameter matrix W independent of the layer depth. The attention score is computed as:

$$\alpha_{i,j} = \text{softmax}(e_{i,j}) = \frac{\exp(e_{i,j})}{\sum_{k \in \mathscr{V}_i \cup \{i\}} \exp(e_{i,k})}, \tag{7.21}$$

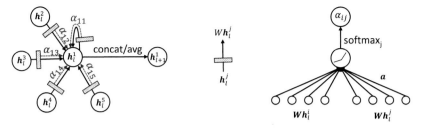

Fig. 7.5: Left: An illustration of graph convolution with single head attentions by node 1 on its neighborhood. Middle: The linear transformation with a shared parameter matrix. Right: The attention mechanism employed in (Veličković et al, 2018).

where self-attention mechanism is applied,

$$e_{i,j} = \text{LeakyReLU}(\mathbf{a}^\top [W\mathbf{h}_l^i \| W\mathbf{h}_l^j]), \tag{7.22}$$

where $\|$ denotes vector concatenation. In general, the attention mechanism can also be denoted as $e_{i,j} = attn(\mathbf{h}_l^i, \mathbf{h}_l^j)$. Therefore, the attention mechanism is a single-layer neural network parameterized by a weight vector $\mathbf{a}$. The attention score $\alpha_{i,j}$ shows the importance of node j to node i.

The above mechanism could also be extended with multi-head attention. Specifically, K independent attention mechanisms are executed in parallel, and the results are concatenated:

$$\mathbf{h}_{l+1}^i = \|_{k=1}^K \sigma\big(\sum_{j \in \mathcal{V}_i \cup \{i\}} \alpha_{i,j}^k W^k \mathbf{h}_l^j \big), \tag{7.23}$$

where $\alpha_{i,j}^k$ is the normalized attention score in the k-th attention mechanism, and W^k is the corresponding parameter matrix.

Besides learning node embeddings, we could also apply attention mechanisms to learn a low-dimensional embedding for the whole graph (Ling et al, 2021). Suppose we are working on an information retrieval problem. Given a set of graphs $\{\mathcal{G}_m\}$, $1 \le m \le M$, and a query q, we want to return the graphs that are most relevant to the query. The embedding of each graph $\mathcal{G}_m$ with respect to q could be computed using the attention mechanism. In the first step, we could apply normal GNN propagation rules as introduced in Equation 7.2, to obtain the embeddings of nodes inside each graph. Let $\mathbf{q}$ denote the embedding of the query, and $\mathbf{h}^{i,m}$ denote the embedding of node i in a graph $\mathcal{G}_m$. The embedding of graph $\mathcal{G}_m$ with respect to the query can be computed as:

$$\mathbf{h}_{\mathcal{G}_m}^q = \frac{1}{|\mathcal{G}_m|} \sum_{i=1}^{|\mathcal{G}_m|} \alpha_{i,q} \mathbf{h}^{i,m} \tag{7.24}$$

where $\alpha_{i,q} = attn(\mathbf{h}^{i,m}, \mathbf{q})$ is the attention score, and $attn()$ is a certain attention function. Finally, $\mathbf{h}_{\mathcal{G}_m}^q$ can be used to compute the similarity of $\mathcal{G}_m$ to the query in the graph retrieval task.

7.3.1.2 Attention Models for Heterogeneous Graphs

A heterogeneous network is a network with multiple types of nodes, links, and even attributes. The structural heterogeneity and rich semantic information bring challenges for designing graph neural networks to fuse information.

Definition 7.4. A *heterogeneous graph* is defined as $\mathscr{G} = (\mathscr{V}, \mathscr{E}, \phi, \psi)$, where $\mathscr{V}$ is the set of node objects and $\mathscr{E}$ is the set of edges. Each node $v \in \mathscr{V}$ is associated with a node type $\phi(v)$, and each edge $(i, j) \in \mathscr{E}$ is associated with an edge type $\psi((i, j))$.

We introduce how the challenge in embedding could be tackled using Heterogeneous graph Attention Network (HAN) (Wang et al, 2019m). Different from traditional GNNs, information propagation on HAN is conducted based on meta-paths.

Definition 7.5. A *meta-path* Φ is defined as a path with the form $v_{i_1} \xrightarrow{r_1} v_{i_2} \xrightarrow{r_2} \cdots \xrightarrow{r_{l-1}} v_{i_l}$, abbreviated as $v_{i_1} v_{i_2} \cdots v_{i_l}$ with a composite relation $r_1 \circ r_2 \circ \cdots \circ r_{l-1}$.

To learn the embedding of node i, we propagate the embeddings from its neighbors within the meta-path. The set of neighbor nodes is denoted as $\mathscr{V}_i^{\Phi}$. Considering that different types of nodes have different feature spaces, a node embedding is first projected to the same space $\mathbf{h}^{j'} = M_{\phi_i} \mathbf{h}^j$. Here M_{ϕ_i} is the transformation matrix for node type ϕ_i. The attention mechanism in HAN is similar to GAT, except that we need to consider the type of meta-path that is currently sampled. Specifically,

$$\mathbf{z}^{i,\Phi} = \sigma\Big(\sum_{j \in \mathscr{V}_i^{\Phi}} \alpha_{i,j}^{\Phi} \mathbf{h}^{j'} \Big), \tag{7.25}$$

where the normalized attention score is

$$\alpha_{i,j}^{\Phi} = \text{softmax}(e_{i,j}^{\Phi}) = \text{softmax}(attn(\mathbf{h}^{i'}, \mathbf{h}^{j'}; \Phi)). \tag{7.26}$$

Given a set of meta-paths $\{\Phi_1, ..., \Phi_P\}$, we can obtain a group of node embeddings denoted as $\{\mathbf{z}^{i,\Phi_1}, ..., \mathbf{z}^{i,\Phi_P}\}$. To fuse embeddings across different meta-paths, another attention algorithm is applied. The fused embedding is computed as:

$$\mathbf{z}^i = \sum_{p=1}^{P} \beta_{\Phi_p} \mathbf{z}^{i,\Phi_p}, \tag{7.27}$$

where the normalized attention score is

$$\beta_{\Phi_p} = \text{softmax}(w_{\Phi_p}) = \text{softmax}\Big(\frac{1}{|\mathscr{V}|} \sum_{i \in \mathscr{V}} \mathbf{q}^{\top} \cdot \text{MLP}(\mathbf{z}^{i,\Phi_p}) \Big), \tag{7.28}$$

where $\mathbf{q}$ is a learnable semantic vector. $\text{MLP}(\cdot)$ denotes a one-layer MLP module. w_{Φ_p} can be explained as the importance of the meta-path Φ_p. Besides modeling heterogeneous types of nodes and edges, HetGNN (Zhang et al, 2019b) extends the discussion by considering heterogeneity in node attributes (e.g., images, texts, categorical features).

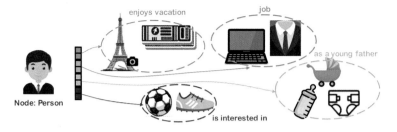

Fig. 7.6: Using multiple embeddings to represent the interests of a user. Each embedding segment corresponds to one aspect in data (Liu et al, 2019a).

7.3.2 Disentangled Representation Learning on Graphs

Traditional representation learning is limited in interpretability due to the opacity of the representation space. Different from manual feature engineering where the meaning of each resultant feature dimension is specified, the meaning of each dimension of the representation space is unknown. Representation learning on graphs also suffers from this limitation. To tackle this issue, several approaches have been proposed to enable assigning concrete meanings to different representation dimensions, thus improving the interpretability of representation learning on graphs.

7.3.2.1 Is A Single Vector Enough?

Many existing representation learning methods on graphs focus on learning a single embedding for each node. However, for those scenarios where some nodes have multiple facets, is a single vector enough to represent each node? Solving such a problem is of great practical value for applications such as recommender systems, where users could have multiple interests. In this case, we could use multiple embeddings to represent each user, and each embedding corresponds to one interest. An example is shown in Fig. 7.6. Specifically, if $\mathbf{h}^i \in \mathbb{R}^D$ denotes the embedding of node i, then $\mathbf{h}^i = [\mathbf{h}^{i,1}; \mathbf{h}^{i,2}; ...; \mathbf{h}^{i,K}]$, where $\mathbf{h}^{i,k} \in \mathbb{R}^{D/K}$ is the embedding for the k-th facet. There are two challenges in learning disentangled representations, i.e., how to discover the K facets, and how to distinguish the update of different embeddings during the training process. The facets could be discovered in an unsupervised manner by using clustering, where each cluster represents a facet. In the following parts, we introduce several approaches for learning disentangled node embeddings on graphs.

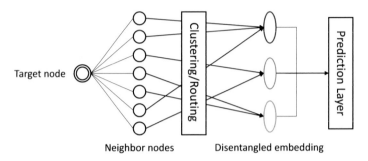

Fig. 7.7: The high-level idea of learning the disentangled node embedding for a target node by using clustering or dynamic routing.

7.3.2.2 Prototypes-Based Soft-Cluster Assignment

We discuss the techniques in the context of recommender system design. Facets that represent item types are discovered as we learn user and item embeddings. Here we assume that each item only has one facet, while each user could still have multiple facets. The embedding of item t is simply $\mathbf{h}^t$, while the embedding of user u is $\mathbf{h}^u = [\mathbf{h}^{u,1}; \mathbf{h}^{u,2}; ...; \mathbf{h}^{u,K}]$. Each item t is associated with a one-hot vector $\mathbf{c}_t = [c_{t,1}, c_{t,2}, ..., c_{t,K}]$, where $c_{t,k} = 1$ if t belongs to facet k, and $c_{t,k} = 0$ otherwise. Besides node embeddings, we also need to learn a set of *prototype* embeddings $\{\mathbf{m}^k\}_{k=1}^K$. The one-hot vector is drawn from the categorical distribution as below:

$$\mathbf{c}_t \sim \text{categorical}(\text{softmax}([s_{t,1}, s_{t,2}, .., s_{t,K}])), \quad s_{t,k} = cos(\mathbf{h}^t, \mathbf{m}^k)/\tau, \quad (7.29)$$

where τ is a hyper-parameter that scales the cosine similarity. Then, the probability of observing an edge (u,t) is

$$p(t|u, \mathbf{c}_t) \propto \sum_{k=1}^K c_{t,k} \cdot \text{similarity}(\mathbf{h}^t, \mathbf{h}^{u,k}). \quad (7.30)$$

Besides the fundamental learning process introduced above, the variational autoencoder framework could also be applied to regularize the learning process (Ma et al, 2019c). The item embeddings and prototype embeddings are jointly updated until convergence. The embedding of each user $\mathbf{h}^u$ is determined by aggregating the embeddings of interacted items, where $\mathbf{h}^{u,k}$ collects embeddings from items that also belong to facet k. In the learning process, the cluster discovery, node-cluster assignments, and embedding learning are jointly conducted.

7.3.2.3 Dynamic Routing Based Clustering

The idea of using dynamic routing for disentangled node representation learning is motivated by the Capsule Network (Sabour et al, 2017). There are two layers of capsules, i.e., low-level capsules and high-level capsules. Given a user u, the set of items that he has interacted with is denoted as $\mathcal{V}_u$. The set of low-level capsules is $\{\mathbf{c}_i^l\}$, $i \in \mathcal{V}_u$, so each capsule is the embedding of an interacted item. The set of high-level capsules is $\{\mathbf{c}_k^h\}$, $1 \leq k \leq K$, where $\mathbf{c}_k^h$ represents the user's k-th interest.

The routing logit value $b_{i,k}$ between low-level capsule i and high-level capsule k is computed as:

$$b_{i,k} = (\mathbf{c}_k^h)^\top S \mathbf{c}_i^l, \tag{7.31}$$

where S is the bilinear mapping matrix. Then, the intermediate embedding for high-level capsule k is computed as a weighted sum of low-level capsules,

$$\mathbf{z}_k^h = \sum_{i \in \mathcal{V}_u} w_{i,k} S \mathbf{c}_i^l,$$

$$w_{i,k} = \frac{\exp(b_{i,k})}{\sum_{k'=1}^{K} \exp(b_{i,k'})} \tag{7.32}$$

so $w_{i,k}$ can be seen as the attention weights connecting the two capsules. Finally, a "squash" function is applied to obtain the embedding of high-level capsules:

$$\mathbf{c}_k^h = \text{squash}(\mathbf{z}_k^h) = \frac{\|\mathbf{z}_k^h\|^2}{1 + \|\mathbf{z}_k^h\|^2} \frac{\mathbf{z}_k^h}{\|\mathbf{z}_k^h\|^2}. \tag{7.33}$$

The above steps constitute one iteration of dynamic routing. The routing process is usually repeated for several iterations to converge. When the routing finishes, the high-level capsules can be used to represent the user u with multiple interests, to be fed into subsequent network modules for inference (Li et al, 2019b), as shown in Fig. 7.7.

7.4 Evaluation of Graph Neural Networks Explanations

In this section, we introduce the setting for evaluating GNN explanations. This includes the *datasets* that are commonly used for constructing and explaining GNNs, as well as the *metrics* that evaluate different aspects of explanations.

7.4.1 Benchmark Datasets

As more approaches have been proposed for explaining GNNs, a variety of datasets have been used to assess their effectiveness. As such a research direction is still in the initial stage of development, a universally accepted benchmark dataset, such as the COCO dataset for image object detection, has not yet been proposed. Here we list a number of datasets that have been used for developing GNN explanation approaches, including synthetic datasets and real-world datasets.

7.4.1.1 Synthetic Datasets

It is difficult to evaluate explanations because there are no ground truths in datasets to compare with. A strategy to mitigate this problem is to use synthetic datasets. In this case, motifs designed by humans could be added to data to play the role as ground truths, and these motifs are assumed to be relevant to the learning task. Some synthetic graph datasets are listed as below.

- **BA-Shapes** (Ying et al, 2019): A Barabási-Albert graph with 300 nodes, to which 80 house-shaped motifs are attached randomly. It is then further augmented by adding 10% random edges.
- **BA-Community** (Ying et al, 2019): A graph consists of two BA-Shapes, with node features in different BA-Shapes following different normal distributions to distinguish them.
- **Tree-Cycle** (Ying et al, 2019): A graph based on an eight-level balance tree, to which 80 hexagonal motifs are attached randomly to the tree.
- **Tree-Grid** (Ying et al, 2019): A graph similar to Tree-Cycle, but with 80 3-by-3 grid motifs instead of the hexagonal motifs.
- **Noisy BA-Community, Noisy Tree-Cycle, Noisy Tree-Grid** (Lin et al, 2020a): These four datasets are obtained by adding 40 important and 10 unimportant node features to the corresponding datasets list above. This design can help to test a method's ability to identify important node features.
- **BA-2Motifs** (Luo et al, 2020): A dataset contains 800 independent graphs that are obtained by adding either a pentagon motif or a house motif to the base BA graph. This dataset is designed for graph classification task while previous ones are for node classification task.

7.4.1.2 Real-World Datasets

Some examples of real-world graph datasets are listed as below.

- **MUTAG** (Debnath et al, 1991): A dataset consisting of 4,337 molecule graphs that are labeled mutagenic or non-mutagenic. The nodes and edges in a graph represent the atoms and chemical bonds. Related studies have shown that molecules with carbon rings and Nitro group (NO_2) may lead to mutagenic effects. Also, there are several other molecule datasets, such as **BBBP**, **BACE** and **TOX21** (Pope et al, 2019).
- **REDDIT-BINARY** (Yanardag and Vishwanathan, 2015): A online-discussion interaction dataset. It contains 2,000 graphs, and each of them is labeled as a

question-answer based or a discussion based community. The nodes and edges represent the users and their interactions, respectively.

- **Delaney Solubility** (Delaney, 2004): A molecule dataset with 1,127 molecule graphs, and their labels are the water-octanol partition coefficient. This dataset is usually for graph regression tasks.
- **Bitcoin-Alpha, Bitcoin-OTC** (Kumar et al, 2016): Two trust-weighted signed networks. Each of them consists of a graph whose nodes are accounts trading on the Bitcoin-Alpha or Bitcoin-OTC platform. The nodes are labeled trustworthy or not according to other members' ratings.
- **MNIST SuperPixel-Graph** (Dwivedi et al, 2020): An image dataset in the form of graphs. Each sample is a graph converted from the corresponding image in the MNIST dataset. Every node is a super-pixel that represents the intensity of corresponding region.

7.4.2 Evaluation Metrics

An appropriate evaluation metric is crucial for methods comparison. Explanation visualization such as heat-map, due to its intuitiveness, has been widely used in explanation for image and text data. However, it loses this advantage since graph data is not intuitive to understand. Only experts with the domain knowledge can make judgment. In this section, we introduce several commonly-used metrics.

- **Accuracy** is only appropriate for datasets with ground truth. The synthetic datasets usually contain the ground truth that is defined by the rule they are constructed. For example, in molecule datasets, the molecule with NO_2 and carbon ring is mutagenic. Considering that carbon ring also occurs in non-mutagenic molecule, the NO_2 group is considered as ground truth. F1 score and ROC-AUC are commonly used accuracy metrics. The limitation of the accuracy metrics is that it is unknown whether the GNN model makes predictions in the same way as humans (i.e., whether the pre-defined ground truth is really valid).
- **Fidelity** (Pope et al, 2019) follows the intuition that removing the truly important features will significantly decrease the model performance. Formally, fidelity is defined as:

$$fidelity = \frac{1}{N} \sum_{i=1}^{N} \left(f^{y_i} \left(\mathscr{G}_i \right) - f^{y_i} \left(\mathscr{G}_i \setminus \mathscr{G}_i' \right) \right) \tag{7.34}$$

where f is the output function target model. $\mathscr{G}_i$ is the i-th graph, $\mathscr{G}_i'$ is the explanation for it, and $\mathscr{G}_i \setminus \mathscr{G}_i'$ represents the perturbed i-th graph in which the identified explanation is removed.

- **Contrastivity** (Pope et al, 2019) uses Hamming distance to measure the differences between two explanations. These two explanations correspond to the model's prediction of one instance for different classes. It is assumed that models would highlight different features when making predictions for different classes. The higher the contrastivity, the better the performance of the interpreter.

- **Sparsity** (Pope et al, 2019) is calculated as the ratio of explanation graph size to input graph size. In some cases, explanations are encouraged to be sparse, because a good explanation should include only the essential features as far as possible and discard the irrelevant ones.
- **Stability** (Sanchez-Lengeling et al, 2020) measures the performance gap of the interpreter before and after adding noise to the explanation. It suggests that a good explanation should be robust to slight changes in the input that do not affect the model's prediction.

7.5 Future Directions

Interpretation on graph neural networks is an emerging domain. There are still many challenges to be tackled. In this section, we list several future directions towards improving the interpretability of graph neural networks.

First, some online applications require real-time responses from models and algorithms. It thus puts forward high requirements on the efficiency of explanation methods. However, many GNN explanation methods conduct sampling or highly iterative algorithms to obtain the results, which is time-consuming. Therefore, one future research direction is how to develop more efficient explanation algorithms without significantly sacrificing explanation precision.

Second, although more and more methods have been developed for interpreting GNN models, how to utilize interpretation towards identifying GNN model defects and improving model properties is still rarely discussed in existing work. Will GNN models be largely affected by adversarial attacks or backdoor attacks? Can interpretation help us to tackle these issues? How to improve GNN models if they have been found to be biased or untrustworthy?

Third, besides attention methods and disentangled representation learning, are there other modeling or training paradigms that could also improve GNN interpretability? In the interpretable machine learning domain, some researchers are interested in providing causal relations between variables, while some others prefer using logic rules for reasoning. Therefore, how to bring causality into GNN learning, or how to use incorporate logic reasoning into GNN inference, may be an interesting direction to explore.

Fourth, most existing efforts on interpretable machine learning have been devoted to get more accurate interpretation, while the human experience aspect is usually overlooked. For end-users, friendly interpretation can promote user experience, and gain their trust to the system. For domain experts without machine learning background, an intuitive interface helps integrate them into the system improvement loop. Therefore, another possible direction is how to incorporate human-computer interaction (HCI) to show explanation in a more user-friendly format, or how to design better human-computer interfaces to facilitate user interactions with the model.

Acknowledgements The work is, in part, supported by NSF (#IIS-1900990, #IIS-1718840, #IIS-1750074). The views and conclusions contained in this paper are those of the authors and should not be interpreted as representing any funding agencies.

Editor's Notes: Similar to the general trend in the machine learning domain, explainability has been ever more widely recognized as an important metric for graph neural networks in addition to those well recognized before such as effectiveness (Chapter 4), complexity (Chapter 5), efficiency (Chapter 6), and robustness (Chapter 8). Explainability can not only broadly influence technique development (e.g., Chapters 9-18) by informing model developers of useful model details, but also could benefit domain experts in various application domains (e.g., Chapters 19-27) by providing them with explanations of predictions.

Chapter 8
Graph Neural Networks: Adversarial Robustness

Stephan Günnemann

Abstract Graph neural networks have achieved impressive results in various graph learning tasks and they have found their way into many applications such as molecular property prediction, cancer classification, fraud detection, or knowledge graph reasoning. With the increasing number of GNN models deployed in scientific applications, safety-critical environments, or decision-making contexts involving humans, it is crucial to ensure their reliability. In this chapter, we provide an overview of the current research on adversarial robustness of GNNs. We introduce the unique challenges and opportunities that come along with the graph setting and give an overview of works showing the limitations of classic GNNs via adversarial example generation. Building upon these insights we introduce and categorize methods that provide provable robustness guarantees for graph neural networks as well as principles for improving robustness of GNNs. We conclude with a discussion of proper evaluation practices taking robustness into account.

8.1 Motivation

The success story of graph neural networks is astonishing. Within a few years, they have become a core component of many deep learning applications. Nowadays they are used in scientific applications such as drug design or medical diagnoses, are integrated in human-centered applications like fake news detection in social media, get applied in decision-making tasks, and even are studied in safety-critical environments like autonomous driving. What unites these domains is their crucial need for reliable results; misleading predictions are not only unfortunate but indeed might lead to dramatic consequences – from false conclusions drawn in science to harm for people. However, can we really trust the predictions resulting from graph neural

Stephan Günnemann
Department of Informatics, Technical University of Munich, e-mail: guennemann@in.tum.de

© The Author(s), under exclusive license to Springer Nature Singapore Pte Ltd. 2022   149
L. Wu et al. (eds.), *Graph Neural Networks: Foundations, Frontiers, and Applications*,
https://doi.org/10.1007/978-981-16-6054-2_8

networks? What happens when the underlying data is corrupted or even becomes deliberately manipulated?

Indeed, the vulnerability of classic machine learning models to (deliberate) perturbations of the data is well known (Goodfellow et al, 2015): even only slight changes of the input can lead to wrong predictions. Such instances, for humans nearly indistinguishable from the original input yet wrongly classified, are also known as *adversarial examples*. One of the most well-known and alarming examples is an image of a stop sign, which is classified as a speed limit sign by a neural network with only very subtle changes to the input; though, for us as humans it still clearly looks like a stop sign (Eykholt et al, 2018). Examples like these illustrate how machine learning models can dramatically fail in the presence of adversarial perturbations. Consequently, adopting machine learning for safety-critical or scientific application domains is still problematic. To address this shortcoming, many researchers have started to analyze the robustness of models in domains like images, natural language, or speech. Only recently, however, GNNs have come into focus. Here, the first work studying GNNs' robustness (Zügner et al, 2018) investigates one of the most prominent tasks, node-level classification, and demonstrated the susceptibility of GNNs to adversarial perturbations as well (see Figure 8.1). Since then, the field of adversarial robustness on graphs has been rapidly expanding, with many works studying diverse tasks and models, and exploring ways to make GNNs more robust.

Fig. 8.1 The upper left graph is the original input. On the right is the graph after performing a small change (e.g. adding an edge or changing some node attributes). The lower part illustrates the predicted classes for each node obtained from a GNN. Is it possible to change the predictions? Are GNNs robust?

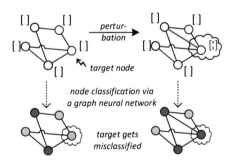

To some degree it is surprising that graphs were not in the focus even earlier. Corrupted data and adversaries are common in many domains where graphs are analyzed, e.g., social media and e-commerce systems. Take for example a GNN-based model for detecting fake news in a social network (Monti et al, 2019; Shu et al, 2020). Adversaries have a strong incentive to fool the system in order to avoid being detected. Similarly, in credit scoring systems, fraudsters try to disguise themselves by creating fake connections. Thus, robustness is an important concern for graph-based learning.

It is important to highlight, though, that adversarial robustness is not only a topic in light of security concerns, where intentional changes, potentially crafted by humans, are used to try to fool the predictions. Instead, adversarial robustness considers worst-case scenarios in general. Especially in safety-critical or scientific applications where reliability is key, understanding the robustness of GNNs to worst-case noise is important, as nature itself might be the adversary. The construction of gene interaction networks, for example, often leads to corrupted graphs containing spurious edges (Tian et al, 2017). Thus, to make sure that graph neural networks work reliably in all these scenarios, we need to investigate robustness under *worst-case/adversarial* corruptions of the data.

Moreover, non-robustness of GNNs shows conceptual gaps: while neural networks are hypothesized to learn meaningful representations that capture the semantics of the domain and task, a non-robust model clearly violates this property. Since the small changes leading to an adversarial example do not alter the meaning, a reasonable representation should also not change the prediction. Thus, understanding adversarial robustness means understanding generalization performance.

Unique Challenges in the Graph Domain

In contrast to other application domains of deep learning, robustness analysis for graphs is especially challenging for multiple reasons:

1. Complex perturbation space: Changes can manifest in various ways including perturbations in the graph structure and the node attributes, leading to a vast space to explore. Importantly, unlike other fields this often means operating in a *discrete data domain* such as adding or removing edges, leading to hard discrete optimization problems as we will see later.
2. Interdependent data: The core feature of GNNs is to exploit the interdependence between instances, for example, in the form of message passing or graph convolution. Perturbations to the graph structure change the message passing scheme, modifying how learned representation are propagated. Specifically, changes to one part of the graph, e.g. one node, might affect many other instances.
3. Notion of similarity: We expect GNN models to be robust to small changes in the graph. If the graphs are almost indistinguishable, the predictions should be the same. However, defining the notion of similarity between graphs itself is a hard problem and unlike, e.g., images, manual inspection by a human is not practical.

Given these challenges, in the following Section 8.2 we first introduce the principle of adversarial attacks on GNNs and highlight some non-robustness results. In Section 8.3, we give an overview of robustness certificates, providing ways for proving the reliability of predictions, followed by Section 8.4 where approaches for improving GNNs' robustness are introduced. We conclude in Section 8.5 with discussing aspects of proper evaluation.

8.2 Limitations of Graph Neural Networks: Adversarial Examples

To understand the (non-)robustness of GNNs, we can try to construct worst-case perturbations – finding a *small* change of the data, which in consequence leads to a *strong* change in the GNN's output. This is also known as performing an adversarial attack and the resulting perturbed data is often called an adversarial example.[1] While random perturbations of the data often have minor effect, specific perturbations, in contrast, can be dramatic. Accordingly, an attack is often phrased as an optimization problem with the goal to *find* a perturbation of the data which maximizes some attack objective (e.g., maximize the predicted probability of some incorrect class).

8.2.1 Categorization of Adversarial Attacks

Before providing a general definition of adversarial attacks, it is helpful to distinguish two very different notions, called *poisoning vs. evasion scenarios*. The difference lies in the stage of the learning process in which the data perturbation is performed. In a poisoning scenario, the perturbation is injected *before* the training of the model; the perturbed data, thus, also affects the learning and the final model we obtain. In contrast, an evasion scenario assumes the model to be given, i.e., already trained and fixed, and the perturbation is applied to *future* data during the application/test phase of the GNN. It is worth to highlight, that for the frequently considered *transductive learning setting* of GNNs – where we have *no* future test data, but only the given (un)labeled data – a poisoning scenario is the more natural choice. Though, in principle any combination of learning (transductive vs. inductive) and attack scenario (poisoning vs. evasion) is worth to be studied.

Given this basic distinction, performing a poisoning adversarial attack can be generally formulated as a bi-level optimization problem

$$\max_{\hat{\mathscr{G}} \in \Phi(\mathscr{G})} \mathscr{O}_{\text{atk}}(f_{\theta^*}(\hat{\mathscr{G}})) \quad s.t. \quad \theta^* = \arg\min_{\theta} \mathscr{L}_{\text{train}}(f_{\theta}(\hat{\mathscr{G}})) \tag{8.1}$$

Here $\Phi(\mathscr{G})$ denotes the set of all graphs we are treating as indistinguishable to the given graph $\mathscr{G}$ at hand, and $\hat{\mathscr{G}}$ denotes a specific perturbed graph from this set. For example, $\Phi(\mathscr{G})$ could capture all graphs which differ from $\mathscr{G}$ in at most ten edges or in a few node attributes. The attacker's goal is to find a graph $\hat{\mathscr{G}}$ that, when passed through the GNN f_{θ^*}, maximizes a specific objective $\mathscr{O}_{\text{atk}}$, e.g., increasing the predicated probability of a certain class for a specific node. Importantly, in a poisoning setting, the weights θ^* of the GNN are not fixed but learned based on the perturbed data, leading to the inner optimization problem that corresponds to the usual training procedure on the (now perturbed) graph. That is, θ^* is obtained

---

[1] Again it is worth highlighting that such 'attacks' are not always due to human adversaries. Thus, the terms 'change' or 'perturbation' might be better suited and have a more neutral connotation.

by minimizing some training loss $\mathcal{L}_{\text{train}}$ on the graph $\hat{\mathcal{G}}$. This nested optimization makes the problem specifically hard.

To define an evasion attack, the above equation can simply be changed by assuming the parameter θ^* to be fixed. Often it is assumed to be given by minimizing the training loss w.r.t. the given graph $\mathcal{G}$ (i.e. $\theta^* = \arg\min_\theta \mathcal{L}_{\text{train}}(f_\theta(\mathcal{G}))$). This makes the above scenario a single-level optimization problem.

This general form of an attack enables us to provide a categorization along different aspects and illustrates the space to explore for robustness characteristics of GNNs in general. While this taxonomy is general, for ease of understanding, it helps to think about an intentional attacker.

Aspect 1: Property under Investigation (Attacker's Goal)

What is the robustness property we want to analyze? For example, do we want to understand how robust the classification of an individual node is? Will it change when perturbing the data? The property under investigation is modeled via $\mathcal{O}_{\text{atk}}$. It intuitively represents the attacker's goal. If $\mathcal{O}_{\text{atk}}$ for example measures the difference between a node's ground truth label and the currently predicted one, maximizing this difference in Eq. equation 8.1 tries to enforce a misclassification.

The attacker's goal is highly task-dependent. The majority of existing works has focused on the robustness of node-level classification based on GNNs, where we have to distinguish two scenarios. Works such as (Zügner et al, 2018; Dai et al, 2018a; Wang and Gong, 2019; Wu et al, 2019b; Chen et al, 2020f; Wang et al, 2020c) investigate how the prediction of an *individual* target node changes under perturbations – also called *local* attack. In contrast, Zügner and Günnemann (2019); Wu et al (2019b); Liu et al (2019c); Ma et al (2020b); Geisler et al (2021); Sun et al (2020d) have investigated how the overall performance on an *entire set* of nodes can drop – called a *global* attack.[2] This seemingly subtle difference between the two scenarios is crucial: In the latter case one has to find a single perturbed graph $\hat{\mathcal{G}} \in \Phi(\mathcal{G})$ which simultaneously changes many predictions, taking into account that all node-level predictions are indeed done jointly based on one input. In the former case, for each individual target node v_i a different perturbation $\hat{\mathcal{G}}_i \in \Phi(\mathcal{G})$ can be selected. Both views are reasonable; they simply model different aspects.

Beyond node-level classification, further works have investigated robustness of graph-level classification (Chen et al, 2020j), link prediction (Chen et al, 2020h; Lin et al, 2020d), and node embeddings (Bojchevski and Günnemann, 2019; Zhang et al, 2019e). The last one is worth mentioning since it targets an unsupervised learning setting, aiming to be task-agnostic. Unlike the other examples, the goal is not to target one specific task but to perturb the quality of the embeddings in general such that one or multiple downstream tasks are hindered. Since it is not known a priori which tasks (classification, link prediction, etc.) will be performed based on the

---

[2] Local attacks have also been called targeted attacks, while global ones untargeted. Since this, however, leads to a name clash with categorizations used in other communities (Carlini and Wagner, 2017) we decided to use local/global here.

node embeddings, defining the objective $\mathcal{O}_{\text{atk}}$ is challenging. As a proxy measure, Bojchevski and Günnemann (2019) for example uses the training loss itself, setting $\mathcal{O}_{\text{atk}} = \mathscr{L}_{\text{train}}$.

Aspect 2: The Perturbation Space (Attacker's Capabilities)

What changes are allowed to the original graph? What do we expect the perturbations to look like? For example, do we want to understand how deleting a few edges influences the prediction? The space of perturbations under consideration is modeled via $\Phi(\mathcal{G})$. It intuitively represents the attacker's capabilities; what and how much they are able to manipulate. The complexity of the perturbation space for graphs represents one of the biggest differences to classical robustness studies and stretches along two dimensions.

(1) What can be changed? Unique to the graph domain are perturbations of the graph structure. In this regard, most publications have studied the scenarios of removing or adding edges to the graph (Dai et al, 2018a; Wang and Gong, 2019; Zügner et al, 2018; Zügner and Günnemann, 2019; Bojchevski and Günnemann, 2019; Zhang et al, 2019e; Zügner et al, 2018; Tang et al, 2020b; Chen et al, 2020f; Chang et al, 2020b; Ma et al, 2020b; Geisler et al, 2021). Focusing on the node level, some works (Wang et al, 2020c; Sun et al, 2020d; Geisler et al, 2021) have considered adding or removing nodes from the graph. Beyond the graph structure, GNN robustness has also been explored for changes to the node attributes (Zügner et al, 2018; Wu et al, 2019b; Takahashi, 2019) and the labels used in semi-supervised node classification (Zhang et al, 2020b).

An intriguing aspect of graphs is to investigate how the interdepenence of instances plays a role in robustness. Due to the message passing scheme, changes to one node might affect (potentially many) other nodes. Often, for example, a node's prediction depends on its k-hop neighborhood, intuitively representing the node's receptive field. Thus, it is not only important what type of change can be performed but also where in the graph this can happen. Consider for example Figure 8.1: to analyze whether the prediction for the highlighted node can change, we are not limited to perturbing the node's own attributes and its incident edges but we can also achieve our aim by perturbing other nodes. Indeed, this reflects real world scenarios much better since it is likely that an attacker has access to a few nodes only, and not to the entire data or the target node itself. Put simply, we also have to consider which nodes can be perturbed. Multiple works (Zügner et al, 2018; Zhang et al, 2019e; Takahashi, 2019) investigate what they call *indirect attacks* (or sometimes influencer attacks), specifically analyzing how an individual node's prediction can change when only perturbing other parts of the graph while leaving the target node untouched.

(2) How much can be changed? Typically, adversarial examples are designed to be nearly indistinguishable to the original input, e.g., changing the pixel values of an image so that it stays visually the same. Unlike image data, where this can easily be verified by manual inspection, this is much more challenging in the graph setting.

Technically, the set of perturbations can be defined based on any graph distance function D measuring the (dis)similarity between graphs. All graphs similar to the given graph $\mathscr{G}$ then define the set $\Phi(\mathscr{G}) = \{\hat{\mathscr{G}} \in \mathbb{G} \mid D(\mathscr{G}, \hat{\mathscr{G}}) \leq \Delta\}$, where $\mathbb{G}$ denotes the space of all potential graphs and Δ the largest acceptable distance.

Defining what are suitable graph distance functions is in itself a challenging task. Beyond that, computing these distances and using them within the optimization problem of Eq. equation 8.1 might be computationally intractable (think, e.g., about the graph edit distance which itself is NP-hard to compute). Therefore, existing works have mainly focused on so called budget constraints, limiting the *number of changes* allowed to be performed. Technically, such budgets correspond to the L_0 pseudo-norm between the clean and perturbed data, e.g., relating to the graphs' adjacency matrix A or its node attributes X.[3] To enable more fine-grained control, often such budget constraints are used locally per node (e.g., limiting the maximal number of edge deletions per node; Δ_i^{loc}) as well as globally (e.g., limiting the overall number of edge deletions; Δ^{glob}). For example

$$\Phi(\mathscr{G}) = \{\hat{\mathscr{G}} = (\hat{A}, \hat{X}) \in \mathbb{G} \mid ||A - \hat{A}||_0 \leq \Delta^{\text{glob}} \wedge \forall i : ||A_i - \hat{A}_i||_0 \leq \Delta_i^{\text{loc}} \wedge X = \hat{X}\}, \tag{8.2}$$

where the graphs $\mathscr{G} = (A, X)$ and $\hat{\mathscr{G}} = (\hat{A}, \hat{X})$ are assumed to have the same size and the node attributes, X resp. $\hat{X}$, to stay unchanged; A_i denotes the ith row of A.

Beyond these budget constraints, it might be useful to preserve further characteristics of the data. In particular for real-world networks many patterns such as specific degree distributions, large clustering coefficients, low diameter, and more are known to hold (Chakrabarti and Faloutsos, 2006). If two graphs show very different patterns, it is easy to tell them apart – and a different prediction could be expected. Therefore, in (Zügner et al, 2018; Zügner and Günnemann, 2019; Lin et al, 2020d) only perturbed graphs are considered which follow similar power-law behavior in the degree distribution. Similarly, one can impose constraints on the attributes considering, e.g., the co-occurrence of specific values.

Aspect 3: Available Information (Attacker's Knowledge)

What information is available to find a harmful perturbation? What is the attacker's knowledge about the system? Considering a human-like adversary, the more knowledge is available, the stronger are the potential attacks.

In general, we have to distinguish between knowledge about the data/graph and knowledge about the model. For the first, either the full graph could be known or only parts of it as, e.g., investigated in (Zügner et al, 2018; Dai et al, 2018a; Chang et al, 2020b; Ma et al, 2020b). While for worst-case analysis we often assume that the attacker has full knowledge, for practical scenarios it is indeed realistic to assume that an attacker only observes subsets of the data. For supervised learning settings,

---

[3] This is a similar approach to image data, where often we take a certain radius as measured by, e.g., an L_p norm around the original input as the allowed perturbation set, assuming that for small radii the semantic meaning of the input does not change.

the ground-truth labels of the target node(s) could additionally be hidden from the attacker. The knowledge about the model includes many aspects such as knowledge about the used GNN architecture, the model's weights, or whether only the output predictions or the gradients are known. Given all these variations, the most common ones are white-box settings, where full information is available, and black-box settings, which usually mean that only the graph and potentially the predicted outputs are available.

Among the three aspects above, the attacker's knowledge seems to be the one which most strongly links to human-like adversaries. It should be highlighted, though, that worst-case perturbations in general are best reflected by the fully white-box setting, making it the preferred choice for strong robustness results. If a model performs robustly in a white-box setting, it will also be robust under the limited scenarios. Moreover, as we will see in Section 8.2.2.1, the transferability of attacks implies that knowledge about the model is not really required.

Aspect 4: The Algorithmic View

Besides the above categorization that focuses on the properties of the attack, another, more technical, view can be taken by considering the algorithmic approach how the (bi-level) optimization problem is solved. In the discussion of the perturbation space we have seen that graph perturbations often relate to the addition/removal of edges or nodes — these are discrete decisions, making Eq. equation 8.1 a discrete optimization problem. This is in stark contrast to other data domains where infinitesimal changes are possible. Thus besides adapting gradient-based approximations, various other techniques can be used to tackle Eq. equation 8.1 for GNNs such as reinforcement learning (Sun et al, 2020d; Dai et al, 2018a) or spectral approximations (Bojchevski and Günnemann, 2019; Chang et al, 2020b). Moreover, the attacker's knowledge has also implications on the algorithmic choice. In a black-box setting where, e.g., only the input and output are observed, we cannot use the true GNN f_θ to compute gradients but have to use other principles like first learning some surrogate model.

8.2.2 The Effect of Perturbations and Some Insights

The above categorization shows that various kinds of adversarial perturbations under different scenarios can be investigated. Summarizing the different results obtained in the literature so far, the trend is clear: standard GNNs trained in the standard way are not robust. In the following, we given an overview of some key insights.

Figure 8.2 illustrates one of the results of the method Nettack as introduced in (Zügner et al, 2018). Here, *local attacks* in an evasion setting focusing on graph structure perturbations are analyzed for a GCN (Kipf and Welling, 2017b). The figure shows the classification margin, i.e., the difference between the predicted

Fig. 8.2 Performing local structure attacks on a GCN model and the Cora ML data with the Nettack (Zügner et al, 2018) approach. If a node is below the dashed line it is misclassified w.r.t. the ground truth label. As shown, almost any node's prediction can be changed.

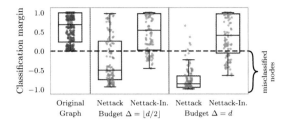

probability of the node's true class minus the one of the second highest class. The left column shows the results for the unperturbed graph where most nodes are correctly classified as illustrated by the predominantly positive classification margin. The second column shows the result after perturbing the graph based on the perturbation found by Nettack using a global budget of $\Delta = \lfloor d_v/2 \rfloor$ and making sure that no singletons occur where d_v is the degree of the node v under attack. Clearly, the GCN model is not robust: almost every node's prediction can be changed. Moreover, the third column shows the impact of indirect attacks. Recall that in these scenarios the performed perturbations cannot happen at the node we aim to misclassify. Even in this setting, a large fraction of nodes is vulnerable. The last two columns show results for an increased budget of $\Delta = d_v$. Not surprisingly, the impact of the attack becomes even more pronounced.

Considering *global attacks* in the poisoning setting similar behavior can be observed. For example, when studying the effect of node additions, the work (Sun et al, 2020d) reports a relative drop in accuracy by up to 7 percentage points with a budget of 1% of additional nodes, without changing the connectivity between existing nodes. For changes to the edge structure, the work (Zügner and Günnemann, 2019) reports performance drops on the test sets by around 6 to 16 percentage points when perturbing 5% of the edges. Noteworthy, on one dataset, these perturbations lead to a GNN obtaining *worse* performance than a logistic regression baseline operating only on the node attributes, i.e., ignoring the graph altogether becomes the better choice.

The following observation from (Zügner and Günnemann, 2019) is important to highlight: One core factor for the obtained lower performance on the perturbed graphs are indeed the learned GNN weights. When using the weights $\theta_{\hat{\mathscr{G}}}$ trained on the perturbed graph $\hat{\mathscr{G}}$ obtained by the poisoning attack, not only the performance on $\hat{\mathscr{G}}$ is low but even the performance on the unperturbed graph $\mathscr{G}$ suffers dramatically. Likewise, when applying weights $\theta_{\mathscr{G}}$ trained on the unperturbed graph $\mathscr{G}$ to the graph $\hat{\mathscr{G}}$, the classification accuracy barely changes. Thus, the poisoning attack performed in (Zügner and Günnemann, 2019) indeed derails the training procedure, i.e., leads to 'bad' weights. This result emphasizes the importance of the training procedure for the performance of graph models. If we are able to find appropriate weights, even perturbed data might be handled more robustly. We will encounter this aspect again in Section 8.4.2.

8.2.2.1 Transferability and Patterns

An interesting question to investigate is the adversarial examples' transferability. Transferability relates to the fact that a harmful perturbation for one model (e.g. a GCN) is also harmful for another model (e.g. GAT (Veličković et al, 2018)). Thus, one can simply reuse one perturbation to fool many models. The transferability of GNN attacks has been investigated in multiple works (Zügner et al, 2018; Zügner and Günnemann, 2019; Lin et al, 2020d; Chen et al, 2020f) and seems to hold across many models. For example, local attacks computed based on Nettack's GCN-like surrogate model in an evasion scenario are also harmful for the original GCN and the Column Network (Pham et al, 2017) model; for evasion and poisoning alike. Interestingly, the performance gets detoriated even for unsupervised node embeddings such as DeepWalk (Perozzi et al, 2014), combined with a subsequent logistic regression to obtain predictions.

The wide transferability of adversarial perturbations could be an indicator that they follow general patterns. There seems to be some systematic change of the graph which hinders many GNN models to perform well. If we can find out what makes, for example, an edge insertion a strong adversarial change, we can use this knowledge to detect adversarial attacks and/or make graph neural networks more robust (see Section 8.4). However, it is yet still not fully understood what makes these adversarial attacks harmful to a variety of models.

In (Zhang et al, 2019b) the predicted categorical distributions over classes for perturbed and unperturbed instances after performing a Nettack attack has been analyzed. Inspecting the average KL-divergence of the predicted categorical distributions of a node and its neighbors, perturbed nodes seem to show higher divergences, i.e., the attacks appear to be aiming to violate the homophily assumption in the graph. Relatedly, Wu et al (2019b) compared the Jaccard similarity between adjacent node's attributes and noticed a change in distribution from the clean and perturbed graph. The work (Zügner et al, 2020) investigated various graph properties, including aspects such as node degree, closeness centrality, PageRank (Brin and Page, 1998) scores, or attribute similarity. They focused on structure attacks using Nettack, allowing only edge insertions and deletions to the target node.

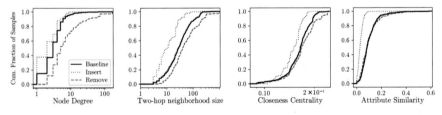

Fig. 8.3: Cumulative distributions of properties of nodes connected to (*Insert*) or disconnected from (*Remove*) the target node by the Nettack method. *Baseline* is the distribution in the entire graph.

Figure 8.3 compares the distribution of such a property (e.g. node degree) when considering all nodes of the unperturbed graph with the distribution of the property when considering only the nodes incident to the inserted/removed adversarial edges. The comparison indicates a statistically significant difference between the distributions. For example, in Figure 8.3 (left) we can see that the Nettack method tends to connect a target node to low-degree nodes. This could be due to the degree-normalization performed in GCN, where low-degree nodes have a higher weight (i.e., influence) on the aggregation of a node. Likewise, considering nodes incident to edges removed by the adversary we can observe that the Nettack method tends to disconnect high-degree nodes from the target node. In Figure 8.3 (second and third plot) we can see that the attack tends to connect the target node with peripheral nodes, as evidenced by small two-hop neighborhood size and low closeness centrality of the adversarially connected nodes. In Figure 8.3 (right) we can see that the adversary tends to connect a target node to other nodes which have dissimilar attributes. As also shown in other works, the adversary appears to try to counter the homophily property in the graph – which is not surprising, since the GNN has likely learned to partly infer a node's class based on its neighbors.

To understand whether such detected patterns are universal, they can be used to design attack principles itself — indeed, this even leads to black-box attacks since the analyzed properties usually relate to the graph only and not the GNN. In (Zügner et al, 2020) a prediction model was learned estimating the potential impact of a perturbation on unseen graphs using the above mentioned properties as input features. While this often resulted in finding effective adversarial perturbations, thus, highlighting the generality of the regularities uncovered, the attack performance was not on par with the original Nettack attack. Similarly, in (Ma et al, 2020b) PageRank-like scores have been used to identify potential harmful perturbations.

8.2.3 Discussion and Future Directions

The aspects along which adversarial attacks on graphs can be studied allow for a huge variety of scenarios. Only a few of them have been thoroughly investigated in the literature. One important aspect to consider, for example, is that in real applications the cost of perturbations differ: while changing node attributes might be relatively easy, injecting edges might be harder. Thus, designing improved perturbation spaces can make the attack scenarios more realistic and better captures the robustness properties one might want to ensure. Moreover, many different data domains such as knowledge graphs or temporal graphs need to be investigated.

Importantly, while first steps have been made to understand the patterns that makes these perturbations harmful, a clear understanding with a sound theoretical backing is still missing. In this regard, it is also worth repeating that all these studies have focused on analyzing perturbations obtained by Nettack; other attacks might potentially lead to very different patterns. This also implies that exploiting the resulting patterns to design more robust GNNs (see Section 8.4.1) is not necessarily a good solution. Moreover, finding reliable patterns also requires more research on how to compute adversarial perturbations in a scalable way (Wang and Gong, 2019; Geisler et al, 2021), since such patterns might be more pronounced on larger graphs.

8.3 Provable Robustness: Certificates for Graph Neural Networks

Adversarial attack approaches are heuristics to highlight potential vulnerabilities of a GNN. However, they do not provide formal guarantees on the reliability of the methods. In particular, an *unsuccessful* attack does *not* imply the robustness of the GNN. It might just be that the attack approach could simply not find an/the adversarial example since it does not solve Eq. equation 8.1 exactly. Attacks, when successful, only provide results about non-robustness. For a safe use of GNNs, however, we need the opposite: we need principles for provable robustness. These methods provide so called *robustness certificates*, giving formal guarantees that no perturbation regarding a specific perturbation model $\Phi(\mathcal{G})$ will change the prediction.

Considering, for example, the task of node-level classification, the problem these certification approaches are aiming to solve is: Given a graph $\mathcal{G}$, a perturbation set $\Phi(\mathcal{G})$, and a GNN f_θ. Verify that the predicted class for node v stays the same for all $\hat{\mathcal{G}} \in \Phi(\mathcal{G})$. If this holds, we say that v is *certifiably robust* w.r.t. $\Phi(\mathcal{G})$.

Only few robustness certificates so far have been proposed for GNNs. They can mainly be categorized into two principles: model-specific and model-agnostic.

8.3.1 Model-Specific Certificates

Model-specific certificates are designed for a specific class of GNN models (e.g., 2-layer GCNs) and a specific task such as node-level classification. A common theme is to phrase certification as a constrained optimization problem: Recall that in a classification task, the final prediction is usually obtained by taking the class with the largest predicted probability or logit. Let $c^* = \arg\max_{c \in \mathcal{C}} f_\theta(\mathcal{G})_c$ denote the predicted class[4] obtained on the unperturbed graph $\mathcal{G}$, where $\mathcal{C}$ is the set of classes and $f_\theta(\mathcal{G})_c$ denotes the logit obtained for class c. This specifically implies, that the margin $f_\theta(\mathcal{G})_{c^*} - f_\theta(\mathcal{G})_c$ between class c^* and any other class c is positive.

A particularly useful quantity for robustness certification is the *worst-case margin*, i.e., the smallest margin possible under any perturbed data $\hat{\mathcal{G}}$:

$$\hat{m}(c^*, c) = \min_{\hat{\mathcal{G}} \in \Phi(\mathcal{G})} [f_\theta(\hat{\mathcal{G}})_{c^*} - f_\theta(\hat{\mathcal{G}})_c] \qquad (8.3)$$

---

[4] This could either be the predicted class for a specific target node v in case of node-level classification; or for the entire graph in case of graph-level classification. We drop the dependency on v since it is not relevant for the discussion. For simplicity, we assume the maximizer c^* to be unique.

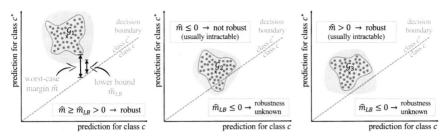

Fig. 8.4: Obtaining robustness certificates via the worst-case margin: The prediction obtained from the unperturbed graph $\mathscr{G}_i$ is illustrated with a cross, while the predictions for the perturbed graphs $\Phi(\mathscr{G}_i)$ are illustrated around it. The worst-case margin measures the shortest distance to the decision boundary. If it is positive (see $\mathscr{G}_1$), all predictions are on the same side of the boundary; robustness holds. If it is negative (see $\mathscr{G}_2$), some predictions cross the decision boundary; the class prediction will change under perturbations, meaning the model is not robust. When using lower bounds — the shaded regions in the figure — robustness is ensured for positive values (see $\mathscr{G}_1$) since the exact worst-case margin can only be larger. If the lower bound becomes negative, no statement can be made (see $\mathscr{G}_2$ and $\mathscr{G}_3$; robustness unknown). Both $\mathscr{G}_2$ and $\mathscr{G}_3$ have a negative lower bound, while the (not tractable to compute) exact worst-case margin differs in sign.

If this term is positive, c can never be the predicted class for node v. And if the worst-case margin $\hat{m}(c^*, c)$ stays positive for all $c \neq c^*$, the prediction is certifiably robust since the logit for class c^* is always the largest – for all perturbed graphs in $\Phi(\mathscr{G})$. This idea is illustrated in Figure 8.4.

As shown, obtaining a certificate means solving the (constrained) optimization problem in Eq. equation 8.3 for every class c. Not surprisingly, however, solving this optimization problem is usually intractable – for similar reasons as computing adversarial attacks is hard. So how can we obtain certificates? Just heuristically solving Eq. equation 8.3 is not helpful since we aim for guarantees.

Lower Bounds on the Worst-Case Margin

The core idea is to obtain tractable lower bounds on the worst-case margin. That is, we aim to find functions $\hat{m}_{LB}$ that ensure $\hat{m}_{LB}(c^*, c) \leq \hat{m}(c^*, c)$ and are more efficient to compute. One solution is to consider relaxations of the original constrained minimization problem, replacing, for example, the model's nonlinearities and hard discreteness constraints via their convex relaxation. For example, instead of requiring that an edge is perturbed or not, indicated by the variables $e \in \{0, 1\}$, we can use $e \in [0, 1]$. Intuitively, using such relaxations leads to supersets of the actually reachable predictions, as visualized in Figure 8.4 with the shaded regions.

Overall, if the lower bound $\hat{m}_{LB}$ stays positive, the robustness certificate holds — since $\hat{m}$ is positive by transitivity as well. This is shown in Figure 8.4 for graph $\mathcal{G}_1$. If $\hat{m}_{LB}$ is negative, no statement can be made since it is only a lower bound of the original worst-case margin $\hat{m}$, which thus can be positive or negative. Compare the two graphs $\mathcal{G}_2$ and $\mathcal{G}_3$ in Figure 8.4: While both have a negative lower bound (i.e., both shaded regions cross the decision boundary), their actual worst-case margins $\hat{m}$ differ. Only for graph $\mathcal{G}_2$ the actually reachable predictions (which are not efficiently computable) also cross the decision boundary. Thus, if the lower bound is negative, the actual robustness remains unknown – similar to an unsuccessful attack, where it remains unclear whether the model is actually non-robust or the attack simply not strong enough. Therefore, besides being efficient to compute, the function $\hat{m}_{LB}$ should be as close as possible to $\hat{m}$ to avoid cases where no answer can be given despite the model being robust.

The above idea, using convex relaxations of the model's nonlinearities and the admissible perturbations, is used in the works (Zügner and Günnemann, 2019; Zügner and Günnemann, 2020) for the class of GCNs and node-level classification. In (Zügner and Günnemann, 2019), the authors consider perturbations to the node attributes and obtain lower bounds via a relaxation to a linear program. The work (Zügner and Günnemann, 2020) considers perturbations in the form of edge deletions and reduces the problem to a jointly constrained bilinear program. Similarly, also using convex relaxations, Jin et al (2020a) has proposed certificates for graph-level classification under edge perturbations using GCNs. Beyond GCNs, model-specific certificates for edge perturbations have also been devised for the class of GNNs using PageRank diffusion (Bojchevski and Günnemann, 2019), which includes label/feature propagation and (A)PPNP (Klicpera et al, 2019a). The core idea of (Bojchevski and Günnemann, 2019) is to treat the problem as a PageRank optimization task which subsequently can be expressed as a Markov decision process. Using this connection one can indeed show that in scenarios where only local budgets are used (see Section 8.2; Eq. equation 8.2) the derived certificates are exact, i.e., no lower bound, while we can still compute them in polynomial time w.r.t. the graph size. In general, all models above consider local and global budget constraints on the number of changes.

Besides providing certificates, being able to efficiently compute (a differentiable lower bound on) the worst-case margin as in Eq. equation 8.3 also enables to improve GNN robustness by incorporating the margin during training, i.e. aiming to make it positive for all nodes. We will discuss this in detail in Section 8.4.2.

Overall, a strong advantage of model-specific certificates is their explicit consideration of the GNN model structure within the margin computation. However, the white-box nature of these certificates is simultaneously their limitation: The proposed certificates capture only a subset of the existing GNN models and any GNN yet to be developed likely requires a new certification technique as well. This limitation is tackled by model-agnostic certificates.

8.3.2 Model-Agnostic Certificates

Model-agnostic certificates treat the machine learning model as a black-box. For example, the work (Bojchevski et al, 2020a) provides certificates for any classifier operating on discrete data, including GNNs. Most importantly, it is sufficient to consider *only* the output of the classifier for different samples to obtain the certificate. This is precisely what makes it particularly appealing for certifying GNNs since it allows us to sidestep a complex analysis of the message-passing dynamics and the non-linear interactions between the nodes. So far, model-agnostic certificates are mainly based on the idea of randomized smoothing (Lecuyer et al, 2019; Cohen et al, 2019), originally proposed for continuous data. To handle graphs, extensions to discrete data have been proposed.

The core idea is to base the certificate on a *smoothed classifier*, which aggregates the output of the original (base) GNN when applied to randomly perturbed versions of the input graph $\mathcal{G}$. For example, the smoothed classifier might report the most likely (majority) class on these randomized samples. While different variants of this approach are possible, we provide one intuitive setting in the following to convey the main idea.

Let $f : \mathbb{G} \to \mathcal{C}$ denote a function (e.g., a GNN) that takes a graph $\mathcal{G} \in \mathbb{G}$ as input and predicts a single class $f(\mathcal{G}) = c \in \mathcal{C}$ as output, e.g. a node's prediction. Let τ be a smoothing distribution, also called randomization scheme, that adds random noise to the input graph. For example, τ might randomly add Bernoulli noise to the adjacency matrix of $\mathcal{G}$, corresponding to randomly adding or deleting edges. Technically, τ assigns probability mass/density $\Pr(\tau(\mathcal{G}) = \mathcal{X})$ to each graph $\mathcal{X} \in \mathbb{G}$. We can construct a *smoothed* (ensemble) classifier g from the *base* classifier f as follows:

$$g(\mathcal{G}) = \arg\max_{c \in \mathcal{C}} \Pr(f(\tau(\mathcal{G})) = c) \tag{8.4}$$

In other words, $g(\mathcal{G})$ returns the most likely class obtained by first randomly perturbing the graph $\mathcal{G}$ using τ and then classifying the resulting graphs $\tau(\mathcal{G})$ with the base classifier f.

As in Section 8.3.1, the goal is to assess whether the prediction does not change under perturbations: denoting with $c^* = g(\mathcal{G})$ the class predicted by the smoothed classifier on $\mathcal{G}$, we want $g(\hat{\mathcal{G}}) = c^*$ for all $\hat{\mathcal{G}} \in \Phi(\mathcal{G})$. Considering for simplicity the case of binary classification, this is equivalent to ensure that $\Pr(f(\tau(\hat{\mathcal{G}})) = c^*) \geq 0.5$ for all $\hat{\mathcal{G}} \in \Phi(\mathcal{G})$; or short: $\min_{\hat{\mathcal{G}} \in \Phi(\mathcal{G})} \Pr(f(\tau(\hat{\mathcal{G}})) = c^*) \geq 0.5$.

Since, unsurprisingly, the term is intractable to compute, we refer again to a lower bound to obtain the certificate:

$$\min_{\hat{\mathcal{G}} \in \Phi(\mathcal{G})} \min_{h \in \mathcal{H}_f} \Pr(h(\tau(\hat{\mathcal{G}})) = c^*) \leq \min_{\hat{\mathcal{G}} \in \Phi(\mathcal{G})} \Pr(f(\tau(\hat{\mathcal{G}})) = c^*) \tag{8.5}$$

Here, $\mathcal{H}_f$ is the set of *all* classifiers sharing some properties with f, e.g., often that the smoothed classifier based on h and f would return the same probability for $\mathcal{G}$, i.e., $\Pr(h(\tau(\mathcal{G})) = c^*) = \Pr(f(\tau(\mathcal{G})) = c^*)$. Since $f \in \mathcal{H}_f$, the inequality holds trivially. Accordingly, if the left hand side of Eq. equation 8.5 is larger than 0.5, also the right hand side is guaranteed to be so, implying that $\mathcal{G}$ would be certifiably robust.

What does Eq. equation 8.5 intuitively mean? It aims to find a base classifier h which minimizes the probability that the perturbed sample $\hat{\mathcal{G}}$ is assigned to class c^*. Thus, h represents a kind of *worst-case base classifier* which, when used within the smoothed classifier, tries to obtain a different prediction for $\hat{\mathcal{G}}$. If even this worst-case base classifier leads to certifiable robustness (left hand side of Eq. equation 8.5 larger than 0.5), then surely the actual base classifier at hand has well.

The most important part to make this all useful, however, is the following: given a set of classifiers $\mathcal{H}_f$, finding the worst-case classifier h and minimizing over the perturbation model $\Phi(\mathcal{G})$ is often tractable. In some cases, the optima can even be calculated in closed-form. This shows some interesting relation to the previous section: There, the intractable minimization over $\Phi(\mathcal{G})$ in Eq. equation 8.3 was replaced by some tractable lower bound, e.g., via relaxations. Now, by finding a worst-case classifier h we not only obtain a lower bound but minimization over $\Phi(\mathcal{G})$ becomes often also immediately tractable. Note, however, that in Section 8.3.1 we obtain a certificate for the base classifier f, while here we obtain a certificate for the smoothed classifier g.

Putting Model-Agnostic Certificates into Practice

As said, given a set of classifiers $\mathcal{H}_f$, finding the worst-case classifier h and minimizing over the perturbation model $\Phi(\mathcal{G})$ is often tractable. The main computational challenge in practice lies in determining $\mathcal{H}_f$. Let's consider our previous example where we enforced all classifiers h to ensure $\Pr(h(\tau(\mathcal{G})) = c^*) = \Pr(f(\tau(\mathcal{G})) = c^*)$. To determine $\mathcal{H}_f$, one needs to compute $\Pr(f(\tau(\mathcal{G})) = c^*)$. Clearly, doing this exactly is again usually intractable. Instead, the probability can be estimated using sampling. To ensure a tight approximation, the base classifier has to be fed a large number of samples from the smoothing distribution. This becomes increasingly expensive as the size and complexity of the GNN model increases. Furthermore, the resulting estimates only hold with a certain probability. Accordingly, also the derived guarantees have the same probability, i.e., one obtains only *probabilistic robustness certificates*. Despite these practical limitations, randomized smoothing has become widely popular, as it is often still more efficient than model-specific certificates.

This general idea of model-agnostic certificates has been investigated for discrete data in (Lee et al, 2019a; Dvijotham et al, 2020; Bojchevski et al, 2020a; Jia et al, 2020), with the latter two focusing also on graph-related tasks. In (Jia et al, 2020), the authors investigate the robustness of community detection. In (Bojchevski et al, 2020a), the main focus is on node-level and graph-level classification w.r.t. graph structure and/or attribute perturbations under global budget constraints. Specifically, Bojchevski et al (2020a) overcomes critical limitations of the other approaches in two regards: it explicitly accounts for sparsity in the data as present in many graphs, and it obtains strong certificates with a dramatically reduced computational complexity. Both aspects are core to making certification useful and possible for graph data. Since the approach of (Bojchevski et al, 2020a) is agnostic to the underlying classifier – it can be used as long as the input is discrete – it has been applied to various GNNs including GCN, GAT, (A)PPNP (Klicpera et al, 2019a), RGCN (Zhu et al, 2019a), and Soft Medoid (Geisler et al, 2020) as well as node-level and graph-level classification.

8.3.3 Advanced Certification and Discussion

Research on robustness certificates for GNNs is still in a very early stage. As we have seen in Section 8.2, the space of attacks is vast with different properties to study and perturbation models to consider. The methods discussed above cover only a few of these scenarios.

One step forward to more powerful certificates is the work of (Schuchardt et al, 2021). Like in local attacks to individual nodes, existing robustness certificates aim to certify each prediction independently. Thus, they assume that an adversary can use different perturbed inputs to attack different predictions. Alternatively, and similar to a global attack, the work (Schuchardt et al, 2021) introduces *collective robustness certificates* which compute the number of predictions which are *simultaneously* guaranteed to remain stable under perturbation. That is, it exploits the fact that a GNN simultaneously outputs multiple predictions based on a single shared input. Given a fixed perturbation budget, using this idea, the number of certifiable predictions can be increased by orders of magnitudes compared to certifying each prediction independently. The work, however, can not handle perturbation models with edge additions. As mentioned before, both views – local and global – are reasonable and it depends on the application which robustness guarantee is more relevant.

To cover the full spectrum of GNN applications, surely further certificates for other scenarios and tasks are required. Specifically, so far, all certificates assume an evasion attack scenario. It is also worth repeating that in the randomized smoothing approaches discussed above, we are actually certifying the smoothed (ensemble) classifier, and not the underlying base classifier. From a practitioner's point of view this means that obtaining a single prediction always requires to feed a large amount of samples through the GNN, leading to a scalability bottleneck which needs to be tackled in the future.

8.4 Improving Robustness of Graph Neural Networks

As we have established, standard GNNs trained in the usual way are not robust to even small changes to the graph, thus, using them in sensitive and critical applications might be risky. Certificates can provide us guarantees about their performance.

However, as a consequence of the non-robustness, the certificates rarely hold for standard models, i.e., only few predictions can be certified. To tackle this limitation, methods aiming to improve robustness have been investigated, i.e. making the models less susceptible to perturbations.[5] In this regard, three broad, not mutually exclusive, categories can be identified.

8.4.1 Improving the Graph

One seemingly clear direction to improve robustness is to remove perturbations from the data, i.e., to revert the performed malicious changes and obtain a more 'clean' graph. While this may sound simple, the inherent challenge is that adversarial perturbations are usually designed to be imperceptible, which makes their identification difficult. Still, as seen in Section 8.2.2.1, some patterns might be present.

Works such as (Zhang et al, 2019b) exploit this idea to perform a 'cleaning' of the graph before it is used as input to the GNN, relying on observations that, for example, the predicted class distribution changes for attacked nodes. Similarly, for attributed graphs, Wu et al (2019b) removes potential adversarial edges based on the Jaccard similary between the nodes' attributes. Such pre-processing steps are not limited to be 'attack detection' approaches that try to spot individual suspicious nodes are edges; they can also be thought of as a kind of denoising. Indeed, the work (Entezari et al, 2020) analyzed that perturbations performed by Nettack affect mainly the high-rank (low-valued) singular components of the graph's adjacency matrix. Thus, to improve robustness they compute a low-rank approximation of the graph which aims to remove the (adversarial) noise in a pre-processing procedure. The limitation is that the resulting graph becomes dense. Overall, such graph cleaning can be used in poisoning as well as evasion scenarios. Note, though, that an approach that has shown to perform well in one scenario, does not imply the success in another.

More generally, while these approaches have shown to be effective in specific scenarios, one has to be aware of one crucial limitation: the exploited patterns are often based on specific attacks like Nettack. Thus, the resulting detections might be limited to certain perturbations and potentially do not generalize to other scenarios.

Improving the graph is not restricted to happen before the training or the inference step, i.e. we do not need to follow a sequential approach of first cleaning and then learning a prediction model. Instead, the cleaning can be interwoven with the learning approach itself. Intuitively speaking, in order to minimize the corresponding training loss, one jointly learns the GNN parameters and also how to clean the graph itself. The benefit of this joint learning approach is that the specific model and task at hand can be taken into account, while the conditions enforced on the clean graph can be rather weak, e.g., only requiring that perturbations should be sparse.

---

[5] In some works, such approaches are called *(heuristic) defenses* to highlight their increased resilience to attacks. Similarly, some works use the term *provable defense* when referring to certificates since they provably prevent attacks to be harmful that are within a certified set $\Phi(\mathscr{G})$.

Fig. 8.5 Illustration of robust training: The classifier corresponding to the orange/solid decision boundary is not robust to perturbations in $\Phi_1(\mathcal{G})$: some graphs cross the boundary and, thus, are assigned a different class. The classifier obtained through robust training (blue/dashed), assigns the same class to all graphs in $\Phi_1(\mathcal{G})$: it is robust w.r.t. $\Phi_1(\mathcal{G})$ – but not $\Phi_2(\mathcal{G})$.

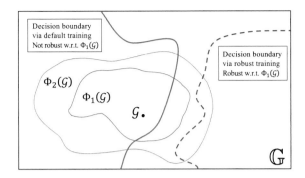

Interestingly, even before the rise of graph neural networks, such joint approaches have been investigated, e.g., in (Bojchevski et al, 2017) to improve the robustness of spectral embeddings. For GNNs, such graph structure learning has been proposed in (Jin et al, 2020e; Luo et al, 2021) where certain properties like low-rank graph structure and attribute similarity are used to define how the clean graph should preferably look like.

8.4.2 Improving the Training Procedure

As discussed in Section 8.2.2, one further reason for the non-robustness of GNNs are the parameters/weights learned during training. Weights resulting from standard training often lead to models that do not generalize well to slightly perturbed data. This is illustrated in Figure 8.5 with the orange/solid decision boundary. Note that the figure shows the input space, i.e., the space of all graphs $\mathbb{G}$; this is in contrast to Figure 8.4 which shows the predicted probabilities. If we were able to improve our training procedure to find 'better' parameters – taking into account that the data is or might become potentially perturbed – the robustness of our model would improve as well. This is illustrated in Figure 8.5 with the blue/dashed decision boundary. There, all perturbed graphs from $\Phi_1(\mathcal{G})$ get the same prediction. As seen, in this regard robustness links to the generalization performance of prediction models in general.

8.4.2.1 Robust Training

Robust training refers to training procedures that aim at producing models that are robust to adversarial (and/or other) perturbations. The common theme is to optimize a *worst-case loss* (also called robust loss), i.e. the loss achieved under the worst-case perturbation. Technically, the training objective becomes:

$$\theta^* = \arg\min_{\theta} \max_{\hat{\mathscr{G}} \in \Phi(\mathscr{G})} \mathscr{L}_{\text{train}}(f_\theta(\hat{\mathscr{G}})) \tag{8.6}$$

where f_θ is the GNN with its trainable weights. As shown, we do not evaluate the loss at the unperturbed graph but instead use the loss achieved in the worst case (compare this to the standard training where we simply minimize $\mathscr{L}_{\text{train}}(f_\theta(\mathscr{G}))$). The weights are steered to obtain low loss under these worst scenarios as well, thus obtaining better generalization.

Not surprisingly, solving Eq. equation 8.6 is usually not tractable for the same reasons as finding attacks and certificates is hard: we have to solve a discrete, highly complex (minmax) optimization problem. In particular, for training, e.g., via gradient based approaches, we also need to compute the gradient w.r.t. the inner maximization. Thus, for feasibility, one usually has to refer to various surrogate objectives, substituting the worst-case loss and the resulting gradient by simpler ones.

Data Augmentation during Training

In this regard, the most naïve approach is to randomly draw samples from the perturbation set $\Phi(\mathscr{G})$ during each training iteration. That is, during training the loss and the gradient are computed w.r.t. these randomly perturbed samples; with different samples drawn in each training iteration. If the perturbation set, for example, contains graphs where up to x edge deletions are admissible, we would randomly create graphs with up to x edges dropped out. Such edge dropout has been analyzed in various works but does not improve adversarial robustness substantially (Dai et al, 2018a; Zügner and Günnemann, 2020); a possible explanation is that the random samples simply do not represent the worst-case perturbations well.

Thus, more common is the approach of *adversarial training* (Xu et al, 2019c; Feng et al, 2019a; Chen et al, 2020i). Here, we do not randomly sample from the perturbation set, but in each training iteration we create adversarial examples $\hat{\mathscr{G}}$ and subsequently compute the gradient w.r.t. these. As these samples are expected to lead to a higher loss, the result of the inner max-operation in Eq. equation 8.6 is much better approximated. Instead of perturbing the input graph, the work (Jin and Zhang, 2019) has investigated a robust training scheme which perturbs the latent embeddings.

It is interesting to note that adversarial training in its standard form requires labeled data since the attack aims to steer towards an incorrect prediction. In the typical transductive graph-learning tasks, however, large amounts of unlabeled data are available. As a solution, virtual adversarial training has also been investigated (Deng et al, 2019; Sun et al, 2020d), operating on the unlabeled data as well. Intuitively, it treats the currently obtained predictions on the unperturbed graph as the ground truth, making it a kind of self-supervised learning. The predictions on the perturbed data should not deviate from the clean predictions, thus enforcing smoothness.

Using (virtual) adversarial training has empirically shown some improvements in robustness, but not consistently. In particular, to well approximate the max term in the robust loss of Eq. equation 8.6, we need powerful adversarial attacks, which are typically costly to compute for graphs (see Section 8.2). Since here attacks need to be computed in every training iteration, the training process is slowed down substantially.

Beyond Data Augmentation - Certificate-Based Loss Functions

At the end of the day, the techniques above perform a costly data augmentation during training, i.e., they use altered versions of the graph. Besides being computationally expensive, there is no guarantee that the adversarial examples are indeed good proxies for the max term in Eq. equation 8.6. An alternative approach, e.g., followed by (Zügner and Günnemann, 2019; Bojchevski and Günnemann, 2019) relies on the idea of certification as discussed previously. Recall that these techniques compute a lower bound $\hat{m}_{LB}$ on the worst-case margin. If it is positive, the prediction is robust for this node/graph. Thus, the lower bound itself acts like a robustness loss $\mathscr{L}_{rob}$, for example instantiated as a hinge loss: $\max(0, \delta - \hat{m}_{LB})$. If the lower-bound is above δ, then the loss is zero; if it is smaller, a penalty occurs. Combining this loss function with, e.g., the usual cross-entropy loss, forces the model not only to obtain good classification performance but also robustness.

Crucially, $\mathscr{L}_{rob}$ and, thus, the lower bound need to be differentiable since we need to compute gradients for training. This, indeed, might be challenging since usually the lower bound itself is still an optimization problem. While in some special cases the optimization problem is directly differentiable (Bojchevski and Günnemann, 2019), another general idea is to relate to the principle of duality. Recall that the worst-case margin $\hat{m}$ (or a potential corresponding lower bound $\hat{m}_{LB}$) is the result of a (primal) *minimization* problem (see Eq. equation 8.3). Based on the principle of duality, the result of the dual *maximization* problem provides, as required, a lower bound to this value. Even more, *any* feasible solution of the dual problem provides a lower bound on the optimal solution. Thus, we actually do not need to solve the dual program. Instead, it is sufficient to compute the objective function of the dual at any single feasible point to obtain an (even lower, thus looser) lower bound; no optimization is required and computing gradients often becomes straightforward. This principle of duality has been used in (Zügner and Günnemann, 2019) to perform robust training in an efficient way.

8.4.2.2 Further Training Principles

Robust training is not the only way to obtain 'better' GNN weights. In (Tang et al, 2020b), for example, the idea of transfer learning (besides further architecture changes; see next section) is exploited. Instead of purely training on a perturbed target graph, the method adopts clean graphs with artificially injected perturbations to first learn suitable GNN weights. These weights are later transferred and fine-tuned to the actual graph at hand. The work (Chen et al, 2020i) exploits smoothing distillation where one trains on predicted soft labels instead of ground-truth labels to enhance robustness. The work (Jin et al, 2019b) argues that graph powering enhances robustness and proposes to minimize the loss not only on the original graph but on a set of graphs consisting of the different graph powers. Lastly, the authors of (You et al, 2021) use a contrastive learning framework using different (graph) data augmentations. Albeit adversarial robustness is not their focus, they report increased adversarial robustness against the attacks of (Dai et al, 2018a). In general, changing the loss function or regularization terms leads to different training, though the effects on robustness for GNNs are not fully understood yet.

8.4.3 Improving the Graph Neural Networks' Architecture

The final category of methods improving robustness is concerned with designing novel GNN architectures itself. Architecture engineering is one core component of neural network research in general, with many advancements in the last years. While traditionally focusing on improving prediction performance, a likewise important property becomes the methods' robustness – both being potentially opposing goals.

8.4.3.1 Adaptively Down-Weighting Edges

Inspired by the idea of graph cleaning as discussed before, a natural idea is to enhance the GNN by mechanisms to reduce the impact of perturbed edges. An obvious choice for this are edge attention principles. However, it is a false conclusion to assume that standard attention-based GNNs like GAT are immediately suitable for this task. Indeed, as shown in (Tang et al, 2020b; Zhu et al, 2019a) such models are non-robust. The problem is that these models still assume clean data to be given; they are not aware that the graph might be perturbed.

Thus, other attention approaches try to incorporate more information in the process. In (Tang et al, 2020b) the attention mechanism is enhanced by taking clean graphs into account for which perturbations have been artificially injected. Since now ground truth information is available (i.e., which edges are harmful), the attention can try to learn down-weighing these while retaining the non-perturbed ones. An alternative idea is used in (Zhu et al, 2019a). Here, the representations of each node in each layer are no longer represented as vectors but as Gaussian distribution. They hypothesize that attacked nodes tend to have large variances, thus using this information within the attention scores. Further attention mechanism considering, e.g., the model and data uncertainty or the neighboring nodes' similarity have been proposed in (Feng et al, 2021; Zhang and Zitnik, 2020).

An alternative to edge attention is to enhance the aggregation used in message passing. In a GNN message passing step, a node's embedding is updated by aggregating over its neighbors' embeddings. In this regard, adversarially inserted edges add additional data points to the aggregation and therefore perturb the output of the message passing step. Aggregation functions such as sum, weighted mean, or the max operation used in standard GNNs can be arbitrarily distorted by only a single outlier. Thus, inspired by the principle of robust statistics, the work (Geisler et al, 2020) proposes to replace the usual GNN's aggregation function with a differentiable version of the Medoid, a provably robust aggregation operation. The idea of enhancing the robustness of the aggregation function used during message passing has further been investigated in (Wang et al, 2020o; Zhang and Lu, 2020).

Overall, all these methods down-weight the relevance of edges, with one crucial difference to the methods discussed in Section 8.4.1: they are adaptive in the sense that the relevance of each edge might vary between, e.g., the different layers of the GNN. Thus, an edge might be excluded/down-weighted in the first layer but included in the second one, depending on the learned intermediate representation. This allows a more fine-grained handling of perturbations. In contrast, the approaches in Section 8.4.1 derive a single cleaned graph that is used in the entire GNN.

8.4.3.2 Further Approaches

Many further ideas to improve robustness have been proposed, which do not all entirely fit into the before mentioned categories. For example, in (Shanthamallu et al, 2021) a surrogate classifier is trained which does not access the graph structure but is aimed to be aligned with the predictions of the GNN, both being jointly trained. Since the final predictor is not using the graph but only the node's attributes, higher robustness to structure perturbations is hypothesized. The work (Miller et al, 2019) proposes to select the training data in specific ways to increase robustness, and Wu et al (2020d) uses the principle of information bottleneck, an information theoretic approach to learn representations balancing expressiveness and robustness. Finally, also randomized smoothing (Section 8.3.2) can be interpreted as a technique to improve adversarial robustness by using an ensemble of predictors on randomized inputs.

8.4.4 Discussion and Future Directions

Considering the current state of research, a surprising observation is that robustness to graph structure perturbations is not well achieved via adversarial training. This is in stark contrast to, e.g., the image domain where robust training (in the form of adversarial training) can be considered one of the highly suitable techniques to improves robustness (Tramer et al, 2020). Focusing on perturbations of the node attributes, in contrast, robust training indeed performs very well as shown in (Zügner and Günnemann, 2019). Surprisingly, such robust training (targeting attributes) also improves robustness under graph structure perturbations (Zügner and Günnemann, 2020) – and, even more, outperforms several adversarial training strategies performing edge dropout. The question remains if structure perturbations have special property that diminishes the effect of robust training or whether the generated adversarial perturbations are not capturing the worst-case; showcasing again the hardness of the problem. This might also explain why the majority of works have focused on principles of weighting/filtering out edges.

In this regard, it is again important to remember that all approaches are typically designed with a specific perturbation model $\Phi(\mathcal{G})$ in mind. Indeed, downweighting/filtering edges implicitly assumes that adversarial edges had been added to the graph. Adversarial edge deletions, in contrast, would require to identify potential edges to (re)add. This quickly becomes intractable due to the large number of possible edges and has not been investigated so far. Moreover, only a few methods so far have provided theoretical guarantees on the methods' robustness behavior.

8.5 Proper Evaluation in the View of Robustness

Progress in the field of GNN robustness requires sound evaluation of the proposed techniques. Importantly, we have to be aware of the potential trade-off between prediction performance (e.g., accuracy) and robustness. For example, we can easily obtain a highly robust classification model by simply always predicting the same class. Clearly, such a model has no use at all. Thus, the evaluation always involves two aspects: (1) Evaluation of the prediction performance. For this, one can simply refer to the established evaluation metrics such as accuracy, precision, recall, or similar, as known for the various supervised and unsupervised learning tasks. (2) Evaluation of the robustness performance.

Perturbation set and radius. Regarding the latter, the first noteworthy point is that robustness always links to a specific perturbation set $\Phi(.)$ that defines the perturbations the model should be robust to. To enable a proper evaluation, existing works therefore usually define some parametric form of the perturbation set, e.g., denoted $\Phi_r(\mathcal{G})$ where r is the maximal number of changes – the budget – we are allowed to perform (e.g., maximal number of edges to add). The variable r is often referred to as the *radius*. This is because the budget usually coincides with a certain maximal norm/distance we are willing to accept between graph $\mathcal{G}$ and perturbed ones. A generalization of the above form to consider multiple budgets/radii is straightforward. Varying the radius enables us to analyze the robustness behavior of the models in detail. Depending on the radius, different robustness results are expected. Specifically, for a large radius low robustness is expected – or even desired – and accordingly, the evaluation should also include these cases showing the limits of the models.

Recall that using the methods discussed in Section 8.2 and Section 8.3 together, we are able to obtain one of the following answers about a prediction's robustness: **(R)** It is robust; the certificate holds since, e.g., the lower bound on the margin is positive. **(NR)** It is non-robust; we are able to find an adversarial example. **(U)** Unknown; no statement possible since, e.g., the lower bound is negative but the attack was not successful either.

Figure 8.6 shows such an example analysis providing insights about the robustness properties of a GCN in detail. Here, local attacks and certificates are computed on standard (left) and robustly (right) trained GCNs for the task of node classification. As the result shows, robust training indeed increases the robustness of a GCN with fewer attacks being successful and more nodes being certifiable.

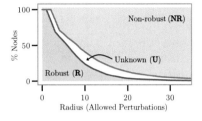 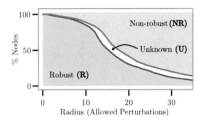

Fig. 8.6: Share of nodes which are provably robust (blue; **R**), non-robust via adversarial example construction (orange; **NR**), or whose robustness is unknown ("gap"; **U**), for increasing perturbation radii. For a given radius, the shares of **(R)+(NR)+(U)**= 100%. **Left**: Standard training; **Right**: robust training as proposed in (Zügner and Günnemann, 2019). Citeseer data and perturbations of node attributes.

It is worth highlighting that case **(U)** – the white gap in Figure 8.6 – occurs only due to the algorithmic inability to solve the attack/certificate problems exactly. Thus, case **(U)** does not give a clear indication about the GNN's robustness but rather about the performance of the attack/certificate.[6] Given this set-up, in the following we distinguish between two evaluation directions, which are reflected in frequently used measures.

Empirical Robustness Evaluation

In an empirical robustness evaluation, we perform an attack on the graph and observe the effects. Common measures are:

- The *drop in performance of the downstream task* (e.g., node classification accuracy), monitoring its decrease after the attack. This metric is typically used in combination with global attacks where a single perturbation is considered that aims to jointly change multiple predictions (see Section 8.2.1, Aspect 1).

---

[6] A large gap indicates that the attacks/certificates are rather loose. The gap might become smaller when improved attacks/certificates become available. Thus, attacks/certificates itself can be evaluated by analyzing the size of the gap since it shows what the maximal possible improvement in either direction is (e.g., the true share of robust predictions can never exceed 100%-NR for a specific radius).

- The *attack success rate*, measuring how many predictions were successfully changed by the attack(s). This simply corresponds to the case **(NR)**, the orange region shown in Fig 8.6. This metric is typically used in combination with local attacks where for each prediction a different perturbation can be used. Naturally, the local attacks' success rate is higher than the overall performance drop due to the flexibility in picking different perturbations.
- In the case of classification, the *classification margin*, i.e., the difference between the predicted probability of the 'true' class minus the second-highest class, and its drop after the attack. See again Figure 8.2 for an example.

The crucial limitation of this evaluation is its dependence on a specific attack approach. The power of the attack strongly affects the result. Indeed, it can be regarded as an *optimistic evaluation* of robustness since a non-successful attack is treated as seemingly robust. However, the conclusion is dangerous since a GNN might only perform well for one type of attack but not another. Thus, the above metrics rather evaluate the power of the attack but only weakly the robustness of the model. *Interpreting the results has to be done with care.* Consequently, when referring to empirical robustness evaluation, it is imperative to use multiple different and powerful attack approaches. Indeed, as also discussed in (Tramer et al, 2020), each robustification principle should come with its own specifically suited attack method (also called adaptive attack) to showcase its limitations.

Provable Robustness Evaluation

A potentially more suitable direction to analyze the robustness behavior of GNNs is to consider provable robustness. As discussed above, case **(U)** corresponds to unclear predictions for which no robustness statement can be given. Since we care about worst-case robustness, we have to assume that these predictions are non-robust as well. In short: **(NR)** and **(U)** should be rare, while case **(R)** should dominate: the number of certifiably robust predictions. Given this idea, the following evaluation metrics are often considered:

- *Certified ratio:* It corresponds to the number of predictions that can be certified as robust given a specific radius r in relation to the number of all predictions. Again take note whether for each prediction a different perturbation can be chosen from $\Phi_r(\mathcal{G})$ (local) or only a joint single one (global). Clearly, the global certified ratio is necessarily (and often significantly) larger than the local one.
- *Certified correctness:* In cases like classification, a prediction can be correct or incorrect. If it is correct and can be certified as well, the prediction is called certified correct. The other, highly undesired, extreme are predictions that are certified incorrect; they are very reliably misclassified.
- *Certified performance:* Based on the idea of certified correct predictions we can also derive a certified version of the original performance metrics, e.g., certified accuracy. Here, only those predictions are treated as correct for the metric if they are 'certified correct'. All other predictions, either incorrect or non-certifiable

are treated as wrong. The certified performance gives a provable lower bound on the performance of the GNN under any admissible perturbation w.r.t. the current perturbation set $\Phi_r(\mathcal{G})$ and the given data.

- *Certified radius:* While the above metrics assume a fixed $\Phi_r(\mathcal{G})$, i.e., a fixed radius r, we can also take another view. For a specific prediction, the largest radius r^* for which the prediction can still be certified as robust is called its certified radius. Given the certified radius of a single prediction, one can easily calculate the *average certifiable radius* over multiple predictions.

Fig. 8.7 Certified ratio of the smoothed classifier obtained from different GNN models using the certificate of (Bojchevski et al, 2020a) where $\Phi_r(\mathcal{G})$ consists of edge deletion perturbations. The model-agnostic nature of the certificate allows to compare the robustness across models.

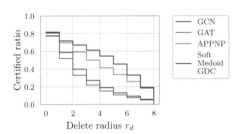

Figure 8.7 shows the certified ratio for different GNN architectures for the task of node-classification when perturbing the graph structure. The smoothed classifier uses 10,000 randomly drawn graphs and the probabilistic certification is based on a confidence level of $\alpha = 0.05$ analogously to the set-up in (Geisler et al, 2020). Since local attacks are considered, the certified ratio is naturally rather low. Still, as shown, there is a significant difference between the models' robustness performance.

Provable robustness evaluation provides strong guarantees in the sense that the evaluation is more pessimistic. E.g. if the certified ratio is high, we know that the actual GNN can only be better. Note again, however, that we still also implicitly evaluate the certificate; with new certificates the result might become even better. Also recall that certificates based on randomized smoothing (Section 8.3.2), evaluate the robustness of the smoothed classifier, thus, not providing guarantees for the base classifier itself. Still, a robust prediction of the smoothed classifier entails that the base classifier predicts the respective class with a high probability w.r.t. the randomization scheme.

As it becomes apparent, evaluating robustness is more complex than evaluating usual prediction performance. To achieve a detailed understanding of the robustness properties of GNNs it is thus helpful to analyze all aspects introduced above.

8.6 Summary

Along with the increasing relevance of graph neural networks in various application domains, comes also an increasing demand to ensure their reliability. In this regard, adversarial robustness plays a central role since perturbed data is omnipresent. As we have seen, standard GNN architectures and training principles – as predominantly used in today's applications – lead to non-robust models, with all the undesired consequences included. However, there is hope: First, various principles to improve robustness of GNNs have started to emerge. The obtained results are already promising giving a first indication that improved robustness can be achieved without giving up too much of the GNNs' prediction performance. Second, robustness certificates provide us ways to even assess certain robustness properties in a formal way. That is, one does not need to rely on heuristics but instead obtains guarantees of the GNN's behavior. In all these directions, one has just started to explore the vast possibilities and many challenges still need to be tackled. Thus, in the upcoming years, various further insights can be expected, pursuing one common goal: to continue the success story of graph neural networks by enabling their reliable use in even sensitive and safety-critical domains.

Acknowledgements

A special thanks to my amazing PhD students Aleksandar Bojchevski, Simon Geisler, Jan Schuchardt, and Daniel Zügner who not only provided valuable feedback to this article but also made many of the research results in this field possible.

Editor's Notes: Adversarial Robustness is one of the hottest topics in Machine Learning/Deep Learning today. This wave of research starts from the robustness of Convolutional Neural Networks in computer vision domain and has rapidly influenced other ML/DL network architectures in other applications domains like NLP and Graphs. Adversarial robustness of Graph Neural Networks is a very important research area, which has a fundamental impact on many other learning tasks, including graph classification task (Chapter 9), link prediction (Chapter 10), graph generation-related tasks (Chapter 11 and Chapter 12), graph matching networks (Chapter 13), and so on. Some chapters (like Chapter 14) can be treated one of potential ways to help alleviate the effect of adversarial robustness by learning a graph structure beyond its intrinsic graph structure.

Part III
Frontiers of Graph Neural Networks

Chapter 9
Graph Neural Networks: Graph Classification

Christopher Morris

Abstract Recently, graph neural networks emerged as the leading machine learning architecture for supervised learning with graph and relational input. This chapter gives an overview of GNNs for graph classification, i.e., GNNs that learn a graph-level output. Since GNNs compute node-level representations, pooling layers, i.e., layers that learn graph-level representations from node-level representations, are crucial components for successful graph classification. Hence, we give a thorough overview of pooling layers. Further, we overview recent research in understanding GNN's limitations for graph classification and progress in overcoming them. Finally, we survey some graph classification applications of GNNs and overview benchmark datasets for empirical evaluation.

9.1 Introduction

Graph-structured data is ubiquitous across application domains ranging from chemo- and bioinformatics (Barabasi and Oltvai, 2004; Stokes et al, 2020) to image (Simonovsky and Komodakis, 2017) and social network analysis (Easley et al, 2012). To develop successful (supervised) machine learning models in these domains, we need techniques to exploit the graph structure's rich information and the feature information within nodes and edges. In recent years, numerous approaches have been proposed for (supervised) machine learning with graphs—most notably, approaches based on *graph kernels* (Kriege et al, 2020) and, more recently, using graph neural networks (GNNs), see (Chami et al, 2020; Wu et al, 2021d) for a general overview. Graph kernels work by predefining a fixed set of features, following a two-step feature extraction and learning task approach. They first compute a vectorial representation of the graph based on predefined features, e.g., small subgraphs, random

Christopher Morris
CERC in Data Science for Real-Time Decision-Making, Polytechnique Montréal, e-mail: chris@christophermorris.info

walks, neighborhood information, or a positive semi-definite kernel matrix reflecting pairwise graph similarities. The resulting features or the kernel matrix are then plugged into a learning algorithm such as a Support Vector Machine. Hence, they rely on human-made feature engineering.

GNNs promise that they possibly offer better adaption to the learning task at hand by learning feature extraction and downstream tasks in an end-to-end fashion. One of the most prominent tasks for GNNs is graph classification or regression, i.e., predicting the class labels or target values of a set of graphs, such as properties of chemical molecules (Wu et al, 2018). Since GNNs learn vectorial representations of nodes, or node-level representations, for successful graph classification, the *pooling layer*, i.e., a layer that learns a graph-level from node-level representations, is crucial. This pooling layer aims to learn, based on the node-level representations, a vectorial representation that captures the graph structure as a whole. Ideally, one wants a graph-level representation that captures local patterns, their interaction, and global patterns. However, the optimal representation should adapt to the given data distribution. What is more, GNNs for graph classification have recently successfully been applied to an extensive range of application areas, the most promising being in pharmaceutical drug research; see (Gaudelet et al, 2020) for a survey. Other important application areas include fields such as material science (Xie and Grossman, 201f8), process engineering (Schweidtmann et al, 2020), and combinatorial optimization (Cappart et al, 2021), some of which we also survey here.

In the following, we give an overview of GNNs for graph classification. Starting from the mid-nineties' classic works, we survey modern works from the current deep learning era, followed by an in-depth review of recent pooling layers.

Before GNNs emerged as the leading architecture for graph classification, research focused on kernel-based algorithms, so-called graph kernels, which work by predefining a set of features. Starting from the early 2000s, researchers proposed a plethora of graph kernels, based on graph features such as shortest-paths (Borgwardt et al, 2005), random walks (Kang et al, 2012; Sugiyama and Borgwardt, 2015; Zhang et al, 2018i), local neighborhood information (Shervashidze et al, 2011a; Costa and De Grave, 2010; Morris et al, 2017, 2020b), and matchings (Fröhlich et al, 2005; Woźnica et al, 2010; Kriege and Mutzel, 2012; Johansson and Dubhashi, 2015; Kriege et al, 2016; Nikolentzos et al, 2017); see (Kriege et al, 2020; Borgwardt et al, 2020) for thorough surveys. For a thorough survey on GNNs, e.g., see (Hamilton et al, 2017b; Wu et al, 2021d; Chami et al, 2020).

9.2 Graph neural networks for graph classification: Classic works and modern architectures

In the following, we survey classic and modern works of GNNs for graph classification. GNNs layers for graph classification date back to at least the mid-nineties in chemoinformatics. For example, Kireev (1995) derived GNN-like neural architectures to predict chemical molecule properties. The work of (Merkwirth and

Lengauer, 2005) had a similar aim. Gori et al (2005); Scarselli et al (2008) proposed the original GNN architecture, introducing the general formulation that was later reintroduced and refined in (Gilmer et al, 2017) by deriving the general *message-passing* formulation, most modern GNN architectures can be expressed in, see Section 9.2.1.

We divide our overview of modern GNN layers for graph classification into *spatial approaches*, i.e., ones that are purely based on the graph structure by aggregating local information around each node, and *spectral approaches*, i.e., ones that rely on extracting information from the graph's spectrum. Although this division is somewhat arbitrary, we stick to it due to historical reasons. Due to the large body of different GNN layers, we cannot offer a complete survey but focus on representative and influential works.

9.2.1 Spatial approaches

One of the earliest *modern*, spatial GNN architectures for graph classification was presented in (Duvenaud et al, 2015b), focusing on the prediction of chemical molecules' properties. Specifically, the authors propose to design a differentiable variant of the well-known Extended Connectivity Fingerprint (ECFP) (Rogers and Hahn, 2010) from chemoinformatics, which works similar to the computation of the WL feature vector. For the computation of their GNN layer, denoted *Neural Graph Fingerprints*, Duvenaud et al (2015b) first initialize the feature vector $\mathbf{f}^0(v)$ of each node v with features of the corresponding atom, e.g., a one hot-encoding representing the atom type. In each iteration or layer t, they compute a feature representation $\mathbf{f}^t(v)$ for node v as

$$\mathbf{f}^t(v) = \mathbf{f}^{t-1}(v) + \sum_{w \in N(v)} \mathbf{f}^{t-1}(w),$$

followed by the application of a one-layer perceptron. Here, $N(v)$ denotes the *neighborhood* of node v, i.e., $N(v) = \{w \in \mathcal{V} \mid (v,w) \in \mathcal{E}\}$. Since the ECFP usually computes sparse feature vectors for small molecules, they apply a linear layer followed by a softmax function, i.e.,

$$\mathbf{f}^{t(v)} = \text{softmax}(\mathbf{f}^t(v) \cdot H^t),$$

which they interpret as a sparsification layer, where H^t is the parameter matrix of the linear layer. The final pooled graph-level representation is computed by summing over all layers' features, and the resulting feature is fed into an MLP for the downstream regression and classification. The above GNN layer is compared to the ECFP on molecular regression datasets showing good performance.

Dai et al (2016) introduced a simple GNN layer inspired by mean-field inference. Concretely, given a graph $\mathcal{G}$, the feature $\mathbf{f}^t(v)$ for node v at layer t is computed as

$$\mathbf{f}^t(v) = \sigma(\mathbf{f}^{t-1}(v) \cdot W_1 + \sum_{w \in N(v)} \mathbf{f}^{t-1}(w) \cdot W_2), \tag{9.1}$$

where W_1 and W_2 are parameter matrices in $\mathbb{R}^{d \times d}$, which are shared across layers, and $\sigma(\cdot)$ is a component-wise non-linearity. The above layer is evaluated on standard, small-scale benchmark datasets (Kersting et al, 2016) showing good performance, similar to classical kernel approaches. Lei et al (2017a) proposed a similar layer and showed a connection to kernel approaches by deriving the corresponding kernel space of the learned graph embeddings.

To explicitly support edge labels, e.g., chemical bonds, Simonovsky and Komodakis (2017) introduced *Edge-Conditioned Convolution*, where a feature for node v is represented as

$$\mathbf{f}^t(v) = \frac{1}{|N(v)|} \sum_{w \in N(v)} F^l(l(w,v), W(l)) \cdot \mathbf{f}^{t-1}(w) + \mathbf{b}^l.$$

Here $l(w,v)$ is the feature (or label) of the edge shared by the nodes v and w. Moreover, $F^l: \mathbb{R}^s \to \mathbb{R}^{d_t \times d_{t-1}}$ is a function, where s denotes the number of components of the edge features and d_t and d_{t-1} denotes the number of components of the features of layer t and $(t-1)$, respectively, mapping the edge feature to a matrix in $\mathbb{R}^{d_t \times d_{t-1}}$. Further, the function F^l is parameterized by the matrix W, conditioned on the edge feature l. Finally, $\mathbf{b}^l$ is a bias term, again conditioned on the edge feature l. The above layer is applied to graph classification tasks on small-scale, standard benchmark datasets (Kersting et al, 2016), and point cloud data from the computer vision.

Building on (Scarselli et al, 2008), Gilmer et al (2017) introduced a general *message-passing* framework, unifying most of the proposed GNN architectures sofar. Specifically, Gilmer et al (2017) replaced the inner sum defined over the neighborhood in the above equations by a general permutation-invariant, differentiable function, e.g., a neural network, and substituted the outer sum over the previous and the neighborhood feature representation, e.g., by a column-wise vector concatenation or LSTM-style update step. Thus, in full generality a new feature $\mathbf{f}^t(v)$ is computed as

$$f_{\text{merge}}^{W_1}\left(\mathbf{f}^{t-1}(v), f_{\text{aggr}}^{W_2}\left(\{\{\mathbf{f}^{t-1}(w) \mid w \in N(v)\}\}\right)\right), \tag{9.2}$$

where $f_{\text{aggr}}^{W_1}$ aggregates over the multi-set of neighborhood features and $f_{\text{merge}}^{W_2}$ merges the node's representation from step $(t-1)$ with the computed neighborhood features. Moreover, it is straighfoward to include edge features as well, e.g., by learning a combined feature representation of the node itself, the neighboring node, and the corresponding edge feature. Gilmer et al (2017) employed the above architecture for regression tasks from quantum chemistry, showing promising performance for regression targets computed by expensive numerical simulations (namely, DFT) (Wu et al, 2018; Ramakrishnan et al, 2014).

Concurrently with (Morris et al, 2020b), Xu et al (2019d) investigated the limits of currently used GNN architectures, showing that their expressiveness is bounded by the WL algorithm, a simple heuristic for the graph isomorphism problem. Specifically, they showed that there does not exist a GNN architecture that can distinguish non-isomorphic graphs that the former algorithm cannot. On the positive side, they proposed the *Graph Isomorphism Network* (GIN) layer and showed that there exists a parameter initialization such that it is as expressive as the WL algorithm. Formally, given a graph $\mathscr{G}$ the feature of node v at layer t is computed as

$$\mathbf{f}^t(v) = \mathrm{MLP}\Big((1+\varepsilon) \cdot \mathbf{f}^{t-1}(v) + \sum_{w \in N(v)} \mathbf{f}^{t-1}(w) \Big), \tag{9.3}$$

where MLP is a standard multi-layer perceptron, and ε is a learnable scalar value. Xu et al (2019d) used standard sum pooling, see below, and achieved good results on standard benchmark datasets compared to other standard GNN layers and kernel approaches Morris et al (2020a).

Xu et al (2018a) investigated how to combine local information at different distances from the target node. Concretely, they investigated different architectural design choices for achieving this, e.g., concatenation, max pooling, and LSTM-style attention, showing mild performance improvements on standard benchmark datasets. Moreover, they drew some connection to random-walk distributions.

Niepert et al (2016) studied neural architectures for graph classification by extracting local patterns from graphs. Starting from each vertex, the approach explores the vertex's k-hop neighborhood, e.g., by using a breadth-first strategy. Using a labeling algorithm, e.g., a centrality index, the vertices in this neighborhood are ordered to transform into a fixed-size vector. Afterwards, a CNN-like neural network followed by an MLP is used to perform the final graph classification. The approach is compared to graph kernel approaches on standard, small-scale benchmark datasets (Kersting et al, 2016) showing promising performance.

Corso et al (2020) investigated the effect and limits of neighborhood aggregation functions. They devised aggregation schemes based on multiple aggregators, e.g., sum, mean, minimum, maximum, and standard deviation, together with so-called *degree scalar*, which combat negative effects due to a different number of neighbors between nodes. Specifically, they introduced the scalar

$$S(d, \alpha) = \left(\frac{\log(d+1)}{\delta} \right)^{\alpha}, \, d > 0, \alpha \in [-1, 1],$$

where

$$\delta = \frac{1}{|\text{train}|} \sum_{i \in \text{train}} \log(d_i + 1),$$

and α is a variable parameter. Here, the set train contains all nodes i in the training set and d_i denotes its degree, resulting in the aggregation function

$$\bigoplus = \underbrace{\begin{bmatrix} I \\ S(D, \alpha = 1) \\ S(D, \alpha = -1) \end{bmatrix}}_{\text{scalers}} \otimes \underbrace{\begin{bmatrix} \mu \\ \sigma \\ \max \\ \min \end{bmatrix}}_{\text{aggregators}}.$$

where $\otimes$ denotes the tensor product. The authors report promising performance over standard aggregation functions on a wide range of standard benchmark datasets, improving over some standard GNN layers.

Vignac et al (2020b) extended the expressivity of GNNs, see also Section 9.4, by using unique node identifiers, generalizing the message-passing scheme proposed by (Gilmer et al, 2017), see Equation (9.2), by computing and passing matrix features instead of vector features. Formally, each node i maintains a matrix U_i in $\mathbb{R}^{n \times c}$, denoted *local context*, where the j-th row contains the vectorial representation of node j of node i. At initialization, each local context U_i is set to $\mathbf{1}$ in $\mathbb{R}^{n \times 1}$, where n denotes the number of nodes in the given graph. Now at each layer l, similar to the above message-passing framework, the local context is updated as

$$U_i^{(l+1)} = u^{(l)}\left(U_i^{(l)}, \tilde{U}_i^{(l)}\right) \in \mathbb{R}^{n \times c_{l+1}} \quad \text{with} \quad \tilde{U}_i^{(l)} = \phi\left(\left\{m^{(l)}(U_i^{(l)}, U_j^{(l)}, y_{ij})\right\}_{j \in N(i)}\right),$$

where $u^{(l)}, m^{(l)}$, and ϕ are update, message, and aggregation functions, respectively, to compute the updated local context, and y_{ij} denotes the edge features shared by node i and j. Moreover, the authors study the expressive power, showing that, in principle, the above layer can distinguish any non-isomorphic pair of graphs and propose more scalable alternative variants of the above architecture. Finally, promising results on standard benchmark datasets are reported.

9.2.2 Spectral approaches

Spectral approaches apply a convolution operator in the spectral domain of the graph's Laplacian matrix, either by directly computing the former's eigendecomposition or by relying on spectral graph theory, see (Chami et al, 2020; Wang et al, 2018a) for more details. Moreover, they have a solid mathematical foundation stemming from signal processing, see, e.g., (Sandryhaila and Moura, 2013; Shuman et al, 2013).

Formally, let $\mathscr{G}$ be an undirected graph on n nodes with adjacency matrix A, then the *graph Laplacian*

$$L = \mathbf{I} - D^{-\frac{1}{2}} A D^{-\frac{1}{2}}$$

of the graph $\mathscr{G}$, where D is the diagonal matrix of node degrees, i.e., $D_{i,i} = \sum_j (A_{i,j})$. Since the graph Laplacian is positive semi-definite, we can factor it as

$$L = U \Lambda U^{\top},$$

where $U = [\mathbf{u_1}, \ldots, \mathbf{u_n}]$ in $\mathbb{R}^{n \times n}$ denotes the matrix of eigenvectors, sorted according to their eigenvalues. Further, the matrix Λ is a diagonal matrix with $\Lambda_{i,i} = \lambda_i$, where λ_i denotes the ith eigenvalue. Let $\mathbf{x}$ in $\mathbb{R}^n$ be a *graph signal*, i.e., a node feature, then the *graph Fourier transform* and its *inverse* for $\mathbf{x}$ is

$$F(\mathbf{x}) = U^{\top} \mathbf{x} \quad \text{and} \quad F^{-1}(\hat{\mathbf{x}}) = U \mathbf{x},$$

respectively, where $\hat{\mathbf{x}} = F(\mathbf{x})$. Hence, formally, the graph Fourier transform is an orthonormal (linear) transform to the space spanned by the basis of the eigenvectors in U; consequently, each element $\mathbf{x} = \sum_i \hat{\mathbf{x}}_i \cdot \mathbf{u}_i$.

Based on this observation, spectrum-based methods generalize convolution (e.g., on grids) to graphs. Thereto, they learn a *convolution filter g*. Formally, this can be expressed as follows:

$$\mathbf{x} * g = U(U^{\top} \mathbf{x} \odot U^{\top} g) = U \cdot \text{diag}(U^{\top} g) \cdot U^{\top} \mathbf{x},$$

where the operator $\cdot$ denotes the elementwise product. If we set $g_\theta = \text{diag}(U^{\top} g)$, the above can be expressed as

$$\mathbf{x} * g_\theta = U g_\theta U^{\top} \mathbf{x}.$$

Then most spectral approaches differ in their implementation of the operator g_θ.

For example, *Spectral Convolutional Neural Networks* (Bruna et al, 2014) set $g_\theta = \Theta_{i,j}^t$, which is a set of learnable parameters. Based on this, they proposed the following spectral GNN layer:

$$H_{\cdot,j}^t = \sigma \left(\sum_{i=1}^{t} U \Theta_{i,j}^t U^{\top} H_{\cdot,i}^{t-1} \right),$$

for j in $\{1, 2, \ldots, t\}$. Here, t is the layer index, H^{t-1} in $\mathbb{R}^{n \times (t-1)}$ is the graph signal, where $H^0 = X$, i.e., the given graph features, and $\Theta_{i,j}^t$ is a diagonal parameter matrix. However, the above layer suffers from a number of drawbacks: The bases of the eigenvectors is not permution invariant, the layer cannot be applied to a graph with a different structure, and the computation of the eigendecomposition is cubic in the number of nodes. Hence, Henaff et al (2015) proposed more scalable variants of the above layer by building on a smoothness notion in the spectral domain, which reduces the numbers of parameters and acts as a regulizer.

To further make the above layer more scalable, Defferrard et al (2016) introduced *Chebyshev Spectral CNNs*, which approximates g_θ by a Chebyshev expansion (Hammond et al, 2011). Namely, they express

$$g_\theta = \sum_{i=0}^{K} \theta_i T_i(\hat{\Lambda}),$$

where $\hat{\Lambda} = 2\Lambda/\lambda_{\max} - \mathbf{I}$, and $\lambda_{\max}$ denotes the largest eigenvalue of the normalized Laplacian $\hat{\Lambda}$. The normalization ensures that the eigenvalues of the Laplacian are in the $[-1, 1]$ real interval, which is required by Chebyshev polynomials. Here, T_i denotes the ith Chebyshev polynomial with $T_1(x) = x$. Alternatively, Levie et al (2019) used Caley polynomials, and show that Chebyshev Spectral CNNs are a special case.

Kipf and Welling (2017b) proposed to make Chebyshev Spectral CNNs more scalable by setting

$$\mathbf{x} * g_\theta = \theta_0 \mathbf{x} - \theta_1 D^{-\frac{1}{2}} A D^{-\frac{1}{2}} \mathbf{x}.$$

Further, they improved the generalization ability of the resulting layer by setting $\theta = \theta_0 = -\theta_1$, resulting in

$$\mathbf{x} * g_\theta = \theta(\mathbf{I} + D^{-\frac{1}{2}} A D^{-\frac{1}{2}}) \mathbf{x}.$$

In fact, the above layer can be understood as a spatial GNN, i.e., it is equivalent to computing a feature

$$\mathbf{f}^t(v) = \sigma \left(\sum_{w \in N(v) \cup v} \frac{1}{\sqrt{d_v d_w}} \mathbf{f}^{t-1}(w) \cdot W \right),$$

for node v in the given graph $\mathscr{G}$, where d_v and d_w denote the degrees of node v and w, respectively. Although the above layer was originally proposed for semi-supervised node classification, it is now one of the most widely used ones and has been applied for tasks such as matrix completion (van den Berg et al, 2018), link prediction (Schlichtkrull et al, 2018), and also as a baseline for graph classification (Ying et al, 2018c).

9.3 Pooling layers: Learning graph-level outputs from node-level outputs

Since GNNs learn vectorial node representations, using them for graph classification requires a pooling layer, enabling going from node to graph-level output. Formally, a pooling layer is a parameterized function that maps a multiset of vectors, i.e., learned node-level representations, to a single vector, i.e., the graph-level representation. Arguably, the simplest of such layers are *sum*, *mean*, and *min* or *max pooling*. That is, given a graph $\mathscr{G}$ and a multiset

$$M = \{\!\{\mathbf{f}(v) \in \mathbb{R}^d \mid v \in \mathscr{V}\}\!\}$$

of node-level representations of nodes in the graph $\mathscr{G}$, sum pooling computes

$$f_{\text{pool}}(\mathscr{G}) = \sum_{\mathbf{f}(v) \in M} \mathbf{f}(v),$$

while mean, min, max pooling take the (component-wise) average, minimum, maximum over the elements in M, respectively. These four simple pooling layers are still used in many published GNN architectures, e.g., see (Duvenaud et al, 2015b). In fact, recent work (Mesquita et al, 2020) showed that more sophisticated layers, e.g., relying on clustering, see below, do not offer any empirical benefits on many real-world datasets, especially those from the molecular domain.

9.3.1 Attention-based pooling layers

Simple attention-based pooling became popular in recent years due to its easy implementation and scalability compared to more sophisticated alternatives; see below. For example, Gilmer et al (2017), see above, used a *seq2seq* architecture for sets (Vinyals et al, 2016) for pooling purposes in their empirical study. Focusing on pooling for GNNs, Lee et al (2019b) introduced the *SAGPool* layer, short for Self-Attention Graph Pooling method for GNNs, using self-attention. Specifically, they computed a *self-attention score* by multiplying the aggregated features of an arbitrary GNN layer by a matrix Θ_{att} in $\mathbb{R}^{d \times 1}$, where d denotes the number of components of the node features. For example, computing the self-attention score $\mathbf{Z}(v)$ for the simple layer of Equation (9.1) equates to

$$\mathbf{Z}(v) = \sigma \left(\mathbf{f}^{t-1}(v) \cdot W_1 + \sum_{w \in N(v)} \mathbf{f}^{t-1}(w) \cdot W_2 \right) \cdot \Theta_{att}.$$

The self-attention score $\mathbf{Z}(v)$ is subsequently used to select the top-k nodes in the graph; similarly, to Cangea et al (2018) and (Gao et al, 2018a), see below, omitting the other nodes, effectively pruning nodes from the graph. Similar attention-based techniques are proposed in (Huang et al, 2019).

9.3.2 Cluster-based pooling layers

The idea of cluster-based pooling layers is to coarsen the graph, i.e., merging similar nodes iteratively. One of the earliest uses has been proposed in (Simonovsky and Komodakis, 2017), see above, where the *Graclus* clustering algorithm (Dhillon et al, 2007) is used. However, one has the note that the algorithm is parameter-free, i.e., it does adapt to the learning task at hand.

The arguably most well-known cluster-based pooling layer is *DiffPool* (Ying et al, 2018c). The idea of DiffPool is to iteratively coarsen the graph by learning a soft clustering of nodes, making the otherwise discrete clustering assignment differentiable. Concretely, at layer t, DiffPool learns a soft cluster assigment S in $[0, 1]^{n_t \times n_{t+1}}$, where n_t and n_{t+1} are the number of nodes at layer t and $(t+1)$, respectively. Each entry $S_{i,j}$ represents the probablity of node i of layer t being clustered into node j of layer $(t+1)$. In each iteration, the matrix S is computed by

$$S = \text{softmax}(\text{GNN}(A_t, F_t)),$$

where A_t and F_t are the adjacency matrix and the feature matrix of the clustered graph at layer t, and the function GNN is an abitrary GNN layer. Finally, in each layer, the adjacency matrix and the feature matrix are updated as

$$A_{t+1} = S^T A_t S \quad \text{and} \quad F_{t+1} = S^T F_t,$$

respectively.

Empirically, the authors show that the DiffPool layer boosts standard GNN layers' performance, e.g., GraphSage (Hamilton et al, 2017b), on standard, small-scale benchmark datasets (Morris et al, 2020a). The downside of the above layer is the added computational cost. The adjacency matrix becomes dense and real-valued after the first pooling layer, leading to a quadratic cost in the number of nodes for each GNN layer's computation. Moreover, the number of clusters has to be chosen in advance, leading to an increase in hyperparameters.

9.3.3 Other pooling layers

Zhang et al (2018g) proposed a pooling layer based on differentiable sorting, denoted *SortPooling*. That is, given the feature matrix F_t of row-wise node features after layer t, SortPooling sorts the rows of F_t in a descending fashion. It truncates the last $n - k$ rows of F_t, or pads with zero rows if $n < k$ for a given graph to unify the graphs' size. Formally, the layer can be written down as

$$F = \text{sort}(F_t) \quad \text{followed by} \quad F_{\text{trunc}} = \text{truncate}(F, k),$$

where the function sort sorts the feature matrix F_t row-wise in a descending fashion, and the functions truncate return the first k of the input matrix. Ties are broken up using the features from previous layers, 1 to $(t-1)$. The resulting tensor F_{trunc} of shape $k \times \sum_{i=1}^{h} d_i$, where d_i denotes the number of features of the ith layer and h the total number of layers, is reshaped into a tensor of size $k(\sum_{i=1}^{h} d_i) \times 1$, row-wise, followed by a standard 1-D convolution with a filter and step size of $\sum_{i=1}^{h} d_i$. Finally, a sequence of max-pooling and 1-D convolutions are applied to identifiy local patterns in the sequence.

Similarly, to combat the high computational cost of some pooling layer, e.g., DiffPool, Cangea et al (2018) introduced a pooling layer dropping $n - \lceil nk \rceil$ nodes of a graph with n nodes in each layer for k in $[0, 1)$. The nodes to be dropped are choosen according to a *projection score* against a learnable vector $\mathbf{p}$. Concretly, they compute the score vector

$$\mathbf{y} = \frac{F_t \cdot \mathbf{p}}{\|\mathbf{p}\|} \quad \text{and} \quad I = \text{top-}k(\mathbf{y}, k),$$

where top-k returns top-k indices from a given vector according to $\mathbf{y}$. Finally, the adjacency A_{t+1} is updated by removing rows and columns that are not in I, while the updated feature matrix

$$F_{t+1} = (F_t \odot \tanh(\mathbf{y})).$$

The authors report slightly lower classification accuracies than the DiffPool layer on most employed datasets while being much faster in computation time. A similar approach was presented in (Gao and Ji, 2019).

To derive more expressive graph representations, Murphy et al (2019c,b) propose *relational pooling*. To increase the expressive power of GNN layers, they average over all permutations of a given graph. Formally, let $\mathscr{G}$ be a graph, then a representation

$$\mathbf{f}(\mathscr{G}) = \frac{1}{|\mathscr{V}|} \sum_{\pi \in \Pi} g(A_{\pi,\pi}, [F_\pi, I_{|V|}]) \tag{9.4}$$

is learned, where Π denotes all possible permutations of the rows and columns of the adjacency matrix of $\mathscr{G}$, g is a permutation-invariant function, and $[\cdot, \cdot]$ denotes column-wise matrix concatenation. Moreover. $A_{\pi,\pi}$ permutes the rows and columns of the adjaceny matrix A according to the permutation π in Π, similarly F_π permutes the rows of the feature matrix F. The author showed that the above architecture is more expressive in terms of distinguishing non-isomorphic graphs than the WL algorithm, and proposed sampling-based techniques to speed up the computation.

Bianchi et al (2020) introduced a pooling layer based on spectral clustering (VON-LUXBURG, 2007). Thereto, they train a GNN together with an MLP, followed by a softmax function, against an approximation of a relaxed version of the k-way normalized Min-cut problem (Shi and Malik, 2000). The resulting cluster assignment matrix S is used in the same way as in Section 9.3.2. The authors evaluated their approach on standard, small-scale benchmark datasets showing promising performance, especially over the DiffPool layer. For another pooling layer based on spectral clustering, see (Ma et al, 2019d).

9.4 Limitations of graph neural networks and higher-order layers for graph classification

In the following, we briefly survey the limitations of GNNs and how their expressive power is upper-bounded by the Weisfeiler-Leman method (Weisfeiler and Leman, 1968; Weisfeiler, 1976; Grohe, 2017). Concretely, a recent line of works by Morris et al (2020b); Xu et al (2019d); Maron et al (2019a) connects the power or expressivity of GNNs to that of the WL algorithm. The results show that GNN architectures generally do not have more power to distinguish between non-isomorphic graphs than the WL. That is, for any graph structure that the WL algorithm cannot distinguish, any possible GNN with any possible choices of parameters will also not be able to distinguish it. On the positive side, the second result states that there is a sequence of parameter initializations such that GNNs have the same power in distinguishing non-isomorphic (sub-)graphs as the WL algorithm, see also Equation (9.3). However, the WL algorithm has many short-comings, see (Arvind et al, 2015; Kiefer et al, 2015), e.g., it cannot distinguish between cycles of different lengths, an important property for chemical molecules, and is not able to distinguish between graphs with different triangle counts, an important property of social networks.

To address this, many recent works tried to build provable more expressive GNNs for graph classification. For example, in (Morris et al, 2020b; Maron et al, 2019b, 2018) the authors proposed *higher-order GNN architectures* that have the same expressive power as the *k-dimensional Weisfeiler-Leman algorithm* (*k*-WL), which is, as *k* grows, a more expressive generalization of the WL algorithm. In the following, we give an overview of such works.

9.4.1 Overcoming limitations

The first GNN architecture that overcame the limitations of the WL algorithm was proposed in (Morris et al, 2020b). Specifically, they introduced so-called *k-GNNs*, which work by learning features over the set of subgraphs on *k* nodes instead of vertices by defining a notion of neighborhood between these subgraphs. Formally, for a given *k*, they consider all *k*-element subsets $[\mathcal{V}]^k$ over $\mathcal{V}$. Let $s = \{s_1, \ldots, s_k\}$ be a *k*-set in $[\mathcal{V}]^k$, then they define the *neighborhood* of *s* as

$$N(s) = \{t \in [\mathcal{V}]^k \mid |s \cap t| = k - 1\}.$$

The *local neighborhood* $N_L(s)$ consists of all *t* in $N(s)$ such that (v, w) in $\mathcal{E}$ for the unique $v \in s \setminus t$ and the unique $w \in t \setminus s$. The *global neighborhood* $N_G(s)$ then is defined as $N(s) \setminus N_L(s)$.

Based on this neighborhood definition, one can generalize most GNN layers for vertex embeddings to more expressive subgraph embeddings. Given a graph $\mathcal{G}$, a feature for a subgraph *s* can be computed as

$$\mathbf{f}_k^t(s) = \sigma\left(\mathbf{f}_k^{t-1}(s) \cdot W_1^t + \sum_{u \in N_L(s) \cup N_G(s)} \mathbf{f}_k^{t-1}(u) \cdot W_2^t\right). \tag{9.5}$$

The authors resort to sum over the local neighborhood in the experiments for better scalability and generalization, showing a significant boost over standard GNNs on a quantum chemistry benchmark dataset (Wu et al, 2018; Ramakrishnan et al, 2014).

The latter approach was refined in (Maron et al, 2019a) and (Morris et al, 2019). Specifically, based on (Maron et al, 2018), Maron et al (2019a) derived an architecture based on standard matrix multiplication that has at least the same power as the 3-WL. Morris et al (2019) proposed a variant of the *k*-WL that, unlike the original algorithm, takes the sparsity of the underlying graph into account. Moreover, they showed that the derived sparse variant is slightly more powerful than the k-WL in distinguishing non-isomorphic graphs and proposed a neural architecture with the same power as the sparse k-WL variant.

An important direction in studying graph representations' expressive power was taken by (Chen et al, 2019f). The authors prove that a graph representation can approximate a function f if and only if it can distinguish all pairs of non-isomorphic graphs $\mathscr{G}$ and $\mathscr{H}$ where $f(\mathscr{G}) \neq f(\mathscr{H})$. With that in mind, they established an equivalence between the set of pairs of graphs a representation can distinguish and the space of functions it can approximate, further introducing a variation of the 2-WL.

Bouritsas et al (2020) enhanced the expressivity of GNNs by annotating node features with subgraph information. Specifically, by fixing a set of predefined, small subgraphs, they annotated each node with their role, formally their automorphism type, in these subgraphs, showing promising performance gains on standard benchmark datasets for graph classification.

Beaini et al (2020) studied how to incorporate directional information into GNNs. Finally, You et al (2021) enhanced GNNs by uniquely coloring central vertices and used two types of message functions to surpass the expressive power of the 1-WL, while Sato et al (2021) and Abboud et al (2020) use random features to achieve the same goal and additionally studied the universality properties of their derived architectures.

9.5 Applications of graph neural networks for graph classification

In the following, we highlight some application areas of GNNs for graph classification, focusing on the molecular domain. One of the most promising applications of GNNs for graph classification is pharmaceutical drug research, see (Gaudelet et al, 2020) for an overview. In this direction, a promising approach was proposed by (Stokes et al, 2020). They used a form of directed message passing neural networks operating on molecular graphs to identify repurposing candidates for antibiotic development. Moreover, they validated their predictions in vivo, proposing suitable repurposing candidates different from know ones.

Schweidtmann et al (2020) used 2-GNNs, see Equation (9.5), to derive GNN models for predicting three fuel ignition quality indicators such as the derived cetane number, the research octane number,and the motor octane number of oxygenated and non-oxygenated hydrocarbons, indicating that the higher-order layers of Equation (9.5) provide significant gains over standard GNNs in the molecular learning domain.

A general principled GNN for the molecular domain, denoted *DimeNet*, was introduced by (Klicpera et al, 2020). By using an edge-based architecture, they computed a message coefficient between atoms based on their relative positioning in 3D space. Concretely, an incoming message to a node is based on the sender's incoming meassage as well as the distance between the atoms and the angles of their atomic bonds. By using this additional information the authors report significant improvements over state-of-the-art GNN models in molecular property prediction tasks .

9.6 Benchmark Datasets

Since most developments for GNNs are driven empirically, i.e., based on evaluations on standard benchmark datasets, meaningful benchmark datasets are crucial for the development of GNNs in the context of graph classification. Hence, the research community has established several widely used repositories for benchmark datasets for graph classification. Two such repositories are worth being highlighted here. First, the *TUDataset* (Morris et al, 2020a) collection contains over 130 datasets provided at `www.graphlearning.io` of various sizes and various areas such as chemistry, biology, and social networks. Moreover, it provides Python-based data loaders and baseline implementations of standards graph kernel and GNNs. Moreover, the datasets can be easily accessed from well-known GNN implementation frameworks such as *Deep Graph Library* (Wang et al, 2019f), *PyTorch Geometric* (Fey and Lenssen, 2019), or *Spektral* (Grattarola and Alippi, 2020). Secondly, the *OGB* (Weihua Hu, 2020) collections contain many large-scale graph classification benchmark datasets, e.g., from chemistry and code analysis with data loaders, prespecified splits, and evaluation protocols. Finally, Wu et al (2018) also provides many large-scale datasets from chemo- and bioinformatics.

9.7 Summary

We gave an overview of GNNs for graph classification. We surveyed classical and modern works in this area, distinguishing between spatial and spectral approaches. Since GNNs compute node-level representations, pooling layers for learning graph-level representations is crucial for successful graph classification. Hence, we surveyed pooling layers based on attention, clustering, and other approaches to pooling. Moreover, we gave an overview of the limitations of GNNs for graph classification and surveyed architectures to overcome these limitations. Finally, we gave an overview of applications of GNNs and benchmark datasets for their evaluation.

Editor's Notes: The success of using GNNs in classification tasks is owing to advanced representation learning (chapter 2) by expressive power of GNNs (chapter 5). And its performance is limited by the scalability (chapter 6), robustness (chapter 8) and transformation capability (chapter 12) of algorithm. As one of the most prominent tasks, one can always face classification in a variety of GNN topic. For example, node classification helps to evaluate performance of AutoML (chapter17) and self-supervised learning (chapter 18) methods of GNNs, graph classification can be token as subpart of adversarial learning in graph generation (chapter 11). Further, there are many promising applications of GNNs in classification, node or edge based ones like urban intelligence (chapter 27), graph based ones like protein and drug prediction (chapter 25).

Chapter 10
Graph Neural Networks: Link Prediction

Muhan Zhang

Abstract Link prediction is an important application of graph neural networks. By predicting missing or future links between pairs of nodes, link prediction is widely used in social networks, citation networks, biological networks, recommender systems, and security, etc. Traditional link prediction methods rely on heuristic node similarity scores, latent embeddings of nodes, or explicit node features. Graph neural network (GNN), as a powerful tool for jointly learning from graph structure and node/edge features, has gradually shown its advantages over traditional methods for link prediction. In this chapter, we discuss GNNs for link prediction. We first introduce the link prediction problem and review traditional link prediction methods. Then, we introduce two popular GNN-based link prediction paradigms, node-based and subgraph-based approaches, and discuss their differences in link representation power. Finally, we review recent theoretical advancements on GNN-based link prediction and provide several future directions.

10.1 Introduction

Link prediction is the problem of predicting the existence of a link between two nodes in a network (Liben-Nowell and Kleinberg, 2007). Given the ubiquitous existence of networks, it has many applications such as friend recommendation in social networks (Adamic and Adar, 2003), co-authorship prediction in citation networks (Shibata et al, 2012), movie recommendation in Netflix (Bennett et al, 2007), protein interaction prediction in biological networks (Qi et al, 2006), drug response prediction (Stanfield et al, 2017), metabolic network reconstruction (Oyetunde et al, 2017), hidden terrorist group identification (Al Hasan and Zaki, 2011), knowledge graph completion (Nickel et al, 2016a), etc.

Muhan Zhang
Institute for Artificial Intelligence, Peking University, e-mail: muhan@pku.edu.cn

© The Author(s), under exclusive license to Springer Nature Singapore Pte Ltd. 2022   195
L. Wu et al. (eds.), *Graph Neural Networks: Foundations, Frontiers, and Applications*,
https://doi.org/10.1007/978-981-16-6054-2_10

Link prediction has many names in different application domains. The term "link prediction" often refers to predicting links in homogeneous graphs, where nodes and links both only have a single type. This is the simplest setting and most link prediction works focus on this setting. Link prediction in bipartite user-item networks is referred to as matrix completion or recommender systems, where nodes have two types (user and item) and links can have multiple types corresponding to different ratings users can give to items. Link prediction in knowledge graphs is often referred to as knowledge graph completion, where each node is a distinct entity and links have multiple types corresponding to different relations between entities. In most cases, a link prediction algorithm designed for the homogeneous graph setting can be easily generalized to heterogeneous graphs (e.g., bipartite graphs and knowledge graphs) by considering heterogeneous node type and relation type information.

There are mainly three types of traditional link prediction methods: heuristic methods, latent-feature methods, and content-based methods. Heuristic methods compute heuristic node similarity scores as the likelihood of links (Liben-Nowell and Kleinberg, 2007). Popular ones include common neighbors (Liben-Nowell and Kleinberg, 2007), Adamic-Adar (Adamic and Adar, 2003), preferential attachment (Barabási and Albert, 1999), and Katz index (Katz, 1953). Latent-feature methods factorize the matrix representations of a network to learn low-dimensional latent representations/embeddings of nodes. Popular network embedding techniques such as DeepWalk (Perozzi et al, 2014), LINE (Tang et al, 2015b) and node2vec (Grover and Leskovec, 2016), are also latent-feature methods because they implicitly factorize some matrix representations of networks too (Qiu et al, 2018). Both heuristic methods and latent-feature methods infer future/missing links leveraging the existing network topology. Content-based methods, on the contrary, leverage explicit node attributes/features rather than the graph structure (Lops et al, 2011). It is shown that combining the graph topology with explicit node features can improve the link prediction performance (Zhao et al, 2017).

By learning from graph topology and node/edge features in a unified way, graph neural networks (GNNs) recently show superior link prediction performance than traditional methods (Kipf and Welling, 2016; Zhang and Chen, 2018b; You et al, 2019; Chami et al, 2019; Li et al, 2020e). There are two popular GNN-based link prediction paradigms: node-based and subgraph-based. Node-based methods first learn a node representation through a GNN, and then aggregate the pairwise node representations as link representations for link prediction. An example is (Variational) Graph AutoEncoder (Kipf and Welling, 2016). Subgraph-based methods first extract a local subgraph around each target link, and then apply a graph-level GNN (with pooling) to each subgraph to learn a subgraph representation, which is used as the target link representation for link prediction. An example is SEAL (Zhang and Chen, 2018b). We introduce these two types of methods separately in Section 10.3.1 and 10.3.2, and discuss their expressive power differences in Section 10.3.3.

To understand GNNs' power for link prediction, several theoretical efforts have been made. The γ-decaying theory (Zhang and Chen, 2018b) unifies existing link prediction heuristics into a single framework and proves their local approximability, which justifies using GNNs to "learn" heuristics from the graph structure instead of using predefined ones. The theoretical analysis of labeling trick (Zhang et al, 2020c) proves that subgraph-based approaches have a higher link representation power than node-based approaches by being able to learn most expressive structural representations of links (Srinivasan and Ribeiro, 2020b) where node-based approaches always fail. We introduce these theories in Section 20.3.

Finally, by analyzing limitations of existing methods, we provide several future directions on GNN-based link prediction in Section 20.4.

10.2 Traditional Link Prediction Methods

In this section, we review traditional link prediction methods. They can be categorized into three classes: heuristic methods, latent-feature methods, and content-based methods.

10.2.1 Heuristic Methods

Heuristic methods use simple yet effective node similarity scores as the likelihood of links (Liben-Nowell and Kleinberg, 2007; Lü and Zhou, 2011). We use x and y to denote the source and target node between which to predict a link. We use $\Gamma(x)$ to denote the set of x's neighbors.

10.2.1.1 Local Heuristics

One simplest heuristic is called **common neighbors** (CN), which counts the number of neighbors two nodes share as a measurement of their likelihood of having a link:

$$f_{\text{CN}}(x,y) = |\Gamma(x) \cap \Gamma(y)|. \tag{10.1}$$

CN is widely used in social network friend recommendation. It assumes that the more common friends two people have, the more likely they themselves are also friends.

Jaccard score measures the proportion of common neighbors instead:

$$f_{\text{Jaccard}}(x,y) = \frac{|\Gamma(x) \cap \Gamma(y)|}{|\Gamma(x) \cup \Gamma(y)|}. \tag{10.2}$$

There is also a famous **preferential attachment** (PA) heuristic (Barabási and Albert, 1999), which uses the product of node degrees to measure the link likelihood:

$$f_{\text{PA}}(x,y) = |\Gamma(x)| \cdot |\Gamma(y)|. \tag{10.3}$$

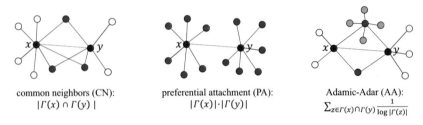

Fig. 10.1: Illustration of three link prediction heuristics: CN, PA and AA.

PA assumes x is more likely to connect to y if y has a high degree. For example, in citation networks, a new paper is more likely to cite those papers which already have a lot of citations. Networks formed by the PA mechanism are called scale-free networks (Barabási and Albert, 1999), which are important subjects in network science.

Existing heuristics can be categorized based on the maximum hop of neighbors needed to calculate the score. CN, Jaccard, and PA are all *first-order heuristics*, because they only involve one-hop neighbors of two target nodes. Next we introduce two *second-order heuristics*.

The **Adamic-Adar** (AA) heuristic (Adamic and Adar, 2003) considers the weight of common neighbors:

$$f_{AA}(x,y) = \sum_{z \in \Gamma(x) \cap \Gamma(y)} \frac{1}{\log |\Gamma(z)|}, \tag{10.4}$$

where a high-degree common neighbor z is weighted less (down-weighted by the reciprocal of $\log |\Gamma(z)|$). The assumption is that a high degree node connecting to both x and y is less informative than a low-degree node.

Resource allocation (RA) (Zhou et al, 2009) uses a more aggressive down-weighting factor:

$$f_{RA}(x,y) = \sum_{z \in \Gamma(x) \cap \Gamma(y)} \frac{1}{|\Gamma(z)|}, \tag{10.5}$$

thus, it favors low-degree common neighbors more.

Both AA and RA are second-order heuristics, as up to two hops of neighbors of x and y are required to compute the score. Both first-order and second-order heuristics are local heuristics, as they can all be computed from a local subgraph around the target link without the need to know the entire network. We illustrate three local heuristics, CN, PA, and AA, in Figure 10.1.

10.2.1.2 Global Heuristics

There are also **high-order heuristics** which require knowing the entire network. Examples include Katz index (Katz, 1953), rooted PageRank (RPR) (Brin and Page, 2012), and SimRank (SR) (Jeh and Widom, 2002).

Katz index uses a weighted sum of all the walks between x and y where a longer walk is discounted more:

$$f_{\text{Katz}}(x,y) = \sum_{l=1}^{\infty} \beta^l |\text{walks}^{\langle l \rangle}(x,y)|. \tag{10.6}$$

Here β is a decaying factor between 0 and 1, and $|\text{walks}^{\langle l \rangle}(x,y)|$ counts the length-l walks between x and y. When we only consider length-2 walks, Katz index reduces to CN.

Rooted PageRank (RPR) is a generalization of PageRank. It first computes the stationary distribution π_x of a random walker starting from x who randomly moves to one of its current neighbors with probability α, or returns to x with probability $1 - \alpha$. Then it uses π_x at node y (denoted by $[\pi_x]_y$) to predict link (x,y). When the network is undirected, a symmetric version of rooted PageRank uses

$$f_{\text{RPR}}(x,y) = [\pi_x]_y + [\pi_y]_x \tag{10.7}$$

to predict the link.

The **SimRank** (SR) score assumes that two nodes are similar if their neighbors are also similar. It is defined in a recursive way: if $x = y$, then $f_{\text{SR}}(x,y) := 1$; otherwise,

$$f_{\text{SR}}(x,y) := \gamma \frac{\sum_{a \in \Gamma(x)} \sum_{b \in \Gamma(y)} f_{\text{SR}}(a,b)}{|\Gamma(x)| \cdot |\Gamma(y)|}, \tag{10.8}$$

where γ is a constant between 0 and 1.

High-order heuristics are global heuristics. By computing node similarity from the entire network, high-order heuristics often have better performance than first-order and second-order heuristics.

10.2.1.3 Summarization

We summarize the eight introduced heuristics in Table 10.1. For more variants of the above heuristics, please refer to (Liben-Nowell and Kleinberg, 2007; Lü and Zhou, 2011). Heuristic methods can be regarded as computing predefined **graph structure features** located in the observed node and edge structures of the network. Although effective in many domains, these handcrafted graph structure features have limited expressivity—they only capture a small subset of all possible structure patterns, and cannot express general graph structure features underlying different networks. Besides, heuristic methods only work well when the network formation mechanism aligns with the heuristic. There may exist networks with complex formation mechanisms which no existing heuristics can capture well. Most heuristics only work for homogeneous graphs.

Table 10.1: Popular heuristics for link prediction

Name	Formula	Order
common neighbors	$\|\Gamma(x) \cap \Gamma(y)\|$	first
Jaccard	$\frac{\|\Gamma(x) \cap \Gamma(y)\|}{\|\Gamma(x) \cup \Gamma(y)\|}$	first
preferential attachment	$\|\Gamma(x)\| \cdot \|\Gamma(y)\|$	first
Adamic-Adar	$\sum_{z \in \Gamma(x) \cap \Gamma(y)} \frac{1}{\log \|\Gamma(z)\|}$	second
resource allocation	$\sum_{z \in \Gamma(x) \cap \Gamma(y)} \frac{1}{\|\Gamma(z)\|}$	second
Katz	$\sum_{l=1}^{\infty} \beta^l \|\text{walks}^{\langle l \rangle}(x,y)\|$	high
rooted PageRank	$[\pi_x]_y + [\pi_y]_x$	high
SimRank	$\gamma \frac{\sum_{a \in \Gamma(x)} \sum_{b \in \Gamma(y)} \text{score}(a,b)}{\|\Gamma(x)\| \cdot \|\Gamma(y)\|}$	high

Notes: $\Gamma(x)$ denotes the neighbor set of vertex x. $\beta < 1$ is a damping factor. $\|\text{walks}^{\langle l \rangle}(x,y)\|$ counts the number of length-l walks between x and y. $[\pi_x]_y$ is the stationary distribution probability of y under the random walk from x with restart, see (Brin and Page, 2012). SimRank score uses a recursive definition.

10.2.2 Latent-Feature Methods

The second class of traditional link prediction methods is called latent-feature methods. In some literature, they are also called latent-factor models or embedding methods. Latent-feature methods compute latent properties or representations of nodes, often obtained by factorizing a specific matrix derived from the network, such as the adjacency matrix and the Laplacian matrix. These latent features of nodes are not explicitly observable—they must be computed from the network through optimization. Latent features are also not interpretable. That is, unlike explicit node features where each feature dimension represents a specific property of nodes, we do not know what each latent feature dimension describes.

10.2.2.1 Matrix Factorization

One most popular latent feature method is matrix factorization (Koren et al, 2009; Ahmed et al, 2013), which is originated from the recommender systems literature. Matrix factorization factorizes the observed adjacency matrix A of the network into the product of a low-rank latent-embedding matrix Z and its transpose. That is, it approximately reconstructs the edge between i and j using their k-dimensional latent embeddings $\mathbf{z}_i$ and $\mathbf{z}_j$:

$$\hat{A}_{i,j} = \mathbf{z}_i^\top \mathbf{z}_j, \tag{10.9}$$

It then minimizes the mean-squared error between the reconstructed adjacency matrix and the true adjacency matrix over the observed edges to learn the latent embeddings:

$$\mathcal{L} = \frac{1}{|\mathcal{E}|} \sum_{(i,j)\in\mathcal{E}} (A_{i,j} - \hat{A}_{i,j})^2. \tag{10.10}$$

Finally, we can predict new links by the inner product between nodes' latent embeddings. Variants of matrix factorization include using powers of A (Cangea et al, 2018) and using general node similarity matrices (Ou et al, 2016) to replace the original adjacency matrix A. If we replace A with the Laplacian matrix L and define the loss as follows:

$$\mathcal{L} = \sum_{(i,j)\in\mathcal{E}} \|\mathbf{z}_i - \mathbf{z}_j\|_2^2, \tag{10.11}$$

then the nontrivial solution to the above are constructed by the eigenvectors corresponding to the k smallest nonzero eigenvalues of L, which recovers the Laplacian eigenmap technique (Belkin and Niyogi, 2002) and the solution to spectral clustering (VONLUXBURG, 2007).

10.2.2.2 Network Embedding

Network embedding methods have gained great popularity in recent years since the pioneering work DeepWalk (Perozzi et al, 2014). These methods learn low-dimensional representations (embeddings) for nodes, often based on training a skip-gram model (Mikolov et al, 2013a) over random-walk-generated node sequences, so that nodes which often appear nearby each other in a random walk (i.e., nodes close in the network) will have similar representations. Then, the pairwise node embeddings are aggregated as link representations for link prediction. Although not explicitly factorizing a matrix, it is shown in (Qiu et al, 2018) that many network embedding methods, including LINE (Tang et al, 2015b), DeepWalk, and node2vec (Grover and Leskovec, 2016), implicitly factorize some matrix representations of the network. Thus, they can also be categorized into latent-feature methods. For example, DeepWalk approximately factorizes:

$$\log \left(\text{vol}(\mathcal{G}) \left(\frac{1}{w} \sum_{r=1}^{w} (D^{-1}A)^r \right) D^{-1} \right) - \log(b), \tag{10.12}$$

where $\text{vol}(\mathcal{G})$ is the sum of node degrees, D is the diagonal degree matrix, w is skip-gram's window size, and b is a constant. As we can see, DeepWalk essentially factorizes the log of some high-order normalized adjacency matrices' sum (up to w). To intuitively understand this, we can think of the random walk as extending a node's neighborhood to w hops away, so that we not only require direct neighbors to have similar embeddings, but also require nodes reachable from each other through w steps of random walk to have similar embeddings.

Similarly, the LINE algorithm (Tang et al, 2015b) in its second-order forms implicitly factorizes:

$$\log \left(\text{vol}(\mathcal{G})(D^{-1}AD^{-1}) \right) - \log(b). \qquad (10.13)$$

Another popular network embedding method, node2vec, which enhances Deep-Walk with negative sampling and biased random walk, is also shown to implicitly factorize a matrix. The matrix does not have a closed form due to the use of second-order (biased) random walk (Qiu et al, 2018).

10.2.2.3 Summarization

We can understand latent-feature methods as extracting low-dimensional node embeddings from the graph structure. Traditional matrix factorization methods use the inner product between node embeddings to predict links. However, we are actually not restricted to inner product. Instead, we can apply a neural network over an arbitrary aggregation of pairwise node embeddings to learn link representations. For example, node2vec (Grover and Leskovec, 2016) provides four symmetric aggregation functions (invariant to the order of two nodes): mean, Hadamard product, absolute difference, and squared difference. If we predict directed links, we can also use non-symmetric aggregation functions, such as concatenation.

Latent-feature methods can take global properties and long-range effects into node representations, because all node pairs are used together to optimize a single objective function, and the final embedding learned for a node can be influenced by all nodes in the same connected component during the optimization. However, latent-feature methods cannot capture structural similarities between nodes Ribeiro et al (2017), i.e., two nodes sharing identical neighborhood structures are not mapped to similar embeddings. Latent-feature methods also need an extremely large dimension to express some simple heuristics (Nickel et al, 2014), making them sometimes have worse performance than heuristic methods. Finally, latent-feature methods are transductive learning methods—the learned node embeddings cannot generalize to new nodes or new networks.

There are many latent-feature methods designed for heterogeneous graphs. For example, the RESCAL model (Nickel et al, 2011) generalizes matrix factorization to multi-relation graphs, which essentially performs a kind of tensor factorization. Metapath2vec (Dong et al, 2017) generalizes node2vec to heterogeneous graphs.

10.2.3 Content-Based Methods

Both heuristic methods and latent-feature methods face the cold-start problem. That is, when a new node joins the network, heuristic methods and latent-feature methods may not be able to predict its links accurately because it has no or only a few existing links with other nodes. In this case, content-based methods might help. Content-based methods leverage explicit content features associated with nodes for link prediction, which have wide applications in recommender systems (Lops et al, 2011). For example, in citation networks, word distributions can be used as content features for papers. In social networks, a user's profile, such as their demographic information and interests, can be used as their content features (however, their friendship information belongs to graph structure features because it is calculated from the graph structure). However, content-based methods usually have worse performance than heuristic and latent-feature methods due to not leveraging the graph structure. Thus, they are usually used together with the other two types of methods (Koren, 2008; Rendle, 2010; Zhao et al, 2017) to enhance the link prediction performance.

10.3 GNN Methods for Link Prediction

In the last section, we have covered three types of traditional link prediction methods. In this section, we will talk about GNN methods for link prediction. GNN methods combine graph structure features and content features by learning them together in a unified way, leveraging the excellent graph representation learning ability of GNNs.

There are mainly two GNN-based link prediction paradigms, node-based and subgraph-based. Node-based methods aggregate the pairwise node representations learned by a GNN as the link representation. Subgraph-based methods extract a local subgraph around each link and use the subgraph representation learned by a GNN as the link representation.

10.3.1 Node-Based Methods

The most straightforward way of using GNNs for link prediction is to treat GNNs as inductive network embedding methods which learn node embeddings from local neighborhood, and then aggregate the pairwise node embeddings of GNNs to construct link representations. We call these methods node-based methods.

10.3.1.1 Graph AutoEncoder

The pioneering work of node-based methods is Graph AutoEncoder (GAE) (Kipf
and Welling, 2016). Given the adjacency matrix A and node feature matrix X of a
graph, GAE (Kipf and Welling, 2016) first uses a GCN (Kipf and Welling, 2017b) to
compute a node representation $\mathbf{z}_i$ for each node i, and then uses $\sigma(\mathbf{z}_i^\top \mathbf{z}_j)$ to predict
link (i, j):

$$\hat{A}_{i,j} = \sigma(\mathbf{z}_i^\top \mathbf{z}_j), \quad \text{where } \mathbf{z}_i = Z_{i,:}, Z = \text{GCN}(X, A) \tag{10.14}$$

where Z is the node representation (embedding) matrix output by the GCN with
the ith row of Z being node i's representation $\mathbf{z}_i$, $\hat{A}_{i,j}$ is the predicted probability for
link (i, j) and σ is the sigmoid function. If X is not given, GAE can use the one-
hot encoding matrix I instead. The model is trained to minimize the cross entropy
between the reconstructed adjacency matrix and the true adjacency matrix:

$$\mathcal{L} = \sum_{i \in \mathcal{V}, j \in \mathcal{V}} (-A_{i,j} \log \hat{A}_{i,j} - (1 - A_{i,j}) \log(1 - \hat{A}_{i,j})). \tag{10.15}$$

In practice, the loss of positive edges ($A_{i,j} = 1$) is up-weighted by k, where k is the
ratio between negative edges ($A_{i,j} = 0$) and positive edges. The purpose is to balance
the positive and negative edges' contribution to the loss. Otherwise, the loss might
be dominated by negative edges due to the sparsity of practical networks.

10.3.1.2 Variational Graph AutoEncoder

The variational version of GAE is called VGAE, or Variational Graph AutoEncoder (Kipf and Welling, 2016). Rather than learning deterministic node embed-
dings $\mathbf{z}_i$, VGAE uses two GCNs to learn the mean $\boldsymbol{\mu}_i$ and variance σ_i^2 of $\mathbf{z}_i$, respectively.

VGAE assumes the adjacency matrix A is generated from the latent node embed-
dings Z through $p(A|Z)$, where Z follows a prior distribution $p(Z)$. Similar to GAE,
VGAE uses an inner-product-based link reconstruction model as $p(A|Z)$:

$$p(A|Z) = \prod_{i \in \mathcal{V}} \prod_{j \in \mathcal{V}} p(A_{i,j}|\mathbf{z}_i, \mathbf{z}_j), \quad \text{where } p(A_{i,j} = 1|\mathbf{z}_i, \mathbf{z}_j) = \sigma(\mathbf{z}_i^\top \mathbf{z}_j). \tag{10.16}$$

And the prior distribution $p(Z)$ takes a standard Normal distribution:

$$p(Z) = \prod_{i \in \mathcal{V}} p(\mathbf{z}_i) = \prod_{i \in \mathcal{V}} \mathcal{N}(\mathbf{z}_i|0, I). \tag{10.17}$$

Given $p(A|Z)$ and $p(Z)$, we may compute the posterior distribution of Z using
Bayes' rule. However, this distribution is often intractable. Thus, given the adjacency matrix A and node feature matrix X, VGAE uses graph neural networks to
approximate the posterior distribution of the node embedding matrix Z:

$$q(Z|X,A) = \prod_{i \in \mathcal{V}} q(\mathbf{z}_i|X,A), \quad \text{where } q(\mathbf{z}_i|X,A) = \mathcal{N}(\mathbf{z}_i|\boldsymbol{\mu}_i, \text{diag}(\boldsymbol{\sigma}_i^2)). \quad (10.18)$$

Here, the mean $\boldsymbol{\mu}_i$ and variance $\boldsymbol{\sigma}_i^2$ of $\mathbf{z}_i$ are given by two GCNs. Then, VGAE maximizes the evidence lower bound to learn the GCN parameters:

$$\mathcal{L} = \mathbb{E}_{q(Z|X,A)}[\log p(A|Z)] - \text{KL}[q(Z|X,A)\|p(Z)], \quad (10.19)$$

where $\text{KL}[q(Z|X,A)\|p(Z)]$ is the Kullback-Leibler divergence between the approximated posterior and the prior distribution of Z. The evidence lower bound is optimized using the reparameterization trick (Kingma and Welling, 2014). Finally, the embedding means $\boldsymbol{\mu}_i$ and $\boldsymbol{\mu}_j$ are used to predict link (i, j) by $\hat{A}_{i,j} = \sigma(\boldsymbol{\mu}_i^\top \boldsymbol{\mu}_j)$.

10.3.1.3 Variants of GAE and VGAE

There are many variants of GAE and VGAE. For example, ARGE (Pan et al, 2018) enhances GAE with an adversarial regularization to regularize the node embeddings to follow a prior distribution. S-VAE (Davidson et al, 2018) replaces the Normal distribution in VGAE with a von Mises-Fisher distribution to model data with a hyperspherical latent structure. MGAE (Wang et al, 2017a) uses a marginalized graph autoencoder to reconstruct node features from corrupted ones through a GCN and applies it to graph clustering.

GAE represents a general class of node-based methods, where a GNN is first used to learn node embeddings and pairwise node embeddings are aggregated to learn link representations. In principle, we can replace the GCN used in GAE/VGAE with any GNN, and replace the inner product $\mathbf{z}_i^\top \mathbf{z}_j$ with any aggregation function over $\{\mathbf{z}_i, \mathbf{z}_j\}$ and feed the aggregated link representation to an MLP to predict the link (i, j). Following this methodology, we can generalize any GNN designed for learning node representations to link prediction. For example, HGCN (Chami et al, 2019) combines hyperbolic graph convolutional neural networks with a Fermi-Dirac decoder for aggregating pairwise node embeddings and outputting link probabilities:

$$p(A_{i,j} = 1|\mathbf{z}_i, \mathbf{z}_j) = [\exp(d(\mathbf{z}_i, \mathbf{z}_j) - r)/t + 1]^{-1}, \quad (10.20)$$

where $d(\cdot, \cdot)$ computes the hyperbolic distance and r, t are hyperparameters.

Position-aware GNN (PGNN) (You et al, 2019) aggregates messages only from some selected anchor nodes during the message passing to capture position information of nodes. Then, the inner product between node embeddings are used to predict links. The PGNN paper also generalizes other GNNs, including GAT (Petar et al, 2018), GIN (Xu et al, 2019d) and GraphSAGE (Hamilton et al, 2017b), to the link prediction setting based on the inner-product decoder.

Many graph neural networks use link prediction as an objective for training node embeddings in an unsupervised manner, despite that their final task is still node classification. For example, after computing the node embeddings, GraphSAGE (Hamilton et al, 2017b) minimize the following objective for each $\mathbf{z}_i$ to encourage connected or nearby nodes to have similar representations:

$$L(\mathbf{z}_i) = -\log\left(\sigma(\mathbf{z}_i^\top \mathbf{z}_j)\right) - k_n \cdot \mathbb{E}_{j' \sim p_n} \log\left(1 - \sigma(\mathbf{z}_i^\top \mathbf{z}_{j'})\right), \tag{10.21}$$

where j is a node co-occurs near i on some fixed-length random walk, p_n is the negative sampling distribution, and k_n is the number of negative samples. If we focus on length-2 random walks, the above loss reduces to a link prediction objective. Compared to the GAE loss in Equation (10.15), the above objective does not consider all $\mathcal{O}(n)$ negative links, but uses negative sampling instead to only consider k_n negative pairs (i, j') for each positive pair (i, j), thus is more suitable for large graphs.

In the context of recommender systems, there are also many node-based methods that can be seen as variants of GAE/VGAE. Monti et al (2017) use GNNs to learn user and item embeddings from their respective nearest-neighbor networks, and use the inner product between user and item embeddings to predict links. Berg et al (2017) propose the graph convolutional matrix completion (GC-MC) model which applies a GNN to the user-item bipartite graph to learn user and item embeddings. They use one-hot encoding of node indices as the input node features, and use the bilinear product between user and item embeddings to predict links. SpectralCF (Zheng et al, 2018a) uses a spectral-GNN on the bipartite graph to learn node embeddings. The PinSage model (Ying et al, 2018b) uses node content features as the input node features, and uses the GraphSAGE (Hamilton et al, 2017b) model to map related items to similar embeddings.

In the context of knowledge graph completion, R-GCN (Relational Graph Convolutional Neural Network) (Schlichtkrull et al, 2018) is one representative node-based method, which considers the relation types by giving different weights to different relation types during the message passing. SACN (Structure-Aware Convolutional Network) (Shang et al, 2019) performs message passing for each relation type's induced subgraphs individually and then uses a weighted sum of node embeddings from different relation types.

10.3.2 Subgraph-Based Methods

Subgraph-based methods extract a local subgraph around each target link and learn a subgraph representation through a GNN for link prediction.

10.3.2.1 The SEAL Framework

The pioneering work of subgraph-based methods is SEAL (Zhang and Chen, 2018b). SEAL first extracts an enclosing subgraph for each target link to predict, and then applies a graph-level GNN (with pooling) to classify whether the subgraph corresponds to link existence. The *enclosing subgraph* around a node set is defined as follows.

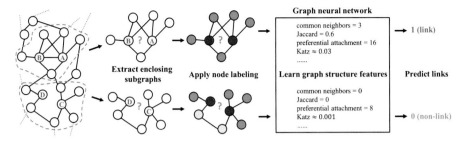

Fig. 10.2: Illustration of the SEAL framework. SEAL first extracts enclosing subgraphs around target links to predict. It then applies a node labeling to the enclosing subgraphs to differentiate nodes of different roles within a subgraph. Finally, the labeled subgraphs are fed into a GNN to learn graph structure features (supervised heuristics) for link prediction.

Definition 10.1. (Enclosing subgraph) For a graph $\mathcal{G} = (\mathcal{V}, \mathcal{E})$, given a set of nodes $S \subseteq \mathcal{V}$, the h-hop enclosing subgraph for S is the subgraph $\mathcal{G}_S^h$ induced from $\mathcal{G}$ by the set of nodes $\cup_{j \in S}\{i \mid d(i, j) \leq h\}$, where $d(i, j)$ is the shortest path distance between nodes i and j.

In other words, the h-hop enclosing subgraph around a node set S contains nodes within h hops of any node in S, as well as all the edges between these nodes. In some literature, it is also called h-hop local/rooted subgraph, or h-hop ego network. In link prediction tasks, the node set S denotes the two nodes between which to predict a link. For example, when predicting the link between x and y, $S = \{x, y\}$ and $\mathcal{G}_{x,y}^h$ denotes the h-hop enclosing subgraph for link (x, y).

The motivation for extracting an enclosing subgraph for each link should be that SEAL aims to automatically learn graph structure features from the network. Observing that all first-order heuristics can be computed from the 1-hop enclosing subgraph around the target link and all second-order heuristics can be computed from the 2-hop enclosing subgraph around the target link, SEAL aims to use a GNN to learn general graph structure features (supervised heuristics) from the extracted h-hop enclosing subgraphs instead of using predefined heuristics.

After extracting the enclosing subgraph $G_{x,y}^h$, the next step is **node labeling**. SEAL applies a Double Radius Node Labeling (DRNL) to give an integer label to each node in the subgraph as its additional feature. The purpose is to use different labels to differentiate nodes of different roles in the enclosing subgraph. For instance, the center nodes x and y are the target nodes between which the target link is located, thus they are different from the rest nodes and should be distinguished. Similarly, nodes at different hops w.r.t. x and y may have different structural importance to the link existence, thus can also be assigned different labels. As discussed in Section 10.4.2, a proper node labeling such as DRNL is crucial for the success of subgraph-based link prediction methods, which makes subgraph-based methods have a higher link representation learning ability than node-based methods.

DRNL works as follows: First, assign label 1 to x and y. Then, for any node i with radius $(d(i,x), d(i,y)) = (1,1)$, assign label 2. Nodes with radius $(1,2)$ or $(2,1)$ get label 3. Nodes with radius $(1,3)$ or $(3,1)$ get 4. Nodes with $(2,2)$ get 5. Nodes with $(1,4)$ or $(4,1)$ get 6. Nodes with $(2,3)$ or $(3,2)$ get 7. So on and so forth. In other words, DRNL iteratively assigns larger labels to nodes with a larger radius w.r.t. the two center nodes.

DRNL satisfies the following criteria: 1) The two target nodes x and y always have the distinct label "1" so that they can be distinguished from the context nodes. 2) Nodes i and j have the same label if and only if their "double radius" are the same, i.e., i and j have the same distances to (x,y). This way, nodes of the same relative positions within the subgraph (described by the double radius $(d(i,x), d(i,y))$) always have the same label.

DRNL has a closed-form solution for directly mapping $(d(i,x), d(i,y))$ to labels:

$$l(i) = 1 + \min(d_x, d_y) + (d/2)[(d/2) + (d\%2) - 1], \qquad (10.22)$$

where $d_x := d(i,x)$, $d_y := d(i,y)$, $d := d_x + d_y$, $(d/2)$ and $(d\%2)$ are the integer quotient and remainder of d divided by 2, respectively. For nodes with $d(i,x) = \infty$ or $d(i,y) = \infty$, DRNL gives them a null label 0.

After getting the DRNL labels, SEAL transforms them into one-hot encoding vectors, or feeds them to an embedding layer to get their embeddings. These new feature vectors are concatenated with the original node content features (if any) to form the new node features. SEAL additionally allows concatenating some pretrained node embeddings such as node2vec embeddings to node features. However, as its experimental results show, adding pretrained node embeddings does not show clear benefits to the final performance (Zhang and Chen, 2018b). Furthermore, adding pretrained node embeddings makes SEAL lose the inductive learning ability.

Finally, SEAL feeds these enclosing subgraphs as well as their new node feature vectors into a graph-level GNN, DGCNN (Zhang et al, 2018g), to learn a graph classification function. The groundtruth of each subgraph is whether the two center nodes really have a link. To train this GNN, SEAL randomly samples N existing links from the network as positive training links, and samples an equal number of unobserved links (random node pairs) as negative training links. After training, SEAL applies the trained GNN to new unobserved node pairs' enclosing subgraphs to predict their links. The entire SEAL framework is illustrated in Figure 10.2. SEAL achieves strong performance for link prediction, demonstrating consistently superior performance than predefined heuristics (Zhang and Chen, 2018b).

10.3.2.2 Variants of SEAL

SEAL inspired many follow-up works. For example, Cai and Ji (2020) propose to use enclosing subgraphs of different scales to learn scale-invariant models. Li et al (2020e) propose Distance Encoding (DE) which generalizes DRNL to node classification and general node set classification problems and theoretically analyzes the power it brings to GNNs. The line graph link prediction (LGLP) model (Cai et al, 2020c) transforms each enclosing subgraph into its line graph and uses the center node embedding in the line graph to predict the original link.

SEAL is also generalized to the bipartite graph link prediction problem of recommender systems (Zhang and Chen, 2019). The model is called Inductive Graph-based Matrix Completion (IGMC). IGMC also samples an enclosing subgraph around each target (user, item) pair, but uses a different node labeling scheme. For each enclosing subgraph, it first gives label 0 and label 1 to the target user and the target item, respectively. The remaining nodes' labels are determined based on both their node types and their distances to the target user and item: if a user-type node's shortest path to reach either the target user or the target item has a length k, it will get a label $2k$; if an item-type node's shortest path to reach the target user or the target item has a length k, it will get a label $2k + 1$. This way, the target nodes can always be distinguished from the context nodes, and users can be distinguished from items (users always have even labels). Furthermore, nodes of different distances to the center nodes can be differentiated, too. Finally, the enclosing subgraphs are fed into a GNN with R-GCN convolution layers to incorporate the edge type information (each edge type corresponds to a different rating). And the output representations of the target user and the target item are concatenated as the link representation to predict the target rating. IGMC is an inductive matrix completion model without relying on any content features, i.e., the model predicts ratings based only on local graph structures, and the learned model can transfer to unseen users/items or new tasks without retraining.

In the context of knowledge graph completion, SEAL is generalized to GraIL (Graph Inductive Learning) (Teru et al, 2020). It also follows the enclosing subgraph extraction, node labeling, and GNN prediction framework. For enclosing subgraph extraction, it extracts the subgraph induced by all the nodes that occur on at least one path of length at most $h + 1$ between the two target nodes. Unlike SEAL, the enclosing subgraph of GraIL does not include those nodes that are only neighbors of one target node but are not neighbors of the other target node. This is because for knowledge graph reasoning, paths connecting two target nodes are of extra importance than dangling nodes. After extracting the enclosing subgraphs, GraIL applies DRNL to label the enclosing subgraphs and uses a variant of R-GCN by enhancing R-GCN with edge attention to output the score for each link to predict.

10.3.3 Comparing Node-Based Methods and Subgraph-Based Methods

At first glance, both node-based methods and subgraph-based methods learn graph structure features around target links based on a GNN. However, as we will show, subgraph-based methods actually have a higher link representation ability than node-based methods due to modeling the associations between two target nodes.

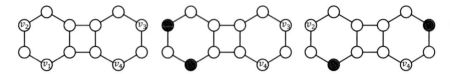

Fig. 10.3: The different link representation ability between node-based methods and subgraph-based methods. In the left graph, nodes v_2 and v_3 are isomorphic; links (v_1,v_2) and (v_4,v_3) are isomorphic; link (v_1,v_2) and link (v_1,v_3) are **not** isomorphic. However, a node-based method cannot differentiate (v_1,v_2) and (v_1,v_3). In the middle graph, when we predict (v_1,v_2), we label these two nodes differently from the rest, so that a GNN is aware of the target link when learning v_1 and v_2's representations. Similarly, when predicting (v_1,v_3), nodes v_1 and v_3 will be labeled differently (shown in the right graph). This way, the representation of v_2 in the left graph will be different from the representation of v_3 in the right graph, enabling GNNs to distinguish (v_1,v_2) and (v_1,v_3).

We first use an example to show node-based methods' limitation for detecting associations between two target nodes. Figure 10.3 left shows a graph we want to perform link prediction on. In this graph, nodes v_2 and v_3 are isomorphic (symmetric to each other), and links (v_1,v_2) and (v_4,v_3) are also isomorphic. However, link (v_1,v_2) and link (v_1,v_3) are **not** isomorphic, as they are not symmetric in the graph. In fact, v_1 is much closer to v_2 than v_3 in the graph, and shares more common neighbors with v_2. Thus, intuitively we do not want to predict (v_1,v_2) and (v_1,v_3) the same. However, because v_2 and v_3 are isomorphic, a node-based method will learn the same node representation for v_2 and v_3 (due to identical neighborhoods). Then, because node-based methods aggregate two node representations as a link representation, they will learn the same link representation for (v_1,v_2) and (v_1,v_3) and subsequently output the same link existence probability for them. This is clearly not what we want.

The root cause of this issue is that node-based methods compute two node representations **independently** of each other, without considering the relative positions and associations between the two nodes. For example, although v_2 and v_3 have different relative positions w.r.t. v_1, a GNN for learning v_2 and v_3's representations is unaware of this difference by treating v_2 and v_3 symmetrically.

With node-based methods, GNNs **cannot even learn to count the common neighbors** between two nodes (which is 1 for (v_1,v_2) and 0 for (v_1,v_3)), one of the most fundamental graph structure features for link prediction. This is still because node-based methods do not consider the other target node when computing one target node's representation. For example, when computing the representation of v_1, node-based methods do not care about which is the other target node—no matter whether the other node has dense connections with it (like v_2) or is far away from it (like v_3), node-based methods will learn the same representation for v_1. The failure to model the associations between two target nodes sometimes results in bad link prediction performance.

Different from node-based methods, subgraph-based methods perform link prediction by extracting an enclosing subgraph around each target link. As we can see, if we extract 1-hop enclosing subgraphs for both (v_1, v_2) and (v_1, v_3), then they are immediately differentiable due to their different enclosing subgraph structures—the enclosing subgraph around (v_1, v_2) is a single connected component, while the enclosing subgraph around (v_1, v_3) is composed of two connected components. Most GNNs can easily assign these two subgraphs different representations.

In addition, the node labeling step in subgraph-based methods also helps model the associations between the two target nodes. For example, let us assume we do not extract enclosing subgraphs, but only apply a node labeling to the original graph. We assume a simplest node labeling which only distinguishes the two target nodes from the rest nodes by assigning label 1 to the two target nodes and label 0 to the rest nodes (we call it *zero-one labeling trick*). Then, when we want to predict link (v_1, v_2), we give v_1, v_2 a different label from those of the rest nodes, as shown by different colors in Figure 10.3 middle. With v_1 and v_2 labeled, when a GNN is computing v_2's representation, it is also "aware" of the source node v_1. And when we want to predict link (v_1, v_3), we will again give v_1, v_3 a different label, as shown in Figure 10.3 right. This way, v_2 and v_3's node representations are no longer the same in the two differently labeled graphs due to the presence of the labeled v_1, and we are able to give different predictions to (v_1, v_2) and (v_1, v_3). This method is called labeling trick (Zhang et al, 2020c). We will discuss it more thoroughly in Section 10.4.2.

10.4 Theory for Link Prediction

In this section, we will introduce some theoretical developments on GNN-based link prediction. For subgraph-based methods, one important motivation is to learn supervised heuristics (graph structure features) from links' neighborhoods. Then, an important question to ask is, how well can GNNs learn existing successful heuristics? The γ-decaying heuristic theory (Zhang and Chen, 2018b) answers this question. In Section 10.3.3, we have seen the limitation of node-based methods for modeling the associations and relationships between two target nodes, and we have also seen that a simple zero-one node labeling can help solve this problem. Why and how can such a simple labeling trick achieve such a better link representation learning ability? What are the general requirements for a node labeling scheme to achieve this ability? The analysis of *labeling trick* answers these questions (Zhang et al, 2020c).

10.4.1 γ-Decaying Heuristic Theory

When using GNNs for link prediction, we want to learn graph structure features useful for predicting links based on message passing. However, it is usually not possible to use very deep message passing layers to aggregate information from the entire network due to the computation complexity introduced by neighbor explosion and the issue of oversmoothing (Li et al, 2018b). This is why node-based methods (such as GAE) only use 1 to 3 message passing layers in practice, and why subgraph-based methods only extract a small 1-hop or 2-hop local enclosing subgraph around each link.

The γ-decaying heuristic theory (Zhang and Chen, 2018b) mainly answers how much structural information useful for link prediction is preserved in local neighborhood of the link, in order to justify applying a GNN only to a local enclosing subgraph in subgraph-based methods. To answer this question, the γ-decaying heuristic theory studies how well can existing link prediction heuristics be approximated from local enclosing subgraphs. If all these existing successful heuristics can be accurately computed or approximated from local enclosing subgraphs, then we are more confident to use a GNN to learn general graph structure features from these local subgraphs.

10.4.1.1 Definition of γ-Decaying Heuristics

Firstly, a direct conclusion from the definition of h-hop enclosing subgraphs (Definition 10.1) is:

Proposition 10.1. *Any h-order heuristic score for (x,y) can be accurately calculated from the h-hop enclosing subgraph $\mathscr{G}_{x,y}^h$ around (x,y).*

For example, a 1-hop enclosing subgraph contains all the information needed to calculate any first-order heuristics, while a 2-hop enclosing subgraph contains all the information needed to calculate any first and second-order heuristics. This indicates that first and second-order heuristics can be learned from local enclosing subgraphs based on an expressive GNN. However, how about high-order heuristics? High-order heuristics usually have better link prediction performance than local ones. To study high-order heuristics' local approximability, the γ-decaying heuristic theory first defines a general formulation of high-order heuristics, namely the γ-*decaying heuristic*.

Definition 10.2. (γ-**decaying heuristic**) A γ-decaying heuristic for link (x,y) has the following form:

$$\mathscr{H}(x,y) = \eta \sum_{l=1}^{\infty} \gamma^l f(x,y,l), \tag{10.23}$$

where γ is a decaying factor between 0 and 1, η is a positive constant or a positive function of γ which is upper bounded by a constant, f is a nonnegative function of x,y,l under the given network, and l can be understood as the iteration number.

Next, it proves that under certain conditions, any γ-decaying heuristic can be approximated from an h-hop enclosing subgraph, and the approximation error decreases at least exponentially with h.

Theorem 10.1. *Given a γ-decaying heuristic $\mathcal{H}(x,y) = \eta \sum_{l=1}^{\infty} \gamma^l f(x,y,l)$, if $f(x,y,l)$ satisfies:*

- *(property 1) $f(x,y,l) \leq \lambda^l$ where $\lambda < \frac{1}{\gamma}$; and*
- *(property 2) $f(x,y,l)$ is calculable from $\mathcal{G}_{x,y}^h$ for $l = 1, 2, \cdots, g(h)$, where $g(h) = ah+b$ with $a, b \in \mathbb{N}$ and $a > 0$,*

then $\mathcal{H}(x,y)$ can be approximated from $\mathcal{G}_{x,y}^h$ and the approximation error decreases at least exponentially with h.

Proof. We can approximate such a γ-decaying heuristic by summing over its first $g(h)$ terms.

$$\widetilde{\mathcal{H}}(x,y) := \eta \sum_{l=1}^{g(h)} \gamma^l f(x,y,l). \tag{10.24}$$

The approximation error can be bounded as follows.

$$|\mathcal{H}(x,y) - \widetilde{\mathcal{H}}(x,y)| = \eta \sum_{l=g(h)+1}^{\infty} \gamma^l f(x,y,l) \leq \eta \sum_{l=ah+b+1}^{\infty} \gamma^l \lambda^l = \eta (\gamma\lambda)^{ah+b+1} (1-\gamma\lambda)^{-1}$$

The above proof indicates that a smaller $\gamma\lambda$ leads to a faster decaying speed and a smaller approximation error. To approximate a γ-decaying heuristic, one just needs to sum its first few terms calculable from an h-hop enclosing subgraph.

Then, a natural question to ask is which existing high-order heuristics belong to γ-decaying heuristics that allow local approximations. Surprisingly, the γ-decaying heuristic theory shows that three most popular high-order heuristics: Katz index, rooted PageRank and SimRank (listed in Table 10.1) are all γ-decaying heuristics which satisfy the properties in Theorem 10.1.

To prove these, we need the following lemma first.

Lemma 10.1. *Any walk between x and y with length $l \leq 2h+1$ is included in $\mathcal{G}_{x,y}^h$.*

Proof. Given any walk $w = \langle x, v_1, \cdots, v_{l-1}, y \rangle$ with length l, we will show that every node v_i is included in $\mathcal{G}_{x,y}^h$. Consider any v_i. Assume $d(v_i, x) \geq h+1$ and $d(v_i, y) \geq h+1$. Then, $2h+1 \geq l = |\langle x, v_1, \cdots, v_i \rangle| + |\langle v_i, \cdots, v_{l-1}, y \rangle| \geq d(v_i, x) + d(v_i, y) \geq 2h+2$, a contradiction. Thus, $d(v_i, x) \leq h$ or $d(v_i, y) \leq h$. By the definition of $\mathcal{G}_{x,y}^h$, v_i must be included in $\mathcal{G}_{x,y}^h$.

Next we present the analysis on Katz, rooted PageRank and SimRank.

10.4.1.2 Katz index

The Katz index (Katz, 1953) for (x,y) is defined as

$$\text{Katz}_{x,y} = \sum_{l=1}^{\infty} \beta^l |\text{walks}^{\langle l \rangle}(x,y)| = \sum_{l=1}^{\infty} \beta^l [A^l]_{x,y}, \tag{10.25}$$

where $\text{walks}^{\langle l \rangle}(x,y)$ is the set of length-l walks between x and y, and A^l is the lth power of the adjacency matrix of the network. Katz index sums over the collection of all walks between x and y where a walk of length l is damped by β^l ($0 < \beta < 1$), giving more weights to shorter walks.

Katz index is directly defined in the form of a γ-decaying heuristic with $\eta = 1, \gamma = \beta$, and $f(x,y,l) = |\text{walks}^{\langle l \rangle}(x,y)|$. According to Lemma 10.1, $|\text{walks}^{\langle l \rangle}(x,y)|$ is calculable from $\mathscr{G}_{x,y}^h$ for $l \leq 2h+1$, thus property 2 in Theorem 10.1 is satisfied. Now we show when property 1 is satisfied.

Proposition 10.2. *For any nodes i,j, $[A^l]_{i,j}$ is bounded by d^l, where d is the maximum node degree of the network.*

Proof. We prove it by induction. When $l = 1$, $A_{i,j} \leq d$ for any (i,j). Thus the base case is correct. Now, assume by induction that $[A^l]_{i,j} \leq d^l$ for any (i,j), we have

$$[A^{l+1}]_{i,j} = \sum_{k=1}^{|V|} [A^l]_{i,k} A_{k,j} \leq d^l \sum_{k=1}^{|V|} A_{k,j} \leq d^l d = d^{l+1}.$$

Taking $\lambda = d$, we can see that whenever $d < 1/\beta$, the Katz index will satisfy property 1 in Theorem 10.1. In practice, the damping factor β is often set to very small values like 5E-4 (Liben-Nowell and Kleinberg, 2007), which implies that Katz can be very well approximated from the h-hop enclosing subgraph.

10.4.1.3 PageRank

The rooted PageRank for node x calculates the stationary distribution of a random walker starting at x, who iteratively moves to a random neighbor of its current position with probability α or returns to x with probability $1 - \alpha$. Let π_x denote the stationary distribution vector. Let $[\pi_x]_i$ denote the probability that the random walker is at node i under the stationary distribution.

Let P be the transition matrix with $P_{i,j} = \frac{1}{|\Gamma(v_j)|}$ if $(i,j) \in E$ and $P_{i,j} = 0$ otherwise. Let $\mathbf{e}_x$ be a vector with the xth element being 1 and others being 0. The stationary distribution satisfies

$$\pi_x = \alpha P \pi_x + (1 - \alpha) \mathbf{e}_x. \tag{10.26}$$

When used for link prediction, the score for (x, y) is given by $[\pi_x]_y$ (or $[\pi_x]_y + [\pi_y]_x$ for symmetry). To show that rooted PageRank is a γ-decaying heuristic, we introduce the *inverse P-distance* theory (Jeh and Widom, 2003), which states that $[\pi_x]_y$ can be equivalently written as follows:

$$[\pi_x]_y = (1 - \alpha) \sum_{w:x \rightsquigarrow y} P[w] \alpha^{\text{len}(w)}, \tag{10.27}$$

where the summation is taken over all walks w starting at x and ending at y (possibly touching x and y multiple times). For a walk $w = \langle v_0, v_1, \cdots, v_k \rangle$, $\text{len}(w) := |\langle v_0, v_1, \cdots, v_k \rangle|$ is the length of the walk. The term $P[w]$ is defined as $\prod_{i=0}^{k-1} \frac{1}{|\Gamma(v_i)|}$, which can be interpreted as the probability of traveling w. Now we have the following theorem.

Theorem 10.2. *The rooted PageRank heuristic is a γ-decaying heuristic which satisfies the properties in Theorem 10.1.*

Proof. We first write $[\pi_x]_y$ in the following form.

$$[\pi_x]_y = (1 - \alpha) \sum_{l=1}^{\infty} \sum_{\substack{w:x \rightsquigarrow y \\ \text{len}(w)=l}} P[w] \alpha^l. \tag{10.28}$$

Defining $f(x, y, l) := \sum_{\substack{w:x \rightsquigarrow y \\ \text{len}(w)=l}} P[w]$ leads to the form of a γ-decaying heuristic. Note that $f(x, y, l)$ is the probability that a random walker starting at x stops at y with exactly l steps, which satisfies $\sum_{z \in V} f(x, z, l) = 1$. Thus, $f(x, y, l) \leq 1 < \frac{1}{\alpha}$ (property 1). According to Lemma 10.1, $f(x, y, l)$ is also calculable from $\mathscr{G}_{x,y}^h$ for $l \leq 2h + 1$ (property 2). ∎

10.4.1.4 SimRank

The SimRank score (Jeh and Widom, 2002) is motivated by the intuition that two nodes are similar if their neighbors are also similar. It is defined in the following recursive way: if $x = y$, then $s(x, y) := 1$; otherwise,

$$s(x, y) := \gamma \frac{\sum_{a \in \Gamma(x)} \sum_{b \in \Gamma(y)} s(a, b)}{|\Gamma(x)| \cdot |\Gamma(y)|} \tag{10.29}$$

where γ is a constant between 0 and 1. According to (Jeh and Widom, 2002), Sim-Rank has an equivalent definition:

$$s(x,y) = \sum_{w:(x,y)\multimap(z,z)} P[w]\gamma^{\,\text{len}(w)}, \qquad (10.30)$$

where $w : (x,y) \multimap (z,z)$ denotes all simultaneous walks such that one walk starts at x, the other walk starts at y, and they first meet at any vertex z. For a simultaneous walk $w = \langle(v_0,u_0),\cdots,(v_k,u_k)\rangle$, $\text{len}(w) = k$ is the length of the walk. The term $P[w]$ is similarly defined as $\prod_{i=0}^{k-1} \frac{1}{|\Gamma(v_i)||\Gamma(u_i)|}$, describing the probability of this walk. Now we have the following theorem.

Theorem 10.3. *SimRank is a γ-decaying heuristic which satisfies the properties in Theorem 10.1.*

Proof. We write $s(x,y)$ as follows.

$$s(x,y) = \sum_{l=1}^{\infty} \sum_{\substack{w:(x,y)\multimap(z,z) \\ \text{len}(w)=l}} P[w]\gamma^{\,l}, \qquad (10.31)$$

Defining $f(x,y,l) := \sum_{\substack{w:(x,y)\multimap(z,z) \\ \text{len}(w)=l}} P[w]$ reveals that SimRank is a γ-decaying heuristic. Note that $f(x,y,l) \leq 1 < \frac{1}{\gamma}$. It is easy to see that $f(x,y,l)$ is also calculable from $\mathscr{G}_{x,y}^h$ for $l \leq h$.

10.4.1.5 Discussion

There exist several other high-order heuristics based on path counting or random walk (Lü and Zhou, 2011) which can be as well incorporated into the γ-decaying heuristic framework. Another interesting finding is that first and second-order heuristics can be unified into this framework too. For example, common neighbors can be seen as a γ-decaying heuristic with $\eta = \gamma = 1$, and $f(x,y,l) = |\Gamma(x) \cap \Gamma(y)|$ for $l = 1$, $f(x,y,l) = 0$ otherwise.

The above results reveal that most existing link prediction heuristics inherently share the same γ-decaying heuristic form, and thus can be effectively approximated from an h-hop enclosing subgraph with exponentially smaller approximation error. The ubiquity of γ-decaying heuristics is not by accident—it implies that a successful link prediction heuristic is better to put exponentially smaller weight on structures far away from the target, as remote parts of the network intuitively make little contribution to link existence. The γ-decaying heuristic theory builds the foundation for learning supervised heuristics from local enclosing subgraphs, as they imply that local enclosing subgraphs already contain enough information to learn good graph structure features for link prediction which is much desired considering learning from the entire network is often infeasible. This motivates the proposition of subgraph-based methods.

To summarize, from small enclosing subgraphs extracted around links, we are able to accurately calculate first and second-order heuristics, and approximate a wide range of high-order heuristics with small errors. Therefore, given a sufficiently expressive GNN, learning from such enclosing subgraphs is expected to achieve performance at least as good as a wide range of heuristics.

10.4.2 Labeling Trick

In Section 10.3.3, we have briefly discussed the difference between node-based methods' and subgraph-based methods' link representation learning abilities. This is formalized into the analysis of *labeling trick* (Zhang et al, 2020c).

10.4.2.1 Structural Representation

We first introduce some preliminary knowledge on *structural representation*, which is a core concept in the analysis of labeling trick.

We define a graph to be $\mathcal{G} = (\mathcal{V}, \mathcal{E}, \mathsf{A})$, where $\mathcal{V} = \{1, 2, \ldots, n\}$ is the set of n vertices, $\mathcal{E} \subseteq \mathcal{V} \times \mathcal{V}$ is the set of edges, and $\mathsf{A} \in \mathbb{R}^{n \times n \times k}$ is a 3-dimensional tensor (we call it adjacency tensor) containing node and edge features. The diagonal components $\mathsf{A}_{i,i,:}$ denote features of node i, and the off-diagonal components $\mathsf{A}_{i,j,:}$ denote features of edge (i, j). We further use $A \in \{0, 1\}^{n \times n}$ to denote the adjacency matrix of $\mathcal{G}$ with $A_{i,j} = 1$ iff $(i, j) \in E$. If there are no node/edge features, we let $\mathsf{A} = A$. Otherwise, A can be regarded as the first slice of A, i.e., $A = \mathsf{A}_{:,:,1}$.

A *permutation* π is a bijective mapping from $\{1, 2, \ldots, n\}$ to $\{1, 2, \ldots, n\}$. Depending on the context, $\pi(i)$ can mean assigning a new index to node $i \in V$, or mapping node i to node $\pi(i)$ of another graph. All $n!$ possible π's constitute the permutation group Π_n. For joint prediction tasks over a set of nodes, we use S to denote the **target node set**. For example, $S = \{i, j\}$ if we want to predict the link between i, j. We define $\pi(S) = \{\pi(i) | i \in S\}$. We further define the permutation of A as $\pi(\mathsf{A})$, where $\pi(\mathsf{A})_{\pi(i),\pi(j),:} = \mathsf{A}_{i,j,:}$.

Next, we define *set isomorphism*, which generalizes graph isomorphism to arbitrary node sets.

Definition 10.3. (Set isomorphism) Given two n-node graphs $\mathcal{G} = (\mathcal{V}, \mathcal{E}, \mathsf{A})$, $\mathcal{G}' = (\mathcal{V}', \mathcal{E}', \mathsf{A}')$, and two node sets $S \subseteq \mathcal{V}$, $S' \subseteq \mathcal{V}'$, we say (S, A) and (S', A') are isomorphic (denoted by $(S, \mathsf{A}) \simeq (S', \mathsf{A}')$) if $\exists \pi \in \Pi_n$ such that $S = \pi(S')$ and $\mathsf{A} = \pi(\mathsf{A}')$.

When $(\mathcal{V}, \mathsf{A}) \simeq (\mathcal{V}', \mathsf{A}')$, we say two graphs $\mathcal{G}$ and $\mathcal{G}'$ are *isomorphic* (abbreviated as $\mathsf{A} \simeq \mathsf{A}'$ because $\mathcal{V} = \pi(\mathcal{V}')$ for any π). Note that set isomorphism is **more strict** than graph isomorphism, because it not only requires graph isomorphism, but also requires that the permutation maps a specific node set S to another node set S'.

In practice, when $S \neq \mathcal{V}$, we are often more concerned with the case of $A = A'$, where we are to find isomorphic node sets **in the same graph** (automorphism). For example, when $S = \{i\}, S' = \{j\}$ and $(i, A) \simeq (j, A)$, we say nodes i and j are isomorphic in graph A (or they have symmetric positions/same structural role within the graph). An example is v_2 and v_3 in Figure 10.3 left.

We say a function f defined over the space of (S, A) is *permutation invariant* (or *invariant* for abbreviation) if $\forall \pi \in \Pi_n$, $f(S, A) = f(\pi(S), \pi(A))$. Similarly, f is *permutation equivariant* if $\forall \pi \in \Pi_n$, $\pi(f(S, A)) = f(\pi(S), \pi(A))$.

Now we define structural representation of a node set, following (Srinivasan and Ribeiro, 2020b; Li et al, 2020e). It assigns a unique representation to each equivalence class of isomorphic node sets.

Definition 10.4. (Most expressive structural representation) Given an invariant function $\Gamma(\cdot)$, $\Gamma(S, A)$ is a most expressive structural representation for (S, A) if $\forall S, A, S', A', \Gamma(S, A) = \Gamma(S', A') \Leftrightarrow (S, A) \simeq (S', A')$.

For simplicity, we will briefly use *structural representation* to denote most expressive structural representation in the rest of this section. We will omit A if it is clear from context. We call $\Gamma(i, A)$ a *structural node representation* for i, and call $\Gamma(\{i, j\}, A)$ a *structural link representation* for (i, j).

Definition 10.4 requires the structural representations of two node sets to be the same if and only if they are isomorphic. That is, isomorphic node sets always have the **same** structural representation, while non-isomorphic node sets always have **different** structural representations. This is in contrast to *positional node embeddings* such as DeepWalk (Perozzi et al, 2014) and matrix factorization (Mnih and Salakhutdinov, 2008), where two isomorphic nodes can have different node embeddings (Ribeiro et al, 2017).

So why do we need structural representations? Formally speaking, Srinivasan and Ribeiro (2020b) prove that any joint prediction task over node sets only requires *most-expressive structural representations* of node sets, which are the same for two node sets if and only if these two node sets are isomorphic. This means, for link prediction tasks, we need to learn the same representation for isomorphic links while discriminating non-isomorphic links by giving them different representations. Intuitively speaking, two links being isomorphic means they should be indistinguishable from any perspective—if one link exists, the other should exist too, and vice versa. Therefore, link prediction ultimately requires finding such a *structural link representation* for node pairs which can uniquely identify link isomorphism classes.

According to Figure 10.3 left, node-based methods that directly aggregate two node representations **cannot** learn such a valid structural link representation because they cannot differentiate non-isomorphic links such as (v_1, v_2) and (v_1, v_3). One may wonder whether using one-hot encoding of node indices as the input node features help node-based methods learn such a structural link representation. Indeed, using node-discriminating features enables node-based methods to learn different representations for (v_1, v_2) and (v_1, v_3) in Figure 10.3 left. However, it also loses GNN's ability to map isomorphic nodes (such as v_2 and v_3) and isomorphic links (such as (v_1, v_2) and (v_4, v_3)) to the same representations, since any two nodes already have different representations from the beginning. This might result in poor generalization ability—two nodes/links may have different final representations even they share identical neighborhoods.

To ease our analysis, we also define a *node-most-expressive GNN*, which gives different representations to all non-isomorphic nodes and gives the same representation to all isomorphic nodes. In other words, a node-most-expressive GNN learns structural node representations.

Definition 10.5. (Node-most-expressive GNN) A GNN is node-most-expressive if it satisfies: $\forall i, A, j, A'$, $\text{GNN}(i, A) = \text{GNN}(j, A') \Leftrightarrow (i, A) \simeq (j, A')$.

Although a polynomial-time implementation of a node-most-expressive GNN is not known, practical GNNs based on message passing can still discriminate almost all non-isomorphic nodes (Babai and Kucera, 1979), thus well approximating its power.

10.4.2.2 Labeling Trick Enables Learning Structural Representations

Now, we are ready to introduce the labeling trick and see how it enables learning structural representations of node sets. As we have seen in Section 10.4.2, a simple zero-one labeling trick can help a GNN distinguish non-isomorphic links such as (v_1, v_2) and (v_1, v_3) in Figure 10.3 left. At the same time, isomorphic links, such as (v_1, v_2) and (v_4, v_3), will still have the same representation, since the zero-one labeled graph for (v_1, v_2) is still symmetric to the zero-one labeled graph for (v_4, v_3). This brings an exclusive advantage over using one-hot encoding of node indices.

Below we give the formal definition of labeling trick, which incorporates the zero-one labeling trick as one specific form.

Definition 10.6. (Labeling trick) Given (S, A), we stack a labeling tensor $L^{(S)} \in \mathbb{R}^{n \times n \times d}$ in the third dimension of A to get a new $A^{(S)} \in \mathbb{R}^{n \times n \times (k+d)}$, where L satisfies: $\forall S, A, S', A', \pi \in \Pi_n$, (1) $L^{(S)} = \pi(L^{(S')}) \Rightarrow S = \pi(S')$, and (2) $S = \pi(S'), A = \pi(A') \Rightarrow L^{(S)} = \pi(L^{(S')})$.

To explain a bit, labeling trick assigns a label vector to each node/edge in graph A, which constitutes the labeling tensor $L^{(S)}$. By concatenating A and $L^{(S)}$, we get the adjacency tensor $A^{(S)}$ of the new labeled graph. By definition we can assign labels to both nodes and edges. For simplicity, here we only consider node labels, i.e., we let off-diagonal components $L_{i,j,:}^{(S)}$ be all zero.

The labeling tensor $L^{(S)}$ should satisfy two conditions in Definition 10.6. The first condition requires the target nodes S to have *distinct labels* from those of the rest nodes, so that S is distinguishable from others. This is because if a permutation π preserving node labels exists between nodes of A and A', then S and S' must have distinct labels to guarantee S' is mapped to S by π. The second condition requires the labeling function to be *permutation equivariant*, i.e., when (S, A) and (S', A') are isomorphic under π, the corresponding nodes $i \in S, j \in S', i = \pi(j)$ must always have the same label. In other words, the labeling should be consistent across different S.

For example, the zero-one labeling is a valid labeling trick by always giving label 1 to nodes in S and 0 otherwise, which is both consistent and S-discriminating. However, an all-one labeling is not a valid labeling trick, because it cannot distinguish the target set S.

Now we introduce the main theorem of labeling trick showing that with a valid labeling trick, a node-most-expressive GNN can learn structural link representations by aggregating its node representations learned from the **labeled** graph.

Theorem 10.4. *Given a node-most-expressive* GNN *and an injective set aggregation function* AGG, *for any* S, A, S', A', $\mathrm{GNN}(S, A^{(S)}) = \mathrm{GNN}(S', A'^{(S')}) \Leftrightarrow (S, A) \simeq (S', A')$, *where* $\mathrm{GNN}(S, A^{(S)}) := \mathrm{AGG}(\{\mathrm{GNN}(i, A^{(S)}) | i \in S\})$.

The proof of the above theorem can be found in Appendix A of (Zhang et al, 2020c). Theorem 10.4 implies that $\mathrm{AGG}(\{\mathrm{GNN}(i, A^{(S)}) | i \in S\})$ is a structural representation for (S, A). Remember that directly aggregating structural node representations learned from the original graph A does not lead to structural link representations. Theorem 10.4 shows that aggregating over the structural node representations learned from the adjacency tensor $A^{(S)}$ of the **labeled graph**, somewhat surprisingly, results in a structural representation for S.

The significance of Theorem 10.4 is that it closes the gap between GNN's node representation nature and link prediction's link representation requirement, which solves the open question raised in (Srinivasan and Ribeiro, 2020b) questioning node-based GNN methods' ability of performing link prediction. Although directly aggregating pairwise node representations learned by GNNs does not lead to structural link representations, combining GNNs with a labeling trick enables learning structural link representations.

It can be easily proved that the zero-one labeling, DRNL and Distance Encoding (DE) (Li et al, 2020e) are all valid labeling tricks. This explains subgraph-based methods' superior empirical performance than node-based methods (Zhang and Chen, 2018b; Zhang et al, 2020c).

10.5 Future Directions

In this section, we introduce several important future directions for link prediction: accelerating subgraph-based methods, designing more powerful labeling tricks, and understanding when to use one-hot features.

10.5.1 Accelerating Subgraph-Based Methods

One important future direction is to accelerate subgraph-based methods. Although subgraph-based methods show superior performance than node-based methods both empirically and theoretically, they also suffer from a huge computation complexity, which prevent them from being deployed in modern recommendation systems. How to accelerate subgraph-based methods is thus an important problem to study.

The extra computation complexity of subgraph-based methods comes from their node labeling step. The reason is that for every link (i, j) to predict, we need to relabel the graph according to (i, j). The same node v will be labeled differently depending on which one is the target link, and will be given a different node representation by the GNN when it appears in different links' labeled graphs. This is different from node-based methods, where we do not relabel the graph and each node only has a single representation.

In other words, for node-based methods, we only need to apply the GNN to the whole graph once to compute a representation for each node, while subgraph-based methods need to repeatedly apply the GNN to differently labeled subgraphs each corresponding to a different link. Thus, when computing link representations, subgraph-based methods require re-applying the GNN for each target link. For a graph with n nodes and m links to predict, node-based methods only need to apply a GNN $\mathcal{O}(n)$ times to get a representation for each node (and then use some simple aggregation function to get link representations), while subgraph-based methods need to apply a GNN $\mathcal{O}(m)$ times for all links. When $m \gg n$, subgraph-based methods have much worse time complexity than node-based methods, which is the price for learning more expressive link representations.

Is it possible to accelerate subgraph-based methods? One possible way is to simplify the enclosing subgraph extraction process and simplify the GNN architecture. For example, we may adopt sampling or random walk when extracting the enclosing subgraphs which might largely reduce the subgraph sizes and avoid hub nodes. It is interesting to study such simplifications' influence on performance. Another possible way is to use distributed and parallel computing techniques. The enclosing subgraph extraction process and the GNN computation on a subgraph are completely independent of each other and are naturally parallelizable. Finally, using multi-stage ranking techniques could also help. Multi-stage ranking will first use some simple methods (such as traditional heuristics) to filter out most unlikely links, and use more powerful methods (such as SEAL) in the later stage to only rank the most promising links and output the final recommendations/predictions.

Either way, solving the scalability issue of subgraph-based methods can be a great contribution to the field. That means we can enjoy the superior link prediction performance of subgraph-based GNN methods without using much more computation resources, which is expected to extend GNNs to more application domains.

10.5.2 Designing More Powerful Labeling Tricks

Another direction is to design more powerful labeling tricks. Definition 10.6 gives a general definition of labeling trick. Although any labeling trick satisfying Definition 10.6 can enable a node-most-expressive GNN to learn structural link representations, the real-world performance of different labeling tricks can vary a lot due to the limited expressive power and depths of practical GNNs. Also, some subtle differences in implementing a labeling trick can also result in large performance differences. For example, given the two target nodes x and y, when computing the distance $d(i,x)$ from a node i to x, DRNL will temporarily mask node y and all its edges, and when computing the distance $d(i,y)$, DRNL will temporarily mask node x and all its edges (Zhang and Chen, 2018b). The reason for this "masking trick" is that DRNL aims to use the pure distance between i and x without the influence of y. If we do not mask y, $d(i,x)$ will be upper bounded by $d(i,y)+d(x,y)$, which obscures the "true distance" between i and x and might hurt the node labels' ability to discriminate structurally-different nodes. As shown in Appendix H of (Zhang et al, 2020c), this masking trick can greatly improve the performance. It is thus interesting to study how to design a more powerful labeling trick (not necessarily based on shortest path distance like DRNL and DE). It should not only distinguish the target nodes, but also assign diverse but generalizable labels to nodes with different roles in the subgraph. A further theoretical analysis on the power of different labeling tricks is also needed.

10.5.3 Understanding When to Use One-Hot Features

Finally, one last important question remaining to be answered is when we should use the original node features and when we should use one-hot encoding features of node indices. Although using one-hot features makes it infeasible to learn structural link representations as discussed in Section 10.4.2, node-based methods using one-hot features show strong performance on dense networks (Zhang et al, 2020c), outperforming subgraph-based methods without using one-hot features by large margins. On the other hand, Kipf and Welling (2017b) show that GAE/VGAE with one-hot features gives worse performance than using original features. Thus, it is interesting to study when to use one-hot features and when to use original features and theoretically understand their representation power differences on networks of different properties. Srinivasan and Ribeiro (2020b) provide a good analysis connecting positional node embeddings (such as DeepWalk) with structural node representations, showing that positional node embeddings can be seen as a sample while the structural node representation can be seen as a distribution. This can serve as a starting point to study the power of GNNs using one-hot encoding features, as GNNs using one-hot encoding features can be seen as combining positional node embeddings with message passing.

Editor's Notes: Link prediction is the problem of predicting the existence of a link between two nodes in a network. Hence the techniques are relevant to graph structure learning (chapter 19), which aims to discover useful graph structure, i.e. links, from data. Scalability property (chapter 6) and expressiveness power theory (chapter 8) play an important role in applying and designing link prediction methods. Link prediction also motivates several downstream tasks in various domains, such as predicting protein-protein and protein-drug interactions (chapter 25), drug development (chapter 24), recommender systems (chapter 19). Besides, predicting links in the complex network, including dynamic graphs (chapter 19), knowledge graphs (chapter 24) and heterogeneous graphs (chapter 26), are also the extension of link prediction tasks.

Chapter 11
Graph Neural Networks: Graph Generation

Renjie Liao

Abstract In this chapter, we first review a few classic probabilistic models for graph generation including the Erdős–Rényi model and the stochastic block model. Then we introduce several representative modern graph generative models that leverage deep learning techniques like graph neural networks, variational auto-encoders, deep auto-regressive models, and generative adversarial networks. At last, we conclude the chapter with a discussion on potential future directions.

11.1 Introduction

The study of graph generation revolves around building probabilistic models over *graphs* which are also called *networks* in many scientific disciplines. This problem has its roots in a branch of mathematics, called *random graph theory* (Bollobás, 2013), which largely lies at the intersection between the probability theory and the graph theory. It is also at the core of a new academic field, called *network science* (Barabási, 2013). Historically, researchers in these fields are often interested in building random graph models (*i.e.*, constructing distributions of graphs using certain parametric families of distributions) and proving the mathematical properties of such models. Albeit being an extremely fruitful and successful research direction that spawns numerous outcomes, these classic models suffer from being too simplistic to capture the complex phenomenon (e.g., highly-clustered, well-connected, scale-free) that appeared in the real-world graphs.

With the advent of powerful deep learning techniques like *graph neural networks*, we can build more expressive probabilistic models of graphs, *i.e.*, the so-called *deep graph generative models*. Such deep models can better capture the complex dependencies within the graph data to generate more realistic graphs and further build accurate predictive models. However, the downside is that these models

Renjie Liao
University of Toronto, e-mail: rjliao@cs.toronto.edu

are often so complicated that we can rarely analyze their properties in a precise manner. The recent practices of these models have demonstrated impressive performances in modeling real-world graphs/networks, e.g., social networks, citation networks, and molecule graphs.

In the following, we first introduce the classic graph generative models in Section 11.2 and then the modern ones that leverage the deep learning techniques in Section 11.3. At last, we conclude the chapter and discuss some promising future directions.

11.2 Classic Graph Generative Models

In this section, we review two popular variants of the classic graph generative models: the Erdős–Rényi model (Erdős and Rényi, 1960) and the stochastic block model (Holland et al, 1983). They are often used as handy baselines in many applications since we have already gained deep understandings of their properties. There are many other graph generative models like the Watts–Strogatz small-world model (Watts and Strogatz, 1998) and the Barabási–Albert (BA) preferential attachment model (Barabási and Albert, 1999). Barabási (2013) provides a thorough survey on these models and other aspects of network science. In the context of machine learning, there are also quite a few non-deep-learning graph generative models like Kronecker graphs (Leskovec et al, 2010). We do not cover these models due to the space limit.

11.2.1 Erdős–Rényi Model

We first explain one of the most well known random graph models, *i.e.*, Erdős–Rényi model (Erdős and Rényi, 1960), named after two Hungarian mathematicians Paul Erdős and Alfréd Rényi. Note that this model has been independently proposed at around the same time by Edgar Gilbert in (Gilbert, 1959). In the following, we first describe the model along with its properties and then discuss its limitations.

11.2.1.1 Model

The Erdős–Rényi model has two closely variants, namely, $G(n,p)$ and $G(n,m)$.

G(n,p) Model In the $G(n,p)$ model, we are given n labeled nodes and generate a graph by randomly connecting an edge linking one node to the other with the probability p, independently from every other edge. In other words, all $\binom{n}{2}$ possible edges have the equal probability p to be included. Therefore, the probability of generating a graph with m edges under this model is as below,

$$p(\text{a graph with } n \text{ nodes and } m \text{ edges}) = p^m(1-p)^{\binom{n}{2}-m}. \tag{11.1}$$

The parameter p controls the "density" of the graph, *i.e.*, a larger value of p makes the graph become more likely to contain more edges. When $p = \frac{1}{2}$, the above probability becomes $\frac{1}{2}\binom{n}{2}$, *i.e.*, all possible $2\binom{n}{2}$ graphs are chosen with equal probability.

Due to the independence of the edges in $G(n, p)$, we can easily derive a few properties from this model.

- The expected number of edges is $\binom{n}{2}p$.
- The degree distribution of any node v is binomial:

$$p(\text{degree}(v) = k) = \binom{n}{k}p^k(1 - p)^{n-1-k} \tag{11.2}$$

- If Np is a constant and $n \to \infty$, the degree distribution of any node v is Poisson:

$$p(\text{degree}(v) = k) = \frac{(np)^k e^{-np}}{k!} \tag{11.3}$$

There is an enormous number of more involved properties of this model that has been proved (e.g., by Erdős and Rényi in the original paper). We list a few others as below.

- If $p > \frac{(1+\varepsilon)\ln n}{n}$, then a graph will almost surely be connected.
- If $p < \frac{(1+\varepsilon)\ln n}{n}$, then a graph will almost surely contain isolated vertices, and thus be disconnected.
- If $Np < 1$, then a graph will almost surely have no connected components of size larger than $O(\log(n))$.

Here almost surely means the probability of the event happens with probability 1 (*i.e.*, the set of possible exceptions has zero measure).

G(n,m) Model In the $G(n, m)$ model, we are given n labeled nodes and generate a graph by uniformly randomly choosing a graph from the set of all graphs with n nodes and m edges, *i.e.*, the probability of choosing each graph is $\binom{\binom{n}{2}}{m}^{-1}$. There are also many important properties associated with the $G(n, m)$ model. In particular, it is interchangeable with the $G(n, p)$ model provided that m is close to $\binom{n}{2}p$ in most investigations. Chapter 2 of (Bollobás and Béla, 2001) provides a comprehensive discussion on the relationship between these two models. The $G(n, p)$ model is more commonly used in practice than the $G(n, m)$ model, partly due to the ease of analysis brought by the independence of the edges.

11.2.1.2 Discussion

As a seminal work in the random graph theory, the Erdős–Rényi model inspires much subsequent work to study and generalize this model. However, the assumptions of this model, e.g., edges are independent and each edge is equally likely to be generated, are too strong to capture the properties of the real-world graphs. For example, the degree distribution of the Erdős–Rényi model has an exponential tail which means we rarely see node degrees span a broad range, e.g., several orders of magnitude. Meanwhile, real-world graphs/networks like the World Wide Web (WWW) are believed to possess a degree distribution that follows a power law, *i.e.*, $p(d) \propto d^{-\gamma}$ where d is the degree and the exponent γ is typically between 2 and 3. Essentially, this means that there are many nodes that have small node degrees, whereas there are a few nodes which have extremely large node degrees (, hubs) in the real-world graphs like WWW. Therefore, many improved models like the scale-free networks (Barabási and Albert, 1999) were later proposed, which fit better to the degree distribution of the real-world graphs.

11.2.2 Stochastic Block Model

Stochastic block models (SBM) are a family of random graphs with clusters of nodes and are often employed as a canonical model for tasks like community detection and clustering. It is proposed independently in a few scientific communities, e.g., machine learning and statistics (Holland et al, 1983), theoretical computer science (Bui et al, 1987), and mathematics (Bollobás et al, 2007). It is arguably the simplest model of a graph with communities/clusters. As a generative model, SBM could provide ground-truth cluster memberships, which in turn could help benchmark and understand different clustering/community detection algorithms. In the following, we first introduce the basics of the model and then discuss its advantages as well as limitations.

11.2.2.1 Model

We start the introduction by denoting the total number of nodes as n and the number of communities/clusters as k. A prior probability vector $\mathbf{p}$ over the k clusters and a $k \times k$ matrix W with entries in $[0, 1]$ are also given. We generate a random graph following the procedure below:

1. For each node, we generate its community label (an integer from $\{1, \cdots, k\}$) by independently sampling from $\mathbf{p}$.
2. For each pair of nodes, denoting their community labels as i and j, we generate an edge by independently sampling with probability $W_{i,j}$.

Basically, the community assignments of a pair of nodes determine the specific entry of W to be used, which in turn indicates how likely we connect this pair of nodes. We denote such a model as SBM$(n, \mathbf{p}, W)$. Note that, if we set $W_{i,j} = q$ for all communities (i, j), then the corresponding SBM degenerates to the Erdős–Rényi model $G(n, q)$.

   In the context of community detection, people are often interested in recovering the community label given a random graph drawn from the SBM model. Denoting the recovered and the ground-truth community labels as $X \in \mathbb{R}^{n \times 1}$ and $Y \in \mathbb{R}^{n \times 1}$, we can define the agreement R between two community labels as,

$$R(X,Y) = \max_{P \in \Pi} \frac{1}{n} \sum_{i=1}^{n} \mathbf{1}\left[X_i = (PY)_i\right], \tag{11.4}$$

where P is a permutation matrix and Π is the set of all permutation matrices. X_i and $(PY)_i$ are the i-th element of X and PY respectively. In short, the agreement considers the best possible reshuffle between two sequences of labels. Depending on the requirement, we could examine the community detection algorithms in the sense of exact recovery (*i.e.*, cluster assignments are exactly recovered almost surely, $p(R(X,Y) = 1) = 1$) or partial recovery (*i.e.*, at most $1 - \varepsilon$ fraction of nodes are mislabeled almost surely, $p(R(X,Y) \geq \varepsilon) = 1$). Researchers have established various conditions under which a particular type of recovery is possible for SBM graphs. For example, for SBMs with $W = \frac{\log(n)Q}{n}$, where Q is a matrix with positive entries and the same size as W, Abbe and Sandon (2015) shows that the exact recovery is possible if and only if the minimum Chernoff-Hellinger divergence between any two columns of $\text{diag}(\mathbf{p})Q$ is no less than 1, where $\text{diag}(\mathbf{p})$ is a diagonal matrix with diagonal entries as $\mathbf{p}$.

11.2.2.2 Discussion

Abbe (2017) provides an up-to-date and comprehensive survey on the SBM and the fundamental limits (from both information-theoretic and computational perspectives) for community detection in the SBM. SBM is a more realistic random graph model for describing graphs with community structures compared to the Erdős–Rényi model. It also spawns many subsequent variants of block models like the mixed membership SBM (Airoldi et al, 2008). However, the estimation of SBMs on real-world graphs is hard since the number of communities is often unknown in advance and some graphs may not exhibit clear community structures.

11.3 Deep Graph Generative Models

In this section, we review several representative deep graph generative models which aim at building probabilistic models of graphs using deep neural networks. Based on the type of deep learning techniques being used, we can roughly divide the current literature into three categories: variational autoencoder (VAEs) (Kingma and Welling, 2014) based methods, deep auto-regressive (Van Oord et al, 2016) methods, and generative adversarial networks (GANs) (Goodfellow et al, 2014b) based methods. We introduce all three model classes in the subsequent sections.

11.3.1 Representing Graphs

We first introduce how a graph is represented in the context of deep graph generative models. Suppose we are given a graph $\mathcal{G} = (\mathcal{V}, \mathcal{E})$ where $\mathcal{V}$ is the set of nodes/vertices and $\mathcal{E}$ is the set of edges. Conditioning on a specific node ordering π, we can represent the graph $\mathcal{G}$ as an adjacency matrix A_π where $A_\pi \in \mathbb{R}^{|\mathcal{V}| \times |\mathcal{V}|}$, where $|\mathcal{V}|$ is the size of set $\mathcal{V}$ (*i.e.*, the number of nodes). The adjacency matrix not only provides a convenient representation of graphs on computers but also offers a natural way to mathematically define a probability distribution over graphs. Here we explicitly write the node ordering π in the subscript to emphasize that the rows and columns of A are arranged according to the π. If we change the node ordering from π to π', the adjacency matrix will be permuted (shuffling rows and columns) accordingly, *i.e.*, $A_{\pi'} = P A_\pi P^\top$, where the permutation matrix P is constructed based on the pair of node orderings (π, π'). In other words, A_π and $A_{\pi'}$ represent the same graph $\mathcal{G}$. Therefore, a graph $\mathcal{G}$ with an adjacency matrix A_π can be equivalently represented as a set of adjacency matrices $\{P A_\pi P^\top | P \in \Pi\}$ where Π is the set of all permutation matrices with size $|\mathcal{V}| \times |\mathcal{V}|$. Note that, depending on the symmetric structures of A_π, there may exist two permutation matrices $P_1, P_2 \in \Pi$ so that $P_1 A_\pi P_1^\top = P_2 A_\pi P_2^\top$. Therefore, we remove such redundancies and keep those uniquely permuted adjacency matrices, denoted as $\mathcal{A} = \{P A_\pi P^\top | P \in \Pi_{\mathcal{G}}\}$. More precisely, $\Pi_{\mathcal{G}}$ is the maximal subset of Π so that $P_1 A_\pi P_1^\top \neq P_2 A_\pi P_2^\top$ holds for any $P_1, P_2 \in \Pi_{\mathcal{G}}$. We add the subscript $\mathcal{G}$ to emphasize that $\Pi_{\mathcal{G}}$ depends on the given graph $\mathcal{G}$. Note that there exists a surjective mapping between Π and $\Pi_{\mathcal{G}}$. For the ease of notations, we will drop the subscript of the node ordering and use $\mathcal{G} \equiv \mathcal{A} = \{P A P^\top | P \in \Pi_{\mathcal{G}}\}$ to represent a graph from now on.

When considering the node features/attributes X, we can denote the graph structured data as $\mathcal{G} \equiv \{(P A P^\top, P X) | P \in \Pi_{\mathcal{G}}\}$[1]. Note that the rows of X are shuffled according to P since each row of X corresponds to a node. In our context, we can assume the maximum number of nodes of all graphs is n. If a graph has fewer nodes than n, we can add dummy nodes (e.g., with all-zero features) which are isolated to other nodes to make the size equal n. Therefore, $X \in \mathbb{R}^{n \times d_X}$ and $A \in \mathbb{R}^{n \times n}$ where d_X is the feature dimension. To simplify the explanation, we do not include the edge feature. But it is straightforward to modify the following models accordingly to incorporate edge features.

11.3.2 Variational Auto-Encoder Methods

Due to the great success of VAEs in image generation (Kingma and Welling, 2014; Rezende et al, 2014), it is natural to extend this framework to graph generation. This

---

[1] Technically, there may exist two permutation matrices $P_1, P_2 \in \Pi$ so that $P_1 A P_1^\top = P_2 A P_2^\top$ and $P_1 X \neq P_2 X$. It thus seems to be necessary to define $\mathcal{G} \equiv \{(P A P^\top, P X) | P \in \Pi\}$. However, as seen later, we are always interested in distributions of node features that are exchangeable over nodes, *i.e.*, $p(P_1 X) = p(P_2 X)$. Therefore, restricting ourselves to $\Pi_{\mathcal{G}}$ is sufficient for our exposition.

idea has been explored from different aspects (Kipf and Welling, 2016; Jin et al, 2018a; Simonovsky and Komodakis, 2018; Liu et al, 2018d; Ma et al, 2018; Grover et al, 2019; Liu et al, 2019b) and is often collectively named as *GraphVAE*. In the following, we first highlight the common framework shared by all these methods and then discuss some important variants.

11.3.2.1 The GraphVAE Family

Similar to vanilla VAEs, every model instance within the GraphVAE family consists of an encoder (*i.e.*, a variational distribution $q_\phi(Z|A,X)$ parameterized by ϕ), a decoder (*i.e.*, a conditional distribution $p_\theta(\mathscr{G}|Z)$ parameterized by θ), and a prior distribution (*i.e.*, a distribution $p(Z)$ typically with fixed parameters). Before introducing individual components, we first describe what the latent variables Z are. In the context of graph generation, we typically assume that each node is associated with a latent vector. Denoting the latent vector of the i-th node as $\mathbf{z}_i$, then $Z \in \mathbb{R}^{n \times d_Z}$ is obtained by stacking $\{\mathbf{z}_i\}$ as row vectors. Such latent vectors should summarize the information of the local subgraphs associated with individual nodes so that we can decode/generate edges based on them. In other words, any pair of latent vectors $(\mathbf{z}_i, \mathbf{z}_j)$ is supposed to be informative to determine whether nodes (i, j) should be connected. We could further introduce edge latent variables $\{\mathbf{z}_{ij}\}$ to enrich the model. Again, we do not consider such an option for simplicity since the underlying modeling technique is roughly the same.

Encoder We first explain how to construct the encoder using a deep neural network. Recall that the input to the encoder is the graph data (A, X). The natural candidate to deal with such data is a graph neural network, e.g., a graph convolutional network (GCN) (Kipf and Welling, 2017b). For example, let us consider a two-layer GCN as below,

$$H = \tilde{A}\sigma(\tilde{A}XW_1)W_2, \tag{11.5}$$

where $H \in \mathbb{R}^{n \times d_H}$ are the node representations (each node is associated with a size-d_H row vector). $\tilde{A} = D^{-\frac{1}{2}}(A+\mathrm{I})D^{-\frac{1}{2}}$ where D is the degree matrix (*i.e.*, a diagonal matrix of which the entries are the row sum of $A + \mathrm{I}$). I is the identity matrix. σ is the nonlinearity which is often chosen to be the rectified linear unit (ReLU) (Nair and Hinton, 2010). $\{W_1, W_2\}$ are the learnable parameters. We can pad a constant to the input feature dimension so that the bias term is absorbed into the weight matrix. We adopt this convention for ease of notation.

Relying on the learned node representations H, we can construct the variational distribution as below,

$$q_\phi(Z|A,X) = \prod_{i=1}^{n} q(\mathbf{z}_i|A,X) \tag{11.6}$$

$$q(\mathbf{z}_i|A,X) = \mathcal{N}(\boldsymbol{\mu}_i, \boldsymbol{\sigma}_i \mathbf{I}) \tag{11.7}$$

$$\boldsymbol{\mu} = \text{MLP}_\mu(H) \tag{11.8}$$

$$\log \boldsymbol{\sigma} = \text{MLP}_\sigma(H). \tag{11.9}$$

Here we typically assume that the variational distribution $q(Z|A,X)$ is conditionally node-wise independent for the tractability consideration. $\boldsymbol{\mu}_i$ and $\boldsymbol{\sigma}_i$ are the i-th rows of $\boldsymbol{\mu}$ and $\boldsymbol{\sigma}$ respectively. The learnable parameters ϕ consist of all parameters of the two multi-layer perceptrons (MLPs) and the above GCN. Although the approximated variational distribution defined in Eq. (11.6) is simple, it possesses a few great properties. First, the probability distribution is invariant w.r.t. the permutation of nodes. Mathematically, it means that given two different permutation matrices $P_1, P_2 \in \Pi$, we have

$$q(P_1 Z|P_1 A P_1^\top, P_1 X) = q(P_2 Z|P_2 A P_2^\top, P_2 X) \tag{11.10}$$

This can be easily verified from the exchangeability of the product of probabilities and the equivariance property of graph neural networks. Second, the neural networks underlying each Gaussian (*i.e.*, "GNN + MLP") are very powerful so that the conditional distributions are expressive in capturing the uncertainty of latent variables. Third, this encoder is computationally cheaper than those which consider the dependencies among different $\{\mathbf{z}_i\}$ (e.g., an autoregressive encoder). It thus provides a solid baseline for investigating whether a more powerful encoder is needed in a given problem.

Prior Similar to most VAEs, GraphVAEs often adopt a prior that is fixed during the learning. For example, a common choice is an node-independent Gaussian as below,

$$p(Z) = \prod_{i=1}^{n} p(\mathbf{z}_i) \tag{11.11}$$

$$p(\mathbf{z}_i) = \mathcal{N}(0, \mathbf{I}). \tag{11.12}$$

Again, we could replace this fixed prior with more powerful ones like an autoregressive model at the cost of more computation and/or a time-consuming pre-training stage. But this prior serves as a good starting point to benchmark more complicated alternatives, e.g., the normalizing flow based one in (Liu et al, 2019b).

Decoder The aim of a decoder in graph generative models is to construct a probability distribution over the graph and its feature/attributes conditioned on the latent variables, *i.e.*, $p(\mathcal{G}|Z)$. However, as we discussed previously, we need to consider all possible node orderings (each corresponds to a permuted adjacency matrix) which leaves the graph unchanged, *i.e.*,

$$p(\mathcal{G}|Z) = \sum_{P \in \Pi_\mathcal{G}} p(PAP^\top, PX|Z). \tag{11.13}$$

Recall that $\Pi_{\mathcal{G}}$ is the maximal subset of the set of all possible permutation matrices Π so that $P_1 A_\pi P_1^\top \neq P_2 A_\pi P_2^\top$ holds for any $P_1, P_2 \in \Pi_{\mathcal{G}}$. To build such a decoder, we first construct a probability distribution over adjacency matrix and node feature matrix. For example, we show a popular and simple construction (Kipf and Welling, 2016) as below,

$$p(A, X|Z) = \prod_{i,j} p(A_{ij}|Z) \prod_{i=1}^{n} p(\mathbf{x}_i|Z) \tag{11.14}$$

$$p(A_{ij}|Z) = \text{Bernoulli}(\Theta_{ij}) \tag{11.15}$$

$$p(\mathbf{x}_i|Z) = \mathcal{N}(\tilde{\boldsymbol{\mu}}_i, \tilde{\boldsymbol{\sigma}}_i) \tag{11.16}$$

$$\Theta_{ij} = \text{MLP}_\Theta([\mathbf{z}_i \| \mathbf{z}_j]) \tag{11.17}$$

$$\tilde{\boldsymbol{\mu}}_i = \text{MLP}_{\tilde{\mu}}(\mathbf{z}_i) \tag{11.18}$$

$$\tilde{\boldsymbol{\sigma}}_i = \text{MLP}_{\tilde{\sigma}}(\mathbf{z}_i), \tag{11.19}$$

where we adopt an edge-independent Bernoulli distribution over edges and node-wise independent Gaussian distribution over node features. $[\mathbf{z}_i \| \mathbf{z}_j]$ means concatenating $\mathbf{z}_i$ and $\mathbf{z}_j$. $\mathbf{x}_i$ is the i-th row of node feature matrix X. The first product term in Eq. (11.14) sums over all n^2 possible edges. The learnable parameters consist of those of three MLPs. This decoder is simple yet powerful. However, given the latent variables Z, the decoder is not permutation invariant in general, *i.e.*, for any two different permutation matrices P_1 and P_2,

$$p(P_1 A P_1^\top, P_1 X|Z) \neq p(P_2 A P_2^\top, P_2 X|Z). \tag{11.20}$$

Note that there are corner cases so that $p(P_1 A P_1^\top, P_1 X|Z) = p(P_2 A P_2^\top, P_2 X|Z)$ holds. For example, if an adjacency matrix A has certain symmetries, there could exist a pair of (P_1, P_2) so that $P_1 A P_1^\top = P_2 A P_2^\top$. But this does not hold for all pairs of (P_1, P_2). As a second example, if all Θ_{ij} are the same for all (i, j), all $\tilde{\boldsymbol{\mu}}_i$ are the same for all i, and all $\tilde{\boldsymbol{\sigma}}_i$ are the same for all i, then for any two permutation matrices (P_1, P_2), we have $p(P_1 A P_1^\top, P_1 X|Z) = p(P_2 A P_2^\top, P_2 X|Z)$. Nevertheless, these two cases happen rarely in practice.

Equipped with the distribution in Eq. (11.14), we can evaluate the terms on the right hand side of Eq. (11.13). However, the number of permutation matrices in $\Pi_{\mathcal{G}}$ can be as large as $n!$ which makes the exact evaluation computationally prohibitive. There are a few ways in the literature to approximate it. For example, we can just use the maximum term as below,

$$p(\mathcal{G}|Z) = \sum_{P \in \Pi_{\mathcal{G}}} p(PAP^\top, PX|Z) \approx \max_{P \in \Pi_{\mathcal{G}}} p(PAP^\top, PX|Z). \tag{11.21}$$

Unfortunately, this maximization problem can be interpreted as an integer quadratic programming which is itself a hard optimization problem. To approximately solve the matching problem, Simonovsky and Komodakis (2018) exploit a relaxed max-pooling matching solver (Cho et al, 2014b). On the other hand, there are some canonical node orderings in certain applications. For example, the simplified molecular-input line-entry system (SMILES) string (Weininger, 1988) provides a sequential ordering of atoms (nodes) of molecule graphs in chemistry. Based on the canonical node ordering, we can construct the corresponding permutation $\tilde{P}$ and simply approximate the conditional probability as,

$$p(\mathcal{G}|Z) = \sum_{P \in \Pi_{\mathcal{G}}} p(PAP^{\top}, PX|Z) \approx p(\tilde{P}A\tilde{P}^{\top}, \tilde{P}X|Z). \tag{11.22}$$

Objective The training objective of GraphVAE is similar to regular VAEs, *i.e.*, the evidence lower bound (ELBO),

$$\max_{\theta, \phi} \quad \mathbb{E}_{q_{\phi}(Z|A,X)}\left[\log p_{\theta}(\mathcal{G}|Z)\right] - \mathrm{KL}(q_{\phi}(Z|A,X)\|p(Z)) \tag{11.23}$$

To learn the encoder and the decoder, we need to sample from the encoder to approximate the expectation in Eq. (11.23) and leverage the reparameterization trick (Kingma and Welling, 2014) to back-propagate the gradient.

11.3.2.2 Hierarchical and Constrained GraphVAEs

There are many variants derived from the GraphVAE family mentioned above. We now briefly introduce two important types of variants, *i.e.*, hierarchical GraphVAE (Jin et al, 2018a) and Constrained GraphVAE (Liu et al, 2018d; Ma et al, 2018).

Hierarchical GraphVAEs One representative work of hierarchical GraphVAEs is *Junction Tree VAEs* (Jin et al, 2018a) which aim at modeling the molecule graphs. The key idea is to build a GraphVAE relying on the hierarchical graph representations of molecules. In particular, we first apply the tree decomposition to obtain a junction tree $\mathcal{T}$ from the original molecule graph $\mathcal{G}$. A *junction tree* is a cluster tree (each node is a set of one or more variables of the original graph) with the running intersection property (Barber, 2004). It provides a coarsened representation of the original graph since one node in a junction tree may correspond to a subgraph with several nodes in the original graph. As shown in Figure 11.1, there are two graphs corresponding to two levels, *i.e.*, the original molecule graph $\mathcal{G}$ (1st level) and the decomposed junction tree $\mathcal{T}$ (2nd level). Since we can efficiently perform tree decomposition to obtain the junction tree, the tree itself is not a latent variable. Jin et al (2018a) propose to use Gated Graph Neural Networks (GGNNs) (Li et al, 2016b) as encoders (one for each level) and construct variational posteriors $q(Z_{\mathcal{G}}|\mathcal{G})$ and $q(Z_{\mathcal{T}}|\mathcal{T})$ as Gaussians. To decode the molecule graph, we need to perform a two-level generation process conditioned on the sampled latent variables $Z_{\mathcal{T}}$ and $Z_{\mathcal{G}}$. A junction tree is first generated by a autoregressive decoder which is again based on GGNNs. Conditioned on the generated tree, Jin et al (2018a) resort to maximum-a-posterior (MAP) formulation to generate the final molecule graph, *i.e.*, finding the compatible subgraphs at each node of the tree so that the overall score (log-likelihood) of the resultant graph (*i.e.*, replacing each node in the tree with the chosen subgraph) is maximized. The whole model can be learned similarly to other

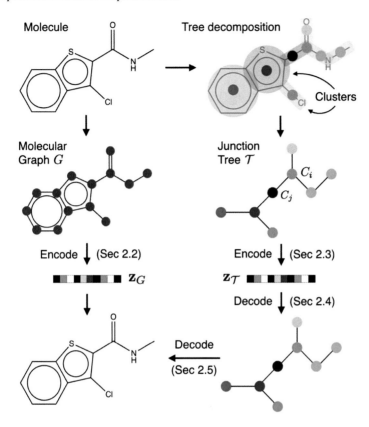

Fig. 11.1: Junction Tree VAEs. The junction tree corresponding to the molecule graph is obtained via the tree decomposition as shown in the top-right. A node/cluster in the junction tree (color-shaded) may correspond to a subgraph in the original molecule graph. Two GNN-based encoders are applied to the molecular graph and junction tree respectively to construct the variational posterior distributions over latent variables $Z_{\mathcal{G}}$ and $Z_{\mathcal{T}}$. During the generation, we first generate the junction tree using an autoregressive decoder and then obtains the final molecule graph via approximately solving a maximum-a-posterior problem. Adapted from Figure 3 of (Jin et al, 2018a).

GraphVAEs. This model provides an interesting extension of GraphVAEs to hierarchical graph generation and demonstrates strong empirical performances. There are other important application-dependent details which greatly improve efficiency. For example, we can build a dictionary of chemically valid subgraphs so that each generation step in the 2nd level decoding generates a subgraph rather than a single node. Nevertheless, the model design largely relies on the efficiency of the chosen junction tree algorithm and certain application-dependent properties. It is unclear how well this model performs on general graphs other than molecules.

Constrained GraphVAEs In many applications of deep graph generative models, certain constraints on the generated graphs are preferred. For example, while generating molecule graphs, the configuration of chemical bonds (edges) must meet the valence criteria of the atoms (nodes). How to ensure the generated graphs satisfy such constraints is a challenging problem. There are generally two types of approaches to overcome it in the context of GraphVAEs. The first type is to design a decoder so that all generated graphs satisfy the constraints by construction. For example, an autoregressive decoder is often adopted as in (Liu et al, 2018d; Dai et al, 2018b). At each step, conditioned on the currently generated graph, the model generates a new node, a new edge, and the node/edge attributes following certain rules, *i.e.*, ruling out invalid options (those would violate the constraints) like what GrammarVAEs (Kusner et al, 2017) do. The other type of approach is to treat the constraints softly. Similar to how constrained optimization problems are converted to unconstrained ones by adding Lagrangians, Ma et al (2018) propose Lagrangian-based regularizers to incorporate constraints like valence constraint for molecule graphs, connectivity constraint, and node compatibility. The benefits of such methods are that the generation could be much simpler and more efficient since we do not need a slow autoregressive decoder. Also, the regularization is only applied during learning and does not bring any overhead in the generation. Of course, the downside is that the generated graph my not exactly satisfy all constraints since the regularization only acts softly in the optimization.

11.3.3 Deep Autoregressive Methods

Deep autoregressive models like PixelRNNs (Van Oord et al, 2016) and PixelCNNs (Oord et al, 2016) have achieved tremendous successes in image modeling. Therefore, it is natural to generalize this type of method to graphs. The shared underlying idea of these autoregressive models is to characterize the graph generation process as a sequential decision-making process and make a new decision at each step conditioning on all previously made decisions. For example, as shown in Figure 11.2, we can first decide whether to add a new node, then decide whether to add a new edge, so on and so forth. If node/edge labels are considered, we can further sample from a categorical distribution at each step to specify such labels. The key question of this class of methods is how to build a probabilistic model so that our current decision depends on all previous historical choices.

11.3.3.1 GNN-based Autoregressive Model

The first GNN-based autoregressive model was proposed in (Li et al, 2018d) of which the high-level idea is exactly the same as shown in Figure 11.2. Suppose at time step $t - 1$, we already generated a partial graph denoted as $\mathscr{G}^{t-1} = (\mathscr{V}^{t-1}, \mathscr{E}^{t-1})$. The corresponding adjacency matrix and node feature matrix are de-

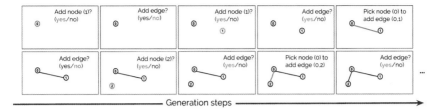

Fig. 11.2: The overview of the deep graph generative model in (Li et al, 2018d). The graph generation is formulated as a sequential decision-making process. At each step of the generation, the model needs to decide: 1) whether add a new node or stop the whole generation; 2) whether add a new edge (one end connected to the new node) or not; 3) which existing node to connect for the new edge. Adapted from Figure 1 of (Li et al, 2018d).

noted as (A^{t-1}, X^{t-1}). At time step t, the model needs to decide: 1) whether we add a new node or we stop the generation (denoting the probability as p_{AddNode}); 2) whether we add an edge that links any existing node to the newly added node (denoting the probability as p_{AddEdge}); 3) choose a existing node to link to the newly added node (denoting the probability as p_{Nodes}). For simplicity, we define p_{AddNode} to be a Bernoulli distribution. We can extend it to a categorical one if node labels/types are considered. p_{AddEdge} is yet another Bernoulli distribution whereas p_{Nodes} is a categorical distribution with size $|\mathcal{V}^{t-1}|$ (*i.e.*, its size will change as the generation goes on).

Message Passing Graph Neural Networks To construct the above probabilities of decisions, we first build a message passing graph neural network (Scarselli et al, 2008; Li et al, 2016b; Gilmer et al, 2017) to learn node representations. The input to the GNN at time step $t-1$ is (A^{t-1}, H^{t-1}) where H^{t-1} is the node representation (one row corresponds to a node). Note that at time 0, since the graph is empty, we need to generate a new node to start. The generation probability p_{AddNode} will be output by the model based on some randomly initialized hidden state. If we model the node labels/types or node features, we can also use them as additional node representations, e.g., concatenating them with rows of H^{t-1}.

The one-step message passing is shown as below,

$$\mathbf{m}_{ij} = f_{\text{Msg}}(\mathbf{h}_i^{t-1}, \mathbf{h}_j^{t-1}) \qquad \forall (i,j) \in \mathcal{E} \qquad (11.24)$$

$$\bar{\mathbf{m}}_i = f_{\text{Agg}}(\{\mathbf{m}_{ij} | \forall j \in \Omega_i\}) \qquad \forall i \in \mathcal{V} \qquad (11.25)$$

$$\tilde{\mathbf{h}}_i^{t-1} = f_{\text{Update}}(\mathbf{h}_i^{t-1}, \bar{\mathbf{m}}_i) \qquad \forall i \in \mathcal{V}, \qquad (11.26)$$

where f_{Msg}, f_{Agg}, and f_{Update} are the message function, the aggregation function, and the node update function respectively. For the message function, we often instantiate f_{Msg} as an MLP. Note that if edge features are considered, one can incorporate them as input to f_{Msg}. f_{Agg} could simply be an average or summation operator. Typical examples of f_{Update} include gated recurrent units (GRUs) (Cho et al, 2014a) and long-short term memory (LSTM) (Hochreiter and Schmidhuber, 1997). $\mathbf{h}_i^{t-1}$ is the input node representation at time step $t-1$. Ω_i denotes the set of neighboring nodes of the node i. $\tilde{\mathbf{h}}_i^{t-1}$ is the updated node representation which serves as the input node representation for the next message passing step. The above message passing process is typically executed for a fixed number of steps, which is tuned as a hyperparameter. Note that the generation step t is different from the message passing step (we deliberately omit its notation to avoid confusion).

Output Probabilities After the message passing process is done, we obtain the new node representations H^t. Now we can construct the aforementioned output probabilities as follows,

$$\mathbf{h}_{\mathcal{G}^{t-1}} = f_{\text{ReadOut}}(H^t) \tag{11.27}$$

$$p_{\text{AddNode}} = \text{Bernoulli}(\sigma(\text{MLP}_{\text{AddNode}}(\mathbf{h}_{\mathcal{G}^{t-1}}))) \tag{11.28}$$

$$p_{\text{AddEdge}} = \text{Bernoulli}(\sigma(\text{MLP}_{\text{AddEdge}}(\mathbf{h}_{\mathcal{G}^{t-1}}, \mathbf{h}_v))) \tag{11.29}$$

$$s_{uv} = \text{MLP}_{\text{Nodes}}(\mathbf{h}_u^t, \mathbf{h}_v) \qquad \forall u \in \mathcal{V}^{t-1} \tag{11.30}$$

$$p_{\text{Nodes}} = \text{Categorical}(\text{softmax}(\mathbf{s})). \tag{11.31}$$

Here we first summarize the graph representation $\mathbf{h}_{\mathcal{G}^{t-1}}$ (a vector) by reading out from the node representation H^t via f_{ReadOut}, which could be an average operator or an attention-based one. Based on $\mathbf{h}_{\mathcal{G}^{t-1}}$, we predict the probability of adding a new node p_{AddNode} where σ is the sigmoid function. If we decide to add a new node by sampling 1 from the Bernoulli distribution p_{AddNode}, we denote the new node as v. We can initialize its representation $\mathbf{h}_v$ as random features by sampling either from $\mathcal{N}(0, I)$ or learned distribution over node type/label if provided. Then we compute similarity scores between every existing node u in $\mathcal{G}^{t-1}$ and v as s_{uv}. $\mathbf{s}$ is the concatenated vector of all similarity scores. Finally, we normalize the scores using softmax to form the categorical distribution from which we sample an existing node to obtain the new edge. By sampling from all these probabilities, we could either stop the generation or obtain a new graph with a new node and/or a new edge. We repeat this procedure by carrying on the node representations along with the generated graphs until the model generates a stop signal from p_{AddNode}.

Training To train the model, we need to maximize the likelihood of the observed graphs. Recall that we need to consider the permutations that leave the graph unchanged as discussed in Section 11.3.2.1. For simplicity, we focus on the adjacency matrix alone following (Li et al, 2018d), i.e., $\mathcal{G} \equiv \{PAP^\top | P \in \Pi_{\mathcal{G}}\}$, where $\Pi_{\mathcal{G}}$ is the maximal subset of Π so that $P_1 A P_1^\top \neq P_2 A P_2^\top$ holds for any $P_1, P_2 \in \Pi_{\mathcal{G}}$. The ideal objective is to maximize the following,

$$\max \quad \log p(\mathcal{G}) \quad \Leftrightarrow \quad \max \quad \log \left(\sum_{P \in \Pi_{\mathcal{G}}} p(PAP^\top) \right). \tag{11.32}$$

Here we omit the variables being optimized, i.e., parameters of models defined in Eq. (11.24) and Eq. (11.27). Note that given a node ordering (corresponding to one specific permutation matrix P), we have a bijection between a sequence of correct decisions and an adjacency matrix. In other words, we can equivalently write $p(PAP^\top)$ as a product of probabilities that are explained in Eq. (11.27). However, the marginalization inside the logarithmic function on the right hand side is intractable due to the nearly factorial size of $\Pi_{\mathcal{G}}$ in practice. Li et al (2018d) propose to randomly sample a few different node orderings as $\tilde{\Pi}_{\mathcal{G}}$ and train the model with following approximated objective,

$$\max \quad \log \left(\sum_{P \in \tilde{\Pi}_{\mathcal{G}}} p(PAP^\top) \right). \tag{11.33}$$

Note that this objective is a strict lower bound of the one in Eq. (11.32). If canonical node orderings like the SMILES ordering for molecule graphs are available, we can also use that to compute the above objective.

Discussion This model formulates the graph generation as a sequential decision-making process and provides a GNN-based autoregressive model to construct probabilities of possible decisions at each step. The overall model design is well-motivated. It also achieves good empirical performances in generating small graphs like molecules (e.g., less than 40 nodes). However, since the model only generates at most one new node and one new edge per step, the total number of generation steps scales with the number of nodes quadratically for dense graphs. It is thus inefficient to generate moderately large graphs (e.g., with a few hundreds of nodes).

11.3.3.2 Graph Recurrent Neural Networks (GraphRNN)

Graph Recurrent Neural Networks (GraphRNN) (You et al, 2018b) is another deep autoregressive model which has a similar sequential decision-making formulation and leverages RNNs to construct the conditional probabilities. We again rely on the adjacency matrix representation of a graph, i.e., $\mathcal{G} \equiv \{PAP^\top | P \in \Pi_{\mathcal{G}}\}$. Before dealing with the permutations, let us assume the node ordering is given so that $P = I$.

A Simple Variant of GraphRNN GraphRNN starts with an autoregressive decomposition of the probability of an adjacency matrix as follows,

$$p(A) = \prod_{t=1}^{n} p(A_t | A_{<t}), \tag{11.34}$$

where A_t is the t-th column of the adjacency matrix A and $A_{<t}$ is a matrix formed by columns $A_1, A_2, \cdots, A_{t-1}$. n is the maximum number of nodes. If a graph has less than n nodes, we pad dummy nodes similarly as discussed in Section 11.3.1. Then we can construct the conditional probability as an edge-independent Bernoulli distribution,

$$p(A_t|A_{<t}) = \text{Bernoulli}(\Theta_t) = \prod_{i=1}^{n} \Theta_{t,i}^{\mathbf{1}[A_{i,t}=1]}(1 - \Theta_{t,i})^{\mathbf{1}[A_{i,t}=0]} \qquad (11.35)$$

$$\Theta_t = f_{\text{out}}(\mathbf{h}_t) \qquad (11.36)$$

$$\mathbf{h}_t = f_{\text{trans}}(\mathbf{h}_{t-1}, A_{t-1}), \qquad (11.37)$$

where Θ_t is a size-n vector of Bernoulli parameters. $\Theta_{t,i}$ denotes its i-th element. $A_{i,t}$ denotes the i-th element of the column vector A_t. f_{out} could be an MLP which takes the hidden state $\mathbf{h}_t$ as input and outputs Θ_t. f_{trans} is the RNN cell function which takes the $(t-1)$-th column of the adjacency matrix A_{t-1} and the hidden state $\mathbf{h}_{t-1}$ as input and outputs the current hidden state $\mathbf{h}_t$. We can use an LSTM or GRU as the RNN cell function. Note that the conditioning on $A_{<t}$ is implemented via the recurrent use of the hidden state in an RNN. The hidden state can be initialized as zeros or randomly sampled from a standard normal distribution. This model variant is very simple and can be easily implemented since it only consists of a few common neural network modules, $i.e.$, an RNN and an MLP.

Full Version of GraphRNN To further improve the model, You et al (2018b) propose a full version of GraphRNN. The idea is to build a hierarchical RNN so that the conditional distribution in Eq. (11.34) becomes more expressive. Specifically, instead of using an edge-independent Bernoulli distribution, we leverage another autoregressive construction to model the dependencies among entries within one column of the adjacency matrix as below,

$$p(A_t|A_{<t}) = \prod_{i=1}^{n} p(A_{i,t}|A_{<i,<t}) \qquad (11.38)$$

$$p(A_{i,t}|A_{<i,<t}) = \text{sigmoid}(g_{\text{out}}(\tilde{\mathbf{h}}_{i,t})) \qquad (11.39)$$

$$\tilde{\mathbf{h}}_{i,t} = g_{\text{trans}}(\tilde{\mathbf{h}}_{i-1,t}, A_{<i,t}) \qquad (11.40)$$

$$\tilde{\mathbf{h}}_{0,t} = \mathbf{h}_t \qquad (11.41)$$

$$\mathbf{h}_t = f_{\text{trans}}(\mathbf{h}_{t-1}, A_{t-1}). \qquad (11.42)$$

Here the bottom RNN cell function f_{trans} still recurrently updates the hidden state to get $\mathbf{h}_t$, thus implementing the conditioning on all previous $t-1$ columns of the adjacency matrix A. To generate individual entries of the t-th column, the top RNN cell function g_{trans} takes its own hidden state $\tilde{\mathbf{h}}_{i-1,t}$ and the already generated t-th column A as input and updates the hidden state as $\tilde{\mathbf{h}}_{i,t}$. The output distribution is a Bernoulli parameterized by the output of an MLP g_{out} which takes $\tilde{\mathbf{h}}_{i,t}$ as input. Note that the initial hidden state $\tilde{\mathbf{h}}_{0,t}$ of the top RNN is set to the hidden state $\mathbf{h}_t$ returned by the bottom RNN.

Objective To train the GraphRNN, we can again resort to the maximum log likelihood similarly to Section 11.3.3.1. We also need to deal with permutations of nodes that leave the graph unchanged. Instead of randomly sampling a few orderings like (Li et al, 2018d), You et al (2018b) propose to use a random-breadth-first-search ordering. The idea is to first randomly sample a node ordering and then pick the first node in this ordering as the root. A breadth-first-search (BFS) algorithm is applied starting from this root node to generate the final node ordering. Let us denote the corresponding permutation matrix as P_{BFS}. The final objective is,

$$\max \quad \log\left(p(P_{\text{BFS}}AP_{\text{BFS}}^\top)\right), \tag{11.43}$$

which is again a strict lower bound of the true log likelihood. Empirical results in (You et al, 2018b) suggest that this random-BFS ordering provides good performances on a few benchmarks.

Discussion The design of the GraphRNN is simple yet effective. The implementation is straightforward since most of the modules are standard. The simple variant is more efficient than the previous GNN-based model (Li et al, 2018d) since it generates multiple edges (corresponding to one column of the adjacency matrix) per step. Moreover, the simple variant performs comparably with the full version in the experiments. Nevertheless, GraphRNN still has certain limitations. For example, RNN highly depends on the node ordering since different node orderings would result in very different hidden states. The sequential ordering could make two nearby (even neighboring) nodes far away in the generation sequence (*i.e.*, far away in the generation time step). Typically, hidden states of an RNN that are far away regarding the generation time step tend to be quite different, thus making it hard for the model to learn that these nearby nodes should be connected. We call this phenomenon the *sequential ordering bias*.

11.3.3.3 Graph Recurrent Attention Networks (GRAN)

Following the line of the work (Li et al, 2018d; You et al, 2018b), Liao et al (2019a) propose the graph recurrent attention networks (GRAN). It is a GNN-based autoregressive model, which greatly improves the previous GNN-based model (Li et al, 2018d) in terms of capacity and efficiency. Furthermore, it alleviates the *sequential ordering bias* of GraphRNN (You et al, 2018b). In the following, we introduce the details of the model.

Model We start with the adjacency matrix representation of graphs, *i.e.*, $\mathscr{G} \equiv \{PAP^\top | P \in \Pi_{\mathscr{G}}\}$. GRAN aims at directly building a probabilistic model over the adjacency matrix similarly to GraphRNN. Again, node/edge features are not of primary interests but can be incorporated without much modification to the model. In particular, from the perspective of modeling the adjacency matrix, the GNN-based autoregressive model in (Li et al, 2018d) generates one entry of the adjacency matrix at a step, whereas GraphRNN (You et al, 2018b) generates one column of entries at a step. GRAN takes a step further along this line by generating a block of columns/rows[2] of the adjacency matrix at a step, which greatly improves the generation speed. Denoting the submatrix with first k rows of the adjacency matrix A as $A_{1:k,:}$, we have the following autoregressive decomposition of the probability,

---

[2] Since we are mainly interested in simple graphs, *i.e.*, unweighted, undirected graphs containing no self-loops or multiple edges, modeling columns or rows makes no difference. We adopt the row-wise notations to follow the original paper.

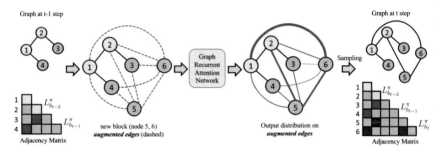

Fig. 11.3: The overview of the graph recurrent attention networks (GRAN). At each step, given an already generated graph, we add a new block of nodes (block size is 2 and color indicates the membership of individual group in the visualization) and augmented edges (dashed lines). Then we apply GRAN to this graph to obtain the output distribution over augmented edges (we show an edge-independent Bernoulli where the line width indicates the probability of generating individual augmented edges). Finally, we sample from the output distribution to obtain a new graph. Adapted from Figure 1 of (Liao et al, 2019a).

$$p(A) = \prod_{t=1}^{\lceil n/k \rceil} p(A_{(t-1)k:tk,:} | A_{:(t-1)k,:}), \tag{11.44}$$

where $A_{:(t-1)k,:}$ indicates the adjacency matrix that has been generated before the t-th step (*i.e.*, $t-1$ blocks with block size k). We use $A_{(t-1)k:tk,:}$ to denote the to-be-generated block at t-th time step. Note that this part is a straightforward generalization to the autoregressive model of GraphRNNs in Eq. (11.34).

To build the condition probability $p(A_{(t-1)k:tk,:} | A_{:(t-1)k,:})$, GRAN leverages a message passing graph neural network. Specifically, denoting the already generated graph before step t (corresponding to $A_{:(t-1)k,:}$) as $\mathcal{G}^{t-1} = (\mathcal{V}^{t-1}, \mathcal{E}^{t-1})$, we first initialize every node representation vector with its corresponding row of the adjacency matrix, *i.e.*, $\mathbf{h}_v = A_{v,:}$ for all $v \leq (t-1)k$. Since we assume the maximum number of nodes is n and pad dummy nodes for graphs with a smaller size, $\mathbf{h}_v$ is of size n. At time step t, we are interested in generating a new block of nodes (corresponding to $A_{(t-1)k:tk,:}$) and their associated edges. For the k new nodes in the t-th block, since their corresponding rows in the adjacency matrix are initially all zeros, we give them an arbitrary ordering from 1 to k and use the one-hot-encoding of the order index as an additional representation to distinguish them, denoting as x_u. We first form a new graph $\tilde{\mathcal{G}}^t = (\mathcal{V}^t, \tilde{\mathcal{E}}^t)$ by connecting the k new nodes to themselves (excluding self-loops) and every other nodes in $\mathcal{G}^{t-1}$. We call such edges as the augmented edges, which are shown as the dashed edges in Figure 11.3. In other words, $\mathcal{V}^t$ is the union of $\mathcal{V}^{t-1}$ and k new nodes whereas $\tilde{\mathcal{E}}^t$ is the union of $\mathcal{E}^{t-1}$ and augmented edges. The core part of GRAN is to construct a probability distribution over such augmented edges from which we can sample a new graph $\mathcal{G}^t$. Note that $\mathcal{G}^t$ has the same set of nodes but potentially fewer edges compared to $\tilde{\mathcal{G}}^t$. To construct the probability, we use a GNN with the following one-step message passing process,

$$\mathbf{m}_{ij} = f_{\mathrm{msg}}(\mathbf{h}_i - \mathbf{h}_j), \qquad \forall (i,j) \in \tilde{\mathscr{E}}^t \qquad (11.45)$$

$$\tilde{\mathbf{h}}_i = [\mathbf{h}_i \| \mathbf{x}_i], \qquad \forall i \in \mathscr{V}^t \qquad (11.46)$$

$$a_{ij} = \mathrm{sigmoid}\left(g_{\mathrm{att}}(\tilde{\mathbf{h}}_i - \tilde{\mathbf{h}}_j)\right), \qquad \forall (i,j) \in \tilde{\mathscr{E}}^t \qquad (11.47)$$

$$\mathbf{h}'_i = \mathrm{GRU}(\mathbf{h}_i, \sum_{j \in \Omega(i)} a_{ij} \mathbf{m}_{ij}), \qquad \forall i \in \mathscr{V}^t \qquad (11.48)$$

where $\mathbf{m}_{ij}$ is the again the message over edge (i,j) and Ω_i is the set of neighboring nodes of node i. The message function f_{msg} and the attention head g_{att} could be MLPs. Note that we set $\mathbf{x}_u$ to zeros for any node u that is in the already generated graph $\mathscr{G}^{t-1}$ since the one-hot-encoding is only used to distinguish those newly added nodes. $[a\|b]$ means concatenating two vectors a and b. The updated node representation $\mathbf{h}'_i$ would serve as the input to the next message passing step. We typically unroll this message passing for a fixed number of steps, which is set as a hyperparameter. Note that the message passing step is independent of the generation step. The attention weights a_{ij} depends on the one-hot-encoding $\mathbf{x}_i$ so that messages on augmented edges could be weighted differently compared to those on edges belonging to $\mathscr{E}^{t-1}$. Based on the final node representations returned by the message passing, we can construct the output distribution is as follows,

$$p(A_{(t-1)k:tk,:} | A_{:(t-1)k,:}) = \sum_{c=1}^{C} \alpha_c \prod_{i=(t-1)k+1}^{tK} \prod_{j=1}^{n} \Theta_{c,i,j} \qquad (11.49)$$

$$\alpha = \mathrm{softmax}\left(\sum_{i=(t-1)k+1}^{tK} \sum_{j=1}^{n} \mathrm{MLP}_\alpha(\mathbf{h}_i^R - \mathbf{h}_j^R) \right) \qquad (11.50)$$

$$\Theta_{c,i,j} = \mathrm{sigmoid}\left(\mathrm{MLP}_\Theta(\mathbf{h}_i^R - \mathbf{h}_j^R) \right). \qquad (11.51)$$

Here we use a mixture of Bernoulli distributions where the mixture coefficients are $\alpha = \{\alpha_1, \cdots, \alpha_C\}$ and the parameters are $\{\Theta_{c,i,j}\}$. Compared to the edge-independent Bernoulli distribution used in the simple variant of GraphRNN, this output distribution can capture dependencies among multiple generated edges. Furthermore, it is more efficient to sample compared to the autoregressive distribution used in the full version of GraphRNN.

Objective To train the model, we also need to deal with permutations in order to maximize the log likelihood. Similar to the strategy used in (Li et al, 2018d; You et al, 2018b), Liao et al (2019a) propose to use a set of canonical orderings, *i.e.*, breadth-first-search (BFS), depth-first-search (DFS), node-degree-descending, node-degree-ascending, and the k-core ordering. In particular, the BFS and the DFS ordering start from the node with the largest node degree. The k-core graph decomposition has been shown to be very useful for modeling cohesive groups in social networks (Seidman, 1983). The k-core of a graph $\mathscr{G}$ is a maximal subgraph that contains nodes of degree k or more. Cores are nested, *i.e.*, i-core belongs to j-core if $i > j$, but they are not necessarily connected subgraphs. Most importantly, the core decomposition, *i.e.*, all cores ranked based on their orders, can be found in linear time (w.r.t. the number of edges) (Batagelj and Zaversnik, 2003). Based on the largest core number per node, we can uniquely determine a partition of all nodes, *i.e.*, disjoint sets of nodes which share the same largest core number. We then assign the core number of each disjoint set by the largest core number of its nodes. Starting from the set with the largest core number, we rank all nodes within the set in node degree descending order. Then we move to the second largest core and so on to obtain the final ordering of all nodes. We call this core descending ordering as *k-core node ordering*.

Our final training objective is,

$$\max \quad \log \left(\sum_{P \in \tilde{\Pi}_{\mathcal{G}}} p(PAP^{\top}) \right). \tag{11.52}$$

where $\tilde{\Pi}_{\mathcal{G}}$ is the set of permutation matrices corresponding to the above node orderings. This is again a strict lower bound of the true log likelihood.

Discussion GRAN improves the previous GNN-based autoregressive model (Li et al, 2018d) and GraphRNN (You et al, 2018b) in the following ways. First, it generates a block of rows of the adjacency matrix per step, which is more efficient than generating an entry per step and then generating a row per step. Second, GRAN uses a GNN to construct the conditional probability. This helps alleviate the sequential ordering bias in GraphRNN since GNN is permutation equivariant, *i.e.*, the node ordering would not affect the conditional probability per step. Third, the output distribution in GRAN is more expressive and more efficient for sampling. GRAN outperforms previous deep graph generative models in terms of empirical performances and the sizes of graphs that can be generated (e.g., GRAN can generate graphs up to 5K nodes). Nevertheless, GRAN still suffers from the fact that the overall model depends on the particular choices of node orderings. It may be hard to find good orderings in certain applications. How to build an order-invariant deep graph generative model would be an interesting open question.

11.3.4 Generative Adversarial Methods

In this part, we review a few methods (De Cao and Kipf, 2018; Bojchevski et al, 2018; You et al, 2018a) that apply the idea of generative adversarial networks (GAN) (Goodfellow et al, 2014b) in the context of graph generation. Based on how a graph is represented during training, we roughly divide them into two categories: adjacency matrix based and random walks based methods. In the following, we explain these two types of methods in detail.

11.3.4.1 Adjacency Matrix Based GAN

MolGAN (De Cao and Kipf, 2018) and graph convolutional policy network (GCPN) (You et al, 2018a) propose a similar GAN-based framework to generate molecule graphs that satisfy certain chemical properties. Here the graph data is represented slightly different from previous sections since one needs to specify both node types (*i.e.*, atom types) and edge types (*i.e.*, chemical bond types). We denote the adjacency matrix[3] as $A \in \mathbb{R}^{N \times N \times Y}$ where Y is the number of chemical bond types. Basically, one slice along the 3rd dimension of A gives an adjacency matrix that characterizes the connectivities among atoms under a specific chemical bond type. We denote the node type as $X \in \mathbb{R}^{N \times T}$ where T is the number of atom types. The goal is to generate (A, X) so that it is similar to observed molecule graphs and possesses certain desirable properties.

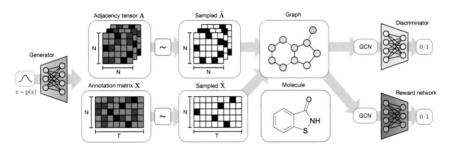

Fig. 11.4: The overview of the MolGAN. We first draw a latent variable $Z \sim p(Z)$ and feed it to a generator which produces a probabilistic (continuous) adjacency matrix A and a probabilistic (continuous) node type matrix X. Then we draw a discrete adjacency matrix $\tilde{A} \sim A$ and a discrete node type matrix $\tilde{X} \sim X$, which together specify a molecule graph. During training, we simultaneously feed the generated graph to a discriminator and a reward network to obtain the adversarial loss (measuring how similar the generated and the observed graphs are) and the negative reward (measuring how likely the generated graphs satisfy the certain chemical constraints). Adapted from Figure 2 of (De Cao and Kipf, 2018).

Model We now explain the details of MolGAN and then highlight the difference between GCPN and MolGAN. Similar to regular GANs, MolGAN consists of a generator $\bar{\mathcal{G}}_\theta(Z)$ and a discriminator $\mathcal{D}_\phi(A, X)$. To ensure the generated samples satisfy desirable chemical properties, MolGAN adopts an additional reward network $\mathcal{R}_\psi(A, X)$. The overall pipeline of MolGAN is illustrated in Figure 11.4.

To generate a molecule graph, we first sample a latent variable $Z \in \mathbb{R}^d$ from some prior, e.g., $Z \sim \mathcal{N}(0, I)$. Then we use an MLP to directly map the sampled Z to a continuous adjacency matrix A and a continuous node type matrix X. The continuous version of the graph data has a natural probabilistic interpretation, *i.e.*, $A_{i,j,c}$

---

[3] Note that A is actually a tensor. We slightly abuse the terminology here to ease the exposition.

means the probability of connecting the atom i and the atom j using the chemical bond type c, whereas $X_{i,t}$ means the probability of assigning the t-th atom type to the i-atom. One can sample a discrete graph data $(\tilde{A}, \tilde{X})$ from the continuous version, i.e., $\tilde{A} \sim A$ and $\tilde{X} \sim X$. This sampling procedure can be implemented using the Gumbel softmax (Jang et al, 2017; Maddison et al, 2017). The discrete adjacency matrix $\tilde{A}$ along with the discrete node type $\tilde{X}$ specify a molecule graph and complete the generation process.

To evaluate how similar the generated graphs and the observed graphs are, we need to build a discriminator. Since we are dealing with graphs, the natural candidate for a discriminator is a graph neural network, e.g., a graph convolutional network (GCN) (Kipf and Welling, 2017b). In particular, we use a variant of GCN (Schlichtkrull et al, 2018) to incorporate multiple edge types. One such graph convolutional layer is shown as below,

$$\mathbf{h}'_i = \tanh\left(f_s(\mathbf{h}_i, \mathbf{x}_i) + \sum_{j=1}^{N} \sum_{y=1}^{Y} \frac{\tilde{A}_{i,j,y}}{|\Omega_i|} f_y(\mathbf{h}_j, \mathbf{x}_i) \right), \tag{11.53}$$

where $\mathbf{h}_i$ and $\mathbf{h}'_i$ are the input and the output node representations of the graph convolutional layer. Ω_i is the set of neighboring nodes of the node i. $\mathbf{x}_i$ is the i-th row of X, i.e., the node type vector of the node i. f_s and f_y are linear transformation functions that are to be learned. After stacking this type of graph convolution for multiple layers, we can readout the graph representation using the following attention-weighted aggregation,

$$\mathbf{h}_{\mathscr{G}} = \tanh\left(\sum_{v \in \mathscr{V}} \text{sigmoid}\left(\text{MLP}_{\text{att}}(\mathbf{h}_v, \mathbf{x}_v)\right) \odot \tanh\left(\text{MLP}(\mathbf{h}_v, \mathbf{x}_v)\right) \right), \tag{11.54}$$

where $\mathbf{h}_v$ is the node representation returned by the top graph convolutional layer. Note that MLP_{att} and MLP are two different instances of MLPs. $\odot$ means element-wise product. We can use the graph representation vector $\mathbf{h}_{\mathscr{G}}$ to compute the discriminator score $\mathscr{D}_\phi(A, X)$, i.e., the probability of classifying a graph as positive (i.e., coming from the data distribution).

Objective Originally, GANs learn the model by performing the minimax optimization as below,

$$\min_\theta \max_\phi \quad \mathbb{E}_{A, X \sim p_{\text{data}}(A, X)}[\log \mathscr{D}_\phi(A, X)] + \mathbb{E}_{Z \sim p(Z)}[\log\left(1 - \mathscr{D}_\phi(\tilde{\mathscr{G}}_\theta(Z))\right)], \tag{11.55}$$

where the generator aims at fooling the discriminator and the discriminator aims at correctly classifying the generated samples and the observed samples. To address certain issues in training GANs such as the mode collapse and the instability, Wasserstein GAN (WGAN) (Arjovsky et al, 2017) and its improved version (Gulrajani et al, 2017) have been proposed. MolGAN follows the improved WGAN and uses the following objective to train the discriminator $\mathscr{D}_\phi(A, X)$,

$$\max_{\phi} \quad \sum_{i=1}^{B} -\mathscr{D}_{\phi}(A^{(i)}, X^{(i)}) + \mathscr{D}_{\phi}(\bar{\mathscr{G}}_{\theta}(Z^{(i)})) + \alpha \left(\|\nabla_{\hat{A}^{(i)}, \hat{X}^{(i)}} \mathscr{D}_{\phi}(\hat{A}^{(i)}, \hat{X}^{(i)})\| - 1 \right)^{2},$$

$$(11.56)$$

where B is the mini-batch size, $Z^{(i)}$ is the i-th sample drawn from the prior, $A^{(i)}, X^{(i)}$ are the i-th graph data drawn from the data distribution, and $\hat{A}^{(i)}, \hat{X}^{(i)}$ are their linear combinations, i.e., $(\hat{A}^{(i)}, \hat{X}^{(i)}) = \varepsilon(A^{(i)}, X^{(i)}) + (1 - \varepsilon)\bar{\mathscr{G}}_{\theta}(Z^{(i)})$, $\varepsilon \sim \mathscr{U}(0,1)$. The squared term on the right-hand side penalizes the gradient of the discriminator so that the training becomes more stable. α is a weighting term to balance the regularization and the objective. Moreover, fixing the discriminator, we train the generator $\bar{\mathscr{G}}_{\theta}(A, X)$ by adding the additional constraint-dependent reward,

$$\min_{\theta} \quad \sum_{i=1}^{B} \lambda \mathscr{D}_{\phi}(\bar{\mathscr{G}}_{\theta}(Z^{(i)})) + (1 - \lambda)\mathscr{L}_{\text{RL}}(\bar{\mathscr{G}}_{\theta}(Z^{(i)})), \qquad (11.57)$$

where $\mathscr{L}_{\text{RL}}$ is the negative reward returned by the reward network $\mathscr{R}_{\psi}$ and λ is the weighting hyperparameter to regulate the trade-off between two losses. The reward could be some non-differentiable quantities that characterize the chemical properties of the generated molecules, e.g., how likely the generated molecule is to be soluble in water. To learn the model with the non-differentiable reward, the deep deterministic policy gradient (DDPG) (Lillicrap et al, 2015) is used. The architecture of the reward network is the same as the discriminator, i.e., a GCN. It is pre-trained by minimizing the squared error between the predicted reward given by $\mathscr{R}_{\psi}$ and an external software which produces a property score per molecule. The pre-training is necessary since the external software is typically slow and could significantly delay the training if it is included in the whole training framework.

Discussion MolGAN demonstrates strong empirical performances on a large chemical database called QM9 (Ramakrishnan et al, 2014). Similar to other GANs, the model is likelihood-free and can thus enjoy more flexible and powerful generators. More importantly, although the generator still depends on the node ordering, the discriminator and the reward networks are order (permutation) invariant since they are built from GNNs. Interestingly enough, graph convolutional policy network (GCPN) (You et al, 2018a) solves the same problem using a similar approach. GCPN has a similar GAN-type of objective and some additional domain-specific rewards that capture the chemical properties of the molecules. It also learns both a generator and a discriminator. However, they do not use a reward network to speed up the reward computation. To deal with the learning of non-differentiable reward, GCPN leverages the proximal policy optimization (PPO) (Schulman et al, 2017) method, which empirically performs better than the vanilla policy gradient method. Another important difference is that GCPN generates the adjacency matrix in an entry-by-entry autoregressive fashion so that the dependencies among multiple generated edges are captured whereas MolGAN generates all entries of the adjacency matrix in parallel conditioned on the latent variable. GCPN also achieves impressive empirical results on another large chemical database called ZINC (Irwin et al, 2012). Nevertheless, there are still limitations with the above models. The discrete gradient estimators (e.g., the policy gradient type of methods) could have large variances, which may slow down the training. Since the domain-specific rewards are non-differentiable and may be time-consuming to obtain, learning a neural network based approximated reward function like what MolGAN does is appealing. However, as reported in MolGAN, pre-training seems to be crucial to make the whole training successful. More exploration along the line of learning a reward function would be beneficial to simplify the whole training pipeline. On the other hand, both methods use some variant of GCNs as the discriminator, which is shown to be insufficient in distinguishing certain graphs[4] (Xu et al, 2019d). Therefore, exploring more powerful discriminators like the Lanczos network (Liao et al, 2019b) that exploits the spectrum of the graph Laplacian as the input feature would be promising to further improve the performance of the above methods.

11.3.4.2 Random Walk Based GAN

In contrast to previous methods, NetGAN (Bojchevski et al, 2018) resorts to the random walk based representations of graphs. The key idea is to map a graph to a set of random walks and learn a generator and a discriminator in the space of random walks. The generator should generate random walks that are similar to those sampled from the observed graphs, whereas the discriminator should correctly distinguish whether a random walk comes from the data distribution or the implicit distribution corresponding to the generator.

Model We start by sampling a set of random walks with fixed length T from the given graph $\mathscr{G}$ using the biased second order random walk sampling strategy described in (Grover and Leskovec, 2016). We denote a random walk as a sequence $(v_1, \cdots, v_T)$ where v_i represents one node in $\mathscr{G}$. Note that a random walk may contain duplicate nodes since it could revisit one node multiple times during the sampling. We again assume the maximum number of nodes for any graph is N. For any node v_i, we use the one-hot-encoding vector as its node feature. In other words, we can view a random walk with a sequence along with its features. Therefore, similar to language models, it is natural to use an RNN as the generator for generating such random walks. NetGAN exploits an LSTM as the generator of which the initial hidden state $\mathbf{h}_0$ and the memory c_0 are computed by feeding a randomly sampled latent vector (drawn from $\mathscr{N}(0, \mathbf{I})$) to two separate MLPs. Then the LSTM generator predicts a categorical distribution over all possible nodes and then samples a node. The one-hot-encoding of the node index is treated as the node representation and fed to the LSTM generator as the input for the next step. We unroll this LSTM for T steps to obtain the final length-T random walk. For the discriminator, we can use another LSTM, which takes a random walk as input and predicts the probability that a given random walk is sampled from the data distribution. The model is trained with the same objective as the improved WGAN (Gulrajani et al, 2017).

---

[4] For example, a GCN can not distinguish two triangles versus a six node circle (both have the same number of nodes and every node has exactly two neighbors) assuming all individual node features are identical.

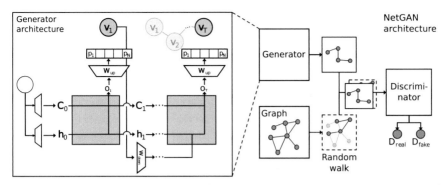

Fig. 11.5: The overview of the NetGAN. We first draw a random vector from a fixed prior $\mathcal{N}(0, \mathbf{I})$ and initialized the memory c_0 and the hidden state $\mathbf{h}_0$ of an LSTM. Then the LSTM generator generates which node to visit per step and is unrolled for a fixed number of steps T. The one-hot-encoding of node index is fed to the LSTM as the input for the next step. The discriminator is another LSTM which performs a binary classification to predict if a given random walk is sampled from a data distribution. Adapted from Figure 2 of (Bojchevski et al, 2018).

After training the LSTM generator, we are capable of generating random walks. However, we need an additional step to construct a graph from a set of generated random walks. The strategy used by NetGAN is as follows. First, we count the edges that appeared in the set of random walks to obtain a scoring matrix S, which has the same size as the adjacency matrix. The (i, j)-th entry of the score matrix $S_{i,j}$ indicates how many times edge (i, j) appears in the set of generated random walks. Second, for each node i, we sample a neighbor according to the probability $\frac{S_{i,j}}{\sum_v S_{i,v}}$. We repeat the sampling until node i has at least one neighbor connected and skip if the edge has already been generated. At last, for any edge (i, j), we perform sampling without replacement according to the probability $\frac{S_{i,j}}{\sum_{u,v} S_{u,v}}$ until the maximum number of edges is reached.

Discussion The random walk based representations for graphs are novel in the context of deep graph generative models. Moreover, they could be more scalable than the adjacency matrix representation since we are not bound by the quadratic (w.r.t. the number of nodes) complexity. The core modules of the NetGAN are LSTMs which are efficient in handling sequences and easy to be implemented. Nevertheless, the graph construction from a set of generated random walks seems to be ad-hoc. There is no theoretical guarantee on how accurate the proposed construction method is. It may require a large number of sampled random walks in order to generate a graph with good qualities.

11.4 Summary

In this chapter, we review a few classic graph generative models and some modern ones which are constructed based on deep neural networks. From the perspectives of the model capacity and the empirical performances, e.g., how good the model can fit observed data, deep graph generative models significantly outperform their classic counterparts. For example, they could generate molecule graphs which are both chemically valid and similar to observed ones in terms of certain graph statistics.

Although we have already made impressive progress in recent years, deep generative models are still in the early stage. Moving forward, there are at least two main challenges. First, how can we scale these models so that they can handle real-world graphs like large scale social networks and WWW? It requires not only more computational resources but also more algorithmic improvements. For example, building a hierarchical graph generative model would be one promising direction to boost efficiency and scale. Second, how can we effectively add domain-specific constraints and/or conditioning on some input information? This question is important since many real-world applications require the graph generation to be conditioned on some inputs (e.g., scene graph generations conditioned on input images). Many graphs in practice come with certain constraints (e.g., chemical validity in the molecule generation).

Editor's Notes: Deep learning-based graph generation can be considered as a downstream task of graph representation learning, where the learned representations are usually enforced to follow some probabilistic assumptions. Hence the techniques in this topic widely enjoy the relevant properties and theories introduced in the previous chapters, such as scalability (Chapter 6), expressiveness power (Chapter 5), and robustness (Chapter 8). Graph generation also further motivates its downstream tasks in various interesting, important, yet usually challenging areas such as drug discovery (see Chapter 24), protein analysis (see Chapter 25), and program synthesis (see Chapter 22).

Chapter 12
Graph Neural Networks: Graph Transformation

Xiaojie Guo, Shiyu Wang, Liang Zhao

Abstract Many problems regarding structured predictions are encountered in the process of "transforming" a graph in the source domain into another graph in target domain, which requires to learn a transformation mapping from the source to target domains. For example, it is important to study how structural connectivity influences functional connectivity in brain networks and traffic networks. It is also common to study how a protein (e.g., a network of atoms) folds, from its primary structure to tertiary structure. In this chapter, we focus on the transformation problem that involves graphs in the domain of deep graph neural networks. First, the problem of graph transformation in the domain of graph neural networks are formalized in Section 12.1. Considering the entities that are being transformed during the transformation process, the graph transformation problem is further divided into four categories, namely node-level transformation, edge-level transformation, node-edge co-transformation, as well as other graph-involved transformations (e.g., sequence-to-graph transformation and context-to-graph transformation), which are discussed in Section 12.2 to Section 12.5, respectively. In each subsection, the definition of each category and their unique challenges are provided. Then, several representative graph transformation models that address the challenges from different aspects for each category are introduced.

Xiaojie Guo
Department of Information Science and Technology, , George Mason University, e-mail: xguo7@gmu.edu

Shiyu Wang
Department of Computer Science, Emory University, e-mail: shiyu.wang@emory.edu

Liang Zhao
Department of Computer Science, Emory University, e-mail: liang.zhao@emory.edu

12.1 Problem Formulation of Graph Transformation

Many problems regarding structured predictions are encountered in the process of "translating" an input data (e.g., images, texts) into a corresponding output data, which is to learn a translation mapping from the input domain to the target domain. For example, many problems in computer vision can be seen as a "translation" from an input image into a corresponding output image. Similar applications can also be found in language translation, where sentences (sequences of words) in one language are translated into corresponding sentences in another language. Such a generic translation problem, which is important yet has been extremely difficult in nature, has attracted rapidly increasing attention in recent years. The conventional data transformation problem typically considers the data under special topology. For example, an image is a type of grid where each pixel is a node and each node has connections to its spatial neighbors. Texts are typically considered as sequences where each node is a word and an edge exists between two contextual words. Both grids and sequences are special types of graphs. In many practical applications, it is required to work on data with more flexible structures than grids and sequences, and hence more powerful translation techniques are required in order to handle more generic graph-structured data. Thus, there emerges a new problem named deep graph transformation, the goal of which is to learn the mapping from the graph in the input domain to the graph in the target domain. The mathematical problem formulation of the graph is provided in detail as below.

A graph is defined as $\mathscr{G}(\mathscr{V}, \mathscr{E}, F, E)$, where $\mathscr{V}$ is the set of N nodes, and $\mathscr{E} \subseteq \mathscr{V} \times \mathscr{V}$ is the set of M edges. $e_{i,j} \in \mathscr{E}$ is an edge connecting nodes $v_i, v_j \in \mathscr{V}$. A graph can be described in matrix or tensor using its (weighted) adjacency matrix A. If the graph has node attributes and edge attributes, there are node attribute matrix $F \in \mathbb{R}^{N \times L}$ where D is the number of node attributes, and edge attribute tensor $E \in \mathbb{R}^{N \times N \times K}$ where K is the number of edge attributes. L is the dimension of node attributes, and K is the dimension of edge attributes. Based on the definition of graph, we define the input graphs from the source domain as $\mathscr{G}_S$ and the output graphs from the target domain as $\mathscr{G}_S \to \mathscr{G}_T$ (Guo et al, 2019c).

Considering the entities that are being transformed during the transformation process, the graph transformation problem is further divided into three categories, namely (1) node-level transformation, where only nodes and nodes attributes can change during translation process; (2) edge-level transformation, where only topology or edge attributes can change during translation process; (3) node-edge co-transformation where both nodes and edges can change during translation process. There are also some other transformations involving graphs, including sequence-to-graph transformation, graph-to-sequence transformation and context-to-graph transformation. Although they can be absorbed into the above three types if regarding sequences as a special case of graphs, we want to separate them out because they may usually attract different research communities.

12.2 Node-level Transformation

12.2.1 Definition of Node-level Transformation

Node-level transformation aims to generate or predict the node attributes or node category of the target graph conditioning on the input graph. It can also be regarded as a node prediction problem with stochasticity. It requires the node set $\mathcal{V}$ or node attributes F to change while the graph edge set and edge attributes are fixed during the transformation namely $\mathcal{G}_S(\mathcal{V}_S, \mathcal{E}, F_S, E) \rightarrow \mathcal{G}_T(\mathcal{V}_T, \mathcal{E}, F_T, E)$. Node transformation has a wide range of real-world applications, such as predicting future states of a system in the physical domain based on the fixed relations (e.g. gravitational forces) (Battaglia et al, 2016) among nodes and the traffic speed forecasting on the road networks (Yu et al, 2018a; Li et al, 2018e). Existing works adopt different frameworks to model the transformation process.

Generally speaking, the straightforward way in dealing with the node translation problem is to regard it as the node prediction problem and utilize the conventional GNNs as encoder to learn the node embedding. Then, based on the node embedding, we can predict the node attributes of the target graphs. While solving the node transformation problem in specific domains, there come various unique requirements, such as considering the spatial and temporal patterns in the traffic flow prediction task. Thus, in this section, we focus on introducing three typical node transformation models in dealing with problems in different areas.

12.2.2 Interaction Networks

Battaglia et al (2016) proposed the interaction network in the task of reasoning about objects, relations, and physics, which is central to human intelligence, and a key goal of artificial intelligence. Many physical problems, such as predicting what will happen next in physical environments or inferring underlying properties of complex scenes, are challenging because their elements are composed and can influence each other as a whole system. It is impossible to solve such problems by considering each object and relation separately. Thus, the node transformation problem can help deal with this task via modeling the interactions and dynamics of elements in a complex system. To deal with the node transformation problem that is formalized in this scenario, an interaction network (IN) is proposed, which combines two main powerful approaches: structured models, simulation, and deep learning. Structured models are operated as the main component based on the GNNs to exploit the knowledge of relations among objects. The simulation part is an effective method for approximating dynamical systems, predicting how the elements in a complex system are influenced by interactions with one another, and by the dynamics of the system.

The overall complex system can be represented as an attributed, directed multigraph $\mathcal{G}$, where each node represents an object and the edge represents the relationship between two objects, e.g., a fixed object attached by a spring to a freely moving mass. To predict the dynamics of a single node (i.e., object), there is an object-centric function, $\mathbf{h}_i^{t+1} = f_O(\mathbf{h}_i^t)$ with the object's state $\mathbf{h}_t$ at time t of the object v_i as the inputs and a future state $\mathbf{h}_i^{t+1}$ at next time step as outputs. Assuming two objects have one directed relationship, the first object v_i influences the second object v_j via their interaction. The effect or influence of this interaction, $\mathbf{e}_{i,j}^{t+1}$ is predicted by a relation-centric function, f_R, with the object states as well as attributes of their relationship as inputs. The object updating process is then written as:

$$\mathbf{e}_{i,j}^{t+1} = f_R(\mathbf{h}_i^t, \mathbf{h}_j^t, \mathbf{r}_i); \quad \mathbf{h}_i^{t+1} = f_O(\mathbf{h}_i^t, \mathbf{e}_{i,j}^{t+1}), \tag{12.1}$$

where $\mathbf{r}_i$ refers to the interaction effects that node v_i receives.

It worth noting that the above operations are for an attributed, directed multi-graph because the edges/ relations can have attributes, and there can be multiple distinct relations between two objects (e.g., rigid and magnetic interactions). In summary, at each step, the interaction effects generated from each relationship is calculated and then an aggregation function is utilized to summarize all the interactions effects on the relevant objects and update the states of each object.

An IN applies the same f_R and f_O to every target nodes, respectively, which makes their relational and object reasoning able to handle variable numbers of arbitrarily ordered objects and relations (i.e., graphs with variables sizes). But one additional constraint must be satisfied to maintain this: the aggregation function must be commutative and associative over the objects and relations, for example summation as aggregation function satisfies this, but division would not.

The IN can be included in the framework of Message Passing Neural Network (MPNN), with the message passing process, aggregation process, and node updating process. However, different from MPNN models which focus on binary relations (i.e., there is one edge per pair of nodes), IN can also handle hyper-graph, where the edges can correspond to n-th order relations by combining n nodes ($n \geq 2$). The IN has shown a strong ability to learn accurate physical simulations and generalize their training to novel systems with different numbers and configurations of objects and relations. They could also learn to infer abstract properties of physical systems, such as potential energy. The IN implementation is the first learnable physics engine that can scale up to real-world problems, and is a promising template for new AI approaches to reasoning about other physical and mechanical systems, scene understanding, social perception, hierarchical planning, and analogical reasoning.

12.2.3 Spatio-Temporal Convolution Recurrent Neural Networks

Spatio-temporal forecasting is a crucial task for a learning system that operates in a dynamic environment. It has a wide range of applications from autonomous vehicles operations, to energy and smart grid optimization, to logistics and supply chain management. The traffic forecasting on road networks, the core component of the intelligent transportation systems, can be formalized as a node transformation problem, the goal of which is to predict the future traffic speeds (i.e., node attributes) of a sensor network (i.e., graph) given historic traffic speeds (i.e., history node attributes). This type of node transformation is unique and challenging due to the complex spatio-temporal dependencies in a series of graphs and inherent difficulty in the long term forecasting. To deal with this, each pair-wise spatial correlation between traffic sensors is represented using a directed graph whose nodes are sensors and edge weights denote proximity between the sensor pairs measured by the road network distance. Then the dynamics of the traffic flow is modeled as a diffusion process and the diffusion convolution operation is utilized to capture the spatial dependency. The overall Diffusion Convolutional Recurrent Neural Network (DCRNN) integrates diffusion convolution, the sequence to sequence architecture and the scheduled sampling technique.

Denote the node information (e.g., traffic flow) observed on a graph $\mathcal{G}$ as a graph signal F and let F^t represent the graph signal observed at time t, the temporal node transformation problem aims to learn a mapping from T' historical graph signals to future T graph signals as: $[F^{t-T'+1}, ..., F^t; \mathcal{G}] \rightarrow [F^{t+1}, ..., F^{t+T}; \mathcal{G}]$. The spatial dependency is modeled by relating node information to a diffusion process, which is characterized by a random walk on $\mathcal{G}$ with restart probability $\alpha \in [0,1]$ and a state transition matrix $D_O^{-1}W$. Here D_O is the out-degree diagonal matrix, and $\mathbf{1}$. After many time steps, such Markov process converges to a stationary distribution $P \in \mathbb{R}^{N \times N}$ whose i-th row represents the likelihood of diffusion from node v_i. Thus, a diffusion convolutional layer can be defined as

$$H_{:,q} = f(\sum_{p=1}^{P} F_{:,p} \star_{\mathcal{G}} f_{\Theta_{p,q,:,:}}), \quad q \in \{1, ..., Q\} \tag{12.2}$$

where the diffusion convolution operation is defined as

$$F_{:,p} \star_{\mathcal{G}} f_\theta = \sum_{k=0}^{K-1} (\phi_{k,1}(D_O^{-1}W)^k + \phi_{k,2}(D_I^{-1}W^T)^k)F_{:,p}, \quad p \in \{1, ..., P\} \tag{12.3}$$

Here the D_O and D_I refer to the out-degree and in-degree diagonal matrix respectively. P and Q refer to the feature dimension of the input and output node features at each diffusion convolution layer. The diffusion convolution is defined on both directed and undirected graphs. When applied to undirected graphs, the existing graph convolution neural networks (GCN) can be considered as a special case of diffusion convolution network.

To deal with the temporal dependency during the node transformation process, the recurrent neural networks (RNN) or Gated Recurrent Unit (GRU) can be leveraged. For example, by replacing the matrix multiplications in GRU with the diffusion convolution, the Diffusion Convolutional Gated Recurrent Unit (DCGRU) is defined as

$$\mathbf{r}^t = \sigma(\Theta_r \star_{\mathscr{G}} [F^t, H^{t-1}] + \mathbf{b}_r^t) \qquad (12.4)$$

$$\mathbf{u}^t = \sigma(\Theta_u \star_{\mathscr{G}} [F^t, H^{t-1}] + \mathbf{b}_u^t)$$

$$C^t = \mathbf{Tanh}(\sigma(\Theta_c \star_{\mathscr{G}} [F^t, (\mathbf{r}^t \odot H^{t-1})] + \mathbf{b}_c^t))$$

$$H^{t-1} = \mathbf{u}^t \odot H^{t-1} + (1 - \mathbf{u}^t) \odot C^t,$$

where X^t and H^t denote the input and output of all the nodes at time t, $\mathbf{r}^t$ and $\mathbf{u}^t$ are reset gate and update gate at time t, respectively. $\star_{\mathscr{G}}$ denotes the diffusion convolution defined in equation 12.3. $\Theta_r, \Theta_u, \Theta_c$ are parameters for the corresponding filters in the diffusion network.

Another typical spatio-temporal graph convolution network for spatial-temporal node transformation is proposed by (Yu et al, 2018a). This model comprises several spatio-temporal convolutional blocks, which are a combination of graph convolutional layers and convolutional sequence learning layers, to model spatial and temporal dependencies. Specifically, the framework consists of two spatio-temporal convolutional blocks (ST-Conv blocks) and a fully-connected output layer in the end. Each ST-Conv block contains two temporal gated convolution layers and one spatial graph convolution layer in the middle. The residual connection and bottleneck strategy are applied inside each block. The input sequence of node information is uniformly processed by ST-Conv blocks to explore spatial and temporal dependencies coherently. Comprehensive features are integrated by an output layer to generate the final prediction. In contrast to the above mentioned DCGRU, this model is built completely from convolutional structures to capture both spatial and temporal patterns without any recurrent neural network; each block is specially designed to uniformly process structured data with residual connection and bottleneck strategy inside.

12.3 Edge-level Transformation

12.3.1 Definition of Edge-level Transformation

Edge-level transformation aims to generate the graph topology and edge attributes of the target graph conditioning on the input graph. It requires the edge set $\mathscr{E}$ and edge attributes E to change while the graph node set and node attributes are fixed during the transformation: $\mathscr{T}: \mathscr{G}_S(\mathscr{V}, \mathscr{E}_S, F, E_S) \rightarrow \mathscr{G}_T(\mathscr{V}, \mathscr{E}_T, F, E_T)$. Edge transformation has a wide range of real-world applications, such as modeling chemical reactions (You et al, 2018a), protein folding (Anand and Huang, 2018) and malware cybernetwork synthesis (Guo et al, 2018b). For example, in social networks where people are the nodes and their contacts are the edges, the contact graph among them varies dramatically across different situations. For example, when the people are organizing a riot, it is expected that the contact graph to become denser and several special "hubs" (e.g., key players) may appear. Hence, accurately predicting the contact network in a target situation is highly beneficial to situational awareness and resource allocation.

Numerous efforts have been contributed to edge-level graph transformation. Here we introduce three typical methods in modelling the edge-level graph transformation problem, including graph transformation generative adversarial networks (GT-GAN), multi-scale graph transformation networks (Misc-GAN), and graph transformation policy networks (CTPN).

12.3.2 Graph Transformation Generative Adversarial Networks

Generative Adversarial Network (GANs) is an alternative method for generation problems. It is designed based on a game theory scenario called the min-max game, where a discriminator and a generator compete against each other. The generator generates data from stochastic noise, and the discriminator tries to tell whether it is real (coming from a training set) or fabricated (from the generator). The absolute difference between carefully calculated rewards from both networks is minimized so that both networks learn simultaneously as they try to outperform each other. GANs can be extended to a conditional model if both the generator and discriminator are conditioned on some extra auxiliary information, such as class labels or data from other modalities. Conditional GANs is realized by feeding the conditional information into the both the discriminator and generator as additional input layer. In this scenario, when the conditional information is a graph, the conditional GANs can be utilized to handle graph transformation problem to learn the mapping from the conditional graph (i.e., input graph) to the target graph (i.e., output graph). Here, we introduce two typical edge-level graph transformation techniques that are based on Conditional GANs.

A novel Graph-Translation-Generative Adversarial Networks (GT-GAN) proposed by (Guo et al, 2018b) can successfully implement and learn the mapping from the input to target graphs. GT-GAN consists of a graph translator $\mathcal{T}$ and a conditional graph discriminator $\mathcal{D}$. The graph translator $\mathcal{T}$ is trained to produce target graphs that cannot be distinguished from "real" ones by our conditional graph discriminator $\mathcal{D}$. Specifically, the generated target graph $\mathcal{G}_{T'} = \mathcal{T}(\mathcal{G}_S, U)$ cannot be distinguished from the real one, $\mathcal{G}_T$, based on the current input graph $\mathcal{G}_S$. U refers to the random noises. $\mathcal{T}$ and $\mathcal{D}$ undergo an adversarial training process based on input and target graphs by solving the following loss function:

$$\mathcal{L}(\mathcal{T}, \mathcal{D}) = \mathbb{E}_{\mathcal{G}_S, \mathcal{G}_T \sim \mathscr{S}}[\log \mathcal{D}(\mathcal{G}_T | \mathcal{G}_S)] \tag{12.5}$$
$$+ \mathbb{E}_{\mathcal{G}_S \sim \mathscr{S}}[\log(1 - \mathcal{D}(\mathcal{T}(\mathcal{G}_S, U) | \mathcal{G}_S))],$$

where $\mathscr{S}$ refers to the dataset. $\mathcal{T}$ tries to minimize this objective while an adversarial $\mathcal{D}$ tries to maximize it, i.e. $\mathcal{T}^* = \arg\min_{\mathcal{T}} \max_{\mathcal{D}} \mathcal{L}(\mathcal{T}, \mathcal{D})$. The graph translator includes two parts: graph encoder and graph decoder. A graph convolution neural net (Kawahara et al, 2017) is extended to serve as the graph encoder in order to embed the input graph into node-level representations, while a new graph deconvolution net is designed as the decoder to generate the target graph. Specifically, the encoder consists of edge-to-edge and edge-to-node convolution layers, which first extract latent edge-level representations and then node-level representations $\{H_i\}_{i=1}^{N}$, where $H_i \in \mathbb{R}^L$ refers to the latent representation of node v_i. The decoder consists of node-to-edge and edge-to-edge deconvolution layers to first get each edge representation $\hat{E}_{i,j}$ based on H_i and H_j, and then finally get edge attribute tensor E based on $\hat{E}$. Based on the graph deconvolution above, it is possible to utilize skips to link the extracted edge latent representations of each layer in the graph encoder with those in the graph decoder.

Specifically, in the graph translator, the output of the l-th "edge deconvolution" layer in the decoder is concatenated with the output of the l-th "edge convolution" layer in the encoder to form joint two channels of feature maps, which are then input into the $(l+1)$-th deconvolution layer. It is worth noting that one key factor for effective translation is the design of a symmetrical encoder-decoder pair, where the graph deconvolution is a mirrored reversed way from graph convolution. This allows skip-connections to directly translate different level's edge information at each layer.

The graph discriminator is utilized to distinguish between the "translated" target graph and the "real" ones based on the input graphs, as this helps to train the generator in an adversarial way. Technically, this requires the discriminator to accept two graphs simultaneously as inputs (a real target graph and an input graph or a generated graph and an input graph) and classify the two graphs as either related or not. Thus, a conditional graph discriminator (CGD) that leverages the same graph convolution layers in the encoder is utilized for the graph classification. Specifically, the input and target graphs are both ingested by the CGD and stacked into a tensor, which can be considered a 2-channel input. After obtaining the node representations, the graph-level embedding is computed by summing these node embeddings. Finally, a softmax layer is implemented to distinguish the input graph-pair from the real graph or generated graph.

To further handle the situation when the pairing information of the input and the output is not available, Gao et al (2018b) proposes an Unpaired Graph Translation Generative Adversarial Nets (UGT-GAN) based on Cycle-GAN (Zhu et al, 2017) and incorporate the same encoder and deconder in GT-GAN to handle the unpaired graph transformation problems. The cycle consistency loss is utilized and generalized into graph cycle consistency loss for unpaired graph translation. Specifically, graph cycle consistency adds an opposite direction translator from target to source domain $\mathscr{T}_r : \mathscr{G}_T \to \mathscr{G}_S$ by training the mappings for both directions simultaneously, and adding a cycle consistency loss that encourages $\mathscr{T}_r(\mathscr{T}(\mathscr{G}_S)) \approx \mathscr{G}_S$ and $\mathscr{T}(\mathscr{T}_r(\mathscr{G}_T)) \approx \mathscr{G}_T$. Combining this loss with adversarial losses on domains $\mathscr{G}_T$ and $\mathscr{G}_S$ yields the full objective for unpaired graph translation.

12.3.3 Multi-scale Graph Transformation Networks

Many real-world networks typically exhibit hierarchical distribution over graph communities. For instance, given an author collaborative network, research groups of well-established and closely collaborated researchers could be identified by the existing graph clustering methods in the lower-level granularity. While, from a coarser level, we may find that these research groups constitute large-scale communities, which correspond to various research topics or subjects. Thus, it is necessary to capture the hierarchical community structures over the graphs for edge-level graph transformation problem. Here, we introduce a graph generation model for learning the distribution of the graphs, which, however, is formalized as a edge-level graph transformation problem.

Based on GANs, a multi-scale graph generative model, Misc-GAN, can be utilized to model the underlying distribution of graph structures at different levels of granularity. Inspired by the success of deep generative models in image translation, a cycle-consistent adversarial network (CycleGAN) (Zhu et al, 2017) is adopted to learn the graph structure distribution and then generate a synthetic coarse graph at each granularity level. Thus, the graph generation task can be realized by "transferring" the hierarchical distribution from the graphs in the source domain to a unique graph in the target domain.

In this framework, the input graph is characterized as several coarse-grained graphs by aggregating the strongly coupled nodes with a small algebraic distance to form coarser nodes. Overall, the framework can be separated into three stages. First, the coarse-grained graphs at K levels of granularity are constructed from the input graph adjacent matrix A_S. The adjacent matrix of the coarse-grained graph $A_S^{(k)} \in \mathbb{R}^{N^{(k)} \times N^{(k)}}$ at the k-th layer is defined as follows:

$$A_S^{(k)} = P^{(k-1)^\top} ... P^{(1)^\top} A_S P^{(1)} ... P^{(k-1)}, \tag{12.6}$$

where $A_S^{(0)} = A_S$ and $P^{(k)} \in \mathbb{R}^{N^{(k)} \times N^{(k)}}$ is a coarse-grained operator for the kth level and $N^{(k)}$ refers to the number of nodes of the coarse-grained graph at level k. In the next stage, each coarse-grained graph at each level k will be reconstructed back into a fine graph adjacent matrix $A_T^{(k)} \in \mathbb{R}^{N^{(k)} \times N^{(k)}}$ as follows:

$$A_T^{(k)} = R^{(1)^\top} ... R^{(k-1)^\top} A_S^{(k)} R^{(k-1)} ... R^{(1)}, \tag{12.7}$$

where $R^{(k)} \in \mathbb{R}^{N^{(k)} \times N^{(k)}}$ is the reconstruction operator for the kth level. Thus all the reconstructed fine graphs at each layer are on the same scale. Finally, these graphs are aggregated into a unique one by a linear function to get the final adjacent matrix as follows: $A_T = \sum_{k=1}^{K} w^k A_T^{(k)} + b^k \mathbf{I}$, where $w^k \in \mathbb{R}$ and $b^k \in \mathbb{R}$ are weights and bias.

12.3.4 Graph Transformation Policy Networks

Beyond the general framework for edge-level transformation problem, it is necessary to deal with some domain-specific problems which may need to incorporate some domain knowledge or information into transformation process. For example, the chemical reaction product prediction problem is a typical edge-level transformation problem, where the input reactant and reagent molecules can be jointly represented as input graphs, and the process of generating product molecules (i.e., output graphs) from reactant molecules can be formulated as a set of edge-level graph transformations. Formalizing the chemical reaction product prediction problem as a edge-level transformation problem is beneficial due to two reasons: (1) it can capture and utilize the molecular graph structure patterns of the input reactants and reagents(i.e., atom pairs with changing connectivity); and (2) it can automatically choose from these reactivity patterns a correct set of reaction triples to generate the desired products.

Do et al (2019) proposed a Graph Transformation Policy Network (GTPN), a novel generic method that combines the strengths of graph neural networks and reinforcement learning, to learn reactions directly from data with minimal chemical knowledge. The GTPN originally aims to generate the output graph by formalizing the graph transformation process as a Markov decision process and modifying the input source graph through several iterations. From the perspective of chemical reaction side, the process of reaction product prediction can be formulated as predicting a set of bond changes given the reactant and reagent molecules as input. A bond change is characterized by the atom pair that holds the bond (where is the change) and the new bond type (what is the change).

Mathematically, given a graph of reactant molecule as input graph, $\mathscr{G}_S$, they predict a set of reaction triples which transforms $\mathscr{G}_S$ into a graph of product molecule $\mathscr{G}_T$. This process is modeled as a sequence consisting of tuples like $(\zeta^t, v_i^t, v_j^t, b^t)$ where v_i^t and v_j^t are the selected nodes from node set at step t whose connection needs to be modified, b^t is the new edge type of (v_i^t, v_j^t) and ζ^t is a binary signal that indicates the end of the sequence. Generally, at every step of the forward pass, GTPN performs seven major steps: 1) computing the atom representation vectors through message passing neural network (MPNN); 2) computing the most possible K reaction atom pairs; 3) predicting the continuation signal ζ^t; 4) predicting the reaction atom pair (v_i^t, v_j^t); 5) predicting a new bond type b^t of this atom pair; 6) updating the atom representations; and 7) updating the recurrent state.

Specifically, the above iterative process of edge-level transformation is formulated as a Markov Decision Process (MDP) characterized by a tuple $(\mathscr{S}, \mathscr{A}, f_P, f_R, \Gamma)$, where $\mathscr{S}$ is a set of states, $\mathscr{A}$ is a set of actions, f_P is a state transition function, f_R is a reward function, and Γ is a discount factor. Thus, the overall model is optimized via the reinforcement learning. Specifically, a state $s^t \in \mathscr{S}$ is a immediate graph that is generated at the step t, and s^0 refers to the input graph. An action $a^t \in \mathscr{A}$ performed at step t is represented as a tuple $(\zeta^t, (v_i^t, v_j^t, b^t))$. The action is composed of three consecutive sub-actions: predicting ζ^t, (v_i^t, v_j^t) and b^t respectively. In the state transition part, If $\zeta^t = 1$, the current graph $\mathscr{G}^t$ is modified based on the reaction triple (v_i^t, v_j^t, b^t) to generate a new intermediate graph $\mathscr{G}^{t+1}$. Regarding the reward, both immediate rewards and delayed rewards are utilized to encourage the model to learn the optimal policy faster. At every step t, if the model predicts $(\zeta^t, (v_i^t, v_j^t, b^t))$ correctly, it will receive a positive reward for each correct sub-action. Otherwise, a negative reward is given. After the prediction process has terminated, if the generated products are exactly the same as the ground-truth products, a positive delayed reward is also given, otherwise a negative reward.

Different from the encoder-decoder frameworks of GT-GAN, GTPN is a typical example of reinforcement learning-based graph transformation network, where the target graph is generated by making modifications on the input graphs in a iterative way. Reinforcement learning (RL) is a commonly used framework for learning controlling policies and generation process by a computer algorithm, the so-called agent, through interacting with its environment. The nature of reinforcement learning methods (i.e.,a sequential generation process) make it a suitable framework for graph transformation problems which sometime requires the step-by-step edits on the input graphs to generate the final target output graphs.

12.4 Node-Edge Co-Transformation

12.4.1 Definition of Node-Edge Co-Transformation

Node-edge co-transformation (NECT) aims to generate node and edge attributes of the target graph conditioned on those of the input graph. It requires that both nodes and edges can vary during the transformation process between the source graph and the target graph as follows: $\mathscr{G}_S(\mathscr{V}_S, \mathscr{E}_S, F_S, E_S) \rightarrow \mathscr{G}_T(\mathscr{V}_T, \mathscr{E}_T, F_T, E_T)$. There are two categories of techniques used to assimilate the input graph to generate the target graph embedding-based and editing-based.

Embedding-based NECT usually encodes the source graph into latent representations using an encoder that contains higher-level information on the input graph which can then be decoded into the target graph by a decoder (Jin et al, 2020c, 2018c; Kaluza et al, 2018; Maziarka et al, 2020b; Sun and Li, 2019). These methods are usually based on either conditional VAEs (Sohn et al, 2015) or conditional GANs (Mirza and Osindero, 2014). Three main techniques will be introduced in this section, including junction-tree variational auto-encoder, molecule cycle-consistent adversarial networks and directed acyclic graph transformation networks.

12.4.1.1 Junction-tree Variational Auto-encoder Transformer

The goal of molecule optimization, which is one of the important molecule generation problems, is to optimize the properties of a given molecule by transforming it into a novel output molecule with optimized properties. The molecule optimization problem is typically formalized as a NECT problem where the input graph refers to the initial molecule and the output graph refers to the optimized molecule. Both the node and edge attributes can change during the transformation process.

The Junction-tree Variational Auto-encoder (JT-VAE) is motivated by the key challenge of molecule optimization in the domain of drug design, which is to find target molecules with the desired chemical properties (Jin et al, 2018a). In terms of the model architecture, JT-VAE extends the VAE (Kingma and Welling, 2014) to molecular graphs by introducing a suitable encoder and a matching decoder. Under JT-VAE, each molecule is interpreted as being formalized from subgraphs chosen from a dictionary of valid components. These components serve as building blocks when encoding a molecule into a vector representation and decoding latent vectors back into optimized molecular graphs. The dictionary of components, such as rings, bonds and individual atoms, is large enough to ensure that a given molecule can be covered by overlapping clusters without forming cluster cycles. In general, JT-VAE generates molecular graphs in two phases, by first generating a tree-structured scaffold over chemical substructures and then combining them into a molecule with a graph message-passing network.

The latent representation of the input graph $\mathscr{G}$ is encoded by a graph message-passing network (Dai et al, 2016; Gilmer et al, 2017). Here, let $\mathbf{x}_v$ denote the feature vector of the vertex v, involving properties of the vertex such as the atom type and valence. Similarly, each edge $(u,v) \in \mathscr{E}$ has a feature vector $\mathbf{x}_{vu}$ indicating its bond type. Two hidden vectors v_{uv} and v_{vu} denote the message from u to v and vice versa. In the encoder, messages are exchanged via loopy belief propagation:

$$\mathbf{v}_{uv}^{(t)} = \tau(W_1^g \mathbf{x}_u + W_2^g \mathbf{x}_{uv} + W_3^g \sum_{w \in N(u) \backslash v} \mathbf{v}_{wu}^{(t-1)}), \qquad (12.8)$$

where $\mathbf{v}_{uv}^t$ is the message computed in the t-th iteration, initialized with $\mathbf{v}_{uv}^{(0)} = 0$, $\tau(\cdot)$ is the ReLU function, W_1^g, W_2^g and W_3^g are weights, and $N(u)$ denotes the neighbors of u. Then, after T iterations, the latent vector of each vertex is generated capturing its local graphical structure:

$$\mathbf{h}_u = \tau(U_1^g \mathbf{x}_u + \sum_{v \in N(u)} U_2^g \mathbf{v}_{vu}^{(T)}), \qquad (12.9)$$

where U_1^g and U_2^g are weights. The final graph representation is $\mathbf{h}_{\mathscr{G}} = \sum_i \mathbf{h}_i / |\mathscr{V}|$, where $|\mathscr{V}|$ is the number of nodes in the graph. The corresponding latent variable $\mathbf{z}_G$ can be sampled from $\mathscr{N}(\mathbf{z}_G; \mu_{\mathscr{G}}, \sigma_{\mathscr{G}}^2)$ and $\mu_{\mathscr{G}}$ and $\sigma_{\mathscr{G}}^2$ can be calculated from $h_{\mathscr{G}}$ via two separate affine layers.

A junction tree can be represented as $(\mathscr{V}, \mathscr{E}, \mathscr{X})$ whose node set is $\mathscr{V} = (C_1, ..., C_n)$ and edge set is $\mathscr{E} = (E_1, ..., E_n)$. This junction tree is labeled by the label dictionary $\mathscr{X}$. Similar to the graph representation, each cluster C_i is represented by a one-hot $\mathbf{x}_i$ and each edge (C_i, C_j) corresponds to two message vectors $\mathbf{v}_{ij}$ and $\mathbf{v}_{ji}$. An arbitrary leaf node is picked as the root and messages are propagated in two phases:

$$s_{ij} = \sum_{k \in N(i) \setminus j} v_{ki} \tag{12.10}$$

$$z_{ij} = \sigma(W^z x_i + U^z s_{ij} + b^z)$$

$$r_{ki} = \sigma(W^r x_i + U^r v_{ki} + b^r)$$

$$\tilde{v}_{ij} = tanh(W x_i + U \sum_{k \in N(i) \setminus j} r_{ki} \odot v_{ki})$$

$$v_{ij} = (1 - z_{ij}) \odot s_{ij} + z_{ij} \odot \tilde{v}_{ij}.$$

h_i, the latent representation of node v_i can now be calculated:

$$h_i = \tau(W^o x_i + \sum_{k \in N(u)} U^o v_{ki}) \tag{12.11}$$

The final tree representation is $h_{\mathscr{T}_{\mathscr{G}}} = h_{root}$. $z_{\mathscr{T}_{\mathscr{G}}}$ is sampled in a similar way as in the encoding process.

Under the JT-VAE framework, the junction tree is decoded from $z_{\mathscr{T}_{\mathscr{G}}}$ using a tree-structured decoder that traverses the tree from the root and generates nodes in their depth-first order. During this process, a node receives information from other nodes, and this information is propagated through message vectors h_{ij}. Formally, let $\tilde{\mathscr{E}} = \{(i_1, j_1), ..., (i_m, j_m)\}$ be the set of edges traversed over the junction tree $(\mathscr{V}, \mathscr{E})$, where $m = 2|\mathscr{E}|$ because each edge is traversed in both directions. The model visits node i_t at time t. Let $\tilde{\mathscr{E}}_t$ be the first t edges in $\tilde{\mathscr{E}}$. The message is updated as $h_{i_t, j_t} = GRU(x_{i_t}, \{h_{k,i_t}\}_{(k,i_t) \in \tilde{\mathscr{E}}_t, k \neq j_t})$, where x_{i_t} corresponds to the node features. The decoder first makes a prediction regarding whether the node i_t still has children to be generated, in which the probability is calculated as:

$$p_t = \sigma(u^d \cdot \tau(W_1^d x_{i_t} + W_2^d z_{\mathscr{T}_{\mathscr{G}}} + W_3^d \sum_{(k,i_t) \in \tilde{\mathscr{E}}_t} h_{k,i_t})), \tag{12.12}$$

where u^d, W_1^d, W_2^d and W_3^d are weights. Then, when a child node j is generated from its parent i, its node label is predicted with:

$$q_j = softmax(U^l \cdot \tau(W_1^l z_{\mathscr{T}_{\mathscr{G}}} + W_2^l h_{ij})), \tag{12.13}$$

where U^l, W_1^l and W_2^l are weights and q_j is a distribution over label dictionary $\mathscr{X}$.

The final step of the model is to reproduce a molecular graph $\mathscr{G}$ to represent the predicted junction tree $(\mathscr{V}, \hat{\mathscr{E}})$ by assembling the subgraphs together into the final molecular graph. Let $\mathscr{G}(\mathscr{T}_{\mathscr{G}})$ be a set of graphs corresponding to the junction tree $\mathscr{T}_{\mathscr{G}}$. Decoding graph $\hat{\mathscr{G}}$ from the junction tree $\hat{\mathscr{T}}_{\mathscr{G}} = (\mathscr{V}, \hat{\mathscr{E}})$ is a structured prediction:

$$\hat{\mathscr{G}} = \arg \max_{\mathscr{G}' = \mathscr{G}(\hat{\mathscr{T}}_{\mathscr{G}})} f^a(\mathscr{G}'), \tag{12.14}$$

where f^a is a scoring function over candidate graphs. The decoder starts by sampling the assembly of the root and its neighbors according to their scores, then proceeds to assemble the neighbors and associated clusters. In terms of scoring the realization of each neighborhood, let $\mathscr{G}_i$ be the subgraph resulting from a particular merging of cluster C_i in the tree with its neighbors $C_j, j \in N_{\hat{\mathscr{T}}_{\mathscr{G}}}(i)$. $\mathscr{G}_i$ is scored as a candidate subgraph by first deriving a vector representation $h_{\mathscr{G}_i}$, and $f_i^a(\mathscr{G}_i) = h_{\mathscr{G}_i} \cdot z_{\mathscr{G}}$ is the subgraph score. For atoms in $\mathscr{G}_i$, let $\alpha_v = i$ if $v \in C_i$ and $\alpha_v = j$ if $v \in C_j \backslash C_i$ to mark the position of atoms in the junction tree and retrieve messages $\hat{\mathbf{m}}_{i,j}$, summarizing the subtree under i along the edge (i, j) obtained by the tree encoder. Then the neural messages can be obtained and aggregated similarly to the encoding step with parameters:

$$\mu_{uv}^{(t)} = \tau(W_1^a \mathbf{x}_u + W_2^a \mathbf{x}_{uv} + W_3^a \hat{\mu}_{uv}^{(t-1)}) \tag{12.15}$$

$$\tilde{\mu}_{uv}^{(t-1)} = \begin{cases} \sum_{w \in N(u) \backslash v} \mu_{wu}^{(t-1)} & \alpha_u = \alpha_v \\ \hat{m}_{\alpha_u, \alpha_v} + \sum_{w \in N(u) \backslash v} \mu_{wu}^{(t-1)} & \alpha_u \neq \alpha_v, \end{cases}$$

where W_1^a, W_2^a and W_3^a are weights.

12.4.1.2 Molecule Cycle-Consistent Adversarial Networks

Cycle-consistent adversarial networks, an alternative to achieve embedding-based NECT, were originally developed to achieve image-to-image transformations. The aim here is to learn to transform an image from a source domain to a target domain in the absence of paired examples by using an adversarial loss. To promote the chemical compound design process, this idea has been borrowed for graph transformation. For instance, Molecule Cycle-Consistent Adversarial Networks (Mol-CycleGAN) have been proposed to generate optimized compounds with high structural similarity to the originals (Maziarka et al, 2020b). Given a molecule $\mathscr{G}_X$ with the desired molecular properties, Mol-CycleGAN aims to train a model to perform the transformation $G : \mathscr{G}_X \to \mathscr{G}_Y$ and then use this model to optimize the molecules. Here $\mathscr{G}_Y$ is the set of molecules without the desired molecular properties. In order to represent the sets $\mathscr{G}_X$ and $\mathscr{G}_Y$, this model requires a reversible embedding that allows both the encoding and decoding of molecules. To achieve this, JT-VAE is employed to provide the latent space during the training process, during which the distance between molecules required to calculate the loss function can be defined directly. Each molecule is represented as a point in latent space, assigned based on the mean of the variational encoding distribution.

For the implementation, the sets $\mathscr{G}_X$ and $\mathscr{G}_Y$ must be defined (e.g., inactive/active molecules), after which the mapping functions $G : \mathscr{G}_X \to \mathscr{G}_Y$ and $F : \mathscr{G}_Y \to \mathscr{G}_X$ are introduced. The discriminators D_X and D_Y are proposed to force generators F and G to generate samples from a distribution close to the distributions of $\mathscr{G}_X$ and $\mathscr{G}_Y$. For this process, F, G, D_X and D_Y are modeled by neural networks. This approach to molecule optimization is designed to (1) take a prior molecule x with no specified features from set $\mathscr{G}_X$ and compute its latent space embedding; (2) use generative neural network G to obtain the embedding of molecule $G(x)$ that has this feature but is also similar to the original molecule x; and (3) decode the latent space coordinates given by $G(x)$ to obtain the optimized molecule.

The loss function to train Mol-CycleGAN is:

$$L(G,F,D_X,D_Y) = L_{GAN}(G,D_Y,\mathscr{G}_X,\mathscr{G}_Y) + L_{GAN}(F,D_X,\mathscr{G}_Y,\mathscr{G}_X) \quad (12.16)$$
$$+ \lambda_1 L_{cyc}(G,F) + \lambda_2 L_{identity}(G,F),$$

and $G^*, F^* = \arg\min_{G,F} \max_{D_X,D_Y} L(G,F,D_X,D_Y)$. The adversarial loss is utilized:

$$L_{GAN}(G,D_Y,\mathscr{G}_X,\mathscr{G}_Y) = \frac{1}{2}\mathbb{E}_{y\sim p_{data}^{\mathscr{G}_Y}}\left[(D_Y(y)-1)^2\right] \quad (12.17)$$
$$+ \frac{1}{2}\mathbb{E}_{x\sim p_{data}^{\mathscr{G}_X}}\left[D_Y(G(x))^2\right],$$

which ensures that the generator G (and F) generates samples from a distribution close to the distribution of $\mathscr{G}_Y$ (or $\mathscr{G}_X$), denoted by $p_{data}^{\mathscr{G}_Y}$ (or $p_{data}^{\mathscr{G}_X}$). The cycle consistency loss

$$L_{cyc}(G,F) = \mathbb{E}_{y\sim p_{data}^{\mathscr{G}_Y}}\left[\|G(F(y))-y\|_1\right] \quad (12.18)$$
$$+ \mathbb{E}_{x\sim p_{data}^{\mathscr{G}_X}}\left[\|F(G(x))-x\|_1\right],$$

reduces the space available to the possible mapping functions such that for a molecule x from set $\mathscr{G}_X$, the GAN cycle constrains the output to a molecule similar to x. The inclusion of the cyclic component acts as a regularization factor, making the model more robust. Finally, to ensure that the generated molecule is close to the original, identity mapping loss is employed:

$$L_{identity}(G,F) = \mathbb{E}_{y\sim p_{data}^{\mathscr{G}_Y}}\left[\|F(y)-y\|_1\right] \quad (12.19)$$
$$+ \mathbb{E}_{x\sim p_{data}^{\mathscr{G}_X}}\left[\|G(x)-x\|_1\right],$$

which further reduces the space available to the possible mapping functions and prevents the model from generating molecules that lay far away from the starting molecule in the latent space of JT-VAE.

12.4.1.3 Directed Acyclic Graph Transformation Networks

Another example of embedding-based NECT is a neural model for learning deep functions on the space of directed acyclic graphs (DAGs) (Kaluza et al, 2018). Mathematically, the neural methodologies developed to handle graph-structured data can be regarded as function approximation frameworks where both the domain and the range of the target function can be graph spaces. In the area of interest here, the embedding and synthesis methodologies are gathered into a single unified framework such that functions can be learned from one graph space onto another graph space without the need to impose a strong assumption of independence between the embedding and generative process. Note that only functions in DAG space are considered here. A general encoder-decoder framework for learning functions from one DAG space onto another has been developed.

Here, RNN is employed to model the function F, denoted as D2DRNN. Specifically, the model consists of an encoder E_α with model parameters α that compute a fixed-size embedding of the input graph $\mathscr{G}_{in}$, and a decoder D_β with parameters β, using the embedding as input and producing the output graph $\hat{\mathscr{G}}_{out}$. Alternatively, the DAG-function can be defined as $F(\mathscr{G}_{in}) := D_\beta(E_\alpha(\mathscr{G}_{in}))$.

The encoder is borrowed from the deep-gated DAG recursive neural network (DG-DAGRNN) (Amizadeh et al, 2018), which generalizes the stacked recurrent neural networks (RNNs) on sequences to DAG structures. Each layer of DG-DAGRNN consists of gated recurrent units (GRUs) (Cho et al, 2014a), which are repeated for each node $v_i \in \mathscr{G}_{in}$. The GRU corresponding to node v contains an aggregated representation of the hidden states of the units regarding its predecessors $\pi(v)$. For an aggregation function A:

$$\mathbf{h}_v = GRU(\mathbf{x}_v, \mathbf{h}'_v), \text{ where } \mathbf{v}' = \mathbf{A}(\{\mathbf{h_u}|\mathbf{u} \in \pi(\mathbf{v})\}). \tag{12.20}$$

Since the ordering of the nodes is defined by the topological sort of $\mathscr{G}_{in}$, all the hidden states $\mathbf{h}_v$ can be computed with a single forward pass along a layer of DG-DAGRNN. The encoder contains multiple layers, each of which passes hidden states to the recurrent units in the subsequent layer corresponding to the same node.

The encoder outputs an embedding $H_{in} = E_\alpha(\mathscr{G}_{in})$, which serves as the input of the DAG decoder. The decoder follows the local-based node-sequential generation style. Specifically, first, the number of nodes of the target graph is predicted by a multilayer perceptron (MLP) with a Poisson regressor output layer, which takes the input graph embedding H_{in} and outputs the mean of a Poisson distribution describing the output graph. Whether it is necessary to add an edge e_{u,v_n} for all the nodes $u \in \{v_1, ..., v_{n-1}\}$ already in the graph is determined by a module of MLP. Since the output nodes are generated in their topological order, the edges are directed from the nodes added earlier to the nodes added later. For each node v, the hidden state $\mathbf{h}_v$ is calculated using a similar mechanism to that used in the encoder, after which they are aggregated and fed to a GRU. The other input for the GRU consists of the aggregated states of all the sink nodes generated so far. For the first node, the hidden state is initialized based on the encoder's output. Then, the output node features are generated based on its hidden state using another module of MLP. Finally, once the last node has been generated, the edges are introduced with probability 1 for sinks in the graph to ensure a connected graph with only one sink node as an output.

12.4.2 Editing-based Node-Edge Co-Transformation

Unlike the encoder-decoder framework, modification-based NECT directly modifies the input graph iteratively to generate the target graphs (Guo et al, 2019c; You et al, 2018a; Zhou et al, 2019c). Two methods are generally used to edit the source graph. One employs a reinforcement-learning agent to sequentially modify the source graph based on a formulated Markov decision process (You et al, 2018a;

Zhou et al, 2019c). The modification at each step is selected from a defined action set that includes "add node", "add edge", "remove bonds" and so on. Another is to update the nodes and edges from the source graph synchronously in a one-shot manner through the MPNN using several iterations (Guo et al, 2019c).

12.4.2.1 Graph Convolutional Policy Networks

Motivated by the large size of chemical space, which can be an issue when designing molecular structures, graph convolutional policy networks (GCPNs) serve as useful general graph convolutional network-based models for goal-directed graph generation through reinforcement learning (RL) (You et al, 2018a). In this model, the generation process can be guided towards the specific desired objectives, while restricting the output space based on underlying chemical rules. To achieve goal-directed generation, three strategies, namely graph representation, reinforcement learning, and adversarial trainings are adopted. In GCPN, molecules are represented as molecular graphs, and partially generated molecular graphs can be interpreted as substructures. GCTN is designed as an RL agent which operates within a chemistry-aware graph generation environment. A molecule is successively constructed by either connecting a new substructure or atom to an existing molecular graph by adding a bond. GCPN is trained to optimize the domain-specific properties of the source molecule by applying a policy gradient to optimize it via a reward composed of molecular property objectives and adversarial loss; it acts in an environment which incorporates domain-specific rules. The adversarial loss is provided by a GCN-based discriminator trained jointly on a dataset of example molecules.

An iterative graph generation process is designed and formulated as a general decision process $M = (\mathscr{S}, \mathscr{A}, P, R, \gamma)$, where $\mathscr{S} = \{s_i\}$ is the set of states that comprises all possible intermediate and final graphs. $\mathscr{A} = (a_i)$ is the set of actions that describe the modifications made to the current graph during each iteration, P represents the transition dynamics that specify the possible outcomes of carrying out an action $p(s_{t+1}|s_t, ..., s_0, a_t)$, $R(s_t) = r_t$ is a reward function that specifies the reward after reaching state s_t and γ is the discount factor. The graph generation process can now be formulated as $(s_0, a_0, r_0, ..., s_n, a_n, r_n)$, and the modification of the graph at each time can be described as a state transition distribution: $p(s_{t+1}|s_t, ..., s_0) = \sum_{a_t} p(a_t|s_t, ..., s_0) p(s_{t+1}|s_t, ..., s_0, a_t)$, where $p(a_t|s_t, ..., s_0)$ is represented as a policy network π_θ. Note that in this process, the state transition dynamics are designed to satisfy the Markov property $p(s_{t+1}|s_t, ...s_0) = p(s_{t+1}|s_t)$.

In this model, a distinct, fixed-dimension, homogeneous action space is defined and amenable to reinforcement learning, where an action is analogous to link prediction. Specifically, a set of scaffold subgraphs $\{C_1, ..., C_s\}$ is first defined based on the source graph, thus serving as a subgraph vocabulary that contains the subgraphs to be added into the target graph during graph generation. Define $C = \cup_{i=1}^s C_i$. Given the modified graph $\mathscr{G}_t$ at step t, the corresponding extended graph can be defined as $\mathscr{G}_t \cup C$. Under this definition, an action can either correspond to connecting a new subgraph C_i to a node in $\mathscr{G}_t$ or connecting existing nodes within graph $\mathscr{G}_t$. GAN is also employed to define the adversarial rewards to ensure that generated molecules do indeed resemble the originals.

Node embedding is achieved by message passing over each edge type for L layers through GCN. At the l-th layer of GCN, messages from different edge types are aggregated to calculate the node embedding $H^{(l+1)} \in \mathbb{R}^{(n+c) \times k}$ of the next layer, where n and c are the sizes of $\mathcal{G}_t$ and C, respectively, and k is the embedding dimension:

$$H^{(l+1)} = AGG(ReLU(\{\hat{D}_i^{-\frac{1}{2}} \hat{E}_i \hat{D}_i^{-\frac{1}{2}} H^{(l)} W_i^{(l)}\}, \forall i \in (1,...,b))). \qquad (12.21)$$

E_i is the ith slice of the edge-conditioned adjacency tensor E, and $\hat{E}_i = E_i + \mathbf{I}$; $\hat{D}_i = \sum_k \hat{E}_{ijk}$ and $W_i^{(l)}$ is the weight matrix for the ith edge type. AGG denotes one of the aggregation functions from $\{MEAN, MAX, SUM, CONTACT\}$.

The link prediction-based action a_t ensures each component samples from a prediction distribution governed by the equations below:

$$a_t = CONCAT(a_{first}, a_{second}, a_{edge}, a_{stop}) \qquad (12.22)$$

$$f_{first}(s_t) = softmax(m_f(X)), \qquad a_{first} \sim f_{first}(s_t) \in \{0,1\}^n \quad (12.23)$$
$$f_{second}(s_t) = softmax(m_s(X_{a_{first}}, X)), \quad a_{second} \sim f_{second}(s_t) \in \{0,1\}^{n+c}$$
$$f_{edge}(s_t) = softmax(m_e(X_{a_{first}}, X)), \quad a_{edge} \sim f_{edge}(s_t) \in \{0,1\}^b$$
$$f_{stop}(s_t) = softmax(m_t(AGG(X))), \qquad a_{stop} \sim f_{stop}(s_t) \in \{0,1\}$$

Here m_f, m_s, m_e and m_f denote MLP modules.

12.4.2.2 Molecule Deep Q-networks Transformer

In addition to GCPN, molecule deep Q-networks (MolDQN) has also been developed for molecule optimization under the node-edge co-transformation problem utilizing an editing-based style. This combines domain knowledge of chemistry with state-of-the-art reinforcement learning techniques (double Q-learning and randomized value functions) (Zhou et al, 2019c). In this field, traditional methods usually employ policy gradients to generate graph representations of molecules, but these suffer from high variance when estimating the gradient (Gu et al, 2016). In comparison, MolDQN is based on value function learning, which is usually more stable and sample efficient. MolDQN also avoids the need for expert pretraining on some datasets, which may lead to lower variance but limits the search space considerably.

In the framework proposed here, modifications of molecules are directly defined to ensure 100% chemical validity. Modification or optimization is performed in a step-wise fashion, where each step belongs to one of the following three categories: (1) atom addition, (2) bond addition, and (3) bond removal. Because the molecule generated depends solely on the molecule being changed and the modification made, the optimization process can be formulated as a Markov decision process (MDP).

Specifically, when performing the action *atom addition*, an empty set of atoms $\mathcal{V}_T$ for the target molecule graph is first defined. Then, a valid action is defined as adding an atom in $\mathcal{V}_T$ and also a bond between the added atom and the original molecule wherever possible. When performing the action *bond addition*, a bond is added between two atoms in $\mathcal{V}_T$. If there is no existing bond between the two atoms, the actions between them can consist of adding a single, double or triple bond. If there is already a bond, this action changes the bond type by increasing the index of the bond type by one or two. When performing the action *bond removal*, the valid bond removal action set is defined as the actions that decrease the bond type index of an existing bond. Possible transitions include: (1) Triple bond → {Double, Single, No} bond, (2) Double bond → {Single, No} bond, and (3) Single bond → {No} bond.

Based on the molecule modification MDP defined above, RL aims to find a policy π that chooses an action for each state that maximizes future rewards. Then, the decision is made by finding the action a for a state s to maximize the Q function:

$$Q^\pi(s,a) = Q^\pi(m,t,a) = \mathbb{E}_\pi[\sum_{n=t}^{T} r_n], \tag{12.24}$$

where r_n is the reward at step n. The optimal policy can therefore be defined as $\pi^*(s) = \arg\max_a Q^{\pi^*}(s,a)$. A neural network is adopted to approximate $Q(s,a,\theta)$, and can be trained by minimizing the loss function:

$$l(\theta) = \mathbb{E}[f_l(y_t - Q(s_t,a_t;\theta))], \tag{12.25}$$

where $y_t = r_t + \max_a Q(s_{t+1},a;\theta)$ is the target value and f_l is the Huber loss:

$$f_l(x) = \begin{cases} \frac{1}{2}x^2 & \text{if } |x| < 1 \\ |x| - \frac{1}{2} & \text{otherwise} \end{cases} \tag{12.26}$$

In a real-world setting, it is usually desirable for several different properties to be optimized at the same time. Under the multi-objective RL setting, the environment will return a vector of rewards at each step t with one reward for each objective. A "scalar" reward framework is applied to achieve multi-objective optimization, with the introduction of a user defined weight vector $\mathbf{w} = [w_1, w_2, ..., w_k]^\top \in \mathbb{R}^k$. The reward is calculated as:

$$r_{s,t} = \mathbf{w}^\top \vec{\mathbf{r}_t} = \sum_{i=1}^{k} w_i r_{i,t}. \tag{12.27}$$

The objective of MDP is to maximize the cumulative scalarized reward.

The Q-learning model (Mnih et al, 2015) is implemented here, incorporating the improvements gained using double Q-learning (Van Hasselt et al, 2016), with a deep neural network being used to approximate the Q-function. The input molecule is converted to a vector, by taking the form of a Morgan fingerprint (Rogers and Hahn, 2010) with the radius of 3 and length of 2048. The number of steps remaining in the episode is concatenated to the vector and a four-layer fully-connected network with hidden state size of [1024, 512, 128, 32] and ReLU activation is used as the architecture.

12.4.2.3 Node-Edge Co-evolving Deep Graph Translator

To overcome a number of challenges including, but not limited to, the mutually dependent translation of the node and edge attributes, asynchronous and iterative changes in the node and edge attributes during graph translation, and the difficulty of discovering and enforcing the correct consistency between node attributes and graph spectra, the Node-Edge Co-evolving Deep Graph Translator (NEC-DGT) has been developed to achieve so-called multi-attributed graph translation and proven to be a generalization of the existing topology translation models (Guo et al, 2019c). This is a node-edge co-evolving deep graph translator that edits the source graph iteratively through a generation process similar to the MPNN-based adjacency-based one-shot method for unconditional deep graph generation, with the main difference being that it takes the graph in the source domain as input rather than the initialized graph (Guo et al, 2019c).

NEC-DGT employs a multi-block translation architecture to learn the distribution of the graphs in the target domain, conditioning on the input graphs and contextual information. Specifically, the inputs are the node and graph attributes, and the model outputs are the generated graphs' node and edge attributes after several blocks. A skip-connection architecture is implemented across the different blocks to handle the asynchronous properties of different blocks, ensuring the final translated results fully utilize various combinations of blocks' information. The following loss function is minimized in the work:

$$\mathscr{L}_{\mathscr{T}} = \mathscr{L}(\mathscr{T}(\mathscr{G}(E_0, F_0), C), \mathscr{G}(E', F')), \tag{12.28}$$

where C corresponds to the contextual information vector, E_0, E' corresponds to the edge attribute tensors of the input and target graphs, respectively, and F_0, F' corresponds to the node attribute tensors of the input and target graphs, respectively.

To jointly handle the various interactions among the nodes and edges, the respective translation paths are considered for each block. For example, in the node translation path, *edges-to-nodes* and *nodes-to-nodes* interactions are considered in the generation of node attributes. Similarly, "node to edges" and "edges-to-edges" are considered in the generation of edge attributes.

The frequency domain properties of the graph are learned, by which the interactions between node and edge attributes are jointly regularized utilizing a nonparametric graph Laplacian. Also, shared patterns among the generated nodes and edges in different blocks are enforced through regularization. Then, the regularization term is

$$\mathscr{R}(\mathscr{G}(E, F)) = \sum_{s=0}^{S} \mathscr{R}_{\theta}(\mathscr{G}(E_S, F_S)) + \mathscr{R}_{\theta}, \tag{12.29}$$

where S corresponds to the number of blocks and θ refers to the overall parameters in the spectral graph regularization. $\mathscr{G}(E_S, F_S)$ is the generated target graph, where E_S is the generated edge attributes tensor and F_S is the node attributes matrix. Then the total loss function is

$$\tilde{\mathscr{L}} = \mathscr{L}(T(\mathscr{G}(E_0, F_0), C), \mathscr{G}(E', F')) + \beta \mathscr{R}(G(E, F)). \tag{12.30}$$

The model is trained by minimizing the MSE of E_S with E', F_S with F', enforced by the regularization. $T(\cdot)$ is the mapping from the input graph to the target graph learned from the multi-attributed graph translation.

The transformation process is modeled by several stages with each stage generating an immediate graph. Specifically, for each stage t, there are two options: node translation paths and edge translation paths. In the node translation path, an MLP-based influence-function is used to calculate the influence $I_i^{(t)}$ on each node v_i from its neighboring nodes. Another MLP-based updating-function is used to update the node attribute as $F_i^{(t)}$ with the input of influence $I_i^{(t)}$. The edge translation path is constructed in the same way as the node translation path, with each edge being generated by the influence from its adjacent edges.

12.5 Other Graph-based Transformations

12.5.1 Sequence-to-Graph Transformation

A deep sequence-to-graph transformation aims to generate a target graph G_T conditioned on an input sequence X. This problem is often seen in domains such as NLP (Chen et al, 2018a; Wang et al, 2018g) and time series mining (Liu et al, 2015; Yang et al, 2020c).

Existing methods (Chen et al, 2018a; Wang et al, 2018g) handle the semantic parsing task by transforming a sequence-to-graph problem into a sequence-to-sequence problem and utilizing the classical RNN-based encoder-decoder model to learn this mapping. A neural semantic parsing approach, named Sequence-to-Action, models semantic parsing as an end-to-end semantic graph generation process (Chen et al, 2018a). Given a sentence $X = \{x_1, ..., x_m\}$, the Sequence-to-Action model generate a sequence of actions $Y = \{y_1, .., y_m\}$ when constructing the correct semantic graph. A semantic graph consists of nodes (including variables, entities, and types) and edges (semantic relationships), with universal operations (e.g., argmax, argmin, count, sum, and not). To generate a semantic graph, six types of actions are defined: *Add Variable Node, Add Entity Node, Add Type Node, Add Edge, Operation Function* and *Argument Action*. In this way, the generated parse tree is represented as a sequence, and the sequence-to-graph problem is transformed into a sequence-to-sequence problem. The attention-based sequence-to-sequence RNN model with an encoder and decoder can be utilized, where the encoder converts the input sequence X to a sequence of context sensitive vectors $\{\mathbf{b}_1, ..., \mathbf{b}_m\}$ using a bidirectional RNN and a classical attention-based decoder generates action sequence Y based on the context sensitive vectors (Bahdanau et al, 2015). The generation of a parse tree as a sequence of actions is represented (Wang et al, 2018g) and concepts from the Stack-LSTM neural parsing model are borrowed, producing two non-trivial improvements, Bi-LSTM subtraction and incremental tree-LSTM, that improve the process of learning a sequence-to-sequence mapping (Dyer et al, 2015).

Other methods have also been developed to handle the problem of Time Series Conditional Graph Generation (Liu et al, 2015; Yang et al, 2020c): given an input multivariate time series, the aim is to infer a target relation graph to model the underlying interrelationship between the time series and each node. A novel model of time series conditioned graph generation-generative adversarial networks (TSGG-GAN) for time series conditioned graph generation has been proposed that explores the use of GANs in a conditional setting (Yang et al, 2020c). Specifically, the generator in a TSGG-GAN adopts a variant of recurrent neural networks known as simple recurrent units (SRU) (Lei et al, 2017b) to extract essential information from the time series, and uses an MLP to generate the directed weighted graph.

12.5.2 Graph-to-Sequence Transformation

A number of graph-to-sequence encoder-decoder models have been proposed to handle rich and complex data structures, which are hard for sequence-to-sequence methods to handle (Gao et al, 2019c; Bastings et al, 2017; Beck et al, 2018; Song et al, 2018; Xu et al, 2018c). A graph-to-sequence model typically employs a graph-neural-network-based (GNN-based) encoder and an RNN/Transformer-based decoder, with most being designed to tackle tasks such as natural language generation (NLG), which is an important task in NLP (YILMAZ et al, 2020). Graph-to-sequence models have the ability to capture the rich structural information of the input and can also be applied to arbitrary graph-structured data.

Early graph-to-sequence methods and their follow-up works (Bastings et al, 2017; Damonte and Cohen, 2019; Guo et al, 2019e; Marcheggiani et al, 2018; Xu et al, 2020b,d; Zhang et al, 2020d,c) have mainly used a graph convolutional network (GCN) (Kipf and Welling, 2017b) as the graph encoder, probably because GCN was the first widely used GNN model that sparked this new wave of research on GNNs and their applications. Early GNN variants, such as GCN, were not originally designed to encode information on the edge type and so cannot be directly applied to the encoding of multi-relational graphs in NLP. Later on, more graph transformer models (Cai and Lam, 2020; Jin and Gildea, 2020; Koncel-Kedziorski et al, 2019) were introduced to the graph-to-sequence architecture to handle these multi-relational graphs. These graph transformer models generally function by either replacing the self-attention network in the original transformer with a masked self-attention network, or explicitly incorporating edge embeddings into the self-attention network.

Because edge direction in an NLP graph often encodes critical information regarding semantic meanings, capturing bidirectional information in the text is helpful and has been widely explored in works such as BiLSTM and BERT (Devlin et al, 2019). Some attention has also been devoted to extending the existing GNN models to handle directed graphs. For example, separate model parameters can be introduced for different edge directions (e.g., incoming/outgoing/self-loop edges) when conducting neighborhood aggregation (Guo et al, 2019e; Marcheggiani et al, 2018; Song et al, 2018). A BiLSTM-like strategy has also been proposed to learn the node embeddings of each direction independently using two separate GNN encoders and then concatenating the two embeddings for each node to obtain the final node embeddings (Xu et al, 2018b,c,d).

In the field of NLP, graphs are usually multi-relational, where the edge type information is vital for the prediction. Similar to the bidirectional graph encoder introduced above, separate model parameters for different edge types are considered when encoding edge type information with GNNs (Chen et al, 2018e; Ghosal et al, 2020; Schlichtkrull et al, 2018). However, usually the total number of edge types is large, leading to non-negligible scalability issues for the above strategies. This problem can be tackled by converting a multi-relational graph to a Levi graph (Levi, 1942), which is bipartite. To create a Levi graph, all the edges in the original graph are treated as new nodes and new edges are added to connect the original nodes and new nodes.

Apart from NLP, graph-to-sequence transformation has been employed in other fields, for example when modeling complex transitions of an individual user's activities among different healthcare subforums over time and learning how this is related to his various health conditions (Gao et al, 2019c). By formulating the transition of user activities as a dynamic graph with multi-attributed nodes, the health stage inference is formalized as a dynamic graph-to-sequence learning problem and, hence, a dynamic graph-to-sequence neural network architecture (DynGraph2Seq) has been proposed (Gao et al, 2019c). This model contains a dynamic graph encoder and an interpretable sequence decoder. In the same work, a dynamic graph hierarchical attention mechanism capable of capturing entire both time-level and node-level attention is also proposed, providing model transparency throughout the whole inference process.

12.5.3 Context-to-Graph Transformation

Deep graph generation conditioning on semantic context aims to generate the target graph $\mathcal{G}_T$ conditioning on an input semantic context that is usually represented in the form of additional meta-features. The semantic context can refer to the category, label, modality, or any additional information that can be intuitively represented as a vector C. The main issue here is to decide where to concatenate or embed the condition representation into the generation process. As a summary, the conditioning information can be added in terms of one or more of the following modules: (1)

the node state initialization module, (2) the message passing process for MPNN-based decoding, and (3) the conditional distribution parameterization for sequential generating.

A novel unified model of graph variational generative adversarial nets has been proposed, where the conditioning semantic context is input into the node state initialization module (Yang et al, 2019a). Specifically, the generation process begins by modeling the embedding Z_i of each node with the separate latent distributions, after which a conditional graph VAE (CGVAE) can be directly constructed by concatenating the condition vector C to each node's latent representation Z_i to obtain the updated node latent representation $\hat{Z}_i$. Thus, the distribution of the individual edge $\mathcal{E}_{i,j}$ is assumed to be a Bernoulli distribution, which is parameterized by the value $\hat{\mathcal{E}}_{i,j}$ and calculated as $\hat{\mathcal{E}}_{i,j} = \text{Sigmoid}(f(\hat{Z}_i)^\top f(\hat{Z}_j))$, where $f(\cdot)$ is constructed using a few fully connected layers. A conditional deep graph generative model that adds the semantic context information into the initialized latent representations Z_i at the beginning of the decoding process has also been proposed (Li et al, 2018d).

Other researchers have added the context information C into the message passing module as part of its MPNN-based decoding process (Li et al, 2018f). Specifically, the decoding process is parameterized as a Markov process and the graph is generated by iteratively refining and updating the initialized graph. At each step t, an action is conducted based on the current node's hidden states $H^t = \{\mathbf{h}_1^t, ..., \mathbf{h}_N^t\}$. To calculate $\mathbf{h}_i^t \in \mathbb{R}^l$ (l denotes the length of the representation) for node v_i in the intermediate graph $\mathcal{G}_t$ after each updating of the graph, a message passing network is utilized with node message propagation. Thus, the context information $C \in \mathbb{R}^k$ is added to the operation of the MPNN layer as follows:

$$\mathbf{h}_i^t = W\mathbf{h}_i^{t-1} + \Phi \sum_{v_j \in N(v_j)} \mathbf{h}_j^{t-1} + \Theta C, \qquad (12.31)$$

where $W \in \mathbb{R}^{l \times l}$, $\Theta \in \mathbb{R}^{l \times l}$ and $\Phi \in \mathbb{R}^{k \times l}$ are all learnable weights vectors and k denotes the length of the semantic context vector.

Semantic context has also been considered as one of the inputs for calculating the conditional distribution parameter at each step during the sequential generating process (Jonas, 2019). The aim here is to solve the molecule inverse problem by inferring the chemical structure conditioning on the formula and spectra of a molecule, which provides a distinguishable fingerprint of its bond structure. The problem is framed as an MDP and molecules are constructed incrementally one bond at a time based on a deep neural network, where they learn to imitate a "subisomorphic oracle" that knows whether the generated bonds are correct. The context information (e.g., spectra) is applied in two places. The process begins with an empty edge set $\mathcal{E}_0$ that is sequentially updated to $\mathcal{E}_k$ at each step k by adding an edge sampled from $p(e_{i,j}|\mathcal{E}_{k-1}, \mathcal{V}, C)$. $\mathcal{V}$ denotes the node set that is defined in the given molecular formula. The edge set keeps updating until the existing edges satisfy all the valence constraints of a molecule. The resulting edge set $\mathcal{E}_K$ then serves as the candidate graph. For a given spectrum C, the process is repeated T times, generating T (potentially different) candidate structures, $\{\mathcal{E}_K^{(i)}\}_{i=1}^T$. Then based on a spectral prediction function $f(\cdot)$, the quality of these candidate structures are evaluated by measuring how close their predicted spectra are to the condition spectrum C. Finally, the optimal generated graph is selected according to $\operatorname*{argmin}_{i} \| f(\mathscr{E}_K^{(i)}) - C \|_2$.

12.6 Summary

In this chapter, we introduce the definitions and techniques for the transformation problem that involves graphs in the domain of deep graph neural networks. We provide a formal definition of the general deep graph transformation problem as well as its four sub-problems, namely node-level transformation, edge-level transformation, node-edge co-transformation, as well as other graph-involved transformations (e.g., sequence-to-graph transformation and context-to-graph transformation). For each sub-problem, its unique challenges and several representative methods are introduced. As an emerging research domain, there are still many open problems to be solved for future exploration, including but not limited to: (1) **Improved scalability**. Existing deep graph transformation models typically have super-linear time complexity to the number of nodes and cannot scale well to large networks. Consequentially, most existing works merely focus on small graphs, typically with dozens to thousands of nodes. It is difficult for them to handle many real-world networks with millions to billions of nodes, such as the internet of things, biological neuronal networks, and social networks. (2) **Applications in NLP**. As more and more GNN-based works have advanced the development of NLP, graph transformation is naturally a good fit for addressing some NLP tasks, such as information extraction and semantic parsing. For example, information extraction can be formalized into a graph-to-graph problem where the input graph is the dependency graph and the output graph is the information graph. (3) **Explainable graph transformation**. When we learn the underlying distribution of the generated target graphs, learning interpretable representations of graph that expose semantic meaning is very important. For example, it is highly beneficial if we could identify which latent variable(s) control(s) which specific properties (e.g., molecule mass) of the target graphs (e.g., molecules). Thus, investigations on the explainable graph transformation process are critical yet unexplored.

> **Editor's Notes**: Graph transformation is deemed very relevant to graph generation (see Chapter 11) and can be considered as an extension of the latter. In many real-world applications, one is usually required to generate graphs with some condition or control from the users. For example, one may want to generate molecules under some targeted properties (see Chapters 24 and 25) or programs under some function (see Chapter 22). In addition, graph-to-graph transformation also has a connection to link prediction (Chapter 10) and node classification (Chapter 4), though the former could be more challenging since it typically requires simultaneous node-edge prediction, and possibly also comes with the consideration of stochasticity.

Chapter 13
Graph Neural Networks: Graph Matching

Xiang Ling, Lingfei Wu, Chunming Wu and Shouling Ji

Abstract The problem of graph matching that tries to establish some kind of structural correspondence between a pair of graph-structured objects is one of the key challenges in a variety of real-world applications. In general, the graph matching problem can be classified into two categories: i) the classic graph matching problem which finds an optimal node-to-node correspondence between nodes of a pair of input graphs and ii) the graph similarity problem which computes a similarity metric between two graphs. While recent years have witnessed the great success of GNNs in learning node representations of graphs, there is an increasing interest in exploring GNNs for the graph matching problem in an end-to-end manner. This chapter focuses on the state of the art of graph matching models based on GNNs. We start by introducing some backgrounds of the graph matching problem. Then, for each category of graph matching problem, we provide a formal definition and discuss state-of-the-art GNN-based models for both the classic graph matching problem and the graph similarity problem, respectively. Finally, this chapter is concluded by pointing out some possible future research directions.

Xiang Ling
Department College of Computer Science and Technology, Zhejiang University, e-mail: lingxiang@zju.edu.cn

Lingfei Wu
JD.COM Silicon Valley Research Center, e-mail: lwu@email.wm.edu

Chunming Wu
Department College of Computer Science and Technology, Zhejiang University, e-mail: wuchunming@zju.edu.cn

Shouling Ji
Department College of Computer Science and Technology, Zhejiang University, e-mail: sji@zju.edu.cn

13.1 Introduction

As graphs are natural and ubiquitous representations for describing sophisticated data structures, the problem of graph matching that tries to establish some kind of structural correspondence between two input graph-structured objects. The graph matching problem is one of the key challenges in a variety of research fields, such as computer vision (Vento and Foggia, 2013), bioinformatics (Elmsallati et al, 2016), cheminformatics (Koch et al, 2019; Bai et al, 2019b), computer security (Hu et al, 2009; Wang et al, 2019i), source/binary code analysis (Li et al, 2019h; Ling et al, 2021), and social network analysis (Kazemi et al, 2015), to name just as few. In particular, recent research advances in graph matching have been closely involved in many real-world applications in the field of computer vision, including visual tracking (Cai et al, 2014; Wang and Ling, 2017), action recognition (Guo et al, 2018a), pose estimation (Cao et al, 2017, 2019), *etc*. In addition to the study in computer vision, graph matching also serves as an important foundation of many other graph-based research tasks, *e.g.*, node and graph classification tasks (Richiardi et al, 2013; Bai et al, 2019c; Ok, 2020), graph generation tasks (You et al, 2018b; Ok, 2020), *etc*.

In a broad sense, according to different goals of graph matching in a wide variety of real-world applications, the general graph matching problem can be classified into two categories (Yan et al, 2016) as follows. The first category is the *classic graph matching problem* (Loiola et al, 2007; Yan et al, 2020a) that tries to establish the node-to-node correspondence (and/or even edge-to-edge correspondence) between the pair of input graphs. The second category is the *graph similarity problem* (Bunke, 1997; Riesen, 2015; Ma et al, 2019a) with the purpose of computing a similarity score between two input graphs. Both categories have the same inputs (*i.e.*, a pair of input graphs) but with different outputs, whereby the output of the first category is mainly formulated as a correspondence *matrix* while the output of the second category is usually expressed as a similarity *scalar*. From the perspective of outputs, the second graph similarity problem can be viewed as a special case of the first graph matching problem, as the similarity scalar reflects a more coarse-grained correspondence representation of graph matching than the correspondence matrix.

Generally, both categories of the graph matching problem are known to be NP-hard (Loiola et al, 2007; Yan et al, 2020a; Bunke, 1997; Riesen, 2015; Ma et al, 2019a), making both problems computationally infeasible for exact and optimum solutions in large-scale and real-world settings. Given the great importance and inherent difficulty of the graph matching problem, it has been heavily investigated in theory and practice and a huge number of approximate algorithms based on theoretical/empirical knowledge of experts have been proposed to find sub-optimal solutions in an acceptable time. Interested readers are referred to (Loiola et al, 2007; Yan et al, 2016; Foggia et al, 2014; Riesen, 2015) for a more extensive review, as these approximation methods are beyond the scope of this chapter. Unfortunately, despite various approximation methods have been devoted to resolving the graph matching problem for the past decades, it still suffers from the issue of poor scalability as well as the issue of heavy reliance on expert knowledge, and thus remains as a challenging and significant research problem for many practitioners.

More recently, GNNs that attempt to adapt deep learning from image to non-euclidean data (*i.e.*, graphs) have received unprecedented attention to learn informative representation (*e.g.*, node or (sub)graph, *etc.*) of graph-structured data in an end-to-end manner (Kipf and Welling, 2017b; Wu et al, 2021d; Rong et al, 2020c). Hereafter, a surge of GNN models have been presented for learning effective node embeddings for downstream tasks, such as node classification tasks (Hamilton et al, 2017a; Veličković et al, 2018; Chen et al, 2020m), graph classification tasks (Ying et al, 2018c; Ma et al, 2019d; Gao and Ji, 2019), graph generation tasks (Simonovsky and Komodakis, 2018; Samanta et al, 2019; You et al, 2018b) as so on. The great success of GNN-based models on these application tasks demonstrates that GNN is a powerful class of deep learning model to better learn the graph representation for downstream tasks.

Encouraged by the great success of GNN-based models obtained from many other graph-related tasks, many researchers have started to adopt GNNs for the graph matching problem and a large number of GNN-based models have been proposed to improve the matching accuracy and efficiency (Zanfir and Sminchisescu, 2018; Rolínek et al, 2020; Wang et al, 2019g; Jiang et al, 2019a; Fey et al, 2020; Yu et al, 2020; Wang et al, 2020j; Bai et al, 2018, 2020b, 2019b; Xiu et al, 2020; Ling et al, 2020; Zhang, 2020; Wang et al, 2020f; Li et al, 2019h; Wang et al, 2019i). During the training stage, these models try to learn a mapping between the pair of input graphs and the ground-truth correspondence in a supervised learning and thus are more time-efficient during the inference stage than traditional approximation methods. In this chapter, we walk through the recent advances and developments of graph matching models based on GNNs. Particularly, we focus on how to incorporate GNNs into the framework of graph matching/similarity learning and try to provide a systematic introduction and review of state-of-the-art GNN-based methods for both categories of the graph matching problem (*i.e.*, the classic graph matching problem in Section 13.2 and the graph similarity problem in Section 13.3, respectively).

13.2 Graph Matching Learning

In this section, we start by introducing the first category of the graph matching problem, *i.e.*, the classic graph matching problem[1], and provide a formal definition of the graph matching problem. Subsequently, we will focus discussion on state-of-the-art graph matching models based on deep learning as well as more advanced GNNs in the literature.

---

[1] For simplicity, we represent the classic graph matching problem as the graph matching problem in the following sections of this chapter.

13.2.1 Problem Definition

A graph of size n (*i.e.*, numbers of nodes) can be represented as $\mathscr{G} = (\mathscr{V}, \mathscr{E}, A, X, E)$, in which $\mathscr{V} = \{v_1, \cdots, v_n\}$ denotes the set of nodes (also known as *vertices*), $\mathscr{E} \subseteq \mathscr{V} \times \mathscr{V}$ denotes the set of edges, $A \in \{0, 1\}^{n \times n}$ denotes the adjacency matrix, $X \in \mathbb{R}^{n \times \cdot}$ denotes the initial feature matrix of nodes, and $E \in \mathbb{R}^{n \times n \times \cdot}$ denotes an optional initial feature matrix of edges.

The purpose of the graph matching problem is to find an optimal node-to-node correspondence between two input graphs, *i.e.*, $\mathscr{G}^{(1)}$ and $\mathscr{G}^{(2)}$. Without loss of generality, we consider the graph matching problem whose two input graphs of equal size[2]. In particular, we provide a formal definition of the graph matching problem in Definition 13.1 as follows and give an example illustration of the node-to-node correspondence in Fig. 13.1.

Definition 13.1 (Graph Matching Problem). Given a pair of input graphs $\mathscr{G}^{(1)} = (\mathscr{V}^{(1)}, \mathscr{E}^{(1)}, A^{(1)}, X^{(1)}, E^{(1)})$ and $\mathscr{G}^{(2)} = (\mathscr{V}^{(2)}, \mathscr{E}^{(2)}, A^{(2)}, X^{(2)}, E^{(2)})$ of equal size n, the graph matching problem is to find a node-to-node correspondence matrix $S \in \{0, 1\}^{n \times n}$ (*i.e.*, also called *assignment matrix* and *permutation matrix*) between the two graphs $\mathscr{G}^{(1)}$ and $\mathscr{G}^{(2)}$. Each element $S_{i,a} = 1$ if and only if the node $v_i \in \mathscr{V}^{(1)}$ in $\mathscr{G}^{(1)}$ corresponds to the node $v_a \in \mathscr{V}^{(2)}$ in $\mathscr{G}^{(2)}$.

Intuitively, the resulting correspondence matrix S represents the possibility of establishing a matching relation between any pair of nodes in two graphs. The graph matching problem is known to be NP-hard and has been investigated by formulating it as a quadratic assignment problem (QAP) (Loiola et al, 2007; Yan et al, 2016). We adopt the general form of Lawler's QAP (Lawler, 1963) with constraints as follows since it has been widely adopted in literature.

Fig. 13.1 An example illustration of the graph matching problem with two input graphs, *i.e.*, the left graph $\mathscr{G}^{(1)}$ and the right graph $\mathscr{G}^{(2)}$ to be matched. The red dotted lines represent the node-to-node correspondences between the two graphs.

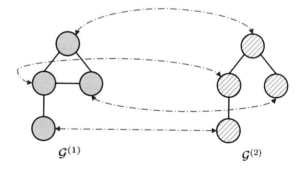

---

[2] For simplicity, we assume that a pair of input graphs in the graph matching problem have the same number of nodes, but we can extend the problem to a pair of graphs with different number of nodes via adding dummy nodes, which is commonly adopted by graph matching literature Krishnapuram et al (2004).

$$\mathbf{s}^* = \arg\max_{\mathbf{s}} \mathbf{s}^\top K \mathbf{s}$$

$$\text{s.t. } S\mathbf{1}_n = \mathbf{1}_n \ \& \ S^\top \mathbf{1}_n = \mathbf{1}_n \tag{13.1}$$

where $\mathbf{s} = \text{vec}(S) \in \{0,1\}^{n^2}$ is the column-wise vectorized version of the assignment matrix S and $\mathbf{1}_n$ is a column vector of length n whose elements are equal to 1. Particularly, $K \in \mathbb{R}^{n^2 \times n^2}$ is the corresponding second-order affinity matrix in which each element $K_{ij,ab}$ measures how well every pair of nodes $(v_i, v_j) \in \mathcal{V}^{(1)} \times \mathcal{V}^{(1)}$ matches $(v_a, v_b) \in \mathcal{V}^{(2)} \times \mathcal{V}^{(2)}$ and can be defined as follows (Zhou and De la Torre, 2012).

$$K_{\text{ind}(i,j),\text{ind}(a,b)} = \begin{cases} c_{ia} & \text{if } i = j \text{ and } a = b, \\ d_{ijab} & \text{else if } A_{i,j}^{(1)} A_{a,b}^{(2)} > 0, \\ 0 & \text{otherwise.} \end{cases} \tag{13.2}$$

where $\text{ind}(\cdot, \cdot)$ is a bijection function that maps a pair of nodes to an integer index, the diagonal element (i.e., c_{ia}) encodes the node-to-node (i.e., first-order) affinity between the node $v_i \in \mathcal{V}^{(1)}$ and the node $v_a \in \mathcal{V}^{(2)}$, and the off-diagonal element (i.e., d_{ijab}) encodes the edge-to-edge (i.e., second-order) affinity between the edge $(v_i, v_j) \in \mathcal{E}^{(1)}$ and the edge $(v_a, v_b) \in \mathcal{E}^{(2)}$.

Another important aspect for the formulation in Equation (13.1) is the constraint, i.e., $S\mathbf{1}_n = \mathbf{1}_n$ and $S^\top \mathbf{1}_n = \mathbf{1}_n$. It demands that the matching output of the graph matching problem, i.e., the correspondence matrix $S \in \{0,1\}^{n \times n}$, should be strictly constrained as a **doubly-stochastic matrix**. Formally the correspondence matrix S is a doubly-stochastic matrix if the summation of each column and each row of it is 1. That is, $\forall i, \sum_j S_{i,j} = 1$ and $\forall j, \sum_i S_{i,j} = 1$. Therefore, the resulting correspondence matrix of the graph matching problem should satisfy the requirement of the doubly-stochastic matrix.

In general, the main challenge in optimizing and solving Equation (13.1) lies in how to model the affinity model as well as how to optimize with the constraint for solutions. Traditional methods mostly utilize pre-defined affinity models with limited capacity (e.g., Gaussian kernel with Euclid distance Cho et al (2010)) and resort to different heuristic optimizations (e.g., graduated assignment (Gold and Rangarajan, 1996), spectral method (Leordeanu and Hebert, 2005), random walk (Cho et al, 2010), etc.). However, such traditional methods suffer from poor scalability and inferior performance for large-scale settings as well as a broad of application scenarios (Yan et al, 2020a). Recently, studies on the graph matching are starting to explore the high capacity of deep learning models, which achieve state-of-the-are performance. In the following subsections, we will first give a brief introduction of deep learning based graph matching models and then discuss state-of-the-art graph matching models based on GNNs.

13.2.2 Deep Learning based Models

Aiming at increasing the matching performance, extensive research interest in leveraging high capacity of deep learning models to solve the problem of graph matching has been ignited since Zanfir and Sminchisescu (2018), which introduces an end-to-end deep learning framework for the graph matching problem for the first time and receives the best paper honorable mention award in CVPR 2018[3].

Deep Graph Matching. In (Zanfir and Sminchisescu, 2018), Zanfir and Sminchisescu first relax the graph matching problem of Equation (13.1) with the ℓ_2 constraint as follows.

$$\mathbf{s}^* = \arg\max_{\mathbf{s}} \mathbf{s}^\top K \mathbf{s}$$
$$\text{s.t. } \|\mathbf{s}\|_2 = 1 \tag{13.3}$$

To solve the problem, they attempt to introduce deep learning techniques to the graph matching and propose an end-to-end training framework with standard differentiable backpropagation and optimization algorithms. The proposed deep graph matching framework first uses the existing pre-trained CNN model (*i.e.*, VGG-16 (Simonyan and Zisserman, 2014b)) to extract node features (*i.e.*, $U^{(1)}$ and $U^{(2)} \in \mathbb{R}^{n \times d}$) and edge features (*i.e.*, $F^{(1)} \in \mathbb{R}^{p \times 2d}$ and $F^{(2)} \in \mathbb{R}^{q \times 2d}$) from the pair of input images in the scenario of computer vision applications. In particular, $F^{(1)}$ and $F^{(2)}$ are row-wise edge feature matrices with p and q as the number of edges in each graph, respectively. As each edge attribute is the concatenation of the start and the end node, the dimension of edge attribute is double $2d$ the dimension of node.

Next, based on extracted node/edge features, it builds the graph matching affinity matrix K via a novel factorization method of graph matching (Zhou and De la Torre, 2012) as follows.

$$\begin{aligned}
K &= \lceil \text{vec}(K_p) \rfloor + (G_2 \otimes G_1) \lceil \text{vec}(K_e) \rfloor (H_2 \otimes H_1)^\top \\
&= \Big\lceil \text{vec}(U^{(1)} U^{(2)\top}) \Big\rfloor + (G_2 \otimes G_1) \Big\lceil \text{vec}(F^{(1)} \Lambda F^{(2)}) \Big\rfloor (H_2 \otimes H_1)^\top
\end{aligned} \tag{13.4}$$

where $\lceil X \rfloor$ denotes a diagonal matrix whose diagonal elements are all X; $\otimes$ denotes the Kronecker product; G_i and H_i ($i = \{1,2\}$) are the node-edge incidence matrices that are recovered from the adjacency matrices $A^{(i)}$, *i.e.*, $A^{(i)} = G_i H_i^\top$ ($i = \{1,2\}$); $K_p \in \mathbb{R}^{n \times n}$ encodes the node-to-node similarity and is directly obtained from the product of two node feature matrices, *i.e.*, $K_p = U^{(1)} U^{(2)\top}$; $K_e \in \mathbb{R}^{p \times q}$ encodes the edge-to-edge similarity and is calculated by $K_e = F^{(1)} \Lambda F^{(2)}$. It is worth to note that $\Lambda \in \mathbb{R}^{2d \times 2d}$ is a learnable parameter matrix and thus the built graph matching affinity matrix K in Equation (13.4) is a learnable affinity model.

Then, with the spectral matching technique (Leordeanu and Hebert, 2005), the graph matching problem is translated into computing the leading eigenvector $\mathbf{s}^*$ which can be approximated by the power iteration algorithm as follows.

---

[3] https://www.thecvf.com/?page_id=413

$$\mathbf{s}_{k+1} = \frac{K\mathbf{s}_k}{\|K\mathbf{s}_k\|_2} \tag{13.5}$$

in which $\mathbf{s}$ is initialized with $\mathbf{s}_0 = \mathbf{1}$ and K is computed from Equation (13.4). It is also worth to note that the spectral graph matching solver in Equation (13.5) is differentiable but un-learnable. Because the resulting $\mathbf{s}_{k+1}$ is not a doubly-stochastic matrix, it employs a bi-stochastic normalization layer to iteratively normalize the matrix by columns and rows over and over again.

Finally, the whole graph matching model is trained in an end-to-end fashion with a displacement loss $\mathscr{L}_{disp}$ which operates the difference between predicted displacement and the ground-truth displacement.

$$\mathscr{L}_{disp} = \sum_{i=0}^{n} \sqrt{\|\mathbf{d}_i - \mathbf{d}_i^{gt}\|_2 + \varepsilon} \quad \text{and} \quad \mathbf{d}_i = \sum_{v_a \in \mathscr{V}^{(2)}} (S_{i,a} P_a^{(2)}) - P_i^{(1)} \tag{13.6}$$

where $P^{(1)}$ and $P^{(2)}$ are coordinates of nodes in both images; the vector of $\mathbf{d}_i$ measures the pixel offset; $\mathbf{d}_i^{gt}$ is the corresponding ground-truth; and ε is a small value for robust penalty.

Deep Graph Matching via Black-box Combinatorial Solver. Motivated by advances in incorporating a combinatorial optimization solver into a neural network (Pogancic et al, 2020), Rolínek et al (2020) propose an end-to-end neural network which seamlessly embeds a black-box combinatorial solver, namely BB-GM, for the graph matching problem. To be specific, given two cost vectors (i.e., $\mathbf{c}^v \in \mathbb{R}^{n^2}$ and $\mathbf{c}^e \in \mathbb{R}^{|\mathscr{E}^{(1)}||\mathscr{E}^{(2)}|}$) for both node-to-node and edge-to-edge correspondences, the graph matching problem is formulated as follows.

$$\text{GM}(\mathbf{c}^v, \mathbf{c}^e) = \underset{(\mathbf{s}^v, \mathbf{s}^e) \in \text{Adm}(\mathscr{G}^{(1)}, \mathscr{G}^{(2)})}{\arg\min} \{\mathbf{c}^v \cdot \mathbf{s}^v + \mathbf{c}^e \cdot \mathbf{s}^e\} \tag{13.7}$$

where GM denotes the black-box combinatorial solver; $\mathbf{s}^v \in \{0,1\}^{n^2}$ is the indicator vector of matched nodes; $\mathbf{s}^e \in \{0,1\}^{|\mathscr{E}^{(1)}||\mathscr{E}^{(2)}|}$ is the indicator vector of matched edges; $\text{Adm}(\mathscr{G}^{(1)}, \mathscr{G}^{(2)})$ represents a set of all possible matching results between $\mathscr{G}^{(1)}$ and $\mathscr{G}^{(2)}$.

By the formulation, the core of the graph matching problem is to construct the two cost vectors $\mathbf{c}^v$ and $\mathbf{c}^e$. Therefore, BB-GM first employs a pre-trained VGG-16 model to extract node embeddings and learns edge embeddings via SplineCNN (Fey et al, 2018). Then, based on the learned node embeddings, $\mathbf{c}^v$ is computed by a weighted inner product similarity between the pair of node embeddings between two graphs, along with a learnable neural network based on the graph-level feature vector. Similarly, $\mathbf{c}^e$ is also computed by a weighted inner product similarity between the pair of edge embeddings between two graphs, along with the same neural network.

13.2.3 Graph Neural Network based Models

More recently, GNNs have started to be studied to deal with the graph matching problem. This is because GNNs bring about new opportunities on the tasks over graph-like data and further improve the model capability taking structural information of graphs into account. Besides, GNNs can be easily incorporated with other deep learning architectures (*e.g.*, CNN, RNN, MLP, *etc.*) and thus provide an end-to-end learning framework for the graph matching problem.

Cross-graph Affinity based Graph Matching. Wang et al (2019g) claim that it is the first work that employs GNNs for deep graph matching learning (as least in computer vision). By exploiting the highly efficient learning capabilities of GNNs that can update the node embeddings with the structural affinity information between two graphs, the graph matching problem, *i.e.*, the quadratic assignment problem, is translated into a linear assignment problem that can be easily solved.

In particular, the authors present the cross-graph affinity based graph matching model with the permutation loss, namely PCA-GM. PCA-GM consists of three steps. First, to enhance learned node embeddings of individual graph with a standard message-passing network (*i.e.*, intra-graph convolution network), PCA-GM further updates node embeddings with an extra cross-graph convolution network, *i.e.*, CrossGConv which not only aggregates the information from local neighbors, but also incorporates the information from the similar nodes in the other graph. Fig. 13.2 illustrates an intuitive comparison between the intra-graph convolution network and the cross-graph convolution network formulated as follows.

$$
\begin{aligned}
H^{(1)(k)} &= \text{CrossGConv}\left(\hat{S}, H^{(1)(k-1)}, H^{(2)(k-1)}\right) \\
H^{(2)(k)} &= \text{CrossGConv}\left(\hat{S}^{\top}, H^{(2)(k-1)}, H^{(1)(k-1)}\right)
\end{aligned}
\tag{13.8}
$$

where $H^{(1)(k)}$ and $H^{(2)(k)}$ are the k-layer node embeddings for the graph $\mathscr{G}^{(1)}$ and $\mathscr{G}^{(2)}$; k denotes the k-th iteration; $\hat{S}$ denotes the predicted assignment matrix which is computed from shallower node embedding layers; and the initial embeddings,

Fig. 13.2 For one node in the left graph $\mathscr{G}^{(1)}$, the intra-graph convolution network only operates on its own graph, *i.e.*, the purple solid lines in $\mathscr{G}^{(1)}$. However, the cross-graph convolution network operates on both its own graph (*i.e.*, the purple solid lines in $\mathscr{G}^{(1)}$) as well as the other graph (*i.e.*, blued dashed lines from all nodes in $\mathscr{G}^{(2)}$ to the node in $\mathscr{G}^{(1)}$).

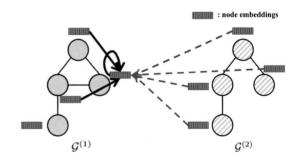

i.e., $H^{(1)(0)}$ and $H^{(2)(0)}$, are extracted via a pre-trained VGG-16 network in line with Zanfir and Sminchisescu (2018).

Second, based on the resulting node embeddings $\widetilde{H}^{(1)}$ and $\widetilde{H}^{(2)}$ for both graphs, PCA-GM computes the node-to-node assignment matrix S by a bi-linear mapping followed by an exponential function as follows.

$$\widetilde{S} = \exp \left(\frac{\widetilde{H}^{(1)} \Theta \widetilde{H}^{(2)\top}}{\tau} \right) \tag{13.9}$$

where Θ denotes the learnable parameter matrix for the assignment matrix learning and $\tau > 0$ is a hyper-parameter. As the obtained $\widetilde{S} \in \mathbb{R}^{n \times n}$ does not satisfy the constraint of the doubly-stochastic matrix, PCA-GM uses the Sinkhorn (Adams and Zemel, 2011) operation for the relaxed linear assignment problem because it is fully differentiable and has been proven effective for the final graph matching prediction.

$$S = \text{Sinkhorn}(\widetilde{S}) \tag{13.10}$$

Finally, PCA-GM adopts the combinatorial permutation loss that computes the cross entropy loss between the final predicted permutation S and ground truth permutation S^{gt} for supervised graph matching learning.

$$\mathscr{L}_{perm} = - \sum_{v_i \in \mathscr{V}^{(1)}, v_a \in \mathscr{V}^{(2)}} S_{i,a}^{gt} \log(S_{i,a}) + (1 - S_{i,a}^{gt}) \log(1 - S_{i,a}) \tag{13.11}$$

Experiment results in (Wang et al, 2019g) demonstrated that graph matching models with the permutation loss outperform that with the displacement loss in Equation (13.6).

Graph Learning–Matching Network. Most prior studies on the graph matching problem rely on established graphs with fixed structure information, *i.e.*, the edge set with or without attributes. Differently, Jiang et al (2019a) present a graph learning-matching network, namely GLMNet, which incorporates the graph structure learning (*i.e.*, learning the graph structure information) into the general graph matching learning to build a unified end-to-end model architecture. To be specific, based on the pair of node feature matrices $X^{(l)} = \{\mathbf{x}_1^{(l)}, \cdots, \mathbf{x}_n^{(l)}\}$ ($l = \{1, 2\}$), GLMNet attempts to learn a pair of optimal graph adjacency matrices $A^{(l)}$ ($l = \{1, 2\}$) for better serving for the latter graph matching learning and each element is computed as follows.

$$A_{i,j}^{(l)} = \phi(\mathbf{x}_i^{(l)}, \mathbf{x}_j^{(l)}; \theta) = \frac{\exp(\sigma(\theta^\top [\mathbf{x}_i^{(l)}, \mathbf{x}_j^{(l)}]))}{\sum_{j=1}^{n} \exp(\sigma(\theta^\top [\mathbf{x}_i^{(l)}, \mathbf{x}_j^{(l)}]))}, \quad l = \{1, 2\} \tag{13.12}$$

where σ is the activation function, *e.g.*, ReLU; $[\cdot, \cdot]$ denotes the concatenation operation; and θ denotes the trainable parameter for the graph structure learning which is shared for both input graphs.

Following PCA-GM (Wang et al, 2019g), GLMNet also explores a series of graph convolution modules to learn informative node embeddings of both input graphs for the latter affinity matrix learning and matching prediction. Based on the obtained $A^{(l)}$ and $X^{(l)}$ ($l = \{1, 2\}$), GLMNet employs the graph smoothing convolution layer (Kipf and Welling, 2017b), the cross-graph convolution layer Wang et al (2019g) and the graph sharpening convolution layer (*i.e.*, defined as the counterpart of Laplacian smoothing in (Kipf and Welling, 2017b)) to further learn and update their node embeddings *i.e.*, $\widetilde{X}^{(l)}$ ($l = \{1, 2\}$). After that, GLMNet directly computes the node-to-node assignment matrix S by Equations (13.9) and (13.10), which is exactly the same as PCA-GM (Wang et al, 2019g) does.

In addition to the permutation cross entropy loss $\mathscr{L}_{perm}$ defined in Equation (13.11), GLMNet adds an extra constraint regularized loss $\mathscr{L}_{con}$ for better satisfying the permutation constraint, *i.e.*, $\mathscr{L} = \mathscr{L}_{perm} + \lambda \mathscr{L}_{con}$ with $\lambda > 0$, in which $\mathscr{L}_{con}$ is defined as follows.

$$\mathscr{L}_{con} = \sum_{v_i, v_j \in \mathscr{V}^{(1)}} \sum_{v_a, v_b \in \mathscr{V}^{(2)}} U_{ij,ab} S_{i,a} S_{j,b}$$

$$U_{ij,ab} = \begin{cases} 1 & \text{if } i = j, a \neq b \text{ or } i \neq j, a = b; \\ 0 & \text{otherwise.} \end{cases} \tag{13.13}$$

where $U \in \mathbb{R}^{n^2 \times n^2}$ represents the conflict relationships of all matches and the optimum correspondence S means $\sum_{v_i, v_j \in \mathscr{V}^{(1)}} \sum_{v_a, v_b \in \mathscr{V}^{(2)}} U_{ij,ab} S_{i,a} S_{j,b} = 0$.

Deep Graph Matching with Consensus. In (Fey et al, 2020), Fey et al also employ GNNs to learn the graph correspondence as previous work, but additionally introduce a *neighborhood consensus* Rocco et al (2018) to further refine the learned correspondence matrix. Firstly, they use common GNN models along with the Sinkhorn operation to compute an initial correspondence matrix S^0 as follows. Ψ_{θ_1} denotes the shared GNN model for both graphs.

$$H^{(l)} = \Psi_{\theta_1}(X^{(l)}, A^{(l)}, E^{(l)}), \ l = \{1, 2\}$$
$$S^0 = \text{Sinkhorn}(H^{(1)} H^{(2)\top}) \tag{13.14}$$

Then, to reach a neighborhood consensus between the pair of matched nodes, they refine the initial correspondence matrix S^0 via another trainable GNN model (*i.e.*, Ψ_{θ_2}) and an MLP model (*i.e.*, ϕ_{θ_3}).

$$O^{(1)} = \Psi_{\theta_2}(I_n, A^{(1)}, E^{(1)})$$
$$O^{(2)} = \Psi_{\theta_2}(S^{k\top} I_n, A^{(2)}, E^{(2)})$$
$$S_{i,a}^{k+1} = \text{Sinkhorn}\left(S_{i,a}^k + \phi_{\theta_3}(\mathbf{o}_i^{(1)} - \mathbf{o}_a^{(2)})\right) \tag{13.15}$$

where I_n is the identify matrix and $\mathbf{o}_i^{(1)} - \mathbf{o}_a^{(2)}$ is computed as the neighborhood consensus between the node pair $(v_i, v_a) \in \mathscr{V}^{(1)} \times \mathscr{V}^{(2)}$ between two graphs (*e.g.*, $\mathbf{o}_i^{(1)} - \mathbf{o}_a^{(2)} \neq \mathbf{0}$ means a false matching over the neighborhoods of v_i and v_j). Finally,

S^K is obtained after K iterations and the final loss function incorporates both feature matching loss and neighborhood consensus loss, *i.e.*, $\mathcal{L} = \mathcal{L}^{init} + \mathcal{L}^{refine}$.

$$
\begin{aligned}
\mathcal{L}^{init} &= - \sum_{v_i \in \mathcal{V}^{(1)}} \log\left(S^0_{i,\pi_{gt}(i)}\right) \\
\mathcal{L}^{refine} &= - \sum_{v_i \in \mathcal{V}^{(1)}} \log\left(S^K_{i,\pi_{gt}(i)}\right)
\end{aligned}
\tag{13.16}
$$

where $\pi_{gt}(i)$ denotes the ground truth correspondence.

Deep Graph Matching with Hungarian Attention. Yu et al (2020) present an end-to-end deep learning model which is almost identical to Wang et al (2019g), including a graph embedding layer based on GNNs, an affinity learning layer (*i.e.*, Equations (13.9) and (13.10)), and the permutation loss (*i.e.*, Equation (13.11)). However, they improve the model with two main contributing aspects. The first aspect is adopting a novel node/edge embedding operation (*i.e.*, CIE) to replace the commonly used GCN operation that simply updates node embeddings while ignores the rich edge attributes. Since the edge information provides a crucial role in determining the graph matching result, CIE updates both node and edge embedding simultaneously by a channel-wise updating function in a multi-head fashion. Interested readers are referred to Section 3.2 in (Yu et al, 2020). Another aspect is a novel loss function. As the previously used permutation loss is prone to overfitting, the authors devise a novel loss function that introduces a Hungarian attention Z into the permutation loss as follows.

$$
Z = \text{Attention}(\text{Hungarian}(S), S^{gt})
$$

$$
\mathcal{L}_{hung} = - \sum_{v_i \in \mathcal{V}^{(1)}, v_a \in \mathcal{V}^{(2)}} Z_{i,a}\left(S^{gt}_{i,a} \log(\mathbf{S}_{i,a}) + (1 - \mathbf{S}^{gt}_{i,a}) \log(1 - \mathbf{S}_{i,a})\right)
\tag{13.17}
$$

where Hungarian denotes a black-box Hungarian algorithm and the role of Z is like a mask that attempts to focus more on those mismatched node pairs and focus less on node pairs that are matched exactly.

Graph Matching with Assignment Graph. Differently, Wang et al (2020j) reformulate the graph matching problem as the problem of selecting reliable nodes in the constructed *assignment graph* (Cho et al, 2010) in which each node represents a potential node-to-node correspondence. The formal definition of assignment graph is given in Definition 13.2 and one example is illustrated in Fig. 13.3.

Definition 13.2 (Assignment Graph). Given two graphs $\mathcal{G}^{(1)} = (\mathcal{V}^{(1)}, \mathcal{E}^{(1)}, X^{(1)}, E^{(1)})$ and $\mathcal{G}^{(2)} = (\mathcal{V}^{(2)}, \mathcal{E}^{(2)}, X^{(2)}, E^{(2)})$, an *assignment graph* $\mathcal{G}^{(A)} = (\mathcal{V}^{(A)}, \mathcal{E}^{(A)}, X^{(A)}, E^{(A)})$ is constructed as follows. $\mathcal{G}^{(A)}$ takes each candidate correspondence $(v_i^{(1)}, v_a^{(2)}) \in \mathcal{V}^{(1)} \times \mathcal{V}^{(2)}$ between two graphs as a node $v_{ia} \in \mathcal{V}^{(A)}$ and link an edge between a pair of nodes $v_{ia}^{(A)}, v_{jb}^{(A)} \in \mathcal{V}^{(A)}$ (*i.e.*, $(v_{ia}^{(A)}, v_{jb}^{(A)}) \in \mathcal{E}^{(A)}$) if and only if both edges *i.e.*, $(v_i^{(1)}, v_j^{(1)}) \in \mathcal{E}^{(1)}$ and $(v_a^{(2)}, v_b^{(2)}) \in \mathcal{E}^{(2)}$, exist in its original graph. Optionally, for

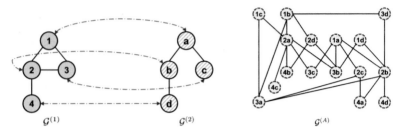

Fig. 13.3: Example illustration of building an assignment graph $\mathscr{G}^{(A)}$ from the pair of graphs $\mathscr{G}^{(1)}$ and $\mathscr{G}^{(2)}$.

node attributes $X^{(A)}$ and edge attributes $E^{(A)}$, each of them could be obtained by concatenating attributes of the pair of nodes or edges in the original graph, respectively.

With the constructed assignment graph $\mathscr{G}^{(A)}$, the reformulated problem of selecting reliable nodes in $\mathscr{G}^{(A)}$ is quite similar to binary node classification tasks Kipf and Welling (2017b) that classify nodes into positive or negative (*i.e.*, meaning matched or un-matched). To solve the problem, the authors propose a fully learnable model based on GNNs which takes the $\mathscr{G}^{(A)}$ as input, iteratively learns node embeddings over graph structural information and predicts a label for each node in $\mathscr{G}^{(A)}$ as output. Besides, the model is trained with a similar loss function to (Jiang et al, 2019a).

13.3 Graph Similarity Learning

In this section, we will first introduce the second category of the general graph matching problem – the graph similarity problem. Then, we will provide an extensive discussion and analysis of state-of-the-art graph similarity learning models based on GNNs.

13.3.1 Problem Definition

Learning a similarity metric between an arbitrary pair of graph-structured objects is one of the fundamental problems in a variety of applications, ranging from similar graph searching in databases (Yan and Han, 2002), to binary function analysis (Li et al, 2019h), unknown malware detection (Wang et al, 2019i), semantic code retrieval (Ling et al, 2021), *etc*. According to different application backgrounds, the similarity metric can be defined by different measures of structural similarity, such as graph edit distance (GED) (Riesen, 2015), maximum common subgraph (MCS) (Bunke, 1997; Bai et al, 2020c), or even more coarse binary similarity (*i.e.*, similar or not) (Ling et al, 2021). As GED is equivalent to the problem of MCS under a fitness function (Bunke, 1997), in this section, we mainly consider the GED computation and focus more on state-of-the-art graph similarity learning models based on GNNs.

Basically, the graph similarity problem intends to compute a similarity score between a pair of graphs, which indicates how similar the pair of graphs is. In the following Definition 13.3, the general graph similarity problem is defined.

Definition 13.3 (Graph Similarity Problem). Given two input graphs $\mathscr{G}^{(1)}$ and $\mathscr{G}^{(2)}$, the purpose of graph similarity problem is to produce a similarity score s between $\mathscr{G}^{(1)}$ and $\mathscr{G}^{(2)}$. In line with the notations in Section 13.2.1, the $\mathscr{G}^{(1)} = (\mathscr{V}^{(1)}, \mathscr{E}^{(1)}, A^{(1)}, X^{(1)})$ is represented as set of n nodes $v_i \in \mathscr{V}^{(1)}$ with a feature matrix $X^{(1)} \in \mathbb{R}^{n \times d}$, edges $(v_i, v_j) \in \mathscr{E}^{(1)}$ formulating an adjacency matrix $A^{(1)}$. Similarly, $\mathscr{G}^{(2)} = (\mathscr{V}^{(2)}, \mathscr{E}^{(2)}, A^{(2)}, X^{(2)})$ is represented as set of m nodes $v_a \in \mathscr{V}^{(2)}$ with a feature matrix $X^{(2)} \in \mathbb{R}^{m \times d}$, edges $(v_a, v_b) \in \mathscr{E}^{(2)}$ formulating an adjacency matrix $A^{(2)}$.

For the similarity score s, if $s \in \mathbb{R}$, the graph similarity problem can be considered as the *graph-graph regression tasks*. On the other hand, if $s \in \{-1, 1\}$, the problem can be considered as the *graph-graph classification tasks*.

Particularly, the computation of GED (Riesen, 2015; Bai et al, 2019b) (sometimes normalized in $[0, 1]$) is a typical case of graph-graph regression tasks. To be specific, GED is formulated as the cost of the shortest sequence of edit operations over nodes or edges which have to undertake to transform one graph into another graph, in which an edit operation can be an insertion or a deletion of a node or an edge. In Fig. 13.4, We give an illustration of GED computation.

Similar to the classic graph matching problem, the computation of GED is also a well-studied NP-hard problem. Although there is a rich body of work (Hart et al, 1968; Zeng et al, 2009; Riesen et al, 2007) that attempts to find sub-optimal solutions in polynomial time via a variety of heuristics (Riesen et al, 2007; Riesen, 2015), these heuristic methods still suffer from the poor scalability (*e.g.*, large search space or excessive memory) and heavy reliance on expert knowledge (*e.g.*, various heuristics based on different application cases). Currently, learning-based models which incorporate GNNs into an end-to-end learning framework for graph similarity learning are gradually becoming more and more available, demonstrating the

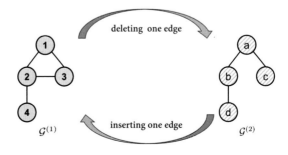

Fig. 13.4 Illustration of computing the GED score between $\mathscr{G}^{(1)}$ and $\mathscr{G}^{(2)}$. Since $\mathscr{G}^{(1)}$ can be transform into in $\mathscr{G}^{(2)}$ by deleting the edge (v_2, v_3) or $\mathscr{G}^{(2)}$ can be transformed into in $\mathscr{G}^{(1)}$ by inserting the edge (v_b, v_c), the GED between two graphs is 1.

superiority by traditional heuristic methods in both effectiveness and efficiency. In two following subsections, we will discuss state-of-the-art GNN-based graph similarity models for graph-graph regression tasks and graph-graph classification tasks, respectively.

13.3.2 Graph-Graph Regression Tasks

As mentioned above, the graph-graph regression task refers to computing a similarity score between a pair of graphs and we focus on the graph similarity learning on GED in this subsection.

Graph Similarity Learning with Convolutional Set Matching. Aiming at accelerating the graph similarity computation while preserving a good performance, Bai et al (2018) first turn the computation of GED into a learning problem rather than approximation methods with combinatorial search, and then propose an end-to-end framework, namely GSimCNN, for the graph similarity learning. For GSimCNN in (Bai et al, 2018) (or GraphSim in (Bai et al, 2020b)[4]), it is probably the first work that applies both GNNs and CNNs for the task of GED computation and consists of three steps in general. First, GSimCNN employs multiple layers of standard GCNs to generate the node embedding vector for each node in the pair of graphs. Second, in each layer of GCNs, GSimCNN uses the BFS node-ordering scheme (You et al, 2018b) to re-order the node embeddings and compute the inner product between the re-ordered node embeddings in two graphs to generate a node-to-node similarity matrix. Finally, after padding or resizing resulting node-to-node similarity matrices into square matrices, the authors transform the task of graph similarity computation into an image processing problem and explore standard CNNs and MLPs for the final graph similarity prediction. GSimCNN is trained with a mean squared error loss function based on predicted similarity scores and the corresponding ground-truth scores.

Graph Similarity Learning with Graph-Level Interaction. Soon after, Bai et al present another GNN-based model, called SimGNN, for graph similarity learning. In SimGNN, it takes not only node-level interactions but also graph-level interactions as considerations for jointly learning the graph similarity score. For the node-level similarity between two graphs, it first adopts a similar approach like GSimCNN to generate the node-to-node similarity matrix, and then extract a histogram feature vector from the matrix as the node-level comparison information. For the graph-level similarity between two graphs, SimGNN first employs a simple graph pooling model via an attention mechanism to generate one graph-level embedding vector for each graph ($\mathbf{h}_{\mathscr{G}(1)}$ and $\mathbf{h}_{\mathscr{G}(2)}$) and then adopts a trainable neural tensor network (NTN) (Socher et al, 2013) to model the relationship between the two graph-

---

[4] It seems that the model architecture of GSimCNN in (Bai et al, 2018) is the same as that of GraphSim in (Bai et al, 2020b), which evaluates the model with additional datasets and similarity metrics (*i.e.*, both GED and MCS).

level embedding vectors as follows.

$$\text{NTN}(\mathbf{h}_{\mathscr{G}(1)}, \mathbf{h}_{\mathscr{G}(2)}) = \sigma\left(\mathbf{h}_{\mathscr{G}(1)}^{\top} W^{[1:K]} \mathbf{h}_{\mathscr{G}(2)} + V\left[\begin{matrix}\mathbf{h}_{\mathscr{G}(1)} \\ \mathbf{h}_{\mathscr{G}(2)}\end{matrix}\right] + \mathbf{b}\right) \qquad (13.18)$$

where σ is the activation function and $[\ \dot{}\ ]$ denotes the concatenation operation. In addition, $W^{[1:K]}$, V and $\mathbf{b}$ are parameters in NTN to be learned and K is a hyper-parameter which determines the length of the graph-level similarity vector calculated by NTN. Finally, to compute the similarity score between two graphs, SimGNN concatenates two similarity vectors from the node level and the graph level along with a small MLP network for prediction.

Graph Similarity Learning based on Hierarchical Clustering. In (Xiu et al, 2020), Xiu et al argue that if two graphs are similar, their corresponding compact graphs should be similar with each other and conversely if two graphs are dissimilar, their corresponding compact graphs should also be dissimilar. They believe that, for the input pair of graphs, different views in regard to different pairs of compact graphs can provide different scales of similarity information between two input graphs and thus benefit the graph similarity computation. To this end, a hierarchical graph matching network (HGMN) (Xiu et al, 2020) is presented to learn the graph similarity from a multi-scale view. Concretely, HGMN first employs multiple stages of hierarchical graph clustering to successively generate more compact graphs with initial node embeddings to provide a multi-scale view of differences between two graphs for subsequent model learning. Then, with the pairs of compact graphs in different stages, HGMN computes the final graph similarity score by adopting a GraphSim-like model (Bai et al, 2020b), including node embeddings update via GCNs, similarity matrices generation and prediction via CNNs. However, in order to ensure permutation invariance of generated similarity matrices, HGMN devises a different node-ordering scheme based on earth mover distance(EMD) (Rubner et al, 1998) rather than BFS node-order method in (Bai et al, 2020b). According to the EMD distance, HGMN first aligns nodes for both input graphs in each stage and then produces the corresponding similarity matrix in the aligned order.

Graph Similarity Learning with Node-Graph Interaction. To learn richer interaction features between a pair of input graphs for computing the graph similarity in an end-to-end fashion, Ling et al propose a multi-level graph matching network (MGMN) (Ling et al, 2020) which consists of a siamese graph neural network (SGNN) and a novel node-graph matching network (NGMN). To learn graph-level interactions between two graphs, SGNN first utilizes a multi-layer of GCNs with the siamese network to generate node embeddings $H^{(l)} = \{\mathbf{h}_i^{(l)}\}_{i=1}^{\{n,m\}} \in \mathbb{R}^{\{n,m\} \times d}$ for all nodes in graph $G^{(l)}$, $l = \{1, 2\}$ and then aggregates a corresponding graph-level embedding vector for each graph. On the other hand, to learn cross-level interaction features between two graphs, NGMN further employs a node-graph matching layer to update node embeddings with learned cross-level interactions between node embeddings of a graph and a corresponding graph-level embedding of the other whole graph. Taking a node $v_i \in \mathcal{V}^{(1)}$ in $\mathcal{G}^{(1)}$ as an example, NGMN first computes an attentive graph-level embedding vector $\mathbf{h}_{G^{(2)}}^{i,att}$ for $\mathcal{G}^{(2)}$ by weighted averaging all node embeddings in $\mathcal{G}^{(2)}$ based on the corresponding cross-graph attention coefficient towards v_i as follows.

$$\mathbf{h}_{G^{(2)}}^{i,att} = \sum_{v_j \in \mathcal{V}^{(2)}} \alpha_{i,j} \mathbf{h}_j^{(2)}, \text{ where } \alpha_{i,j} = \text{cosine}(\mathbf{h}_i^{(1)}, \mathbf{h}_j^{(2)}) \; \forall v_j \in \mathcal{V}^{(2)} \tag{13.19}$$

where *att* in the superscript of $\mathbf{h}_{G^{(2)}}^{i,att}$ means it is an attentive graph-level embedding vector of $G^{(2)}$ in terms of the node v_i in $\mathcal{G}^{(1)}$.

Then, to update the node embedding of v_i with cross-graph interactions, NGMN learns similarity feature vector between the node embedding (*i.e.*, $\mathbf{h}_i^{(1)}$) and the attentive graph-level embedding vector (*i.e.*, $\mathbf{h}_{G^{(2)}}^{i,att}$) via a multi-perspective matching function. After performing the above node-graph matching layer over all nodes for both graphs, NGMN aggregates a corresponding graph-level embedding vector for each graph. The full model MGMN concatenates the two aggregated graph-level embeddings from both SGNN and NGMN for each graph and feed those concatenated embeddings into a final small prediction network for the graph similarity computation.

Graph Similarity Learning based on GRAPH-BERT. As previous studies on the graph similarity learning are mostly trained in a supervised manner and cannot guarantee the basic properties (*e.g.*, triangle inequality) of the graph similarity metric like GED, Zhang introduces a novel training framework of GB-DISTANCE (Zhang, 2020) based on GRAPH-BERT (Zhang et al, 2020a). First, GB-DISTANCE adapts the pre-trained GRAPH-BERT model to update node embeddings and further aggregate a graph-level representation embedding of vector $\mathbf{h}_{\mathcal{G}^{(i)}}$ for the graph $\mathcal{G}^{(i)}$. Then, GB-DISTANCE computes the graph similarity $d_{i,j}$ between the pair of graphs $(\mathcal{G}^{(i)}, \mathcal{G}^{(j)})$ with several fully connected layers as follows.

$$d(\mathcal{G}^{(i)}, \mathcal{G}^{(j)}) = 1 - \exp\left(-\text{FC}\left((\mathbf{h}_{\mathcal{G}^{(i)}} - \mathbf{h}_{\mathcal{G}^{(j)}}) **2\right)\right) \tag{13.20}$$

where FC denotes the employed fully connected layers and $(\cdot) **2$ denotes the element-wise square of the input vector. In (Zhang, 2020), GB-DISTANCE considers a scenario that inputs a set of m graphs (*i.e.*, $\{\mathcal{G}^{(i)}\}_{i=1}^m$) and outputs the similarity between any pair of graphs, *i.e.*, a similarity matrix $D = \{D_{i,j}\}_{i,j=1}^{i,j=m} = \{d(\mathcal{G}^{(i)}, \mathcal{G}^{(j)})\}_{i,j=1}^{i,j=m} \in \mathbb{R}^{m \times m}$, and formulates the graph similarity problem in a supervised or semi-supervised settings as follows.

$$\min \|M \odot (D - \hat{D})\|_p \text{ with } M_{i,j} = \begin{cases} 1 & \text{if } D_{i,j} \text{ is labeled} \\ \alpha & \text{if } D_{i,j} \text{ is unlabeled} \wedge i \neq j \\ \beta & \text{if } i = j \end{cases} \tag{13.21}$$

$$\text{s.t.} \, D_{i,j} \leq D_{i,k} + D_{k,j}, \forall i, j, k \in \{1, \cdots, m\}$$

where $\|\cdot\|_p$ denotes the L_p norm; $\hat{D}$ denotes the ground-truth similarity matrix; M is a mask matrix for the semi-supervised learning with two hyper-parameters α and β; the constraint of $D_{i,j} \leq D_{i,k} + D_{k,j}$, $\forall i,j,k \in \{1,\cdots,m\}$ tries to ensure the triangle inequality of graph similarity metrics. To optimize the model with such constraints, GB-DISTANCE devises a two-phase training algorithm with the constrained metric refining methods.

Graph Similarity Computation based on A*. It is obviously observed that all these aforementioned approaches directly compute the GED similarity score between two graphs, however, failing to produce the edit path, which can explicitly express the sequence of edit operations for transforming one graph into the other graph. To output the edit path like the traditional A* (Hart et al, 1968; Riesen et al, 2007) algorithm, Wang et al propose a graph similarity learning model GENN-A* (Wang et al, 2020f) which incorporates the existing solution of A* with a learnable GENN model based on GNNs. A* (Hart et al, 1968; Riesen et al, 2007) is a tree-searching algorithm which explores the space of all possible node/edge mappings between two graphs as an ordered search tree and further expands successors of a node p in the search tree by the minimum induced edit cost $g(p) + h(p)$, in which $g(p)$ is the cost of current partial edit path induced so far and $h(p)$ is the estimated cost of edit path between the remaining un-matched sub-graphs. Because of the poor scalability of A*, GENN-A* thus replaces the heuristics with a learning-based model (*i.e.*, GENN) to predict $h(p)$. GENN is almost the same as SimGNN (Bai et al, 2019b) with the removal of the histogram module and is used to predict a normalized GED score $s(p) \in (0,1)$ between the remaining un-matched sub-graphs. After that, the $h(p)$ is obtained as follows where $\hat{n}$ and $\hat{m}$ denote the numbers of nodes of the un-matched sub-graphs.

$$h(p) = -0.5(\hat{n} + \hat{m})\log(s(p)) \tag{13.22}$$

13.3.3 Graph-Graph Classification Tasks

In addition to the computation of GED, learning a binary label $s \in \{-1,1\}$ (*i.e.*, similar or not) between a pair of graphs can be view as a task of the graph-graph classification learning[5] and has been widely studied in many real-world applications, including binary code analysis, source code analysis, malware detection, *etc.*

Graph Similarity Learning via Cross-graph Matching. In the scenario of detecting whether two binary functions are similar or not, Li et al present a message-passing based graph matching network (GMN) (Li et al, 2019h) to learn a similarity label between the two control-flow graphs (CFGs) which represent two input binary functions. In particular, GMN employs a similar cross-graph matching network

---

[5] The termed graph-graph classification learning is totally different from the general graph classification task (Ying et al, 2018c; Ma et al, 2019d) that only predicts a label for one input graph rather than a pair of input graphs.

based on standard message-passing GNNs to iteratively generate more discrimina-tive node embeddings (*e.g.*, $H^{(l)} = \{\mathbf{h}_i^{(l)}\}_{v_i \in \mathscr{V}^{(l)}}$, $l = \{1, 2\}$) for two input graphs. Intuitively, it updates the node embeddings of one input graphs by incorporating the attentive association information of another through a soft attention, which is similar to the cross-graph convolution network introduced in Equation (13.8) and Fig. 13.2. Subsequently, in order to calculate the similarity score, GMN adopts an aggrega-tion operation (Li et al, 2016b) as follows to output a graph-level embedding vector (*i.e.*, $\mathbf{h}_{G^{(l)}}$, $l = \{1, 2\}$) for each graph and applies an existing similarity function for the final similarity prediction, *i.e.*, $s(\mathbf{h}_{G^{(1)}}, \mathbf{h}_{G^{(2)}}) = f_s(\mathbf{h}_{G^{(1)}}, \mathbf{h}_{G^{(2)}})$, where f_s can be an arbitrary existing similarity function such as Euclidean, cosine or Hamming similarity function.

$$\mathbf{h}_{G^{(l)}} = \text{MLP}_{\theta 1} \left(\sum_{v_i \in \mathscr{V}^{(l)}} \sigma\big(\text{MLP}_{\theta 2}(\mathbf{h}_i^{(l)}) \big) \odot \text{MLP}_{\theta 3}(\mathbf{h}_i^{(l)}) \right), \ l = \{1, 2\} \qquad (13.23)$$

where σ denotes the activation function; $\odot$ denotes the element-wise multiplication operation; $\text{MLP}_{\theta 1}$, $\text{MLP}_{\theta 2}$, $\text{MLP}_{\theta 3}$ are MLP networks to be trained. Based on dif-ferent supervisions of training samples (*e.g.*, the ground-truth binary label between two graphs or relative similarity among three graphs), GMN adopts two margin-based loss functions, *i.e.*, the pair loss function and the triplet loss function. As for different similarity functions f_s employed, the formulation of the corresponding loss function is quite different. Thus, we refer interested readers for the loss functions to (Li et al, 2019h).

Graph Similarity Learning on Heterogeneous Graphs. Motivated by ever-growing malware threats, a heterogeneous graph matching network (MatchGNet) frame-work (Wang et al, 2019i) is proposed for unknown malware detection. To better represent programs (*e.g.*, benign or malicious) in enterprise systems and capture in-teraction relationships between system entities (*e.g.*, files, processes, sockets, *etc.*), a heterogeneous invariant graph is constructed for each program. Therefore, the mal-ware detection problem is equivalent to detecting whether two representation graphs (*i.e.*, the graph of the input program and the graph of the existing benign program) are similar or not. Due to the heterogeneity of the invariant graph, MatchGNet em-ploys a hierarchical attention graph neural encoder (HAGNE)-based GNN to learn a graph-level embedding vector for each program. Particularly, HAGNE first identi-fies path-relevant sets of neighbors via meta-paths (Sun et al, 2011) and then updates node embeddings by aggregating the entities under each path-relevant neighbor set. The graph-level embedding over all the meta-paths is computed by a weighted sum-marization of all embeddings of meta-paths. Finally, MatchGNet directly calculates the cosine similarity between the two graph-level embedding vector as the final pre-dicted label for malware detection.

13.4 Summary

In this chapter, we have introduced the general graph matching learning, whereby objective functions are formulated for establishing an optimal node-to-node correspondence matrix between two graphs for the classic graph matching problem and computing a similarity metric between two graphs for the graph similarity problem, respectively. In particular, we have thoroughly analyzed and discussed state-of-the-art GNN-based graph matching models and graph similarity models. In the future, for better graph matching learning, some directions we believe are requiring more efforts:

- **Fined-grained cross-graph features.** For the graph matching problem which inputs the pair of graphs, interaction features between two graphs are fundamental and key features in both the graph matching learning and the graph similarity learning. Although several existing methods (Li et al, 2019h; Ling et al, 2020) have been devoted to learning interacted features between two graphs for better representation learning, these models have caused non-negligible extra computational overhead. Better fined-grained cross-graph feature learning with efficient algorithms could make a new state of the art.
- **Semi-supervised learning and un-supervised learning.** Because of the complexity of graphs in the real-world application scenarios, it is common to train the model in a semi-supervised setting or even in an un-supervised setting. Making full use of relationships between existing graphs and, if possible, the other data that is not directly relevant to the graph matching problem could further promote the development of graph matching/similarity learning in more practical applications.
- **Vulnerability and robustness.** Although adversarial attacks have been extensively studied for image classification tasks (Goodfellow et al, 2015; Ling et al, 2019) and node/graph classification tasks (Zügner et al, 2018; Dai et al, 2018a), there is currently only one preliminary work (Zhang et al, 2020f) that studies adversarial attacks on the graph matching problem. Therefore, studying the vulnerability of the state-of-the-art graph matching/similarity models and further building more robust models is a highly challenging problem.

> **Editor's Notes:** Graph Matching Networks is an emerging research topic recently and have drawn significant number of interests in both research community and industrial community due to its broad range of application domains such as computer vision (Chapter 20), Natural Language Processing (Chapter 21), Program Analysis (Chapter 22), Anomaly Detection (Chapter 26). Graph Matching Networks is built on graph node representation learning (Chapter 4) but focuses more on the interaction of two graphs from low-level nodes to high-level graphs. It has tight connection with link prediction (Chapter 10) and self-supervised learning (Chapter 18), where graph matching could be formulated as one of the sub-tasks for these graph learning tasks. Obviously, adversarial robustness (Chapter 8) could have direct impact of graph matching networks, which has recently been extensively studied as well.

Chapter 14
Graph Neural Networks: Graph Structure Learning

Yu Chen and Lingfei Wu

Abstract Due to the excellent expressive power of Graph Neural Networks (GNNs) on modeling graph-structure data, GNNs have achieved great success in various applications such as Natural Language Processing, Computer Vision, recommender systems, drug discovery and so on. However, the great success of GNNs relies on the quality and availability of graph-structured data which can either be noisy or unavailable. The problem of graph structure learning aims to discover useful graph structures from data, which can help solve the above issue. This chapter attempts to provide a comprehensive introduction of graph structure learning through the lens of both traditional machine learning and GNNs. After reading this chapter, readers will learn how this problem has been tackled from different perspectives, for different purposes, via different techniques, as well as its great potential when combined with GNNs. Readers will also learn promising future directions in this research area.

14.1 Introduction

Recent years have seen a significantly increasing amount of interest in Graph Neural Networks (GNNs) (Kipf and Welling, 2017b; Bronstein et al, 2017; Gilmer et al, 2017; Hamilton et al, 2017b; Li et al, 2016b) with a wide range of applications in Natural Language Processing (Bastings et al, 2017; Chen et al, 2020p), Computer Vision (Norcliffe-Brown et al, 2018), recommender systems (Ying et al, 2018b), drug discovery (You et al, 2018a) and so on. GNN's powerful ability in learning expressive graph representations relies on the quality and availability of graph-structured data. However, this poses some challenges for graph representation

Yu Chen
Facebook AI, e-mail: hugochan2013@gmail.com

Lingfei Wu
JD.COM Silicon Valley Research Center, e-mail: lwu@email.wm.edu

learning with GNNs. On the one hand, in some scenarios where the graph structure is already available, most of the GNN-based approaches assume that the given graph topology is perfect, which does not necessarily hold true because i) the real-word graph topology is often noisy or incomplete due to the inevitably error-prone data measurement or collection; and ii) the intrinsic graph topology might merely represent physical connections (e.g the chemical bonds in molecule), and fail to capture abstract or implicit relationships among vertices which can be beneficial for certain downstream prediction task. On the other hand, in many real-world applications such as those in Natural Language Processing or Computer Vision, the graph representation of the data (e.g., text graph for textual data or scene graph for images) might be unavailable. Early practice of GNNs (Bastings et al, 2017; Xu et al, 2018d) heavily relied on manual graph construction which requires extensive human effort and domain expertise for obtaining a reasonably performant graph topology during the data preprocessing stage.

In order to tackle the above challenges, graph structure learning aims to discover useful graph structures from data for better graph representation learning with GNNs. Recent attempts (Chen et al, 2020m,o; Liu et al, 2021; Franceschi et al, 2019; Ma et al, 2019b; Elinas et al, 2020; Velickovic et al, 2020; Johnson et al, 2020) focus on joint learning of graph structures and representations without resorting to human effort or domain expertise. Different sets of techniques have been developed for learning discrete graph structures and weighted graph structures for GNNs. More broadly speaking, graph structure learning has been widely studied in the literature of traditional machine learning in both unsupervised learning and supervised learning settings (Kalofolias, 2016; Kumar et al, 2019a; Berger et al, 2020; Bojchevski et al, 2017; Zheng et al, 2018b; Yu et al, 2019a; Li et al, 2020a). Besides, graph structure learning is also closely related to important problems such as graph generation (You et al, 2018a; Shi et al, 2019a), graph adversarial defenses (Zhang and Zitnik, 2020; Entezari et al, 2020; Jin et al, 2020a,e) and transformer models (Vaswani et al, 2017).

This chapter is organized as follows. We will first introduce how graph structure learning has been studied in the literature of traditional machine learning, prior to the recent surge of GNNs (section 14.2). We will introduce existing works on both unsupervised graph structure learning (section 14.2.1) and supervised graph structure learning (section 14.2.2). Readers will later see how some of the introduced techniques originally developed for traditional graph structure learning have been revisited and improve graph structure learning for GNNs. Then we will move to our main focus of this chapter which is graph structure learning for GNNs in section 14.3. This part will cover various topics including joint graph structure and representation learning for both unweighted and weighted graphs (section 14.3.1), and the connections to other problems such as graph generation, graph adversarial defenses and transformers (section 14.3.2). We will highlight some future directions in section 24.5 including robust graph structure learning, scalable graph structure learning, graph structure learning for heterogeneous graphs, and transferable graph structure learning. We will summarize this chapter in section 14.5.

14.2 Traditional Graph Structure Learning

Graph structure learning has been widely studied from different perspectives in the literature of traditional machine learning, prior to the recent surge of Graph Neural Networks. Before we move to the recent achievements of graph structure learning in the field of Graph Neural Networks, which is the main focus of this chapter, in this section, we will first examine this challenging problem through the lens of traditional machine learning.

14.2.1 Unsupervised Graph Structure Learning

The task of unsupervised graph structure learning aims to directly learn a graph structure from a set of data points in an unsupervised manner. The learned graph structure may be later consumed by subsequent machine learning methods for various prediction tasks. The most important benefit of this kind of approaches is that they do not require labeled data such as ground-truth graph structures for supervision, which could be expensive to obtain. However, because the graph structure learning process does not consider any particular downstream prediction task on the data, the learned graph structure might be sub-optimal for the downstream task.

14.2.1.1 Graph Structure Learning from Smooth Signals

Graph structure learning has been extensively studied in the literature of Graph Signal Processing (GSP). It is often referred to as the graph learning problem in the literature whose goal is to learn the topological structure from smooth signals defined on the graph in an unsupervised manner. These graph learning techniques (Jebara et al, 2009; Lake and Tenenbaum, 2010; Kalofolias, 2016; Kumar et al, 2019a; Kang et al, 2019; Kumar et al, 2020; Bai et al, 2020a) typically operate by solving an optimization problem with certain prior constraints on the properties (e.g., smoothness, sparsity) of graphs. Here, we introduce some representative prior constraints defined on graphs which have been widely used for solving the graph learning problem.

Before introducing the specific graph learning techniques, we first provide the formal definition of a graph and graph signals. Consider a graph $\mathscr{G} = \{\mathscr{V}, \mathscr{E}\}$ with the vertex set $\mathscr{V}$ of cardinality n and edge set $\mathscr{E}$, its adjacency matrix $A \in \mathbb{R}^{n \times n}$ governs its topological structure where $A_{i,j} > 0$ indicates there is an edge connecting vertex i and j and $A_{i,j}$ is the edge weight. Given an adjacency matrix A, we can further obtain the graph Laplacian matrix $L = D - A$ where $D_{i,i} = \sum_j A_{i,j}$ is the degree matrix whose off-diagonal entries are all zero.

A graph signal is defined as a function that assigns a scalar value to each vertex of a graph. We can further define multi-channel signals $X \in \mathbb{R}^{n \times d}$ on a graph that assigns a d dimensional vector to each vertex, and each column of the feature matrix

X can be considered as a graph signal. Let $X_i \in \mathbb{R}^d$ denote the graph signal defined on the i-th vertex.

Fitness. Early works (Wang and Zhang, 2007; Daitch et al, 2009) on graph learning utilized the neighborhood information of each data point for graph construction by assuming that each data point can be optimally reconstructed using a linear combination of its neighbors. Wang and Zhang (2007) proposed to learn a graph with normalized degrees by minimizing the following objective,

$$\sum_i ||X_i - \sum_j A_{i,j} X_j||^2 \tag{14.1}$$

where $\sum_j A_{i,j} = 1$, $A_{i,j} \geq 0$.

Similarly, Daitch et al (2009) proposed to minimize a measure of fitness that computes a weighted sum of the squared distance from each vertex to the weighted average of its neighbors, formulated as follows:

$$\sum_i ||D_{i,i} X_i - \sum_j A_{i,j} X_j||^2 = ||LX||_F^2 \tag{14.2}$$

where $||M||_F = (\sum_{i,j} M_{i,j}^2)^{1/2}$ is the Frobenius norm.

Smoothness. Smoothness is another widely adopted assumption on natural graph signals. Given a set of graph signals $X \in \mathbb{R}^{n \times d}$ defined on an undirected weighted graph with an adjacency matrix $A \in \mathbb{R}^{n \times n}$, the smoothness of the graph signals is usually measured by the Dirichlet energy (Belkin and Niyogi, 2002),

$$\Omega(A, X) = \frac{1}{2} \sum_{i,j} A_{i,j} ||X_i - X_j||^2 = \mathrm{tr}(X^\top LX) \tag{14.3}$$

where L is the Laplacian matrix and $\mathrm{tr}(\cdot)$ denotes the trace of a matrix. Lake and Tenenbaum (2010); Kalofolias (2016) proposed to learn a graph by minimizing $\Omega(A, X)$ which forces neighboring vertices to have similar features, thus enforcing graph signals to change smoothly on the learned graph. Notably, solely minimizing the above smoothness loss can lead to the trivial solution $A = 0$.

Connectivity and Sparsity. In order to avoid the trivial solution caused by solely minimizing the smoothness loss, Kalofolias (2016) imposed additional constraints on the learned graph,

$$-\alpha \vec{1}^\top \log(A\vec{1}) + \beta ||A||_F^2 \tag{14.4}$$

where the first term penalizes the formation of disconnected graphs via the logarithmic barrier, and the second term controls sparsity by penalizing large degrees due to the first term. Note that $\vec{1}$ denotes the all-ones vector. As a result, this improves the overall connectivity of the graph, without compromising sparsity.

Similarly, Dong et al (2016) proposed to solve the following optimization problem:

$$\min_{L\in\mathbb{R}^{n\times n}, Y\in\mathbb{R}^{n\times p}} ||X - Y||_F^2 + \alpha\,\mathrm{tr}(Y^\top LY) + \beta\,||L||_F^2$$

$$\text{s.t.} \quad \mathrm{tr}(L) = n,$$

$$L_{i,j} = L_{j,i} \le 0, \quad i \ne j, \qquad\qquad (14.5)$$

$$L \cdot \vec{1} = \vec{0}$$

which is equivalent to finding jointly the graph Laplacian L and Y (i.e., a "noise-less" version of the zero-mean observation X), such that Y is close to X, and in the meantime Y is smooth on the sparse graph. Note that the first constraint acts as a normalization factor and permits to avoid trivial solutions, and the second and third constraints guarantee that the learned L is a valid Laplacian matrix that is positive semidefinite.

Ying et al (2020a) aimed to learn a sparse graph under Laplacian constrained Gaussian graphical model, and proposed a nonconvex penalized maximum likelihood method by solving a sequence of weighted l1-norm regularized sub-problems. Maretic et al (2017) proposed to learn a sparse graph signal model by alternating between a signal sparse coding and a graph update step.

In order to reduce the computational complexity of solving the optimization problem, many approximation techniques (Daitch et al, 2009; Kalofolias and Per-raudin, 2019; Berger et al, 2020) have been explored. Dong et al (2019) provided a good literature review on learning graphs from data from a GSP perspective.

14.2.1.2 Spectral Clustering via Graph Structure Learning

Graph structure learning has also been studied in the field of clustering analysis. For example, in order to improve the robustness of spectral clustering methods for noisy input data, Bojchevski et al (2017) assumed that the observed graph A can be decomposed into the corrupted graph A^c and the good (i.e., clean) graph A^g, and it is beneficial to only perform the spectral clustering on the clean graph. They hence proposed to jointly perform the spectral clustering and the decomposition of the observed graph, and adopted a highly efficient block coordinate-descent (alternating) optimization scheme to approximate the objective function. Huang et al (2019b) proposed a multi-view learning model which simultaneously conducts multi-view clustering and learns similarity relationships between data points in kernel spaces.

14.2.2 Supervised Graph Structure Learning

The task of supervised graph structure learning aims to learn a graph structure from data in a supervised manner. They may or may not consider a particular downstream prediction task during the model training phase.

14.2.2.1 Relational Inference for Interacting Systems

Relational inference for interacting systems aims to study how objects in complex systems interact. Early works considered a fixed or fully-connected interaction graph (Battaglia et al, 2016; van Steenkiste et al, 2018) while modeling the interaction dynamics among objects. Sukhbaatar et al (2016) proposed a neural model to learn continuous communication among a dynamically changing set of agents where the communication graph changes over time as agents move, enter and exit the environment. Recent efforts (Kipf et al, 2018; Li et al, 2020a) have been made to simultaneously infer the latent interaction graph and model the interaction dynamics. Kipf et al (2018) proposed a variational autoencoder (VAE) (Kingma and Welling, 2014) based approach which learns to infer the interaction graph structure and model the interaction dynamics among physical objects simultaneously from their observed trajectories in an unsupervised manner. The discrete latent code of VAE represents edge connections of the latent interaction graph, and both the encoder and decoder take the form of a GNN to model the interaction dynamics among objects. Because the latent distribution of VAE is discrete, the authors adopted a continuous relaxation in order to use the reparameterization trick (Kingma et al, 2014). While Kipf et al (2018) focused on inferring a static interaction graph, Li et al (2020a) designed a dynamic mechanism to evolve the latent interaction graph adaptively over time. A Gated Recurrent Unit (GRU) (Cho et al, 2014a) was applied to capture the history information and adjust the prior interaction graph.

14.2.2.2 Structure Learning in Bayesian Networks

A Bayesian network (BN) is a Probabilistic Graphical Model (PGM) which encodes conditional dependencies between random variables via a directed acyclic graph (DAG), where each random variable is represented as a node in DAG. The problem of learning the BN structure is important yet challenging in Bayesian networks research. Most existing works on BN learning focus on score-based learning of DAGs, and aim to find a DAG with the maximal score where a score indicates how well any candidate DAG is supported by the observed data (and any prior knowledge). Early works treat BN learning as a combinatorial optimization problem which is NP-hard due to the intractable search space of DAGs scaling superexponentially with the number of nodes. Some efficient methods have been proposed for exact BN learning via dynamic programming (Koivisto and Sood, 2004; Silander and Myllymäki, 2006) or integer programming (Jaakkola et al, 2010; Cussens, 2011). Recently, Zheng et al (2018b) proposed to formulate the traditional combinatorial optimization problem into a purely continuous optimization problem over real matrices with a smooth equality constraint ensuring acyclicity of the graph. The resulting problem can hence be efficiently solved by standard numerical algorithms. A follow-up work (Yu et al, 2019a) leveraged the expressive power of GNNs, and proposed a variational autoencoder (VAE) based deep generative model with a variant of the structural constraint to learn the DAG. The VAE was parameterized by a GNN that can naturally handle both discrete and vector-valued random variables.

14.3 Graph Structure Learning for Graph Neural Networks

Graph structure learning has recently been revisited in the field of GNNs so as to handle the scenarios where the graph-structured data is noisy or unavailable. Recent attempts in this line of research mainly focus on joint learning of graph structures and representations without resorting to human effort or domain expertise. fig. 14.1 shows the overview of graph structure learning for GNNs. Besides, we see several important problems being actively studied (including graph generation, graph adversarial defenses and transformer models) in recent years which are closely related to graph structure learning for GNNs. We will discuss their connections and differences in this section.

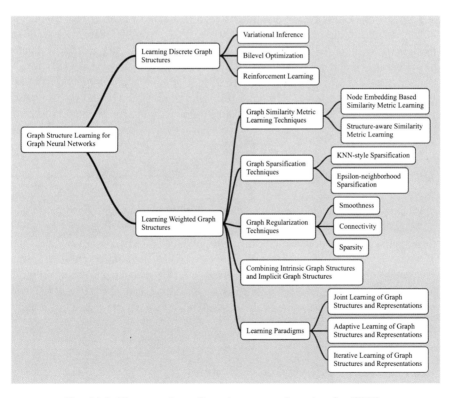

Fig. 14.1: The overview of graph structure learning for GNNs.

14.3.1 Joint Graph Structure and Representation Learning

In recent practice of Gnns, joint graph structure and representation learning has drawn a growing attention. This line of research aims to jointly optimize the graph structure and GNN parameters toward the downstream prediction task in an end-to-end manner, and can be roughly categorized into two groups: learning discrete graph structures and learning weighted adjacency matrices. The first kind of approaches (Chen et al, 2018e; Ma et al, 2019b; Zhang et al, 2019d; Elinas et al, 2020; Pal et al, 2020; Stanic et al, 2021; Franceschi et al, 2019; Kazi et al, 2020) operate by sampling a discrete graph structure (i.e., corresponding to a binary adjacency matrix) from the learned probabilistic adjacency matrix, and then feeding the graph to a subsequent GNN in order to obtain the task prediction. Because the sampling operation breaks the differentiability of the whole learning system, techniques such as variational inference (Hoffman et al, 2013) or Reinforcement Learning (Williams, 1992) are applied to optimize the learning system. Considering that discrete graph structure learning often has the optimization difficulty introduced by the non-differentiable sampling operation and it is hence difficult to learn weights on edges, the other kind of approaches (Chen et al, 2020m; Li et al, 2018c; Chen et al, 2020o; Huang et al, 2020a; Liu et al, 2019b, 2021; Norcliffe-Brown et al, 2018) focuses on learning the weighted (and usually sparse) adjacency matrix associated to a weighted graph which will be later consumed by a subsequent GNN for the prediction task. We will discuss these two types of approaches in great detail next. Before discussing different techniques for joint graph structure and representation learning, let's first formulate the joint graph structure and representation learning problem.

14.3.1.1 Problem Formulation

Let the graph $\mathscr{G} = (\mathscr{V}, \mathscr{E})$ be represented as a set of n nodes $v_i \in \mathscr{V}$ with an initial node feature matrix $X \in \mathbb{R}^{d \times n}$, and a set of m edges $(v_i, v_j) \in \mathscr{E}$ (binary or weighted) formulating an initial noisy adjacency matrix $A^{(0)} \in \mathbb{R}^{n \times n}$. Given a noisy graph input $\mathscr{G} := \{A^{(0)}, X\}$ or only a node feature matrix $X \in \mathbb{R}^{d \times n}$, the joint graph structure and representation learning problem we consider aims to produce an optimized graph $\mathscr{G}^* := \{A^{(*)}, X\}$ and its corresponding node embeddings $Z = f(\mathscr{G}^*, \theta) \in \mathbb{R}^{h \times n}$, with respect to certain downstream prediction task. Here, we denote f as a GNN and θ as its model parameters.

14.3.1.2 Learning Discrete Graph Structures

In order to deal with the issue of uncertainty on graphs, many of the existing works on learning discrete graph structures regard the graph structure as a random variable where a discrete graph structure can be sampled from certain probabilistic adjacency matrix. They usually leverage various techniques such as variational inference (Chen et al, 2018e; Ma et al, 2019b; Zhang et al, 2019d; Elinas et al, 2020; Pal et al, 2020; Stanic et al, 2021), bilevel optimization (Franceschi et al, 2019), and Reinforcement Learning (Kazi et al, 2020) to jointly optimize the graph structure and GNN parameters. Notably, they are often limited to the transductive learning setting where the node features and graph structure are fully observed during both the training and inference stages. In this section, we introduce some representative works on this topic and show how they approach the problem from different perspectives.

Franceschi et al (2019) proposed to jointly learn a discrete probability distribution on the edges of the graph and the parameters of GNNs by treating the task as a bilevel optimization problem Colson et al (2007), formulated as,

$$
\min_{\vec{\theta} \in \mathscr{H}_N} \mathbb{E}_{A \sim \mathrm{Ber}(\vec{\theta})}[F(w_\theta, A)]
$$
$$
\text{such that } w_\theta = \mathrm{argmin}_w \mathbb{E}_{A \sim \mathrm{Ber}(\vec{\theta})}[L(w, A)]
$$
$$(14.6)$$

where $\overline{\mathscr{H}}_N$ denotes the convex hull of the set of all adjacency matrices for N nodes, and $L(w, A)$ and $F(w_\theta, A)$ are both task-specific loss functions measuring the difference between GNN predictions and ground-truth labels which are computed on a training set and validation set, respectively. Each edge (i.e., node pair) of the graph is independently modeled as a Bernoulli random variable, and an adjacency matrix $A \sim \mathrm{Ber}(\vec{\theta})$ can thus be sampled from the graph structure distribution parameterized by $\vec{\theta}$. The outer objective (i.e., the first objective) aims to find an optimal discrete graph structure given a GCN and the inner objective (i.e., the second objective) aims to find the optimal parameters w_θ of a GCN given a graph. The authors approximately solved the above challenging bilevel problem with hypergradient descent.

Considering that real-word graphs are often noisy, Ma et al (2019b) viewed the node features, graph structure and node labels as random variables, and modeled the joint distribution of them with a flexible generative model for the graph-based semi-supervised learning problem. Inspired by random graph models from the network science field (Newman, 2010), they assumed that the graph is generated based on node features and labels, and thus factored the joint distribution as the following:

$$
p(X, Y, G) = p_{\vec{\theta}}(G | X, Y) p_{\vec{\theta}}(Y | X) p(X) \tag{14.7}
$$

where X, Y and G are random variables corresponding to the node features, labels and graph structure, and $\vec{\theta}$ are learnable model parameters. Note that the conditional probabilities $p_{\vec{\theta}(G|X,Y)}$ and $p_{\vec{\theta}(Y|X)}$ can be any flexible parametric families of distributions as long as they are differentiable almost everywhere w.r.t. $\vec{\theta}$. In the paper, $p_{\vec{\theta}(G|X,Y)}$ is instantiated with either latent space model (LSM) (Hoff et al, 2002) or stochastic block models (SBM) (Holland et al, 1983). During the inference stage, in order to infer the missing node labels denoted as Y_{miss}, the authors leveraged the recent advances in scalable variational inference (Kingma and Welling, 2014; Kingma et al, 2014) to approximate the posterior distribution $p_{\vec{\theta}(Y_{\mathrm{miss}}|X, Y_{\mathrm{obs}}, G)}$ via a recognition model $q_{\vec{\phi}(Y_{\mathrm{miss}}|X, Y_{\mathrm{obs}}, G)}$ parameterized by $\vec{\phi}$ where Y_{obs} denotes the observed node labels. In the paper, $q_{\vec{\phi}(Y_{\text{miss}}|X,Y_{\text{obs}},G)}$ is instantiated with a GNN. The model parameters $\vec{\theta}$ and $\vec{\phi}$ are jointly optimized by maximizing the Evidence Lower Bound (Bishop, 2006) of the observed data (Y_{obs}, G) conditioned on X.

Elinas et al (2020) aimed to maximize the posterior over the binary adjacency matrix given the observed data (i.e., node features X and observed node labels Y^o), formulated as,

$$p(A|X,Y^o) \propto p_{\vec{\theta}(Y^o|X,A)p(A)} \tag{14.8}$$

where $p_{\vec{\theta}(Y^o|X,A)}$ is a conditional likelihood which can be further factorized following the conditional independence assumption,

$$p_{\vec{\theta}(Y^o|X,A)} = \prod_{y_i \in Y^o} p_{\vec{\theta}(y_i|X,A)}$$
$$p_{\vec{\theta}(y_i|X,A)} = \text{Cat}(y_i|\vec{\pi}_i) \tag{14.9}$$

where $\text{Cat}(y_i|\vec{\pi}_i)$ denotes a categorical distribution, and is the i-th row of a probability matrix $\Pi \in \mathbb{R}^{N \times C}$ modeled by a GCN, namely, $\Pi = \text{GCN}(X,A,\vec{\theta})$. As for the prior distribution over the graph $p(A)$, the authors considered the following form,

$$p(A) = \prod_{i,j} p(A_{i,j})$$
$$p(A_{i,j}) = \text{Bern}(A_{i,j}|\rho^o_{i,j}) \tag{14.10}$$

where $\text{Bern}(A_{i,j}|\rho^o_{i,j})$ is a Bernoulli distribution over the adjacency matrix $A_{i,j}$ with parameter $\rho^o_{i,j}$. In the paper, $\rho^o_{i,j} = \rho_1 \overline{A}_{i,j} + \rho_2(1 - \overline{A}_{i,j})$ was constructed to encode the degree of belief on the absence and presence of observed links with hyperparameters $0 < \rho_1, \rho_2 < 0$. Note that $\overline{A}_{i,j}$ is the observed graph structure which can potentially be perturbed. If there is no input graph available, a KNN graph can be employed. Given the above formulations, the authors developed a stochastic variational inference algorithm by leveraging the reparameterization trick (Kingma et al, 2014) and Concrete distributions techniques (Maddison et al, 2017; Jang et al, 2017) to optimize the graph posterior $p(A|X,Y^o)$ and the GCN parameters $\vec{\theta}$ jointly.

Kazi et al (2020) designed a probabilistic graph generator whose underlying probability distribution is computed based on pair-wise node similarity, formulated as,

$$p_{i,j} = e^{-t||X_i - X_j||} \tag{14.11}$$

where t is a temperature parameter, and X_i is the node embedding of node v_i. Given the above edge probability distribution, they adopted the Gumbel-Top-k trick (Kool et al, 2019) to sample an unweighted KNN graph which would be fed into a GNN-based prediction network. Note that the sampling operation breaks the differentiability of the model, the authors thus exploited Reinforcement Learning to reward edges involved in a correct classification and penalize edges which led to misclassification.

14.3.1.3 Learning Weighted Graph Structures

Unlike the kind of graph structure learning approaches focusing on learning a discrete graph structure (i.e., binary adjacency matrix) for the GNN, there is a class of approaches instead focusing on learning a weighted graph structure (i.e., weighted adjacency matrix). In comparison with learning a discrete graph structure, learning a weighted graph structure has several advantages. Firstly, optimizing a weighted adjacency matrix is much more tractable than optimizing a binary adjacency matrix because the former can be easily achieved by SGD techniques (Bottou, 1998) or even convex optimization techniques (Boyd et al, 2004) while the later often has to resort to more challenging techniques such as variational inference (Hoffman et al, 2013), Reinforcement Learning (Williams, 1992) and combinatorial optimization techniques (Korte et al, 2011) due to its non-differentiability. Secondly, a weighted adjacency matrix is able to encode richer information on edges compared to a binary adjacency matrix, which could benefit the subsequent graph representation learning. For example, the widely used Graph Attention Network (GAT) (Veličković et al, 2018) essentially aims to learn edge weights for the input binary adjacency matrix which benefit the subsequent message passing operations. In this subsection, we will first introduce some common graph similarity metric learning techniques as well as graph sparsification techniques widely used in existing works for learning a sparse weighted graph by considering pair-wise node similarity in the embedding space. Some representative graph regularization techniques will be later introduced for controlling the quality of the learned graph structure. We will then discuss the importance of combining both of the intrinsic graph structures and learned implicit graph structures for better learning performance. Finally, we will cover some important learning paradigms for the joint learning of graph structures and graph representations that have been successfully adopted by existing works.

Graph Similarity Metric Learning Techniques

As introduced in section 14.2.1.1, prior works on unsupervised graph structure learning from smooth signals also aim to learn a weighted adjacency matrix from data. Nevertheless, they are incapable of handling inductive learning setting where there are unseen graphs or nodes in the inference phase. This is because they often learn by directly optimizing an adjacency matrix based on certain prior constraints on the graph properties. Many works on discrete graph structure learning (section 14.3.1.2) have trouble conducting inductive learning as well on account of the similar reason.

Inspired by the success of attention-based techniques (Vaswani et al, 2017; Veličković et al, 2018) for modeling relationships among objects, many recent works in the literature cast graph structure learning as similarity metric learning defined upon the node embedding space assuming that the node attributes more or less contain useful information for inferring the implicit topological structure of the graph. One biggest advantage of this strategy is that the learned similarity metric function can be later applied to an unseen set of node embeddings to infer a graph structure, thus enabling inductive graph structure learning.

For data deployed in non-Euclidean domains such as graph data, the Euclidean distance is not necessarily the optimal metric for measuring node similarity. Common options for metric learning include cosine similarity (Nguyen and Bai, 2010), radial basis function (RBF) kernel (Yeung and Chang, 2007) and attention mechanisms (Bahdanau et al, 2015; Vaswani et al, 2017). In general, according to the types of raw information sources needed, we group the similarity metric learning functions into two categories: *Node Embedding Based Similarity Metric Learning* and *Structure-aware Similarity Metric Learning*. Next, we will introduce some representative metric learning functions from both categories which have been successfully adopted in prior works on graph structure learning for GNNs.

Node Embedding Based Similarity Metric Learning

Node embedding based similarity metric learning functions are designed to learn a pair-wise node similarity matrix based on node embeddings which ideally encode important semantic meanings of the nodes for graph structure learning.

Attention-based Similarity Metric Functions Most similarity metric functions proposed so far are based on the attention mechanism Bahdanau et al (2015); Vaswani et al (2017). Norcliffe-Brown et al (2018) adopted a simple metric function which computes the dot product between any pair of node embeddings (eq. (14.12)). Given its limited learning capacity, it might have difficulty learning an optimal graph structure.

$$S_{i,j} = \vec{v}_i^\top \vec{v}_j \qquad (14.12)$$

where $S \in \mathbb{R}^{n \times n}$ is a node similarity matrix, and $\vec{v}_i$ is the vector representation of node v_i.

To enrich the learning capacity of dot product, Chen et al (2020n) proposed a modified dot product by introducing learnable parameters, formulated as follows:

$$S_{i,j} = (\vec{v}_i \odot \vec{u})^\top \vec{v}_j \qquad (14.13)$$

where $\odot$ denotes element-wise multiplication, and $\vec{u}$ is a non-negative trainable weight vector which learns to highlight different dimensions of the node embeddings. Note that the output similarity matrix S is asymmetric.

Chen et al (2020o) proposed a more expressive version of dot product by introducing a weight matrix, formulated as follows:

$$S_{i,j} = \text{ReLU}(W\vec{v}_i)^\top \text{ReLU}(W\vec{v}_j) \qquad (14.14)$$

where W is a $d \times d$ weight matrix, and $\text{ReLU}(x) = \max(0,x)$ is a rectified linear unit (ReLU) (Nair and Hinton, 2010) which is used here to enforce the sparsity of the output similarity matrix.

Similar to (Chen et al, 2020o), On et al (2020) introduced a learnable mapping function to node embeddings before computing the dot product, and applied a ReLU function to enforce sparsity, formulated as follows:

$$S_{i,j} = \text{ReLU}(f(\vec{v}_i)^\top f(\vec{v}_j)) \tag{14.15}$$

where $f : \mathbb{R} \to \mathbb{R}$ is a single-layer feed-forward network without non-linear activation.

Besides using ReLU to enforce sparsity, Yang et al (2018c) applied the square operation to stabilize training, and the row-normalization operation to obtain a normalized similarity matrix, formulated as follows:

$$S_{i,j} = \frac{(\text{ReLU}((W_1\vec{v}_i)^\top W_2\vec{v}_j + b))^2}{\sum_k (\text{ReLU}((W_1\vec{v}_k)^\top W_2\vec{v}_j + b))^2} \tag{14.16}$$

where W_1 and W_2 are $d \times d$ weight matrices, and b is a scalar parameter.

Unlike Chen et al (2020o) that applied the same linear transformation to node embeddings, Huang et al (2020a) applied different linear transformations to the two node embeddings when computing the pair-wise node similarity, formulated as follows:

$$S_{i,j} = \text{softmax}((W_1\vec{v}_i)^\top W_2\vec{v}_j) \tag{14.17}$$

where W_1 and W_2 are $d \times d$ weight matrices, and $\text{softmax}(\vec{z})_i = \frac{e^{z_i}}{\sum_j e^{z_j}}$ is applied to obtain a row-normalized similarity matrix.

Velickovic et al (2020) aimed at graph structure learning in a temporal setting where the implicit graph structure to be learned changes over time. At each time step t, they first computed the pair-wise node similarity $a_{i,j}^{(t)}$ using the same attention mechanism as in (Huang et al, 2020a), and based on that, they further obtained an "aggregated" adjacency matrix $S_{i,j}^{(t)}$ by deriving a new edge for node i by choosing node j with the maximal $\bar{a}_{ij}$. The whole process is formulated as follows:

$$
\begin{aligned}
a_{i,j}^{(t)} &= \text{softmax}((W_1\vec{v}_i^{(t)})^\top W_2\vec{v}_j^{(t)}) \\
\widetilde{S}_{i,j}^{(t)} &= \mu_i^{(t)}\widetilde{S}_{i,j}^{(t-1)} + (1 - \mu_i^{(t)})\mathbb{I}_{j=\text{argmax}_k(a_{i,k}^{(t)})} \\
S_{i,j}^{(t)} &= \widetilde{S}_{i,j}^{(t)} \vee \widetilde{S}_{j,i}^{(t)}
\end{aligned}
\tag{14.18}
$$

where $\mu_i^{(t)}$ is a learnable binary gating mask, $\vee$ denotes logical disjunction between the two operands to enforce symmetry, and W_1 and W_2 are $d \times d$ weight matrices. Because the argmax operation makes the whole learning system non-differentiable, the authors provided the ground-truth graph structures for supervision at each time step.

Cosine-based Similarity Metric Functions Chen et al (2020m) proposed a multi-head weighted cosine similarity function which aims at capturing pair-wise node similarity from multiple perspectives, formulated as follows:

$$S^p_{i,j} = \cos(\vec{w}_p \odot \vec{v}_i, \vec{w}_p \odot \vec{v}_j)$$
$$S_{i,j} = \frac{1}{m} \sum_{p=1}^{m} S^p_{ij} \tag{14.19}$$

where $\vec{w}_p$ is a learnable weight vector associated to the p-th perspective, and has the same dimension as the node embeddings. Intuitively, $S^p_{i,j}$ computes the pair-wise cosine similarity for the p-th perspective where each perspective considers one part of the semantics captured in the embeddings. Moreover, as observed in (Vaswani et al, 2017; Veličković et al, 2018), employing multi-head learners is able to stabilize the learning process and increase the learning capacity.

Kernel-based Similarity Metric Functions Besides attention-based and cosine-based similarity metric functions, researchers also explored to apply kernel-based metric functions for graph structure learning. Li et al (2018c) applied a Gaussian kernel to the distance between any pair of node embeddings, formulated as follows:

$$d(\vec{v}_i, \vec{v}_j) = \sqrt{(\vec{v}_i - \vec{v}_j)^\top M (\vec{v}_i - \vec{v}_j)}$$
$$S(\vec{v}_i, \vec{v}_j) = \frac{-d(\vec{v}_i, \vec{v}_j)}{2\sigma^2} \tag{14.20}$$

where σ is a scalar hyperparameter which determines the width of the Gaussian kernel, and $d(\vec{v}_i, \vec{v}_j)$ computes the Mahalanobis distance between the two node embeddings $\vec{v}_i$ and $\vec{v}_j$. Notably, M is the covariance matrix of the node embeddings distribution if we assume all the node embeddings of the graph are drawn from the same distribution. If we set $M = I$, the Mahalanobis distance reduces to the Euclidean distance. To make M a symmetric and positive semi-definite matrix, the authors let $M = WW^\top$ where W is a $d \times d$ learnable weight matrix. We can also regard W as the transform basis to the space where we measure the Euclidean distance between two vectors.

Similarly, Henaff et al (2015) first computed the Euclidean distance between any pair of node embeddings, and then applied a Gaussian Kernel or a self-tuning diffusion kernel (Zelnik-Manor and Perona, 2004), formulated as follows:

$$d(\vec{v}_i, \vec{v}_j) = \sqrt{(\vec{v}_i - \vec{v}_j)^\top (\vec{v}_i - \vec{v}_j)}$$
$$S(\vec{v}_i, \vec{v}_j) = \frac{-d(\vec{v}_i, \vec{v}_j)}{\sigma^2}$$
$$S_{\text{local}}(\vec{v}_i, \vec{v}_j) = \frac{-d(\vec{v}_i, \vec{v}_j)}{\sigma_i \sigma_j} \tag{14.21}$$

where $S_{\text{local}}(\vec{v}_i, \vec{v}_j)$ defines a self-tuning diffusion kernel whose variance is locally adapted around each node. Specifically, σ_i is computed as the distance $d(\vec{v}_i, \vec{v}_{i_k})$ corresponding to the k-th nearest neighbor i_k of node i.

Structure-aware Similarity Metric Learning

When learning implicit graph structures from data, it might be beneficial to utilize the intrinsic graph structures as well if they are available.

Utilizing Intrinsic Edge Embeddings for Similarity Metric Learning Inspired by recent works on structure-aware transformers (Zhu et al, 2019b; Cai and Lam, 2020) which brought the intrinsic graph structure to the self-attention mechanism in the transformer architecture, some works designed structure-aware similarity metric functions which additionally consider the edge embeddings of the intrinsic graph. Liu et al (2019b) introduced a structure-aware attention mechanism as the following:

$$S_{i,j}^l = \text{softmax}(\vec{u}^\top \tanh(W[\vec{h}_i^l, \vec{h}_j^l, \vec{v}_i, \vec{v}_j, \vec{e}_{i,j}])) \qquad (14.22)$$

where $\vec{v}_i$ denotes the node attributes for node i, $\vec{e}_{i,j}$ represents the edge attributes between node i and j, $\vec{h}_i^l$ is the vector representation of node i in the l-th GNN layer, and $\vec{u}$ and W are trainable weight vector and weight matrix, respectively.

Similarly, Liu et al (2021) proposed a structure-aware global attention mechanism for learning pair-wise node similarity, formulated as follows,

$$S_{i,j} = \frac{\text{ReLU}(W^Q \vec{v}_i)^\top (\text{ReLU}(W^K \vec{v}_i) + \text{ReLU}(W^R \vec{e}_{i,j}))}{\sqrt{d}} \qquad (14.23)$$

where $\vec{e}_{i,j} \in \mathbb{R}^{d_e}$ is the embedding of the edge connecting node i and j, $W^Q, W^K \in \mathbb{R}^{d \times d_v}$, $W^R \in \mathbb{R}^{d \times d_e}$ are learnable weight matrices, and d, d_v and d_e are the dimensions of hidden vectors, node embeddings and edge embeddings, respectively.

Utilizing Intrinsic Edge Connectivity Information for Similarity Metric Learning In the case where only the edge connectivity information is available in the intrinsic graph, Jiang et al (2019b) proposed a masked attention mechanism for graph structure learning, formulated as follows,

$$S_{i,j} = \frac{A_{i,j} \exp(\text{ReLU}(\vec{u}^\top |\vec{v}_i - \vec{v}_j|))}{\sum_k A_{i,k} \exp(\text{ReLU}(\vec{u}^\top |\vec{v}_i - \vec{v}_k|))} \qquad (14.24)$$

where $A_{i,j}$ is the adjacency matrix of the intrinsic graph and $\vec{u}$ is a weight vector with the same dimension as node embeddings $\vec{v}_i$. This idea of using masked attention to incorporate the initial graph topology shares the same spirit with the GAT (Veličković et al, 2018) model.

Graph Sparsification Techniques

The aforementioned similarity metric learning functions all return a weighted adjacency matrix associated to a fully-connected graph. A fully-connected graph is not only computationally expensive but also might introduce noise such as unimportant edges. In real-word applications, most graph structures are much more sparse. Therefore, it can be beneficial to explicitly enforce sparsity to the learned graph structure. Besides applying the ReLU function in the similarity metric functions (Chen et al, 2020o; On et al, 2020; Yang et al, 2018c; Liu et al, 2021; Jiang et al, 2019b), various graph sparsification techniques have been adopted to enhance the sparsity of the learned graph structure.

Norcliffe-Brown et al (2018); Klicpera et al (2019b); Chen et al (2020o,n); Yu et al (2021a) adopted a KNN style sparsification operation to obtain a sparse adjacency matrix from the node similarity matrix computed by the similarity metric learning function, formulated as follows:

$$A_{i,:} = \text{topk}(S_{i,:}) \tag{14.25}$$

where topk is a KNN-style operation. Specifically, for each node, only the K nearest neighbors (including itself) and the associated similarity scores are kept, and the remaining similarity scores are masked off.

Klicpera et al (2019b); Chen et al (2020m) enforced a sparse adjacency matrix by considering only the ε-neighborhood for each node, formulated as follows:

$$A_{i,j} = \begin{cases} S_{i,j} & S_{i,j} > \varepsilon \\ 0 & \text{otherwise} \end{cases} \tag{14.26}$$

where those elements in S which are smaller than a non-negative threshold ε are all masked off (i.e., set to zero).

Graph Regularization Techniques

As discussed earlier, many works in the field of Graph Signal Processing typically learn the graph structure from data by directly optimizing the adjacency matrix to minimize the constraints defined based on certain graph properties, without considering any downstream tasks. On the contrary, many works on graph structure learning for GNNs aim to optimize a similarity metric learning function (for learning graph structures) toward the downstream prediction task. However, they do not explicitly enforce the learned graph structure to have some common properties (e.g., smoothness) presented in real-word graphs.

Chen et al (2020m) proposed to optimize the graph structures by minimizing a hybrid loss function combining both the task prediction loss and the graph regularization loss. They explored three types of graph regularization losses which pose constrains on the smoothness, connectivity and sparsity of the learned graph.

Smoothness The smoothness property assumes neighboring nodes to have similar features.

$$\Omega(A, X) = \frac{1}{2n^2} \sum_{i,j} A_{i,j} \|X_i - X_j\|^2 = \frac{1}{n^2} \text{tr}(X^\top L X) \tag{14.27}$$

where $\text{tr}(\cdot)$ denotes the trace of a matrix, $L = D - A$ is the graph Laplacian, and $D_{i,i} = \sum_j A_{i,j}$ is the degree matrix. As can be seen, minimizing $\Omega(A, X)$ forces adjacent nodes to have similar features, thus enforcing smoothness of the graph signals on the graph associated with A. However, solely minimizing the smoothness loss will result in the trivial solution $A = 0$. We might also want to pose other constraints to the graph.

Connectivity The following equation penalizes the formation of disconnected graphs via the logarithmic barrier.

$$\frac{-1}{n}\vec{1}^{\top}\log(A\vec{1}) \tag{14.28}$$

where n is the number of nodes.

Sparsity The following equation controls sparsity by penalizing large degrees.

$$\frac{1}{n^2}||A||_F^2 \tag{14.29}$$

where $||\cdot||_F$ denotes the Frobenius norm of a matrix.

In practice, solely minimizing one type of graph regularization losses might not be desirable. For instance, solely minimizing the smoothness loss will result in the trivial solution $A = 0$. Therefore, it could be beneficial to balance the trade-off among different types of desired graph properties by computing a linear combination of the various graph regularization losses, formulated as follows:

$$\frac{\alpha}{n^2}\mathrm{tr}(X^{\top}LX) + \frac{-\beta}{n}\vec{1}^{\top}\log(A\vec{1}) + \frac{\gamma}{n^2}||A||_F^2 \tag{14.30}$$

where α, β and γ are all non-negative hyperparameters for controlling the smoothness, connectivity and sparsity of the learned graph.

Besides the above graph regularization techniques, other prior assumptions such as neighboring nodes tend to share the same label (Yang et al, 2019c) and learned implicit adjacency matrix should be close to the intrinsic adjacency matrix (Jiang et al, 2019b) have been adopted in the literature.

Combining Intrinsic Graph Structures and Implicit Graph Structures

Recall that one of the most important motivations for graph structure learning is that the intrinsic graph structure (if it is available) might be error-prone (e.g., noisy or incomplete) and sub-optimal for the downstream prediction task. However, the intrinsic graph typically still carries rich and useful information regarding the optimal graph structure for the downstream task. Hence, it could be harmful to totally discard the intrinsic graph structure.

A few recent works (Li et al, 2018c; Chen et al, 2020m; Liu et al, 2021) proposed to combine the learned implicit graph structure with the intrinsic graph structure for better downstream prediction performance. The rationales are as follows. First of all, they assume that the optimized graph structure is potentially a "shift" (e.g., substructures) from the intrinsic graph structure, and the similarity metric function is intended to learn such a "shift" which is supplementary to the intrinsic graph structure. Secondly, incorporating the intrinsic graph structure can help accelerate the training process and increase the training stability considering there is no prior knowledge on the similarity metric, the trainable parameters are randomly initialized, and thus it may take long to converge.

Different ways for combining intrinsic and implicit graph structures have been proposed. For instance, Li et al (2018c); Chen et al (2020m) proposed to compute a linear combination of the normalized graph Laplacian of the intrinsic graph structure and the normalized adjacency matrix of the implicit graph structure, formulated as follows:

$$\widetilde{A} = \lambda L^{(0)} + (1 - \lambda) f(A) \qquad (14.31)$$

where $L^{(0)}$ is the normalized graph Laplacian matrix, $f(A)$ is the normalized adjacency matrix associated to the learned implicit graph structure, and λ is a hyperparameter controlling the trade-off between the intrinsic and implicit graph structures. Note that $f : \mathbb{R}^{n \times n} \to \mathbb{R}^{n \times n}$ can be arbitrary normalization operations such as graph Laplacian operation and row-normalization operation. Liu et al (2021) proposed a hybrid message passing mechanism for GNNs which fuses the two aggregated node vectors from the intrinsic graph and the learned implicit graph, respectively, and then feed the fused vector to a GRU (Cho et al, 2014a) to update node embeddings.

Learning Paradigms

Most existing methods for graph structure learning for GNNs consist of two key learning components: graph structure learning (i.e., similarity metric learning) and graph representation learning (i.e., GNN module), and the ultimate goal is to learn the optimized graph structures and representations with respect to certain downstream prediction task. How to optimize the two separate learning components toward the same ultimate goal becomes an important question?

Joint Learning of Graph Structures and Representations

The most straightforward strategy is to jointly optimize the whole learning system in an end-to-end manner toward the downstream prediction task which provides certain form of supervision, as illustrated in fig. 14.2. Jiang et al (2019b); Yang et al (2019c); Chen et al (2020m) designed a hybrid loss function combining both the task prediction loss and the graph regularization loss, namely, $\mathscr{L} = \mathscr{L}_{\text{pred}} + \mathscr{L}_{\mathscr{G}}$. The aim of introducing the graph regularization loss is to bring some prior knowledge to the graph properties (e.g., smoothness, sparsity) as we discussed above so as to enforce learning more meaningful graph structures and alleviate the potential overfitting issue.

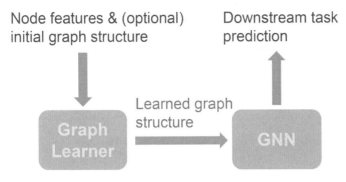

Fig. 14.2: Joint learning paradigm.

Adaptive Learning of Graph Structures and Representations

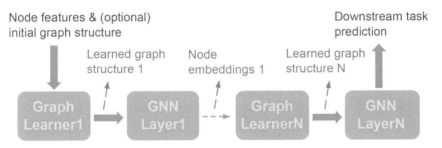

Fig. 14.3: Adaptive learning paradigm.

It is common practice to sequentially stack multiple GNN layers so as to capture long-range dependencies in a graph. As a result, the graph representations updated by one GNN layer will be consumed by the next GNN layer as the initial graph representations. Since input graph representations to each GNN layer are transformed by the previous GNN layer, one may naturally think whether the input graph structure to each GNN layer should be adaptively adjusted to reflect the changes of the graph representations, as illustrated in fig. 14.3. One such example is the GAT (Veličković et al, 2018) model which adatptively reweights the importance of neighboring node embeddings by applying the self-attention mechanism to the previously updated node embeddings when performing neighborhood aggregation at each GAT layer. However, the GAT model does not update the connectivity information of the intrinsic graph. In the literature of graph structure learning for GNNs, some methods (Yang et al, 2018c; Liu et al, 2019b; Huang et al, 2020a; Saire and Ramírez Rivera, 2019) also operate by adaptively learning a graph structure for every GNN layer based on the updated graph representations produced by the previous GNN layer. And the whole learning system is usually jointly optimized in an end-to-end manner toward the downstream prediction task.

Iterative Learning of Graph Structures and Representations

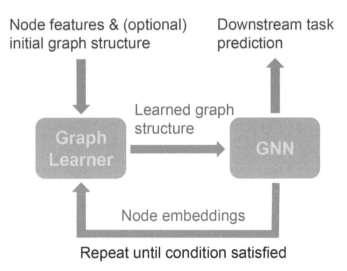

Fig. 14.4: Iterative learning paradigm.

Both of aforementioned joint learning and adaptive learning paradigms aim to learn a graph structure by applying a similarity metric function to the graph representations in a one-shot effort. Even though the adaptive learning paradigm aims to learn a graph structure at each GNN layer based on the updated graph representations, the graph structure learning procedure at each GNN layer is still one-shot. One big limitation of such a one-shot graph structure learning paradigm is that the quality of the learned graph structure heavily relies on the quality of the graph representations. Most existing methods assume that raw node features capture a good amount of information about the graph topology, which unfortunately is not always the case. Thus, it can be challenging to learn good implicit graph structures from the raw node features which do not contain adequate information about the graph topology.

Chen et al (2020m) proposed a novel end-to-end graph learning framework, dubbed as IDGL, for jointly and iteratively learning graph structures and representations. As illustrated in fig. 14.4, the IDGL framework operates by learning a better graph structure based on better graph representations, and in the meantime, learning better graph representations based on a better graph structure in an iterative manner. More specifically, the IDGL framework iteratively searches for an implicit graph structure that augments the intrinsic graph structure (if not available, a KNN graph is used) which is optimized for the downstream prediction task. And this iterative learning procedure dynamically stops when the learned graph structure approaches close enough to the optimized graph (with respect to the downstream task) according to certain stopping criterion (i.e., the difference between learned adjacency matrices at consecutive iterations are smaller than certain threshold). At each iteration, a hybrid loss combining both the task prediction loss and the graph regularization loss is added to the overall loss. After all iterations, the overall loss is back-propagated through all previous iterations to update model parameters.

This iterative learning paradigm for repeatedly refining the graph structure and graph representations has a few advantages. On the one hand, even when the raw node features do not contain adequate information for learning implicit relationships among nodes, the node embeddings learned by the graph representation learning component could ideally provide useful information for learning a better graph structure because these node embeddings are optimized toward the downstream task. On the other hand, the newly learned graph structure could be a better graph input for the graph representation learning component to learn better node embeddings.

14.3.2 Connections to Other Problems

Graph structure learning for GNNs has interesting connections to a few important problems. Thinking about these connections might spur further research in those areas.

14.3.2.1 Graph Structure Learning as Graph Generation

The task of graph generation focuses on generating realistic and meaningful graphs. The early works of graph generation formalized the problem as a stochastic generation process, and proposed various random graph models for generating a pre-selected family of graphs such as ER graphs (Erdős and Rényi, 1959), small-world networks (Watts and Strogatz, 1998), and scale-free graphs (Albert and Barabási, 2002). However, these approaches typically make certain simplified and carefully-designed apriori assumptions on graph properties, and thus in general have limited modeling capacity on complex graph structures. Recent attempts focus on building deep generative models for graphs by leveraging RNN You et al (2018b), VAE (Jin et al, 2018a), GAN (Wang et al, 2018a), flow-based techniques (Shi et al, 2019a) and other specially designed models (You et al, 2018a). And GNNs are usually adopted by these models as a powerful graph encoder.

Even though the graph generation task and the graph structure learning task both focus on learning graphs from data, they have essentially different goals and methodologies. Firstly, the graph generation task aims to generate new graphs where both nodes and edges are added to together construct a meaningful graph. However the graph structure learning task aims to learn a graph structure given a set of node attributes. Secondly, generative models for graphs typically operate by learning the distribution from the observed set of graphs, and generating more realistic graphs by sampling from the learned graph distribution. But graph structure learning methods typically operate by learning the pair-wise relationships among the given set of nodes, and based on that, building the graph topology. It will be an interesting research direction to study how the two tasks can help each other.

14.3.2.2 Graph Structure Learning for Graph Adversarial Defenses

Recent studies (Dai et al, 2018a; Zügner et al, 2018) have shown that GNNs are vulnerable to carefully-crafted perturbations (a.k.a adversarial attacks), e.g., small deliberate perturbations in graph structures and node/edge attributes. Researchers working on building robust GNNs found graph structure learning a powerful tool against topology attacks. Given an initial graph whose topology might become unreliable because of adversarial attacks, they leveraged graph structure learning techniques to recover the intrinsic graph topology from the poisoned graph.

For instance, assuming that adversarial attacks are likely to violate some intrinsic graph properties (e.g., low-rank and sparsity), Jin et al (2020e) proposed to jointly learn the GNN model and the "clean" graph structure from the perturbed graph by optimizing some hybrid loss combining both the task prediction loss and the graph regularization loss. In order to restore the structure of the perturbed graph, Zhang and Zitnik (2020) designed a message-passing scheme that can detect fake edges, block them and then attend to true, unperturbed edges. In order to address the noise brought by the task-irrelevant information on real-life large graphs, Zheng et al (2020b) introduced a supervised graph sparsification technique to remove potentially task-irrelevant edges from input graphs. Chen et al (2020d) proposed a Label-Aware GCN (LAGCN) framework which can refine the graph structure (i.e., filtering distracting neighbors and adding valuable neighbors for each node) before the training of GCN.

There are many connections between graph adversarial defenses and graph structure learning. On the one hand, graph structure learning is partially motivated by improving potentially error-prone (e.g., noisy or incomplete) input graphs for GNNs, which share the similar spirit with graph adversarial defenses. On the other hand, the task of graph adversarial defenses can benefit from graph structure learning techniques as evidenced by some recent works.

However, there is a key difference between their problem settings. The graph adversarial defenses task deals with the setting where the initial graph structure is available, but potentially poisoned by adversarial attacks. And the graph structure learning task aims to handle both the scenarios where the input graph structure is available or unavailable. Even when the input graph structure is available, one can still improve it by "denoising" the graph structure or augmenting the graph structure with an implicit graph structure which captures implicit relationships among nodes.

14.3.2.3 Understanding Transformers from a Graph Learning Perspective

Transformer models (Vaswani et al, 2017) have been widely used as a powerful alternative to Recurrent Neural Networks, especially in the Natural Language Processing field. Recent studies (Choi et al, 2020) have shown the close connection between transformer models and GNNs. By nature, transformer models aim to learn a self-attention matrix between every pair of objects, which can be thought as an adjacency matrix associated with a fully-connected graph containing each object as a node. Therefore, one can claim that transformer models also perform some sort of joint graph structure and representation learning, even though these models typically do not consider any initial graph topology and do not control the quality of the learned fully-connected graph. Recently, many variants of the so-called graph transformers (Zhu et al, 2019b; Yao et al, 2020; Koncel-Kedziorski et al, 2019; Wang et al, 2020k; Cai and Lam, 2020) have been developed to combine the benefits of both GNNs and transformers.

14.4 Future Directions

In this section, we will introduce some advanced topics of graph structure learning for GNNs and highlight some promising future directions.

14.4.1 Robust Graph Structure Learning

Although one of the major motivations of developing graph structure learning techniques for GNNs is to handle noisy or incomplete input graphs, robustness does not lie in the heart of most existing graph structure learning techniques. Most of existing works did not evaluate the robustness of their approaches to noisy initial graphs. Recent works showed that random edge addition or deletion attacks significantly downgraded the downstream task performance (Franceschi et al, 2019; Chen et al, 2020m). Moreover, most existing works admit that the initial graph structure (if provided) might be noisy and thus unreliable for graph representation learning, but they still assume that node features are reliable for graph structure learning, which is often not true in real-world scenarios. Therefore, it is challenging yet rewarding to explore robust graph structure learning techniques for data with noisy initial graph structures and noisy node attributes.

14.4.2 Scalable Graph Structure Learning

Most existing graph structure learning techniques need to model the pair-wise relationships among all the nodes in order to discover the hidden graph structure. Therefore, their time complexity is at least $O(n^2)$ where n is the number of graph nodes. This can be very expensive and even intractable for large-scale graphs (e.g., social networks) in real word. Recently, Chen et al (2020m) proposed a scalable graph structure learning approach by leveraging the anchor-based approximation technique to avoid explicitly computing the pair-wise node similarity, and achieved linear complexity in both computational time and memory consumption with respect to the number of graph nodes. In order to improve the scalability of transformer models, different kinds of approximation techniques have also been developed in recent works (Tsai et al, 2019; Katharopoulos et al, 2020; Choromanski et al, 2021; Peng et al, 2021; Shen et al, 2021; Wang et al, 2020g). Considering the close connections between graph structure learning for GNNs and transformers, we believe there are many opportunities in building scalable graph structure learning techniques for GNNs.

14.4.3 Graph Structure Learning for Heterogeneous Graphs

Most existing graph structure learning works focus on learning homogeneous graph structures from data. In comparison with homogeneous graphs, heterogeneous graphs are able to carry on richer information on node types and edge types, and occur frequently in real-world graph-related applications. Graph structure learning for heterogeneous graphs is supposed to be more challenging because more types of information (e.g., node types, edge types) are expected to be learned from data. Some recent attempts (Yun et al, 2019; Zhao et al, 2021) have been made to learn graph structures from heterogeneous graphs.

14.5 Summary

In this chapter, we explored and discussed graph structure learning from multiple perspectives. We first reviewed the existing works on graph structure learning in the literature of traditional machine learning, including both unsupervised graph structure learning and supervised graph structure learning. As for unsupervised graph structure learning, we mainly looked into some representative techniques developed from the Graph Signal Processing community. We also introduced some recent works on clustering analysis that leveraged graph structure learning techniques. As for supervised graph structure learning, we introduced how this problem was studied in the research on modeling interacting systems and Bayesian Networks. The main focus of this chapter is on introducing recent advances in graph structure learning for GNNs. We motivated graph structure learning in the GNN field by discussing the scenarios where the graph-structured data is noisy or unavailable. We then moved on to introduce recent research progress in joint graph structure and representation learning, including learning discrete graph structures and learning weighted graph structures. The connections and differences between graph structure learning and other important problems such as graph generation, graph adversarial defenses and transformer models were also discussed. We then highlighted several remaining challenges and future directions in the research of graph structure learning for GNNs.

Editor's Notes: Graph Structure Learning is a fast-emerging research topic and have seen a significant number of interests in recent years. The key idea is to learn an optimized graph structure in order to generate a better node representation (Chapter 4) and a more robust node representation (Chapter 8). Obviously, the graph structure learning could be expensive if the common pair-wise learning approach is adopted and thus the scalability issue could be a real major concern (Chapter 6). Meanwhile, it has tight connection with graph generation (Chapter 11) and self-supervised learning (Chapter 18), since they all consider partially how to modify/leverage graph structure. This chapter can be applicable to a broad range of application domains such as recommendation system (Chapter 19), computer vision (Chapter 20), Natural Language Processing (Chapter 21), Program Analysis (Chapter 22), and so on.

Chapter 15
Dynamic Graph Neural Networks

Seyed Mehran Kazemi

Abstract The world around us is composed of entities that interact and form relations with each other. This makes graphs an essential data representation and a
crucial building-block for machine learning applications; the nodes of the graph
correspond to entities and the edges correspond to interactions and relations. The
entities and relations may evolve; e.g., new entities may appear, entity properties
may change, and new relations may be formed between two entities. This gives rise
to dynamic graphs. In applications where dynamic graphs arise, there often exists
important information within the evolution of the graph, and modeling and exploiting such information is crucial in achieving high predictive performance. In this
chapter, we characterize various categories of dynamic graph modeling problems.
Then we describe some of the prominent extensions of graph neural networks to dynamic graphs that have been proposed in the literature. We conclude by reviewing
three notable applications of dynamic graph neural networks namely skeleton-based
human activity recognition, traffic forecasting, and temporal knowledge graph completion.

15.1 Introduction

Traditionally, machine learning models were developed to make predictions about
entities (or objects or examples) given only their features and irrespective of their
connections with the other entities in the data. Examples of such prediction tasks
include predicting the political party a social network user supports given their other
features, predicting the topic of a publication given its text, predicting the type of
the object in an image given the image pixels, and predicting the traffic in a road (or
road segment) given historical traffic data in that road.

Seyed Mehran Kazemi
Borealis AI, e-mail: `mehran.kazemi@borealisai.com`

In many applications, there exist relationships between the entities that can be exploited to make better predictions about them. As a few examples, social network users that are close friends or family members are more likely to support the same political party, two publications by the same author are more likely to have the same topic, two images taken from the same website (or uploaded to social media by the same user) are more likely to have similar objects in them, and two roads that are connected are more likely to have similar traffic volumes. The data for these applications can be represented in the form of a graph where nodes correspond to entities and edges correspond to the relationships between these entities.

Graphs arise naturally in many real-world applications including recommender systems, biology, social networks, ontologies, knowledge graphs, and computational finance. In some domains the graph is static, i.e. the graph structure and the node features are fixed over time. In other domains, the graph changes over time. In a social network, for example, new edges are added when people make new friends, existing edges are removed when people stop being friends, and node features change as people change their attributes, e.g., when they change their career assuming that career is one of the node features. In this chapter, we focus on the domains where the graph is dynamic and changes over time.

In applications where dynamic graphs arise, modeling the evolution of the graph is often crucial in making accurate predictions. Over the years, several classes of machine learning models have been developed that capture the structure and the evolution of dynamic graphs. Among these classes, extensions of graph neural networks (GNNs) (Scarselli et al, 2008; Kipf and Welling, 2017b) to dynamic graphs have recently found success in several domains and they have become one of the essential tools in the machine learning toolbox. In this chapter, we review the GNN approaches for dynamic graphs and provide several application domains where dynamic GNNs have provided striking results. The chapter is not meant to be a full survey of the literature but rather a description of the common techniques for applying GNNs to dynamic graphs. For a comprehensive survey of representation learning approaches for dynamic graphs we refer the reader to (Kazemi et al, 2020), and for a more specialized survey of GNN-based approaches to dynamic graphs we refer the reader to (Skarding et al, 2020).

The rest of the chapter is organized as follows. In Section 15.2, we define the notation that will be used throughout the chapter and provide the necessary background to follow the rest of the chapter. In Section 15.3, we describe different types of dynamic graphs and different prediction problems on these graphs. In Section 15.4, we review several approaches for applying GNNs on dynamic graphs. In Section 15.5, we review some of the applications of dynamic GNNs. Finally, Section 15.6 summarizes and concludes the chapter.

15.2 Background and Notation

In this section, we define our notation and provide the background required to follow the rest of the chapter.

We use lowercase letters z to denote scalars, bold lowercase letters $\mathbf{z}$ to denote vectors and uppercase letters $\mathbf{Z}$ to denote matrices. z_i denotes the i element of $\mathbf{z}$, $\mathbf{Z}_i$ denotes a column vector corresponding to the i row of $\mathbf{Z}$, and $\mathbf{Z}_{i,j}$ denotes the element at the i row and j column of $\mathbf{Z}$. $\mathbf{z}$ denotes the transpose of $\mathbf{z}$ and $\mathbf{Z}$ denotes the transpose of $\mathbf{Z}$. $(\mathbf{z}\mathbf{z}') \in \mathbb{R}^{d+d'}$ corresponds to the concatenation of $\mathbf{z} \in \mathbb{R}^d$ and $\mathbf{z}' \in \mathbb{R}^{d'}$. We use to represent an identity matrix. We use $\odot$ to denote element-wise (Hadamard) product. We represent a sequence as $[e_1, e_2, \ldots, e_k]$ and a set as $\{e_1, e_2, \ldots, e_k\}$ where e_is represent the elements in the sequence or set.

In this chapter, we mainly consider attributed graphs. We represent an *attributed graph* as $G = (V, A, X)$ where $V = \{v_1, v_2, \ldots, v_n\}$ is the set of vertices (aka nodes), $n = |V|$ denotes the number of nodes, $A \in \mathbb{R}^{n \times n}$ is an *adjacency matrix*, and $X \in \mathbb{R}^{n \times d}$ is a feature matrix where X_i represents the features associated with the i node v_i and d denotes the number of features. If there exists no edge between v_i and v_j, then $A_{i,j} = 0$; otherwise, $A_{i,j} \in \mathbb{R}_+$ represents the weight of the edge where $\mathbb{R}_+$ represents positive real numbers.

If G is *unweighted*, then the range of A is $\{0, 1\}$ (i.e. $A \in \{0, 1\}^{n \times n}$). G is *undirected* if the edges have no directions; it is *directed* if the edges have directions. For an undirected graph, A is symmetric (i.e. $A = A$). For each edge $A_{i,j} > 0$ of a directed graph, we call v_i the source and v_j the target of the edge. If G is *multi-relational* with a set $R = \{r_1, \ldots, r_m\}$ of relations, then the graph has m adjacency matrices where the i adjacency matrix represents the existence of the i relation r_i between the nodes.

15.2.1 Graph Neural Networks

In this chapter, we use the term *Graph Neural Network (GNN)* to refer to the general class of neural networks that operate on graphs through message-passing between the nodes. Here, we provide a brief description of GNNs.

Let $G = (V, A, X)$ be a static attributed graph. A GNN is a function $f : \mathbb{R}^{n \times n} \times \mathbb{R}^{n \times d} \to \mathbb{R}^{n \times d'}$ that takes G (or more specifically A and X) as input and provides as output a matrix $Z \in \mathbb{R}^{n \times d'}$ where $Z_i \in \mathbb{R}^{d'}$ corresponds to a hidden representation for the i node v_i. This hidden representation is called the *node embedding*. Providing a node embedding for each node v_i can be viewed as dimensionality reduction where the information from v_i's initial features as well as the information from its connectivity to other nodes and the features of these nodes are captured in a vector Z_i. This vector can be used to make informed predictions about v_i. In what follows, we describe two example GNNs namely *graph convolutions networks* and *graph attention networks* for undirected graphs.

Graph Convolutional Networks: Graph convolutional networks (GCNs) (Kipf and Welling, 2017b) stack multiple layers of graph convolution. The l layer of GCN for an undirected graph $G = (V, A, X)$ can be formulated as follows:

$$Z^{(l)} = \sigma(D^{-\frac{1}{2}} \tilde{A} D^{-\frac{1}{2}} Z^{(l-1)} W^{(l)}) \tag{15.1}$$

where $\tilde{A} = A+$ corresponds to the adjacency matrix with self-loops, D is a diagonal degree matrix with $D_{i,i} = \tilde{A}_i \mathbf{1}$ (**1** represents a column vector of ones) and $D_{i,j} = 0$ for $i \neq j$, $D^{-\frac{1}{2}} \tilde{A} D^{-\frac{1}{2}}$ corresponds to a row and column normalization of $\tilde{A}$, $Z^{(l)} \in \mathbb{R}^{n \times d^{(l)}}$ and $Z^{(l-1)} \in \mathbb{R}^{n \times d^{(l-1)}}$ represent the node embeddings in layer l and $(l-1)$ respectively with $Z^{(0)} = X$, $W^{(l)} \in \mathbb{R}^{d^{(l-1)} \times d^{(l)}}$ represents the weight matrix at layer l, and σ is an activation function.

The l layer of a GCN model can be described in terms of the following steps. First, it applies a linear projection to the node embeddings $Z^{(l-1)}$ using the weight matrix $W^{(l)}$, then for each node v_i it computes a weighted sum of the projected embeddings of v_i and its neighbors where the weights for the weighted sum are specified according to $D^{-\frac{1}{2}} \tilde{A} D^{-\frac{1}{2}}$, and finally it applies a non-linearity to the weighted sums and updates the node embeddings. Notice that in a L-layer GCN, the embedding for each node is computed based on its L-hop neighborhood (i.e. based on the nodes that are at most L hops away from it).

Graph Attention Networks: Instead of fixing the weights when computing a weighted sum of the neighbors, attention-based GNNs replace $D^{-\frac{1}{2}} \tilde{A} D^{-\frac{1}{2}}$ in equation 15.1 with an attention matrix $\hat{A}^{(l)} \in \mathbb{R}^{n \times n}$ such that:

$$Z^{(l)} = \sigma(\hat{A}^{(l)} Z^{(l-1)} W^{(l)}) \tag{15.2}$$

$$\hat{A}_{i,j}^{(l)} = \frac{E_{i,j}^{(l)}}{\sum_k E_{i,k}^{(l)}}, \quad E_{i,j}^{(l)} = \tilde{A}_{i,j} \exp\left(\alpha(Z_i^{(l-1)}, Z_j^{(l-1)}; \theta^{(l)})\right) \tag{15.3}$$

where $\alpha : \mathbb{R}^{d^{(l-1)}} \times \mathbb{R}^{d^{(l-1)}} \to \mathbb{R}$ is a function with parameters $\theta^{(l)}$ that computes *attention weights* for pairs of nodes. Here, $\tilde{A}$ acts as a mask that ensures $E_{i,j}^{(l)} = 0$ (and consequently $\hat{A}_{i,j}^{(l)} = 0$) if v_i and v_j are not connected. The exp function in the computation of $E_{i,j}^{(l)}$ and the normalization $\frac{E_{i,j}^{(l)}}{\sum_k E_{i,k}^{(l)}}$ correspond to a (masked) softmax function of the attention weights. Different attention-based GNNs can be constructed with different choices of α. In graph attention networks (GATs) (Veličković et al, 2018), $\theta^{(l)} \in \mathbb{R}^{2d^{(l)}}$ and α is defined as follows:

$$\alpha(Z_i^{(l-1)}, Z_j^{(l-1)}; \theta^{(l)}) = \sigma\left(\theta^{(l)}(W^{(l)} Z_i^{(l-1)} \,\|\, W^{(l)} Z_j^{(l-1)})\right) \tag{15.4}$$

where σ is an activation function. The formulation in equation 15.2 corresponds to a *single-head* attention-based GNN. A *multi-head* attention-based GNN computes multiple attention matrices $\hat{A}^{(l,1)}, \ldots, \hat{A}^{(l,\beta)}$ using equation 15.3 but with different weights $\theta^{(l,1)}, \ldots, \theta^{(l,\beta)}$ and $W^{(l,1)}, \ldots, W^{(l,\beta)}$ and then replaces equation 15.2 with:

$$Z^{(l)} = \sigma(\hat{A}^{(l,1)} Z^{(l-1)} W^{(l,1)} \| \cdots \| \hat{A}^{(l,\beta)} Z^{(l-1)} W^{(l,\beta)}) \qquad (15.5)$$

where β is the number of heads. Each head may learn to aggregate the neighbors differently and extract different information.

15.2.2 Sequence Models

Over the years, several models have been proposed that operate on sequences. In this chapter, we are mainly interested in neural sequence models that take as input a sequence $[x^{(1)}, x^{(2)}, \ldots, x^{(\tau)}]$ of observations where $x^{(t)} \in \mathbb{R}^d$ for all $t \in \{1, \ldots, \tau\}$, and produce as output hidden representations $[h^{(1)}, h^{(2)}, \ldots, h^{(\tau)}]$ where $h^{(t)} \in \mathbb{R}^{d'}$ for all $t \in \{1, \ldots, \tau\}$. Here, τ represents the length of the sequence or the timestamp for the last element in the sequence. Each hidden representation $h^{(t)}$ is a *sequence embedding* capturing information from the first t observations. Providing a sequence embedding for a given sequence can be viewed as dimensionality reduction where the information from the first t observations in the sequence is captured in a single vector $h^{(t)}$ which can be used to make informed predictions about the sequence. In what follows, we describe *recurrent neural networks*, *Transformers*, and *convolutional neural networks* for sequence modeling.

Recurrent Neural Networks: Recurrent neural networks (RNNs) (Elman, 1990) and its variants have achieved impressive results on a range of sequence modeling problems. The core principle of the RNN is that its output is a function of the current data point as well as a representation of the previous inputs. Vanilla RNNs consume the input sequence one by one and provides embeddings using the following equation (applied sequentially for t in $[1, \ldots, \tau]$):

$$h^{(t)} = RNN(x^{(t)}, h^{(t-1)}) = \sigma(W^{(i)} x^{(t)} + W^{(h)} h^{(t-1)} + b) \qquad (15.6)$$

where $W^{(\cdot)}$s and b are the model parameters, $h^{(t)}$ is the hidden state corresponding to the embedding of the first t observations, and $x^{(t)}$ is the t observation. One may initialize $h^{(0)} = 0$, where 0 is a vector of 0s, or let $h^{(0)}$ be learned during training. Training vanilla RNNs is typically difficult due to gradient vanishing and exploding.

Long short term memory (LSTMs) (Hochreiter and Schmidhuber, 1997) (and *gated recurrent units (GRUs)* (Cho et al, 2014a)) alleviate the training problem of vanilla RNNs through gating mechanism and additive operations. An LSTM model consumes the input sequence one by one and provides embeddings using the following equations:

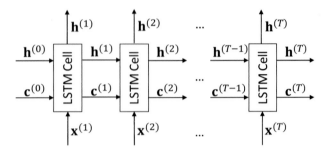

Fig. 15.1: An LSTM model taking as input a sequence $x^{(1)}, x^{(2)}, \ldots, x^{(\tau)}$ and producing hidden representations $h^{(1)}, h^{(2)}, \ldots, h^{(\tau)}$ as output. Equations 15.7-15.11 describe the operations in *LSTM Cells*.

$$i^{(t)} = \sigma \left(W^{(ii)} x^{(t)} + W^{(ih)} h^{(t-1)} + b^{(i)} \right) \tag{15.7}$$

$$f^{(t)} = \sigma \left(W^{(fi)} x^{(t)} + W^{(fh)} h^{(t-1)} + b^{(f)} \right) \tag{15.8}$$

$$c^{(t)} = f^{(t)} \odot c^{(t-1)} + i^{(t)} \odot Tanh \left(W^{(ci)} x^{(t)} + W^{(ch)} h^{(t-1)} + b^{(c)} \right) \tag{15.9}$$

$$o^{(t)} = \sigma \left(W^{(oi)} x^{(t)} + W^{(oh)} h^{(t-1)} + b^{(o)} \right) \tag{15.10}$$

$$h^{(t)} = o^{(t)} \odot Tanh \left(c^{(t)} \right) \tag{15.11}$$

Here $i^{(t)}$, $f^{(t)}$, and $o^{(t)}$ represent the input, forget and output gates respectively, $c^{(t)}$ is the memory cell, $h^{(t)}$ is the hidden state corresponding to the embedding of the sequence until t observation, σ is an activation function (typically Sigmoid), *Tanh* represents the hyperbolic tangent function, and $W^{(..)}$s and $b^{(.)}$s are weight matrices and vectors. Similar to vanilla RNNs, one may initialize $h^{(0)} = c^{(0)} = 0$ or let them be vectors with learnable parameters. Figure 15.1 shows an overview of an LSTM model.

A bidirectional RNN (BiRNN) (Schuster and Paliwal, 1997) is a combination of two RNNs one consuming the input sequence $[x^{(1)}, x^{(2)}, \ldots, x^{(\tau)}]$ in the forward direction and producing hidden representations $[\overrightarrow{h}^{(1)}, \overrightarrow{h}^{(2)}, \ldots, \overrightarrow{h}^{(\tau)}]$ as output, and the other consuming the input sequence backwards (i.e. $[x^{(\tau)}, x^{(\tau-1)}, \ldots, x^{(1)}]$) and producing hidden representations $[\overleftarrow{h}^{(\tau)}, \overleftarrow{h}^{(\tau-1)}, \ldots, \overleftarrow{h}^{(1)}]$ as output. These two hidden representations are then concatenated producing a single hidden representation $h^{(t)} = (\overrightarrow{h}^{(t)} \overleftarrow{h}^{(t)})$. Note that in RNNs, $h^{(t)}$ is computed only based on observations at or before t whereas in BiRNNs, $h^{(t)}$ is computed based on observations at, before, or after t. BiLSTMs Graves et al (2005) are a specific version of BiRNNs where the RNN is an LSTM.

Transformers: Consuming the input sequence one by one makes RNNs not amenable to parallelization. It also makes capturing long-range dependencies difficult. To solve these issues, the *Transformer* model Vaswani et al (2017) allows processing a sequence as a whole. The central operation in Transformer models is the self-attention mechanism. Let $H^{(l-1)}$ be an embedding matrix in layer $(l-1)$ such that its t row $H_t^{(l-1)}$ represents the embedding of the first t observations. The self-attention mechanism at each layer l can be described similar to equation 15.2 and equation 15.3 for attention-based GNNs by defining $\tilde{A}$ in equation 15.3 as a lower triangular matrix where $\tilde{A}_{i,j} = 1$ if $i \leq j$ and $\tilde{A}_{i,j} = 0$ otherwise, replacing $Z^{(l)}$ and $Z^{(l-1)}$ with $H^{(l)}$ and $H^{(l-1)}$, and defining the α function in equation 15.3 as follows:

$$\alpha(H_t^{(l-1)}, H_{t'}^{(l-1)}; \theta^{(l)}) = \frac{Q_t K_{t'}}{\sqrt{d^{(k)}}}, Q = W^{(l,Q)} H^{(l-1)}, K = W^{(l,K)} H^{(l-1)}$$

(15.12)

where $\theta^l = \{W^{(l,Q)}, W^{(l,K)}\}$ are the weights with $W^{(l,Q)}, W^{(l,K)} \in \mathbb{R}^{d^{(l-1)} \times d^{(k)}}$. The matrices Q and K are called the *query* and *key* matrices[1]. Q_t and $K_{t'}$ represent column vectors corresponding to the t and t' th row of Q and K, respectively. After L layers, the hidden representations $H^{(L)}$ contain the sequence embeddings with $H_t^{(L)}$ corresponding to the embedding of the first t observations (denoted as $h^{(t)}$ for RNNs). The lower-triangular matrix $\tilde{A}$ ensures that the embedding $H_t^{(L)}$ is computed based only on the observations at and before the t observation. One may define $\tilde{A}$ as a matrix of all 1s to allow $H_t^{(L)}$ to be computed based on the observations at, before, and after the t observation (similar to BiRNNs).

In equation 15.12, the embeddings are updated based on an aggregation of the embeddings from the previous timestamps, but the order of these embeddings is not modeled explicitly. To enable taking the order into account, the embeddings in the Transformer model are initialized as $H_t^{(0)} = x^{(t)} + p^{(t)}$ or $H_t^{(0)} = (x^{(t)} \| p^{(t)})$ where $H_t^{(0)}$ is the t row of $H^{(0)}$, $x^{(t)}$ is the t observation, and $p^{(t)}$ is a positional encoding capturing information about the position of the observation in the sequence. In the original work, the positional encodings are defined as follows:

$$p_{2i}^{(t)} = \sin(t/10000^{2i/d}), \quad p_{2i+1}^{(t)} = \sin(t/10000^{2i/d} + \pi/2)$$

(15.13)

Note that $p^{(t)}$ is constant and does not change during training.

Convolutional Neural Networks: Convolutional neural networks (CNNs) (Le Cun et al, 1989) have revolutionized many computer vision applications. Originally, CNNs were proposed for 2D signals such as images. They were later used for 1D signals such as sequences and time-series. Here, we describe 1D CNNs. We start with describing 1D convolutions. Let $H \in \mathbb{R}^{n \times d}$ be a matrix and $F \in \mathbb{R}^{u \times d}$ be a convolution filter. Applying the filter F on H produces a vector $h' \in \mathbb{R}^{n-u+1}$ as follows:

$$h_i' = \sum_{j=1}^{u} \sum_{k=1}^{d} H_{i+j-1,k} F_{j,k}$$

(15.14)

---

[1] For readers familiar with Transformers, in our description the *values* matrix corresponds to the multiplication of the embedding matrix with the weight matrix $W^{(l)}$ in equation 15.2.

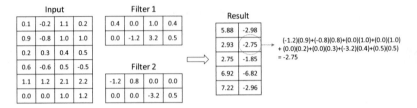

Fig. 15.2: An example of a 1D convolution operation with two convolution filters.

It is also possible to produce a vector $h' \in \mathbb{R}^n$ (i.e. a vector whose dimension is the same as the first dimension of H) by padding H with zeros. Having d' convolution filters, one can generate d' vectors as in equation 15.14 and stack them to generate a matrix $H' \in \mathbb{R}^{(n-u+1) \times d'}$ (or $H' \in \mathbb{R}^{n \times d'}$). Figure 15.2 provides an example of 1D convolution.

The 1D convolution operation in equation 15.14 is the main building block of the 1D CNNs. Similar to equation 15.12, let us assume $H^{(l-1)}$ represents the embeddings in the l layer with $H_t^{(0)} = x^{(t)}$ where $H_t^{(0)}$ represents the t row of $H^{(0)}$ and $x^{(t)}$ is the t observation. 1D CNN models apply multiple convolution filters to $H^{(l-1)}$ as described above and produce a matrix to which activation and (sometimes) pooling operations are applied to produce $H^{(l)}$. The convolution filters are the learnable parameters of the model. Hereafter, we use the term CNN to refer to the general family of 1D convolutional neural networks.

15.2.3 Encoder-Decoder Framework and Model Training

A deep neural network model can typically be decomposed into an encoder and a decoder module. The encoder module takes the input and provides vector-representation (or embeddings), and the decoder module takes the embeddings and provides predictions. The GNNs and sequence models described in Sections 15.2.1 and 15.2.2 correspond to the encoder modules of a full model; they provide node embeddings Z and sequence embeddings H, respectively. The decoder is typically task-specific. As an example, for a node classification task, the decoder can be a feed-forward neural network applied on a node embedding Z_i provided by the encoder, followed by a softmax function. Such a decoder provides as output a vector $\hat{y} \in \mathbb{R}^{|C|}$ where C represents the classes, $|C|$ represents the number of classes, and $\hat{y}_j$ shows the probability of the node belonging to the j class. A similar decoder can be used for sequence classification. As another example, for a link prediction problem, the decoder can take as input the embeddings for two nodes, take the sigmoid of a dot-product of the two node embeddings, and use the produced number as the probability of an edge existing between the two nodes.

The parameters of a model are learned through optimization by minimizing a task-specific loss function. For a classification task, for instance, we typically assume having access to a set of ground-truth labels Y where $Y_{i,j} = 1$ if the i example belongs to the j class and $Y_{i,j} = 0$ otherwise. We learn the parameters of the model by minimizing (e.g., using stochastic gradient descent) the cross entropy loss defined as follows:

$$L = -\frac{1}{|Y_{i,j}|} \sum_i \sum_j Y_{i,j} log(\hat{Y}_{i,j}) \qquad (15.15)$$

where $|Y_{i,j}|$ denotes the number of rows in $Y_{i,j}$ corresponding to the number of labeled examples, and $\hat{Y}_{i,j}$ is the probability of the i example belonging to the j class according to the model. For other tasks, one may use other appropriate loss functions.

15.3 Categories of Dynamic Graphs

Different applications give rise to different types of dynamic graphs and different prediction problems. Before commencing the model development, it is crucial to identify the type of dynamic graph and its static and evolving parts, and have a clear understanding of the prediction problem. In what follows, we describe some general categories of dynamic graphs, their evolution types, and some common prediction problems for them.

15.3.1 Discrete vs. Continues

As pointed out in (Kazemi et al, 2020), dynamic graphs can be divided into discrete-time and continuous-time categories. Here, we describe the two categories and point out how discrete-time can be considered a specific case of continuous-time dynamic graphs.

A discrete-time dynamic graph (DTDG) is a sequence $[G^{(1)}, G^{(2)}, \ldots, G^{(\tau)}]$ of graph snapshots where each $G^{(t)} = (V^{(t)}, A^{(t)}, X^{(t)})$ has vertices $V^{(t)}$, adjacency matrix $A^{(t)}$ and feature matrix $X^{(t)}$. DTDGs mainly appear in applications where (sensory) data is captured at regularly-spaced intervals.

Example 15.1. Figure 15.3 shows three snapshots of an example DTDG. In the first snapshot, there are three nodes. In the next snapshot, a new node v_4 is added and a connection is formed between this node and v_2. Furthermore, the features of v_1 are updated. In the third snapshot, a new edge has been added between v_3 and v_4.

A special type of DTDGs is the spatio-temporal graphs where a set of entities are spatially (i.e. in terms of closeness in space) and temporally correlated and data is captured at regularly-spaced intervals. An example of such a spatio-temporal graph is traffic data in a city or a region where traffic statistics at each road are computed at regularly-spaced intervals; the traffic at a particular road at time t is correlated with

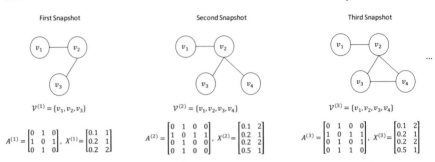

Fig. 15.3: Three snapshots of an example DTDG. In the first snapshot, there are 3 nodes. In the second snapshot, a new node v_4 is added and a connection is formed between this node and v_2. Moreover, the features of v_1 are updated. In the third snapshot, a new edge has been added between v_3 and v_4.

the traffic at the roads connected to it at time t (spatial correlation) as well as the traffic at these roads and the ones connected to it at previous timestamps (temporal correlation). In this example, the nodes in each $G^{(t)}$ may represent roads (or road segments), the adjacency matrix $A^{(t)}$ may represent how the roads are connected, and the feature matrix $X^{(t)}$ may represent the traffic statistics in each road at time t.

A *continuous-time dynamic graph (CTDG)* is a pair $(G^{(t_0)}, O)$ where $G^{(t_0)} = (V^{(t_0)}, A^{(t_0)}, X^{(t_0)})$ is a static graph[2] representing an initial state at time t_0 and O is a sequence of temporal observations/events. Each observation is a tuple of the form *(event type, event, timestamp)* where *event type* can be a node or edge addition, node or edge deletion, node feature update, etc., *event* represents the actual event that happened, and *timestamp* is the time at which the event occurred.

Example 15.2. An example of a CTDG is a pair $(G^{(t_0)}, O)$ where $G^{(t_0)}$ is the graph in the first snapshot of Figure 15.3 and the observations are as follows:

$$O = [(add\ node, v_4, 20\text{-}05\text{-}2020), (add\ edge, (v_2, v_4), 21\text{-}05\text{-}2020),$$
$$(Feature\ update, (v_1, [0.1, 2]), 28\text{-}05\text{-}2020), (add\ edge, (v_3, v_4), 04\text{-}06\text{-}2020)]$$

where, e.g., *(add node, v_4, 20-05-2020)* is an observation corresponding to a new node v_4 being added to the graph at time 20-05-2020.

At any point $t \geq t_0$ in time, a snapshot $G^{(t)}$ (corresponding to a static graph) can be obtained from a CTDG by updating $G^{(t_0)}$ sequentially according to the observations O that occurred before (or at) time t. In some cases, multiple edges may have been added between two nodes giving rise to multi-graphs; one may aggregate the edges to convert the multi-graph into a simple graph if required. Therefore, a DTDG can be viewed as a special case of a CTDG where only some regularly spaced snapshots of the CTDG are available.

---

[2] Note that we can have $V^{(t_0)} = \{\}$ corresponding to a graph with no nodes. We can also have $A^{(t_0)}_{i,j} = 0$ for all i, j corresponding to a graph with no edges.

Example 15.3. For the CTDG in Example 15.2, assume $t_0 = $ 01-05-2020 and we only observe the state of the graph on the first day of each month (01-05-2020, 01-06-2020 and 01-07-2020 for this example). In this case, the CTDG will reduce to the DTDG snapshots in Figure 15.3.

15.3.2 Types of Evolution

For both DTDGs and CTDGs, various parts of the graph may change and evolve. Here, we describe some of the main types of evolution. As a running example, we use a dynamic graph corresponding to a social network where the nodes represent *users* and the edges represent connections such as *friendship*.

Node addition/deletion: In our running example, new users may join the platform resulting in new nodes being added to the graph, and some users may leave the platform resulting in some nodes being removed from the graph.

Feature update: Users may have multiple features such as *age, country of residence, occupation*, etc. These features may change over time as users become older, move to a new country, or change their occupation.

Edge addition/deletion: As time goes by, some users become friends resulting in new edges and some people stop being friends resulting in some edges being removed from the graph. As pointed out in (Trivedi et al, 2019), the observations corresponding to events between two nodes may be categorized into *association* and *communication* events. The former corresponds to events that lead to structural changes in the graph and result in a long-lasting flow of information between the nodes (e.g., the formation of new friendships in social networks). The latter corresponds to events that result in a temporary flow of information between nodes (e.g., the exchange of messages in a social network). These two event categories typically evolve at different rates and one may model them differently, especially in applications where they are both present.

Edge weight updates: The adjacency matrix corresponding to the friendships may be weighted where the weights represent the strength of the friendships (e.g., computed based on the duration of friendship or other features). In this case, the strength of the friendships may change over time resulting in edge weight updates.

Relation updates: The edges between the users may be labeled where the label indicates the type of the connection, e.g., *friendship, engagement*, and *siblings*. In this case, the relation between two users may change over time (e.g., it may change from *friendship* to *engagement*). One may see relation update as a special case of edge evolution where one edge is deleted and another edge is added (e.g., the *friendship* edge is removed and an *engagement* edge is added).

15.3.3 Prediction Problems, Interpolation, and Extrapolation

We review four types of prediction problems for dynamic graphs: node classification/regression, graph classification, link prediction, and time prediction. Some of these problems can be studied under two settings: interpolation and extrapolation. They can also be studied under a transductive or inductive prediction setting. In what follows, we will describe each prediction problem. We let be a (discrete-time or continuous-time) dynamic graph containing information in a time interval $[t_0, \tau]$.

Node classification/regression: Let $V^{(t)} = \{v_1, \ldots, v_n\}$ represent the nodes in at time t. Node classification at time t is the problem of classifying a node $v_i \in V^{(t)}$ into a predefined set of classes C. Node regression at time t is the problem of predicting a continuous feature for a node $v_i \in V^{(t)}$. In the extrapolation setting, we make predictions about a future state (i.e. $t \geq \tau$) and the predictions are made based on the observations before or at t (e.g., forecasting the weather for the upcoming days). In the interpolation setting, $t_0 \leq t \leq \tau$ and the predictions are made based on all the observations (e.g., filling the missing values).

Graph classification: Let $\{1, 2, \ldots, k\}$ be a set of dynamic graphs. Graph classification is the problem of classifying each dynamic graph i into a predefined set of classes C.

Link prediction: Link prediction is the problem of predicting new links between the nodes of a dynamic graph. In the case of interpolation, the goal is to predict if there was an edge between two nodes v_i and v_j at timestamp $t_0 \leq t \leq \tau$ (or a time interval between t_0 and τ), assuming that v_i and v_j are in at time t. The interpolation problem is also known as the *completion* problem and can be used to predict missing links. In the case of extrapolation, the goal is to predict if there is going to be an edge between two nodes v_i and v_j at a timestamp $t > \tau$ (or a time interval after τ) assuming that v_i and v_j are in the at time τ.

Time prediction: Time prediction is the problem of predicting when an event happened or when it will happen. In the case of interpolation (sometimes called *temporal scoping*), the goal is to predict the time $t_0 \leq t \leq \tau$ when an event occurred (e.g., when two nodes v_i and v_j started or ended their connection). In the extrapolation case (sometimes called *time to event prediction*), the goal is to predict the time $t > \tau$ when an event will happen (e.g., when a connection will be formed between v_i and v_j).

Transductive vs. Inductive: The above problem definitions for node classification/regression, link prediction, and time prediction correspond to a transductive setting in which at the test time, predictions are to be made for entities already observed during training. In the inductive setting, information about previously unseen entities (or entirely new graphs) is provided at the test time and predictions are to be made for these entities (see (Hamilton et al, 2017b; Xu et al, 2020a; Albooyeh et al, 2020) for examples). The graph classification task is inductive by nature as it requires making predictions for previously unseen graphs at the test time.

15.4 Modeling Dynamic Graphs with Graph Neural Networks

In Section 15.2.1, we described how applying a GNN on a static graph G provides an embedding matrix $Z \in \mathbb{R}^{n \times d'}$ where n is the number of nodes, d' is the embedding dimension, and Z_i represents the embedding for the i entity v_i and can be used to make predictions about it. For dynamic graphs, we wish to extend GNNs to obtain embeddings $Z^{(t)} \in \mathbb{R}^{n_t \times d'}$ for any timestamp t, where n_t is the number of nodes in the graph at time t and $Z_i^{(t)}$ captures the information about the i entity at time t. In this section, we review several such extensions of GNNs. We mainly describe the encoder part of the models for dynamic graphs as the decoder and the loss functions can be defined similarly to Section 15.2.3.

15.4.1 Conversion to Static Graphs

A simple but sometimes effective approach for applying GNNs on dynamic graphs is to first convert the dynamic graph into a static graph and then apply a GNN on the resulting static graph. The main benefits of this approach include simplicity as well as enabling the use of a wealth of GNN models and techniques for static graphs. One disadvantage with this approach, however, is the potential loss of information. In what follows, we describe two conversion approaches.

Temporal aggregation: We start with describing temporal aggregation for a particular type of dynamic graphs and then explain how it extends to more general cases. Consider a DTDG $[G^{(1)}, G^{(2)}, \ldots, G^{(\tau)}]$ where each $G^{(t)} = (V^{(t)}, A^{(t)}, X^{(t)})$ such that $V^{(1)} = \cdots = V^{(\tau)} = V$ and $X^{(1)} = \cdots = X^{(\tau)} = X$ (i.e. the nodes and their features are fixed over time and only the adjacency matrix evolves). Note that in this case, the adjacency matrices have the same shape. One way to convert this DTDG into a static graph is through a weighted aggregation of the adjacency matrices as follows:

$$A^{(agg)} = \sum_{t=1}^{\tau} \phi(t, \tau) A^{(t)} \tag{15.16}$$

where $\phi : \mathbb{R} \times \mathbb{R} \to \mathbb{R}$ provides the weight for the t adjacency matrix as a function of t and τ. For extrapolation problems, a common choice for ϕ is $\phi(t, \tau) = \exp(-\theta(\tau - t))$ corresponding to exponentially decaying the importance of the older adjacency matrices (Yao et al, 2016). Here, θ is a hyperparameter controlling how fast the importance decays. For interpolation problems where a prediction is to be made for a timestamp $1 \leq t' \leq \tau$, one may define the function as $\phi(t, t') = \exp(-\theta|t' - t|)$ corresponding to exponentially decaying the importance of the adjacency matrices as they move further away from t'. Through this aggregation, one can convert the DTDG above into a static graph $G = (V, A^{(agg)}, X)$ and subsequently apply a static GNN model on it to make predictions. It is important to note that the aggregated adjacency matrix is weighted (i.e. $A^{(agg)} \in \mathbb{R}^{n \times n}$) so one can only use the GNN models that can handle weighted graphs.

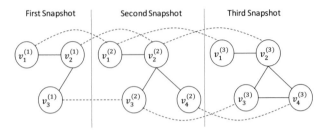

Fig. 15.4: An example of converting a DTDG into a static graph through temporal unrolling. Solid lines represent the edges in the graph at different timestamps and dashed lines represent the added edges. In this example, each node is connected to the node corresponding to the same entity only in the previous timestamp (i.e. $\omega = 1$).

In the case where node features also evolve, one may use a similar aggregation as in equation 15.16 and compute $X^{(agg)}$ based on $[X^{(1)}, X^{(2)}, \ldots, X^{(\tau)}]$. In the case where nodes are added and removed, one possible way of aggregation is as follows. Let $V^{(s)} = \{v \mid v \in V^{(1)} \cup \cdots \cup V^{(\tau)}\}$ represent the set of all the nodes that existed throughout time. We can expand every $A^{(t)}$ to a matrix in $\mathbb{R}^{|V^{(s)}| \times |V^{(s)}|}$ where the values for the rows and columns corresponding to any node $v \notin V^{(t)}$ are all 0s. The feature vectors can be expanded similarly. Then, equation 15.16 can be applied on the expanded adjacency and feature matrices. A similar aggregation can be done for CTDGs by first converting it into a DTDG (see Section 15.3.1) and then applying equation 15.16.

Example 15.4. Consider a DTDG with the three snapshots in Figure 15.3. We let $V^{(s)} = \{v_1, v_2, v_3, v_4\}$, add a row and a column of zeros to $A^{(1)}$, and add a row of zeros to $X^{(1)}$. Then, we use equation 15.16 with some value of θ to compute $A^{(agg)}$ and $X^{(agg)}$. Then we apply a GNN on the aggregated graph.

Temporal unrolling: Another way of converting a dynamic graph into a static graph is unrolling the dynamic graph and connecting the nodes corresponding to the same object across time. Consider a DTDG $[G^{(1)}, G^{(2)}, \ldots, G^{(\tau)}]$ and let $G^{(t)} = (V^{(t)}, A^{(t)}, X^{(t)})$ for $t \in \{1, \ldots, \tau\}$. Let $G^{(s)} = (V^{(s)}, A^{(s)}, X^{(s)})$ represent the static graph to be generated from the DTDG. We let $V^{(s)} = \{v^{(t)} \mid v \in V^{(t)}, t \in \{1, \ldots, \tau\}\}$. That is, every node $v \in V^{(t)}$ at every timestamp $t \in \{1, \ldots, \tau\}$ becomes a new node named $v^{(t)}$ in $V^{(s)}$ (so $|V^{(s)}| = \sum_{t=1}^{\tau} |V^{(t)}|$). Note that this is different from the way we constructed $V^{(s)}$ for temporal aggregation: here every node at every timestamp becomes a node in $V^{(s)}$ whereas in temporal aggregation we took a union of the nodes across timestamps. For every node $v^{(t)} \in V^{(s)}$, we let the features of $v^{(t)}$ in $X^{(s)}$ to be the same as its features in $X^{(t)}$. If two nodes $v_i, v_j \in V^{(t)}$ are connected according to $A^{(t)}$, we connect the corresponding nodes in $A^{(s)}$. We also connect each node $v^{(t)}$ to $v^{(t')}$ for $t' \in \{max(1, t - \omega), \ldots, t - 1\}$ so a node corresponding to an entity at time t becomes connected to the nodes corresponding to the same entity at the previous ω timestamps, where ω is a hyperparameter. One may assign different weights to these temporal edges in $\boldsymbol{A}^{(s)}$ based on the difference between t and t' (e.g., exponentially decaying the weight). Having constructed the static graph $G^{(s)}$, one may apply a GNN model on it and, e.g., use the resulting embedding for $v^{(t)}$s (i.e. the nodes corresponding to the t timestamp of the DTDG) to make predictions about the nodes.

Example 15.5. Figure 15.4 provides an example of temporal unrolling for the DTDG in Figure 15.3 with $\omega = 1$. The graph has 11 nodes overall and so $\boldsymbol{A}^{(s)} \in \mathbb{R}^{11 \times 11}$. The node features are set according to the ones in Figure 15.3, e.g., the feature values for $v_1^{(2)}$ are 0.1 and 2.

15.4.2 Graph Neural Networks for DTDGs

One natural way of developing models for DTDGs is by combining GNNs with sequence models; the GNN captures the information within the node connections and the sequence model captures the information within their evolution. A large number of the works on dynamic graphs in the literature follow this approach – see, e.g., (Seo et al, 2018; Manessi et al, 2020; Xu et al, 2019a). Here, we describe some generic ways of combining GNNs with sequence models.

GNN-RNN: Let be a DTDG with a sequence $[G^{(1)}, \ldots, G^{(\tau)}]$ of snapshots where $G^{(t)} = (V^{(t)}, \boldsymbol{A}^{(t)}, \boldsymbol{X}^{(t)})$ for each $t \in \{1, \ldots, \tau\}$. Suppose we want to obtain node embeddings at some time $t \leq \tau$ based on the observations at or before t. For simplicity, let us assume $V^{(1)} = V^{(2)} = \cdots = V^{(\tau)} = V$, i.e. the nodes are the same throughout time (in cases where the nodes change, one may use a similar strategy as in Example 15.4).

We can apply a GNN to each of the $G^{(t)}$s and obtain a hidden representation matrix $\boldsymbol{Z}^{(t)}$ whose rows correspond to node embeddings. Then, for the i node v_i, we obtain a sequence of embeddings $[\boldsymbol{Z}_i^{(1)}, \boldsymbol{Z}_i^{(2)}, \ldots, \boldsymbol{Z}_i^{(\tau)}]$. These embeddings do not yet contain temporal information. To incorporate the temporal aspect of the DTDG into the embeddings and obtain a temporal embedding for v_i at time t, we can feed the sequence $[\boldsymbol{Z}_i^{(1)}, \boldsymbol{Z}_i^{(2)}, \ldots, \boldsymbol{Z}_i^{(t)}]$ into an RNN model defined in equation 27.1 by replacing $\boldsymbol{x}^{(t)}$ with $\boldsymbol{Z}_i^{(t)}$ and using the hidden representation of the RNN model as the temporal node embedding for v_i. The temporal embedding for other nodes can be obtained similarly by feeding their sequence of embeddings produced by the GNN model to the same RNN model. The following formulae describe a variant of the GNN-RNN model where the GNN is a GCN (defined in equation 15.1), the RNN is an LSTM model, and the LSTM operations are applied to all nodes embeddings at the same time (the formulae are applied sequentially for t in $[1, 2, \ldots, \tau]$).

$$Z^{(t)} = GCN(X^{(t)}, A^{(t)}) \tag{15.17}$$

$$I^{(t)} = \sigma\left(Z^{(t)}W^{(ii)} + H^{(t-1)}W^{(ih)} + b^{(i)}\right) \tag{15.18}$$

$$F^{(t)} = \sigma\left(Z^{(t)}W^{(fi)} + H^{(t-1)}W^{(fh)} + b^{(f)}\right) \tag{15.19}$$

$$C^{(t)} = F^{(t)} \odot C^{(t-1)} + I^{(t)} \odot Tanh\left(Z^{(t)}W^{(ci)} + H^{(t-1)}W^{(ch)} + b^{(c)}\right) \tag{15.20}$$

$$O^{(t)} = \sigma\left(Z^{(t)}W^{(oi)} + H^{(t-1)}W^{(oh)} + b^{(o)}\right) \tag{15.21}$$

$$H^{(t)} = O^{(t)} \odot Tanh\left(C^{(t)}\right) \tag{15.22}$$

where, similar to equations 15.7-15.11, $I^{(t)}$, $F^{(t)}$, and $O^{(t)}$ represent the input, forget and output gates for the nodes respectively, $C^{(t)}$ is the memory cell, $H^{(t)}$ is the hidden state corresponding to the node embeddings for the first t observation, and $W^{(..)}$s and $b^{(.)}$s are weight matrices and vectors. In the above formulae, when we add a matrix $Z^{(t)}W^{(.i)} + H^{(t-1)}W^{(.h)}$ with a bias vector $b^{(.)}$, we assume the bias vector $b^{(.)}$ as added to every row of the matrix. $H^{(0)}$ and $C^{(0)}$ can be initialized with zeros or learned from the data. $H^{(t)}$ corresponds to the temporal node embeddings at time t and can be used to make predictions about them. We can summarize the equations above into:

$$Z^{(t)} = GCN(X^{(t)}, A^{(t)}) \tag{15.23}$$

$$H^{(t)}, C^{(t)} = LSTM(Z^{(t)}, H^{(t-1)}, C^{(t-1)}) \tag{15.24}$$

In a similar way, one can construct other variations of the GNN-RNN model such as GCN-GRU, GAT-LSTM, GAT-RNN, etc. Figure 15.5 provides an overview of the GCN-LSTM model.

RNN-GNN: In cases where the graph structure is fixed through time (i.e. $A^{(1)} = \cdots = A^{(\tau)} = A$) and only node features change, instead of first applying a GNN model and then applying a sequence model to obtain temporal node embeddings, one may apply the sequence model first to capture the temporal evolution of the node features and then apply a GNN model to capture the correlations between the nodes. We can create different variations of this generic model by using different GNN and sequence models (e.g., LSTM-GCN, LSTM-GAT, GRU-GCN, etc.). The formulation for a LSTM-GCN model is as follows:

$$H^{(t)}, C^{(t)} = LSTM(X^{(t)}, H^{(t-1)}, C^{(t-1)}) \tag{15.25}$$

$$Z^{(t)} = GCN(H^{(t)}, A) \tag{15.26}$$

with $Z^{(t)}$ containing the temporal node embeddings at time t. Note that RNN-GNN is only appropriate if the the adjacency matrix is fixed over time; otherwise, RNN-GNN fails to capture the information within the evolution of the graph structure.

GNN-BiRNN and BiRNN-GNN: In the case of GNN-RNN and RNN-GNN, the obtained node embeddings $H^{(t)}$ contain information about the observations at

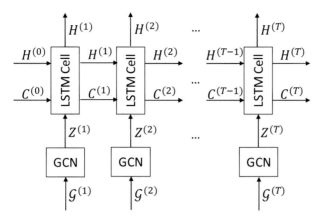

Fig. 15.5: The GCN-LSTM model taking a sequence $G^{(1)}, G^{(2)}, \ldots, G^{(\tau)}$ as input and producing hidden representations $\boldsymbol{H}^{(1)}, \boldsymbol{H}^{(2)}, \ldots, \boldsymbol{H}^{(\tau)}$ as output. The operations in *LSTM Cells* are described in equations 15.18-15.22. The GCN modules have shared parameters.

or before time t. This is appropriate for extrapolation problems. For interpolation problems (e.g., when we want to predict missing links between edges at a timestamp $t \leq \tau$), however, we may want to use the observations before, at, or after time t. One possible way of achieving this is by combining a GNN with a BiRNN so that the BiRNN provides information from not only the observations at or before time t but also after time t.

GNN-Transformer: Combining GNNs with Transformers can be done in a similar way as in GNN-RNNs. We apply a GNN to each of the $G^{(t)}$s and obtain a hidden representation matrix $\boldsymbol{Z}^{(t)}$ whose rows correspond to node embeddings. Then for the i entity v_i, we create a matrix $\boldsymbol{H}^{(0,i)}$ such that $\boldsymbol{H}_t^{(0,i)} = \boldsymbol{Z}_i^{(t)} + \boldsymbol{p}^{(t)}$ (or $\boldsymbol{H}_t^{(0,i)} = \boldsymbol{Z}_i^{(t)} \boldsymbol{p}^{(t)}$) where $\boldsymbol{p}^{(t)}$ is the positional encoding vector for position t. That is, the t row of $\boldsymbol{H}^{(0,i)}$ contains the embedding $\boldsymbol{Z}_i^{(t)}$ of v_i obtained by applying the GCN model on $G^{(t)}$, plus the positional encoding. The 0 superscript in $\boldsymbol{H}^{(0,i)}$ shows that $\boldsymbol{H}^{(0,i)}$ corresponds to the input of a Transformer model in the 0 layer. Once we have $\boldsymbol{H}^{(0,i)}$, we can apply an L-layer Transformer model (see equations 15.2, 15.3 and 15.12) to obtain $\boldsymbol{H}^{(L,i)}$ where $\boldsymbol{H}_t^{(L,i)}$ corresponds to the temporal embedding of v_i at time t. For extrapolation, the matrix $\tilde{\boldsymbol{A}}$ in equation 15.3 is a lower triangular matrix with $\tilde{\boldsymbol{A}}_{i,j} = 1$ if $i \leq j$ and 0 otherwise; for interpolation, $\tilde{\boldsymbol{A}}$ is a matrix of all 1s. The GCN-Transformer variant of the GNN-Transformer model can be described using the following equations:

$$Z^{(t)} = GCN(X^{(t)}, A^{(t)}) \; for \; t \in \{1, 2, \dots, \tau\} \tag{15.27}$$

$$H_t^{(0,i)} = Z_i^{(t)} + p^{(t)} \; for \; t \in \{1, 2, \dots, \tau\}, \; i \in \{1, 2, \dots, |V|\} \tag{15.28}$$

$$H^{(L,i)} = Transformer(H^{(0,i)}, \tilde{A}) \; for \; i \in \{1, 2, \dots, |V|\} \tag{15.29}$$

GNN-CNN: In a similar way as GNN-RNN and GNN-Transformer, one can combine GNNs with CNNs where the GNN provides $[Z^{(1)}, Z^{(2)}, \dots, Z^{(t)}]$, then the embeddings $[Z_i^{(1)}, Z_i^{(2)}, \dots, Z_i^{(t)}]$ for each node v_i are stacked into a matrix $H^{(0,i)}$ similar to the GNN-Transformer model, and then a 1D CNN model is applied on $H^{(0,i)}$ (see Section 15.2.2) to provide the final node embeddings.

Creating Deeper Models: Consider the GCN-LSTM model in Figure 15.5. The output of the GCN module is a sequence $[Z^{(1)}, Z^{(2)}, \dots, Z^{(\tau)}]$ and the outputs of the LSTM module is a sequence of hidden representation matrices $[H^{(1)}, H^{(2)}, \dots, H^{(\tau)}]$ Let us call the output of the GCN module as $[Z^{(1,1)}, Z^{(1,2)}, \dots, Z^{(1,\tau)}]$ and the output of the LSTM module as $[H^{(1,1)}, H^{(1,2)}, \dots, H^{(1,\tau)}]$ where the added superscript 1 indicates that these are the hidden representations created at layer 1. One may consider each $H^{(1,t)}$ as the new node features for the nodes in $G^{(t)}$ and run a GCN module (with separate parameters from the initial GCN) again to obtain $[Z^{(2,1)}, Z^{(2,2)}, \dots, Z^{(2,\tau)}]$. Then, another LSTM module may operate on these matrices to produce $[H^{(2,1)}, H^{(2,2)}, \dots, H^{(2,\tau)}]$. Stacking L of these GCN-LSTM blocks produces $[H^{(L,1)}, H^{(L,2)}, \dots, H^{(L,\tau)}]$ as output. These hidden matrices can then be used for making predictions about the nodes. The l layer of this model can be formulated as below (the formulae are applied sequentially for t in $[1, \dots, \tau]$):

$$Z^{(l,t)} = GCN(H^{(l-1,t)}, A^{(t)}) \tag{15.30}$$

$$H^{(l,t)}, C^{(l,t)} = LSTM(Z^{(l,t)}, H^{(l,t-1)}, C^{(l,t-1)}) \tag{15.31}$$

where $H^{(0,t)} = X^{(t)}$ for $t \in \{1, \dots, \tau\}$. The above two equations define what is called a *GCN-LSTM block*. Other blocks can be constructed using similar combinations.

15.4.3 Graph Neural Networks for CTDGs

Recently, developing models that operate on CTDGs without converting them to DTDGs (or converting them to static graphs) has been the subject of several studies. One class of models for CTDGs is based on extensions of the sequence models described in Section 15.2.2, especially RNNs. The general idea behind these models is to consume the observations sequentially and update the embedding of a node whenever a new observation is made about that node (or, in some works, about one of its neighbors). Before describing GNN-based approaches for CTDGs, we briefly describe some of the RNN-based models for CTDGs.

Consider a CTDG with $G^{(t_0)} = (V^{(t_0)}, A^{(t_0)}, X^{(t_0)})$ with $A_{i,j}^{(t_0)} = 0$ for all i, j (i.e. no initial edges) and observations O whose only type is edge additions. Since the

only observation types are edge additions, for this CTDG, the nodes and their features are fixed over time. Let $\boldsymbol{Z}^{(t-)}$ represent the node embeddings right before time
t (initially, $\boldsymbol{Z}^{(t_0)} = \boldsymbol{X}^{(t_0)}$ or $\boldsymbol{Z}^{(t_0)} = \boldsymbol{X}^{(t_0)} \boldsymbol{W}$ where $\boldsymbol{W}$ is a weight matrix with learnable parameters). Upon making an observation $(AddEdge, (v_i, v_j), t)$ corresponding
to a new directed edge between two nodes $v_i, v_j \in V$, the model developed in (Kumar
et al, 2019b) updates the embeddings for v_i and v_j as follows:

$$\boldsymbol{Z}_i^{(t)} = RNN_{source}((\boldsymbol{Z}_j^{(t-)} \parallel \Delta t_i \parallel \boldsymbol{f}), \boldsymbol{Z}_i^{(t-)}) \tag{15.32}$$

$$\boldsymbol{Z}_j^{(t)} = RNN_{target}((\boldsymbol{Z}_i^{(t-)} \parallel \Delta t_j \parallel \boldsymbol{f}), \boldsymbol{Z}_j^{(t-)}) \tag{15.33}$$

where RNN_{source} and RNN_{target} are two RNNs with different weights[3], Δt_i and Δt_j
represent the time elapsed since v_i's and v_j's previous interactions respectively[4], $\boldsymbol{f}$
represents a vector of features corresponding to edge features (if any), $\parallel$ indicates
concatenation, and $\boldsymbol{Z}_i^{(t)}$ and $\boldsymbol{Z}_j^{(t)}$ represent the updated embeddings at time t. The
first RNN takes as input a new observation $(\boldsymbol{Z}_j^{(t-)} \parallel \Delta t_i \parallel \boldsymbol{f})$ and the previous
hidden state of a node $\boldsymbol{Z}_i^{(t-)}$ and provides an updated representation (similarly for
the second RNN). Besides learning a temporal embedding $\boldsymbol{Z}^{(t)}$ as described above,
in (Kumar et al, 2019b) another embedding vector is also learned for each entity
that is fixed over time and captures the static features of the nodes. The two embeddings are then concatenated to produce the final embedding that is used for making
predictions.

In Trivedi et al (2017), a similar strategy is followed to develop a model for
CTDGs with multi-relational graphs in which two custom RNNs update the node
embeddings for the source and target nodes once a new labeled edge is observed
between them. In Trivedi et al (2019), a model is developed that is similar to
the above models but closer in nature to GNNs. Upon making an observation
$(AddEdge, (v_i, v_j), t)$, the node embedding for v_i is updated as follows (and similarly for v_j):

$$\boldsymbol{Z}_i^{(t)} = RNN((z_{\mathcal{N}(v_j)} \Delta t_i), \boldsymbol{Z}_i^{(t-)}) \tag{15.34}$$

where $z_{\mathcal{N}(v_j)}$ is an embedding that is computed based on a custom attention-
weighted aggregation of the embeddings of v_j and its neighbors at time t, and Δt_i is
defined similarly as in equation 15.32. Unlike equation 15.32 where the RNN updates the embedding of v_i based on the embedding of v_j alone, in equation 15.34
the embedding of v_i is updated based on an aggregation of the embeddings from the
first-order neighborhood of v_j which makes it close in nature to GNNs.

Many of the existing RNN-based approaches for CTDGs only compute the node
embeddings based on their immediate neighboring nodes (or nodes that are 1-hop

---

[3] The reason for using two RNNs is to allow the source and target nodes of a directed graph to be
updated differently upon making the observation $(AddEdge, (v_i, v_j), t)$. If the graph is undirected,
one may use a single RNN.
[4] If this is the first interaction of v_i (or v_j), then Δt_i (or Δt_j) can be the time elapsed since t_0.

away from them) and do not take into account the nodes that are multi-hops away. We now describe a GNN-based model for CTDGs named *temporal graph attention networks (TGAT)* and developed in (Xu et al, 2020a) that computes node embeddings based on the k-hop neighborhood of the nodes (i.e. based on the nodes that are at most k hops away). Being a GNN-based model, TGAT can learn embeddings for new nodes that are added to a graph and can be used for inductive settings where at the test time, predictions are to be made for previously unseen nodes.

Similar to the Transformer model, TGAT removes the recurrence and instead relies on self-attention and an extension of positional encoding to continuous time encoding named Time2Vec. In Time2Vec (Kazemi et al, 2019), time t (or a delta of time as in equation 15.32 and equation 15.34) is represented as a vector $z^{(t)}$ defined as follows:

$$z_i^{(t)} = \begin{cases} \omega_i t + \varphi_i, & \text{if } i = 0. \\ \sin(\omega_i t + \varphi_i), & \text{if } 1 \leq i \leq k. \end{cases} \tag{15.35}$$

where ω and φ are vectors with learnable parameters. TGAT uses a specific case of Time2Vec where the linear term is removed and the parameters φ are fixed to 0s and $\frac{\pi}{2}$s similar to equation 15.13. We refer the reader to Kazemi et al (2019); Xu et al (2020a) for theoretical and practical motivations of such a time encoding.

Now we describe how TGAT computes node embeddings. For a node v_i and timestamp t, let $\mathcal{N}_i^{(t)}$ represent the set of nodes that interacted with v_i at or before time t and the timestamps for the interaction. Each element of $\mathcal{N}_i^{(t)}$ is of the form (v_j, t_k) where $t_k \leq t$. The l layer of TGAT computes the embedding $h^{(t,l,i)}$ for v_i at time t in layer l using the following steps:

1. For any node v_i, $h^{(t,0,i)}$ (corresponding to the embedding of v_i in the 0 layer in time t) is assumed to be equal to X_i for any value of t.
2. A matrix $K^{(t,l,i)}$ with $|\mathcal{N}_i^{(t)}|$ rows is created such that for each $(v_j, t_k) \in \mathcal{N}_i^{(t)}$, $K^{(t,l,i)}$ has a row $(h^{(t_k,l-1,j)} \,||\, z^{(t-t_k)})$ where $h^{(t_k,l-1,j)}$ corresponds to the embedding of v_j in layer $(l-1)$ at the time t_k of its interaction with v_i and $z^{(t-t_k)}$ is an encoding for the delta time $(t - t_k)$ as in equation 15.35. Note that each $h^{(t_k,l-1,j)}$ is computed recursively using the same steps outlined here.
3. A vector $q^{(t,l,i)}$ is computed as $(h^{(t,l-1,i)} z^{(0)})$ where $h^{(t,l-1,i)}$ is the embedding of v_i at time t in layer $(l-1)$ and $z^{(0)}$ is an encoding for a delta of time equal to 0 as in equation 15.35.
4. $q^{(t,l,i)}$ is used to determine how much v_i should attend to each row of $K^{(t,l,i)}$ corresponding to the representation of its neighbors[5]. Attention weights $a^{(t,l,i)}$ are computed using equation 15.12 where the j element of $a^{(t,l,i)}$ is computed as $a_j^{(t,l,i)} = \alpha(q^{(t,l,i)}, K_j^{(t,l,i)}; \theta^{(l)})$.
5. Having the attention weights, a representation $\tilde{h}^{(t,l,i)}$ is computed for v_i using equation 15.2 where the attention matrix $\hat{A}^{(l)}$ is replaced with the attention vector $a^{(t,l,i)}$.

---

[5] For simplicity, here we describe a single-head attention-based GNN version of TGAT; in the original work, a multi-head version is used (see equation 15.5 for details.)

6. Finally, $h^{(t,l,i)} = FF^{(l)}(h^{(t,l-1,i)}\tilde{h}^{(t,l,i)})$ computes the representation for node v_i at time t in layer l where $FF^{(l)}$ is a feed-forward neural network in layer l.

An L-layer TGAT model computes node embeddings based on the L-hop neighborhood of a node.

Suppose we run a 2-layer TGAT model on a temporal graph where v_i interacted with v_j at time $t_1 < t$ and v_j interacted with v_k at time $t_2 < t_1$. The embedding $h^{(t,2,i)}$ is computed based on the embedding $h^{(t_1,1,j)}$ which is itself computed based on the embedding $h^{(t_2,0,k)}$. Since we are now at 0 layer, $h^{(t_2,0,k)}$ in TGAT is approximated with X_k thus ignoring the interactions v_k has had before time t_2. This may be suboptimal if v_k has had important interactions before t_2 as these interactions are not reflected on $h^{(t_1,1,j)}$ and hence not reflected on $h^{(t,2,i)}$. In (Rossi et al, 2020), this problem is remedied by using a recurrent model (similar to those introduced at the beginning of this subsection) that provides node embeddings at any time based on their previous local interactions, and initializing $h^{(t,0,i)}$s with these embeddings.

15.5 Applications

In this chapter, we provide some examples of real-world problems that have been formulated as predictions over dynamic graphs and modeled using GNNs. In particular, we review applications in computer vision, traffic forecasting, and knowledge graphs. This is by no means a comprehensive list; other application domains include recommendation systems Song et al (2019a), physical simulation of object trajectories Kipf et al (2018), social network analysis Min et al (2021), automated software bug triaging Wu et al (2021a), and many more.

15.5.1 Skeleton-based Human Activity Recognition

Human activity recognition from videos is a well-studied problem in computer vision with several applications. Given a video of a human, the goal is to classify the activity performed by the human in the video into a pre-defined set of classes such as *walking*, *running*, *dancing*, etc. One possible approach for this problem is to make predictions based on the human body skeleton as the skeleton conveys important information for human action recognition. In this subsection, we provide a dynamic graph formulation of this problem and a modeling approach based mainly on (a simplified version of) the approach of (Yan et al, 2018a).

Let us begin with formulating the skeleton-based activity recognition problem as reasoning over a dynamic graph. A video is a sequence of frames and each frame can be converted into a set of n nodes corresponding to the key points in the skeleton using computer vision techniques (see, e.g., (Cao et al, 2017)). These n nodes each have a feature vector representing their (2D or 3D) coordinates in the image frame. The human body specifies how these key points are connected to each other. With

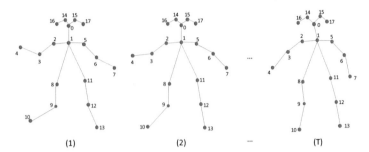

Fig. 15.6: The human skeleton represented as a graph for each snapshot of a video. The nodes represent the key points and the edges represent connections between these key points. The t graph corresponds to the human skeleton obtained from the t frame of a video.

this description, we can formulate the problem as reasoning over a DTDG consisting of a sequence $[G^{(1)}, G^{(2)}, \ldots, G^{(\tau)}]$ of graphs where each $G^{(t)} = (V^{(t)}, A^{(t)}, X^{(t)})$ corresponds to the t frame of a video with $V^{(t)}$ representing the set of key points in the t frame, $A^{(t)}$ representing their connections, and $X^{(t)}$ representing their features. An example is provided in Figure 15.6. One may notice that $V^{(1)} = \cdots = V^{(\tau)} = V$ and $A^{(1)} = \cdots = A^{(\tau)} = A$, i.e. the nodes and the adjacency matrices remain fixed throughout the sequence because they correspond to the key points and how they are connected in the human body. For instance, in the graphs of Figure 15.6, the node numbered as 3 is always connected to the nodes numbered as 2 and 4. The feature matrices $X^{(t)}$, however, keep changing as the coordinates of the key points change in different frames. The activity recognition can now be cast as classifying a dynamic graph into a set of predefined classes C.

The approach employed in (Yan et al, 2018a) is to convert the above DTDG into a static graph through temporal unrolling (see Section 15.4.1). In the static graph, the node corresponding to a key point at time t is connected to other key points at time t according to the human body (or, in other words, according to $A^{(t)}$) as well as the nodes representing the same key point and its neighbors in the previous ω timestamps. Once a static graph is constructed, a GNN can be applied to obtain embeddings for every joint at every timestamp. Since activity recognition corresponds to graph classification in this formulation, the decoder may consist of a (max, mean, or another type of) pooling layer on the node embeddings to obtain a graph embedding followed by a feed-forward network and a softmax layer to make class predictions.

In the l layer of the GNN in (Yan et al, 2018a), the adjacency matrix is multiplied element-wise to a mask matrix $M^{(l)}$ with learnable parameters (i.e. $A \odot M^{(l)}$ is used as the adjacency matrix). $M^{(l)}$ can be considered a data-independent attention map that learns weights for the edges in A. The goal of $M^{(l)}$ is to learn which connections are more important for activity recognition. Multiplying by $M^{(l)}$ only allows for changing the weight of the edges in A but it cannot add new edges. Connecting the key points according to the human body may arguably not be the best choice as, e.g., the connection between the hands is important in recognizing the *clapping* activity. In (Li et al, 2019e), the adjacency is summed with two other matrices $B^{(l)}$ and $C^{(l)}$ (i.e. $A + B^{(l)} + C^{(l)}$ is used as the adjacency) where $B^{(l)}$ is a data-independent attention matrix similar to $M^{(l)}$ and $C^{(l)}$ is a data-dependent attention matrix. Adding two matrices $B^{(l)}$ and $C^{(l)}$ to A allows for not only changing the edge weights in A but also adding new edges.

Instead of converting the dynamic graph to a static graph through temporal unrolling and applying a GNN on the static graph as in the previous two works, in Shi et al (2019b), (among other changes) a GNN-CNN model is used. One can use other combinations of a GNN and a sequence model (e.g., GNN-RNN) to obtain embeddings for joints at different timestamps. Note that activity recognition is not an extrapolation problem (i.e. the goal is not to predict the future based on the past). Therefore, to obtain the joint embeddings at time t, one may use information not only from $G^{(t')}$ where $t' \leq t$ but also from timestamps $t' > t$. This can be done by using, e.g., a GNN-BiRNN model (see Section 15.4.2).

15.5.2 Traffic Forecasting

For urban traffic control, traffic forecasting plays a paramount role. To predict the future traffic of a road, one needs to consider two important factors: spatial dependence and temporal dependence. The traffics in different roads are spatially dependent on each other as future traffic in one road depends on the traffic in the roads that are connected to it. The spatial dependence is a function of the topology of the road networks. There is also temporal dependence for each road because the traffic volume on a road at any time depends on the traffic volume at the previous times. There are also periodic patterns as, e.g., the traffic in a road may be similar at the same times of the day or at the same times of the week.

Early approaches for traffic forecasting mainly focused on temporal dependencies and ignored the spatial dependencies (Fu et al, 2016). Later approaches aimed at capturing spatial dependencies using convolutional neural networks (CNNs) (Yu et al, 2017b), but CNNs are typically restricted to grid structures. To enable capturing both spatial and temporal dependencies, several recent works have formulated traffic forecasting as reasoning over a dynamic graph (DTDGs in particular).

We first start by formulating traffic forecasting as a reasoning problem over a dynamic graph. One possible formulation is to consider a node for each road segment and connect two nodes if their corresponding road segments intersect with each other. The node features are the traffic flow variables (e.g., speed, volume, and density). The edges can be directed, e.g., to show the flow of the traffic in one-way roads, or undirected, showing that traffic flows in both directions. The structure of the graph can also change over time as, e.g., some road segments or some intersections may get (temporarily) closed. One may record the traffic flow variables and the state of the roads and intersections at regularly-spaced time intervals resulting in a DTDG. Alternatively, one may record the variables at different (asynchronous) time intervals resulting in a CTDG. The prediction problem is a node regression problem as we require to predict the traffic flow for the nodes, and it is an extrapolation problem as we need to predict the future state of the flow. The problem can be studied under a transductive setting where a model is trained based on the traffic data in a region and tested for making predictions about the same region. It can also be studied under an inductive setting where a model is trained based on the traffic data in multiple regions and is tested on new regions.

In (Zhao et al, 2019c), a model is proposed for transductive traffic forecasting in which the problem is formulated as reasoning over a DTDG with a sequence $[G^{(1)}, G^{(2)}, \ldots, G^{(\tau)}]$ of snapshots. The graph structure is considered to be fixed (i.e. no changes in road or intersection conditions) but the node features, corresponding to traffic flow features, change over time. The proposed model is a GCN-GRU model (see Section 15.4.2) where the GCN captures the spatial dependencies and the GRU captures the temporal dependencies. At any time t, the model provides a hidden representation matrix $\boldsymbol{H}^{(t)}$ based on the information at or before t; the rows of this matrix correspond to the node embeddings. These embeddings can then be used to make predictions about the traffic flow in the next timestamp(s). Assuming $\hat{\boldsymbol{Y}}^{(t+1)}$ represents the predictions for the next timestamp and $\boldsymbol{Y}^{(t+1)}$ represents the ground truth, the model is trained by minimizing an L2-regularized sum of the absolute errors $||\hat{\boldsymbol{Y}}^{(t+1)} - \boldsymbol{Y}^{(t+1)}||$.

As explained in Section 15.2.2, RNN-based models (e.g., the GCN-GRU model above) typically require sequential computations and are not amenable to parallelization. In (Yu et al, 2018a), the temporal dependencies are captured using CNNs instead of RNNs. The proposed model contains multiple blocks of CNN-GNN-CNN where the GNN is a generalization of GCNs to multi-dimensional tensors and the CNNs are gated.

The two works described so far consider the adjacency matrix to be fixed in different timestamps. As explained earlier, however, the adjacency matrix may change over time, e.g., due to accidents and roadblocks. In (Diao et al, 2019), the change in the adjacency matrix is taken into account through estimating the change in the topology of the roads based on the short-term traffic data.

15.5.3 Temporal Knowledge Graph Completion

Knowledge graphs (KGs) are databases of facts. A KG contains a set of facts in the form of triples (v_i, r_j, v_k) where v_i and v_k are called the subject and object entities and r_j is a relation. A KG can be viewed as a directed multi-relational graph with nodes $V = \{v_1, \ldots, v_n\}$, relations $R = \{r_1, \ldots, r_m\}$, and m adjacency matrices where the j adjacency matrix corresponds to the relations of type r_j between the nodes according to the triples.

A *temporal knowledge graph (TKG)* contains a set of temporal facts. Each fact may be associated with a single timestamp indicating the time when the event specified by the fact occurred, or a time interval indicating the start and end timestamps.

The facts with a single timestamp typically represent communication events and the facts with a time interval typically represent associative events (see Section 15.3.2)[6]. Here, we focus on facts with a single timestamp for which a TKG can be defined as a set of quadruples of the form (v_i, r_j, v_k, t) where t indicates the time when (v_i, r_j, v_k) occurred. Depending on the granularity of the timestamps, one may think of a TKG as a DTDG or a CTDG.

TKG completion is the problem of learning models based on the existing temporal facts in a TKG to answer queries of the type $(v_i, r_j, ?, t)$ (or $(?, r_j, v_k, t)$) where the correct answer is an entity $v \in V$ such that (v_i, r_j, v, t) (or (v, r_j, v_k, t)) has not been observed during training. It is mainly an interpolation problem as queries are to be answered at a timestamp t based on the past, present, and future facts. Currently, the majority of the models for TKG completion are not based on GNNs (e.g., see (Goel et al, 2020; García-Durán et al, 2018; Dasgupta et al, 2018; Lacroix et al, 2020)). Here, we describe a GNN-based approach that is mainly based on the work in (Wu et al, 2020b).

Since TKGs correspond to multi-relational graphs, to develop a GNN-based model that operates on a TKG we first need a relational GNN. Here, we describe a model named *relational graph convolution network (RGCN)* (Schlichtkrull et al, 2018) but other relational GNN models can also be used (see, e.g., (Vashishth et al, 2020)). Whereas GCN projects all neighbors of a node using the same weight matrix (see Section 15.2.1), RGCN applies relation-specific projections. Let $\hat{R}$ be a a set of relations that includes every relation in $R = \{r_1, \dots, r_m\}$ as well as a self-loop relation r_0 where each node has the relation r_0 only with itself. As is common in directed graphs (see, e.g., (Marcheggiani and Titov, 2017)) and specially for multi-relational graphs (see, e.g., (Kazemi and Poole, 2018)), for each relation $r_j \in R$ we also add an auxiliary relation r_j^{-1} to $\hat{R}$ where v_i has relation r_j^{-1} with v_k if and only if v_k has relation r_j with v_i. The l layer of an RGCN model can then be described as follows:

$$Z^{(l)} = \sigma\Big(\sum_{r \in \hat{R}} D^{(r)^{-1}} A^{(r)} Z^{(l-1)} W^{(l,r)} \Big)$$ (15.36)

where $A^{(r)} \in \mathbb{R}^{n \times n}$ represents the adjacency matrix corresponding to relation r, $D^{(r)}$ is the degree matrix of $A^{(r)}$ with $D_{i,i}^{(r)}$ representing the number of incoming relations of type r for the i node, $D^{(r)^{-1}}$ is a normalization term[7], $W^{(l,r)}$ is a relation-specific weight matrix for layer l, $Z^{(l-1)}$ represents the node embeddings in the (l-1) layer, and $Z^{(l)}$ represents the updated node embeddings in the l layer. If initial features X are provided as input, $Z^{(0)}$ can be set to X. Otherwise, $Z^{(0)}$ can either be set as 1-hot encodings where $Z_i^{(0)}$ is a vector whose elements are all zeros except in the

---

[6] This, however, is not always true as one may break a fact such as $(v_i, LivedIn, v_j)$ with a time interval $[2010, 2015]$ (meaning from 2010 until 2015) into a fact $(v_i, StartedLivingIn, v_j)$ with a timestamp of 2010 and another fact $(v_i, EndedLivingIn, v_j)$ with a timestamp of 2015.

[7] One needs to handle the cases where $D_{i,i}^{(r)} = 0$ to avoid numerical issues.

i position where it is 1, or it can be randomly initialized and then learned from the data.

In (Wu et al, 2020b), a TKG is formulated as a DTDG consisting of a sequence of snapshots $[G^{(1)}, G^{(2)}, \ldots, G^{(\tau)}]$ of multi-relational graphs. Each $G^{(t)}$ contains the same set of entities V and relations R (corresponding to all the entities and relations in the TKG) and contains the triples (v_i, r_j, v_k, t) from the TKG that occurred at time t. Then, RGCN-BiGRU and RGCN-Transformer models are developed (see Section 15.4.2) that operate on the DTDG formulation of the TKG where the RGCN model provides the node embeddings at every timestamp and the BiGRU and Transformer models aggregate the temporal information. Note that in each $G^{(t)}$ there may be several nodes with no incoming and outgoing edges (and also no features since TKGs typically do not have node features). RGCN does not learn a representation for these nodes as there exists no information about them in $G^{(t)}$. To handle this, special BiGRU and Transformer models are developed in (Wu et al, 2020b) that handle missing values.

The RGCN-BiGRU and RGCN-Transformer models provide node embeddings $\boldsymbol{H}^{(t)}$ at any timestamp t. To answer a query such as $(v_i, r_j, ?, t)$, one can compute the plausibility score of (v_i, r_j, v_k, t) for every $v_k \in V$ and select the entity that achieves the highest score. A common approach to find the score for an entity v_k for the above query is to use the TransE decoder Bordes et al (2013) according to which the score is $-\|\boldsymbol{H}_i^{(t)} + \boldsymbol{R}_j - \boldsymbol{H}_k^{(t)}\|$ where $\boldsymbol{H}_i^{(t)}$ and $\boldsymbol{H}_k^{(t)}$ correspond to the node embeddings for v_i and v_k at time t (provided by the RGCN) and $\boldsymbol{R}$ is a matrix with learnable parameters which has $m = |R|$ rows each corresponding to an embedding for a relation. TransE and its extensions are known to make unrealistic assumptions about the types and properties of the relations Kazemi and Poole (2018), so, alternatively, one may use other decoders that has been developed within the knowledge graph embedding community (e.g., the models in (Kazemi and Poole, 2018; Trouillon et al, 2016)).

When the timestamps in the TKG are discrete and there are not many of them, one can use a similar approach as above to answer queries of the form $(v_i, r_j, v_k, ?)$ through finding the score for every t in the set of discrete timestamps and selecting the one that achieves the highest score (see, e.g., (Leblay and Chekol, 2018)). Time prediction for TKGs has been also studied in an extrapolation setting where the goal is to predict when an event is going to happen in the future. This has been mainly done using temporal point processes as decoders (see, e.g., (Trivedi et al, 2017, 2019)).

15.6 Summary

Graph-based techniques are emerging as leading approaches in the industry for application domains with relational information. Among these techniques, graph neural networks (GNNs) are currently among the top-performing approaches. While GNNs and other graph-based techniques were initially developed mainly for static graphs, extending these approaches to dynamic graphs has been the subject of several recent studies and has found success in several important areas. In this chapter, we reviewed the techniques for applying GNNs to dynamic graphs. We also reviewed some of the applications of dynamic GNNs in different domains including computer vision, traffic forecasting, and knowledge graphs.

Editor's Notes: In the universe, the only thing unchanged is "change" itself, so do networks. Hence extending techniques for simple, static networks to those for dynamic ones is inevitably the trend while this domain is progressing. While there is a fast-increasing research body for dynamic networks in recent years, much more efforts are needed in order for substantial progress in the key issues such as scalability and validity discussed in Chapter 5 and other chapters. Extensions of the techniques in Chapters 9-18 are also needed. Many real-world applications radically speaking, requires to consider dynamic network, such as recommender system (Chapter 19) and urban intelligence (Chapter 27). So they could also benefit from the technique advancement toward dynamic networks.

Chapter 16
Heterogeneous Graph Neural Networks

Chuan Shi

Abstract Heterogeneous graphs (HGs) also called heterogeneous information networks (HINs) have become ubiquitous in real-world scenarios. Recently, employing graph neural networks (GNNs) to heterogeneous graphs, known as heterogeneous graph neural networks (HGNNs) which aim to learn embedding in low-dimensional space while preserving heterogeneous structure and semantic for downstream tasks, has drawn considerable attention. This chapter will first give a brief review of the recent development on HG embedding, then introduce typical methods from the perspective of shallow and deep models, especially HGNNs. Finally, it will point out future research directions for HGNNs.

16.1 Introduction to HGNNs

Heterogeneous graphs (HGs) (Sun and Han, 2013), which compose different types of entities and relations, also known as heterogeneous information networks (HINs), are ubiquitous in real-world scenarios, ranging from bibliographic networks, social networks to recommender systems. For example, as shown in Fig. 16.1 (a), a bibliographic network can be represented by a HG, which consists of four types of entities (author, paper, venue, and term) and three types of relations (author-write-paper, paper-contain-term and conference-publish-paper); and these basic relations can be further derived for more complex semantics (e.g., author-write-paper-contain-item). It has been well recognized that HG is a powerful model that embraces rich semantic and structural information. Therefore, researches on HG have been experiencing tremendous growth in data mining and machine learning, many of which have successful applications such as recommendation (Shi et al, 2018a; Hu et al, 2018a), text

Chuan Shi
School of Computer Science, Beijing University of Posts and Telecommunications, e-mail: shichuan@bupt.edu.cn

analysis (Linmei et al, 2019; Hu et al, 2020a), and cybersecurity (Hu et al, 2019b; Hou et al, 2017).

Due to the ubiquity of HGs, how to learn embedding of HGs is a key research problem in various graph analysis applications, e.g., node/graph classification (Dong et al, 2017; Fu et al, 2017), and node clustering (Li et al, 2019g). Traditionally, matrix factorization methods (Newman, 2006b) generate latent features in HGs. However, the computational cost of decomposing a large-scale matrix is usually very expensive, and also suffers from its statistical performance drawback (Shi et al, 2016; Cui et al, 2018). To address this challenge, heterogeneous graph embedding, aiming to learn a function that maps input space into lower-dimensional space while preserving heterogeneous structure and semantic, has drawn considerable attention in recent years.

Although there have been ample studies of embedding technology on homogeneous graphs (Cui et al, 2018) which consist of only one type of nodes and edges, these techniques cannot be directly applicable to HGs due to heterogeneity. Specifically, (1) the structure in HGs is usually semantic dependent, e.g., meta-path structure (Dong et al, 2017) can be very different when considering different types of relations; (2) different types of nodes and edges have different attributes located in different feature spaces; (3) HGs are usually application dependent, which may need sufficient domain knowledge for meta-path/meta-graph selection.

To tackle the above issues, various HG embedding methods have been proposed (Chen et al, 2018b; Hu et al, 2019a; Dong et al, 2017; Fu et al, 2017; Wang et al, 2019m; Shi et al, 2018a; Wang et al, 2020n). From the technical perspective, we divide the widely used models in HG embedding into two categories: shallow models and deep models. In summary, shallow models initialize the node embeddings randomly, then learn the node embeddings through optimizing some well-designed objective functions to preserve heterogeneous structures and semantics. Deep model aims to use deep neural networks (DNNs) to learn embedding from node attributes or interactions, where heterogeneous graph neural networks (HGNNs) stand out and will be the focus of this chapter. And there have demonstrated the success of HG embedding techniques deployed in real-world applications including recommender systems (Shi et al, 2018a; Hu et al, 2018a; Wang et al, 2020n), malware detection systems (Hou et al, 2017; Fan et al, 2018; Ye et al, 2019a), and healthcare systems (Cao et al, 2020; Hosseini et al, 2018).

The remainder of this chapter is organized as follows. In Sect. 27.1, we first introduce basic concepts in HGs, then discuss unique challenges of HG embedding due to the heterogeneity and give a brief review of the recent development on HG embedding. In Sect. 24.2 and 20.3, we categorize and introduce HG embedding in details according to the shallow and deep models. In Sect. 20.4, we further review pros and cons of the models introduced above. Finally, Sect. 20.5 forecasts the future research directions for HGNNs.

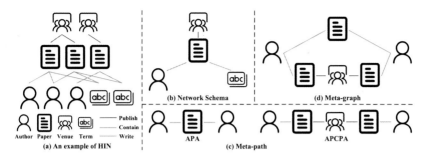

Fig. 16.1: An illustrative example of a heterogeneous graph (Wang et al, 2020l). (a) A bibliographic graph including four types of entities (i.e., author, paper, venue and term) and three types of relations (i.e., publish, contain and write). (b) Network schema of the bibliographic graph. (c) Two meta-paths (i.e., author-paper-author and paper-term-paper). (d) A meta-graph used in the bibliographic graph.

16.1.1 Basic Concepts of Heterogeneous Graphs

In this section, we will first formally introduce basic concepts in HGs and illustrate the symbols used throughout this chapter. HG is a graph consisting of different types of entities (i.e., nodes) and/or different types of relations (i.e., edges), which can be defined as follows.

Definition 16.1. Heterogeneous Graph (or Heterogeneous Information Network) (Sun and Han, 2013). A HG is defined as a graph $\mathscr{G} = \{\mathscr{V}, \mathscr{E}\}$, in which $\mathscr{V}$ and $\mathscr{E}$ represent the node set and the edge set, respectively. Each node $v \in \mathscr{V}$ and each edge $e \in \mathscr{E}$ are associated with their mapping function $\phi(v) : \mathscr{V} \to \mathscr{A}$ and $\varphi(e) : \mathscr{E} \to \mathscr{R}$. $\mathscr{A}$ and $\mathscr{R}$ denote the node type set and edge type set, respectively, where $|\mathscr{A}| + |\mathscr{R}| > 2$. The **network schema** for $\mathscr{G}$ is defined as $\mathscr{S} = (\mathscr{A}, \mathscr{R})$, which can be seen as a meta template of a heterogeneous graph $\mathscr{G} = \{\mathscr{V}, \mathscr{E}\}$ with the node type mapping function $\phi(v) : \mathscr{V} \to \mathscr{A}$ and the edge type mapping function $\varphi(e) : \mathscr{E} \to \mathscr{R}$. The network schema is a graph defined over node types $\mathscr{A}$, with edges as relation types from $\mathscr{R}$.

HG not only provides graph structure of data association, but also portrays higher-level semantics. An example of HG is illustrated in Fig. 16.1 (a), which consists of four node types (author, paper, venue, and term) and three edge types (author-write-paper, paper-contain-term, and conference-publish-paper), and Fig. 16.1 (b) illustrates the network schema. To formulate semantics of higher-order relationships among entities, meta-path (Sun et al, 2011) is further proposed whose definition is given below.

Definition 16.2. Meta-path (Sun et al, 2011). A meta-path p is based on network schema $\mathscr{S}$, which is denoted as $p = N_1 \xrightarrow{R_1} N_2 \xrightarrow{R_2} \cdots \xrightarrow{R_l} N_{l+1}$ (simplified to

$N_1N_2\cdots N_{l+1}$) with node types $N_1, N_2, \cdots, N_{l+1} \in \mathcal{N}$ and edge types $R_1, R_2, \cdots R_l \in \mathcal{R}$.

Note that different meta-paths describe semantic relationships in different views. For example, the meta-path APA indicates the co-author relationship and $APCPA$ represents the co-conference relation. Both of them can be used to formulate the relatedness over authors. Although meta-path can be used to depict the relatedness over entities, it fails to capture a more complex relationship, such as motifs (Milo et al, 2002). To address this challenge, meta-graph (Huang et al, 2016b) is proposed to use a directed acyclic graph of entity and relation types to capture more complex relationships between entities, defined as follows.

Definition 16.3. Meta-graph (Huang et al, 2016b). A meta-graph $\mathcal{T}$ can be seen as a directed acyclic graph (DAG) composed of multiple meta-paths with common nodes. Formally, meta-graph is defined as $\mathcal{T} = (\mathcal{V}_{\mathcal{T}}, \mathcal{E}_{\mathcal{T}})$, where $\mathcal{V}_{\mathcal{T}}$ is a set of nodes and $\mathcal{E}_{\mathcal{T}}$ is a set of edges. For any node $v \in \mathcal{V}_{\mathcal{T}}, \phi(v) \in \mathcal{A}$; for any edge $e \in \mathcal{E}_{\mathcal{T}}, \varphi(e) \in \mathcal{R}$.

An example meta-graph is shown in Fig. 16.1 (d), which can be regarded as the combination of meta-path APA and $APCPA$, reflecting high-order similarity of two nodes. Note that a meta-graph can be symmetric or asymmetric (Zhang et al, 2020g). To learn embeddings of HG, we formalize the problem of heterogeneous graph embedding.

Definition 16.4. Heterogeneous Graph Embedding (Shi et al, 2016). Heterogeneous graph embedding aims to learn a function $\Phi : \mathcal{V} \to \mathbb{R}^d$ that embeds the nodes $v \in \mathcal{V}$ in HG into low-dimensional Euclidean space with $d \ll |\mathcal{V}|$.

16.1.2 Challenges of HG Embedding

Different from homogeneous graph embedding (Cui et al, 2018), where the basic problem is preserving structure and property in node embedding (Cui et al, 2018). Due to the heterogeneity, HG embedding imposes more challenges, which are illustrated below.

Complex Structure (the complex HG structure caused by multiple types of nodes and edges). In a homogeneous graph, the fundamental structure can be considered as first-order, second-order, and even higher-order structures (Tang et al, 2015b). All these structures are well defined and have good intuition. However, the structure in HGs will dramatically change depending on the selected relations. Let's still take the academic graph in Fig. 16.1 (a) as an example, the neighbors of one paper will be authors with the "write" relation; while with "contain" relation, the neighbors become terms. Complicating things further, the combination of these relations, which can be considered as higher-order structures in HGs, will result in different and more complicated structures. Therefore, how to efficiently and effectively preserve these complex structures is of great challenge in HG embedding,

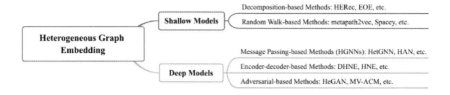

Fig. 16.2: Heterogeneous graph embedding tree classification diagram.

while current efforts have been made towards the meta-path structure (Dong et al, 2017) and meta-graph structure (Zhang et al, 2018b).

Heterogeneous Attributes (the fusion problem caused by the heterogeneity of attributes). Since nodes and edges in a homogeneous graph have the same type, each dimension of the node or edge attributes has the same meaning. In this situation, node can directly fuse attributes of its neighbors. However, in HGs, the attributes of different types of nodes and edges may have different meanings (Zhang et al, 2019b; Wang et al, 2019m). For example, the attributes of author can be research fields, while paper may use keywords as attributes. Therefore, how to overcome the heterogeneity of attributes and effectively fuse the attributes of neighbors poses another challenge in HG embedding.

Application Dependent. HG is closely related to the real-world applications, while many practical problems remain unsolved. For example, constructing an appropriate HG may require sufficient domain knowledge in a real-world application. Also, meta-path and/or meta-graph are widely used to capture the structure of HGs. However, unlike homogeneous graph, where the structure (e.g., the first-order and second-order structure) is well defined, meta-path selection may also need prior knowledge. Furthermore, to better facilitate the real-world applications, we usually need to elaborately encode side information (e.g., node attributes) (Wang et al, 2019m; Zhang et al, 2019b) or more advanced domain knowledge (Shi et al, 2018a; Chen and Sun, 2017) to HG embedding process.

16.1.3 Brief Overview of Current Development

Most of early works on graph data are based on high-dimensional sparse vectors for matrix analysis. However, the sparsity of the graph in reality and its growing scale have created serious challenges for such methods. A more effective way is to map nodes to latent space and use low-dimensional vectors to represent them. Therefore, they can be more flexibly applied to different data mining tasks, i.e., graph embedding.

There has been a lot of works dedicated to homogeneous graph embedding (Cui et al, 2018). These works are mainly based on deep models and combined with graph properties to learn embeddings of nodes or edges. For instance, DeepWalk (Perozzi et al, 2014) combines random walk and skip-gram model; LINE (Tang et al, 2015b) utilizes first-order and second-order similarity to learn distinguished node embedding for large-scale graphs; SDNE (Wang et al, 2016) uses deep auto-encoders to extract non-linear characteristics of graph structure. In addition to structural information, many methods further use the content of nodes or other auxiliary information (such as text, images, and tags) to learn more accurate and meaningful node embeddings. Some survey papers comprehensively summarize the work in this area (Cui et al, 2018; Hamilton et al, 2017c).

Due to the heterogeneity, embedding techniques for homogeneous graphs cannot be directly applicable to HGs. Therefore, researchers have begun to explore HG embedding methods, which emerge in recent years but develop rapidly. From the technical perspective, we summarize the widely used techniques (or models) in HG embedding, which can be generally divided into two categories: shallow models and deep models, as shown in Fig. 16.2. Specifically, shallows model mainly rely on meta-paths to simplify the complex structure of HGs, which can be classified into decomposition-based and random walk-based. Decomposition-based techniques Chen et al (2018b); Xu et al (2017b); Shi et al (2018b,c); Matsuno and Murata (2018); Tang et al (2015a); Gui et al (2016) decompose complex heterogeneous structure into several simpler homogeneous structures; while random walk-based (Dong et al, 2017; Hussein et al, 2018) methods utilize meta-path-guided random walk to preserve specific first-order and high-order structures. In order to take full advantage of heterogeneous structures and attributes, deep models are three-fold: message passing-based (HGNNs), encoder-decoder-based and adversarial-based methods. Message passing mechanism, i.e., the core idea of graph neural networks (GNNs), seamlessly integrates structure and attribute information. HGNNs inherit the message passing mechanism and design suitable aggregation functions to capture rich semantic in HGs (Wang et al, 2019m; Fu et al, 2020; Hong et al, 2020b; Zhang et al, 2019b; Cen et al, 2019; Zhao et al, 2020b; Zhu et al, 2019d; Schlichtkrull et al, 2018). The remaining encoder-decoder-based (Tu et al, 2018; Chang et al, 2015; Zhang et al, 2019c; Chen and Sun, 2017) and adversarial-based (Hu et al, 2018a; Zhao et al, 2020c) techniques employ encoder-decoder framework or adversarial learning to preserve complex attribute and structural information of HGs. In the following sections, we will introduce representative works of their subcategories in detail and compare their pros and cons.

16.2 Shallow Models

Early HG embedding methods focus on employing shallow models. They first initialize node embeddings randomly, then learn node embeddings through optimizing some well-designed objective functions. We divide the shallow model into two categories: decomposition-based and random walk-based.

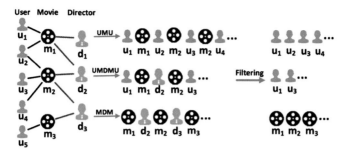

Fig. 16.3: An illustrative example of the proposed meta-path-guided random walk in HERec (Shi et al, 2018a). HERec first perform random walks guided by some selected meta-paths, then filter node sequences not with the user type or item type.

16.2.1 Decomposition-based Methods

To cope with the challenges brought by heterogeneity, decomposition-based techniques (Chen et al, 2018b; Xu et al, 2017b; Shi et al, 2018b,c; Matsuno and Murata, 2018; Tang et al, 2015a; Gui et al, 2016) decompose HG into several simpler subgraphs and preserve the proximity of nodes in each sub-graph, finally merge the information to achieve the effect of divide and conquer.

Specifically, HERec (Shi et al, 2018a) aims to learn embeddings of users and items under different meta-paths and fuses them for recommendation. It first finds the co-occurrence of users and items based on the meta-path-guided random walks on user-item HG, as shown in Fig. 16.3. Then it uses node2vec (Grover and Leskovec, 2016) to learn preliminary embeddings from the co-occurrence sequences of users and items. Because embeddings under different meta-paths contain different semantic information, for better recommendation performance, HERec designs a fusion function to unify the multiple embeddings:

$$g(\mathbf{h}_u^p) = \frac{1}{|P|} \sum_{p=1}^{P} (W^p \mathbf{h}_u^p + \mathbf{b}^p), \qquad (16.1)$$

where $\mathbf{h}_u^p$ is the embedding of user node u in meta-path p. P denotes the set of meta-paths. The fusion of item embeddings is similar to users. Finally, a prediction layer is used to predict the items that users prefer. HERec optimizes the graph embedding and recommendation objective jointly.

As another example, EOE is proposed to learn embeddings for coupled HGs, which consist of two different but related subgraphs. It divides the edges in HG into intra-graph edges and inter-graph edges. Intra-graph edge connects two nodes with the same type, and inter-graph edge connects two nodes with different types. To capture the heterogeneity in inter-graph edge, EOE (Xu et al, 2017b) uses the relation-specific matrix M_r to calculate the similarity between two nodes, which can be formulated as:

Fig. 16.4 The architecture of metapath2vec (Dong et al, 2017). Node sequence is generated under the meta-path PAP. It projects the embedding of the center node, e.g., p_2 into latent space and maximizes the probability of its meta-path-based context nodes, e.g., p_1, p_3, a_1 and a_2, appearing.

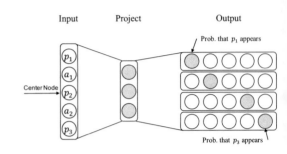

$$S_r(v_i, v_j) = \frac{1}{1 + \exp\left\{-\mathbf{h}_i^\top M_r \mathbf{h}_j\right\}}. \qquad (16.2)$$

Similarly, PME (Chen et al, 2018b) decomposes HG into some bipartite graphs according to the types of edges and projects each bipartite graph into a relation-specific semantic space. PTE (Tang et al, 2015a) divides the documents into word-word graph, word-document graph and word-label graph. Then it uses LINE (Tang et al, 2015b) to learn the shared node embeddings for each sub-graph. HEBE (Gui et al, 2016) samples a series of subgraphs from a HG and preserves the proximity between the center node and its subgraph.

The above-mentioned two-step framework of decomposition and fusion, as a transition product from homogeneous networks to HGs, is often used in the early attempt of HG embedding. Later, researchers gradually realized that extracting homogeneous graphs from HGs would irreversibly lose information carried by heterogeneous neighbors, and began to explore HG embedding methods that truly adapted to heterogeneous structure.

16.2.2 Random Walk-based Methods

Random walk, which generates some node sequences in a graph, is often used to describe the reachability between nodes. Therefore, it is widely used in graph representation learning to sample neighbor relationships of nodes and capture local structure in the graph (Grover and Leskovec, 2016). In homogeneous graphs, the node type is single and random walk can walk along any path. While in HGs, due to the type constraints of nodes and edges, meta-path-guided random walk is usually adopted, so that the generated node sequence contains not only the structural information, but also the semantic information. Through preserving the node sequence structure, node embedding can preserve both first-order and high-order proximity (Dong et al, 2017). A representative work is metapath2vec (Dong et al, 2017), which uses meta-path-guided random walk to capture semantic information of two nodes, e.g., the co-author relationship in academic graph as shown in Fig. 16.4.

Metapath2vec (Dong et al, 2017) mainly uses meta-path-guided random walk to generate heterogeneous node sequences with rich semantic, then it designs a heterogeneous skip-gram technique to preserve the proximity between node v and its context nodes, i.e., neighbors in the random walk sequences:

$$\arg\max_{\theta} \sum_{v \in \mathcal{V}} \sum_{t \in \mathcal{N}} \sum_{c_t \in C_t(v)} \log p(c_t|v; \theta), \tag{16.3}$$

where $C_t(v)$ represents the context nodes of node v with type t. $p(c_t|v; \theta)$ denotes the heterogeneous similarity function on node v and its context neighbors c_t:

$$p(c_t|v; \theta) = \frac{e^{\mathbf{h}_{c_t} \cdot \mathbf{h}_v}}{\sum_{\tilde{v} \in \mathcal{V}} e^{\mathbf{h}_{\tilde{v}} \cdot \mathbf{h}_v}}. \tag{16.4}$$

From the diagram shown in Fig. 16.4, Eq. (16.4) needs to calculate similarity between center node and its neighbors. Then Mikolov et al (2013b) introduces a negative sampling strategy to reduce the computation. Hence, Eq. (16.4) can be approximated as:

$$\log \sigma(\mathbf{h}_{c_t} \cdot \mathbf{h}_v) + \sum_{q=1}^{Q} \mathbb{E}_{\tilde{v}^q \sim P(\tilde{v})} \left[\log \sigma \left(-\mathbf{h}_{\tilde{v}^q} \cdot \mathbf{h}_v \right) \right], \tag{16.5}$$

where $\sigma(\cdot)$ is the sigmoid function, and $P(\tilde{v})$ is the distribution in which the negative node $\tilde{v}^q$ is sampled for Q times. Through the strategy of negative sampling, the time complexity is greatly reduced. However, when choosing the negative samples, metapath2vec does not consider the types of nodes, i.e., different types of nodes are from the same distribution $P(\tilde{v})$. Thus it further designs metapath2vec++, which samples negative nodes of the same type as the central node, i.e., $\tilde{v}_t^q \sim P(\tilde{v}_t)$. The formulation can be rewritten as:

$$\log \sigma(\mathbf{h}_{c_t} \cdot \mathbf{h}_v) + \sum_{q=1}^{Q} \mathbb{E}_{\tilde{v}_t^q \sim P(\tilde{v}_t)} \left[\log \sigma \left(-\mathbf{h}_{\tilde{v}_t^q} \cdot \mathbf{h}_v \right) \right]. \tag{16.6}$$

After minimizing the objective function, metapath2vec and metapath2vec++ can capture both structural information and semantic information effectively and efficiently.

Based on metapath2vec, a series of variants have been proposed. Spacey (He et al, 2019) designs a heterogeneous spacey random walk to unify different metapaths with a second-order hyper-matrix to control transition probability among different node types. JUST (Hussein et al, 2018) proposes a random walk method with Jump and Stay strategies, which can flexibly choose to change or maintain the type of the next node in the random walk without meta-path. BHIN2vec (Lee et al, 2019e) proposes an extended skip-gram technique to balance the various types of relations. It treats heterogeneous graph embedding as multiple relation-based tasks, and balances the influence of different relations on node embeddings by adjusting the training ratio of different tasks. HHNE (Wang et al, 2019n) conducts meta-path-guided random walk in hyperbolic space (Helgason, 1979), where the similarity between nodes can be measured using hyperbolic distance. In this way, some properties of

HGs, e.g., hierarchical and power-law structure, can be naturally reflected in learned node embeddings.

16.3 Deep Models

In recent years, deep neural networks (DNNs) have achieved great success in the fields of computer vision and natural language processing. Some works have also begun to use deep models to learn embedding from node attributes or interactions among nodes in HGs. Compared with shallow models, deep models can better capture the non-linear relationship, which can be roughly divided into three categories: message passing-based, encoder-decoder-based and adversarial-based.

16.3.1 Message Passing-based Methods (HGNNs)

Graph neural networks (GNNs) have emerged recently. Its core idea is the message passing mechanism, which aggregates neighborhood information and transmits it as messages to neighbor nodes. Different from GNNs that can directly fuse attributes of neighbors to update node embeddings, due to different types of nodes and edges, HGNNs need to overcome the heterogeneity of attributes and design effective fusion methods to utilize neighborhood information. Therefore, the key component is to design a suitable aggregation function, which can capture semantic and structural information of HGs (Wang et al, 2019m; Fu et al, 2020; Hong et al, 2020b; Zhang et al, 2019b; Cen et al, 2019; Zhao et al, 2020b; Zhu et al, 2019d; Schlichtkrull et al, 2018).

Unsupervised HGNNs. Unsupervised HGNNs aim to learn node embeddings with good generalization. To this end, they always utilize interactions among different types of attributes to capture the potential commonalities. HetGNN (Zhang et al, 2019b) is the representative work of unsupervised HGNNs. It consists of three parts: content aggregation, neighbor aggregation, and type aggregation. Content aggregation is designed to learn fused embeddings from different node contents, such as images, text, or attributes:

$$f_1(v) = \frac{\sum_{i \in C_v}[\overrightarrow{LSTM}\{\mathscr{FC}(\mathbf{h}_i)\} \oplus \overleftarrow{LSTM}\{\mathscr{FC}(\mathbf{h}_i)\}]}{|C_v|}, \quad (16.7)$$

where C_v is the type of node v's attributes. $\mathbf{h}_i$ is the i-th attributes of node v. A bi-directional Long Short-Term Memory (Bi-LSTM) (Huang et al, 2015) is used to fuse the embeddings learned by multiple attribute encoder $\mathscr{FC}$. Neighbor aggregation aims to aggregate the nodes with same type by using a Bi-LSTM to capture the position information:

$$f_2^t(v) = \frac{\sum_{v' \in N_t(v)} [\overrightarrow{LSTM}\{f_1(v')\} \oplus \overleftarrow{LSTM}\{f_1(v')\}]}{|N_t(v)|}, \tag{16.8}$$

where $N_t(v)$ is the first-order neighbors of node v with type t. Type aggregation uses an attention mechanism to mix the embeddings of different types and produces the final node embeddings.

$$\mathbf{h}_v = \alpha^{v,v} f_1(v) + \sum_{t \in O_v} \alpha^{v,t} f_2^t(v). \tag{16.9}$$

where $\mathbf{h}_v$ is the final embedding of node v, and O_v denotes the set of node types. Finally, a heterogeneous skip-gram loss is used as the unsupervised graph context loss to update node embeddings. Through these three aggregation methods, HetGNN can preserve the heterogeneity of both graph structures and node attributes.

Other unsupervised methods capture either heterogeneity of node attributes or heterogeneity of graph structures. HNE (Chang et al, 2015) is proposed to learn embeddings for the cross-model data in HGs, but it ignores the various types of edges. SHNE (Zhang et al, 2019c) focuses on capturing semantic information of nodes by designing a deep semantic encoder with gated recurrent units (GRU) (Chung et al, 2014). Although it uses heterogeneous skip-gram to preserve the heterogeneity of graph, SHNE is designed specifically for text data. Cen proposes GATNE (Cen et al, 2019), which aims to learn node embeddings in multiplex graph, i.e., a heterogeneous graph with different types of edges. Compared with HetGNN, GATNE pays more attention to distinguishing different edge relationships between node pairs.

Semi-supervised HGNNs. Different from unsupervised HGNNs, semi-supervised HGNNs aim to learn task-specific node embeddings in an end-to-end manner. For this reason, they prefer to use the attention mechanism to capture the most relevant structural and attribute information to the task. Wang (Wang et al, 2019m) propose heterogeneous graph attention network (HAN), which uses a hierarchical attention mechanism to capture both node and semantic importance. The architecture of HAN is shown in Fig. 16.5.

It consists of three parts: node-level attention, semantic-level attention, and prediction. Node-level attention aims to utilize the self-attention mechanism (Vaswani et al, 2017) to learn importances of neighbors in a certain meta-path:

$$\alpha_{ij}^m = \frac{\exp(\sigma(\mathbf{a}_m^{\mathrm{T}} \cdot [\mathbf{h}_i' \| \mathbf{h}_j']))}{\sum_{k \in \mathcal{N}_i^m} \exp(\sigma(\mathbf{a}_m^{\mathrm{T}} \cdot [\mathbf{h}_i' \| \mathbf{h}_k']))}, \tag{16.10}$$

where $\mathcal{N}_i^m$ is the neighbors of node v_i in meta-path m, α_{ij}^m is the weight of node v_j to node v_i under meta-path m. The node-level aggregation is defined as:

$$\mathbf{h}_i^m = \sigma \left(\sum_{j \in \mathcal{N}_i^m} \alpha_{ij}^m \cdot \mathbf{h}_j \right), \tag{16.11}$$

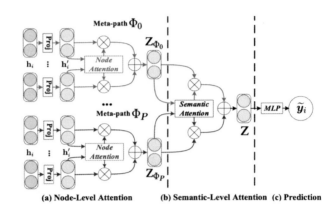

(a) Node-Level Attention   (b) Semantic-Level Attention (c) Prediction

Fig. 16.5: The architecture of HAN (Wang et al, 2019m). The whole model can be divided into three parts: Node-Level Attention aims to learn the importance of neighbors' features. Semantic-Level Attention aims to learn the importance of different meta-paths. Prediction layer utilizes the labeled nodes to update node embeddings.

where $\mathbf{h}_i^m$ denotes the learned embedding of node i based on meta-path m. Because different meta-paths capture different semantic information of HG, a semantic-level attention mechanism is designed to calculated the importance of meta-paths. Given a set of meta-paths $\{m_0, m_1, \cdots, m_P\}$, after feeding node features into node-level attention, it has P semantic-specific node embeddings $\{H_{m_0}, H_{m_1}, \cdots, H_{m_P}\}$. To effectively aggregate different semantic embeddings, HAN designs a semantic-level attention mechanism:

$$w_{m_i} = \frac{1}{|\mathcal{V}|} \sum_{i \in \mathcal{V}} \mathbf{q}^{\mathrm{T}} \cdot \tanh(W \cdot \mathbf{h}_i^m + \mathbf{b}), \tag{16.12}$$

where $W \in \mathbb{R}^{d' \times d}$ and $\mathbf{b} \in \mathbb{R}^{d' \times 1}$ denote the weight matrix and bias of the MLP, respectively. $\mathbf{q} \in \mathbb{R}^{d' \times 1}$ is the semantic-level attention vector. In order to prevent the node embeddings from being too large, HAN uses the softmax function to normalize w_{m_i}. Hence, the semantic-level aggregation is defined as:

$$H = \sum_{i=1}^{P} \beta_{m_i} \cdot H_{m_i}, \tag{16.13}$$

where β_{m_i} denotes the normalized w_{m_i}, which represents the semantic importance. $H \in \mathbb{R}^{N \times d}$ denotes the final node embeddings. Finally, a task-specific layer is used to fine-tune node embeddings with a small number of labels and the embeddings H can be used in downstream tasks, such as node clustering and link prediction. HAN is the first to extend GNNs to the heterogeneous graph and design a hierarchical attention mechanism, which can capture both structural and semantic information.

Subsequently, a series of attention-based HGNNs was proposed (Fu et al, 2020; Hong et al, 2020b; Hu et al, 2020e). MAGNN (Fu et al, 2020) designs intra-metapath aggregation and inter-metapath aggregation. The former samples some meta-path instances surrounding the target node and uses an attention layer to learn the importance of different instances, and the latter aims to learn the importance of different meta-paths. HetSANN (Hong et al, 2020b) and HGT (Hu et al, 2020e) treat one type of node as query to calculate the importance of other types of nodes around it, through which the method can not only capture interactions among different types of nodes, but also assign different weights to neighbors during aggregation.

In addition, there are some HGNNs that focus on other issues. NSHE (Zhao et al, 2020b) proposes to incorporate network schema, instead of meta-path, in aggregating neighborhood information. GTN (Yun et al, 2019) aims to automatically identify the useful meta-paths and high-order edges in the process of learning node embeddings. RSHN (Zhu et al, 2019d) uses both original node graph and coarsened line graph to design a relation-structure aware HGNN. RGCN (Schlichtkrull et al, 2018) uses multiple weight matrices to project the node embeddings into different relation spaces, thus capturing the heterogeneity of the graph.

Compared with shallow models, HGNNs have an obvious advantage that they have the ability of inductive learning, i.e., learning embeddings for out-of-sample nodes. Besides, HGNNs need smaller memory space because they only need to store model parameters. These two reasons are important for the real-world applications. However, they still suffer from the huge time costing in inferencing and retraining.

16.3.2 Encoder-decoder-based Methods

Encoder-decoder-based techniques aim to employ some neural networks as encoder to learn embedding from node attributes and design a decoder to preserve some properties of the graphs (Tu et al, 2018; Chang et al, 2015; Zhang et al, 2019c; Chen and Sun, 2017; Zhang et al, 2018a; Park et al, 2019).

For example, DHNE (Tu et al, 2018) proposes hyper-path-based random walk to preserve both structural information and indecomposability of hyper-graphs. Specifically, it designs a novel deep model to produce a non-linear tuple-wise similarity function while capturing the local and global structures of a given HG. As shown in Fig. 16.6, taking a hyperedge with three nodes a, b and c as an example. The first layer of DHNE is an autoencoder, which is used to learn latent embeddings and preserve the second-order structures of graph (Tang et al, 2015b). The second layer is a fully connected layer with embedding concatenated:

$$L = \sigma(W_a \mathbf{h}_a \oplus W_b \mathbf{h}_b \oplus W_c \mathbf{h}_c), \tag{16.14}$$

where L denotes the embedding of the hyperedge; $\mathbf{h}_a, \mathbf{h}_b$ and $\mathbf{h}_c \in \mathbb{R}^{d \times 1}$ are the embeddings of node a, b and c learn by the autoencoder. W_a, W_b and $W_c \in \mathbb{R}^{d' \times d}$ are the transformation matrices for different node types. Finally, the third layer is used

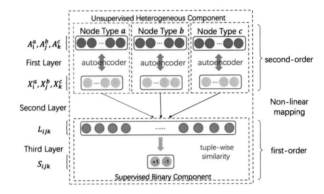

Fig. 16.6: The framework of DHNE (Tu et al, 2018). DHNE learns embeddings for nodes in heterogeneous hypernetworks, which can simultaneously address inde-composable hyperedges while preserving rich structural information.

to calculate the indecomposability of hyperedge:

$$\mathscr{S} = \sigma(W \cdot L + \mathbf{b}), \qquad (16.15)$$

where $\mathscr{S}$ denote the indecomposability of hyperedge; $W \in \mathbb{R}^{1 \times 3d'}$ and $\mathbf{b} \in \mathbb{R}^{1 \times 1}$ are the weight matrix and bias, respectively. A higher value of $\mathscr{S}$ means these nodes are from the existing hyperedges, otherwise it should be small.

Similarly, HNE (Chang et al, 2015) focuses on multi-modal heterogeneous graph. It uses CNN and autoencoder to learn embedding from images and texts, respectively. Then it uses the embedding to predict whether there is an edge between the images and texts. Camel (Zhang et al, 2018a) uses GRU as an encoder to learn paper embedding from the abstracts. A skip-gram objective function is used to preserve the local structures of the graphs.

16.3.3 Adversarial-based Methods

Adversarial-based techniques utilize the game between generator and discriminator to learn robust node embedding. In homogeneous graph, the adversarial-based techniques only consider the structural information, for example, GraphGAN (Wang et al, 2018a) uses Breadth First Search when generating virtual nodes. In a HG, the discriminator and generator are designed to be relation-aware, which captures the rich semantics on HGs. HeGAN (Hu et al, 2018a) is the first to use GAN in HG embedding. It incorporates the multiple relations into the generator and discriminator, so that the heterogeneity of a given graph can be considered.

As shown in Fig. 16.7 (c), HeGAN mainly consists of two competing players, the discriminator and the generator. Given a node, the generator attempts to produce

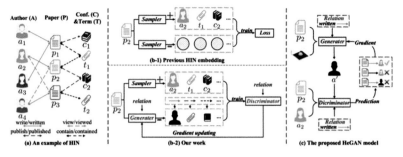

Fig. 16.7: Overview of HeGAN (Hu et al, 2018a). (a) A toy HG for bibliographic data. (b) Comparison between HeGAN and previous works. (c) The framework of HeGAN for adversarial learning on HGs.

fake samples associated with the given node to feed into the discriminator, whereas the discriminator tries to improve its parameterization to separate the fake samples from the real ones actually connected to the given node. The better trained discriminator would then force the generator to produce better fake samples, and the process is repeated. During such iterations, both the generator and discriminator receive mutual, positive reinforcement. While this setup may appear similar to previous efforts (Cai et al, 2018c; Dai et al, 2018c; Pan et al, 2018) on GAN-based network embedding, HeGAN employs two major novelties to address the challenges of adversarial learning on HINs.

First, existing studies only leverage GAN to distinguish whether a node is real or fake *w.r.t.* structural connections to a given node, without accounting for the heterogeneity in HINs. For example, given a paper p_2, they treat nodes a_2, a_4 as real, whereas a_1, a_3 are fake simply based on the topology of the HIN shown in Fig. 16.7 (a). However, a_2 and a_4 are connected to p_2 for different reasons: a_2 writes p_2 and a_4 only views p_2. Thus, they miss out on valuable semantics carried by HGs, unable to differentiate a_2 and a_4 even though they play distinct semantic roles. Given a paper p_2 as well as a relation, say, write/written, HeGAN introduces a relation-aware discriminator to tell apart a_2 and a_4. Formally, relation-aware discriminator $C(\mathbf{e}_v \mid u, r; \theta^C)$ evaluates the connectivity between the pair of nodes u and v *w.r.t.* a relation r:

$$C(\mathbf{e}_v \mid u, r; \theta^C) = \frac{1}{1 + \exp(-\mathbf{e}_u^{C\top} M_r^C \mathbf{e}_v)}, \quad (16.16)$$

where $\mathbf{e}_v \in \mathbb{R}^{d \times 1}$ is the input embedding of the sample v, $\mathbf{e}_u \in \mathbb{R}^{d \times 1}$ is the learnable embedding of node u, and $M_r^C \in \mathbb{R}^{d \times d}$ is a learnable relation matrix for relation r.

Second, existing studies are limited in sample generation in both effectiveness and efficiency. They typically model the distribution of nodes using some form of softmax over all nodes in the original graph. In terms of effectiveness, their fake samples are constrained to the nodes in the graph, whereas the most representative fake samples may fall "in between" the existing nodes in the embedding space. For example, given a paper p_2, they can only choose fake samples from $\mathcal{V}$, such as

a_1 and a_3. However, both may not be adequately similar to real samples such as a_2. Towards a better sample generation, we introduce a generalized generator that can produce latent nodes such as a' shown in Fig. 16.7 (c), where it is possible that $a' \notin \mathcal{V}$. In particular, the generalized generator leverage the following Gaussian distribution:

$$\mathcal{N}(\mathbf{e}_u^{G^\top} M_r^G, \sigma^2 I), \tag{16.17}$$

where $\mathbf{e}_u^G \in \mathbb{R}^{d \times 1}$ and $M_r^G \in \mathbb{R}^{d \times d}$ denote the node embedding of $u \in \mathcal{V}$ and the relation matrix of $r \in \mathcal{R}$ for the generator.

Except for HeGAN, MV-ACM (Zhao et al, 2020c) uses GAN to generate the complementary views by computing the similarity of nodes in different views. Overall, adversarial-based methods prefer to utilize the negative samples to enhance the robustness of embeddings. But the choice of negative samples has a huge influence on the performance, thus leading higher variances.

16.4 Review

Based on the above representative work of the shallow and deep models, it can be found that the shallow models mainly focus on the structure of HGs, and rarely use additional information such as attributes. One of the possible reasons is that shallow models are hard to depict the relationship between additional and structural information. The learning ability of DNNs supports modeling of this complex relationship. For example, message passing-based techniques are good at encoding structures and attributes simultaneously, and integrate different semantic information. Compared with message passing-based techniques, encoder-decoder-based techniques are weak in fusing information due to the lack of messaging mechanism. But they are more flexible to introduce different objective functions through different decoders. Adversarial-based methods prefer to utilize the negative samples to enhance the robustness of embeddings. But the choice of negative samples has a huge influence on the performance, thus leading higher variances (Hu et al, 2019a).

However, shallow and deep models each have their own pros and cons. Shallow models lack non-linear representation capability, but are efficient and easy to parallelize. Specially, the complexity of random walk technique consists of two parts: random walk and skip-gram, both of which are linear with the number of nodes. Decomposition technique needs to divide HGs into sub-graphs according to the type of edges, so the complexity is linear with the number of edges, which is higher than random walk. Deep models have stronger representation capability, but they are easier to fit noise and have higher time and space complexity. Additionally, the cumbersome hyperparameter adjustment of deep models is also criticized. But with the popularity of deep learning, deep models, especially HGNNs, have become the main research direction in HG embedding.

16.5 Future Directions

HGNNs have made great progress in recent years, which clearly shows that it is a powerful and promising graph analysis paradigm. In this section, we discuss additional issues/challenges and explore a series of possible future research directions.

16.5.1 Structures and Properties Preservation

The basic success of HGNNs builds on the HG structure preservation. This also motivates many HGNNs to exploit different HG structures, where the most typical one is meta-path (Dong et al, 2017; Shi et al, 2016). Following this line, meta-graph structure is naturally considered (Zhang et al, 2018b). However, HG is far more than these structures. Selecting the most appropriate meta-path is still very challenging in the real world. An improper meta-path will fundamentally hinder the performance of HGNNs. Whether we can explore other techniques, e.g., motif (Zhao et al, 2019a; Huang et al, 2016b) or network schema (Zhao et al, 2020b) to capture HG structure is worth pursuing. Moreover, if we rethink the goal of traditional graph embedding, i.e., replacing structure information with the distance/similarity in a metric space, a research direction to explore is whether we can design HGNNs which can naturally learn such distance/similarity rather than using pre-defined meta-path/meta-graph.

As mentioned before, many current HGNNs mainly take the structures into account. However, some properties, which usually provide additional useful information to model HGs, have not been fully considered. One typical property is the dynamics of HG, i.e., a real-world HG always evolves over time. Despite that the incremental learning on dynamic HG is proposed (Wang et al, 2020m), dynamic heterogeneous graph embedding is still facing big challenges. For example, Bian et al (2019) is only proposed with a shallow model, which greatly limits its embedding ability. How can we learn dynamic heterogeneous graph embedding in HGNNs framework is worth pursuing. The other property is the uncertainty of HG, i.e., the generation of HG is usually multi-faceted and the node in a HG contains different semantics. Traditionally, learning a vector embedding usually cannot well capture such uncertainty. Gaussian distribution may innately represent the uncertainty property (Kipf and Welling, 2016; Zhu et al, 2018), which is largely ignored by current HGNNs. This suggests a huge potential direction for improving HGNNs.

16.5.2 Deeper Exploration

We have witnessed the great success and large impact of GNNs, where most of the existing GNNs are proposed for homogeneous graph (Kipf and Welling, 2017b; Veličković et al, 2018). Recently, HGNNs have attracted considerable attention (Wang et al, 2019m; Zhang et al, 2019b; Fu et al, 2020; Cen et al, 2019).

One natural question may arise that what is the essential difference between GNNs and HGNNs. More theoretical analysis on HGNNs is seriously lacking. For example, it is well accepted that the GNNs suffer from over-smoothing problem (Li et al, 2018b), so will HGNNs also have such a problem? If the answer is yes, what factor causes the over-smoothing problem in HGNNs since they usually contain multiple aggregation strategies (Wang et al, 2019m; Zhang et al, 2019b).

In addition to theoretical analysis, new technique design is also important. One of the most important directions is the self-supervised learning. It uses the pre-text tasks to train neural networks, thus reducing the dependence on manual labels (Liu et al, 2020f). Considering the actual demand that label is insufficient, self-supervised learning can greatly benefit the unsupervised and semi-supervised learning, and has shown remarkable performance on homogeneous graph embedding (Veličković et al, 2018; Sun et al, 2020c). Therefore, exploring self-supervised learning on HGNNs is expected to further facilitate the development of this area.

Another important direction is the pre-training of HGNNs (Hu et al, 2020d; Qiu et al, 2020a). Nowadays, HGNNs are designed independently, i.e., the proposed method usually works well for certain tasks, but the transfer ability across different tasks is ill-considered. When dealing with a new HG or task, we have to train HGNNs from scratch, which is time-consuming and requires a large amount of labels. In this situation, if there is a well pre-trained HGNN with strong generalization that can be fine-tuned with few labels, the time and label consumption can be reduced.

16.5.3 Reliability

Except for properties and techniques in HGs, we are also concerned about ethical issues in HGNNs, such as fairness, robustness, and interpretability. Considering that most methods are black boxes, making HGNNa reliable is an important future work.

Fairness. The embeddings learned by methods are sometimes highly related to certain attributes, e.g., age or gender, which may amplify societal stereotypes in the prediction results (Du et al, 2020). Therefore, learning fair or de-biased embeddings is an important research direction. There are some researches on the fairness of homogeneous graph embedding (Bose and Hamilton, 2019; Rahman et al, 2019). However, the fairness of HGNNs is still an unsolved problem, which is an important research direction in the future.

Robustness. Also, the robustness of HGNNs, especially the adversarial attacking, is always an important problem (Madry et al, 2017). Since many real-world applications are built based on HGs, the robustness of HGNNs becomes an urgent yet unsolved problem. What is the weakness of HGNNs and how to enhance it to improve the robustness need to be further studied.

Interpretability. Moreover, in some risk-aware scenarios, e.g., fraud detection (Hu et al, 2019b) and bio-medicine (Cao et al, 2020), the explanation of models or embeddings is important. A significant advantage of HG is that it contains rich semantics, which may provide eminent insight to promote the explanation of HGNNs. Besides, the emerging disentangled learning (Siddharth et al, 2017; Ma et al, 2019c), which divides the embedding into different latent spaces to improve the interpretability, can also be considered.

16.5.4 Applications

Many HG-based applications have stepped into the era of graph embedding. There have demonstrated the strong performance of HGNNs on E-commerce and cyber-security. Exploring more capacity of HGNNs on other areas holds great potential in the future. For example, in software engineering area, there are complex relations among test sample, requisition form, and problem form, which can be naturally modeled as HGs. Therefore, HGNNs are expected to open up broad prospects for these new areas and become a promising analytical tool. Another area is the biological system, which can also be naturally modeled as a HG. A typical biological system contains many types of objects, e.g., Gene Expression, Chemical, Phenotype, and Microbe. There are also multiple relations between Gene Expression and Phenotype (Tsuyuzaki and Nikaido, 2017). HG structure has been applied to biological system as an analytical tool, implying that HGNNs are expected to provide more promising results.

In addition, since the complexity of HGNNs are relatively large and the techniques are difficult to parallelize, it is difficult to apply the existing HGNNs to large-scale industrial scenarios. For example, the number of nodes in E-commerce recommendation may reach one billion (Zhao et al, 2019b). Therefore, successful technique deployment in various applications while resolving the scalability and efficiency challenges will be very promising.

> **Editor's Notes**: The concept of the heterogeneous graph is essentially originated from the data mining domain. Although heterogeneous graphs can usually be formulated as attributed graphs (Chapter 4), the research focus of the former is typically the frequent combinatorial patterns of node types in a subgraph (usually a path). Heterogeneous graphs represent a wide range of real-world applications which usually consist of multiple, heterogeneous data sources. For example, in recommender systems introduced in Chapter 19, we have both the "user" node and "item" node as well as higher-order patterns formed by multi-node types. Similarly, molecules and proteins as well as many networks in Natural Language Processing and Program Analysis can also be considered as heterogeneous graphs (see Chapters 21,22,24,25).

Chapter 17
Graph Neural Networks: AutoML

Kaixiong Zhou, Zirui Liu, Keyu Duan and Xia Hu

Abstract Graph neural networks (Gnns) are efficient deep learning tools to analyze networked data. Being widely applied in graph analysis tasks, the rapid evolution of GNNs has led to a growing number of novel architectures. In practice, both neural architecture construction and training hyperparameter tuning are crucial to the node representation learning and the final model performance. However, as the graph data characteristics vary significantly in the real-world systems, given a specific scenario, rich human expertise and tremendous laborious trials are required to identify a suitable GNN architecture and training hyperparameters. Recently, automated machine learning (AutoML) has shown its potential in finding the optimal solutions automatically for machine learning applications. While releasing the burden of the manual tuning process, AutoML could guarantee access of the optimal solution without extensive expert experience. Motivated from the previous successes of AutoML, there have been some preliminary automated GNN (AutoGNN) frameworks developed to tackle the problems of GNN neural architecture search (GNN-NAS) and training hyperparameter tuning. This chapter presents a comprehensive and up-to-date review of AutoGNN in terms of two perspectives, namely search space and search algorithm. Specifically, we mainly focus on the GNN-NAS problem and present the

Kaixiong Zhou
Department of Computer Science and Engineering, Texas A&M University, e-mail: zkxiong@tamu.edu

Zirui Liu
Department of Computer Science and Engineering, Texas A&M University, e-mail: tradigrada@tamu.edu

Keyu Duan
Department of Computer Science and Engineering, Texas A&M University, e-mail: k.duan@tamu.edu

Xia Hu
Department of Computer Science and Engineering, Texas A&M University, e-mail: hu@cse.tamu.edu

state-of-the-art techniques in these two perspectives. We further discuss the open problems related to the existing methods for future research.

17.1 Background

Graph neural networks (GNNs) have made substantial progress in integrating deep learning approaches to analyze graph-structured data collected from various domains, such as social networks (Ying et al, 2018b; Huang et al, 2019d; Monti et al, 2017; He et al, 2020), academic networks (Yang et al, 2016b; Kipf and Welling, 2017b; Gao et al, 2018a), and biochemical modular graphs (Zitnik and Leskovec, 2017; Aynaz Taheri, 2018; Gilmer et al, 2017; Jiang and Balaprakash, 2020). Following the common message passing strategy, GNNs apply spatial graph convolutional layer to learn a node's embedding representation via aggregating the representations of its neighbors and combining them to the node itself. A GNN architecture is then constructed by the stacking of multiple such layers and their inter-layer skip connections, where the elementary operations of a layer (e.g., aggregation & combination functions) and the concrete inter-layer connections are specified specifically in each design. To adapt to different real-world applications, a variety of GNN architectures have been explored, including GCN (Kipf and Welling, 2017b), GraphSAGE (Hamilton et al, 2017b), GAT (Veličković et al, 2018), SGC (Wu et al, 2019a), JKNet (Xu et al, 2018a), and GCNII (Chen et al, 2020l). They vary in how to aggregate the neighborhood information (e.g., mean aggregation in GCN versus neighbor attention learning in GAT) and the choices of skip connections (e.g., none connection in GCN versus initial connection in GCNII).

Despite the significant success of GNNs, their empirical implementations are usually accompanied with careful architecture engineering and training hyperparameter tuning, aiming to adapt to the different types of graph-structured data. Based on the researcher's prior knowledge and trial-and-error tuning processes, a GNN architecture is instantiated from its model space specifically and evaluated in each graph analysis task. For example, considering the underlying model Graph-SAGE (Hamilton et al, 2017b), the various-size architectures determined by the different hidden units are applied respectively for citation networks and protein-protein interaction graphs. Furthermore, the optimal skip connection mechanisms in JKNet architectures (Xu et al, 2018a) vary with the real-world tasks. Except the architecture engineering, the training hyperparameters play important roles in the final model performance, including learning rate, weight decay, and epoch numbers. In the open repositories, their hyperparameters are manually manipulated to get the desired model performances. The tedious selections of GNN architectures and training hyperparameters not only burden data scientists, but also make it difficult for beginners to access the high-performance solutions quickly for their tasks on hand.

Automated machine learning (AutoML) has emerged as a prevailing research to liberate the community from the time-consuming manual tuning processes (Chen et al, 2021). Given any task and based on the predefined search space, AutoML aims at automatically optimizing the machine learning solutions (or denoted with the term designs), including neural architecture search (NAS) and automated hyperparameter tuning (AutoHPT). While NAS targets the optimization of architecture-related parameters (e.g., the layer number and hidden units), AutoHPT indicates the selections of training-related parameters (e.g., the learning rate and weight decay). They are the sub-fields of AutoML. It has been widely reported that the novel neural architectures discovered by NAS outperform the human-designed ones in many machine learning applications, including image classification (Zoph and Le, 2016; Zoph et al, 2018; Liu et al, 2017b; Pham et al, 2018; Jin et al, 2019a; Luo et al, 2018; Liu et al, 2018b,c; Xie et al, 2019a; Kandasamy et al, 2018), semantic image segmentation (Chenxi Liu, 2019), and image generation (Wang and Huan, 2019; Gong et al, 2019). Dating back to 1900's (Kohavi and John, 1995), it has been commonly acknowledged that AutoHPT could improve over the default training setting (Feurer and Hutter, 2019; Chen et al, 2021). Motivated by the previous successful applications of AutoML, there have been some recent efforts on conjoining the researches of AutoML and GNNs (Gao et al, 2020b; Zhou et al, 2019a; You et al, 2020a; Ding et al, 2020a; Zhao et al, 2020a,g; Nunes and Pappa, 2020; Li and King, 2020; Shi et al, 2020; Jiang and Balaprakash, 2020). They generally define the automated GNN (AutoGNN) as an optimization problem and formulate their own working pipelines from three perspectives, as shown in Figure 17.1, the search space, search algorithm, and performance estimation strategy. The search space consists of a large volume of candidate designs, including GNN architectures and the training hyperparameters. On top of the search space, several heuristic search algorithms are proposed to solve the NP-complete optimization problem by iteratively approximating the well-performing designs, including random search (You et al, 2020a). The objective of performance estimation is to accurately estimate the task performance of every candidate design explored at each step. Once the search progress terminates, the best neural architecture accompanied with the suitable training hyperparameters is returned to be evaluated on the downstream machine learning task.

In this chapter, we will organize the existing efforts and illustrate AutoGNN framework with the following sections: notations, problem definition, and challenges of AutoGNN (in Sections 17.1.1, 17.1.2, and 17.1.3), search space (in Section 17.2), and search algorithm (in Section 17.3). We then present the open problems for future research in Section 17.4. Specially, since the community's interests mainly focus on discovering the powerful GNN architecture, we pay more attentions to GNN-NAS in this chapter.

17.1.1 Notations of AutoGNN

Following the previous expressions (You et al, 2020a), we use the term "design" to refer to an available solution of the optimization problem in AutoGNN. A design consists of a concrete GNN architecture and a specific set of training hy-

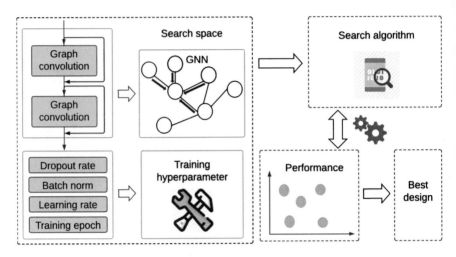

Fig. 17.1: Illustration of a general framework for AutoGNN. The search space consists of plenty of designs, including GNN architectures and the training hyperparameters. At each step, the search algorithm samples a candidate design from the search space and estimates its model performance on the downstream task. Once the search progress terminates, the best design accompanied with the highest performance on the validation set is returned and exploited for the real-world system.

perparameters. Specifically, the design is characterized by multiple dimensions, including architecture dimensions (e.g., the layer number, skip connections, aggregation, and combination functions) and hyperparameter dimensions (e.g., the learning rate and weight decay). Along each design dimension, there is a series of different elementary options provided to support the automated architecture engineering or training hyperparameter tuning. For example, we could have candidates $\{SUM, MEAN, MAX\}$ at the aggregation function dimension, and use $\{1e-4, 5e-4, 1e-3, 5e-3, 0.01, 0.1\}$ at the learning rate dimension. Given the series of candidate options along each dimension, the search space in AutoGNN is constructed by Cartesian product of all the design dimensions. A design is instantiated by assigning concrete values to these dimensions, such as a GNN architecture with the aggregation function of MEAN and learning rate of 1e-3. Note that GNN-NAS and AutoHPT explore in the search spaces consisted of expansive GNN architectures and hyperparameter combinations, respectively; AutoGNN optimizes in a more comprehensive search space containing both of them.

17.1.2 Problem Definition of AutoGNN

Before diving into detailed techniques, we examine the essence of AutoGNN by formally defining its optimization problem. To be specific, let $\mathscr{F}$ be the search space. Let $\mathscr{D}_{\text{train}}$ and $\mathscr{D}_{\text{valid}}$ be the training and validation sets, respectively. Let M be the performance evaluation metric of a design in any given graph analysis task, e.g., F1 score or accuracy in the node classification task. The objective of AutoGNN is to find the optimal design $f^* \in \mathscr{F}$ in terms of M evaluated on the validation set $\mathscr{D}_{\text{valid}}$. Formally, AutoGNN requires solving the following bi-level optimization problem:

$$
\begin{aligned}
f^* &= \text{argmax}_{f \in \mathscr{F}} \ M(f(\theta^*); D_{\text{valid}}), \\
\text{s.t.} \quad \theta^* &= \text{argmin}_\theta \ L(f(\theta); D_{\text{train}}).
\end{aligned}
\tag{17.1}
$$

where θ^* denotes the optimized trainable weights of design f and L denotes the loss function. For each design, AutoGNN will first optimize its associated weights θ by minimizing the loss on the training set through gradient descent, and then evaluates it on the validation set to decide whether this design is the optimal one. By solving the above optimization problem, AutoGNN automates the architecture engineering and training hyperparameter tuning procedure, and pushes GNN designs to examine a broad scope of candidate solutions. However, it is well known that such the bi-level optimization problem is NP-complete (Chen et al, 2021), thereby it would be extremely time-consuming for searching and evaluating the well-performing designs on large graphs with massive nodes and edges. Fortunately, there have been some heuristic search techniques proposed to locate the local optimal design (e.g., CNN or RNN architecture) as close as possible to the global one in the applications of image classification and natural language processing, including reinforcement learning (RL) (Zoph and Le, 2016; Zoph et al, 2018; Pham et al, 2018; Cai et al, 2018a; Baker et al, 2016), evolutionary methods (Liu et al, 2017b; Real et al, 2017; Miikkulainen et al, 2019; Xie and Yuille, 2017; Real et al, 2019), and Bayesian optimization (Jin et al, 2019a). They iteratively explore the next design and update the search algorithm based on the performance feedback of the new design, in order to move toward the global optimal solution. Compared with the previous efforts, the characteristics of AutoGNN problem could be viewed from two aspects: the search space and search algorithms tailored to identify the optimal design of GNN. In the following sections, we list the challenge details and the existing AutoGNN work.

17.1.3 Challenges in AutoGNN

The direct application of existing AutoML frameworks to automate GNN designs is non-trivial, due to the two major challenges as follows.

First, the search space of AutoGNN is significantly different from the ones in the AutoML literature. Taking NAS applied in discovering CNN architectures (Zoph and Le, 2016) as an example, the search space of convolution operation is mainly specified by the convolutional kernel size. In contrast, considering the message-passing based graph convolution, the search space of spatial graph convolution is constructed by multiple key architecture dimensions, including aggregation, combination, and embedding activation functions. With the growing number of GNN model variants, it is important to formulate a good search space being both expressive and compact. On the one hand, the search space should cover the important architecture dimensions to subsume the existing human-designed architectures and adapt to a series of diverse graph analysis tasks. On the other hand, the search space should be compact by excluding the non-general dimensions and incorporating modest ranges of options along each dimension, in order to save the search time cost.

Second, the search algorithm should be tailored to discover the well-performing design efficiently based on the special search space in AutoGNN. The search controller determines how to iteratively explore the search space and update the search algorithm according to the performance feedbacks of sampled designs. A good controller needs to balance the trade-off between exploration and exploitation during the search progress, in order to avoid the premature sub-optimal region and quickly discover the well-performing designs, respectively. However, the previous search algorithms may be inefficient to the application of GNN-NAS. Specially, one of the key properties in GNN architectures is that the model performance may vary significantly with a slight modification along an architecture dimension. For example, it has been theoretically and empirically demonstrated that the graph classification accuracy could be improved by simply replacing the max pooling with summation in the aggregation function dimension of GNN (Xu et al, 2019d). The previous RL-based methods sample and evaluate the whole architecture at each search step. It would be hard for the search algorithms to learn the following relationship towards exploring better GNN: which part of the architecture dimension modifications improves or degrades the model performance. Another challenging problem is the surge of new graph analysis tasks, which requires huge computation resources to optimize GNN architectures. Instead of searching the optimal GNN from scratch, it is crucial to transfer the well-performing architectures discovered before to the new task to save the expensive computation cost.

17.2 Search Space

In this section, we summarize the search spaces in literature. As shown in Figure 17.2, the search spaces of designs in AutoGNN are differentiated according to GNN architectures and training hyperparameters, whose details are listed as below.

17.2.1 Architecture Search Space

Considering the existing AutoGNN frameworks (Gao et al, 2020b; Zhou et al, 2019a), GNN model is commonly implemented based on the spatial graph convolution mechanism. To be specific, the spatial graph convolution takes the input graph as a computation graph and learns node embeddings by passing messages along edges. A node embedding is updated recursively by aggregating the embedding representations of its neighbors and combining them to the node itself. Formally, the k-th spatial graph convolutional layer of GNN could be expressed as:

$$
\begin{aligned}
\mathbf{h}_i^{(k)} &= \text{AGGREGATE}(\{a_{ij}^{(k)} W^{(k)} \mathbf{x}_j^{(k-1)} : j \in \mathcal{N}(i)\}), \\
\mathbf{x}_i^{(k)} &= \text{ACT}(\text{COMBINE}(W^{(k)} \mathbf{x}_i^{(k-1)}, \mathbf{h}_i^{(k)})).
\end{aligned}
\tag{17.2}
$$

$\mathbf{x}_i^{(k)}$ denotes the embedding vector of node v_i at the k-th layer. $\mathcal{N}(i)$ denotes the set of neighbors adjacent to node v_i. $W^{(k)}$ denotes the trainable weight matrix used to project node embeddings. $a_{ij}^{(k)}$ denotes the message-passing weight along edge connecting nodes v_i and v_j, which is determined by normalized graph adjacency matrix or learned from attention mechanism. Function AGGREGATE, such as mean, max, and sum pooling, is used to aggregate neighbor representations. Function COMBINE is used to combine neighbor embedding $\mathbf{h}_i^{(k)}$ as well as node embedding $\mathbf{x}_i^{(k-1)}$ from the last layer. Finally, function ACT (e.g., ReLU) is used to add non-linearity to the embedding learning.

As shown in Figure 17.2, GNN architecture consists of several graph convolutional layers defined in Eq. equation 17.2, and may incorporate skip connection between any two arbitrary layers similar to residual CNN (He et al, 2016a). Following the previous definitions in NAS, we use the term "micro-architecture" to represent a graph convolutional layer, including the specifications of hidden units and graph convolutional functions; we use the term "macro-architecture" to represent network topology, including the choices of layer depth, inter-layer skip connections, and pre/post-processing layers. The architecture search space contains a large volume of diverse GNN architectures, which could be categorized into the search spaces of micro-architectures as well as macro-architectures.

17.2.1.1 Micro-architecture Search Space

According to Eq. equation 17.2 and as shown in Figure 17.2, the micro-architecture of a graph convolutional layer is characterized by the following five architecture dimensions:

- **Hidden units:** Trainable matrix $W^{(k)} \in \mathbb{R}^{d^{(k-1)} \times d^{(k)}}$ maps node embeddings to a new space and learns to extract the informative features. $d^{(k)}$ is the number of hidden units and plays key role in the task performance. In the GNN-NAS

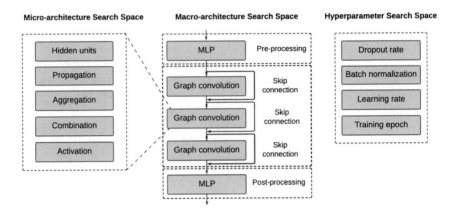

Fig. 17.2: Illustration of a comprehensive search space, which consists of micro-architecture, macro-architecture, and training hyperpameter search spaces. Each space is characterized by multiple dimensions, such as hidden units, propagation function, etc, in the micro-architecture search space. Each dimension provides a series of candidate options, and the search space is constructed by Cartesian product of all its dimensions. A discrete point in the comprehensive search space represents a specific design, which adopts one option at each dimension.

frameworks of GraphNAS (Gao et al, 2020b) and AGNN (Zhou et al, 2019a), $d^{(k)}$ is usually selected from set $\{4, 8, 16, 32, 64, 128, 256\}$.

- **Propagation function:** It determines the message-passing weight $a_{ij}^{(k)}$ to specify how node embeddings are propagated upon the input graph structure. In a wide variety of GNN models (Kipf and Welling, 2017b; Wu et al, 2019a; Hamilton et al, 2017b; Ding et al, 2020a), $a_{ij}^{(k)}$ is defined by the corresponding element from the normalized adjacency matrix: $\tilde{D}^{-\frac{1}{2}}\tilde{A}\tilde{D}^{-\frac{1}{2}}$ or $\tilde{D}^{-1}\tilde{A}$, where $\tilde{A}$ is the self-loop graph adjacency matrix and $\tilde{D}$ is its degree matrix, respectively. Note that the real-world graph-structured data could be both complex and noisy (Lee et al, 2019c), which leads to the inefficient neighbor aggregation. GAT (Veličković et al, 2018) applies attention mechanism to compute $a_{ij}^{(k)}$ to attend on relevant neighbors. Based on the existing GNN-NAS frameworks (Gao et al, 2020b; Zhou et al, 2019a; Ding et al, 2020a), we list the common choices of propagation functions in Table 17.1.
- **Aggregation function:** Depending on the input graph structure, a proper application of aggregation function is important to learn the informative neighbor distribution (Xu et al, 2019d). For example, a mean pooling function takes the average of neighbors, while a max pooling only preserves the significant one. The aggregation function is usually selected from set $\{\text{SUM}, \text{MEAN}, \text{MAX}\}$.
- **Combination function:** It is used to combine neighbor embedding $\mathbf{h}_i^{(k)}$ and projected embedding $W^{(k)}\mathbf{x}_i^{(k-1)}$ of the node itself. Examples of combination

function include sum and multiple layer perceptron (MLP), etc. While the sum operation simply adds the two embeddings, MLP further applies linear mapping based upon the summation or concatenation of these two embeddings.

- **Activation function:** The candidate activation function is usually selected from {Sigmoid, Tanh, ReLU, Linear, Softplus, LeakyReLU, ReLU6, ELU}.

Given the above five architecture dimensions and their associated candidate options, the micro-architecture search space is constructed by their Cartesian product. Each discrete point in the micro-architecture search space corresponds to a concrete micro-architecture, e.g., a graph convolutional layer with {Hidden units: 64, Propagation function: GAT, aggregation function: SUM, combination function: MLP, Activation function: ReLU}. By providing the extensive candidate options along each dimension, the micro-architecture search space covers most of layer implementations in the state-of-the-art models, such as Chebyshev (Defferrard et al, 2016), GCN (Kipf and Welling, 2017b), GAT (Veličković et al, 2018), and LGCN (Gao et al, 2018a).

Table 17.1: Propagation function candidates to compute weight $a_{ij}^{(k)}$ if nodes v_i and v_j are connected; otherwise $a_{ij}^{(k)} = 0$. Symbol $||$ denotes the concatenation operation, $\mathbf{a}$, $\mathbf{a}_l$ and $\mathbf{a}_r$ denote trainable vectors, and $W_G^{(k)}$ is a trainable matrix.

Propagation Types	Propagation functions	Equations
Normalized adjacency	$\tilde{A}$	1
	$\tilde{D}^{-\frac{1}{2}}\tilde{A}\tilde{D}^{-\frac{1}{2}}$	$\frac{1}{\sqrt{|\mathcal{N}(i)||\mathcal{N}(j)|}}$
	$\tilde{D}^{-1}\tilde{A}$	$\frac{1}{|\mathcal{N}(i)|}$
Attention mechanism	GAT	$\text{LeakyReLU}(\mathbf{a}^\top(W^{(k)}\mathbf{x}_i^{(k-1)}||W^{(k)}\mathbf{x}_j^{(k-1)}))$
	SYM-GAT	$a_{ij}^{(k)} + a_{ji}^{(k)}$ based on GAT
	COS	$\mathbf{a}^\top(W^{(k)}\mathbf{x}_i^{(k-1)}||W^{(k)}\mathbf{x}_j^{(k-1)})$
	LINEAR	$\tanh(\mathbf{a}_l^\top W^{(k)}\mathbf{x}_i^{(k-1)} + \mathbf{a}_r^\top W^{(k)}\mathbf{x}_j^{(k-1)})$
	GERE-LINEAR	$W_G^{(k)}\tanh(W^{(k)}\mathbf{x}_i^{(k-1)} + W^{(k)}\mathbf{x}_j^{(k-1)})$

17.2.1.2 Macro-architecture Search Space

Besides the micro-architecture, another architectural level of GNN is its macro-architecture as shown in Figure 17.2, i.e., the network topology. The macro-architecture of GNN specifies the numbers of graph convolutional layers as well as pre/post-processing layers, and the choices of skip connections (You et al, 2020a; Li et al, 2018b, 2019c). We list the details of these four architecture dimensions in the following.

- **Graph convolutional layer depth:** The direct stacking of multiple layers is commonly adopted to improve the reception fields of nodes. Let l_{gc} denote the number of graph convolutional layers. l_{gc} is usually selected from range $[2, 10]$.
- **Pre-processing layer depth:** In real-world applications, the length of nodes' input features may be too large and leads to costly computation in hidden feature learning. The feature pre-processing is included in search space (You et al, 2020a) for the first time and conducted by MLP, whose layer number is denoted as l_{pre}. l_{pre} is sampled from candidates $\{0, 1, 2, 3\}$.
- **Post-processing layer depth:** Similarly, the post-processing layers of MLP are applied to project hidden embeddings into task-specific space, e.g., the embedding space with dimensions the same as class labels in the node classification task. Let l_{post} denote the layer number with examples $\{0, 1, 2, 3\}$.
- **Skip connections:** Following the residual deep CNNs in computer vision and the recent deep GNNs, skip connections have been incorporated in the search space of GNN-NAS frameworks (You et al, 2020a; Zhao et al, 2020g,a). To be specific, at layer l, the embeddings of up to $l - 1$ previous layers could be sampled and combined to the current layer's output, leading to 2^{k-1} possible decisions at layer k. For the prior node embeddings that are connected to the current output, there have been a series of candidate options developed to combine them, namely $\{SUM, CAT, MAX, LSTM\}$. Specially, option SUM, CAT or MAX adds, concatenates or element-wisely max pools these connected embeddings. LSTM uses an attention mechanism to compute the importance score of each layer, and then obtain the weighted average of the connected embeddings (Xu et al, 2018a).

The entire architecture space is constructed by Cartesian product of the micro and macro-architecture search spaces, which is totally characterized by the nine architecture dimensions. It could be extremely huge and comprehensive to subsume the recent residual GNN models, such as JKNet (Xu et al, 2018a) and deeperGCN (Li et al, 2018b).

17.2.2 Training Hyperparameter Search Space

The training hyperparameters have significant impacts on the task performances of GNN architectures, and have been explored in AutoGNN frameworks (You et al, 2020a; Shi et al, 2020). We summarize four important dimensions of training hyperparameters in the following and show them in Figure 17.2.

- **Dropout rate:** At the beginning of each graph convolutional layer or pre/post-processing layer, a proper dropout rate is crucial to avoid the over-fitting issue. The widely-used examples are $\{False, 0.05, 0.1, 0.2, 0.3, 0.4, 0.5, 0.6\}$.
- **Batch normalization:** It is applied after graph convolutional layer or pre/post-processing layer to normalize node embeddings of the whole graph or a batch (Zhou et al, 2020d; Zhao and Akoglu, 2019; Ioffe and Szegedy, 2015). The candidate

normalization techniques include $\{$False, BatchNorm $(Ioffe\ and\ Szegedy,\ 2015)$, PairNorm $(Zhao\ and\ Akoglu,\ 2019)$, DGN $(Zhou\ et\ al,\ 2020d)$, NodeNorm $(Zhou\ et\ al,\ 2020c)$, GraphNorm $(Cai\ et\ al,\ 2020d)\}$.

- **Learning rate**: While a larger learning rate leads to a premature suboptimal solution, a smaller one will make the optimization process converge slowly. The candidate learning rates are $\{$1e-4, 5e-4, 1e-3, 5e-3, 0.01, 0.1$\}$.
- **Training epoch**: According to the common practice (You et al, 2020a; Kipf and Welling, 2017b), the training epoch examples are $\{100, 200, 400, 500, 1000\}$.

17.2.3 Efficient Search Space

Given the micro-architecture, macro-architecture, and training hyperparameters search spaces, in the practical systems, the applied search space is formulated by Cartesian product of any combination of them. Although a large search space subsumes the diverse GNN architectures and training environments to adapt to the different graph analysis tasks, it would be time-consuming to explore the optimal design. To make the search progress efficient, there are two mainstream simplifying search spaces applied in the existing AutoGNN frameworks.

- **Focus on GNN-NAS:** Instead of fully tuning the training hyperparameters, most of AutoGNN (or GNN-NAS) frameworks (Gao et al, 2020b; Zhou et al, 2019a; Zhao et al, 2020a,g; Ding et al, 2020a; Nunes and Pappa, 2020; Li and King, 2020; Jiang and Balaprakash, 2020) focus on tackling the problem of discovering the well-performing GNN architectures. Comparing with AutoHPT, it is commonly acknowledged that a novel architecture discovered from GNN-NAS is more important and challenging to the research community, which could motivate the data scientist to improve GNN model paradigms in the future. In GNN-NAS, the search space is thus reduced to the one containing only the neural architecture variants.
- **Simplify architecture search space:** Even in GNN-NAS, the plenty of architecture dimensions and their associated candidate options still make the search space complex. Based on the prior knowledge about the impacts of different modules on model performances, one would prefer to explore only along the crucial architecture dimensions in the practical systems. For example, it is found that the simplified search space (Zhao et al, 2020a) characterized by aggregation function and skip connections could generate the high-performance GNN architectures comparable to ones from the comprehensive search spaces (Gao et al, 2020b; Zhou et al, 2019a). Specially, since the decision cardinality of skip connections increases exponentially with layers, the simplified search space even only explores the skip connections in the last layer similar to JKNet (Xu et al, 2018a). In another simplified search space, the model-specific architecture dimensions are excluded and pre-defined based on expert experiences, including the hidden units, propagation function, and combination function.

17.3 Search Algorithms

Many different search strategies can be used to explore the search space in AutoGNN, including random search, evolutionary methods, RL, and differentiable search methods. In this section, we will introduce the basic concepts of these search algorithms and how to utilize them to explore candidate designs.

17.3.1 Random Search

Given a search space, random search randomly samples the various designs with equal probability. The random search is the most basic approach, yet it is quite effective in practice. In addition to serve as a baseline in AutoGNN works (Zhou et al, 2019a; Gao et al, 2020b), random search is the standard benchmark for comparing the effectiveness of different candidate options along a dimension in the search space (You et al, 2020a). Specially, suppose the dimension to be evaluated is batch normalization, whose candidate examples are given by {False, BatchNorm}. To comprehensively compare the effectiveness of these two options, a series of diverse designs are randomly sampled from the search space, where the batch normalization is reset to False and BatchNorm in each design, respectively. Each pair of designs (referred to Normalization=False and Normalization=BatchNorm) are compared in terms of their model performances on a downstream graph analysis task. It is found that the designs with Normalization=BatchNorm generally rank higher than the others, which indicates the benefit of including BatchNorm in the model design.

17.3.2 Evolutionary Search

Evolutionary methods evolve a population of designs, i.e., the set of different GNN architectures and training hyperparameters. In every evolution step, at least one design from the population is sampled and serves as a parent to generate a new child design by applying mutations to it. In the context of AutoGNN, the design mutations are local operations, such as changing the aggregation function from MAX to SUM, altering the hidden units, and altering a specific training hyperparameter. After training the child design, its performance is evaluated on the validation set. The superior design will be added to the population. Specifically, Shi et al (2020) proposes to select two parent designs and then crossover them along some dimensions. To generate the diverse child designs, Shi et al (2020) further mutates the above crossover designs.

17.3.3 Reinforcement Learning Based Search

RL (Silver et al, 2014; Sutton and Barto, 2018) is a learning paradigm concerned with how agents ought to take actions in an environment to maximize the reward. In the context of AutoGNN, the agent is the so-called "controller", which tries to generate promising designs. The generation of design can be regarded as the controller's action. The controller's reward is often defined as the model performance of generated design on the validation set, such as validation accuracy for the node classification task. The controller is trained in a loop as shown in Figure 17.3: the controller first samples a candidate design and trains it to convergence to measure its performance on the task of desire. Note that the controller is usually realized by RNN, which generates the design of GNN architecture and training hyperparameters as a string of variable strength. The controller then uses the performance as a guiding signal to update itself toward finding the more promising design in the future search progress.

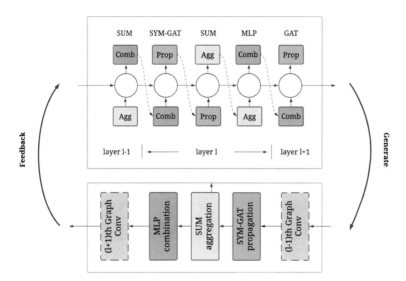

Fig. 17.3: A illustration of reinforcement learning based search algorithm. The controller (upper block) generates a GNN architecture (lower block) and tests it on the validation dataset. By treating the architecture as a string with variable length, the controller usually applies RNN to sequentially sample options in the different dimensions (e.g., combination, aggregation, and propagation functions) to formulate the final GNN architecture. The validation performance is then used as feedback to train the controller. Note that the architecture dimensions here are just used for the illustration purpose. Please refer to Section 17.2 for a complete introduction of the search space.

The existing RL-based AutoGNN frameworks target at the sub-field problem of GNN-NAS. Generally, in RL-based GNN-NAS, there are two sets of trainable parameters: the parameters of the controller, denoted by ω, and the parameters of a GNN architecture, denoted by θ. The training procedure consists of two interleaving phases, which alternatively solves the bi-level optimization problem as shown in Eq. equation 17.1. The first phase trains θ on the training data set $\mathscr{D}_{train}$ with a fixed number of epochs using standard back-propagation. The second phase trains ω to learn to sample high-performance GNN architectures evaluated on the validation set $\mathscr{D}_{valid}$. These two phases are alternated during the training. Specifically, in the first phase, the controller proposes a GNN architecture f and performs gradient descent on θ to minimize the loss function $\mathscr{L}(f(\theta); \mathscr{D}_{train})$, which is computed on the batches of training data. In the second phase, the optimized parameter θ^* is fixed to update the controller parameters ω, aiming to maximize the expected reward:

$$\omega^* = \text{argmax}_\omega \, \mathbb{E}_{f \sim \pi(f;\omega)}[\mathscr{R}(f(\theta^*); \mathscr{D}_{valid})]. \tag{17.3}$$

Here, $\pi(f;\omega)$ is the controller's policy parameterized by ω to sample and generate GNN architecture f. The reward $\mathscr{R}(f(\theta^*); \mathscr{D}_{valid})$ is the model performance defined by the task of desire, such as the accuracy for the node classification task. Furthermore, the reward is computed on the validation set, rather than on the training set, to encourage the controller to select architectures that generalize well. In most of the existing work, the gradient of the expected reward $\mathbb{E}_{f \sim \pi(f;\omega)}[\mathscr{R}(f(\theta^*); \mathscr{D}_{valid})]$ with respect to ω is computed using REINFORCE rule (Sutton et al, 2000).

Considering GNN-NAS efforts in literature, RL-based search algorithms differ in how they represent and train the controller. GraphNAS uses an RNN controller to sequentially sample from the multiple architecture dimensions and generate a string that encodes a GNN architecture (Gao et al, 2020b). Based on the expected reward signaling the quality of the whole architecture, the RNN controller has to optimize the sampling policies along all the dimensions. AGNN (Zhou et al, 2019a) is motivated by an observation that the minor modification to an architecture dimension can lead to abrupt change in performance. For example, the graph classification accuracy of GNN may be significantly improved by only changing the choice of aggregation function from MAX to SUM (Xu et al, 2019d). Based on this observation, AGNN proposes a more efficient controller consisted of a series of RNN sub-controllers, each corresponding to an independent architecture dimension. At each step, AGNN only applies one of the RNN sub-controllers to sample new options from the corresponding dimension, and uses these options to mutate the best architecture found so far. By evaluating such a slightly-mutated design, the RNN sub-controller can exclude the noises generated from the other architecture dimension modifications, and better trains the sampling policy of its own dimension.

17.3.4 Differentiable Search

There are several candidate options along each architecture dimension. For example, for the aggregation function at a particular layer, we have the option of applying either a SUM, a MEAN, or a MAX pooling. The common search approaches in GNN-NAS, such as random search, evolutionary algorithms, and RL-based search methods, treat selecting the best option as a black-box optimization problem over a discrete domain. At each search step, they sample and evaluate a single architecture from the discrete architecture search space. However, such the search process towards well-performing GNNs will be very time-consuming since the number of possible models is extremely large. Differentiable search algorithms relax the discrete search space to be continuous, which can be optimized efficiently by gradient descent. Specifically, for each architecture dimension, the differentiable search algorithms usually relax the hard choice from the candidate set into a continuous distribution, where each option is assigned with a probability. One example for illustrating the differentiable search along the aggregation function dimension is shown in Figure 17.4. At the k-th layer, the node embedding output of aggregation function can be decomposed and expressed as:

$$
\mathbf{h}_i^{(k)} = \begin{cases} \sum_m \alpha_m o_m(\mathbf{x}_j^{(k-1)} : j \in \mathcal{N}(i) \cup \{i\}), \\ \text{or} \\ \alpha_m o_m(\mathbf{x}_j^{(k-1)} : j \in \mathcal{N}(i) \cup \{i\}), \ m \sim p(\alpha_m), \end{cases} \tag{17.4}
$$
$$
\text{s.t.} \ \sum_m \alpha_m = 1.
$$

o_m represents the m-th aggregation function option, and α_m is the sampling probability associated with the corresponding option. The probability distribution along a dimension is regularized to have the sum of one. The architecture distribution is then formulated by the union probability distribution of all the dimensions. At each search step, as shown in Eq.equation 17.4 (with the example of the aggregation function dimension), the real operation of a dimension in a new architecture could be generated by two different ways: weighted option combination and option sampling. For the case of weighted option combination, the real operation is represented by the weighted average of all candidate options. For the other case, the real operation is instead sampled from the probability distribution $p(\alpha_m)$ of the corresponding architecture dimension. In both cases, the adopted options are scaled by their sampling probabilities to support the architecture distribution optimization by gradient descent. The architecture distribution is then updated directly by backpropagating the training loss at each training step. During the testing, the discrete architecture can be obtained by retaining the strongest candidate with the highest probability α_m along each dimension. In contrast to black-box optimization, gradient-based optimization is significantly more data efficient, and hence greatly speeds up the search process.

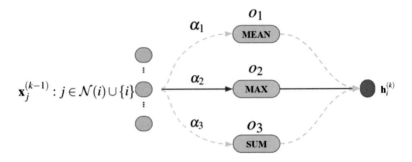

Fig. 17.4: One example for illustrating the differentiable search for the aggregation function. At a search step, the aggregation function is given by the weighted combination of the three candidates, or instead realized by one sampled option (e.g., MAX scaled with probability α_2). Once the search progress terminates, the option with the highest probability (e.g., MAX with solid arrow) is used in the final architecture to be evaluated on testing set.

Compared with RL-based search, differentiable search based algorithm is less popular in the GNN-NAS literature. PDNAS (Zhao et al, 2020g) relaxes the discrete search space into a continuous one by employing the Gumbel-sigmoid, enabling optimization via gradient descent. POSE focuses on searching the propagation function, whose discrete search space is relaxed by a softmax approximation.

17.3.5 Efficient Performance Estimation

To solve the bi-level optimization problem of AutoGNN, all the above search algorithms share a common two-stage working pipeline: sampling a new design and adjusting the search algorithm based on the performance estimation of the new design at each step. Once the search progress terminates, the optimal design with the highest model performance will be treated as the desired solution to the concerned optimization problem. Therefore, an accurate performance estimation strategy is crucial to AutoGNN framework. The simplest way of performance estimation is to perform a standard training for each generated design, and then obtain the model performance on the split validation set. However, such an intuitive strategy is computationally expensive given the long search progress and massive graph datasets.

Parameter sharing is one of the efficient strategies to reduce the cost of performance estimation, which avoids training from scratch for each design. Parameter sharing is first proposed in ENAS (Pham et al, 2018) to force all designs to share weights to improve efficiency. A new design could be immediately estimated by reusing the weights well trained before. However, such a strategy cannot be directly adopted in GNN-NAS since the GNN architectures in search space may have weights with different dimensions or shapes. To tackle the challenge, recent work modified the parameter sharing strategy to customize for GNNs. GraphNAS (Gao et al, 2020b) categorizes and stores the optimized weights based on their shapes, and applies the one with the same shape to the new design. After parameter sharing, AGNN (Zhou et al, 2019a) further uses a few training epochs to fully adapt the transferred weights to the new design. In the differentiable GNN-NAS frameworks, the parameter sharing is conducted naturally between GNN architectures sharing the common computation options (Zhao et al, 2020g; Ding et al, 2020a).

17.4 Future Directions

We have reviewed various search spaces and search algorithms. Although some initial AutoGNN efforts have been paid, compared with the rapid development of AutoML in computer vision, AutoGNN is still in the preliminary research stage. In this section, we discuss several future directions, especially for research on GNN-NAS.

- **Search space.** The design of architecture search space is the most important portion in GNN-NAS framework. An appropriate search space should be comprehensive by covering the key architecture dimensions and their state-of-the-art primitive options to guarantee the performance of searched architecture for any given task. Besides, the search space should be compact by incorporating a moderate number of powerful options to make the search progress efficient. However, most of the existing architecture search spaces are constructed based on vanilla GCN and GAT, failing to consider the recent GNN developments. For example, graph pooling (Ying et al, 2018c; Gao and Ji, 2019; Lee et al, 2019b; Zhou et al, 2020e) has attracted increasing research interests to enable encoding the graph structures hierarchically. Based on the wide variety of pooling algorithms, the corresponding hierarchical GNN architectures gradually shrink the graph size and enhance the neighborhood reception field, empirically improving the downstream graph analysis tasks. Furthermore, a series of novel graph convolution mechanisms have been proposed from different perspectives, such as neighbor-sampling methods to accelerate computation (Hamilton et al, 2017b; Chen et al, 2018c; Zeng et al, 2020a), and PageRank based graph convolutions to extend neighborhood size (Klicpera et al, 2019a,a; Bojchevski et al, 2020b). With the development in GNN community, it is crucial to update the search space to subsume the state-of-the-art models.
- **Deep graph neural networks.** All the existing search spaces are implemented with shallow GNN architectures, i.e., the number of graph convolutional layers $l_{gc} \leq 10$. Unlike the widely adopted deep neural networks (e.g., CNNs and transformers) in computer vision and natural language processing, GNN architectures are usually limited with less than 3 layers (Kipf and Welling, 2017b; Veličković et al, 2018). As the layer number increases, the node representations will converge to indistinguishable vectors due to the recursive neighborhood aggregation and non-linear activation (Li et al, 2018b; Oono and Suzuki, 2020). Such phenomenon is recognized as the over-smoothing issue (NT and Maehara,

2019), which prevents the construction of deep GNNs from modeling the dependencies to high-order neighbors. Recently, many efforts have been proposed to relieve the over-smoothing issue and construct deep GNNs, including embedding normalization (Zhao and Akoglu, 2019; Zhou et al, 2020d; Ioffe and Szegedy, 2015), residual connection (Li et al, 2019c, 2018b; Chen et al, 2020l; Klicpera et al, 2019a), and random data augmentation (Rong et al, 2020b; Feng et al, 2020). However, most of them only achieve comparable or even worse performance compared to their corresponding shallow models. By incorporating these new techniques into the search space, GNN-NAS could effectively combine them and identify the novel deep GNN model, which unleashes the deep learning power for graph analytics.

- **Applications to emerging graph analysis tasks.** One limitation of GNN-NAS frameworks in literature is that they are usually evaluated on a few benchmark datasets, such as Cora, Citeseer, and Pubmed for node classification (Yang et al, 2016b). However, the graph-structured data is ubiquitous, and the novel graph analysis tasks are always emerging in real-world applications, such as property prediction of biochemical molecules (i.e., graph classification) (Zitnik and Leskovec, 2017; Aynaz Taheri, 2018; Gilmer et al, 2017; Jiang and Balaprakash, 2020), item/friend recommendation in social networks (i.e., link prediction) (Ying et al, 2018b; Monti et al, 2017; He et al, 2020), and circuit design (i.e., graph generation) (Wang et al, 2020b; Li et al, 2020h; Zhang et al, 2019d). The surge of novel tasks poses significant challenges for the future search of well-performing architectures in GNN-NAS, due to the diverse data characteristics and objectives of tasks and the expensive searching cost. On one hand, since the new tasks may do not resemble any of the existing benchmarks, the search space has to be re-constructed by considering their specific data characteristics. For example, in the knowledge graph with informative edge attributes, the micro-architecture search space needs to incorporate edge-aware graph convolutional layers to guarantee a desired model performance (Schlichtkrull et al, 2018; Shang et al, 2019). On the other hand, if the new tasks are similar to the existing ones, the search algorithms could re-exploit the best architectures discovered before to accelerate the search progress in the new tasks. For example, one can simply initialize the search progress with these sophisticated architectures and uses several epochs to explore the potentially good ones within a small region. Especially for the massive graphs with a large volume of nodes and edges, the reuse of well-performing architectures from similar tasks could significantly save the computation cost. The research challenge is how to quantify the similarities between the different graph-structured data.

Acknowledgements

This work is, in part, supported by NSF (#IIS-1750074 and #IIS-1718840). The views, opinions, and/or findings contained in this paper are those of the authors and should not be interpreted as representing any funding agencies.

Editor's Notes: Automated graph neural networks introduce automated machine learning to tackle the problem of GNN neural architecture search and hyperparameter search. Hence, this chapter is orthogonal to most of the other chapters in this book, which generally depend on expert experience to design specific models and tune hyperparameters. Neural architecture search space contains the components of manually designed models, such as kinds of aggregators introduced in chapter 4 and chapter 5. Automated graph neural networks support common graph analysis tasks, such as node classification (chapter 4), graph classification (chapter 9), and link prediction (chapter 10).

Chapter 18
Graph Neural Networks: Self-supervised Learning

Yu Wang, Wei Jin, and Tyler Derr

Abstract Although deep learning has achieved state-of-the-art performance across numerous domains, these models generally require large annotated datasets to reach their full potential and avoid overfitting. However, obtaining such datasets can have high associated costs or even be impossible to procure. Self-supervised learning (SSL) seeks to create and utilize specific pretext tasks on unlabeled data to aid in alleviating this fundamental limitation of deep learning models. Although initially applied in the image and text domains, recent interest has been in leveraging SSL in the graph domain to improve the performance of graph neural networks (GNNs). For node-level tasks, GNNs can inherently incorporate unlabeled node data through the neighborhood aggregation unlike in the image or text domains; but they can still benefit by applying novel pretext tasks to encode richer information and numerous such methods have recently been developed. For GNNs solving graph-level tasks, applying SSL methods is more aligned with other traditional domains, but still presents unique challenges and has been the focus of a few works. In this chapter, we summarize recent developments in applying SSL to GNNs categorizing them via the different training strategies and types of data used to construct their pretext tasks, and finally discuss open challenges for future directions.

Yu Wang
Department of Electrical Engineering and Computer Science, Vanderbilt University, e-mail: yu.wang.1@vanderbilt.edu

Wei Jin
Department of Computer Science and Engineering, Michigan State University, e-mail: jinwei2@msu.edu

Tyler Derr
Department of Electrical Engineering and Computer Science, Vanderbilt University, e-mail: tyler.derr@vanderbilt.edu

© The Author(s), under exclusive license to Springer Nature Singapore Pte Ltd. 2022
L. Wu et al. (eds.), *Graph Neural Networks: Foundations, Frontiers, and Applications*,
https://doi.org/10.1007/978-981-16-6054-2_18

18.1 Introduction

Recent years have witnessed the great success of applying deep learning in numerous fields. However, the superior performance of deep learning heavily depends on the quality of the supervision provided by the labeled data and collecting a large amount of high-quality labeled data tends to be time-intensive and resource-expensive (Hu et al, 2020c; Zitnik and Leskovec, 2017). Therefore, to alleviate the demand for massive labeled data and provide sufficient supervision, self-supervised learning (SSL) has been introduced. Specifically, SSL designs domain-specific pretext tasks that leverage extra supervision from unlabeled data to train deep learning models and learn better representations for downstream tasks. In computer vision, various pretext tasks have been studied, e.g., predicting relative locations of image patches (Noroozi and Favaro, 2016) and identifying augmented images generated from image processing techniques such as cropping, rotating and resizing (Shorten and Khoshgoftaar, 2019). In natural language processing, self-supervised learning has also been heavily utilized, e.g., predicting the masked word in BERT (Devlin et al, 2019).

Simultaneously, graph representation learning has emerged as a powerful strategy for analyzing graph-structured data over the past few years (Hamilton, 2020). As the generalization of deep learning to the graph domain, Graph Neural Networks (GNNs) has become one promising paradigm due to their efficiency and strong performance in real-world applications (You et al, 2021; Zitnik and Leskovec, 2017). However, the vanilla GNN model (i.e., Graph Convolutional Network (Kipf and Welling, 2017b)) and even more advanced existing GNNs (Hamilton et al, 2017b; Xu et al, 2019d, 2018a) are mostly established in a semi-supervised or supervised manner, which still requires high-cost label annotation. Additionally, these GNN models may not take full advantage of the abundant information in unlabeled data, such as the graph topology and node attributes. Hence, SSL can be naturally harnessed for GNNs to gain additional supervision and thoroughly exploit the information in the unlabeled data.

Compared with grid-based data such as images or text (Zhang et al, 2020e), graph-structured data is far more complex due to its highly irregular topology, involved intrinsic interactions and abundant domain-specific semantics (Wu et al, 2021d). Different from images and text where the entire structure represents a single entity or expresses a single semantic meaning, each node in the graph is an individual instance with its own features and positioned in its own local context. Furthermore, these individual instances are inherently related with each other, which forms diverse local structures that encode even more complex information to be discovered and analyzed. While such complexity engenders tremendous challenges in analyzing graph-structured data, the substantial and diverse information contained in the node features, node labels, local/global graph structures, and their interactions and combinations provide golden opportunities to design self-supervised pretext tasks.

Embracing the challenges and opportunities to study self-supervised learning in GNNs, the works (Hu et al, 2020c, 2019c; Jin et al, 2020d; You et al, 2020c) have been the first research that systematically design and compare different self-supervised pretext tasks in GNNs. For example, the works (Hu et al, 2019c; You et al, 2020c) design pretext tasks to encode the topological properties of a node such as centrality, clustering coefficient, and its graph partitioning assignment, or to encode the attributes of a node such as individual features and clustering assignments in embeddings output by GNNs. The work (Jin et al, 2020d) designs pretext tasks to align the pairwise feature similarity or the topological distance between two nodes in the graph with the closeness of two nodes in the embedding space. Apart from the supervision information employed in creating pretext tasks, designing effective training strategies and selecting reasonable loss functions are another crucial components in incorporating SSL into GNNs. Two frequently used training strategies that equip GNNs with SSL are 1) pre-training GNNs through completing pretext task(s) and then fine-tuning the GNNs on downstream task(s), and 2) jointly training GNNs on both pretext and downstream tasks (Jin et al, 2020d; You et al, 2020c). There are also few works (Chen et al, 2020c; Sun et al, 2020c) applying the idea of self-training in incorporating SSL into GNNs. In addition, loss functions are selected to be tailored for purposes of specific pretext tasks, which includes classification-based tasks (cross-entropy loss), regression-based tasks (mean squared error loss) and contrastive-based tasks (contrastive loss).

In view of the substantial progress made in the field of graph neural networks and the significant potential of self-supervised learning, this chapter aims to present a systematic and comprehensive review on applying self-supervised learning into graph neural networks. The rest of the chapter is organized as follows. Section 18.2 first introduces self-supervised learning and pretext tasks, and then summarizes frequently used self-supervised methods from the image and text domains. In Section 18.3, we introduce the training strategies that are used to incorporate SSL into GNNs and categorize the pretext tasks that have been developed for GNNs. Section 18.4 and 18.5 present detailed summaries of numerous representative SSL methods that have been developed for node-level and graph-level pretext tasks. Thereafter, in Section 18.6 we discuss representative SSL methods that are developed using both node-level and graph-level supervision, which we refer to as node-graph-level pretext tasks. Section 18.7 collects and reinforces the major results and the insightful discoveries in prior sections. Concluding remarks and future forecasts on the development of SSL in GNNs are provided in Section 18.8.

18.2 Self-supervised Learning

Supervised learning is the machine learning task of training a model that maps an input to an output based on the ground-truth input-output pairs provided by a labeled dataset. Good performance of supervised learning requires a decent amount of labeled data (especially when using deep learning models), which are expensive to manually collect. Conversely, self-supervised learning generates supervisory signals from unlabeled data and then trains the model based on the generated supervisory signals. The task used for training the model based on the generative signal is referred to as the pretext task. In comparison, the task whose ultimate performance we care about the most and expect our model to solve is referred to as the downstream task. To guarantee the performance benefits from self-supervised learning, pretext tasks should be carefully designed such that completing them encourages the model to have the similar or complementary understanding as completing downstream tasks. Self-supervised learning initially originated to solve tasks in image and text domains. The following part focuses on introducing self-supervised learning in these two fields with the specific emphasis on different pretext tasks.

In computer vision (CV), many ideas have been proposed for self-supervised representation learning on image data. A common example is that we expect that small distortion on an image does not affect its original semantic meaning or geometric forms. The idea to create surrogate training datasets with unlabeled image patches by first sampling patches from different images at varying positions and then distorting patches by applying a variety of random transformations are proposed in (Dosovitskiy et al, 2014). The pretext task is to discriminate between patches distorted from the same image or from different images. Rotation of an entire image is another effective and inexpensive way to modify an input image without changing semantic content (Gidaris et al, 2018). Each input image is first rotated by a multiple of 90 degrees at random. The model is then trained to predict which rotation has been applied. However, instead of performing pretext tasks on an entire image, the local patches could also be extracted to construct the pretext tasks. Examples of methods using this technique include predicting the relative position between two random patches from one image (Doersch et al, 2015) and designing a jigsaw puzzle game to place nine shuffled patches back to the original locations (Noroozi and Favaro, 2016). More pretext tasks such as colorization, autoencoder, and contrastive predictive coding have also been introduced and effectively utilized (Oord et al, 2018; Vincent et al, 2008; Zhang et al, 2016d).

While computer vision has achieved amazing progress on self-supervised learning in recent years, self-supervised learning has been heavily utilized in natural language processing (NLP) research for quite a while. Word2vec (Mikolov et al, 2013b) is the first work that popularized the SSL ideas in the NLP field. Center word prediction and neighbor word prediction are two pretext tasks in Word2vec where the model is given a small chunk of the text and asked to predict the center word in that text or vice versa. BERT (Devlin et al, 2019) is another famous pre-trained model in NLP where two pretest tasks are to recover randomly masked words in a text or to classify whether two sentences can come one after another or not. Similar works have also been introduced, such as having the pretext task classify whether a pair of sentences are in the correct order (Lan et al, 2020), or a pretext task that first randomly shuffles the ordering of sentences and then seeks to recover the original ordering (Lewis et al, 2020).

Compared with the difficulty of data acquisition encountered in image and text domains, machine learning in the graph domain faces even more challenges in acquiring high-quality labeled data. For example, for molecular graphs it can be extremely expensive to perform the necessary laboratory experiments to label some molecules (Rong et al, 2020a), and in a social network obtaining ground-truth labels for individual users may require large-scale surveys or be unable to be released due to privacy agreements/concerns (Chen et al, 2020a). Therefore, the success achieved by applying SSL in CV and NLP naturally leads the question as to whether SSL can be effectively applied in the graph domain. Given that graph neural network is among the most powerful paradigms for graph representation learning, in following sections we will mainly focus on introducing self-supervised learning within the framework of graph neural networks and highlighting/summarizing these recent advancements.

18.3 Applying SSL to Graph Neural Networks: Categorizing Training Strategies, Loss Functions and Pretext Tasks

When seeking to apply self-supervised learning to GNNs, the major decisions to be made are how to construct the pretext tasks, which includes what information to leverage from the unlabeled data, what loss function to use, and what training strategy to use for effectively improving the GNN's performance. Hence, in this section we will first mathematically formalize the graph neural network with self-supervised learning and then discuss each of the above. More specifically, we will introduce three training strategies, three loss functions that are frequently employed in the current literature, and categorize current state-of-the-art pretext tasks for GNNs based on the type of information they leverage for constructing the pretext task.

Given an undirected attributed graph $\mathscr{G} = \{\mathscr{V}, \mathscr{E}, X\}$, where $\mathscr{V} = \{v_1, ..., v_{|\mathscr{V}|}\}$ represents the vertex set with $|\mathscr{V}|$ vertices, $\mathscr{E}$ represents the edge set and $e_{ij} = (v_i, v_j)$ is an edge between node v_i and v_j, $X \in \mathbb{R}^{|\mathscr{V}| \times d}$ represents the feature matrix and $\mathbf{x}_i = X[i,:]^\top \in \mathbb{R}^d$ is the d-dimensional feature vector of the node v_i. $A \in \mathbb{R}^{|\mathscr{V}| \times |\mathscr{V}|}$ is the adjacency matrix where $A_{ij} = 1$ if $e_{ij} \in \mathscr{E}$ and $A_{ij} = 0$ if $e_{ij} \notin \mathscr{E}$. We denote any GNN-based feature extractor as $f_\theta : \mathbb{R}^{|\mathscr{V}| \times d} \times \mathbb{R}^{|\mathscr{V}| \times |\mathscr{V}|} \to \mathbb{R}^{|\mathscr{V}| \times d'}$ parametrized by θ, which takes any node feature matrix X and the graph adjacency matrix A and outputs the d'-dimensional representation for each node $Z_{\text{GNN}} = f_\theta(X, A) \in \mathbb{R}^{|\mathscr{V}| \times d'}$, which is further fed into any permutation invariant function READOUT : $\mathbb{R}^{|\mathscr{V}| \times d'} \to \mathbb{R}^{d'}$ to obtain the graph embeddings $\mathbf{z}_{\text{GNN}, \mathscr{G}} = \text{READOUT}(f_\theta(X, A)) \in \mathbb{R}^{d'}$. More specifically, we note that here θ represents the parameters encoded in the corresponding network architectures of the GNN (Hamilton et al, 2017b; Kipf and Welling, 2017b; Petar et al, 2018; Xu et al, 2019d, 2018a). Considering the transductive semi-supervised tasks where we are provided with the labeled node set $\mathscr{V}_l \subset \mathscr{V}$, the labeled graph $\mathscr{G}$, the associated node label matrix $Y_{\text{sup}} \in \mathbb{R}^{|\mathscr{V}_l| \times l}$, and the graph label $\mathbf{y}_{\text{sup}, \mathscr{G}} \in \mathbb{R}^l$ with label dimension l, we aim to classify nodes and graphs. The node and graph representations output by GNNs are firstly processed by the extra adaptation layer $h_{\theta_{\text{sup}}}$ parametrized by the supervised adaptation parameter θ_{sup} to obtain the predicted l-dimensional node label $Z_{\text{sup}} \in \mathbb{R}^{|\mathscr{V}| \times l}$ and graph label $\mathbf{z}_{\text{sup}, \mathscr{G}} \in \mathbb{R}^l$ by Eq. equation 18.1-equation 18.2. Then the model parameters θ in GNN-based extractor f_θ and the parameters θ_{sup} in adaptation layer $h_{\theta_{\text{sup}}}$ are learned by optimizing the supervised loss calculated between the output/predicted label and the true label for labeled nodes and the labeled graph, which can be formulated as:

$$Z_{\text{sup}} = h_{\theta_{\text{sup}}}(f_\theta(X,A)) \tag{18.1}$$

$$\mathbf{z}_{\text{sup},\mathscr{G}} = h_{\theta_{\text{sup}}}(\text{READOUT}(f_\theta(X,A))) \tag{18.2}$$

$$\theta^*, \theta_{\text{sup}}^* = \arg\min_{\theta,\theta_{\text{sup}}} \mathscr{L}_{\text{sup}}(\theta,\theta_{\text{sup}}) = \begin{cases} \underbrace{\arg\min_{\theta,\theta_{\text{sup}}} \frac{1}{|\mathcal{V}_l|} \sum_{v_i \in \mathcal{V}_l} \ell_{\text{sup}}(\mathbf{z}_{\text{sup},i}, \mathbf{y}_{\text{sup},i})}_{\text{Node supervised task}} \\ \underbrace{\arg\min_{\theta,\theta_{\text{sup}}} \ell_{\text{sup}}(\mathbf{z}_{\text{sup},\mathscr{G}}, \mathbf{y}_{\text{sup},\mathscr{G}})}_{\text{Graph supervised task}} \end{cases}, \tag{18.3}$$

where $\mathscr{L}_{\text{sup}}$ is the total supervised loss function and ℓ_{sup} is the supervised loss function for each example, $\mathbf{y}_{\text{sup},i} = Y_{\text{sup}}[i,:]^\top$ indicates the true label for node v_i in node supervised task and $\mathbf{y}_{\text{sup},\mathscr{G}}$ indicates the true label for graph $\mathscr{G}$ in graph supervised task. Their corresponding predicted label distributions are denoted as $\mathbf{z}_{\text{sup},i} = Z_{\text{sup}}[i,:]^\top$ and $\mathbf{z}_{\text{sup},\mathscr{G}}$. $\theta, \theta_{\text{sup}}$ are parameters to be optimized for any GNN model and the extra adaptation layer for the supervised downstream task, respectively. Note that for ease of notation, we assume the above graph supervised task is operated only on one graph but the above framework can be easily adapted to supervised tasks on multiple graphs.

18.3.1 Training Strategies

In this chapter, we view SSL as the process of designing a specific pretext task and learning the model on the pretext task. In this sense, SSL can either be used as unsupervised pre-training or be integrated with semi-supervised learning.

The model capability of extracting features for completing pretext and downstream tasks is improved through optimizing the model parameters $\theta, \theta_{\text{ssl}}$, and θ_{sup}, where θ_{ssl} denotes the parameters of the adaptation layer for the pretext task. Inspired by relevant discussions (Hu et al, 2019c; Jin et al, 2020d; Sun et al, 2020c; You et al, 2020b,c), we summarize three possible training strategies that are popular in the literature to train GNNs in the self-supervised setting as self-training, pre-training with fine-tuning, and joint training.

18.3.1.1 Self-training

Self-training is a strategy that leverages the supervision information in the training process generated by the model itself (Li et al, 2018b; Riloff, 1996). A typical self-training pipeline begins with first training the model over the labeled data, then generating pseudo labels to unlabeled samples that have highly confident predictions, and including them into the labeled data in the next round of training. In this way, the pretext task is the same as the downstream task by utilizing the pseudo labels for some of the originally unlabeled data. A detailed overview is presented in Fig. 18.1 where the prediction results are re-utilized to augment the training data in the next iteration as done in (Sun et al, 2020c).

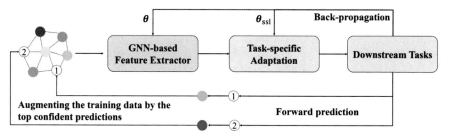

Fig. 18.1: An overview of GNNs with SSL using self-training.

18.3.1.2 Pre-training and Fine-tuning

A common strategy to utilize features learned from completing pretext tasks includes applying the optimized parameters from self-supervision as initialization for fine-tuning in downstream tasks. This strategy consists of two stages: pre-training on the self-supervised pretext tasks and fine-tuning on the downstream tasks. The overview of this two-stage optimization strategy is given in Fig. 18.2.

The whole model consists of one shared GNN-based feature extractor and two adaptation modules, one for the pretext task and one for the downstream task. In the pre-training process, the model is trained with the self-supervised pretext task(s) as:

$$Z_{ssl} = h_{\theta_{ssl}}(f_\theta(X,A)), \tag{18.4}$$

$$\mathbf{z}_{ssl,\mathscr{G}} = h_{\theta_{ssl}}(\text{READOUT}(f_\theta(X,A))), \tag{18.5}$$

$$\theta^*, \theta^*_{\text{ssl}} = \arg\min_{\theta,\theta_{\text{ssl}}} \mathscr{L}_{\text{ssl}}(\theta, \theta_{\text{ssl}}) = \begin{cases} \underbrace{\arg\min_{\theta,\theta_{\text{ssl}}} \frac{1}{|\mathscr{V}|} \sum_{v_i \in \mathscr{V}} \ell_{\text{ssl}}(\mathbf{z}_{\text{ssl},i}, \mathbf{y}_{\text{ssl},i})}_{\text{Node pretext tasks}} \\ \underbrace{\arg\min_{\theta,\theta_{\text{ssl}}} \ell_{\text{ssl}}(\mathbf{z}_{\text{ssl},\mathscr{G}}, \mathbf{y}_{\text{ssl},\mathscr{G}})}_{\text{Graph pretext tasks}} \end{cases}, \quad (18.6)$$

where θ_{ssl} denotes the parameters of the adaptation layer $h_{\theta_{\text{ssl}}}$ for the pretext tasks, ℓ_{ssl} is the self-supervised loss function for each example, and $\mathscr{L}_{\text{ssl}}$ is the total loss function of completing the self-supervised task. In node pretext tasks, $\mathbf{z}_{\text{ssl},i} = Z_{\text{ssl}}[i,:]^\top$ and $\mathbf{y}_{\text{ssl},i} = Y_{\text{ssl}}[i,:]^\top$, which are the self-supervised predicted and true label(s) for the node v_i, respectively. In graph pretext tasks, $\mathbf{z}_{\text{ssl},\mathscr{G}}$ and $\mathbf{y}_{\text{ssl},\mathscr{G}}$ are the self-supervised predicted and true label(s) for the graph $\mathscr{G}$, respectively. Then, in the fine-tuning process, the feature extractor f_θ is trained by completing downstream tasks in Eq. equation 18.1-equation 18.3 with the pre-trained θ^* as the initialization. Note that to utilize the pre-trained node/graph representations the fine-tuning process can also be replaced by training a linear classifier (e.g., Logistic Regression (Peng et al, 2020; Veličković et al, 2019; You et al, 2020b; Zhu et al, 2020c)).

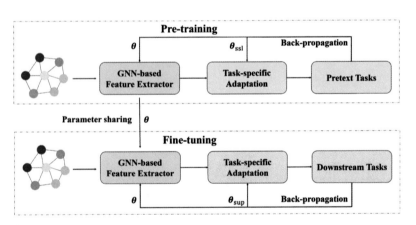

Fig. 18.2: An overview of GNNs with SSL using pre-training and fine-tuning.

18.3.1.3 Joint Training

Another natural idea to harness self-supervised learning for graph neural networks is to combine losses of completing pretext task(s) and downstream task(s) and jointly train the model. The overview of the joint training is shown in Fig. 18.3.

The joint training consists of two components: feature extraction by a GNN and adaption processes for both the pretext tasks and downstream tasks. In the feature extraction process, a GNN takes the graph adjacency matrix A and the feature matrix X as input and outputs the node embeddings Z_{GNN} and/or graph embeddings $\mathbf{z}_{GNN,\mathcal{G}}$. In the adaptation procedure, the extracted node and graph embeddings are further transformed to complete pretext and downstream tasks via $h_{\theta_{ssl}}$ and $h_{\theta_{sup}}$, respectively. We then jointly optimize the pretext and downstream task losses as:

$$Z_{sup} = h_{\theta_{sup}}(f_\theta(X,A)), \quad Z_{ssl} = h_{\theta_{ssl}}(f_\theta(X,A)), \tag{18.7}$$

$$\mathbf{z}_{sup,\mathcal{G}} = h_{\theta_{sup}}(\text{READOUT}(f_\theta(X,A))), \quad \mathbf{z}_{ssl,\mathcal{G}} = h_{\theta_{ssl}}(\text{READOUT}(f_\theta(X,A))), \tag{18.8}$$

$$\theta^*, \theta^*_{sup}, \theta^*_{ssl} = \begin{cases} \underbrace{\arg\min_{\theta,\theta_{sup},\theta_{ssl}} \frac{1}{|\mathcal{V}|} \sum_{v_i \in \mathcal{V}} \left(\alpha_1 \ell_{sup}(\mathbf{z}_{sup,i}, \mathbf{y}_{sup,i}) + \alpha_2 \ell_{ssl}(\mathbf{z}_{ssl,i}, \mathbf{y}_{ssl,i}) \right)}_{\text{Node pretext tasks}} \\ \underbrace{\arg\min_{\theta,\theta_{sup},\theta_{ssl}} \alpha_1 \ell_{sup}(\mathbf{z}_{sup,\mathcal{G}}, \mathbf{y}_{sup,\mathcal{G}}) + \alpha_2 \ell_{ssl}(\mathbf{z}_{ssl,\mathcal{G}}, \mathbf{y}_{ssl,\mathcal{G}})}_{\text{Graph pretext tasks}} \end{cases},$$

$$\tag{18.9}$$

where $\alpha_1, \alpha_2 \in \mathbb{R} > 0$ are the weights for combining the supervised loss ℓ_{sup} and the self-supervised loss ℓ_{ssl}.

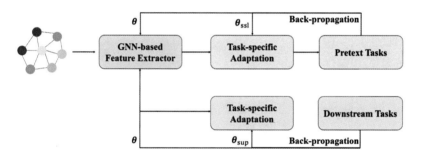

Fig. 18.3: An overview of GNNs with SSL using joint training.

18.3.2 Loss Functions

A loss function is used to evaluate the performance of how well the algorithm models the data. Generally, in GNNs with self-supervised learning, the loss function for the pretext task has three forms, which are classification loss, regression loss and contrastive learning loss. Note that the loss functions we discuss here are only for the pretext tasks rather than downstream tasks.

18.3.2.1 Classification and Regression Loss

In completing classification-based pretext tasks such as node clustering where node embeddings are expected to encode the assignment information of the clusters, the objective for the pretext is to minimize the following loss function:

$$
\mathscr{L}_{ssl} = \begin{cases} \underbrace{\frac{1}{|\mathcal{V}|} \sum_{v_i \in \mathcal{V}} \ell_{CE}(\mathbf{z}_{ssl,i}, \mathbf{y}_{ssl,i}) = -\frac{1}{|\mathcal{V}|} \sum_{v_i \in \mathcal{V}} \sum_{j=1}^{L} \mathbb{1}(\mathbf{y}_{ssl,ij} = 1) \log(\tilde{\mathbf{z}}_{ssl,ij})}_{\text{Node pretext tasks}} \\ \underbrace{\ell_{CE}(\mathbf{z}_{ssl,\mathscr{G}}, \mathbf{y}_{ssl,\mathscr{G}}) = -\sum_{j=1}^{L} \mathbb{1}(\mathbf{y}_{ssl,\mathscr{G}j} = 1) \log(\tilde{\mathbf{z}}_{ssl,\mathscr{G}j})}_{\text{Graph pretext tasks}} \end{cases},
$$

(18.10)

where ℓ_{CE} indicates the cross entropy function, $\mathbf{z}_{ssl,i}$ and $\mathbf{z}_{ssl,\mathscr{G}}$ represents the predicted label distribution of node v_i and graph $\mathscr{G}$ for the pretext task, and their corresponding class probability distribution $\tilde{\mathbf{z}}_{ssl,i}$ and $\tilde{\mathbf{z}}_{ssl,\mathscr{G}}$ are calculated by softmax normalization, respectively. For example, $\tilde{\mathbf{z}}_{ssl,ij}$ is the probability of node v_i belonging to class j. Since every node v_i has its own pseudo label (i.e., $\mathbf{y}_{ssl,i}$) in completing pretext tasks, we can consider all the nodes $\mathcal{V}$ in the graph compared to only the labeled set of nodes $\mathcal{V}_l$ as before in downstream tasks.

In completing regression-based pretext tasks, such as feature completion, the mean squared error loss is typically used as the loss function:

$$
\mathscr{L}_{ssl} = \begin{cases} \underbrace{\frac{1}{|\mathcal{V}|} \sum_{v_i \in \mathcal{V}} \ell_{MSE}(\mathbf{z}_{ssl,i}, \mathbf{y}_{ssl,i}) = \frac{1}{|\mathcal{V}|} \sum_{v_i \in \mathcal{V}} ||\mathbf{z}_{ssl,i} - \mathbf{y}_{ssl,i}||^2}_{\text{Node pretext tasks}} \\ \underbrace{\ell_{MSE}(\mathbf{z}_{ssl,\mathscr{G}}, \mathbf{y}_{ssl,\mathscr{G}}) = ||\mathbf{z}_{ssl,\mathscr{G}} - \mathbf{y}_{ssl,\mathscr{G}}||^2}_{\text{Graph pretext tasks}} \end{cases}, \quad (18.11)
$$

where the objective is minimizing the distance from our learned embedding to $\mathbf{y}_{ssl,i}$ which represents any ground-truth value of node v_i, such as the original attribute in the feature completion or other values of node v_i.

18.3.2.2 Contrastive Learning Loss

Inspired by the significant progress achieved by employing the contrastive learning in natural language processing and computer vision (Le-Khac et al, 2020), recent studies (Hassani and Khasahmadi, 2020; Veličković et al, 2019; You et al, 2020b; Zhu et al, 2020c, 2021) propose similar contrastive frameworks to enable SSL in GNNs. The general goal of contrastive learning in GNNs is to train GNN-based encoders such that the agreement of representations between similar graph instances (e.g., multiple views generated from the same instance) is maximized while the agreement between dissimilar graph instances (e.g., multiple views generated from different instances) is minimized. Such maximization and minimization of agree-

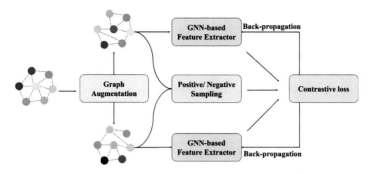

Fig. 18.4: An overview of GNNs with SSL using contrastive learning.

ments between different views of instances is typically formalized as maximizing the mutual information $\mathscr{I}(Z_{\text{ssl}}^1, Z_{\text{ssl}}^2)$ between representations Z_{ssl}^1 and Z_{ssl}^2 under two different views as:

$$\max_{\theta, \theta_{\text{ssl}}} \mathscr{I}(Z_{\text{ssl}}^1, Z_{\text{ssl}}^2), \qquad (18.12)$$

where $Z_{\text{ssl}}^1, Z_{\text{ssl}}^2$ correspond to representations output from any GNN-based encoder followed by an adaptation layer $h_{\theta_{\text{ssl}}}$ under two different graph views $\mathscr{G}^1, \mathscr{G}^2$.

In order to computationally estimate and maximize the mutual information that is originally intractable to be exactly computed in most cases (Belghazi et al, 2018; Gabrié et al, 2019; Paninski, 2003; Xie et al, 2021), multiple estimators to evaluate the lower bounds to the mutual information are derived, including normalized temperature-scaled cross-entropy (NT-Xent) (Chen et al, 2020l), Donsker-Varadhan representation of the KL-divergence (Donsker and Varadhan, 1976), noise-contrastive estimation (InfoNCE) gutmann2010noise, Jensen-Shannon estimator (Nowozin et al, 2016). For simplicity, here we only present one frequently used mutual information estimator NT-Xent, which is formalized as:

$$\mathscr{L}_{\text{ssl}} = \frac{1}{|\mathscr{P}^+|} \sum_{(i,j) \in \mathscr{P}^+} \ell_{\text{NT-Xent}}(Z_{\text{ssl}}^1, Z_{\text{ssl}}^2, \mathscr{P}^-)$$

$$= -\frac{1}{|\mathscr{P}^+|} \sum_{(i,j) \in \mathscr{P}^+} \log \frac{\exp(\mathscr{D}(\mathbf{z}_{\text{ssl},i}^1, \mathbf{z}_{\text{ssl},j}^2))}{\sum_{k \in \{j \cup \mathscr{P}_i^-\}} \exp(\mathscr{D}(\mathbf{z}_{\text{ssl},i}^1, \mathbf{z}_{\text{ssl},k}^2))} \qquad (18.13)$$

where $\mathscr{D}(\mathbf{z}_{\text{ssl},i}^1, \mathbf{z}_{\text{ssl},j}^2)) = \frac{\text{sim}(\mathbf{z}_{\text{ssl},i}^1, \mathbf{z}_{\text{ssl},j}^2)}{\tau}$ is a learnable discriminator parametrized with the similarity function (i.e., cosine similarity) and the temperature factor τ, $\mathscr{P}^+$ represents the set of all pairs of positive samples while $\mathscr{P}^- = \bigcup_{(i,j) \in \mathscr{P}^+} \mathscr{P}_i^-$ represents all sets of negative samples. Especially $\mathscr{P}_i^-$ contains all negative samples of the sample i. Note that we can contrast both node representations, graph representations and node-graph representations under different views. Therefore, $\mathbf{z}_{\text{ssl}}^1$ is not limited to the node embeddings, but could refer to the embeddings of both node and graph under the first graph view $\mathscr{G}^1$. Thus, i, j, k could refer to both node and graph samples.

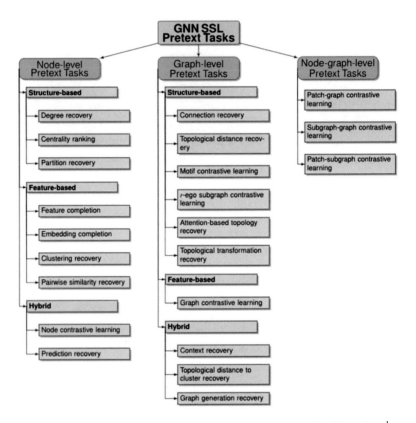

Fig. 18.5: A categorization of pretext tasks in self-supervised learning.[1]

18.3.3 Pretext Tasks

Pretext tasks are constructed by leveraging different types of supervision information coming from different components of graphs. Based on the components that generate the supervision information, pretext tasks that are prevalent in the literature are categorized into node-level, graph-level and node-graph level. In completing node-level and graph-level pretext tasks, three types of information can be leveraged: graph structure, node features, or hybrid, where the latter combines the infor-

---

[1] Additional summary details and the corresponding code links for these methods can be found at https://github.com/NDS-VU/GNN-SSL-Chapter.

mation from node features, graph structure, and even information from the known training labels (as presented in (Jin et al, 2020d)). We summarize the categorization of pretext tasks as a tree where each leaf node represents a specific type of pretext tasks in Fig. 18.5 while also including the corresponding references. In the next three sections, we give detailed explanations about each of these pretext tasks and summarize the majority of existing methods.

18.4 Node-level SSL Pretext Tasks

For node-level pretext tasks, methods have been developed to use easily-accessible data to generate pseudo labels for each node or relationships for each pair of nodes. In this way, the GNNs are then trained to be predictive of the pseudo labels or to keep the equivalence between the node embeddings and the original node relationships.

18.4.1 Structure-based Pretext Tasks

Different nodes have different structure properties in graph topology, which can be measured by the node degree, centrality, node partition, etc. Thus, for structure-based pretext tasks at the node-level, we expect to align node embeddings extracted from the GNNs with their structure properties, in an attempt to ensure this information is preserved while GNNs learn the node embeddings.

Since degree is the most fundamental topological property, Jin et al (2020d) designs the pretext task to recover the node degree from the node embeddings as follows:

$$\mathcal{L}_{\text{ssl}} = \frac{1}{|\mathcal{V}|} \sum_{v_i \in \mathcal{V}} \ell_{\text{MSE}}(\mathbf{z}_{\text{ssl},i}, d_i) \qquad (18.14)$$

where d_i represents the degree of node i and $\mathbf{z}_{\text{ssl},i} = Z_{\text{ssl}}[i,:]^{\top}$ denotes the self-supervised GNN embeddings of node i. It should be noted that this pretext task can be generalized to harness any structural property in the node level.

Node centrality measures the importance of nodes based on their structure roles in the whole graph (Newman, 2018). Hu et al (2019c) designs a pretext task to have GNNs estimate the rank scores of node centrality. The specific centrality measures considered are eigencentrality, betweenness, closeness, and subgraph centrality. For a node pair (u, v) and a centrality score s, with relative order $R^s_{u,v} = 1(s_u > s_v)$ where $R^s_{u,v} = 1$ if $s_u > s_v$ and $R_{u,v} = 0$ if $s_u \leq s_v$, a decoder D^{rank} for centrality score s estimates its rank score by $S_v = D^{rank}_s(\mathbf{z}_{\text{GNN},v})$. The probability of estimated rank order is defined by the sigmoid function $\tilde{R}^s_{u,v} = \frac{\exp(S_u - S_v)}{1 + \exp(S_u - S_v)}$. Then predicting the relative order between pairs of nodes could be formalized as a binary classification problem with the loss:

$$\mathscr{L}_{\text{ssl}} = -\sum_{s}\sum_{u,v\in\mathscr{V}}(R_{u,v}^{s}\log\tilde{R}_{u,v}^{s} + (1-R_{u,v}^{s})\log(1-\tilde{R}_{u,v}^{s})). \qquad (18.15)$$

Different from peer works, Hu et al (2019c) does not consider any node feature but instead extract the node features directly from the graph topology, which includes: (1) degree that defines the local importance of a node; (2) core-number that defines the connectivity of the subgraph around a node; (3) collective influence that defines the neighborhood importance of a node; and (4) local clustering coefficient, which defines the connectivity of 1-hop neighborhood of a node. Then, the four features (after min-max normalization) are concatenated with a nonlinear transformation and fed into the GNN where (Hu et al, 2019c) uses the pretext tasks: centrality ranking, clustering recovery and edge prediction. Another innovative idea in (Hu et al, 2019c) is to choose a fix-tune boundary in the middle layer of GNNs. The GNN blocks below this boundary are fixed, while the ones above the boundary are fine-tuned. For downstream tasks that are closely related to the pre-trained tasks, a higher boundary is used.

Another important node-level structural property is the partition each node belongs after performing a graph partitioning method. In (You et al, 2020c), the pretext task is to train the GNNs to encode the node partition information. Graph partitioning is to partition the nodes of a graph into different groups such that the number of edges between each group is minimized. Given the node set $\mathscr{V}$, the edge set $\mathscr{E}$, and a preset number of partitions $p\in[1,|\mathscr{V}|]$, a graph partitioning algorithm (e.g., (Karypis and Kumar, 1995) as used in (You et al, 2020c)) will output a set of nodes $\{\mathscr{V}_{\text{par}_1},...,\mathscr{V}_{\text{par}_p}|\mathscr{V}_{\text{par}_i}\subset\mathscr{V},i=1,...,p\}$. Then the classification loss is set exactly the same as:

$$\mathscr{L}_{\text{ssl}} = -\frac{1}{|\mathscr{V}|}\sum_{v_i\in\mathscr{V}}\ell_{\text{CE}}(\mathbf{z}_{\text{ssl},i},\mathbf{y}_{\text{ssl},i}) \qquad (18.16)$$

where $\mathbf{z}_{\text{ssl},i}$ denotes the embedding of node v_i and assuming that the partitioning label is a one-hot encoding $\mathbf{y}_{\text{ssl},i}\in\mathbb{R}^p$ with k-th entry as 1 and others as 0 if $v_i\in\mathscr{V}_{\text{par}_k}, i=1,...,|\mathscr{V}|,\exists k\in[1,p]$.

18.4.2 Feature-based Pretext Tasks

Node features are another important information that can be leveraged to provide extra supervision. Since the state-of-the-art GNNs suffer from over-smoothing (Chen et al, 2020c), the original feature information is partially lost after fed into the GNNs. In order to reduce the information loss in node embeddings, the pretext task in (Hu et al, 2020c; Jin et al, 2020d; Manessi and Rozza, 2020; Wang et al, 2017a; You et al, 2020c) is to first mask node features and let the GNN predict those features. More specifically, they randomly mask input node features by replacing them with special mask indicators and then apply GNNs to obtain the corresponding node embeddings. Finally a linear model is applied on top of embeddings to predict the corresponding masked node features. Assuming the set of nodes that are masked is

$\mathcal{V}_m$, then the self-supervised regression loss to reconstruct these masked features is:

$$\mathscr{L}_{ssl} = \frac{1}{|\mathcal{V}_m|} \sum_{v_i \in \mathcal{V}_m} \ell_{MSE}(\mathbf{z}_{ssl,i}, \mathbf{x}_i) \tag{18.17}$$

To handle the high sparsity of the node features, it is beneficial to first perform feature dimensionality reduction on X (such as principal component analysis (PCA) used in (Jin et al, 2020d)). Additionally, instead of reconstructing node features, node embeddings could also be reconstructed from their corrupted version, such as in (Manessi and Rozza, 2020).

Contrary to the graph partitioning where nodes are grouped by the graph topology, in graph clustering the clusters of nodes are discovered based on their features (You et al, 2020c). In this way the pretext task can be designed to recover the node clustering assignment. Given the node set $\mathcal{V}$, the feature matrix X, and a preset number of clusters $p \in [1, |\mathcal{V}|]$ (or without if the clustering algorithm automatically learns the number of clusters) as input, the clustering algorithm will output a set of node clusters $\{\mathcal{V}_{clu_1}, \ldots, \mathcal{V}_{clu_p} | \mathcal{V}_{clu_i} \subset \mathcal{V}, i = 1, ..., p\}$ and assuming for node v_i, the partitioning label is a one-hot encoding $\mathbf{y}_{ssl,i} \in \mathbb{R}^p$ with k-th entry as 1 and others as 0 if $v_i \in \mathcal{V}_{clu_k}, i = 1, ..., |\mathcal{V}|, \exists k \in [1, p]$. Then the loss is the same as Eq. equation 18.16.

Instead of focusing on individual nodes, pretext tasks have also been developed based on the relationship between pairs of nodes (Jin et al, 2021, 2020d). The basic idea is to retain the node pairwise feature similarity in the node embeddings from GNNs. Suppose $\mathcal{T}_s, \mathcal{T}_d$ denote the sets of node pairs having the highest and the lowest similarity:

$$\mathcal{T}_s = \{(v_i, v_j) | \; sim(\mathbf{x}_i, \mathbf{x}_j) \text{ in top-}B \text{ of } \{sim(\mathbf{x}_i, \mathbf{x}_b)\}_{b=1}^B \backslash sim(\mathbf{x}_i, \mathbf{x}_i), \forall v_i \in \mathcal{V}\}, \tag{18.18}$$

$$\mathcal{T}_d = \{(v_i, v_j) | \; sim(\mathbf{x}_i, \mathbf{x}_j) \text{ in bottom-}B \text{ of } \{sim(\mathbf{x}_i, \mathbf{x}_b)\}_{b=1}^B \backslash sim(\mathbf{x}_i, \mathbf{x}_i), \forall v_i \in \mathcal{V}\}, \tag{18.19}$$

where $sim(\mathbf{x}_i, \mathbf{x}_j)$ measures the cosine similarity of features between two nodes v_i, v_j and B is the number of top/bottom pairs selected for each node. Then the pretext task is to optimize the following regression loss:

$$\mathscr{L}_{ssl} = \frac{1}{|\mathcal{T}_s \cup \mathcal{T}_d|} \sum_{(v_i, v_j) \in \mathcal{T}_s \cup \mathcal{T}_d} \ell_{MSE}\big(f_w(|\mathbf{z}_{GNN,i} - \mathbf{z}_{GNN,j}|), sim(\mathbf{x}_i, \mathbf{x}_j)\big), \tag{18.20}$$

where f_w is a function mapping the difference between two node embeddings from GNNs to a scalar representing the similarity between them.

18.4.3 Hybrid Pretext Tasks

Instead of employing only the topology or only the feature information as the extra supervision, some pretext tasks combine them together as the hybrid supervision, or even utilize information from the known training labels.

A contrastive framework for unsupervised graph representation learning, GRACE, where two correlated graph views are generated by randomly performing corruption on attributes (masking node features) and topology (removing or adding graph edges) is proposed in (Zhu et al, 2020c). Then the GNNs are trained using a contrastive loss to maximize the agreement between node embeddings in these two views. In each iteration two graph views $\mathcal{G}^1 = \{A^1, X^1\}$ and $\mathcal{G}^2 = \{A^2, X^2\}$ are generated randomly according to the possible augmentation functions from an input graph $\mathcal{G} = \{A, X\}$.

The objective is to maximize the similarity of the same nodes in different views of the graph while minimizing the similarity of different nodes in the same or different views of the graph. Thus, if we denote the node embeddings in the two views as $Z_{GNN}^1 = f_\theta(X^1, A^1), Z_{GNN}^2 = f_\theta(X^2, A^2)$, then the contrastive NT-Xent loss is:

$$\mathcal{L}_{ssl} = \frac{1}{|\mathscr{P}^+|} \sum_{(v_i^1, v_i^2) \in \mathscr{P}^+} \ell_{NT\text{-}Xent}(Z_{GNN}^1, Z_{GNN}^2, \mathscr{P}^-), \qquad (18.21)$$

where $\mathscr{P}^+$ includes positive pairs of (v_i^1, v_i^2) where v_i^1, v_i^2 correspond to the same node in different views, while $\mathscr{P}^- = \bigcup_{(v_i^1, v_i^2) \in \mathscr{P}^+} \mathscr{P}_{v_i^1}^-$ represents all sets of negative samples with $\mathscr{P}_{v_i^1}^-$ containing nodes different from v_i in the same view (intra-view negative pairs) or the other view (inter-view negative pairs).

More specifically, in the above, the two graph corruptions are removing edges and masking node features. In removing edges, a random masking matrix $M \in \{0,1\}^{|\mathcal{V}| \times |\mathcal{V}|}$ is randomly sampled whose entry is drawn from a Bernoulli distribution $M_{ij} \sim \mathcal{B}(1 - p_r)$ if $A_{ij} = 1$ for the original graph. p_r is the probability of each edge being removed. The resulting matrix can be computed as $A' = A \odot M$ creating the adjacency matrix of graph view $\mathcal{G}'$ from $\mathcal{G}$.

In masking node features, a random vector $\mathbf{m} \in \{0,1\}^d$ is utilized, where each dimension of m is independently drawn from a Bernoulli distribution with probability $1 - p_m$ and d is the dimension of the node features X. Then, the generated node features X' for graph view $\mathcal{G}'$ from $\mathcal{G}$ is computed by:

$$X' = [\mathbf{x}_1 \odot \mathbf{m}; \mathbf{x}_2 \odot \mathbf{m}; \cdots; \mathbf{x}_{|\mathcal{V}|} \odot \mathbf{m}], \qquad (18.22)$$

where $[;]$ is the concatenation operator. Moreover, a modified version of the GRACE is proposed in (Zhu et al, 2021) where the whole contrastive procedure is the same as GRACE except that the graph augmentation is adaptively performed based on the importance of nodes and edges. Specifically, the probability of removing an edge between nodes v_i, v_j should reflect the importance of the edge (v_i, v_j) such that the augmentation function is more likely to corrupt unimportant edges while keeping important connective structures intact in augmented views. Similarly the feature dimensions frequently appearing in influential nodes are seen as important and so are masked with lower probability.

The observation made in (Chen et al, 2020b) that nodes with further topological distance to the labeled nodes are more likely to be misclassified indicates the uneven distribution of the ability of GNNs to embed node features in the whole graph. However, existing graph contrastive learning methods ignore this uneven distribution, which motivates Chen et al (2020b) to propose the distance-wise graph contrastive learning (DwGCL) method that can adaptively augment the graph topology, sample the positive and negative pairs, and maximize the mutual information. The topology information gain (TIG) is calculated based on Group PageRank and node features to describe the task information effectiveness that the node obtains from labeled nodes along the graph topology. By ranking the performance of GNNs on nodes according to their TIG values with/without contrastive learning, it is found that contrastive learning mainly improves the performance on nodes that are topologically far away from the labeled nodes. Based on the above finding, Chen et al (2020b) propose to: 1) perturb the graph topology by augmenting nodes according to their TIG value; 2) sampling the positive and negative pairs considering local/-global topology distance and node embedding distance; and 3) assigning different weights to nodes in the self-supervised loss based on their TIG rankings. Results demonstrate the performance improvement of this distance-wise graph contrastive learning over the typical contrastive learning approach.

Another special supervision information to exploit is the prediction results of the model itself. Sun et al (2020c) leverages the multi-stage training framework to utilize the information of the pseudo labels generated by predictions in the next rounds of training. The multi-stage training algorithm repeatedly adds the most confident predictions of each class to the label set and re-utilizes these pseudo labeled data to train the GNNs. Furthermore, a self-checking mechanism based on Deep-Cluster (Caron et al, 2018) is proposed to guarantee the precision of labeled data. Assuming that the cluster assignment for node v_i is $\mathbf{c}_i \in \{0,1\}^p$ (here the number of clusters is assumed to equal to the number of predefined classes p in the downstream classification task) and the centroid matrix $C \in R^{d' \times p}$ represents the feature of each cluster, then we obtain the cluster assignment $\mathbf{c}_i$ for each node v_i by optimizing:

$$\min_{C} \frac{1}{\mathcal{V}} \sum_{v_i \in \mathcal{V}} \min_{\mathbf{c}_i \in \{0,1\}^p} ||\mathbf{z}_{\text{GNN},i} - C\mathbf{c}_i||_2^2, \quad s.t. \quad \mathbf{c}_i^{\text{T}} \mathbf{1}_p = 1. \quad (18.23)$$

After applying DeepCluster to group nodes into multiple clusters, an aligning mechanism is used to assign nodes in each cluster to their corresponding class defined by downstream tasks. For each cluster $k \in [1, p]$ in unlabeled data, the computation of aligning mechanism is:

$$c^k = \arg\min_{m} ||\kappa_k - \mu_m||^2, \quad (18.24)$$

where μ_m denotes the centroid of class m in labeled data, κ_k denotes the centroid of cluster k in unlabeled data and c^k represents the aligned class that has the closest distance to the centroid κ_k of the cluster k among all centroids of classes in the original labeled data. Note that the self-checking can be directly performed by comparing the distance of each unlabeled node to centroids of classes in labeled data. However, directly checking in this naïve way is very time-consuming.

18.5 Graph-level SSL Pretext Tasks

After having just presented the node-level SSL pretext tasks, in this section we focus on the graph-level SSL pretext tasks where we desire the node embeddings coming from the GNNs to encode information of graph-level properties.

18.5.1 Structure-based Pretext Tasks

As the counterpart of the nodes in the graph, the edges encode abundant information of the graph, which can also be leveraged as an extra supervision to design pretext tasks. The pretext task in (Zhu et al, 2020a) is to recover the graph topology, i.e., predict edges, after randomly removing edges in the graph. After node embeddings $\mathbf{z}_{\mathrm{GNN},i}$ is obtained for each node v_i, the probability of the edge between any pair of nodes v_i, v_j is calculated by their feature similarity as follows:

$$A'_{ij} = \mathrm{sigmoid}(\mathbf{z}_{\mathrm{GNN},i}(\mathbf{z}_{\mathrm{GNN},j})^{\top}), \qquad (18.25)$$

and the weighted cross-entropy loss is used during training, which is defined as:

$$\mathscr{L}_{\mathrm{ssl}} = -\sum_{v_i, v_j \in \mathscr{V}} W(A_{ij} \log A'_{ij}) + (1 - A_{ij}) \log(1 - A'_{ij}), \qquad (18.26)$$

where W is the weight hyperparameter used for balancing two classes; which are node pairs having an edge and node pairs without an edge between them.

As it is known that unclean graph structure usually impedes the applicability of GNNs (Cosmo et al, 2020; Jang et al, 2019). A method that trains the GNNs by downstream supervised tasks based on the cleaned graph structure reconstructed from completing a self-supervised pretext task is introduced in (Fatemi et al, 2021). The self-supervised pretext task aims to train a separate GNN to denoise the corrupted node feature $\hat{X}$ generated by either randomly zeroing some dimensions of the original node feature X when having binary features or by adding independent Gaussian noise when X is continuous. Two methods are used to generate the initial graph adjacency matrix $\tilde{A}$. The first method Full Parametrization (FP) treats every entry in $\tilde{A}$ as a parameter and directly optimizes its $|\mathscr{V}|^2$ parameters by denoising the corrupted feature $\hat{X}$. The second method MLP-kNN considers a mapping function kNN(MLP(X)), where a multilayer perceptron (i.e., MLP($\cdot$)) updates the original node features and kNN($\cdot$) produces a sparse matrix by selecting top-k similar nodes to each node and adds edges between them. Then, the generated initial adjacency matrix $\tilde{A}$ is normalized and symmetrized into a new adjacency matrix A as follows:

$$A = D^{-\frac{1}{2}} \frac{\tilde{P}(\tilde{A}) + \tilde{P}(\tilde{A})^\top}{2} D^{-\frac{1}{2}}, \qquad (18.27)$$

where $\tilde{P}$ is a function with a non-negative range to ensure the positivity of every entry in A. In MLP-kNN method, $\tilde{P}$ is the element-wise ReLU function. However, the ReLU function could result in the gradient flow problem in the FP method, thus the element-wise ELU function followed by an addition of 1 to avoid the problem of gradient flow is used instead. Next, a separate GNN-based encoder takes noisy node features $\hat{X}$ and the new normalized adjacency matrix A as input and output the updated node features $\hat{Z} = \text{GNN}(\hat{X}, A)$. The parameters in FP and MLP-kNN used for generating the initial adjacency matrix $\tilde{A}$ is optimized by:

$$\mathscr{L}_{\text{ssl}} = \frac{1}{|\mathscr{V}_{\text{m}}|} \sum_{v_i \in \mathscr{V}_{\text{m}}} \ell_{\text{MSE}}(\mathbf{x}_i, \hat{\mathbf{z}}_i), \qquad (18.28)$$

where $\hat{\mathbf{z}}_i = \hat{Z}[i,:]^\top$ is the noisy embedding vector of the node v_i obtained by the separate GNN-based encoder. The optimized parameters in FP and MLP-kNN leads to the generation of more cleaned graph adjacency matrix, which in turn results in the better performance in the downstream tasks.

In addition to the graph edges and the adjacency matrix, topological distance between nodes is another important global structural property in graph. The pretext task in (Peng et al, 2020) is to recover the topological distance between nodes. More specifically, they leverage the shortest path length between nodes denoted as p_{ij} between nodes v_i and v_j, but this could be replaced with any other distance measure. Then, they define the set $\mathscr{C}_i^k$ as all the nodes having the shortest path distance of length k from node v_i. More formally, this is defined as:

$$\mathscr{C}_i = \mathscr{C}_i^1 \cup \mathscr{C}_i^2 \cup \cdots \cup \mathscr{C}_i^{\delta_i}, \quad \mathscr{C}_i^k = \{v_j | d_{ij} = k\}, \quad k = 1, 2, \cdots, \delta_i, \qquad (18.29)$$

where δ_i is the upper bound of the hop count from other nodes to v_i, d_{ij} is the length of the path p_{ij}, and $\mathscr{C}_i$ is the union of all the k-hop shortest path neighbor sets C_i^k. Based on these sets, one-hot encodings $\mathbf{d}_{ij} \in \mathbb{R}^{\delta_i}$ are created for pairs of nodes v_i, v_j, where $v_j \in \mathscr{C}_i$, according to their distance d_{ij}. Then, the GNN model is guided to extract node embeddings that encode node topological distance as follows:

$$\mathscr{L}_{\text{ssl}} = \sum_{v_i \in \mathscr{V}} \sum_{v_j \in \mathscr{C}_i} \ell_{\text{CE}}(f_w(|\mathbf{z}_{\text{GNN},i} - \mathbf{z}_{\text{GNN},j}|), \mathbf{d}_{ij}), \qquad (18.30)$$

where f_w is a function mapping the difference between two node embeddings to the probabilities of pairs of nodes belonging to the corresponding category of the topological distance. Since the number of the categories depends on the upper bound of the hop count (topological distance) but precisely determining this upper bound is time-consuming for a big graph, it is assumed that the number of hops (distance) is under control based on small-world phenomenon (Newman, 2018) and is further divided into several major categories that clearly discriminates the dissimilarity and partly tolerates the similarity. Experiments demonstrate that dividing the topological distance into four categories: $\mathscr{C}_i^1, \mathscr{C}_i^2, \mathscr{C}_i^3, \mathscr{C}_i^k (k \geq 4)$ achieves the best performance (i.e., δ_i=4). Another problem is that the number of nodes that are close to the focal node v_i is much less than the nodes that are further away (i.e., the magnitude of $\mathscr{C}_i^{\delta_i}$ will be significantly larger than other sets). To circumvent this imbalance problem, node pairs are sampled with an adaptive ratio.

Network motifs are recurrent and statistically significant subgraphs of a larger graph and (Zhang et al, 2020f) designs a pretext task to train a GNN encoder that can automatically extract graph motifs. The learned motifs are further leveraged to generate informative subgraphs used in graph-subgraph contrastive learning. Firstly, a GNN-based encoder f_θ and a m-slot embedding table $\{\mathbf{m}_1, ..., \mathbf{m}_m\}$ denoting m cluster centers of m motifs are initialized. Then, a node affinity matrix $U \in \mathbb{R}^{|\mathcal{V}| \times |\mathcal{V}|}$ is calculated by softmax normalization on the embedding similarity $\mathscr{D}(\mathbf{z}_{\text{GNN},i}, \mathbf{z}_{\text{GNN},j})$ between nodes i, j as in Eq. equation 18.13. Afterwards, spectral clustering (VON-LUXBURG, 2007) is performed on U to generate different groups, within which $n_\mathscr{G}$ connected components that have more than three nodes are collected as the sampled subgraphs from the graph $\mathscr{G}$ and their embeddings are calculated by applying READOUT function. For each subgraph, its cosine similarity to each of the m motifs is calculated to obtain a similarity metric $S \in \mathbb{R}^{m \times n_\mathscr{G}}$. To produce semantic-meaningful subgraphs that are close to motifs, the top 10% most similar subgraphs to each motif are selected based on the similarity metric S and are collected into a set $\mathscr{G}^{\text{top}}$. The affinity values in U between pairs of nodes in each of these subgraphs are increased by optimizing the loss:

$$\mathscr{L}_1 = -\frac{1}{|\mathscr{G}^{\text{top}}|} \sum_{i=1}^{|\mathscr{G}^{\text{top}}|} \sum_{(v_j, v_k) \in \mathscr{G}_i^{\text{top}}} U[j, k]. \qquad (18.31)$$

The optimization of the above loss forces nodes in motif-like subgraphs to be more likely to be grouped together in spectral clustering, which leads to more subgraph samples aligned with the motifs. Next, the embedding table of motifs is optimized based on the sampled subgraphs. The assignment matrix $Q \in \mathbb{R}^{m \times n_\mathscr{G}}$ is found by maximizing similarities between embeddings and its assigned motif:

$$\max_Q \ Tr(Q^{\mathsf{T}}S) - \frac{1}{\lambda} \sum_{i,j} Q[i,j] \log Q[i,j], \qquad (18.32)$$

where the second term controlled by hyperparameter λ is to avoid all representations collapsing into a single cluster center. After the cluster assignment matrix Q is obtained, the GNN-based encoder and the motif embedding table are trained, which is equivalent to a supervised m-class classification problem with labels Q and the prediction distribution $\widetilde{S}$ obtained by applying a column-wise softmax normalization with temperature τ:

$$\mathcal{L}_2 = -\frac{1}{n_{\mathcal{G}}} \sum_{i=1}^{n_{\mathcal{G}}} \ell_{CE}(\mathbf{q}_i, \tilde{\mathbf{s}}_i), \qquad (18.33)$$

where $\mathbf{q}_i = Q[:,i]$ and $\tilde{\mathbf{s}}_i = \tilde{S}[:,i]$ denote the assignment distribution and predicted distribution for the subgraph i, respectively. Optimizing Eq. equation 18.33 jointly enhances the ability of GNN encoder to extract subgraphs that are similar to motifs and improves the embeddings of motifs. The last step is to train the GNN-based encoder by a classification task where subgraphs are reassigned back to their corresponding graphs. Note that the subgraphs are generated by the Motif-guided extractor, which is more likely to capture higher-level semantic information compared with randomly sampled subgraphs. The whole framework is trained jointly by weighted combining $\mathcal{L}_1, \mathcal{L}_2$ and the contrastive loss.

Aside from the network motifs, other subgraph structures can be leveraged to provide extra supervision in designing pretext tasks. In (Qiu et al, 2020a), an r-ego network for a certain vertex is defined as the subgraph induced by nodes that have shortest path with length shorter than r. Then a random walk with restart is initiated at ego vertex v_i and the subgraph induced by nodes that are visited during the random walk starting at v_i are used as the augmented version of the r-ego network. First, two augmented r-ego networks centered around vertex v_i are obtained by performing the random walk twice (i.e., $\mathcal{G}_i$ and $\mathcal{G}_i^+$), which are defined as a positive pair since they come from the same r-ego network. In comparison, a negative pair corresponds to two subgraphs augmented from different r-ego networks (e.g., one coming from v_i and another coming from v_j resulting in random walk induced subgraphs $\mathcal{G}_i$ and $\mathcal{G}_j$, respectively). Based on the above defined positive and negative subgraph pairs, a contrastive loss is set up to optimize the GNNs as follows:

$$\mathcal{L}_{ssl} = \frac{1}{|\mathcal{P}^+|} \sum_{(\mathcal{G}_i, \mathcal{G}_i^+) \in \mathcal{P}^+} \ell_{NT\text{-Xent}}(Z_{ssl}^1, Z_{ssl}^2, \mathcal{P}^-), \qquad (18.34)$$

where Z_{ssl}^1, Z_{ssl}^2 denotes the GNN-based graph embeddings and specifically here the two different views are the same $Z_{ssl}^1 = Z_{ssl}^2$. $\mathcal{P}^+$ contains positive pairs of subgraphs $(\mathcal{G}_i, \mathcal{G}_i^+)$ sampled by random walk starting at the same ego vertex v_i in the same graph while $P^- = \bigcup_{(\mathcal{G}_i, \mathcal{G}_i^+) \in \mathcal{P}^+} \mathcal{P}_{\mathcal{G}_i}^-$ represents all sets of negative samples. Specifically $\mathcal{P}_{\mathcal{G}_i}^-$ represents subgraphs sampled by random walk starting at either different ego vertex from v_i in $\mathcal{G}$ or directly sampled by random walk in different graphs from $\mathcal{G}$.

Although Graph Attention Network (GAT) (Petar et al, 2018) achieves performance improvements over the original GCN (Kipf and Welling, 2017b), there is little understanding of what graph attention learns. To this end, Kim and Oh (2021) proposes a specific pretext task to leverage the edge information to supervise what graph attention learns:

$$\mathscr{L}_{ssl} = \frac{1}{|\mathscr{E} \cup \mathscr{E}^-|} \sum_{(j,i) \in \mathscr{E} \cup \mathscr{E}^-} 1\big((j,i) \in \mathscr{E}\big) \cdot \log \chi_{ij} + 1\big((j,i) \in \mathscr{E}^-\big) \log(1 - \chi_{ij}),$$
(18.35)

where $\mathscr{E}$ is the set of edges, $\mathscr{E}^-$ is the sampled set of node pairs without edges, and χ_{ij} is the edge probability between node i, j calculated from their embeddings. Based on two primary edge attentions, the GAT attention (shortly as GO) (Petar et al, 2018) and the dot-product attention (shortly as DP) (Luong et al, 2015), two advanced attention mechanisms, SuperGAT$_{SD}$ (Scaled Dot-product, shortly as SD) and SuperGAT$_{MX}$ (Mixed GO and DP, shortly as MX) are proposed:

$$e_{ij,SD} = e_{ij,DP}/\sqrt{F}, \qquad \chi_{ij,SD} = \sigma(e_{ij,SD}), \qquad (18.36)$$

$$e_{ij,MX} = e_{ij,GO} \cdot \sigma(e_{ij,DP}), \qquad \chi_{ij,MX} = \sigma(e_{ij,DP}), \qquad (18.37)$$

where σ denotes the sigmoid function taking the edge weight e_{ij} and calculating the edge probability χ_{ij}. SuperGAT$_{SD}$ divides the dot-product of edge $e_{ij,DP}$ by a square root of dimension as Transformer (Vaswani et al, 2017) to prevent some large values from dominating the entire attention after softmax. SuperGAT$_{MX}$ multiplies GO and DP attention with sigmoid, which is motivated by the gating mechanism of Gated Recurrent Units (GRUs) (Cho et al, 2014a). Since DP attention with the sigmoid denotes the edge probability, multiplying $\sigma(e_{ij,DP})$ in calculating $e_{ij,MX}$ can softly drop neighbors that are not likely linked while implicitly assigning importance to the remaining nodes. $e_{ij,DP}, e_{ij,GO}$ are the weight of edge (i, j) used to calculate the GO and DP attention. Results disclose several insightful discovers including the GO attention learns label-agreement better than DP, whereas DP predicts edge presence better than GO, and the performance of the attention mechanism is not fixed but depends on homophily and average degree of the specific graph.

The topological information can also be generated manually for designing pretext tasks. Gao et al (2021) proposes to encode the transformation information between two different graph topologies in the representations of nodes obtained by GNNs. First, they transform the original graph adjacency matrix A into $\hat{A}$ by randomly adding or removing edges from the original edge set. Then, by feeding the original and transformed graph topology and the node feature matrix into any GNN-based encoder, the feature representation $Z_{GNN}, \hat{Z}_{GNN}$ before and after topology transformation are calculated and their difference $\Delta Z \in \mathbb{R}^{N \times F'}$ is defined as:

$$\Delta Z = \hat{Z}_{GNN} - Z_{GNN} = [\Delta z_{GNN,1}, ..., \Delta z_{GNN,N}]^\top = [\hat{z}_{GNN,1} - z_{GNN,1}, ..., \hat{z}_{GNN,N} - z_{GNN,N}]^\top.$$
(18.38)

Next they predict the topology transformation between node v_i and v_j through the node-wise feature difference ΔZ by constructing the edge representation as:

$$\mathbf{e}_{ij} = \frac{\exp(-(\Delta \mathbf{z}_i - \Delta \mathbf{z}_j) \odot (\Delta \mathbf{z}_i - \Delta \mathbf{z}_j))}{||\exp(-(\Delta \mathbf{z}_i - \Delta \mathbf{z}_j) \odot (\Delta \mathbf{z}_i - \Delta \mathbf{z}_j))||}, \qquad (18.39)$$

where $\odot$ denotes the Hardamard product. This edge representation $\mathbf{e}_{ij}$ is then fed into an MLP for the prediction of the topological transformation, which includes four classes: edge addition, edge deletion, keeping disconnection and keeping connection between each pair of nodes. Thus, the GNN-based encoder is trained by:

$$\mathscr{L}_{\text{ssl}} = \frac{1}{|\mathscr{V}|^2} \sum_{v_i, v_j \in \mathscr{V}} \ell_{\text{CE}}(\text{MLP}(\mathbf{e}_{ij}), \mathbf{t}_{ij}) \tag{18.40}$$

where we denote the topological transformation category between nodes v_i and v_j as one-hot encoding $\mathbf{t}_{ij} \in \mathbb{R}^4$.

18.5.2 Feature-based Pretext Tasks

Typically, graphs do not come with any feature information and here the graph-level features refer to the graph embeddings obtained after applying a pooling layer on all node embeddings from GNNs.

GraphCL (You et al, 2020b) designs the pretext task to first augment graphs by four different augmentations including node dropping, edge perturbation, attribute masking and subgraph extraction and then maximize the mutual information of the graph embeddings between different augmented views generated from the same original graph while also minimizing the mutual information of the graph embeddings between different augmented views generated from different graphs. The graph embeddings Z_{ssl} are obtained through any permutational-invariant READOUT function on node embeddings followed by applying an adaptation layer. Then the mutual information is maximized by optimizing the following NT-Xent contrastive loss:

$$\mathscr{L}_{\text{ssl}} = \frac{1}{|\mathscr{P}^+|} \sum_{(\mathscr{G}_i, \mathscr{G}_j) \in \mathscr{P}^+} \ell_{\text{NT-Xent}}(Z_{\text{ssl}}^1, Z_{\text{ssl}}^2, \mathscr{P}^-), \tag{18.41}$$

where $Z_{\text{ssl}}^1, Z_{\text{ssl}}^2$ represent graph embeddings under two different views. The view could be the original view without any augmentation or the one generated from applying four different augmentations. $\mathscr{P}^+$ contains positive pairs of graphs $(\mathscr{G}_i, \mathscr{G}_j)$ augmented from the same original graph while $\mathscr{P}^- = \bigcup_{(\mathscr{G}_i, \mathscr{G}_j) \in \mathscr{P}^+} \mathscr{P}_{\mathscr{G}_i}^-$ represents all sets of negative samples. Specifically $\mathscr{P}_{\mathscr{G}_i}^-$ contains graphs augmented from the graph different from $\mathscr{G}_i$. Numerical results demonstrate that the augmentation of edge perturbations benefits social networks but hurts biochemical molecules. Applying attribute masking achieves better performance in denser graphs. Node dropping and subgraph extraction are generally beneficial across all datasets.

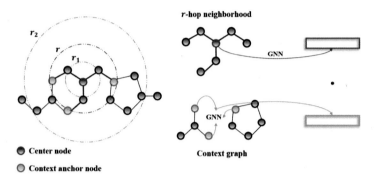

Fig. 18.6: An example of a context and r-neighborhood graph.

18.5.3 Hybrid Pretext Tasks

One way to use the information of the training nodes in designing pretext tasks is developed in (Hu et al, 2020c) where the context concept is raised. The goal of this work is to pre-train a GNN so that it maps nodes appearing in similar graph structure contexts to nearby embeddings. For every node v_i, the r-hop neighborhood of v_i contains all nodes and edges that are at most r-hops away from v_i in the graph. The context graph of v_i is a subgraph between r_1-hops and r_2-hops away from node v_i. It is required that $r_1 < r$ so that some nodes are shared between the neighborhood and the context graph, which is referred to as context anchor nodes. Examples of neighborhood and context graphs are shown in Fig. 18.6. Two GNN encoders are set up: the main GNN encoder is to get the node embedding $\mathbf{z}^r_{\text{GNN},i}$ based on their r-hop neighborhood node features and the context GNN is to get the node embeddings of every other node in the context anchor node set, which are then averaged to get the node context embedding $\mathbf{c}_i$. Then Hu et al (2020c) used negative sampling to jointly learn the main GNN and the context GNN. In the optimization process, positive samples refer to the situation when the center node of the context and the neighborhood graphs is the same while the negative samples refer to the situation when the center nodes of the context and the neighborhood graphs are different. The learning objective is a binary classification of whether a particular neighborhood and a particular context graph have the same center node and the negative likelihood loss is used as follows:

$$\mathscr{L}_{\text{ssl}} = -\left(\frac{1}{|\mathscr{K}|} \sum_{(v_i,v_j)\in\mathscr{K}} (y_i \log(\sigma((\mathbf{z}^r_{\text{GNN},i})^\top \mathbf{c}_j)) + (1-y_i)\log(1-\sigma((\mathbf{z}^r_{\text{GNN},i})^\top \mathbf{c}_j))))$$

(18.42)

where $y_i = 1$ for the positive sample where $i = j$ while $y_i = 0$ for the negative sample where $i \neq j$, with $\mathscr{K}$ denoting the set of positive and negative pairs, and σ is the sigmoid function computing the probability.

Similar idea to employ the context concept in completing pretext tasks is also proposed in (Jin et al, 2020d). Specifically, the context here is defined as:

$$\mathbf{y}_{ic} = \frac{|\Gamma_{\mathcal{V}_l}(v_i,c)| + |\Gamma_{\mathcal{V}_u}(v_i,c)|}{|\Gamma_{\mathcal{V}_l}(v_i)| + |\Gamma_{\mathcal{V}_u}(v_i)|}, c = 1,...,l, \qquad (18.43)$$

where $\mathcal{V}_u$ and $\mathcal{V}_l$ denote the unlabeled and labeled node set, $\Gamma_{\mathcal{V}_u}(v_i)$ denotes the unlabeled nodes that are adjacency to node v_i, $\Gamma_{\mathcal{V}_u}(v_i,c)$ denotes the unlabeled nodes that have been assigned class c and are adjacency to node v_i, $\mathcal{N}_{\mathcal{V}_l}(v_i)$ denotes the labeled nodes that are adjacency to node v_i, $\Gamma_{\mathcal{V}_l}(v_i,c)$ denotes the labeled nodes that are adjacency to node v_i and of class c. To generate labels for the unlabeled nodes so as to calculate the context vector $\mathbf{y}_i$ for each node v_i, label propagation (LP) (ZHU, 2002) or the iterative classification algorithm (ICA) (Neville and Jensen, 2000) is used to construct pseudo labels for unlabeled nodes in $\mathcal{V}_u$. Then the pretext task is approached by optimizing the following loss function:

$$\mathscr{L}_{\text{ssl}} = \frac{1}{|\mathcal{V}|} \sum_{v_i \in \mathcal{V}} \ell_{\text{CE}}(\mathbf{z}_{\text{ssl},i}, \mathbf{y}_i), \qquad (18.44)$$

The main issue of the above pretext task is the error caused by generating labels from LP or ICA. The paper (Jin et al, 2020d) further proposed two methods to improve the above pretext task. The first method is to replace the procedure of assigning labels of unlabeled nodes based on only one method such as LP or ICA with assigning labels by ensembling results from multiple different methods. Their second method treats the initial labeling from LP or ICA as noisy labels, and then leverages an iterative approach (Han et al, 2019) to improve the context vectors, which leads to significant improvements based on this correction phase.

One previous pretext task is to recover the topological distance between nodes. However, calculating the distance of the shortest path for all pairs of nodes even after the sampling is time-consuming. Therefore, Jin (Jin et al, 2020d) replaces the pairwise distance between nodes with the distance between nodes and their corresponding clusters. For each cluster, a fixed set of anchor/center nodes is established. For each node, its distance to this set of anchor nodes is calculated. The pretext task is to extract node features that encode the information of this node2cluster distance. Suppose k clusters are obtained by applying the METIS graph partitioning algorithm (Karypis and Kumar, 1998) and the node with the highest degree is assumed to be the center of the corresponding cluster, then each node v_i will have a cluster distance vector $\mathbf{d}_i \in \mathbb{R}^k$ and the distance-to-cluster pretext task is completed by optimizing:

$$\mathscr{L}_{\text{ssl}} = \frac{1}{|\mathcal{V}|} \sum_{v_i \in \mathcal{V}} \ell_{\text{MSE}}(\mathbf{z}_{\text{ssl},i}, \mathbf{d}_i), \qquad (18.45)$$

Aside from the graph topology and the node features, the distribution of the training nodes and their training labels are another valuable source of information for designing pretext tasks. One of the pretext tasks in (Jin et al, 2020d) is to require the node embeddings output by GNNs to encode the information of the topological distance from any node to the training nodes. Assuming that the total number of classes is p and for class $c \in \{1, ..., p\}$ and the node $v_i \in \mathcal{V}$, the average, minimum and maximum shortest path length from v_i to all labeled nodes in class c is calculated and denoted as $\mathbf{d}_i \in \mathbb{R}^{3p}$, then the objective is to optimize the same regression loss as defined in Eq. equation 18.45

The generating process of networks encodes abundant information for designing pretext tasks. Hu et al (2020d) propose the GPT-GNN framework for generative pre-training of GNNs. This framework performs attribute and edge generation to enable the pre-trained model to capture the inherent dependency between node attributes and graph structure. Assuming that the likelihood over this graph by this GNN model is $p(\mathcal{G}; \theta)$ which represents how the nodes in $\mathcal{G}$ are attributed and connected, GPT-GNN aims to pre-train the GNN model by maximizing the graph likelihood, i.e., $\theta^* = \max_\theta p(\mathcal{G}; \theta)$. Given a permutated order, the log likelihood is factorized autoregressively - generating one node per iteration as:

$$\log p_\theta(X, \mathcal{E}) = \sum_{i=1}^{|\mathcal{V}|} \log p_\theta(\mathbf{x}_i, \mathcal{E}_i | X_{<i}, \mathcal{E}_{<i}) \tag{18.46}$$

For all nodes that are generated before the node i, their attributes $X_{<i}$, and the edges between these nodes $\mathcal{E}_{<i}$ are used to generate a new node v_i, including both its attribute $\mathbf{x}_i$ and its connections with existing nodes $\mathcal{E}_i$. Instead of directly assuming that $\mathbf{x}_i, \mathcal{E}_i$ are independent, they devise a dependency-aware factorization mechanism to maintain the dependency between node attributes and edge existence. The generation process can be decomposed into two coupled parts: (1) generating node attributes given the observed edges, and (2) generating the remaining edges given the observed edges and the generated node attributes. For computing the loss of attribute generation, the generated node feature matrix X is corrupted by masking some dimensions to obtain the corrupted version $\hat{X}^{\text{Attr}}$ and further fed together with the generated edges into GNNs to get the embeddings $\hat{Z}_{\text{GNN}}^{\text{Attr}}$. Then, the decoder $\text{Dec}^{\text{Attr}}(\cdot)$ is specified, which takes $\hat{Z}_{\text{GNN}}^{\text{Attr}}$ as input and outputs the predicted attributes $\text{Dec}^{\text{Attr}}(\hat{Z}_{\text{GNN}}^{\text{Attr}})$. The attribute generation loss is:

$$\mathcal{L}_{\text{ssl}}^{\text{Attr}} = \frac{1}{|\mathcal{V}|} \sum_{v_i \in \mathcal{V}} \ell_{\text{MSE}}(\text{Dec}^{\text{Attr}}(\hat{z}_{\text{GNN},i}^{\text{Attr}}), \mathbf{x}_i), \tag{18.47}$$

where $\hat{z}_{\text{GNN},i}^{\text{Attr}} = \hat{Z}_{\text{GNN}}^{\text{Attr}}[i,:]^\top$ denotes the decoded embedding of node v_i. For computing the loss of edge reconstruction, the original generated node feature matrix X is directly fed together with the generated edges into GNNs to get the embeddings $Z_{\text{GNN}}^{\text{Edge}}$. Then the contrastive NT-Xent loss is calculated:

$$\mathcal{L}_{\text{ssl}}^{\text{Edge}} = \frac{1}{|\mathcal{P}^+|} \sum_{(v_i, v_j) \in \mathcal{P}^+} \ell_{\text{NT-Xent}}(Z_{\text{GNN}}^{\text{Edge}}, Z_{\text{GNN}}^{\text{Edge}}, \mathcal{P}^-), \tag{18.48}$$

where $\mathscr{P}^+$ contains positive pairs of connected nodes (v_i, v_j) while $\mathscr{P}^- = \bigcup_{(v_i, v_j) \in \mathscr{P}^+} \mathscr{P}_{v_i}^-$ represents all sets of negative samples and $P_{v_i}^-$ contains all nodes that are not directly linked with node v_i. Note here two views are set equal, i.e., $Z^1 = Z^2 = Z_{\text{GNN}}^{\text{Edge}}$.

18.6 Node-graph-level SSL Pretext Tasks

All the above pretext tasks are designed based on either the node or the graph level supervision. However, there is another final line of research combining these two sources of supervision to design pretext tasks, which we summarize in this section.

Veličković et al (2019) proposed to maximize the mutual information between representations of high-level graphs and low-level patches. In each iteration, a negative sample $\hat{X}, \hat{A}$ is generated by corrupting the graph through shuffling node features and removing edges. Then a GNN-based encoder is applied to extract node representations Z_{GNN} and $\hat{Z}_{\text{GNN}}$, which are also named as the local patch representations. The local patch representations are further fed into an injective readout function to get the global graph representations $\mathbf{z}_{\text{GNN},\mathscr{G}} = \text{READOUT}(Z_{\text{GNN}})$. Then the mutual information between Z_{GNN} and $\mathbf{z}_{\text{GNN},\mathscr{G}}$ is maximized by minimizing the following loss function:

$$\mathscr{L}_{\text{ssl}} = \frac{1}{|\mathscr{P}^+| + |\mathscr{P}^-|} \Big(\sum_{i=1}^{|\mathscr{P}^+|} \mathbb{E}_{(X,A)} [\log \sigma(\mathbf{z}_{\text{GNN},i}^\top W \mathbf{z}_{\text{GNN},\mathscr{G}})] \tag{18.49}$$

$$+ \sum_{j=1}^{|\mathscr{P}^-|} \mathbb{E}_{(\hat{X},\hat{A})} [\log(1 - \sigma(\tilde{\mathbf{z}}_{\text{GNN},i}^\top W \mathbf{z}_{\text{GNN},\mathscr{G}}))] \Big),$$

where $|\mathscr{P}^+|$ and $|\mathscr{P}^-|$ are the number of the positive and negative pairs, σ stands for any nonlinear activation function and PReLU is used in (Veličković et al, 2019), $\mathbf{z}_{\text{GNN},i}^\top W \mathbf{z}_{\text{GNN},\mathscr{G}}$ calculates the weighted similarity between the patch representation centered at node v_i and the graph representation. A linear classifier is followed up to classify nodes after the above contrastive pretext task.

Similar to (Veličković et al, 2019) where the mutual information between the patch representations and the graph representations is maximized, Hassani and Khasahmadi (2020) proposed another framework of contrasting the node representations of one view and the graph representations of another view. The first view is the original graph and the second view is generated by a graph diffusion matrix. The heat and personalized PageRank (PPR) diffusion matrix are considered, which are:

$$S^{\text{heat}} = \exp(tAD^{-1} - t), \tag{18.50}$$

$$S^{\text{PPR}} = \alpha(\mathbf{I}_n - (1 - \beta)D^{-1/2}AD^{-1/2})^{-1}, \tag{18.51}$$

where β denotes teleport probability, t is the diffusion time, and D is the diagonal degree matrix. After D is obtained, two different GNN encoders followed by a shared projection head are applied on nodes in the original graph adjacency matrix and the generated diffusion matrix to get two different node embeddings Z_{GNN}^1 and Z_{GNN}^2. Two different graph embeddings $\mathbf{z}_{GNN,\mathscr{G}}^1$ and $\mathbf{z}_{GNN,\mathscr{G}}^2$ are further obtained by applying a graph pooling function to the node representations (before the projection head) and followed by another shared projection head. The mutual information between nodes and graphs in different views is maximized through:

$$\mathscr{L}_{ssl} = -\frac{1}{|\mathscr{V}|} \sum_{v_i \in V} (MI(\mathbf{z}_{GNN,i}^1, \mathbf{z}_{GNN,\mathscr{G}}^2) + MI(\mathbf{z}_{GNN,i}^2, \mathbf{z}_{GNN,\mathscr{G}}^1)), \qquad (18.52)$$

where the MI represents the mutual information estimator and four estimators are explored, which are noise-contrastive estimator, Jensen-Shannon estimator, normalized temperature-scaled cross-entropy, and Donsker-Varadhan representation of the KL-divergence. Note that the mutual information in Eq. equation 18.52 is averaged over all graphs in the original work (Hassani and Khasahmadi, 2020). Additionally, their results demonstrate that Jensen-Shannon estimator achieves better results across all graph classification tasks, whereas in the node classification task, noise contrastive estimation achieves better results. They also discover that increasing the number of views does not increase the performance on downstream tasks.

18.7 Discussion

Existing methods employing self-supervision to graph neural networks achieve performance improvements and numerous insightful results are also discovered in the meantime. While most of the self-supervised pretext tasks are helpful for the downstream tasks, there are still a fair proportion of pretext tasks that bring weak improvement or even fail to boost the performance (Gao et al, 2021; Jin et al, 2020d; Manessi and Rozza, 2020; You et al, 2020c). This is either because these pretext tasks are highly unrelated to the primary task, i.e., the encoded features useful for pretext tasks are useless or even harmful (Manessi and Rozza, 2020) for downstream tasks or because the information learned from completing pretext tasks can already be learned from completing downstream tasks by GNNs (Jin et al, 2020d). Besides, the strength of the performance improvement depends on the specific GNN architecture used for completing pretext and downstream tasks. The improvements are more significant for basic GNNs such as GCN, GAT, and GIN while less for more advanced GNNs such as GMNN (You et al, 2020c). Furthermore, one pretext task is not universally the best across multiple datasets (Gao et al, 2021; Manessi and Rozza, 2020). Therefore, whether a self-supervised pretext task helps GNNs in the standard target performance is determined by first whether the dataset allows the GNNs to extract extra feature information through completing pretext tasks, and second whether the extra self-supervised information complement, contradict to or has already been covered by information extracted from existing architecture (You et al, 2020c). Numerous works focus on applying contrastive learning as a form of self-supervised learning (Chen et al, 2020b; Hassani and Khasahmadi, 2020; Veličković et al, 2019; You et al, 2020b; Zhu et al, 2021). Generally they find that while composing different augmentations benefits the performance (You et al, 2020b), increasing the number of views generated from the same graph augmentation technique to more than two cause no further improvement (Hassani and Khasahmadi, 2020), which is different from visual representation learning. Moreover, the beneficial combinations of augmentations are data-specific because of the highly heterogeneous nature of the graph-structured data and harder contrastive tasks are more helpful than overly simple ones (You et al, 2020b). Therefore, designing viable pretext tasks requires domain specific knowledge and should be targeted towards specific types of networks, GNN architectures and downstream tasks.

18.8 Summary

In this chapter, we provided a systemic, categorical and comprehensive overview on the recent works leveraging self-supervised learning in graph neural networks. Despite recent successes achieved by applying self-supervised learning in the text and image domains, self-supervised learning applied to the graph domain, especially for graph neural networks, is still in its emerging stage. Several promising directions could be pursued to further advance this field. First, although a large surge of research focuses on designing effective pretext tasks boosting the performance of graph neural networks, few works focus on visualizing, interpreting and explaining the underlying reason causing such beneficial performance improvements. Deeply understanding the intrinsic mechanism as to why and how SSL helps GNNs could help us design more powerful pretext tasks. Second, similar to the work defining the architectural design space for GNNs to quickly query the best GNN design for a novel task on a novel dataset (You et al, 2020a), we should collect and classify various pretext tasks and create a design space for SSL in GNNs. This allows for transferring the best designs of pretext tasks across different downstream tasks, GNN architectures and datasets. We hope that this chapter can shed some light on the main ideas of applying self-supervised learning to graph neural networks and related applications in order to encourage progress in the field.

Editor's Notes: Although methods introduced in the previous chapter (chapter 4, 5, 6, 15, and 16) have achieved state-of-the-art performances in corresponding tasks, they require large annotated datasets. Self-supervised learning seeks to create and utilize pretext labels on unlabeled data. Pretext tasks are relevant to traditional graph analysis tasks, such as node-level tasks (chapter 4) and graph level tasks (chapter 9), while pretext tasks use pseudo labels. The development of self-supervised GNN is of great significance to domains where labeled data are difficult to obtain, such as drug development (chapter 24). Besides, domains that have accumulated a large number of unlabeled data sets, such as computer vision (chapter 20) and natural language processing (chapter 21), also benefit from self-supervised learning.

Part IV
Broad and Emerging Applications with Graph Neural Networks

Chapter 19
Graph Neural Networks in Modern Recommender Systems

Yunfei Chu, Jiangchao Yao, Chang Zhou and Hongxia Yang

Abstract Graph is an expressive and powerful data structure that is widely applicable, due to its flexibility and effectiveness in modeling and representing graph structure data. It has been more and more popular in various fields, including biology, finance, transportation, social network, among many others. Recommender system, one of the most successful commercial applications of the artificial intelligence, whose user-item interactions can naturally fit into graph structure data, also receives much attention in applying graph neural networks (GNNs). We first summarize the most recent advancements of GNNs, especially in the recommender systems. Then we share our two case studies, dynamic GNN learning and device-cloud collaborative Learning for GNNs. We finalize with discussions regarding the future directions of GNNs in practice.

19.1 Graph Neural Networks for Recommender System in Practice

19.1.1 Introduction

The Introduction of GNNs Graph has a long history originated from the Seven Bridges of Königsberg problem in 1736 (Biggs et al, 1986). It is flexible to model

Yunfei Chu,
DAMO Academy, Alibaba Group, e-mail: `fay.cyf@alibaba-inc.com`

Jiangchao Yao
DAMO Academy, Alibaba Group, e-mail: `jiangchao.yjc@alibaba-inc.com`

Chang Zhou
DAMO Academy, Alibaba Group, e-mail: `ericzhou.zc@alibaba-inc.com`

Hongxia Yang
DAMO Academy, Alibaba Group, e-mail: `yang.yhx@alibaba-inc.com`

complex relationships among individuals, which makes it a ubiquitous data structure widely applied in numerous fields, *e.g.*, biology, finance, transportation, social network, recommender systems.

Despite there are traditional topics of extracting deterministic information in graph theory like shortest path, connected components, local clustering, graph isomorphism, and *etc.*, machine learning applications for graph data focus more on predicting the missing parts or future dynamics. Among these applications, the most typical research problems studied in recent year, are predicting whether there exists or will emerge an edge between two nodes (link prediction), and inferring node-level or graph-level labels (node/graph classification).

The recent progress in deep learning leads to a booming learning paradigm called representation learning, which also becomes the de facto standard in solving graph machine learning problems. The idea of graph representation learning is to encode graph primitives as real-valued vectors in the same metric space, which are then involved in downstream applications. The encoder takes as input the original graph such as node attributes vector and graph adjacency matrix in an end-to-end fashion, rather than traditional methods that require extracting heuristic features such as betweenness centrality, pagerank value, number of closed triangles.

Next, we summarize recent graph node representation techniques in a unified framework and focus only on the link prediction task. We illustrate several representative approaches in recent literature from a node-centric perspective, since the node-centric view can naturally fit into scalable message passing implementations that are originally popular in graph mining community (Malewicz et al, 2010; Y.Low et al, 2012) and then borrowed to GNNs community (Wang et al, 2019f; Zhu et al, 2019c).

For a graph $\mathcal{G} = (\mathcal{V}, \mathcal{E})$ with adjacency matrix A, a standard graph neural network model has the following components.

- An ego-network extractor EGO that extracts a local subgraph around the node v. This local subgraph is also referred to as the receptive field of v which is then used by the node encoder.
- An encoder ENC that maps each node $v \in \mathcal{V}$ into a vector in a metric space R^d. The encoder takes as input the ego-network of v, as well as any node/edge representation in $EGO(v)$. A similarity function is defined on R^d to measure how close two nodes appear to be.
- A learning objective $\mathcal{L}$. We do not discuss node classification here and only focus on unsupervised node representation learning. The objective can be reconstructing the adjacency matrix A, transformations of A, or any sampled form of A and its transformations.

Random Walk-style

Early graph representation learning approaches (Perozzi et al, 2014; Tang et al, 2015b; Cao et al, 2015; Zhou et al, 2017; Ou et al, 2016; Grover and Leskovec, 2016) in deep learning era are inspired by word2vec (Mikolov et al, 2013b), an efficient word embedding method in natural language processing community. These methods do not need any neighborhood for encoding, where EGO plays as an identity mapping. The encoder ENC takes as the node id in the graph and assigns a trainable vector to each node.

The very different part of these methods is the learning objective. Approaches like Deepwalk, LINE, Node2vec use different random walk strategies to create positive node pairs (u, v) as the training example, and estimate the probability of visiting v given u, $p(v|u)$, as a multinomial distribution,

$$p(v|u) = \frac{\exp(sim(u,v))}{\sum_{v'} \exp(sim(u,v'))},$$

where sim is a similarity function. They exploit an approximated Noise Constrained Estimation (NCE) loss (Gutmann and Hyvärinen, 2010), known as skip gram with negative sampling originated in word2vec as the following, to reduce the high computation cost,

$$\log \sigma(sim(u,v)) + k\mathbb{E}_{v' \sim q_{neg}} \log(1 - \sigma(sim(u,v'))).$$

q_{neg} is a proposed negative distribution, which impacts the variation of the optimization target (Yang et al, 2020d). Note that this formula can be also approximated with sampled softmax (Bengio and Senécal, 2008; Jean et al, 2014), which in our experience performs better in top-k recommendation tasks as the node number becomes extremely large (Zhou et al, 2020a).

These learning objectives have connections with traditional node proximity measurements in graph mining community. GraRep (Cao et al, 2015), APP (Zhou et al, 2017) borrows the idea from (Levy and Goldberg, 2014) and point out these random walk based method are equivalent to preserving their corresponding transformations of the adjacency matrix A, such as personalized pagerank.

Matrix Factorization-style

HOPE (Ou et al, 2016) provides a generalized matrix form of other types of node proximity measurement, *e.g.*, katz, adamic-adar, and adopts matrix factorization to learn embedding that preserve these proximity. NetMF (Qiu et al, 2018) unifies several classic graph embedding methods in the framework of matrix factorization, provides connections between the deepwalk-like approaches and the theory of graph Laplacian.

GNN-style

Graph neural network (Kipf and Welling, 2017b; Scarselli et al, 2008) provides an end-to-end semi-supervised learning paradigm that was previously modeled via label propagations. It can also be used to learn node representations in an unsupervised manner like the above graph embedding methods. GNN-like approaches for unsupervised learning, compared to deepwalk-like methods, are more powerful in capturing local structural, *e.g.*, have at most the power of WL-test (Xu et al,

2019d). The downstream link prediction task that requires local-structural aware representation or cooperation with node features may benefit more from GNN-style approaches.

The *EGO* operator collects and constructs the receptive field of each node. For GCN (Kipf and Welling, 2017b), a full k-layer neighborhood is required for each node, making it hard to work for large graphs which usually follow power-law degree distribution. GraphSage (Hamilton et al, 2017b) instead samples a fixed-size neighborhood in each layer, mitigates this problem and can scale to large graphs. LCGNN (Qiu et al, 2021) samples a local cluster around each node by short random-walks with theoretical guarantee.

Then different kinds of Aggregation functions are proposed within this receptive field. GraphSage investigates several neighborhood aggregation alternatives, including mean/max pooling, LSTM. GAT (Veličković et al, 2018) utilizes self-attention to perform the aggregation, which shows stable and superior performance in many graph benchmarks. GIN (Xu et al, 2019d) has a slightly different aggregation function, whose discriminative/representational power is proved to be equal to the power of the WL test. As link prediction task may also consider structural similarity between two nodes besides their distance, this local structural preserving method may achieve good performance for networks that have obvious local structural patterns.

The learning objectives of GNN-style approaches are similar with those in random walk style ones.

Introduction of Modern Recommender System

Recommender system, one of the most successful commercial applications of the artificial intelligence, whose user-item interactions can naturally fit into graph structure data, also receives much attention to applying GNNs. We now give a brief introduction about the problem settings, the classic methods in recommender systems.

The user-item relationships are the most typical form of recommender systems, *e.g.*, news recommendation, e-commerce recommendation, video recommendation. Although recommender systems are eventually optimizing for a complex ecosystem of multi-sided participants (Abdollahpouri et al, 2020), i.e., the users, the platform and the content provider, we only focus on how the platform will maximize the user-side utility in this chapter.

In a user-item recommender system $\mathscr{S}$ with recommender algorithm $\mathscr{A}$, $\mathscr{U}$ is the user set and $\mathscr{I}$ is the item set. At timestamp t, a user $u \in \mathscr{U}$ visits $\mathscr{S}$, a list of items $\mathscr{I}_{u,t}$ is produced by $\mathscr{A}$. u takes positive actions, *e.g.*, click, buy, play, on parts of the items in $\mathscr{I}_{u,t}$, referred to as $\mathscr{I}_{u,t}^+$, while performing the corresponding negative actions on the others, *e.g.*, not click, not buy, not play, referred to as $\mathscr{I}_{u,t}^-$.

The basic data collected from an industrial recommender system, can be described as

$$\mathscr{D}_{\mathscr{S},\mathscr{A}} = \{(t, \mathscr{I}_{u,t}^+, \mathscr{I}_{u,t}^-) | u \in \mathscr{U}, t\}. \tag{19.1}$$

The short-term objective [1] of an algorithm in modern recommender systems, can be summarized as

$$\mathscr{A} = \arg\max_{\mathscr{A}} \sum_{u,t} Utility(\mathscr{I}_{u,t}^+), \tag{19.2}$$

in which the *Utility* function could be considered as maximizing *click through rate*, *GMV*, or a mixture of multiple objectives (Ribeiro et al, 2014; McNee et al, 2006).

A modern commercial recommender system, especially for those with over millions of end-users and items, has adopted a multi-stage modeling pipeline as the tradeoff between the business goals and the efficiency given the constraints of limited computing resources. Different stages have different simplifications of the data organization and objectives, which many research papers do not put in a clear way.

In the following, we first review several simplifications of the industrial recommendation problem setting, that are clean enough for the research community. Then we describe the multi-stage pipeline and the problem in each stage, review classic methods to handle the problem and revisit how GNNs are applied in existing methods, trying to give an objective view about these methods.

Simplifications of the collected data.

- *Impression bias.* The user feedback data generated under algorithm $\mathscr{A}$, has a bias towards estimating the oracle user preference. This critical and unique problem for recommender system, is usually not considered, especially for the early works.
- *Negative feedback.* $|\mathscr{L}_{u,t}^-|$, the number of negative behaviors in one display, is orders of magnitude larger than $|\mathscr{L}_{u,t}^+|$, and very few dataset has collected negative feedback. Most of the well-known papers in the research community ignore those true negative user feedback, instead, they simulate negative feedback by sampling from a proposal distribution, which is not the ground truth and the metrics designed over the simulated feedback may not reveal true performance.
- *Temporal information.* Early studies prefer a static view of recommendation, which eliminates the temporal information of t in the user behavior sequences.

Multi-stage model pipeline in modern recommender systems.

- Retrieval Phase. This phase is also referred to as candidate generation or recall phase. It narrows down the collection of relevant items from billions to hundreds via efficient similarity-based learning, indexing, and searching. To prevent from sticking into dead loops caused by fitting the exposure distribution, retrieval phase has to independently provide sufficient diversity for different downstream purposes or strategies, while retaining the accuracy. As the candidate set is in extremely large size, approaches in the recall phase are usually in the form of point-wise modeling that is simple to build sophisticated index and perform

---

[1] We indicate the short-term objective as the objective in the sense of each request response. Here we do not consider further impacts on the ecosystem brought by an algorithm.

Table 19.1: Data simplifications in different settings

Setting / Phase in Pipeline	Data Simplification
Matrix Completion / Retrieval Phase	$\mathscr{D}_{\mathscr{S}} = \{\mathscr{L}_u^+ \mid u \in \mathscr{U}\}$
Click Through Rate Prediction / Rank Phase	$\mathscr{D}_{\mathscr{S}} = \{(\mathscr{L}_u^+, \mathscr{L}_u^-) \mid u \in \mathscr{U}\}$
Sequential Recommendation / Retrieval Phase	$\mathscr{D}_{\mathscr{S}} = \{(t, \mathscr{L}_{u,t}^+) \mid u \in \mathscr{U}, t\}$

efficient retrieval. The most widely used measurement for this phase is the top-k hit ratio.

- Rank Phase. The problem space is quite different from those in the retrieval phase, since rank phase needs to give precise comparison within a much smaller subspace, instead of recalling as many as good items from the entire item candidates set. Restricted to a small number of candidates, it is capable of exploiting more complex methods over the user-item interaction in acceptable response time.

- Re-rank Phase. Considering the effects studied in the discrete choice model (Train, 1986), the relationships among the displayed items may have significant impacts on the user behavior. This poses opportunities to consider from the combinational optimization perspective, i.e., how to chose a combination of the subset which maximizes the whole utilities of the recommendation list.

The above stages can be adjusted according to different characteristics of the recommendation scenario. For example, if the candidate set is at hundreds or thousands, recall phase is not necessarily required as the computation power is usually enough to cover such rank-all operation at once. The re-rank phase is also not necessary if the item number per request is few.

We summarize in Table 19.1 the different data simplifications made in different problem settings with their corresponding pipeline stages.

19.1.2 Classic Approaches to Predict User-Item Preference

The fundamental ability required by Recommender System is to predict the possibility that a user will take actions on a specific displayed item, which we refer to as the point-wise preference estimation, $p(item|user)$. Now we review several classic approaches in dealing with the cleanest setting of Matrix Completion in Table 19.1. The user-item iteraction matrix perspective of data organization $\mathscr{D}_{\mathscr{S}} = \{\mathscr{L}_u^+ \mid u \in \mathscr{U}\}$ is $M = \{M_{u,i} \mid u \in \mathscr{U}, i \in \mathscr{I}\}$, where each row $M_u = \mathscr{L}_u^+$. The famous Collaborative Filtering methods in recommendation can be categorized into neighborhood-based one and model-based one.

Neighborhood-based Approaches

Item-based collaborative filtering first identifies a set of similar items for each of the items that the user has clicked/purchased/rated, and then recommends top-N items by aggregating the similarities. User-based CF, on the other hand, identifies similar users and then performs aggregation on their clicked items.

The key part in Neighborhood-based Approaches is the definition of the similarity metric. Take item-based CF as an example, top-k heuristic approaches calculate item-item similarity from the user-item interaction matrix M, *e.g.*, pearson correlation, cosine similarity. Storing $|\mathscr{I}|x|\mathscr{I}|$ similarity score pairs is intractable. Instead, to help produce a top-k recommendation list efficiently, neighborhood-based k-nearest-neighbor CF usually memorizes top few similar items for each item, resulting in a sparse similarity matrix C. Despite the heuristics, SLIM (Ning and Karypis, 2011) learns such sparse similarity by reconstructing M via MC with zero diagonal and sparse constraints in C.

One draw back of storing only the sparse similarity is that, it cannot identify less-similar relationships which restricts its downstream applications.

Model-based Approaches

Model-based methods learn similarity functions between user and item by optimizing an objective function. Matrix Factorization, the prior of which is that the user-behavior matrix is low-rank, i.e., all users' tastes can be described by linear combinations of a few style latent factors. The prediction for a user's preference on an item can be calculated as the dot product of the corresponding user and item factor.

19.1.3 Item Recommendation in user-item Recommender Systems: a Bipartite Graph Perspective

The matrix completion setting also has an equivalent form in bipartite graph,

$$\mathscr{G} = (\mathscr{V}, \mathscr{E}), \tag{19.3}$$

where $\mathscr{V} = \mathscr{U} \cup \mathscr{I}$, *i.e.*, the union of the user set $\mathscr{U}$ and the item set $\mathscr{I}$, and $\mathscr{E} = \{(u,i)|i \in \mathscr{I}_u^+, u \in \mathscr{U}\}$, *i.e.*, the collection of the edges between u and his/her clicked i. Then the point-wise user-item preference estimation can be viewed as a link prediction task in this user-item interaction bipartite graph.

Heuristic graph mining approaches, which fall into the category of neighborhood-based CF, are widely used in the retrieval phase. We can calculate user-item similarity by performing graph mining tasks like Common Neighbors, Adar (Adamic and Adar, 2003), Katz (Katz, 1953), Personalized PageRank (Haveliwala, 2002), over the original bipartite graph, or calculate item-item similarity on its induced item-item correlation graph (Zhou et al, 2017; Wang et al, 2018b) which are then used in the final user preference aggregation.

Graph embedding techniques for industrial recommender system are first explored in (Zhou et al, 2017) and its successor with side information support (Wang et al, 2018b). They construct an item correlation graph of billions of edges from user-item click sequences organized by sessions. Then a deepwalk-style graph embedding method is applied to calculate the item representations, which then provides item-item similarities in the retrieval phase. Though it's shown in (Zhou et al, 2017) that embedding based method has advantage in scenarios where the top-k heuristics cannot provide any item-pair similarity, it's still debatable whether the similarity given by graph embedding methods can outperform carefully designed heuristic ones when all the top-k similar item can be retrieved.

We also note that, graph embedding techniques can be regarded as matrix factorization for a transformation of the graph adjacency matrix A, as discussed in earlier sections. That means, theoretically the difference between graph embedding techniques and the basic matrix factorization are their priors, i.e., what matrix is assumed to be the best to factorize. Factorization of the transformations of A indicates to fit an evolved system in the future while traditional MF methods are factorizing the current static system.

Graph neural networks for industrial recommender system are first studied in (Ying et al, 2018b), whose backend model is a variant of GraphSage. PinSage computes the L1 normalized visit counts of nodes during random walks started from a given node v, and the top-k counted nodes are regarded as v's receptive field. Weighted aggregation is performed among the nodes according to their normalized counts. As GraphSage-like approaches do not suffer from too large neighborhood, PinSage is scalable to web-scale recommender system with millions of users and items. It adopts a triplet loss, instead of NCE-variants that are usually used in other papers.

We want to discuss more about the choice of negative examples in representation learning based recommender models, including GNNs, in the retrieval phase. As retrieval phase aims to retrieve the k most relevant items from the entire item space, it's crucial to keep an item's global position far from *all* irrelevant items. In an industrial system with an extremely large candidate set, we find the performance of any representation-based model very sensitive to the choice of negative samples and the loss function. Though there seems a trend in mixing all kinds of hand-crafted hard examples (Ying et al, 2018b; Huang et al, 2020b; Grbovic and Cheng, 2018) in binary cross entropy loss or triplet loss, unfortunately, it has even no theoretical support that can lead us to the right direction. In practice, we find it a good choice to apply sampled softmax (Jean et al, 2014; Bengio and Senécal, 2008), InfoNCE (Zhou et al, 2020a) in the retrieval phase with an extremely large candidate set, where the latter has also an effect of debiasing.

GNNs are a useful tool to incorporate with relational features of user and item. KGCN (Wang et al, 2019e) enhances the item representation by performing aggregations among its corresponding entity neighborhood in a knowledge graph. KGNN-LS (Wang et al, 2019c) further poses a label smoothness assumption, which posits that similar items in the knowledge graph are likely to have similar user preference. It adds a regularization term to help learn such a personalized weighted knowledge graph. KGAT (Wang et al, 2019j) shares a generally similar idea with KGCN. The only main difference is an auxiliary loss for knowledge graph reconstruction.

Despite there are many more paper discussing about how to fuse external knowledge, relationships of other entities, which all argue it's beneficial for downstream recommendation tasks, one should seriously consider whether its system needs such external knowledge or it will introduce more noises than benefits.

19.2 Case Study 1: Dynamic Graph Neural Networks Learning

19.2.1 Dynamic Sequential Graph

In a recommender, we can obtain a list of user-item interaction tuples $\mathscr{E} = \{(u,i,t)\}$ observed in a time window, where the user $u \in \mathscr{U}$ interacts with an item $i \in \mathscr{I}$ associated with a timestamp $t \in \mathbb{R}^+$. For a user $u \in \mathscr{U}$ (or an item $i \in \mathscr{I}$) at time t, we define the 1-depth dynamic sequential subgraph of user u (or item i) at time t as a set of interactions of user u (or item i) before time t in chronological order, denoted by $\mathscr{G}_{u,t}^{(1)} = \{(u,i,\tau)|\tau < t, (u,i,\tau) \in \mathscr{E}\}$ (or $\mathscr{G}_{i,t}^{(1)} = \{(u,i,\tau)|\tau < t, (u,i,\tau) \in \mathscr{E}\}$). Given the k-depth dynamic sequential subgraphs $\mathscr{G}_{i,t}^{(k)}$ for $i \in \mathscr{I}$ (or $\mathscr{G}_{u,t}^{(k)}$ for $u \in \mathscr{U}$), we define the $(k+1)$-depth dynamic sequential subgraph of user u (or item i) at time t as a set of k-depth dynamic sequential subgraphs that user u (or item i) interacts in chronological order with its 1-depth dynamic sequential subgraphs, $\mathscr{G}_{u,t}^{(k+1)} = \{\mathscr{G}_{i,\tau}^{(k)}|\tau < t, (u,i,\tau) \in \mathscr{E}\} \cup \mathscr{G}_{u,t}^{(1)}$ (or $\mathscr{G}_{i,t}^{(k+1)} = \{\mathscr{G}_{u,\tau}^{(k)}|\tau < t, (u,i,\tau) \in \mathscr{E}\} \cup \mathscr{G}_{i,t}^{(1)}$). The illustration of DSG is shown in Figure 19.1. We define the historical behavior sequence of user u (or item i) at time t as a sequence of interacted items (or users) in chronological order, denoted by $\mathscr{S}_{u,t} = \{(i,\tau)|\tau < t, (u,i,\tau) \in \mathscr{E}\}$ (or $\mathscr{S}_{i,t} = \{(u,\tau)|\tau < t, (u,i,\tau) \in \mathscr{E}\}$).

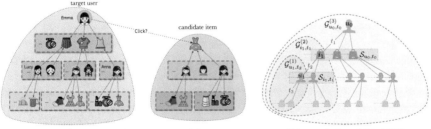

(a) Dynamic sequential graphs in recommendation.   (b) An example of a user's 3-depth DSG.

Fig. 19.1: Illustration of Dynamic Sequential Graph. DSG is a heterogeneous time-evolving dynamic graph combining the high-hop connectivity in graphs and the temporal dependency in sequences. DSG is constructed from bottom to top recursively.

19.2.2 DSGL: Dynamic Sequential Graph Learning

19.2.2.1 Overview

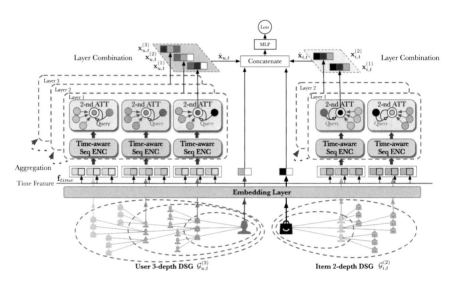

Fig. 19.2: Framework of the proposed DSGL method. DSGL constructs DSGs for the target user u (left) and the candidate item i (right) respectively. Their representations are refined with multiple aggregation layers, each of which consists of a time-aware sequence encoding layer and a second-order graph attention layer. DSGL gets the final representations via layer combination followed by an MLP-based prediction layer. Modules of the same color share the same set of parameters.

Based on the constructed user-item interaction DSG, we propose the edge learning model named Dynamic Sequential Graph Learning (DSGL), as illustrated in Figure 19.2. The basic idea of DSGL is to perform graph convolution iteratively on the DSGs for the target user and the candidate item on their corresponding devices, by aggregating the embeddings of neighbors as the new representation of a target node. The aggregator consists of two parts: (1) the time-aware sequence encoding that encodes the behavior sequence with time information and temporal dependency captured; and (2) the second-order graph attention that activates the related behavior in the sequence to eliminate noisy information. Besides the above two components, we also propose an embedding layer that initializes user, item, and time embeddings, a layer combination module that combines the embeddings of multiple layers to achieve final representations, and a prediction layer that outputs the prediction score.

19.2.2.2 Embedding Layer

There are four groups of inputs in the proposed DSGL: the target user u, the candidate item i, the k-depth DSGs of the target user $\mathscr{G}_{u,t}^k$ and $(k-1)$-depth DSGs of the candidate item $\mathscr{G}_{i,t}^{k-1}$. For each field of discrete features, such as age, gender,category, brand, and ID, we represent it as an embedding matrix. By concatenating all fields of features, we have the node feature of items, denoted by $\mathbf{f}_{item} \in \mathbb{R}^{d_i}$. Similarly, $\mathbf{f}_{user} \in \mathbb{R}^{d_u}$ represents the concatenated embedding vectors of fields in the category of user. As for the interaction timestamp in DSG, we compute the time intervals between the interaction time and its parent interaction time as time decays. Given a historical behavior sequence $\mathscr{S}_{u,t}$ of user u at the timestamp t, each interaction $(u,i,\tau) \in \mathscr{S}_{u,t}$ corresponds to a time decay $\Delta_{(u,i,\tau)} = t - \tau$. Following (Li et al, 2020g), we transform the continuous time decay values to discrete features by mapping them to a series of buckets with the ranges $[b^0, b^1), [b^1, b^2), \ldots, [b^l, b^{l+1})$, where the base b is a hyper-parameter. Then by performing the embedding lookup operation, the time decay embedding can be obtained, denoted by $\mathbf{f}_{time} \in \mathbb{R}^{d_t}$.

19.2.2.3 Time-Aware Sequence Encoding

The nodes at each layer of DSGs are in time order, which reflects the time-varying preference of users as well as the popularity evolution of items. Thus we perform sequence modeling as a part of GNN to capture the dynamics of the interaction sequences. We design a *time-aware sequential encoder* to utilize the time information explicitly. For each interaction (u,i,t), we have the historical behavior sequence $\mathscr{S}_{u,t}$ of user u and $\mathscr{S}_{i,t}$ of item i. For sequence $\mathscr{S}_{u,t}$, by feeding each interacted item along with the time decay in the sequence into the embedding layer, the behavior embedding sequence is formed with the combined feature sequence, as $\{\mathbf{e}_{i,\tau}|(i,\tau) \in \mathscr{S}_{u,t}\}$, where $\mathbf{e}_{i,\tau} = [\mathbf{f}_{item_i}; \mathbf{f}_{time_\tau}] \in \mathbb{R}^{d_i+d_t}$ is the embedding of item i in the sequence. Similarly, for sequence $\mathscr{S}_{i,t}$, we have the embedding sequence as $\{\mathbf{e}_{u,\tau}|(u,\tau) \in \mathscr{S}_{i,t}\}$, where $\mathbf{e}_{u,\tau} = [\mathbf{f}_{user_u}; \mathbf{f}_{time_\tau}] \in \mathbb{R}^{d_u+d_t}$. We take the obtained embedding as the zero-layer of inputs in the time-aware sequence encoder, i.e., $\mathbf{x}_{u,t}^{(0)} = \mathbf{e}_{u,t}$ and $\mathbf{x}_{i,t}^{(0)} = \mathbf{e}_{i,t}$. For ease of notation, we will drop the superscript in the rest of the following two subsections.

In the time-aware sequence encoding, we infer the hidden state of each node in the behavior sequence step by step in a RNN-based manner. Given the behavior sequences $\mathscr{S}_{u,t}$ and $\mathscr{S}_{i,t}$, we represent j-th item's hidden states and inputs in the sequence $\mathscr{S}_{u,t}$ as $\mathbf{h}_{item_j}$ and $\mathbf{x}_{item_j}$, and j-th user's hidden states and inputs in the sequence $\mathscr{S}_{i,t}$ as $\mathbf{h}_{user_j}$ and $\mathbf{x}_{user_j}$. The forward formulas are

$$\mathbf{h}_{item_j} = \mathscr{H}_{item}(\mathbf{h}_{item_{j-1}}, \mathbf{x}_{item_j}); \quad \mathbf{h}_{user_j} = \mathscr{H}_{user}(\mathbf{h}_{user_{j-1}}, \mathbf{x}_{user_j}). \quad (19.4)$$

where $\mathscr{H}_{user}(\cdot, \cdot)$ and $\mathscr{H}_{item}(\cdot, \cdot)$ represent the encoding functions specific to user and item, respectively. We adopt the long short-term memory (LSTM) (Hochreiter and Schmidhuber, 1997) as the encoder instead of the Transformer (Vaswani et al, 2017), since LSTM can utilize time feature to control the information to be propagated with the time decay feature as inputs. After the time-aware sequence encoding, we obtain the corresponding hidden states sequence of historical behavior sequence $\mathscr{S}_{u,t}$ of user u and $\mathscr{S}_{i,t}$ of item i. The time-aware sequence encoding functions can be represented as:

$$
\begin{aligned}
&\text{LSTM}_{item}(\{\mathbf{x}_{i,\tau}|(i,\tau) \in \mathscr{S}_{u,t}\}) = \{\mathbf{h}_{i,\tau}|(i,\tau) \in \mathscr{S}_{u,t}\}; \\
&\text{LSTM}_{user}(\{\mathbf{x}_{u,\tau}|(u,\tau) \in \mathscr{S}_{i,t}\}) = \{\mathbf{h}_{u,\tau}|(u,\tau) \in \mathscr{S}_{i,t}\}.
\end{aligned}
\tag{19.5}
$$

19.2.2.4 Second-Order Graph Attention

In practice, there may exist noisy neighbors, whose interest or audience is irrelevant to the target node. To eliminate the noise brought by the unreliable nodes, we propose an attention mechanism to activate related nodes in the behavior sequence. Traditional graph attention mechanism, like GAT (Veličković et al, 2018), computes attention weights between the central node and the neighbor nodes, which indicate the importance of each neighbor node to the central node. Although they perform well on the node classification task, they may increase noise diffusion for recommendation when there exists an unreliable connection.

To address the above problem, we propose a graph attention mechanism that uses both the parent node of the central node and the central node itself to build the query and takes the neighbor nodes as the key and value. Since we use the parent node of the central node to enhance the expressive power of the query, which is connected to the key node with two hops, we name it *second-order graph attention*. The parent node of the central node can be seen as a complement when the central node is unreliable, thus improving the robustness.

Following the scaled dot-product attention (Vaswani et al, 2017), the attention function is defined as

$$
\text{Attention}(Q,K,V) = \frac{\text{softmax}(QK^{\top})}{\sqrt{d}}V
\tag{19.6}
$$

where Q, K and V represent the query, key and value, respectively, and d is the dimension of K and Q. The multi-head attention is defined as follows:

$$
\text{MultiHead}(Q,K,V) = [\text{head}_1;\text{head}_2;\ldots;\text{head}_h]W_O
\tag{19.7}
$$

$$
\text{head}_i = \text{Attention}(QW_{Q_i},KW_{K_i},VW_{V_i})
\tag{19.8}
$$

where weights W_Q, W_K, W_V and W_O are trained parameters.

Given the behavior hidden states sequence $\{\mathbf{h}_{i,\tau}|(i,\tau) \in \mathscr{S}_{u,t}\}$ and $\{\mathbf{h}_{u,\tau}|(u,\tau) \in \mathscr{S}_{i,t}\}$ after the time-aware sequence encoding, we represents the attention process as:

$$\mathbf{x}_{u,t} = \text{ATT}_{item}(\{\mathbf{h}_{i,\tau}|(i,\tau) \in \mathscr{S}_{u,t}\}); \mathbf{x}_{i,t} = \text{ATT}_{user}(\{\mathbf{h}_{u,\tau}|(u,\tau) \in \mathscr{S}_{i,t}\}). \quad (19.9)$$

19.2.2.5 Aggregation and Layer Combination

The core idea of GCN is to learn representation for nodes by performing convolution over their neighborhood. In DSGL, we stack the time-aware sequence encoding and the second-order graph attention, and the aggregator can be represented as:

$$
\begin{aligned}
\mathbf{x}_{u,t}^{(k+1)} &= \text{ATT}_{item}(\text{LSTM}_{item}(\{\mathbf{x}_{i,t}^{(k)}|i \in \mathscr{S}_{u,t}\})); \\
\mathbf{x}_{i,t}^{(k+1)} &= \text{ATT}_{user}(\text{LSTM}_{user}(\{\mathbf{x}_{u,t}^{(k)}|i \in \mathscr{S}_{i,t}\})).
\end{aligned}
\quad (19.10)
$$

Different from traditional GCN models that use the last layer as the final node representation, inspired by (He et al, 2020), we combine the embeddings obtained at each layer to form the final representation of a user (an item):

$$\hat{\mathbf{x}}_{u,t} = \frac{1}{k_u}\sum_{k=1}^{k_u}\mathbf{x}_{u,t}^{(k)}; \quad \hat{\mathbf{x}}_{i,t} = \frac{1}{k_i}\sum_{k=1}^{k_i}\mathbf{x}_{i,t}^{(k)}, \quad (19.11)$$

where K_u and K_i denote the numbers of DSGL layers for user u and item i, respectively.

19.2.3 Model Prediction

Given an interaction triplet (u,i,t), we can predict the possibility of the user interacting with the item as:

$$\hat{y} = \mathscr{F}(u,i,\mathscr{G}_{u,t}^{(k)},\mathscr{G}_{i,t}^{(k-1)};\Theta) = \text{MLP}([\mathbf{e}_{u,t};\mathbf{e}_{i,t};\hat{\mathbf{x}}_{u,t};\hat{\mathbf{x}}_{i,t}]) \quad (19.12)$$

where $\text{MLP}(\cdot)$ represents the MLP layer and Θ denotes the network parameters. We adopt the cross-entropy loss function:

$$\mathscr{L} = -\sum_{(u,i,t,y)\in\mathscr{D}}[y\log\hat{y} + (1-y)\log(1-\hat{y})] \quad (19.13)$$

where $\mathscr{D}$ is the set of training samples, and $y \in \{0,1\}$ denotes the real label. The algorithm procedure is presented in Algorithm 1.

Algorithm 2 The algorithm of DSGL.

Input:
    The training set $\mathscr{D} = \{(u,i,t,y)\}$; User set $\mathscr{U}$; Item set $\mathscr{I}$; Interaction set $\mathscr{E}$; Depths k_u, k_i;
    Number of epochs E.

Output: Network parameters Θ.

 1: Initialize input feature $\mathbf{f}_{user_u}$ of user $u \in \mathscr{U}$ and $\mathbf{f}_{item_i}$ of item $i \in \mathscr{I}$;
 2: **for** $e \leftarrow 1$ to E **do**
 3:    **for** $(u,i,t,y) \in \mathscr{D}$ **do**
 4:        Construct DSGs $\mathscr{G}_{u,t}^{(k_u)}$, $\mathscr{G}_{i,t}^{(k_i)}$ for user u and item i from $\mathscr{E}$;
 5:        **for** $(v,j,\tau) \in \mathscr{G}_{u,t}^{(k_u)} \bigcup \mathscr{G}_{i,t}^{(k_i)}$ **do**
 6:            Obtain the behavior sequence $\mathscr{S}_{v,\tau}$ and $\mathscr{S}_{j,\tau}$;
 7:            $\mathbf{x}_{v,\tau}^{(0)} \leftarrow \mathbf{e}_{v,\tau}$; $\mathbf{x}_{j,\tau}^{(0)} \leftarrow \mathbf{e}_{j,\tau}$;
 8:            **for** $k \leftarrow 1$ to k_u **do**
 9:                $\mathbf{x}_{v,\tau}^{(k)} \leftarrow \text{ATT}_{item}(\text{LSTM}_{item}(\{\mathbf{x}_{j,\tau}^{(k-1)} | i \in \mathscr{S}_{v,\tau}\}))$;
10:            **end for**
11:            **for** $k \leftarrow 1$ to k_i **do**
12:                $\mathbf{x}_{j,\tau}^{(k)} \leftarrow \text{ATT}_{user}(\text{LSTM}_{user}(\{\mathbf{x}_{v,\tau}^{(k-1)} | i \in \mathscr{S}_{j,\tau}\}))$;
13:            **end for**
14:        **end for**
15:        $\hat{\mathbf{x}}_{u,t} \leftarrow \frac{1}{k_u}\sum_{k=1}^{k_u} \mathbf{x}_{u,t}^{(k)}$; $\hat{\mathbf{x}}_{i,t} \leftarrow \frac{1}{k_i}\sum_{k=1}^{k_i} \mathbf{x}_{i,t}^{(k)}$;
16:        $\hat{y}_{u,i,t} \leftarrow \text{MLP}([\mathbf{e}_{u,t}; \mathbf{e}_{i,t}; \hat{\mathbf{x}}_{u,t}; \hat{\mathbf{x}}_{i,t}])$;
17:        Update the parameters Θ by optimizing Eq.19.13;
18:    **end for**
19: **end for**=0

19.2.4 Experiments and Discussions

We evaluate our methods on the real-world Amazon product datasets[2], and use five subsets. The widely used metrics for the CTR prediction task, i.e., AUC (the area under the ROC curve) and Logloss, are adopted. The compared recommendation methods can be grouped into five categories, including conventional methods (SVD++ (Koren, 2008) and PNN (Qu et al, 2016)), sequential methods with user behaviors (GRU4Rec (Hidasi et al, 2015), CASER (Tang and Wang, 2018), ATRANK (Zhou et al, 2018a) and DIN (Zhou et al, 2018b)), sequential methods with user and item behaviors (Topo-LSTM (Wang et al, 2017b), TIEN (Li et al, 2020g) and DIB (Guo et al, 2019a)), static-graph-based methods (NGCF (Wang et al, 2019k) and LightGCN (He et al, 2020)), and dynamic-graph-based method (SR-GNN (Wu et al, 2019c)).

19.2.4.1 Performance Comparison

To demonstrate the performance of the proposed model, we compare DSGL with the state-of-the-art recommendation methods. We find that DSGL consistently out-

---

[2] http://snap.stanford.edu/data/amazon/productGraph/

performs all other baselines, demonstrating its effectiveness. The sequential models outperform the conventional methods by a large margin, proving the effectiveness of capturing temporal dependency in recommendation. The sequential methods which model both user behaviors and item behaviors outperform the methods that only use the user behavior sequences, which verifies the importance of both user- and item-side behavior information. The performance of the static-graph-based methods, including LightGCN and NGCF, are not competitive. The reasons are two folds. First, these methods ignore the new interactions in the testing set in the inference phase. Second, since they do not model the temporal dependency of interactions, they cannot capture the evolving interests, degrading the performances compared with sequential models. The session-graph-based method SR-GNN outperforms static-graph-based methods, because SR-GNN incorporates all the interacted items before the current moment into graphs dynamically. However, it underperforms the sequential methods. One possible reason could be that the ratio of repeated items in the sequences is low in the Amazon datasets, and the transitions of items are not complex enough to be modeled as graphs.

19.2.4.2 Effectiveness of Graph Structure and Layer Combination

To show the effectiveness of the graph structure and layer combination, we compare the performance of **DSGL** and its variant **DSGL w/o LC** that uses the last layer instead of the combined layer as the final representation w.r.t different numbers of layers. Focusing on DSGL with layer combination, the performance gradually improves with the increase of layers. We attribute the improvement to the collaborative information carried by the second-order and third-order connectivity in the graph structure. Comparing DSGL and DSGL w/o LC, we find that removing the layer combination degrades the performance largely, which demonstrates the effectiveness of layer combination.

19.2.4.3 Effectiveness of Time-Aware Sequence Encoding

In DSGL, we perform time-aware sequence encoding to preserve both the order of behaviors and the time information. Thus, we design ablation experiments to study how the temporal dependency and time information in DSGL contributes to the final performance. To evaluate the role of time information, we test the removal of time feature only of the item bahavior (i.e., **DSGL w/o time in UBH**), of the user behavior (i.e., **DSGL w/o time in IBH**), and of both behaviors (i.e., **DSGL w/o time**). To evaluate the contribution of the behavior order, we test the removal of the sequence encoding module while retaining time information (i.e., **DSGL w/o Seq ENC**) and the removal of the time-aware sequence encoding (i.e., **DSGL w/o TA Seq ENC**). From the comparison, we find that DSGL outperforms DSGL w/o TA Seq ENC by a significant margin, demonstrating the efficacy of the time-aware sequence encoding layer. Comparing DSGL w/o time, DSGL w/o time in UBH and

DSGL w/o time in IBH with the default DSGL, we observe that removing the time information on either user or item behavior side will cause performance degradation. DSGL outperforms DSGL w/o Seq ENC, confirming the importance of temporal dependency carried by the historical behavior sequence.

19.2.4.4 Effectiveness of Second-Order Graph Attention

In DSGL, we propose a second-order graph attention to eliminate noise from unreliable neighbors. To justify its rationality, we explore different choices here. We test the performance without graph attention (i.e., **DSGL w/o ATT**). We also replace the second-order graph attention with the traditional graph attention (i.e., **DSGL-GAT**). Note that the attention function in DSGL-GAT here is the same as the one in DSGL, and the only difference is the query. DSGL-GAT takes the central node as the query. From the results, we have the following observations:

- The best setting in all cases is adopting the second-order graph attention (i.e., the current design of DSGL). Replacing it with GAT drops the performance, demonstrating the effectiveness of second-order attention in activating related neighbors and eliminating the noise from reliable neighbors.
- Removing the attention mechanism (i.e., DSGL w/o ATT), the performance degrades largely, worse than DSGL with traditional graph attention. In some cases, the performance is even not as good as the best baseline. The observation demonstrates the necessity to introduce the attention mechanism in GNN-based recommendation methods due to the inevitable noise in the multi-hop neighborhood.

19.3 Case Study 2: Device-Cloud Collaborative Learning for Graph Neural Networks

19.3.1 The proposed framework

Recently, several works (Sun et al, 2020e; Cai et al, 2020a; Gong et al, 2020; Yang et al, 2019e; Lin et al, 2020e; Niu et al, 2020) have explored the on-device computing advantages in recommender systems. This drives the development of on-device GNNs, e.g., DSGL in the previous section. However, these early works either only consider the cloud modeling, or on-device inference, or the aggregation of the temporal on-device training pieces to handle the privacy constraint. Little has explored the device modeling and the cloud modeling jointly to benefit both sides for GNNs. To bridge this gap, we introduce a Device-Cloud Collaborative Learning framework as shown in Figure 23.2. Given a recommendation dataset $\{(\mathbf{x}_n, y_n)\}_{n=1,\ldots,N}$, we target to learn a GNN-based mapping function $f : \mathbf{x}_n \to y_n$ on the cloud side. Here, $\mathbf{x}_n$ is the graph feature that contains all available candidate features and user context, y_n is the user implicit feedback (click or not) to the corresponding candidate and N is the

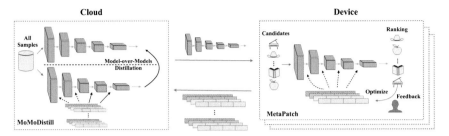

Fig. 19.3: The general DCCL framework for recommendation. The cloud side is responsible to learn the centralized cloud GNN model via the model-over-models distillation from the personalized on-device GNN models. The device receives the cloud GNN model to conduct the on-device personalization. We propose *MoMoDistill* and *MetaPatch* to instantiate each side respectively.

sample number. On the device side, each device (indexed by m) has its own local dataset, $\left\{\left(\mathbf{x}_n^{(m)}, y_n^{(m)}\right)\right\}_{n=1,\dots,N^{(m)}}$. We add a few parameter-efficient patches (Yuan et al, 2020a) to the cloud GNN model f (freezing its parameters on the device side) for each device to build a new GNN $f^{(m)} : \mathbf{x}_n^{(m)} \to y_n^{(m)}$. In the following, we will present the practical challenges in the deployment and our solutions.

19.3.1.1 MetaPatch for On-device Personalization

Although the device hardware has been greatly improved in the recent years, it is still resource-constrained to learn a complete big model on the device. Meanwhile, only finetuning last few layers is performance-limited due to the feature basis of the pretrained layers. Fortunately, some previous works have demonstrated that it is possible to achieve the comparable performance as the whole network finetuning via patch learning (Cai et al, 2020b; Yuan et al, 2020a; Houlsby et al, 2019). Inspired by these works, we insert the model patches on basis of the cloud model f for on-device personalization. Formally, the output of the l-th layer attached with one patch on the m-th device is expressed as

$$f_l^{(m)}(\cdot) = f_l(\cdot) + \mathbf{h}_l^{(m)}(\cdot) \circ f_l(\cdot), \tag{19.14}$$

where LHS of Eq.19.14 is the sum of the original $f_l(\cdot)$ and the patch response of $f_l(\cdot)$. Here, $\mathbf{h}_l^{(m)}(\cdot)$ is the trainable patch function and $\circ$ denotes the function composition that treats the output of the previous function as the input. Note that, the model patch could have different neural architectures. Here, we do not explore its variants but specify the same bottleneck architecture like (Houlsby et al, 2019).

Nevertheless, we empirically find that the parameter space of multiple patches is still relatively too large and easily overfits the sparse local samples. To overcome this issue, we propose *MetaPatch* to reduce the parameter space. It is a kind of meta learning methods to generate parameters (Ha et al, 2017; Jia et al, 2016). Concretely, assume the parameters of each patch are denoted by $\theta_l^{(m)}$ (flatten all parameters in the patch into a vector). Then, we can deduce the following decomposition

$$\theta_l^{(m)} = \Theta_l * \hat{\theta}^{(m)}, \tag{19.15}$$

where Θ_l is the globally shared parameter basis (freezing it on the device and learned in the cloud) and $\hat{\theta}^{(m)}$ is the surrogate tunable parameter vector to generate each patch parameter $\theta_l^{(m)}$ in the device-GNN-model $f^{(m)}$. To facilitate the understanding, we term $\hat{\theta}^{(m)}$ as the metapatch parameter. In this paper, we keep the number of patch parameters is greatly less than that of the metapatch parameters to be learned for personalization. Note that, regarding the pretraining of Θ_l, we leave the discussion in the following section to avoid the clutter, since it is learned on the cloud side. According to Eq. 19.15, we implement the patch parameter generation via the metapatch parameter $\hat{\theta}^{(m)}$ instead of directly learning $\theta^{(m)}$. To learn the metapatch parameter, we can leverage the local dataset to minimize the following loss function.

$$\min_{\hat{\theta}^{(m)}} \ell(y, \tilde{y})\big|_{\tilde{y}=f^{(m)}(\mathbf{x})}, \tag{19.16}$$

where ℓ is the pointwise cross-entropy loss, $f^{(m)}(\cdot) = f_L^{(m)}(\cdot) \circ \cdots f_l^{(m)}(\cdot) \cdots \circ f_1^{(m)}(\cdot)$ and L is the number of total layers. After training the device specific parameter $\hat{\theta}^{(m)}$ by Eq. 19.16, we can use Eq. 19.15 to generate all patches, and then insert them into the cloud GNN model f via Eq. 19.14 to get the final personalized GNN model $f^{(m)}$, which will provide the on-device personalized recommendation.

19.3.1.2 MoMoDistill to Enhance the Cloud Modeling

The conventional incremental training of the centralized cloud model follows the "model-over-data" paradigm. That is, when the new training samples are collected from devices, we directly perform the incremental learning based on the model trained in the early sample collection. The objective is formulated as follows,

$$\min_{W_f} \ell(y, \hat{y})\big|_{\hat{y}=f(\mathbf{x})}, \tag{19.17}$$

where W_f is the network parameter of the cloud GNN model f to be trained. This is an independent perspective without considering the device modeling. However, the on-device personalization actually can be more powerful than the centralized cloud model to handle the corresponding local samples. Thus, the guidance from the on-device models could be a meaningful prior to help the cloud modeling. Inspired by this, we propose a "model-over-models" paradigm to simultaneously learn from data and aggregate the knowledge from on-device models, to enhance the training of the centralized cloud model. Formally, the objective with the distillation procedure on the samples from all devices is defined as,

$$\min_{W_f} \ell(y,\hat{y}) + \beta \ \mathrm{KL}(\tilde{y},\hat{y})\big|_{\hat{y}=f(\mathbf{x}),\tilde{y}=f^{(m)}(\mathbf{x})}, \qquad (19.18)$$

where β is the hyperparameter to balance the distillation and "model-over-data" learning. Note that, the feasibility of the distillation in Eq. 19.18 critically depends on the patch mechanism in the previous section, since it allows us to input the meta-patch parameters like features with only loading the other parameters of $f^{(m)}$ in one time. Otherwise, we will suffer from the engineering issue of reloading numerous checkpoints frequently, which is almost impossible for current frameworks.

In *MetaPatch*, we introduce the global parameter basis $\{\Theta_l\}$ (simplified by Θ) to reduce the parameter space on the device. Regarding its training, we empirically find that coupled learning with W_f easily falls into undesirable local optimal, since they play different roles in terms of their semantics. Therefore, we resort to a progressive optimization strategy, that is, first optimize f based on Eq. 19.18, and then distill the knowledge for the parameter basis Θ with the learned f. For the second step, we design an auxiliary component by considering the heterogeneous characteristics of the metapatches from all devices and the cold-start issue at the beginning. Concretely, given the dataset $\{(x,y,\mathbf{u}^{(I(x))},\hat{\theta}^{(I(x))})\}_{n=1,...,N}$, where I maps the sample index to the device index and $\mathbf{u} \subset x$ is the user profile features (*e.g.*, age, gender, purchase level, etc) of the corresponding device, we define the following auxiliary encoder,

$$U(\hat{\theta},u) = W^{(1)}\tanh(W^{(2)}\hat{\theta} + W^{(3)}\mathbf{u}), \qquad (19.19)$$

where $W^{(1)}$, $W^{(2)}$, $W^{(3)}$ are tunable projection matrices. Here, we use W_e denoting the collection $\{W^{(1)},W^{(2)},W^{(3)}\}$ for simplicity. To learn the global parameter basis, we replace $\hat{\theta}$ by $U(\hat{\theta},\mathbf{u})$ to simulate Eq. 19.15 to generate the model patch, *i.e.*, $\Theta *U(\hat{\theta},\mathbf{u})$, since actually $\hat{\theta}$ is too heterogeneous to be directly used. Then, combining $\Theta *U(\hat{\theta},\mathbf{u})$ with f learned in the first distillation step, we can form a new proxy device model $\hat{f}^{(m)}$ (different from $f^{(m)}$ in the patch generation). Here, we leverage such a proxy $\hat{f}^{(m)}$ to directly distill the knowledge from the true $f^{(m)}$ collected from devices, which optimizes Θ and the parameters of the auxiliary encoder,

$$\min_{(\Theta,\ W_e)} \ell(y,\hat{y}) + \beta \ \mathrm{KL}(\tilde{y},\hat{y})\big|_{\hat{y}=\hat{f}^{(m)}(x),\tilde{y}=f^{(m)}(x)}, \qquad (19.20)$$

Eq. 19.18 and Eq. 19.20 progressively help learn the centralized cloud model and the global parameter basis. We specially term this progressive distillation mechanism as *MoMoDistill* to emphasize our "model-over-models" paradigm different from the conventional "model-over-data" incremental training on the cloud side. Finally, in Algorithm 3, we summarize the complete procedure of DCCL.

Algorithm 3 Device-Cloud Collaborative Learning for GNNs

Pretrain the cloud GNN model f, and then learn the global parameter basis Θ based on Eq. 19.20 by setting $\hat{\theta}$ as 0.

while lifecycle **do**   Send f and Θ to devices.
Device(f, Θ): ▷ *MetaPatch*
1) Accumulate the local data into batches
2) On-device personalization via Eq.19.16
3) If time > threshold: upload personalized GNN model $f^{(m)}$
4) Else: return the step 1).
Recycle all model patches $\{\hat{\theta}^{(m)}\}$.
Cloud($\{\hat{\theta}^{(m)}\}$): ▷ *MoMoDistill*
1) Optimize the cloud GNN model f based on Eq.19.18
2) Learn the parameter basis Θ by Eq.19.20

19.3.2 Experiments and Discussions

To demonstrate the effectiveness of the proposed framework, we conduct a range of experiments on three recommendation datasets Amazon, Movielens-1M and Taobao. Generally, all these three datasets are user interactive history in sequence format, and the last user interacted item is cut out as test sample. For each last interacted item, we randomly sample 100 items that do not appear in the user history. We compare our framework with some classical cloud models, namely, the conventional methods MF (Koren et al, 2009) and FM (Rendle, 2010), deep learning-based methods NeuMF (He et al, 2017b) and DeepFM (Guo et al, 2017), and sequence-based methods SASRec (Kang and McAuley, 2018) and DIN (Zhou et al, 2018b). For the whole experiments, we implement our model on the basis of DIN, where we insert the model patches in the last second fully-connected layer and the first two fully-connected layers after the feature embedding layer. In all comparisons, we term *MetaPatch* as DCCL-e, and *MoMoDistill* as DCCL-m, since the whole framework resembles EM iterations. The default method to compare the baselines is named DCCL, which indicates that it goes through both on-device personalization and the "model-over-models" distillation. The performance are measured by HitRate, NDCG and macro-AUC.

19.3.2.1 How is the performance of DCCL compared with the SOTAs?

To demonstrate the effectiveness of DCCL, we conduct the experiments on Amazon, Movielens and Taobao to compare to a range of baselines. Aligned with the popular experimental settings (He et al, 2017b; Zhou et al, 2018b), the last interactive item of each user on three datasets is left for evaluation and all items before the last one are used for training. For DCCL, we split the training data into two parts on average according to the temporal order: one part is for the pretraining of the backbone (DIN) and the other part is for the training of DCCL. In the experiments, we conduct one-round DCCL-e and DCCL-m. Finally, the DCCL-m is used to compare with the six representative models. We find that the deep learning based methods NeuMF and DeepFM usually outperform the conventional methods MF and FM, and the sequence-based methods SASRec and DIN consistently outperform previous non-sequence-based methods. Our DCCL builds upon on the best baseline DIN and further improves its results. Specifically, DCCL shows about 2% or more improvements in terms of NDCG@10, and at least 1% improvements in terms of HitRate@10 on all three datasets. The performances on both small and large datasets confirm the superiority of our DCCL.

19.3.2.2 Whether on-device personalization benefits to the cloud model?

In this section, we target to demonstrate that how on-device personalization via *MetaPatch* (abbreviated as DCCl-e) can improve the recommendation performance from different levels of users compared with the centralized cloud model. Considering the data scale and the availability of the context information for visualization, only the Taobao dataset is used to conduct this experiment. To validate the performance of DCCL-e in the fine-grained granularity, we sort the users based on their sample numbers and then partition them into 20 groups on average along the sorted user axis (see the statistic of the sample number *w.r.t.* the user in the appendix). After on-device model personalization, we calculate the performance for each group based on the personalized models. Here, the macro-AUC metric is used, which equally treats the users in the group instead of the group AUC in (Zhou et al, 2018b).

We use DIN as baseline and pretrain it on the Taobao Dataset of the first 20 days. Then, we test the model in the data of the remaining 10 days. For DCCL-e, we first pretrain DIN on the Taobao Dataset of the first 10 days, and then insert the patches into the pretrained DIN same as previous settings. Finally, we perform the on-device personalization in the subsequent 10 days. Similarly, we test DCCL-e on the data of the last 10 days. The evaluation is respectively conducted in the 20 groups. According to the results, we find that with the increase of the group index number, the performance approximately decreases. This is because the users in the group of larger indices are more like the long-tailed users based on our partition, and their patterns are easily ignored or even sacrificed by the centralized cloud model. In comparison, DCCL-e shows the consistent improvement over DIN on all groups, and especially can achieve a large improvement in long-tailed user groups.

19.3.2.3 The iterative characteristics of the multi-round DCCL.

To illustrate the convergence property of DCCL, we conduct the experiments on the Taobao dataset in different device-cloud interaction temporal intervals. Concretely, we specify every 2, 5, 10 days interactions between device and cloud, and respectively trace the performance of each round evaluated on the last click of each user. According to the results, we observe that frequent interactions achieve much better performance than the infrequent counterparts. We speculate that, as *MeatPatch* and *MoMoDistill* could promote each other at every round, the advantages in performance have been continuously strengthened with more frequent interactions. However, the side effect is we have to frequently update the on-device models, which may introduce other uncertain crash risks. Thus, in the real-world scenarios, we need to make a trade-off between performance and the interaction interval.

19.3.2.4 Ablation Study of DCCL

For the first study, we given the results of the one-round DCCL on the Taobao dataset and compare with DIN. From the results, we can observe the progressive improvement after DCCL-e and DCCL-m, and DCCL-m acquires more benefit than DCCL-e in terms of the improvement. The revenue behind DCCL-e is *MetaPatch* customizes a personalized model for each user to improve their recommendation experience once new behavior logs are collected on device, without the delayed update from the centralized cloud server. The further improvements from DCCL-m confirm the necessity of *MoMoDistill* to re-calibrate the backbone and the parameter basis in a long term. However, if we conduct the experiments without our two modules, the model performance is as DIN, which is not better than DCCL.

For the second ablation study, we explore the effect of the model patches in different layer junctions. In previous sections, we insert two patches (1st Junction, 2nd Junction) in the two fully-connected layers respectively after the feature embedding layer, and one patch (3rd Junction) to the layer before the last softmax transformation layer. In this experiment, we validate their effectiveness by only keep each of them in one-round DCCL. Compared with the full model, we can find that removing the model patch would decrease the performance. The results suggest the patches in the 1st and 2nd junctions are more effective than the one in the 3rd junction.

19.4 Future Directions

Certainly, we have witnessed the arising trends for GNNs to be applied in various areas. We believe the following directions should be paid more attention for GNNs to have wider impacts in big data areas, especially in search, recommendation or advertisement.

- There is still a lot to understand about GNNs, but there were quite a few important results about how they work (Loukas, 2020; Xu et al, 2019d; Oono and Suzuki, 2020). Future research works of GNNs should balance between technical simplicity, high practical impact, and far-reaching theoretical insights.
- It is also great to see how GNNs can be applied for other real-world tasks (Wei et al, 2019; Wang et al, 2019a; Paliwal et al, 2020; Shi et al, 2019a; Jiang and

Balaprakash, 2020; Chen et al, 2020o). For example, we see applications in fixing bugs in Javascript, game playing, answering IQ-like tests, optimization of TensorFlow computational graphs, molecule generation, and question generation in dialogue systems, among many others.

- It will become popular to see GNNs applied for knowledge graph reasoning (Ren et al, 2020; Ye et al, 2019b). A knowledge graph is a structured way to represent facts where nodes and edges actually bear some semantic meaning, such as the name of the actor or act of playing in movies.
- Recently there are new perspectives on how we should approach learning graph representations, especially considering the balance between local and global information. For example, Deng et al (2020) presents a way to improve running time and accuracy in node classification problem for any unsupervised embedding method. Chen et al (2019c) shows that if one replaces a non-linear neighborhood aggregation function with its linear counterpart, which includes degrees of the neighbors and the propagated graph attributes, then the performance of the model does not decrease. This is aligned with previous statements that many graph data sets are trivial for classification and raises a question of the proper validation framework for this task.
- Algorithmic works of GNNs should be integrated with system design more closely, to empower end-to-end solutions for users to address their scenarios by taking graph into deep learning frameworks. It should allow pluggable operators to adapt to the fast development of GNN community and excels in graph building and sampling. As an independent and portable system, the interfaces of AliGraph (Zhu et al, 2019c) can be integrated with any tensor engine that is used for expressing neural network models. By co-designing the flexible Gremlin like interfaces for both graph query and sampling, users can customize data accessing pattern freely. Moreover, AliGraph also shows excellent performance and scalability.

Editor's Notes: Recommender system is one of the hottest topics in both research and industrial communities due to its huge value in a number of commercial businesses such as Amazon, Facebook, LinkedIn, and so on. Since user-item interactions, user-user interaction and item-item similarity can naturally formulate into graph structure data, various graph representation learning techniques (GNN Methods in Chapter 4, GNN Scalability in Chapter 6, Graph Structure Learning in Chapter 14, Dynamic GNNs in Chapter 15, and Heterogeneous GNNs in Chapter 16) can serve a strong set of algorithmic foundations in applying GNNs for developing an effective and efficient modern recommendation system.

Chapter 20
Graph Neural Networks in Computer Vision

Siliang Tang, Wenqiao Zhang, Zongshen Mu, Kai Shen, Juncheng Li, Jiacheng Li
and Lingfei Wu

Abstract Recently Graph Neural Networks (GNNs) have been incorporated into
many Computer Vision (CV) models. They not only bring performance improvement to many CV-related tasks but also provide more explainable decomposition to
these CV models. This chapter provides a comprehensive overview of how GNNs
are applied to various CV tasks, ranging from single image classification to cross-
media understanding. It also provides a discussion of this rapidly growing field from
a frontier perspective.

Siliang Tang,
College of Computer Science and Technology, Zhejiang University e-mail: `siliang@zju.edu.cn`

Wenqiao Zhang,
College of Computer Science and Technology, Zhejiang University, e-mail: `wenqiaozhang@zju.edu.cn`

Zongshen Mu,
College of Computer Science and Technology, Zhejiang University, e-mail: `zongshen@zju.edu.cn`

Kai Shen,
College of Computer Science and Technology, Zhejiang University, e-mail: `shenkai@zju.edu.cn`

Juncheng Li,
College of Computer Science and Technology, Zhejiang University, e-mail: `junchengli@zju.edu.cn`

Jiacheng Li,
College of Computer Science and Technology, Zhejiang University, e-mail: `lijiacheng@zju.edu.cn`

Lingfei Wu
JD.COM Silicon Valley Research Center, e-mail: `lwu@email.wm.edu`

© The Author(s), under exclusive license to Springer Nature Singapore Pte Ltd. 2022   447
L. Wu et al. (eds.), *Graph Neural Networks: Foundations, Frontiers, and Applications*,
https://doi.org/10.1007/978-981-16-6054-2_20

20.1 Introduction

Recent years have seen great success of Convolutional Neural Network (CNN) in Computer Vision (CV). However, most of these methods lack the fine-grained analysis of relationships among the visual data (e.g., relation visual regions, adjacent video frames). For example, an image can be represented as a spatial map while the regions in an image are often spatially and semantically dependent. Similarly, video can be represented as spatio-temporal graphs, where each node in the graph represents a region of interest in the video and the edges capture relationships between such regions. These edges can describe the relations and capture the interdependence between nodes in the visual data. Such fine-grained dependencies are critical to perceiving, understanding, and reasoning the visual data. Therefore, graph neural networks can be naturally utilized to extract patterns from these graphs to facilitate the corresponding computer vision tasks.

This chapter introduces the graph neural network model in various computer vision tasks, including specific tasks for image, video and cross-media (cross-modal) (Zhuang et al, 2017). For each task, this chapter demonstrates how graph neural networks can be adapted to and improve the aforementioned computer vision tasks with representative algorithms.

Ultimately, to provide a frontier perspective, we also introduce some other distinctive GNN modeling methods and application scenarios on the subfield.

20.2 Representing Vision as Graphs

In this section, we introduce the representation of visual graph $\mathcal{G}^V = \{\mathcal{V}, \mathcal{E}\}$. We focus on how to construct nodes $\mathcal{V} = \{v_1, v_2, ..., v_N\}$ and edges (or relations) $\mathcal{E} = \{e_1, e_2, ..., e_M\}$ in the visual graph.

20.2.1 Visual Node representation

Nodes are essential entities in a graph. There are three kinds of methods to represent the node of the image $\mathbf{X} \in \mathbb{R}^{h \times w \times c}$ or the video $\mathbf{X} \in \mathbb{R}^{f \times h \times w \times c}$, where (h, w) is the resolution of the original image, c is the number of channels, and f is the number of frames.

Firstly, it is possible to split the image or the frame of the video into regular grids referring to Fig. 20.1, each of which is the (p, p) resolution of the image patch (Dosovitskiy et al, 2021; Han et al, 2020). Then each grid servers as the vertex of the visual graph and apply neural networks to get its embedding.

Secondly, some pre-processed structures like Fig. 20.2 can be directly borrowed for vertex representation. For example, by object detection framework like Faster R-CNN (Ren et al, 2015) or YOLO (Heimer et al, 2019), visual regions in the first

Fig. 20.1: Split an image into fixed-size patches and view as vertexes

column of the figure, have been processed and can be thought of as vertexes in the graph. We map different regions to the same dimensional features and feed them to the next training step. Like the middle column of the figure, scene graph generation models (Xu et al, 2017a; Li et al, 2019i) not only achieve visual detection but also aim to parse an image into a semantic graph which consists of objects and their semantic relationships, where it is tractable to get vertexes and edges to deploy downstream tasks in the image or video. In the last one, human joints linked by skeletons naturally form a graph and learn human action patterns (Jain et al, 2016b; Yan et al, 2018a)

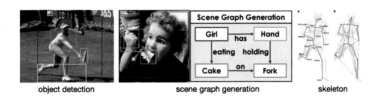

Fig. 20.2: Pre-processed visual graph examples

At last, some works utilize semantic information to represent visual vertexes. Li and Gupta (2018) assigns pixels with similar features to the same vertex, which is soft and likely groups pixels into coherent regions. Pixel features in the group are further aggregated to form a single vertex feature as Fig. 20.3. Using convolutions to learn densely-distributed, low-level patterns, Wu et al (2020a) processes the input image with several convolution blocks and treat these features from various filters as vertexes to learn more sparsely-distributed, higher-order semantic concepts. A point cloud is a set of 3D points recorded by LiDAR scans. Te et al (2018) and Landrieu and Simonovsky (2018) aggregate k-nearest neighbor to form superpoint (or vertex) and build their relations by ConvGNNs to explore the topological structure and 'see' the surrounding environment.

Fig. 20.3: Grouping similar pixels as vertexes (different colors)

20.2.2 Visual Edge representation

Edges depict the relations of nodes and play an important role in graph neural networks.For a 2D image, the nodes in the image can be linked with different spatial relations. For a clip of video stacked by continuous frames, it adds temporal relations between frames besides spatial ones within the frame. On the one hand, these relations can be fixed by predefined rules to train GNNs, referred to as static relations. Learning to learn relations (thought of as dynamic relations) attracts more and more attention on the other hand.

20.2.2.1 Spatial Edges

To capture spatial relations is the key step in the image or video. For static methods, generating scene graphs (Xu et al, 2017a) and human skeletons (Jain et al, 2016b) are natural to choose edges between nodes in the visual graph described in the Fig. 20.2. Recently, some works (Bajaj et al, 2019; Liu et al, 2020g) use fully-connected graph (every vertex is linked with other ones) to model the relations among visual nodes and compute union region of them to represent edge features. Furthermore, self-attention mechanism (Yun et al, 2019; Yang et al, 2019f) are introduced to learn the relations among visual nodes, whose main idea is inspired by transformer (Vaswani et al, 2017) in NLP. When edges are represented, we can choose either spectral-based or spatial-based GNNs for applications (Zhou et al, 2018c; Wu et al, 2021d).

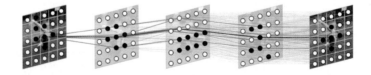

Fig. 20.4: A spatial-temporal graph by extracting nodes from each frame and allowing directed edges between nodes in neighbouring frames

20.2.2.2 Temporal Edges

To understand the video, the model not only builds spatial relations in a frame but also captures temporal connections among frames. A series of methods (Yuan et al, 2017; Shen et al, 2020; Zhang et al, 2020h) compute each node in the current frame with near frames by semantic similarity methods like k-Nearest Neighbors to construct temporal relations among frames. Especially, as you can see in the Fig. 20.4, Jabri et al (2020) represent video as a graph using a Markov chain and learn a random walk among nodes by dynamic adjustment, where nodes are image patches, and edges are affinities (in some feature space) between nodes of neighboring frames. Zhang et al (2020g) use regions as visual vertexes and evaluate the IoU (Intersection of Union) of nodes among frames to represent the weight edges.

20.3 Case Study 1: Image

20.3.1 Object Detection

Object detection is a fundamental and challenging problem in computer vision, which received great and lasting attention in recent years. Given a natural image, the object detection task seeks to locate the visual object instances from certain categories (e.g. humans, animals, or trees). Generally speaking, object detection can be grouped into two categories (Liu et al, 2020b): 1) generic object detection and 2) salient object detection. The first class aims to detect unlimited instances of objects in the digital image and predict their class attributes from some pre-defined categories. The goal of the second type is to detect the most salient instance. In recent years deep learning-based methods have achieved tremendous success in this field, such as Faster-RCNN (Ren et al, 2015), YOLO (Heimer et al, 2019), and etc. Most of the early methods and their follow-ups (Ren et al, 2015; He et al, 2017a) usually adopt the region selection module to extract the region features and predict the active probability for each candidate region. Although they are demonstrated successful, they mostly treat the recognition of each candidate region separately, thus leading to nonnegligible performance drops when facing the nontypical and nonideal occasions, such as heavy long-tail data distributions and plenty of confusing categories (Xu et al, 2019b). The graph neural network (GNN) is introduced to effectively address this troublesome challenge by modeling the correlations between regions explicitly and leveraging them to achieve better performance. In this section, we will present one typical case SGRN (Xu et al, 2019b) to discuss this promising direction.

The SGRN can be simply divided into two modules: 1) sparse graph learner which learns the graph structure explicitly during the training and 2) the spatial-aware graph embedding module which leverages the learned graph structure information and obtains the graph representation. To make it clear, we denote the graph as $\mathcal{G}(\mathcal{V}, \mathcal{E})$, where $\mathcal{V}$ is the vertex set and $\mathcal{E}$ is the edge set. The image is $\mathcal{I}$. And we formulate the regions as $R = \{\mathbf{f}_i\}_{i=1}^{n_\mathcal{I}}, \mathbf{f}_i \in \mathbb{R}^d$ for a specific image $\mathcal{I}$, where d is the region feature's dimension. We will discuss these two parts and omit other details.

Unlike previous attempts in close fields which build category-to-category graph (Dai et al, 2017; Niepert et al, 2016), the SGRN treats the candidate regions R as graph nodes $\mathcal{V}$ and constructs dynamic graph $\mathcal{G}$ on top of them. Technically, they project the region features into the latent space $\mathbf{z}$ by:

$$\mathbf{z}_i = \phi(\mathbf{f}_i) \tag{20.1}$$

where ϕ is the two fully-connected layers with ReLU activation, $\mathbf{z}_i \in \mathbb{R}^l$ and l is the latent dimension.

The region graph is constructed by latent representation $\mathbf{z}$ as follows:

$$S_{i,j} = \mathbf{z}_i \mathbf{z}_j^\top \tag{20.2}$$

where $S \in \mathbb{R}^{n_r \times n_r}$. It is not proper to reserve all relations between region pairs since there are many negative (i.e., background) samples among the region proposals, which may affect the down task's performance. If we use the dense matrix S as the graph adjacency matrix, the graph will be fully-connected, which leads to computation burden or performance drop since most existing GNN methods work worse on fully-connected graphs (Sun et al, 2019). To solve this issue, the SGRN adopt KNN to make the graph sparse (Chen et al, 2020n,o). In other words, for the learned similarity matrix $S_i \in \mathbb{R}^{N_r}$, they only keep the K nearest neighbors (including itself) as well as the associated similarity scores (i.e., they mask off the remaining similarity scores). The learned graph adjacency is denoted as:

$$A = \text{KNN}(S) \tag{20.3}$$

The node's initial embedding is obtained by the pre-trained visual classifier. We omit the details and simply denote it as $X = \{\mathbf{x}_i\}_{i=1}^{n_r}$. The SGRN introduces a spatial-aware graph reasoning module to learn the spatial-aware node embedding. Formally, they introduce a patch of operator adapted by graph convolutional network (GCN) with learnable gaussian kernels, given by:

$$f_k'(i) = \sum_{j \in \mathcal{N}(i)} \omega_k(\mu(i,j)) \mathbf{x}_j A_{i,j} \tag{20.4}$$

where $\mathcal{N}(i)$ denotes the neighborhood of node i, $\mu(i,j)$ is the distance of node i, j calculated by the center of them in a polar coordinate system, and $\omega_k()$ is the k-th gaussian kernel. Then the K kernels' results are concatenated together and projected to the latent space as follows:

$$\mathbf{h}_i = g([f_1'(i); f_2'(i); ...; f_K'(i)]) \tag{20.5}$$

where $g(\cdot)$ denotes the projection with non-linearity. Finally, $\mathbf{h}_i$ is combined with the original visual region feature $\mathbf{f}_i$ to enhance classification and regression performance.

20.3.2 Image Classification

Inspired by the success of deep learning techniques, significant improvement has been made in the image classification field, such as ResNet (He et al, 2016a). However, the CNN-based models are limited in modeling relations between samples. The graph neural network is introduced to image classification, which aims to model the fine-grained region correlations to enhance classification performance (Hong et al, 2020a), combining labeled and unlabeled image instances for semi-supervised image classification (Luo et al, 2016; Satorras and Estrach, 2018). In this section, we will present a typical case for semi-supervised image classification to show the effectiveness of GNN.

We denote the data samples as $(x_i, y_i) \in \mathscr{T}$, where x_i is the image and $y_i \in \mathbb{R}^K$ is the image label. For semi-supervised setting, the $\mathscr{T}$ is divided into labeled part $\mathscr{T}_{labeled}$ and unlabeled part $\mathscr{T}_{unlabeld}$. We assume that there are N_l labeled samples and N_u unlabeled samples, respectively. The proposed GNN is dynamic and multi-layer, which means for each layer, it will learn the graph topology from the previous layer's the node embedding and learn the new embedding on top of it. Thus, we denote the layer number as M and only present the detailed graph construction and graph embedding techniques of layer k. Technically, they construct the graph for the image set and formulate the posterior prediction task as message passing with graph neural network. They cast the samples as graph $\mathscr{G}(\mathscr{V}, \mathscr{E})$, whose nodes set is the image set consisting of both labeled and unlabeled data. The edge set $\mathscr{E}$ is constructed during training.

First, they denote the initial node representation as $X = \{\mathbf{x}_i\}_{i=1}^{n_l+n_u}$ as follows:

$$\mathbf{x_i}^0 = (\phi(x_i), h(y_i)) \tag{20.6}$$

where $\phi()$ is the convolutional neural network and $h()$ is the one-hot label encoding. Note that for unlabeled data, they replace the $h()$ with the uniform distribution over the K-simplex.

Second, the graph topology is learned by current layer's node embedding denoted as $\mathbf{x}^k$. The distance matrix modeling the distance in the embedding space between nodes is denoted as S given by:

$$S_{i,j}^k = \varphi(\mathbf{x}_i, \mathbf{x}_j) \tag{20.7}$$

where φ is a parametrized symmetric function as follows:

$$\varphi(\mathbf{a}, \mathbf{b}) = MLP(abs(\mathbf{a} - \mathbf{b})) \tag{20.8}$$

where $MLP()$ is a multilayer perceptron network and $abs()$ is the absolute function. Then the adjacency matrix A is calculated by normalizing the row of S using softmax operation.

Then a GNN layer is adapted to encode the graph nodes with learned topology A. The GNN layer receives the node embedding matrix $\mathbf{x}^k$ and outputs the aggregated node representation $\mathbf{x}^{k+1}$ as:

$$\mathbf{x}_l^{k+1} = \rho\left(\sum_{B \in A} B\mathbf{x}^k \theta_{B,l}^k\right), l = d_1...d_{k+1} \tag{20.9}$$

where $\{\theta_1^k, ..., \theta_{|A|}^k\}$ are trainable parameters, and $\rho()$ is non-linear activate function (leaky ReLU here).

The graph neural network is effective in modeling the unstructured data's correlation. In this work, the GNN explicitly exploits the relation between samples, especially the labeled and unlabeled data, contributing to few-shot image classification challenges.

20.4 Case Study 2: Video

20.4.1 Video Action Recognition

Action recognition in video is a highly active area of research, which plays a crucial role in video understanding. Given a video as input, the task of action recognition is to recognize the action appearing in the video and predict the action category. Over the past few years, modeling the spatio-temporal nature of video has been the core of research in the field of video understanding and action recognition. Early approaches of activity recognition such as Hand-crafted Improved Dense Trajectory(iDT) (Wang and Schmid, 2013), two-Stream ConvNets (Simonyan and Zisserman, 2014a), C3D (Tran et al, 2015), and I3D (Carreira and Zisserman, 2017) have focused on using spatio-temporal appearance features. To better model longer-term temporal information, researchers also attempted to model the video as an ordered frame sequence using Recurrent Neural Networks (RNNs) (Yue-Hei Ng et al, 2015; Donahue et al, 2015; Li et al, 2017b). However, these conventional deep learning approaches only focus on extracting features from the whole scenes and are unable to model the relationships between different object instances in space and time. For example, to recognize the action in the video corresponds to "opening a book", the temporal dynamics of objects and human-object and object-object interactions are crucial. We need to temporally link book regions across time to capture the shape of the book and how it changes over time.

To capture relations between objects across time, several deep models (Chen et al, 2019d; Herzig et al, 2019; Wang and Gupta, 2018; Wang et al, 2018e) have been recently introduced that represent the video as spatial-temporal graph and leverage recently proposed graph neural networks. These methods take dense object proposals as graph nodes and learn the relations between them. In this section, we take the framework proposed in (Wang and Gupta, 2018) as one example to demonstrate how graph neural networks can be applied to action recognition task.

As illustrated in Fig 20.5, the model takes a long clip of video frames as input and forwards them to a 3D Convolutional Neural Network to get a feature map $I \in \mathbb{R}^{t \times h \times w \times d}$, where t represents the temporal dimension, $h \times w$ represents the spatial dimensions and d represents the channel number. Then the model adopts the Region Proposal Network (RPN) (Ren et al, 2015) to extract the object bounding boxes followed by RoIAlign (He et al, 2017a) extracting d-dimension feature for each object proposal. The output n object proposals aggregated over t frames are corresponding to n nodes in the building graphs. There are mainly two types of graphs: Similarity Graph and Spatial-Temporal Graph.

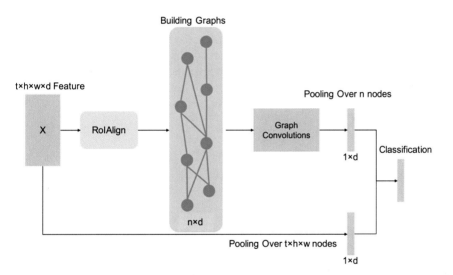

Fig. 20.5: Overview of the GNN-based model for Video Action Recognition.

The similarity graph is constructed to measure the similarity between objects. In this graph, pairs of semantically related objects are connected. Formally, the pairwise similarity between every two nodes can be represented as:

$$F(\mathbf{x}_i, \mathbf{x}_j) = \phi(\mathbf{x}_i)^{\top} \phi'(\mathbf{x}_j) \tag{20.10}$$

where ϕ and ϕ' represent two different transformations of the original features.

After computing the similarity matrix, the normalized edge values A_{ij}^{sim} from node i to node j can be defined as:

$$A_{ij}^{sim} = \frac{expF(\mathbf{x}_i, \mathbf{x}_j)}{\sum_{j=1}^{n} expF(\mathbf{x}_i, \mathbf{x}_j)} \tag{20.11}$$

The spatial-temporal graph is proposed to encode the relative spatial and temporal relations between objects, where objects in nearby locations in space and time are connected together. The normalized edge values of the spatial-temporal graph can be formulated as:

$$A_{ij}^{front} = \frac{\sigma_{ij}}{\sum_{j=1}^{n} \sigma_{ij}} \tag{20.12}$$

where G^{front} represents the forward graph which connects objects from frame t to frame t + 1, and σ_{ij} represents the value of Intersection Over Unions (IoUs) between object i in frame t and object j in frame $t + 1$. The backward graph A^{back} can be computed in a similar way. Then, the Graph Convolutional Networks (GCNs) (Kipf and Welling, 2017b) is applied to update features of each object node. One layer of graph convolutions can be represented as:

$$Z = AXW \tag{20.13}$$

where A represents one of the adjacency matrix (A^{sim}, A^{front}, or A^{back}), X represents the node features, and W is the weight matrix of the GCN.

The updated node features after graph convolutions are forwarded to an average pooling layer to obtain the global graph representation. Then, the graph representation and pooled video representation are concatenated together for video classification.

20.4.2 Temporal Action Localization

Temporal action localization is the task of training a model to predict the boundaries and categories of action instances in untrimmed videos. Most existing methods (Chao et al, 2018; Gao et al, 2017; Lin et al, 2017; Shou et al, 2017, 2016; Zeng et al, 2019) tackle temporal action localization in a two-stage pipeline: they first generate a set of 1D temporal proposals and then perform classification and temporal boundary regression on each proposal individually. However, these methods process each proposal separately, failing to leverage the semantic relations between proposals. To model the proposal-proposal relations in the video, graph neural networks are then adopted to facilitate the recognition of each proposal instance. P-GCN (Zeng et al, 2019) is recently proposed method to exploit the proposal-proposal relations using Graph Convolutional Networks. P-GCN first constructs an action proposal graph, where each proposal is represented as a node and their relations between two proposals as an edge. Then P-GCN performs reasoning over the proposal graph using GCN to model the relations among different proposals and update their representations. Finally, the updated node representations are used to refine their boundaries and classification scores based on the established proposal-proposal dependencies.

20.5 Other Related Work: Cross-media

Graph-structured data widely exists in different modal data (images, videos, texts), and is used in existing cross-media tasks (e.g., *visual caption, visual question answer, cross-media retrieval*). In other words, using of graph structure data and GNN rationally can effectively improve the performance of cross-media tasks.

20.5.1 Visual Caption

Visual caption aims at building a system that automatically generates a natural language description of a given image or video. The problem of image captioning is interesting not only because it has important practical applications, such as helping visually impaired people see, but also because it is regarded as a grand challenge for vision understanding. The typical solutions of visual captioning are inspired by machine translation and equivalent to translating an image to a text. In these methods (Li et al, 2017d; Lu et al, 2017a; Ding et al, 2019b), Convolutional Neural Network (CNN) or Region-based CNN (R-CNN) is usually exploited to encode an image and a decoder of Recurrent Neural Network (RNN) w/ or w/o attention mechanism is utilized to generate the sentence. However, a common issue not fully studied is how visual relationships should be leveraged in view that the mutual correlations or interactions between objects are the natural basis for describing an image.

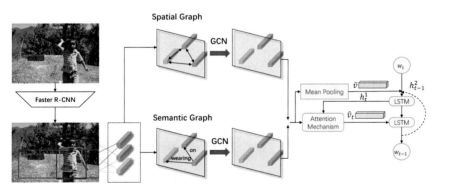

Fig. 20.6: Framework of GCN-LSTM.

In recent years, Yao et al (2018) presented Graph Convolutional Networks plus Long Short-Term Memory (GCN-LSTM) architecture, which explores visual relationship for boosting image captioning. As shown in Fig. 20.6, they study the problem from the viewpoint of modeling mutual interactions between objects/regions to enrich region-level representations that are feed into sentence decoder. Specifically, they build two kinds of visual relationships, i.e., semantic and spatial correlations, on the detected regions, and devised Graph Convolutions on the region-level representations with visual relationships to learn more powerful representations. Such relation-aware region-level representations are then input into attention LSTM for sentence generation.

Then, Yang et al (2019g) presented a novel Scene Graph Auto-Encoder (SGAE) for image captioning. This captioning pipeline contains two step: 1) extracting the scene graph for an image and using GCN to encode the corresponding scene graph, then decoding the sentence by the recoding representation; 2) incorporating the image scene graph to the captioning model. They also use GCNs to encode the visual scene graph . Given the representation of visual scene graph, they introduce joint visual and language memory to choose appropriate representation to generate image description.

20.5.2 Visual Question Answering

Visual Question Answering (VQA) aims at building a system that automatically answers natural language questions about visual information. It is a challenging task that involves mutual understanding and reasoning across different modalities. In the past few years, benefiting from the rapid developments of deep learning, the prevailing image and video question methods (Shah et al, 2019; Zhang et al, 2019g; Yu et al, 2017a) prefer to represent the visual and linguistic modalities in a common latent subspace, use the encoder-decoder framework and attention mechanism, which has made remarkable progress.

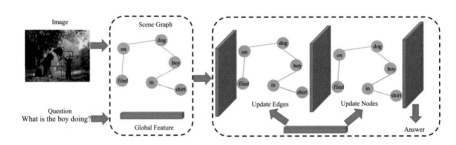

Fig. 20.7: GNN-based Visual QA.

However, the aforementioned methods also have not considered the graph information in the VQA task. Recently, Zhang et al (2019a) investigates an alternative approach inspired by conventional QA systems that operate on knowledge graphs. Specifically, as shown in Fig. 20.7, they investigate the use of scene graphs derived from images, then naturally encode information on graphs and perform structured reasoning for Visual QA. The experimental results demonstrate that scene graphs, even automatically generated by machines, can definitively benefit Visual QA if paired with appropriate models like GNNs. In other words, leveraging scene graphs largely increases the Visual QA accuracy on questions related to counting, object presence and attributes, and multi-object relationships.

Another work (Li et al, 2019d) presents the Relation-aware Graph Attention Network (ReGAT), a novel framework for VQA, to model multi-type object relations with question adaptive attention mechanism. A Faster R-CNN is used to generate a set of object region proposals, and a question encoder is used for question embedding. The convolutional and bounding-box features of each region are then injected into the relation encoder to learn the relation-aware, question-adaptive, region-level representations from the image. These relation-aware visual features and the question embeddings are then fed into a multimodal fusion module to produce a joint representation, which is used in the answer prediction module to generate an answer.

20.5.3 Cross-Media Retrieval

Image-text retrieval task has become a popular cross-media research topic in recent years. It aims to retrieve the most similar samples from the database in another modality. The key challenge here is how to match the cross-modal data by understanding their contents and measuring their semantic similarity. Many approaches (Faghri et al, 2017; Gu et al, 2018; Huang et al, 2017b) have been proposed. They often use global representations or local to express the whole image and sentence. Then, a metric is devised to measure the similarity of a couple of features in different modalities. However, the above methods lose sight of the relationships between objects in multi-modal data, which is also the key point for image-text retrieval.

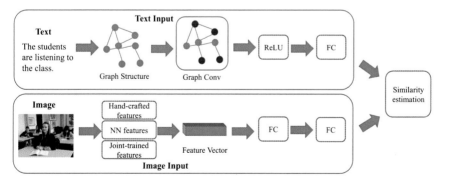

Fig. 20.8: Overview of dual-path neural network for Image-text retrieval.

To utilize the graph data in image and text better, as shown in Fig. 20.8, Yu et al (2018b) proposes a novel cross-modal retrieval model named dual-path neural network with graph convolutional network. This network takes both irregular graph-structured textual representations and regular vector-structured visual representations into consideration to jointly learn coupled feature and common latent semantic space.

In addition, Wang et al (2020i) extract objects and relationships from the image and text to form the visual scene graph and text scene graph, and design a so-called Scene Graph Matching (SGM) model, where two tailored graph encoders encode the visual scene graph and text scene graph into the feature graph. After that, both object-level and relationship-level features are learned in each graph, and the two feature graphs corresponding to two modalities can be finally matched at two levels more plausibly.

20.6 Frontiers for Graph Neural Networks on Computer Vision

In this section, we introduce the frontiers for GNNs on Computer Vision. We focus on the advanced modeling methods of GNN for Computer Vision and their applications in a broader area of the subfield.

20.6.1 Advanced Graph Neural Networks for Computer Vision

The main idea of the GNN modeling method on CV is to represent visual information as a graph. It is common to represent pixels, object bounding boxes, or image frames as nodes and further build a homogeneous graph to model their relations. Despite this kind of method, there are also some new ideas for GNN modeling.

Considering the specific task nature, some works try to represent different forms of visual information in the graph.

- **Person Feature Patches** Yan et al (2019); Yang et al (2020b); Yan et al (2020b) build spatial and temporal graphs for person re-identification (Re-ID). They horizontally partition each person feature map into patches and use the patches as the nodes of the graph. GCN is further used to modeling the relation of body parts across frames.
- **Irregular Clustering Regions** Liu et al (2020h) introduce the bipartite GNN for mammogram mass detection. It first leverages kNN forward mapping to partition an image feature map into irregular regions. Then the features in an irregular region are further integrated as a node. The bipartite node sets are constructed by cross-view images respectively, while the bipartite edge learns to model both inherent cross-view geometric constraints and appearance similarities.

- **NAS Cells** Lin et al (2020c) proposed graph-guided Neural Architecture Search (NAS) algorithms. The proposed models represent an operation cell as a node and apply the GCNs to model the relationship of cells in network architecture search.

20.6.2 Broader Area of Graph Neural Networks on Computer Vision

In this subsection, we introduce some other application scenarios of GNNs on CV, including but not limited to the following:

- **Point Cloud Analysis** Point Cloud Analysis aims to recognize a set of points in a coordinate system. Each point is represented by its three coordinates with some other features. In order to utilize CNN, the early works (Chen et al, 2017; Yan et al, 2018b; Yang et al, 2018a; Zhou and Tuzel, 2018) convert a point cloud to a regular grid such as image and voxel. Recently, a series of works (Chen et al, 2020g; Lin et al, 2020f; Xu et al, 2020e; Shi and Rajkumar, 2020; Shu et al, 2019) use a graph representation to preserve the irregularity of a point cloud. GCN plays a similar role as CNN in image processing for aggregating local information. Chen et al (2020g) develops a hierarchical graph network structure for 3D object detection on point clouds. Lin et al (2020f) proposes a learnable GCN kernel and a 3D graph max pooling with a receptive field of K nearest neighboring nodes. Xu et al (2020e) proposes a Coverage-Aware Grid Query and a Grid Context Aggregation to accelerate 3D scene segmentation. Shi and Rajkumar (2020) designs a Point-GNN with an auto-registration mechanism to detect multiple objects in a single shot.
- **Low Resource Learning** Low-resource learning models the ability of learning from a very small amount of data or transferring from prior. Some works leverage GNN to incorporate structural information for the low-resource image classification. Wang et al (2018f); Kampffmeyer et al (2019) use knowledge graphs as extra information to guide zero-short image classification. Each node corresponds to an object category and takes the word embeddings of nodes as input for predicting the classifier of different categories. Except for the knowledge graph, the similarity between images in the dataset is also helpful for the few-shot learning. Garcia and Bruna (2017); Liu et al (2018e); Kim et al (2019) set up similarity metrics and further modeling the few-shot learning problem as a label propagating or edge-labeling problem.
- **Face Recognition** Wang et al (2019p) formulates the face clustering task as a link prediction problem. It utilizes the GCN to infer the likelihood of linkage between pairs in the face sub-graphs. Yang et al (2019d) proposes a proposal-detection-segmentation framework for face clustering on an affinity graph. Zhang et al (2020b) propose a global-local GCN to perform label cleansing for face recognition.

- **Miscellaneous** We also introduce some distinctive GNN applications on the subfield. Wei et al (2020) proposes a view-GCN to recognizes 3D shape through its projected 2D images. Wald et al (2020) extends the concept of scene graph to the 3D indoor scene. Ulutan et al (2020) leverage GCNs to reason the interactions between humans and objects. Cucurull et al (2019) predicts fashion compatibility between two items by formulating an edge prediction problem. Sun et al (2020b) builds a social behavior graph from a video and uses GNNs to propagate social interaction information for trajectory prediction. Zhang et al (2020i) builds a vision and language relation graph to alleviate the hallucination problem in the grounded video description task.

20.7 Summary

This chapter shows that GNN is a promising and fast-developing research field that offers exciting opportunities in computer vision techniques. Nevertheless, it also presents some challenges. For example, graphs are often related to real scenarios, while the aforementioned GNNs lack interpretability, especially the decision-making problems (e.g., medical diagnostic model) in the computer vision field. However, compared to other black-box models (e.g., CNN), interpretability for graph-based deep learning is even more challenging since graph nodes and edges are often heavily interconnected. Thus, a further direction worth exploring is how to improve the interpretability and robustness of GNN for computer vision tasks.

Editor's Notes: Convolutional Neural Network has achieved huge success in computer vision domain. However, recent years have seen the rise of relational machine learning like GNNs and Transformers to modeling more fine-grained correlations in both images and videos. Certainly, graph structure learning techniques in Chapter 14 becomes very important for constructing an optimized graph from an image or a video and learning node representations on this learnt implicit graph. Dynamic GNNs in Chapter 15 will play an important role when coping with a video. GNN Methods in Chapter 4 and GNN Scalability in Chapter 6 are then another two basic building blocks for the use of GNNs for CV. This chapter is also highly correlated with the Chapter 21 (GNN for NLP) since vision and language is a fast-growing research area and multi-modality data is widely used today.

Chapter 21
Graph Neural Networks in Natural Language Processing

Bang Liu, Lingfei Wu

Abstract Natural language processing (NLP) and understanding aim to read from unformatted text to accomplish different tasks. While word embeddings learned by deep neural networks are widely used, the underlying linguistic and semantic structures of text pieces cannot be fully exploited in these representations. Graph is a natural way to capture the connections between different text pieces, such as entities, sentences, and documents. To overcome the limits in vector space models, researchers combine deep learning models with graph-structured representations for various tasks in NLP and text mining. Such combinations help to make full use of both the structural information in text and the representation learning ability of deep neural networks. In this chapter, we introduce the various graph representations that are extensively used in NLP, and show how different NLP tasks can be tackled from a graph perspective. We summarize recent research works on graph-based NLP, and discuss two case studies related to graph-based text clustering, matching, and multi-hop machine reading comprehension in detail. Finally, we provide a synthesis about the important open problems of this subfield.

21.1 Introduction

Language serves as a cornerstone of human cognition. Enable machines to understand natural language is at the very heart of machine intelligence. Natural language processing (NLP) concerns with the interaction between machines and human languages. It is a critical subfield of computer science, linguistics, and artificial intelligence (AI). Ever since the early research about machine translation in the 1950s

Bang Liu
Department of Computer Science and Operations Research, University of Montreal, e-mail: bang.liu@umontreal.ca

Lingfei Wu
JD.COM Silicon Valley Research Center, e-mail: lwu@email.wm.edu

L. Wu et al. (eds.), *Graph Neural Networks: Foundations, Frontiers, and Applications*,
https://doi.org/10.1007/978-981-16-6054-2_21

until nowadays, NLP has been playing an essential role in the research of machine learning and artificial intelligence.

NLP has a wide range of applications in the life and business of modern society. Critical NLP applications include but not limited to: machine translation applications that aim to translate text or speech from a source language to another target language (e.g., Google Translation, Yandex Translate); chatbots or virtual assistants that conduct an on-line chat conversation with a human agent (e.g., Apple Siri, Microsoft Cortana, Amazon Alexa); search engines for information retrieval (e.g., Google, Baidu, Bing); question answering (QA) and machine reading comprehension in different fields and applications (e.g., open-domain question answering in search engines, medical question answering); knowledge graphs and ontologies that extract and represent knowledge from multi-sources to improve various applications (e.g., DBpedia (Bizer et al, 2009), Google Knowledge Graph); and recommender systems in E-commerce based on text analysis (e.g., E-commerce recommendation in Alibaba and Amazon). Therefore, AI breakthroughs in NLP are big for business.

Two crucial research problems lie at the core of NLP: i) how to represent natural language texts in a format that computers can read; and ii) how to compute based on the input format to understand the input text pieces. We observe that researchers' ideas on representing and modeling text keep evolving during the long history of NLP development.

Up to the 1980s, most NLP systems were symbolic-based. Different text pieces were considered as symbols, and the models for various NLP tasks were implemented based on complex sets of hand-written rules. For example, classic rule-based machine translation (RBMT) involves a host of rules defined by linguists in grammar books. Such systems include Systran, Reverso, Prompt, and LOGOS (Hutchins, 1995). Rule-based approaches with symbolic representations are fast, accurate, and explainable. However, acquiring the rules for different tasks is difficult and needs extensive expert efforts.

Starting in the late 1980s, statistical machine learning algorithms brought revolution to NLP research. In statistical NLP systems, usually a piece of text is considered as a bag of its words, disregarding grammar and even word order but keeping multiplicity (Manning and Schutze, 1999). Many of the notable early successes occurred in machine translation due to statistical models were developed. Statistical systems were able to take advantage of multilingual textual corpora. However, it is hard to model the semantic structure and information of human language by simply considering the text as a bag of words.

Since the early 2010s, the field of NLP has shifted to neural networks and deep learning, where word embeddings techniques such as Word2Vec (Mikolov T, 2013) or GloVe (Pennington et al, 2014) were developed to represent words as fixed vectors. We have also witnessed an increase in end-to-end learning for tasks such as question answering. Besides, by representing text as a sequence of word embedding vectors, different neural network architectures, such as vanilla recurrent neural networks (Pascanu et al, 2013), Long Short-Term Memory (LSTM) networks (Greff et al, 2016), or convolutional neural networks (Dos Santos and Gatti, 2014), were applied to model text. Deep learning has brought a new revolution in NLP, greatly improving the performance of various tasks.

In 2018, Google introduced a neural network-based technique for NLP pre-training called Bidirectional Encoder Representations from Transformers (BERT) (Devlin et al, 2019). This model has enabled many NLP tasks to achieve superhuman performance in different benchmarks and has spawned a series of follow-up studies on pre-training large-scale language models (Qiu et al, 2020b). In such approaches, the representations of words are contextual sensitive vectors. By taking the contextual information into account, we can model the polysemy of words. However, large-scale pre-trained language models require massive consumption of data and computing resources. Besides, existing neural network-based models lack explainability or transparency, which can be a major drawback in health, education, and finance domains.

Along with the evolving history of text representations and computational models, from symbolic representations to contextual-sensitive embeddings, we can see an increase of semantical and structural information in text modeling. A key question is: how to further improve the representation of various text pieces and the computational models for different NLP tasks? We argue that representing text as graphs and applying graph neural networks to NLP applications is a highly promising research direction. Graphs are of great significance to NLP research. The reasons are multi-aspect, which will be illustrated in the following.

First, our world consists of things and the relations between them. The ability to draw logical conclusions about how different things are related to one another, or so-called relational reasoning, is central to both human and machine intelligence. In NLP, understanding human language also requires modeling different text pieces and reasoning over their relations. Graph provides a unified format to represent things and the relations between them. By modeling text as graphs, we can characterize the syntactic and semantic structures of different texts and perform explainable reasoning and inference over such representations.

Second, the structure of languages is intrinsically compositional, hierarchical, and flexible. From corpus to documents, paraphrases, sentences, phrases, and words, different text pieces form a hierarchical semantic structure, in which a higher-level semantic unit (e.g., a sentence) can be further decomposed into more fine-grained units (e.g., phrases and words). Such structural nature of human languages can be characterized by tree structures. Furthermore, due to the flexibility of languages, the same meaning can be expressed in different sentences, such as active and passive voices. However, we can unify the representation of varying sentences by semantic graphs like Abstract Meaning Representation (AMR) (Schneider et al, 2015) to make NLP models more robust.

Last but not least, graphs have always been extensively utilized and formed an essential part of NLP applications ranging from syntax-based machine translation, knowledge graph-based question answering, abstract meaning representation for common sense reasoning tasks, and so on. On the other hand, with the vigorous research on graph neural networks, the recent research trend of combining graph neural networks and NLP has become more and more prosperous. Moreover, by utilizing the general representation ability of graphs, we can incorporate multi-modal information (e.g., images or videos) to NLP, integrating different signals, modeling the world contexts and dynamics, and jointly learning multi-tasks.

In this chapter, we present a brief overview of the status of graphs in NLP. We will introduce and categorize different graph representations adopted and show how NLP tasks can be mapped onto graph-based problems and solved by graph neural network-based approaches in Sec. 21.2. After that, we will discuss two case studies. The first case study in Sec. 21.3 introduces graph-based text clustering and matching for hot events discovery and organization. The second one in Sec. 21.4 presents graph-based multi-hop machine reading comprehension. We then provide a synthesis about the important open problems of this subfield in Sec. 22.7. Finally, we conclude this chapter in Sec. 21.6.

Concurrently, a few very recent survey and tutorials (Wu et al, 2021c,b; Vashishth et al, 2019) aim to comprehensively introduce the historical and modern developments of machine learning (especially deep learning) on graphs for NLP. In addition, a recent released Graph4NLP library [1] is the first and an easy-to-use library at the intersection of Deep Learning on Graphs and Natural Language Processing. It provides both full implementations of state-of-the-art models for data scientists and also flexible interfaces to build customized models for researchers and developers with whole-pipeline support.

21.2 Modeling Text as Graphs

In this section, we will provide an overview of different graph representations in NLP. After that, we will discuss how different NLP tasks can be tackled from a graph perspective.

21.2.1 Graph Representations in Natural Language Processing

Various graph representations have been proposed for text modeling. Based on the different types of graph nodes and edges, a majority of existing works can be generalized into five categories: text graphs, syntactic graphs, semantic graphs, knowledge graphs, and hybrid graphs.

Text graphs use words, sentences, paragraphs, or documents as nodes and establish edges by word co-occurrence, location, or text similarities. Rousseau and Vazirgiannis (2013); Rousseau et al (2015) represented a document as graph-of-word, where nodes represent unique terms and directed edges represent co-occurrences between the terms within a fixed-size sliding window. Wang et al (2011) connected terms with syntactic dependencies. Schenker et al (2003) connected two words by

---

[1] Graph4NLP library can be accessed via this link `https://github.com/graph4ai/graph4nlp`.

a directed edge if one word immediately precedes another word in the document title, body, or link. The edges are categorized by the three different types of linking. Balinsky et al (2011); Mihalcea and Tarau (2004); Erkan and Radev (2004) connected sentences if they near to each other, share at least one common keyword, or the sentence similarity is above a threshold. Page et al (1999) connected web documents by hyperlinks. Putra and Tokunaga (2017) constructed directed graphs of sentences for text coherence evaluation. It utilized sentence similarities as weights and connects sentences with various constraints about sentence similarity or location. Text graphs can be established quickly, but they can not characterize the syntactic or semantic structure of sentences or documents.

Syntactic graphs (or trees) emphasize the syntactical dependencies between words in a sentence. Such structural representations of sentences are achieved by parsing, which constructs the syntactic structure of a sentence according to a formal grammar. Constituency parsing tree and dependency parsing graph are two types of syntactic representations of sentences that use different grammars (Jurafsky, 2000). Based on syntactic analysis, documents can also be structured. For example, Leskovec et al (2004) extracted subject-predicate-object triples from text based on syntactic analysis and merges them to form a directed graph. The graph was further normalized by utilizing WordNet (Miller, 1995) to merge triples belonging to the same semantic pattern.

While syntactic graphs show the grammatical structure of text pieces, semantic graphs aim to represent the meaning being conveyed. A model of semantics could help disambiguate the meaning of a sentence when multiple interpretations are valid. Abstract Meaning Representation (AMR) graphs (Banarescu et al, 2013) are rooted, labeled, directed, acyclic graphs (DAGs), comprising whole sentences. Sentences that are similar in meaning will be assigned the same AMR, even if they are not identically worded. In this way, AMR graphs abstract away from syntactic representations. The nodes in an AMR graph are AMR concepts, which are either English words, PropBank framesets (Kingsbury and Palmer, 2002), or special keywords. The edges are approximately 100 relations, including frame arguments following PropBank conventions, semantic relations, quantities, date-entities, lists, and so on.

Knowledge graphs (KGs) are graphs of data intended to accumulate and convey knowledge of the real world. The nodes of a KG represent entities of interest, and the edges represent relations between these entities (Hogan et al, 2020). Prominent examples of KGs include DBpedia (Bizer et al, 2009), Freebase (Bollacker et al, 2007), Wikidata (Vrandečić and Krötzsch, 2014) and YAGO (Hoffart et al, 2011), covering various domains. KGs are broadly applied for commercial use-cases, such as web search in Bing (Shrivastava, 2017) and Google (Singhal, 2012), commerce recommendation in Airbnb (Chang, 2018) and Amazon (Krishnan, 2018), and social networks like Facebook (Noy et al, 2019) and LinkedIn (He et al, 2016b). There are also graph representations that connect terms in a document to real-world entities or concepts based on KGs such as DBpedia (Bizer et al, 2009) and WordNet (Miller, 1995). For example, Hensman (2004) identifies the semantic roles in a sentence with

WordNet and VerbNet, and combines these semantic roles with a set of syntactic rules to construct a concept graph.

Hybrid graphs contain multiple types of nodes and edges to integrate heterogeneous information. In this way, the various text attributes and relations can be jointly utilized for NLP tasks. Rink et al (2010) utilized sentences as nodes and encodes lexical, syntactic, and semantic relations in edges. Jiang et al (2010) extracted tokens, syntactic structure nodes, semantic nodes and so on from each sentence and link them by different types of edges. Baker and Ellsworth (2017) built a sentence graph based on Frame Semantics and Construction Grammar.

21.2.2 Tackling Natural Language Processing Tasks from a Graph Perspective

Understanding natural language is essentially understanding different textual elements and their relationships. Therefore, we can tackle different NLP tasks from a graph perspective based on the different representations we have introduced. In recent years, many research works apply graph neural networks (Wu et al, 2021d) to solve NLP problems. A majority of them are actually solving the following problems: node classification, link prediction, graph classification, graph matching, community detection, graph-to-text generation, and reasoning over graphs.

For tasks focusing on assigning labels to words or phrases, they can be modeled as node classification. Cetoli et al (2017) showed that dependency trees play a positive role for named entity recognition by using a graph convolutional network (GCN) (Kipf and Welling, 2017b) to boost the results of a bidirectional LSTM. In (Gui et al, 2019), a GNN-based approach was proposed to alleviate the word ambiguity in Chinese NER. Lexicons are used to construct the graph and provide word-level features. Yao et al (2019) proposed a text classification method termed Text Graph Convolutional Networks. It builds a heterogeneous word document graph for a whole corpus and turns document classification into a node classification problem.

In addition to node classification, predicting the relationships between two elements is also an essential problem in NLP research, especially for knowledge graphs. Zhang and Chen (2018b) proposed a novel link prediction framework to simultaneously learn from local enclosing subgraphs, embeddings, and attributes based on graph neural networks. Rossi et al (2021) presented an extensive comparative analysis on link prediction models based on KG embeddings. They found that the graph structural features play paramount effects on the effectiveness of link prediction models. Guo et al (2019d) introduced the Attention Guided Graph Convolutional Networks (AGGCNs) for relation extraction tasks. The model operates directly on the full dependency trees and learns to distill the useful information from them in an end-to-end fashion.

Graph classification techniques are applied to text classification problems to utilize the intrinsic structure of texts. In (Peng et al, 2018), a graph-CNN based deep learning model was proposed for text classification. It first converts texts to graph-of-words and then utilizes graph convolution operations to convolve the word graph. Huang et al (2019a); Zhang et al (2020d) proposed graph-based methods for text classification, where each text owns its structural graph and text level word interactions can be learned.

For NLP tasks involving a pair of text, graph matching techniques can be applied to incorporate the structural information of a text. Liu et al (2019a) proposed the Concept Interaction Graph to represent an article as a graph of concepts. It then matches a pair of articles by comparing the sentences that enclose the same concept node through a series of encoding techniques and aggregate the matching signals through a graph convolutional network. Haghighi et al (2005) represented sentences as directed graphs extracted from a dependency parser and develops a learned graph matching approach to approximating textual entailment. Xu et al (2019e) formulated the KB-alignment task as a graph matching problem, and proposed a graph attention-based approach. It first matches all entities in two KGs, and then jointly models the local matching information to derive a graph-level matching vector.

Community detection provides a means of coarse-graining the complex interactions or relations between nodes, which is suitable for text clustering problems. For example, Liu et al (2017a, 2020a) described a news content organization system at Tencent which discovers events from vast streams of breaking news and evolves news story structures in an online fashion. They constructed a keyword graph and applied community detection over it to perform coarse-grained keyword-based text clustering. After that, they further constructed a document graph for each coarse-grained clusters, and applied community detection again to get fine-grained event-level document clusters.

The task of graph-to-text generation aims at producing sentences that preserve the meaning of input graphs (Song et al, 2020b). Koncel-Kedziorski et al (2019) introduced a graph transforming encoder which can leverage the relational structure of knowledge graphs and generate text from them. Wang et al (2020k); Song et al (2018) proposed graph-to-sequence models (Graph Transformer) to generate natural language texts from AMR graphs. Alon et al (2019a) leveraged the syntactic structure of programming languages to encode source code and generate text.

Last but not least, reasoning over graphs plays a key role in multi-hop question answering (QA), knowledge-based QA, and conversational QA tasks. Ding et al (2019a) presented a framework CogQA to tackle multi-hop machine reading problem at scale. The reasoning process is organized as a cognitive graph, reaching entity-level explainability. Tu et al (2019) represented documents as a heterogeneous graph and employ GNN-based message passing algorithms to accumulate evidence on the proposed graph to solve the multi-hop reading comprehension problem across multiple documents. Fang et al (2020) created a hierarchical graph by constructing nodes on different levels of granularity (questions, paragraphs, sentences, entities), and proposed Hierarchical Graph Network (HGN) for multi-hop QA. Chen et al (2020n) dynamically constructed a question and conversation history aware context graph at each conversation turn and utilized a Recurrent Graph Neural Network and a flow mechanism to capture the conversational flow in a dialog.

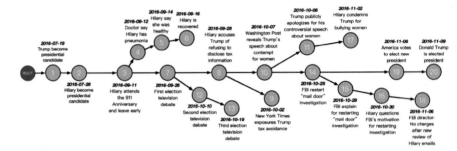

Fig. 21.1: The story tree of "2016 U.S. presidential election". Figure credit: Liu et al (2020a).

In the following, we will present two case studies to illustrate how graphs and graph neural networks can be applied to different NLP tasks with more details.

21.3 Case Study 1: Graph-based Text Clustering and Matching

In this case study, we will describe the Story Forest intelligent news organization system designed for fine-grained hot event discovery and organization from web-scale breaking news (Liu et al, 2017a, 2020a). Story Forest has been deployed in the Tencent QQ Browser, a mobile application that serves more than 110 million daily active users. Specifically, we will see how a number of graph representations are utilized for fine-grained document clustering and document pair matching and how GNN contributes to the system.

21.3.1 Graph-based Clustering for Hot Events Discovery and Organization

In the fast-paced modern society, tremendous volumes of news articles are constantly being generated by different media providers, leading to information explosion. In the meantime, the large quantities of daily news stories that can cover different subjects and contain redundant or overlapping data are becoming increasingly difficult for readers to digest. Many news app users feel that they are overwhelmed by extremely repetitive information about a variety of current hot events while still struggling to get information about the events in which they are genuinely interested. Besides, search engines conduct document retrieval on the basis of user-entered requests. They do not, however, provide users with a natural way to view trending topics or breaking news.

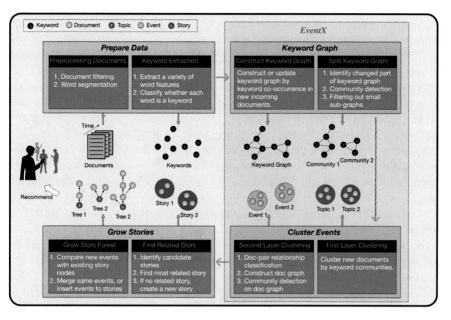

Fig. 21.2: An overview of the system architecture of Story Forest. Figure credit: Liu et al (2020a).

In (Liu et al, 2017a, 2020a), a novel news organization system named Story Forest was proposed to address the aforementioned challenges. The key idea of the Story Forest system is that, instead of providing users a list of web articles based on input queries, it proposes the concept of "event" and "story", and propose to organize tremendous of news articles into story trees to organize and track evolving hot events, revealing the relationships between them and reduce the redundancies. An event is a set of news articles reporting the same piece of real-world breaking news. And a story is a tree of related events that report a series of evolving real-world breaking news.

Figure 21.1 presents an example of a story tree, which showcases the story of "2016 U.S. presidential election". There are 20 nodes in the story tree. Each node indicates an event in the U.S. election in 2016, and each edge represents a temporal development relationship or a logical connection between two breaking news events. For example, event 1 is talking about Trump becomes a presidential candidate, and event 20 says Donald Trump is elected president. The index number on each node represents the event sequence over the timeline. The story tree contains 6 paths, where the main path $1 \rightarrow 20$ captures the process of the presidential election, the branch $3 \rightarrow 6$ describes Hilary's health conditions, the branch $7 \rightarrow 13$ is focusing on the television debates, $14 \rightarrow 18$ are about "mail door" investigation, etc. As we can see, users can easily understand the logic of news reports and learn the key facts quickly by modeling the evolutionary and logical structure of a story into a story tree.

The story trees are constructed from web-scale news articles by the Story Forest system. The system's architecture is shown in Fig. 21.2. It consists primarily of four components: preprocessing, keyword graph construction, clustering documents to events, and growing story trees with events. The overall process is split into eight stages. First, a range of NLP and machine learning tools will be used to process the input news document stream, including document filtering and word segmentation. Then the system extracts keywords, construct/update the co-occurrence graph of keywords, and divide the graph into sub-graphs. After that, it utilizes *EventX*, a graph-based fine-grained clustering algorithm to cluster documents into fine-grained events. Finally, the story trees (formed previously) are updated by either inserting each discovered event into an existing story tree at the right place or creating a new story tree if the event does not belong to any current story.

We can observe from Fig. 21.2 that a variety of text graphs are utilized in the Story Forest system. Specifically, the *EventX* clustering algorithm is based on two types of text graphs: keyword co-occurrence graph and document relationship graph. The keyword co-occurrence graph connects two keywords if they co-occurred for more than n times in a news corpus, where n is a hyperparameter. On the other hand, the document relationship graph connects document pairs based on whether two documents are talking about the same event. Based on such two types of text graphs, EventX can accurately extract fine-grained document clusters, where each cluster contains a set of documents that focus on the same event.

In particular, EventX performs two-layer graph-based clustering to extract events. The first layer performs community detection over the constructed keyword co-occurrence graph to split it into sub-graphs, where each sub-graph the keywords for a specific topic. The intuition for this step is that keywords related to a common topic usually will frequently appear in documents belonging to that topic. For example, documents belonging to the topic "2016 U.S. presidential election" will often mention keywords such as "Donald Trump", "Hillary Clinton", "election", and so on. Therefore, highly correlated keywords will be linked to each other and form dense subgraphs, whereas keywords that are not highly related will have sparse or no links. The goal here is to extract dense keyword subgraphs linked to various topics. After obtaining the keyword subgraphs (or communities), we can assign each document to its most correlated keyword subgraph by calculating their TF-IDF similarity. At this point, we have grouped documents by topics in the first layer clustering.

In the second layer, EventX constructs a document relationship graph for each topic obtained in the first layer. Specifically, a binary classifier will be applied to each pair of documents in a topic to detect whether two documents are talking about the same event. If yes, we connect the pair of documents. In this way, the set of documents in a topic turn into a document relationship graph. After that, the same community detection algorithm in the first layer will be applied to the document relationship graph, splitting it into sub-graphs where each sub-graph now represents a fine-grained event instead of a coarse-grained topic. Since the number of news articles belonging to each topic is significantly less after the first-layer document clustering, the graph-based clustering on the second layer is highly efficient, making it applicable for real-world applications. After extracting fine-grained events, we can

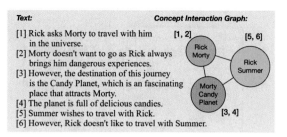

Fig. 21.3 An example to show a piece of text and its Concept Interaction Graph representation. Figure credit: Liu et al (2019a)

update the story trees by inserting an event to its related story or creating a new story tree if it doesn't belong to any existing stories. We refer to (Liu et al, 2020a) for more details about the Story Forest system.

21.3.2 Long Document Matching with Graph Decomposition and Convolution

During the construction of the document relationship graph in the Story Forest system, a fundamental problem is determining whether two news articles are talking about the same event. It is a problem of semantic matching, which is a core research problem that lies at the core of many NLP applications, including search engines, recommender systems, news systems, etc. However, previous research about semantic matching is mainly designed for matching sentence pairs (Wan et al, 2016; Pang et al, 2016), e.g., for paraphrase identification, answer selection in question-answering, and so on. Due to the long length of news articles, such methods are not suitable and do not perform well on document matching (Liu et al, 2019a).

To solve this challenge, Liu et al (2019a) presented a divide-and-conquer strategy to align a pair of documents and shift deep text comprehension away from the currently dominant sequential modeling of language elements and toward a new level of graphical document representation that is better suited to longer articles. Specifically, Liu et al (2019a) proposed the Concept Interaction Graph (CIG) as a way to view a document as a weighted graph of concepts, with each concept node being either a keyword or a group of closely related keywords. Furthermore, two concept nodes will be connected by a weighted edge which indicates their interaction strength.

As a toy example, Fig. 21.3 shows how to convert a document into a Concept Interaction Graph (CIG). First, we extract keywords such as *Rick*, *Morty*, and *Summer* from the document using standard keyword extraction algorithms, e.g., TextRank (Mihalcea and Tarau, 2004). Second, similar to what we have done in the Story Forest system, we can group keywords into sub-graphs by community detection. Each keyword community turns into a "concept" in the document. After extracting concepts, we attach each sentence in the document to its most related concept node by calculating the similarities between a sentence and each concept. In Fig. 21.3, sen-

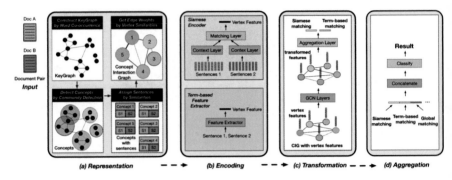

Fig. 21.4: An overview of our approach for constructing the Concept Interaction Graph (CIG) from a pair of documents and classifying it by Graph Convolutional Networks. Figure credit: Liu et al (2019a).

tences 5 and 6 are mainly talking about the relationship between *Rick* and *Summer*, and are thus attached to the concept (*Rick*, *Summer*). Similarly, we can attach other sentences to nodes, decomposing the content of a document into a number of concepts. To construct edges, we represent each node's sentence set as a concatenation of the sentences attached to it and measure the edge weight between any two nodes as the TF-IDF similarity between their sentence sets to create edges that show the correlation between different concepts. An edge will be removed if its weight is below a threshold. For a pair of documents, the process of converting them into a CIG is similar. The only differences are that the keywords are from both documents, and each concept node will have two sets of sentences from the two documents. As a result, we have represented the original document (or document pair) with a graph of key concepts, each with a (or a pair of) sentence subset(s), as well as the interaction topology among them.

The CIG representation of a document pair decomposes its content into multiple parts. Next, we need to match the two documents based on such representation. Fig. 21.4 illustrates the process of matching a pair of long documents. The matching process consists of four steps: a) preprocessing the input document pair and transform it into a CIG; b) matching the sentences from two documents over each node to get local matching features; c) structurally transforming local matching features by graph convolutional layers; and d) aggregating all the local matching features to get the final result.

Specifically, for the local matching on each concept node, the inputs are the two sets of sentences from two documents. As each node only contains a small portion of the document sentences, the long text matching problems transform into short text matching on a number of concept nodes. In (Liu et al, 2019a), two different matching models are utilized: i) similarity-based matching, which calculate a variety of text similarities between two set of sentences; ii) Siamese matching, which utilizes a Siamese neural network (Mueller and Thyagarajan, 2016) to encode the two sentence sets and get a local matching vector. After getting local matching results, the next question is: how to get an overall matching score? Liu et al (2019a) aggregates the local matching vectors into a final matching score for the pair of articles by utilizing the ability of the graph convolutional network filters (Kipf and Welling, 2017b) to capture the patterns exhibited in the CIG at multiple scales. In particular, the local matching vectors of the concept nodes are transformed by multi-layer GCN layers to take the interaction structure between nodes (or concepts in two documents) into consideration. After getting the transformed feature vectors, they are aggregated by mean pooling to get a global matching vector. Finally, the global matching vector will be fed into a classifier (e.g., a feed-forward neural network) to get the final matching label or score. The local matching module, global aggregation module, and the final classification module are trained end-to-end.

In (Liu et al, 2019a), extensive evaluations were performed to test the performance of the proposed approach for document matching. A key discovery made by (Liu et al, 2019a) is that the graph convolution operation significantly improves the performance of matching, demonstrating the effect of applying graph neural networks to the proposed text graph representation. The structural transformation on the matching vectors via GCN can efficiently capture the semantic interactions between sentences, and the transformed matching vectors better capture the semantic distance over each concept node by integrating the information of its neighbor nodes.

21.4 Case Study 2: Graph-based Multi-Hop Reading Comprehension

In this case study, we further introduce how graph neural networks can be applied to machine reading comprehension in NLP. Machine reading comprehension (MRC) aims to teach machines to read and understand unstructured text like a human. It is a challenging task in artificial intelligence and has great potential in various enterprise applications. We will see that by representing text as a graph and applying graph neural networks to it, we can mimic the reasoning process of human beings and achieve significant improvements for MRC tasks.

Suppose we have access to a Wikipedia search engine, which can be utilized to retrieve the introductory paragraph $para[x]$ of an entity x. How can we answer the question "Who is the director of the 2003 film which has scenes in it filmed at the Quality Cafe in Los Angeles?" with the search engine? Naturally, we will start with pay attention to related entities such as "Quality Cafe", look up relevant introductions through Wikipedia, and quickly locate "Old School" and "Gone in 60 Seconds" when it comes to Hollywood movies. By continuing to inquire about the introduction of the two movies, we further found their director. The last step is to determine which director it is. This requires us to analyze the semantics and qualifiers of the sentence. After knowing that the movie is in 2003, we can make the final judgment: "Todd Phillips" is the answer we want. Figure 21.5 illustrates such

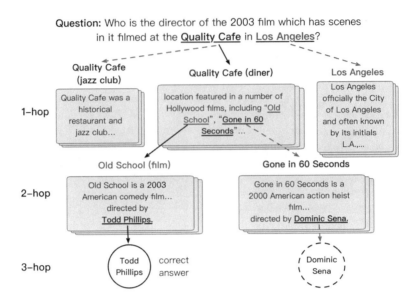

Fig. 21.5: An example of cognitive graph for multi-hop QA. Each *hop node* corresponds to an entity (e.g., "Los Angeles") followed by its introductory paragraph. The circles mean *ans nodes*, answer candidates to the question. Cognitive graph mimics human reasoning process. Edges are built when calling an entity to "mind". The solid black edges are the correct reasoning path. Figure credit: Ding et al (2019a).

process. Answering the aforementioned question requires multi-hop reasoning over different information, that is so-called multi-hop question answering.

In fact, "pay attention to related entities quickly" and "analyze the meaning of sentences for inference" are two different thinking processes. In cognition, the well-known "dual process theory" (Kahneman, 2011) believes that human cognition is divided into two systems. **System 1** is an implicit, unconscious and intuitive thinking system. Its operation relies on experience and association. **System 2** performs explicit, conscious and controllable reasoning process. This system uses knowledge in working memory to perform slow but reliable logical reasoning. System 2 is the embodiment of human advanced intelligence.

Guided by the dual process theory, the Cognitive Graph QA (CogQA) framework was proposed in (Ding et al, 2019a). It adopts a directed graph structure, named cognitive graph, to perform step-by-step deduction and exploration in the cognitive process of multi-hop question answering. Figure 21.5 presents the cognitive graph for answering the previously mentioned question. Denote the graph as $\mathscr{G}$, each node in $\mathscr{G}$ represents an entity or possible answer x, also interchangeably denoted as node x. The solid black edges are the correct reasoning path to answer the question. The cognitive graph is constructed by an extraction module that acts like System 1. It takes the introductory paragraph $para[x]$ of entity x as input, and outputs answer candidates (i.e., *ans nodes*) and useful next-hop entities (i.e., *hop nodes*) from the paragraph. These new nodes gradually expand $\mathscr{G}$, forming an explicit graph structure for System 2 reasoning module. During the expansion of $\mathscr{G}$, the new nodes or existing nodes with new incoming edges bring new *clue* about the answer. Such nodes are referred as *frontier nodes*. For *clue*, it is a form-flexible concept, referring to information from predecessors for guiding System 1 to better extract spans. To perform neural network-based reasoning over $\mathscr{G}$ instead of rule-based, System 1 also summarizes $para[x]$ into an initial hidden representation vector when extracting spans, and System 2 updates all paragraphs' hidden vectors X based on graph structure as reasoning results for downstream prediction.

The procedure of the framework CogQA is as follows. First, the cognitive graph $\mathscr{G}$ is initialized with the entities mentioned in the input question Q, and the entities are marked as initial frontier nodes. After initialization, a node x is popped from frontier nodes, and then a two-stage iterative process is conducted with two models $\mathscr{S}_1$ and $\mathscr{S}_2$ mimicking System 1 and System 2, respectively. In the first stage, the System 1 module in CoQA extracts question-relevant entities, answers candidates from paragraphs, and encodes their semantic information. Extracted entities are organized as a cognitive graph, which resembles the working memory. Specifically, given x, CogQA collects $clues[x, \mathscr{G}]$ from predecessor nodes of x, where the *clues* can be sentences where x is mentioned. It further fetches introductory paragraph $para[x]$ in Wikipedia database $\mathscr{W}$ if any. After that, $\mathscr{S}_1$ generates $sem[x, Q, clues]$, which is the initial X_x (i.e., the embedding of x). If x is a *hop node*, then $\mathscr{S}_1$ finds hop (e.g., entities) and answer spans in $para[x]$. For each hop span y, if $y \notin \mathscr{G}$ and $y \in \mathscr{W}$, then create a a new hop node for y and add it to $\mathscr{G}$. If $y \in \mathscr{G}$ but $edge(x, y) \notin \mathscr{G}$, then add a new edge (x, y) to $\mathscr{G}$ and mark node y as a frontier node, as it needs to be revisited with new information. For each answer span y, a new answer node y and edge (x, y) will be added to $\mathscr{G}$. In the second stage, System 2 conducts the reasoning procedure over the graph and collects clues to guide System 1 to better extract next-hop entities. In particular, the hidden representation X of all paragraphs will be updated by $\mathscr{S}_2$. The above process is iterated until there is no frontier node in the cognitive graph (i.e., all possible answers are found) or the graph is large enough. Then the final answer is chosen with a predictor $\mathscr{F}$ based on the reasoning results X from System 2.

The CogQA framework can be implemented as the system in Fig. 21.6. It utilizes BERT (Devlin et al, 2019) as System 1 and GNN as System 2. For clues $clues[x, \mathscr{G}]$, they are the sentences in paragraphs of x's predecessor nodes, from which x is extracted. We can observe from Fig. 21.6 that the input to BERT is the concatenation of the question, the clues passed from predecessor nodes, and the introductory paragraph of x. Based on these inputs, BERT outputs hop spans and answer spans, as well as uses the output at position 0 as $sem[x, Q, clues]$.

For System 2, CogQA utilizes a variant of GNN to update the hidden representations of all nodes. For each node x, its initial representation $X_x \in \mathbb{R}^h$ is the semantic vector $sem[x, Q, clues]$ from System 1 (i.e., BERT). The updating formula of the GNN layers are as follows:

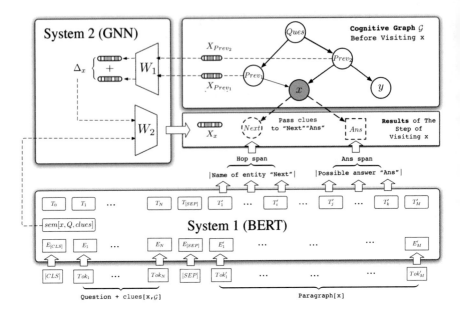

Fig. 21.6: Overview of CogQA implementation. When visiting the node x, System 1 generates new hop and answer nodes based on the $clues[x, \mathcal{G}]$ discovered by System 2. It also creates the inital representation $sem[x, Q, clues]$, based on which the GNN in System 2 updates the hidden representations X_x. Figure credit: Ding et al (2019a).

$$\Delta = \sigma((AD^{-1})^{\top}\sigma(XW_1)) \tag{21.1}$$

$$X' = \sigma(XW_2 + \Delta) \tag{21.2}$$

where X' is the new hidden representations after a propagation step of GNN. $W_1, W_2 \in \mathbb{R}^{h \times h}$ are weight matrices, σ is the activation function. $\Delta \in \mathbb{R}^{n \times h}$ are aggregated vectors passed from neighbors in the propagation. A is the adjacent matrix of $\mathcal{G}$. It is column-normalized to AD^{-1}, where D is the degree matrix of $\mathcal{G}$. By left multiplying the transformed hidden vector $\sigma(XW_1)$ with $(AD^{-1})^{\top}$, the GNN performs a localized spectral filtering. In the iterative step of visiting frontier node x, its hidden representation X_x is updated following the above equations.

Finally, a two-layer fully connected network (FCN) is utilized to serve as predictor $\mathscr{F}$:

$$answer = \underset{\text{answer node } x}{\arg\max} \; \mathscr{F}(X_x) \tag{21.3}$$

In this way, one answer candidate can be selected as the final answer. In the HotpotQA dataset (Yang et al, 2018b), there are also questions that aim to compare a certain property of entity x and y. Such questions are regarded as binary classification with input $X_x - X_y$ and solved by another identical FCNs.

The cognitive graph structure in the CogQA framework offers ordered and entity-level explainability and suits for relational reasoning, owing to the explicit reasoning paths in it. Aside from simple paths, it can also clearly display joint or loopy reasoning processes, where new predecessors might bring new clues about the answer. As we can see, by modeling the context information as a cognitive graph and applying GNN to such representation, we can mimic the dual process of human perception and reasoning and achieve excellent performance on multi-hop machine reading comprehension tasks, as demonstrated in (Ding et al, 2019a).

21.5 Future Directions

Applying graph neural networks to NLP tasks with suitable graph representations for text can bring significant benefits, as we have discussed and shown through the case studies. Although GNNs have achieved outstanding performance in many tasks, including text clustering, classification, generation, machine reading comprehension and so on, there are still numerous open problems to solve at the moment to better understand human language with graph-based representations and models. In particular, here we categorize and discuss the open problems or future directions for graph-based NLP in terms of five aspects: model design of GNNs, data representation learning, multi-task relationship modeling, world model, and learning paradigm.

Although several GNN models are applicable to NLP tasks, only a small subset of them is explored for model design. More advanced GNN models can be utilized or improved to handle the scale, depth, dynamics, heterogeneity, and explainability of natural language texts. First, scaling GNNs to large graphs helps to utilize resources such as large-scale knowledge graphs better. Second, most GNN architectures are shallow, and the performance drops after two to three layers. Design deeper GNNs enables node representation learning with information from larger and more adaptive receptive fields (Liu et al, 2020c). Third, we can utilize dynamic graphs to model the evolving or temporal phenomenons in texts, e.g., the development of stories or events. Correspondingly, dynamic or temporal GNNs (Skarding et al, 2020) can help capture the dynamic nature in specific NLP tasks. Forth, the syntactic, semantic, as well as knowledge graphs in NLP are essentially heterogeneous graphs. Developing heterogeneous GNNs (Wang et al, 2019i; Zhang et al, 2019b) can help better utilizing the various nodes and edge information in text and understanding its semantic. Last but not least, the need for improved explainability, interpretability, and trust of AI systems in general demands principled methodologies. One way is using GNNs as a model of neural-symbolic computing and reasoning (Lamb et al, 2020), as the data structure and reasoning process can be naturally captured by graphs.

For data representations, most existing GNNs can only learn from input when a graph-structure of input data is available. However, real-world graphs are often noisy and incomplete or might not be available at all. Designing effective models and algorithms to automatically learn the relational structure in input data with limited structured inductive biases can efficiently solve this problem. Instead of manually designing specific graph representations of data for different applications, we can enable models to automatically identify the implicit, high-order, or even casual relationships between input data points, and learn the graph structure and representations of inputs. To achieve these, recent research on graph pooling (Lee et al, 2019b), graph transformers (Yun et al, 2019), and hypergraph neural networks (Feng et al, 2019c) can be applied and further explored.

Multi-task learning (MTL) in deep neural networks for NLP has recently received increasing interest as it has the potential to efficiently regularize models and to reduce the need for labeled data (Bingel and Søgaard, 2017). We can marriage the representation power of graph structures with multi-task learning to integrate diverse input data, such as images, text pieces, and knowledge bases, and jointly learn a unified and structured representation for various tasks. Furthermore, we can learn the relationships or correlations between different tasks and exploit the learned relationship for curriculum learning to accelerate the convergence rate for model training. Finally, with the unified graph representation and integration of different data, as well as the joint and curriculum learning of different tasks, NLP or AI systems will gain the ability to continually acquire, fine-tune, and transfer knowledge and skills throughout their lifespan.

Grounded language learning or acquisition (Matuszek, 2018; Hermann et al, 2017) is another trending research topic that aims at learning the meaning of language as it applies to the physical world. Intuitively, language can be better learned when presented and interpreted in the context of the world it pertains to. It has been demonstrated that GNNs can efficiently capture joint dependencies between different elements in the world (Li et al, 2017e). Besides, they can also efficiently utilize the rich information in multiple modalities of the world to help understand the meaning of scene texts (Gao et al, 2020a). Therefore, representing the world or environment with graphs and GNNs to improve the understanding of languages deserves more research endeavors.

Lastly, research about self-supervised pre-training for GNNs is also attracting more attention. Self-supervised representation learning leverages input data itself as supervision and benefits almost all types of downstream tasks (Liu et al, 2020f). Numerous successful self-supervised pre-training strategies, such as BERT (Devlin et al, 2019) and GPT (Radford et al, 2018) have been developed to tackle a variety of language tasks. For graph learning, when task-specific labeled data is extremely scarce, or the graphs in the training set are structurally very different from graphs in the test set, pre-training GNNs can serve as an efficient approach for transfer learning on graph-structured data (Hu et al, 2020c).

21.6 Conclusions

Over the past few years, graph neural networks have become powerful and practical tools for a variety of problems that can be modeled by graphs. In this chapter, we did a comprehensive overview of combining graph representations and graph neural networks in NLP tasks. We introduced the motivation of applying graph representations and GNNs to NLP problems through the developing history of NLP research. After that, we provided a brief overview of various graph representations in NLP, as well as discussed how to tackle different NLP tasks from a graph perspective. To illustrate how graphs and GNNs are applied in NLP applications with more details, we presented two case studies related to graph-based hot event discovery and multi-hop machine reading comprehension. Finally, we categorized and discussed several frontier research and open problems for graph-based NLP.

Editor's Notes: Graph-based methods for Natural Language Processing have been long studied over the last two decades. Indeed, the human language is high-level symbol and thus there are rich hidden structural information beyond the original simple text sequence. In order to make full use of GNNs for NLP, graph structure learning techniques in Chapter 14 and GNN Methods in Chapter 4 serve as the two fundamental building blocks. Meanwhile, GNN Scalability in Chapter 6, Heterogeneous GNNs in Chapter 16, GNN Robustness in Chapter 8, and so on are also highly important for developing an effective and efficient approach with GNNs for various NLP applications. This chapter is also highly correlated with the Chapter 20 (GNN for CV) since vision and language is a fast-growing research area and multi-modality data is widely used today.

Chapter 22
Graph Neural Networks in Program Analysis

Miltiadis Allamanis

Abstract Program analysis aims to determine if a program's behavior complies with some specification. Commonly, program analyses need to be defined and tuned by humans. This is a costly process. Recently, machine learning methods have shown promise for probabilistically realizing a wide range of program analyses. Given the structured nature of programs, and the commonality of graph representations in program analysis, graph neural networks (GNN) offer an elegant way to represent, learn, and reason about programs and are commonly used in machine learning-based program analyses. This chapter discusses the use of GNNs for program analysis, highlighting two practical use cases: variable misuse detection and type inference.

22.1 Introduction

Program analysis is a widely studied area in programming language research that has been an active and lively research domain for decades with many fruitful results. The goal of program analysis is to determine properties of a program with regards to its behavior (Nielson et al, 2015). Traditionally analysis methods aim to provide formal guarantees about some program property e.g., that the output of a function always satisfies some condition, or that a program will always terminate. To provide those guarantees, traditional program analysis relies on rigorous mathematical methods that can deterministically and conclusively prove or disprove a formal statement about a program's behavior.

However, these methods cannot learn to employ coding patterns or probabilistically handle ambiguous information that is abundant in real-life code and is widely used by coders. For example, when a software engineer encounters a variable named

Miltiadis Allamanis
Microsoft Research, e-mail: miallama@microsoft.com

"counter", without any additional context, she/he will conclude with a high probability that this variable is a non-negative integer that enumerates some elements or events. In contrast, a formal program analysis method — having no additional context — will conservatively conclude that "counter" may contain any value.

Machine learning-based program analysis (Section 22.2) aims to provide this human-like ability to learn to reason over ambiguous and partial information at the cost of foregoing the ability to provide (absolute) guarantees. Instead, through learning common coding patterns, such as naming conventions and syntactic idioms, these methods can offer (probabilistic) evidence about aspects of the behavior of a program. This is not to say that machine learning makes traditional program analyses redundant. Instead, machine learning provides a useful weapon in the arsenal of program analysis methodologies.

Graph representations of programs play a central role in program analysis and allow reasoning over the complex structure of programs. Section 22.3 illustrates one such graph representation which we use throughout this and discusses alternatives. We then discuss GNNs which have found a natural fit for machine learning-based program analyses and relate them to other machine learning models (Section 22.4). GNNs allow us to represent, learn, and reason over programs elegantly by integrating the rich, deterministic relationships among program entities with the ability to learn over ambiguous coding patterns. In this , we discuss how to approach two practical static program analyses using GNNs: bug detection (Section 22.5), and probabilistic type inference (Section 22.6). We conclude this (Section 22.7) discussing open challenges and promising new areas of research in the area.

22.2 Machine Learning in Program Analysis

Before discussing program analysis with GNNs, it is important to take a step back and ask where machine learning can help program analysis and why. At a first look these two fields seem incompatible: static program analyses commonly seek guarantees (e.g., a program *never* reaches some state) and dynamic program analyses certify some aspect of a program's execution (e.g., specific inputs yield expected outputs), whereas machine learning models probabilities of events.

At the same time, the burgeoning area of machine learning for code (Allamanis et al, 2018a) has shown that machine learning can be applied to source code across a series of software engineering tasks. The premise is that although code has a deterministic, unambiguous structure, humans write code that contains patterns and ambiguous information (e.g. comments, variable names) that is valuable for understanding its functionality. It is this phenomenon that program analysis can also take advantage of.

There are two broad areas where machine learning can be used in program analysis: learning proof heuristics, and learning static or dynamic program analyses. Commonly static program analyses resort into converting the analysis task into a combinatorial search problem, such as a Boolean satisfiability problem (SAT), or another form of theorem proving. Such problems are known to often be computationally intractable. Machine learning-based methods, such as the work of (Irving et al, 2016) and (Selsam and Bjørner, 2019) have shown the promise that heuristics can be *learned* to guide combinatorial search. Discussing this exciting area of research is out-of-scope for this . Instead, we focus on the static program analysis learning problem.

Conceptually, a specification defines a desired aspect of a program's functionality and can take many forms, from natural language descriptions to formal mathematical constructs. Traditional static program analyses commonly resort to formulating program analyses through rigorous formal methods and dynamic analyses through observations of program executions. However, defining such program analyses is a tedious, manual task that can rarely scale to a wide range of properties and programs. Although it is imperative that formal methods are used for safety-critical applications, there is a wide range of applications that miss on the opportunity to benefit from program analysis. Machine learning-based program analysis aims to address this, but sacrifice the ability to provide guarantees. Specifically, machine learning can help program analyses deal with the two common sources of ambiguities: latent specifications, and ambiguous execution contexts (e.g., due to dynamically loaded code). Program analysis learning commonly takes one of three forms, discussed next.

Specification Tuning where an expert writes a sound program analysis which may yield many false positives (false alarms). Raising a large number of false alarms leads to the analogue of Aesop's "The Boy who Cried Wolf": too many false alarms, lead to true positives getting ignored, diminishing the utility of the analysis. To address this, work such as those of (Raghothaman et al, 2018) and (Mangal et al, 2015) use machine learning methods to "tune" (or post-process) a program analysis by learning which aspects of the formal analysis can be discounted, increasing precision at the cost of recall (soundness).

Specification Inference where a machine learning model is asked to learn to predict a plausible specification from existing code. By making the (reasonable) assumption that most of the code in a codebase complies with some latent specification, machine learning models are asked to infer closed forms of those specifications. The predicted specifications can then be input to traditional program analyses that check if a program satisfies them. Examples of such models are the factor graphs of (Kremenek et al, 2007) for detecting resource leaks, the work of (Livshits et al, 2009) and (Chibotaru et al, 2019) for information flow analysis, the work of (Si et al, 2018) for generating loop invariants, and the work of (Bielik et al, 2017) for synthesizing rule-based static analyzers from examples. The type inference problem discussed in Section 22.6 is also an instance of specification inference.

Weaker specifications — commonly used in dynamic analyses — can also be inferred. For example, Ernst et al (2007) and Hellendoorn et al (2019a) aim to predict invariants (assert statements) by observing the values during execution. Tufano et al (2020) learn to generate unit tests that describe aspects of the code's behavior.

Black Box Analysis Learning where the machine learning model acts as a black box that performs the program analysis and raises warnings but never explicitly formulates a concrete specification. Such forms of program analysis have great flexibility and go beyond what many traditional program analyses can do. However, they often sacrifice explainability and provide no guarantees. Examples of such methods include DeepBugs (Pradel and Sen, 2018), Hoppity (Dinella et al, 2020), and the variable misuse problem (Allamanis et al, 2018b) discussed in Section 22.5.

In Section 22.5 and 22.6, we showcase two learned program analyses using GNNs. However, we first need to discuss how to represent programs as graphs (Section 22.3) and how to process these graphs with GNNs (Section 22.4).

22.3 A Graph Represention of Programs

Many traditional program analysis methods are formulated over graph representations of programs. Examples of such representations include syntax trees, control flow, data flow, program dependence, and call graphs each providing different views of a program. At a high level, programs can be thought as a set of heterogeneous entities that are related through various kinds of relations. This view directly maps a program to a heterogeneous directed graph $\mathcal{G} = (\mathcal{V}, \mathcal{E})$, with each entity being represented as a node and each relationship of type r represented as an edge $(v_i, r, v_j) \in \mathcal{E}$. These graphs resemble knowledge bases with two important differences (1) nodes and edges can be deterministically extracted from source code and other program artifacts (2) there is one graph per program/code snippet.

However, deciding which entities and relations to include in a graph representation of a program is a form of feature engineering and task-dependent. Note that there is no unique or widely accepted method to convert a program into a graph representation; different representations offer trade-offs between expressing various program properties, the size of the graph representation, and the (human and computational) effort required to generate them.

In this section we illustrate one possible program graph representation inspired by (Allamanis et al, 2018b), who model each source code file as a single graph. We discuss other graph representations at the end of this section. Figure 22.1 shows the graph for a hand-crafted synthetic Python code snippet curated to illustrate a few aspects of the graph representation. A high-level explanation of the entities and relations follows; for a detailed overview of the relevant concepts, we refer the reader to programming language literature, such as the compiler textbook of (Aho et al, 2006).

Tokens A program's source code is at its most basic form a string of characters. By construction programming languages can be deterministically tokenized (lexed) into a sequence of tokens (also known as lexemes). Each token can then be represented as a node (white boxes with gray border in Figure 22.1) of "token" type. These

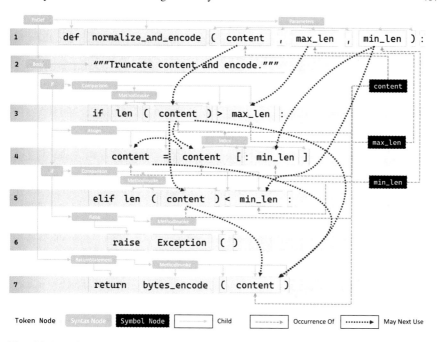

Fig. 22.1: A heterogeneous graph representation of a simple synthetic Python program (some nodes omitted for visual clarity). Source code is represented as a heterogeneous graph with typed nodes and edges (shown at the bottom of the figure). Code is originally made of tokens (token nodes) which can deterministically be parsed into a syntax tree with non-terminal nodes (vertexes). The symbols present in the snippet (e.g. variables) can then be computed (Symbol nodes) and each reference of symbol denoted by an OccurenceOf edge. Finally, dataflow edges can be computed (MayNextUse) to indicate the possible flows of values in the program. Note, the snippet here contains a bug in line 4 (see Section 22.5).

nodes are connected with a NextToken edge (not shown in Figure 22.1) to form a linear chain.

Syntax The sequence of tokens is parsed into a syntax tree. The leafs of the tree are the tokens and all other nodes of the tree are "syntax nodes" (Figure 22.1; grey blue rounded boxes). Using edges of Child type all syntax nodes and tokens are connected to form a tree structure. This stucture provides contextual information about the syntactical role of the tokens, and groups them into expressions and statements; core units in program analysis.

Symbols Next, we introduce "symbol" nodes (Figure 22.1; black boxes with dashed outline). Symbols in Python are the variables, functions, packages that are available at a given scope of a program. Like most compilers and interpreters, after parsing the code, Python creates a symbol table containing all the symbols within each file of code. For each symbol, a node is created. Then, every identifier token (e.g., the content tokens in Figure 22.1) or expression node is connected to the symbol node it refers to. Symbol nodes act as a central point of reference among the uses of variables and are useful for modeling the long-range relationships (e.g., how an object is used).

Data Flow To convey information about the program execution we add data flow edges to the graph (dotted curved lines in Figure 22.1) using an intraprocedural dataflow analysis. Although, the actual data flow within the program during execution is unknown due to the use of branching in loops and if statements, we can add edges indicating all the valid paths that data *may* flow through the program. Take as an example the parameter min_len in Figure 22.1. If the condition in line 3 is true, then min_len will be accessed in line 4, but not in line 5. Conversely, if the condition in line 3 is false, then the program will proceed to line 5, where min_len will be accessed. We denote this information with a MayNextUse edge. This construction resembles a program dependence graph (PDG) used in compilers and conventional program analyses. In contrast to the edges previously discussed, MayNextUse has a different flavor. It does not indicate a deterministic relationship but sketches all possible data flows during execution. Such relationships are central in program analyses where existential or universal properties of programs need to be computed. For example, a program analysis may need to compute that *for all* ($\forall$) possible execution paths some property is true, or that there exists ($\exists$) at least one possible execution with some property.

It is interesting to observe that just using the token nodes and NextToken edges we can (deterministically) compute all other nodes and edges. Compilers do exactly that. Then why introduce those additional nodes and edges and not let a neural network figure them out? Extracting such graph representations is cheap computationally and can be performed using the compiler/interpreter of the programming language without substantial effort. By directly providing this information to machine learning models — such as GNNs — we avoid "spending" model capacity for learning deterministic facts and introduce inductive biases that can help on program analysis tasks.

Alternative Graph Representations So far we presented a simplified graph representation inspired from (Allamanis et al, 2020). However, this is just one possible representation among many, that emphasizes the local aspects of code, such as syntax, and intraprocedural data flow. These aspects will be useful for the tasks discussed in Sections 22.5 and 22.6. Others entities and relationships can be added, in the graph representation of Figure 22.1. For example, Allamanis et al (2018b) use a GuardedBy edge type to indicate that a statement is guarded by a condition (i.e., it is executed only when the condition is true), and Cvitkovic et al (2018) use a SubtokenOf edge to connect tokens to special subtoken nodes indicating that the nodes share a common subtoken (e.g., the tokens max_len and min_len in Figure 22.1 share the len subtoken).

Representations such as the one presented here are *local*, i.e. emphasize the local structure of the code and allow detecting and using fine-grained patterns. Other local representations, such as the one of (Cummins et al, 2020) emphasize the data and control flow removing the rich natural language information in identifiers and comments, which is unnecessary for some compiler program analysis tasks. However, such local representations yield extremely large graphs when representing multiple files and the graphs become too large for current GNN architectures to meaningfully process (e.g., due to very long distances among nodes). Although a single, general graph representation that includes every imaginable entity and relationship would seem useful, existing GNNs would suffer to process the deluge of data. Nevertheless, alternative graph constructions that emphasize different program aspects are found in the literature and provide different trade-offs.

One such representation is the *global* hypergraph representation of (Wei et al, 2019) that emphasizes the inter- and intraprocedural type constraints among expressions in a program, ignoring information about syntactic patterns, control flow, and intraprocedural data flow. This allows processing whole programs (instead of single files; as in the representation of Figure 22.1) in a way that is suitable for predicting type annotations, but misses the opportunity to learn from syntactic and control-flow patterns. For example, it would be hard argue for using this representation for the variable misuse bug detection discussed in Section 22.5.

Another kind of graph representations is the *extrinsic* one defined by (Abdelaziz et al, 2020) who combine syntactic and semantic information of programs with metadata such as documentation and content from question and answer (Q&A) websites. Such representations often de-emphasize aspects of the code structure focusing on other natural language and social elements of software development. Such a representation would be unsuitable for the program analyses of Sections 22.5 and 22.6.

22.4 Graph Neural Networks for Program Graphs

Given the predominance of the graph representations for code, a variety of machine learning techniques has been employed for program analyses over program graphs, well before GNNs got established in the machine learning community. In these methods, we find some of the origins and motivations for GNNs.

One popular approach has been to project the graph into another simpler representation that other machine learning methods can accept as input. Such projections include sequences, trees, and paths. For example, Mir et al (2021) encode the sequences of tokens around each variable usage to predict its type (as in the usecase of Section 22.6). Sequence-based models offer great simplicity and have good computational performance but may miss the opportunity to capture complex structural patterns such as data and control flow.

Another successful representation is the extraction of paths from trees or graphs. For example, Alon et al (2019a) extract a sample of the paths between every two terminal nodes in an abstract syntax tree, which resembles random walk methods (Vishwanathan et al, 2010). Such methods can capture the syntactic information and learn to derive some of code's semantic information. These paths are easy to extract and provide useful features to learn about code. Nevertheless, they are lossy projections of the entities and relations within a program, that a GNN can – in principle – use in full.

Finally, factor graphs, such as conditional random fields (CRF) work directly on graphs. Such models commonly include carefully constructed graphs that capture only the relevant relationships. The most prominent example in program analysis includes the work of Raychev et al (2015) that captures the type constraints among expressions and the names of identifiers. While such models accurately represent entities and relationships, they commonly require manual feature engineering and cannot easily learn "soft" patterns beyond those explicitly modeled.

Graph Neural Networks GNNs rapidly became a valuable tool for learned program analyses given their flexibility to learn from rich patterns and the easiness of combining them with other neural network components. Given a program graph representation, GNNs compute the network embeddings for each node, to be used for downstream tasks, such as those discussed in Section 22.5 and 22.6. First, each entity/node v_i is embedded into a vector representation $\mathbf{n}_{v_i}$. Program graphs have rich and diverse information in their nodes, such as meaningful identifier names (e.g. max_len). To take advantage of the information within each token and symbol node, its string representation is subtokenized (e.g. "max", "len") and each initial node representation $\mathbf{n}_{v_i}$ is computed by pooling the embeddings of the subtokens, i.e., for a node v_i and for sum pooling, the input node representation is computed as

$$\mathbf{n}_{v_i} = \sum_{s \in \text{SUBTOKENIZE}(v_i)} \mathbf{t}_s$$

where $\mathbf{t}_s$ is a learned embedding for a subtoken s. For syntax nodes, their initial state is the embedding of the type of the node. Then, any GNN architecture that can process directed heterogeneous graphs[1] can be used to compute the network embeddings, i.e.,

$$\{\mathbf{h}_{v_i}\} = \text{GNN}\left(\mathcal{G}', \{\mathbf{n}_{v_i}\}\right), \tag{22.1}$$

where the GNN commonly has a fixed number of "layers" (e.g. 8), $\mathcal{G}' = (\mathcal{V}, \mathcal{E} \cup \mathcal{E}_{inv})$, and $\mathcal{E}_{inv}$ is the set of inverse edges of $\mathcal{E}$, i.e., $\mathcal{E}_{inv} = \left\{ (v_j, r^{-1}, v_i), \forall (v_i, r, v_j) \in \mathcal{E} \right\}$ The network embeddings $\{\mathbf{h}_{v_i}\}$ are then the input to a task-specific neural network. We discuss two tasks in the next sections.

---

[1] GGNNs (Li et al, 2016b) have historically been a common option, but other architectures have shown improvements (Brockschmidt, 2020) over plain GGNNs for some tasks.

22.5 Case Study 1: Detecting Variable Misuse Bugs

We now focus on a black box analysis learning problem that utilizes the graph representation discussed in the previous section. Specifically, we discuss the variable misuse task, first introduced by (Allamanis et al, 2018b) but employ the formulation of (Vasic et al, 2018). A variable misuse is the incorrect use of one variable instead of another already in the scope. Figure 22.1 contains such a bug in line 4, where instead of min_len, the max_len variable needs to be used to correctly truncate the content. To tackle this task a model needs to first *localize* (locate) the bug (if one exists) and then suggest a repair.

Such bugs happen frequently, often due to careless copy-paste operations and can often be though as "typos". Karampatsis and Sutton (2020) find that more than 12% of the bugs in a large set of Java codebases are variable misuses, whereas Tarlow et al (2020) find 6% of Java build errors in the Google engineering systems are variable misuses. This is a lower bound, since the Java compiler can only detect variable misuse bugs though its type checker. The author conjectures — from his personal experience — that many more variable misuse bugs arise during code editing and are resolved before being committed to a repository.

Note that this is a black box analysis learning task. No explicit specification of what the user tries to achieve exists. Instead the GNN needs to infer this from common coding patterns, natural language information within comments (like the one in line 2; Figure 22.1) and identifier names (like min, max, and len) to reason about the presence of a likely bug. In Figure 22.1 it is reasonable to assume that the developer's intent is to truncate content to max_len when it exceeds that size (line 4). Thus, the goal of the variable misuse analysis is to (1) localize the bug (if one exists) by pointing to the buggy node (the min_len token in line 4), and (2) suggest a repair (the max_len symbol).

To achieve this, assume that a GNN has computed the network embeddings $\{\mathbf{h}_{v_i}\}$ for all nodes $v_i \in \mathcal{V}$ in the program graph $\mathcal{G}$ (Equation 22.1). Then, let $\mathcal{V}_{vu} \subset \mathcal{V}$ be the set of token nodes that refer to variable usages, such as the min_len token in line 4 (Figure 22.1). First, a localization module aims to pinpoint which variable usage (if any) is a variable misuse. This is implemented as a pointer network (Vinyals et al, 2015) over $\mathcal{V}_{vu} \cup \{\emptyset\}$ where $\emptyset$ denotes the "no bug" event with a learned $\mathbf{h}_\emptyset$ embedding. Then using a (learnable) projection $\mathbf{u}$ and a softmax, we can compute the probability distribution over $\mathcal{V}_{vu}$ and the special "no bug" event,

$$p_{loc}(v_i) = \operatorname*{softmax}_{v_j \in \mathcal{V}_{vu} \cup \{\emptyset\}} \left(\mathbf{u}^\top \mathbf{h}_{v_i} \right). \tag{22.2}$$

In the case of Figure 22.1, a GNN detecting the variable misuse bug in line 4, would assign a high p_{loc} to the node corresponding to the min_len token, which is the location of the variable misuse bug. During (supervised) training the loss is simply the cross-entropy classification loss of the probability of the ground-truth location (Equation 22.2).

```
1 def describe_identity_pool(self, identity_pool_id):
2   identity_pool = self.identity_pools.get(identity_pool_id, None)
3
4   if not identity_pool:
5 -     raise ResourceNotFoundError(identity_pool)
6 +     raise ResourceNotFoundError(identity_pool_id)
7 ...
```

Fig. 22.2: A diff snippet of code with a real-life variable misuse error caught by a GNN-based model in the https://github.com/spulec/moto open-source project.

Repair given the location of a variable misuse bug can also be represented as a pointer network over the nodes of the symbols that are in scope at the variable misuse location v_{bug}. We define $\mathscr{V}_{s@v_{bug}}$ as the set of the symbol nodes of the alternative candidate symbols that are in scope at v_{bug}, except from the symbol node of v_{bug}. In the case of Figure 22.1 and the bug in line 4, $\mathscr{V}_{s@v_{bug}}$ would contain the content and max_len symbol nodes. We can then compute the probability of repairing the localized variable misuse bug with the symbol s_i as

$$p_{rep}(s_i) = \operatorname*{softmax}_{s_j \in \mathscr{V}_{s@v_{bug}}} \left(\mathbf{w}^\top [\mathbf{h}_{v_{bug}}, \mathbf{h}_{s_i}] \right),$$

i.e., the softmax of the concatenation of the node embeddings of v_{bug} and s_i, projected onto a $\mathbf{w}$ (i.e., a linear layer). For the example of Figure. 22.1, $p_{rep}(s_i)$ should be high for the symbol node of max_len, which is the intended repair for the variable misuse bug. Again, in supervised training, we minimize the cross-entropy loss of the probability of the ground-truth repair.

Training When a large dataset of variable misuse bugs and the relevant fixes can be mined, the GNN-based model discussed in this section can be trained in a supervised manner. However, such datasets are hard to collect at the scale that existing deep learning methods require to achieve reasonable performance. Instead work in this area has opted to automatically insert random variable misuse bugs in code scraped from open-source repositories — such as GitHub — and create a corpus of randomly inserted bugs (Vasic et al, 2018; Hellendoorn et al, 2019b). However, the random generation of buggy code needs to be carefully performed. If the randomly introduced bugs are "too obvious", the learned models will not be useful. For example, random bug generators should avoid introducing a variable misuse that causes a variable to be used before it is defined (use-before-def). Although such randomly generated corpora are not entirely representative of real-life bugs, they have been used to train models that can catch real-life bugs.

When evaluating variable misuse models — like those presented in this section — they achieve relatively high accuracy over randomly generated corpora with accuracies of up to 75% (Hellendoorn et al, 2019b). However, in the author's experience for real-life bugs — while some variable misuse bugs are recalled — precision tends to be low making them impractical for deployment. Improving upon this is an important open research problem. Nevertheless, actual bugs have been caught in practice. Figure 22.2 shows such an example caught by a GNN-based variable misuse detector. Here, the developer incorrectly passed identity_pool instead of identity_pool_id as the exception argument when identity_pool was None (no pool with the requested id could be found). The GNN-based black-box analysis seems to have learned to "understand" that it is unlikely that the developer's intention is to pass None to the ResourceNotFoundError constructor and instead suggests that it should be replaced by identity_pool_id. This is without ever formulating a formal specification or creating a symbolic program analysis rule.

22.6 Case Study 2: Predicting Types in Dynamically Typed Languages

Types are one of the most successful innovations in programming languages. Specifically, type annotations are explicit specifications over the valid values a variable can take. When a program *type checks*, we get a formal guarantee that the values of variables will *only* take the values of the annotated type. For example, if a variable has an int annotation, it must contain integers but not strings, floats, etc. Furthermore, types can help coders understand code more easily and software tools such as auto-completion and code navigation to be more precise. However, many programming languages either have to decide to forgo the guarantees provided by types or require their users to explicitly provide type annotations.

To overcome these limitations, specification inference methods can be used to predict plausible type annotations and bring back some of the advantages of typed code. This is especially useful in code with partial contexts (e.g., a standalone snippet of code in a webpage) or optionally typed languages. This section looks into Python, which provides an optional mechanism for defining type annotations. For example, content in Figure 22.1 can be annotated as content: str in line 1 to indicate that the developer expects that it will only contain string values. These annotations can then be used by type checkers, such as mypy (mypy Contributors, 2021) and other developer tools and code editors. This is the probabilistic type inference problem, first proposed by (Raychev et al, 2015). Here we use the GRAPH2CLASS GNN-based formulation of (Allamanis et al, 2020) treating this as a classification task over the symbols of the program similar to (Hellendoorn et al, 2018). Pandi et al (2020) offer an alternative formulation of the problem.

For type checking methods to operate explicit types annotations need to be provided by a user. When those are not present, type checking may not be able to function and provide any guarantees about the program. However, this misses the opportunity to probabilistically reason over the types of the program from other sources of information – such as variable names and comments. Concretely, in the example of Figure 22.1, it would be reasonable to assume that min_len and max_len have an integer type given their names and usage. We can then use this "educated guess" to type check the program and retrieve back some guarantees about the program execution.

Such models can find multiple applications. For example, they can be used in recommendation systems that help developers annotate a code base. They may help developers find incorrect type annotations or allow editors to provide assistive features — such as autocomplete — based on the predicted types. Or they may offer "fuzzy" type checking of a program (Pandi et al, 2020).

At its simplest form, predicting types is a node classification task over the subset of symbol nodes. Let $\mathcal{V}_s$ be the set of nodes of "symbol" type in the heterogeneous graph of a program. Let also, Z be a fixed vocabulary of type annotations, along with a special Any type[2]. We can then use the node embeddings of every node $v \in \mathcal{V}_s$ to predict the possible type of each symbol.

$$p(s_j : \tau) = \operatorname*{softmax}_{\tau' \in Z} \left(E_\tau{}^\top \mathbf{h}_{v_{s_j}} + b_\tau \right),$$

i.e., the inner product of each symbol node embedding with a learnable type embedding E_τ for each type $\tau \in T$ plus a learnable bias b_τ. Training can then be performed by minimizing some classification loss, such as the cross entropy loss, over a corpus of (partially) annotated code.

Type Checking The type prediction problem is a specification inference problem (Section 22.2) and the predicted type annotations can be passed to a standard type checking tool which can verify that the predictions are consistent with the source code's structure (Allamanis et al, 2020) or search for the most likely prediction that is consistent with the program's structure (Pradel et al, 2020). This approach allows to reduce false positives, but does not eliminate them. A trivial example is an identity function def foo(x): return x. A machine learning model may incorrectly deduce that x is a str and that foo returns a str. Although the type checker will consider this prediction type-correct it is hard to justify as correct in practice.

Training The type prediction model discussed in this section can be trained in a supervised fashion. By scraping large corpora of code, such as open-source code found on GitHub[3], we can collect thousands of type-annotated symbols. By stripping those type annotations from the original code and using them as a ground truth a training and validation set can be generated.

Such systems have shown to achieve a reasonably high accuracy (Allamanis et al, 2020) but with some limitations: type annotations are highly structured and sparse. For example Dict[Tuple[int, str], List[bool]] is a valid type annotation that may appear infrequently in code. New user-defined types (classes) will also appear at test time. Thus, treating type annotations as district classes of a classification problem

---

[2] The type Any representing the top of the type lattice and is somewhat analogous to the special UNKNOWN token used in NLP.

[3] Automatically scraped code corpora are known to suffer from a large number of duplicates (Allamanis, 2019). When collecting such corpora special care is needed to remove those duplicates to ensure that the test set is not contaminated with training examples.

```
1  def __init__(
2    self,
3  - embedding_dim: float = 768,
4  - ffn_embedding_dim: float = 3072,
5  - num_attention_heads: float = 8,
6  + embedding_dim: int = 768,
7  + ffn_embedding_dim: int = 3072,
8  + num_attention_heads: int = 8,
9    dropout: float = 0.1,
10   attention_dropout: float = 0.1,
```

Fig. 22.3: A diff snippet from the incorrect type annotation caught by Typilus (Allamanis et al, 2020) in the open-source fairseq library.

is prone to severe class imbalance issues and fails to capture information about the structure within types. Adding new types to the model can be solved by employing meta-learning techniques such as those used in Typilus (Allamanis et al, 2020; Mir et al, 2021), but exploiting the internal structure of types and the rich type hierarchy is still an open research problem.

Applications of type prediction models include suggesting new type annotations to previously un-annotated code but can also be used for other downstream tasks that can exploit information for a probabilistic estimate of the type of some symbol. Additionally, such models can help find incorrect type annotations provided by the users. Figure 22.3 shows such an example from Typilus (Allamanis et al, 2020). Here the neural model "understands" from the parameter names and the usage of the parameters (not shown) that the variables cannot contain floats but instead should contain integers.

22.7 Future Directions

GNNs for program analysis is an exciting interdisciplinary field of research combining ideas of symbolic AI, programming language research, and deep learning with many real-life applications. The overarching goal is to build analyses that can help software engineers build and maintain the software that permeates every aspect of our lives. Still there are many open challenges that need to be addressed to deliver upon this promise.

From a program analysis and programming language perspective a lot of work is needed to bridge the domain expertise of that community to machine learning. What kind of learned program analysis can be useful to coders? How can existing program analyses be improved using learned components? What are the inductive biases that machine learning models need to incorporate to better represent program-related concepts? How should learned program analyses be evaluated amidst the lack of large annotated corpora? Until recently, program analysis research has limited itself to primarily using the formal structure of the program, ignoring ambiguous information in identifiers and code comments. Researching analyses that can better leverage this information may light new and fruitful directions to help coders across many application domains.

Crucially, the question of how to integrate formal aspects of program analyses into the learning process is still an open question. Most specification inference work (e.g. Section 22.6) commonly treats the formal analyses as a separate pre- or postprocessing step. Integrating the two viewpoints more tightly will create better, more robust tools. For example, researching better ways to incorporate (symbolic) constraints, search, and optimization concepts within neural networks and GNNs will allow for better learned program analyses that can learn to better capture program properties.

From a software engineering research additional research is needed for the user experience (UX) of the program analysis results presented to users. Most of the existing machine learning models do not have performance characteristics that allow them to work autonomously. Instead they make probabilistic suggestions and present them to users. Creating or finding the affordances of the developer environment that allow to surface probabilistic observations and communicate the probabilistic nature of machine learning model predictions will significantly help accelerate the use of learned program analyses.

Within the research area of GNNs there are many open research questions. GNNs have shown the ability to learn to replicate some of the algorithms used in common program analysis techniques (Veličković et al, 2019) but with strong supervision. How can complex algorithms be learned with GNNs using just weak supervision? Additionally, existing techniques often lack the representational capabilities of formal methods. Combinatorial concepts found in formal methods, such as sets and lattices lack direct analogues in deep learning. Researching richer combinatorial — and possibly non-parametric — representations will provide valuable tools for learning program analyses.

Finally, common themes in deep learning also arise within this domain:

- The explainability of the decisions and warnings raised by learned program analyses is important to coders who need to understand them and either mark them as false positives or address them appropriately. This is especially important for black-box analyses.
- Traditional program analyses offer explicit guarantees about a program's behavior even within adversarial settings. Machine learning-based program analyses relax many of those guarantees towards reducing false positives or aiming to provide some value beyond the one offered by formal methods (e.g. use ambiguous information). However, this makes these analyses vulnerable to adversarial attacks (Yefet et al, 2020). Retrieving some form of adversarial robustness is still desirable for learned program analyses and is still an open research problem.
- Data efficiency is also an important problem. Most existing GNN-based program analysis methods either make use of relatively large datasets of annotated code (Section 22.6) or use unsupervised/self-supervised proxy objectives (Sec-

tion 22.5). However, many of the desired program analyses do not fit these frameworks and would require at least some form of weak supervision. Pre-training on graphs is one promising direction that could address this problem, but has so far is focused on homogeneous graphs, such as social/citation networks and molecules. However, techniques developed for homogeneous graphs, such as the pre-training objectives used, do *not* transfer well to heterogeneous graphs like those used in program analysis.

- All machine learning models are bound to generate false positive suggestions. However when models provide well-calibrated confidence estimates, suggestions can be accurately filtered to reduce false positives and their confidence better communicated to the users. Researching neural methods that can make accurate and calibrated confidence estimates will allow for greater impact of learned program analyses.

Acknowledgements The author would like to thank Earl T. Barr for useful discussions and feedback on drafts of this chapter.

Editor's Notes: Program analysis is one of the important downstream tasks of graph generation (Chapter 11). The main challenging problem of program analysis lies in graph representation learning (Chapter 2), which integrates the relationships and entities of the program. On basis of these graph representations, heterogeneous GNN (Chapter 16) and other variants can be used to learn the embedding of each node for task-specific neural networks. It has achieved state-of-art performances in bug detection and probabilistic type inference. There are also many emerging problems in program analysis, e.g. explainability (Chapter 7) of decisions and warnings, and adversarial robustness (Chapter 8).

Chapter 23
Graph Neural Networks in Software Mining

Collin McMillan

Abstract Software Mining encompasses a broad range of tasks involving software, such as finding the location of a bug in the source code of a program, generating natural language descriptions of software behavior, and detecting when two programs do basically the same thing. Software tends to have an extremely well-defined structure, due to the linguistic confines of source code and the need for programmers to maintain readability and compatibility when working on large teams. A tradition of graph-based representations of software has therefore proliferated. Meanwhile, advances in software repository maintenance have recently helped create very large datasets of source code. The result is fertile ground for Graph Neural Network representations of software to facilitate a plethora of software mining tasks. This chapter will provide a brief history of these representations, describe typical software mining tasks that benefit from GNNs, demonstrate one of these tasks in detail, and explain the benefits that GNNs can provide. Caveats and recommendations will also be discussed.

23.1 Introduction

Software Mining is broadly defined as any task that seeks to solve a software engineering problem by analyzing the myriad artifacts in projects and their connections (Hassan and Xie, 2010; Kagdi et al, 2007; Zimmermann et al, 2005). Consider the task of writing documentation. A human performing this task may gain comprehension of the software by reading the source code and understanding how different parts of the code interact. Then he or she may write documentation explaining the behavior of the system based on that comprehension. Likewise, if a machine is to automate writing that documentation, the machine must also analyze the software in order to comprehend it. This analysis is often called "Software Mining."

Collin McMillan
Department of Computer Science, University of Notre Dame, e-mail: cmc@nd.edu

While human comprehension of software is a cognitive process that occurs naturally as engineers read and interact with that software (Letovsky, 1987; Maalej et al, 2014), machine comprehension must be formally defined and quantifiable. Typically this boils down to a vectorized representation of each software artifact. For example, each identifier name in a function may be assigned an, e.g., 100-length vector denoting its position in a word embedding space. Then the function may be the average of those vectors for the identifier names it contains. Or it may be the output of a recurrent neural network given those identifier name vectors, or perhaps only the names that occur in particular locations. The point is that machine comprehension of software is often quantifiable as a vectorized representation of the artifacts composing that software.

Evidence is accumulating that Graph Neural Networks are an effective means to obtain these vectorized representations and thus improve machine comprehension of software. There is a long tradition in the Software Engineering research literature of treating software as a graph. Control flow graphs, call graphs, abstract syntax trees, execution path graphs, and many others are frequently the output of both static and dynamic analysis. Meanwhile, advances in software repository management have enabled the creation of datasets covering billions of lines of code. The result is fertile ground for GNNs.

This chapter covers the history and state-of-the-art in representing software as a graph for GNNs, followed by a high-level discussion of current approaches, a detailed look at a specific approach, and caveats for future researchers.

23.2 Modeling Software as a Graph

Software is a high-value target for GNNs partly because software tends to be very highly structured as a graph or set of graphs. Different software mining tasks may take advantage of different graph structures from software. Graph representations of software go far beyond any specific software mining task. Graph representations are baked into the way compilers convert source code into machine code (e.g., parse trees). They are used during linking and dependency resolution (e.g., program dependence graphs). And they have long the basis for many visualization and support tools to help programmers understand large software projects (Gema et al, 2020; Ottenstein and Ottenstein, 1984; Silva, 2012).

When considering how to make use of these different graph structures in software, basically the questions one must ask are: "what are the nodes?" and "what are the edges?" These questions take two forms in software engineering research: a macro- and a micro-level representation. The macro-level representation tends to concern connections among large software artifacts, such as a graph in which every source code file is a node and every dependency among the files is an edge. The micro-level representation, in contrast, tends to include small details, such as a graph in which every token in a function is a node, and every edge is a syntactic link between the nodes, such as are often extracted from an Abstract Syntax Tree.

This section compares and contrasts these representations as they relate to using GNNs for Software Mining tasks.

23.2.1 Macro versus Micro Representations

Graph structures in software may be broadly classified as either macro- or micro-level. In theory, the distinction is superfluous because a micro-level representation may be scaled up to arbitrary size. For example, an entire large program may be represented as one large abstract syntax tree. But in practice, time and space constraints necessitate a separation of macro- and micro-level representations. In a recent collection of Java programs (LeClair and McMillan, 2019), the average number of nodes in the AST of a function is over 120, with at least one edge per node. The average number of functions per program is over 1800, and there are over 28,000 programs in the dataset. The reality is that a micro-level representation of an entire program is often not feasible, so a macro-level representation is introduced to capture the "big picture."

23.2.1.1 Macro-level Representations

A macro-level graph representation of software captures the high-level structure and intent behind a program while avoiding a deep dive into details required to implement that intent. Inspiration for macro-level representations is often drawn from software design documents, such as those formally defined via UML (Braude and Bernstein, 2016; Horton, 1992). An example is a class diagram for an object-oriented program. Each class is a node in the graph. Edges in the graph may variously be dependency, inheritance, realization, composition, among others. Nodes may also have attributes that refer to the member variables and methods of a class.

In practice, selecting a macro-level representation for a software mining task using GNNs tends to be severely constrained by what can actually be obtained from the dataset. Often this constraint precludes the use of behavior-based graphs such as use case diagrams, because proper use case diagrams are rare, and those that are available are usually not in a consistent format. For example, because some engineers might follow different conventions, or only provide these diagrams informally. Software repositories tend to be replete with source code but lack documentation, especially design documentation (Kalliamvakou et al, 2014).

Therefore, by far, the most popular macro-level graph representations tend to be ones that can be extracted directly from source code. A decision often arises related to the degree of granularity, which usually is a choice between packages/directories, classes/files, or methods/functions. The class diagram is relatively easy to locate every class in a software project, then analyze each class to find their dependencies, inheritances, and etc. Package diagrams are similar, having the advantage of quickly providing a very high level view of a program – even large projects may only have a few dozen packages. But a very popular alternative is a function/method call graph, in which each function in a program is a node and each call relationship from one function to another is a directed edge between two nodes. Call graphs are popular within Software Engineering literature because they are relatively easy to extract while giving enough detail for a strong macro-level view of a program without overwhelming data sizes (recall a typical program has around 1800 functions (LeClair and McMillan, 2019)).

23.2.1.2 Micro-level Representations

A micro-level representation describes a portion of the software in great detail. Micro-level representations have been the focus of a majority of research using GNNs for software mining. Allamanis et al (2018b) describe one approach, pointing out that the "backbone of a program graph is the program's abstract syntax tree." However, as mentioned above, it is often not feasible to build a model relying on the entire AST of an entire program. Instead, a typical practice is to generate the AST for small portions of code, such as individual functions. Each function is treated as a graph, independent of all other functions.

The benefit of treating each function as a separate graph is that a GNN model can be trained on each independently. A prediction model of nearly any kind will require independent, self-contained examples. There will be some context about which an output prediction is generated (or against which a sample prediction is used for training). By treating each function as an independent graph, a GNN can be trained using each function as the context. This is a tidy solution in software mining for two reasons. First, many tasks in software mining involve predictions about specific functions, such as whether that function is likely to contain a fault (see the next section). Second, graphs of functions derived from the AST exhibit a community structure. In a typical function, there are many connections among nodes inside the function, but relatively few connections from nodes inside the function to nodes outside the function – the variables, conditionals, loops, and etc., in the code of a function interact closely with each other, while must less frequently referring to something outside the function such as the use of a global variable or call.

One may concoct any number of micro-level representations of software, based on different tokens in the source code and relationships of those tokens. For example, control flow relationships have occasionally been highlighted as often more valuable for comprehension than data dependencies (Dearman et al, 2005; Ko et al, 2006). At other times, method invocations (Mcmillan et al, 2013; Sillito et al, 2008) or signatures (Roehm et al, 2012) are proposed as providing superior information for different software mining tasks. Yet the pattern is that a micro-level representation is generated for many small portions of a software system, and these portions are treated as independent of each other. A GNN can take advantage of these micro-level representations by learning from each one as a different sample.

23.2.2 Combining the Macro- and Micro-level

Macro- and micro-level representations may be combined. One strategy would be to compute both macro- and micro-level representations independently, then concatenate them into one large context matrix. Such a model may be referred to as "dual encoder" (Chidambaram et al, 2019; Yang et al, 2019h) or "cascading" (Wang et al, 2017h) in that they learn two representations of the same object but at different levels of granularity. An alternative would be to use the output of the micro-level representation to seed the macro-level representation, for example, by learning a representation of each function using the AST and then using it as the initial value for the nodes in a function call graph.

23.3 Relevant Software Mining Tasks

Graph neural networks are becoming a staple of research in software mining tasks. The history of deep learning for software mining tasks is chronicled in several surveys (Allamanis et al, 2018a; Lin et al, 2020b; Semasaba et al, 2020; Song et al, 2019b). Allamanis et al (2018a) cast a particularly wide net and broadly classify software mining tasks that rely on neural networks as either "code generational" or "code representational." This classification is based on a big picture view of the models used for these tasks. In a code generational task, the output of the model is source code. Tasks in this category include automatic program repair (Chen et al, 2019e; Dinella et al, 2020; Wang et al, 2018d; Vasic et al, 2018; Yasunaga and Liang, 2020), code completion (Li et al, 2018a; Raychev et al, 2014), and compiler optimization (Brauckmann et al, 2020). These models tend to be trained with large volumes of code vetted somehow to ensure quality, with the aim of learning norms in code that lead to that quality. Then, during inference, the goal is to bring arbitrary code into closer conformance with those norms. For example, a model may be presented with code containing a bug, and that bug may be repaired by changing the code to be more like the model's predictions (which, it is hoped, represent the norms learned in training).

In contrast to code generational tasks are code representational tasks. These tasks use source code primarily as the input to a neural model during training but have a wide variety of outputs. Tasks in this category include code clone detection (Ain et al, 2019; Li et al, 2017c; White et al, 2016), code search (Chen and Zhou, 2018; Sachdev et al, 2018; Zhang et al, 2019f), type prediction (Pradel et al, 2020), and code summarization (Song et al, 2019b). In models designed to solve these tasks, the goal is usually to create a vectorized representation of code, which is then used for a specific task that may only be tangentially related to the code itself. For instance, for source code search, a neural model may be used to project the source code in a large repository into a vector space. Then a different model is used to project a natural language query into the same vector space. The code nearest to the query in the vector space is considered as the search result for that query. Code clone detection is similar: code is projected into a vector space, and very nearby code may be considered a clone in that space.

The use of graph neural networks is ballooning in both categories of software mining tasks. In code generational tasks, the focus tends to be on modifications to a program graph such as an AST that bring that graph into closer conformity with the model's expectations. While some approaches focus on code as a sequence (Chen et al, 2019e), the recent trend has been to recommend graph transformations or highlight non-conforming areas of the graph (Dinella et al, 2020; Yasunaga and Liang, 2020). This is useful in code because a recommendation may relate to code elements that are quite far away from each other, such as the declaration of a variable and a use of that variable. In contrast, in code representational tasks, the focus tends to be on creating ever more complex graph representations of code and then using GNN architectures to exploit that complexity. For example, the first GNN-based approaches tended to use only the AST (LeClair et al, 2020), while newer approaches use attention-based GNNs to emphasize the most important edges out of a multitude that can be extracted from code (Zügner et al, 2021). Despite differences in code generational and representational tasks, the trend in both categories has strongly favored GNNs.

Consider the task of code summarization, which exemplifies the trend towards GNNs. Code summarization is the task of writing natural language descriptions of source code. Typically these descriptions are used in documentation for that source code, e.g., JavaDocs. The evolution of this research area is shown in Figure 23.1. The term "code summarization" was coined around 2010, and several years of active research followed using templated and IR-based solutions. Then around 2017, solutions based on neural networks proliferated. At first, these were essentially seq2seq models in which the encoder sequence is the code and decoder sequence is the description. Starting around 2018, the state-of-the-art moved to linearized AST representations. Graph neural networks were proposed around this time as a better solution (Allamanis et al, 2018b), but it would be another year or more for GNN-based approaches to appear in the literature. GNNs are poised to underpin the state-of-the-art. In the next section, we dive into the details of a GNN-based solution, showing why it works and areas of future growth.

23.4 Example Software Mining Task: Source Code Summarization

This section describes source code summarization as an example software mining task that benefits from GNNs. Source code summarization, as mentioned above, is the task of writing natural language descriptions of source code. The input to a code summarization model includes at least the source code being described, though may also include other details about the software project from which the code originates. The output is the natural language description. This task is considered "code representational" because it primarily relies on a learned representation of code in order to make predictions about the description.

23.4.1 Primer GNN-based Code Summarization

As a primer towards GNN-based code summarization, consider a technique presented by LeClair et al (2020). This model is intended to be a straightforward application of convolutional GNNs in the vein of graph2seq (Xu et al, 2018c).

	IR	M	T	A	S	G
Haiduc et al (2010)	x					
Sridhara et al (2011)		x	x			
Rastkar et al (2011)	x	x	x			
De Lucia et al (2012)	x					
Panichella et al (2012)	x	x				
Moreno et al (2013)	x		x			
Rastkar and Murphy (2013)	x					
McBurney and McMillan (2014)		x	x			
Rodeghero et al (2014)	x					
Rastkar et al (2014)		x				
Cortés-Coy et al (2014)	x					
Moreno et al (2014)	x					
Oda et al (2015)			x			
Abid et al (2015)		x	x			
Iyer et al (2016)			x			
McBurney et al (2016)	x	x				
Zhang et al (2016a)		x	x			
Rodeghero et al (2017)		x				
Fowkes et al (2017)	x					
Badihi and Heydarnoori (2017)		x	x			
Loyola et al (2017)				x		
Lu et al (2017b)				x		
Jiang et al (2017)				x		
Hu et al (2018c)				x		
Hu et al (2018b)				x	x	
Wan et al (2018)				x	x	
Liang and Zhu (2018)				x	x	
Alon et al (2019a,b)				x	x	
Gao et al (2019b)				x		
LeClair et al (2019)				x	x	
Nie et al (2019)				x	x	
Haque et al (2020)				x	x	
Haldar et al (2020)				x	x	
LeClair et al (2020)				x	x	x
Ahmad et al (2020)				x	x	
Zügner et al (2021)				x	x	x
Liu et al (2021)				x	x	x

Table 23.1: Overview of papers on the topic of source code summarization, from the paper to coin the term "code summarization" in 2010 to the following ten years. Note the evolution from IR/template-based solutions to neural models and now to GNN models. Column *IR* indicates if the approach is based on Information Retrieval. *M* indicates manual features/heuristics. *T* indicates templated natural language. *A* indicates Artificial Intelligence (usually Neural Network) solutions. *S* means structural data such as the AST is used (for AI-based models). *G* means a GNN is the primary means of representing that structural data.

23.4.1.1 Model Input / Output

The input to this technique is a micro-level representation of code: it is just the AST of a single subroutine. The nodes in the graph are all nodes in the GNN, whether they are visible to the programmer or not. The only edge type is the parent-child relationship in the AST. Consider the code and example summaries in Example 23.1 and the AST of this code in Figure 23.1. Regarding the Figure 23.1, bold indicates text from source code that is visible to a human reader in the source code file – a depth-first search of the leaf nodes reveals the code sequence. E.g., "public void send guess ..." Non-bold indicates AST nodes that the compiler uses to represent structure. Visible text is preprocessed as it would appear to the model. For example, the name `sendGuess` is split into `send` and `guess`, and both nodes are children of a `name` node, which is a child of `function`. Neither `name` nor `function` is visible to a human reader. The circled areas 1-4 are reference points for discussion in Sections 23.4.1.4 and 23.4.2.

The AST in Figure 23.1 is the only input to the model, from which the model must generate an English description. Technically, the AST is srcml (Collard et al, 2011) preprocessed (e.g., splitting identifies such as `sendGuess` into `send` and `guess`) using community standard procedures (LeClair and McMillan, 2019). The reference output description in Example 23.1 is the actual JavaDoc summary written by a human programmer. The summary labeled "gnn ast" is the prediction from this approach. The summary labeled "flat ast" is the output from an immediate predecessor that used an RNN on a linearization of the AST. The only difference between the GNN and flat AST approach is the structure of the encoder; all other model details are identical. Yet, we note that the GNN-based approach matched the reference exactly, while the flat AST approach matched only a few words. Shortly we will analyze this example to provide intuition about why the model performed so well.

---

summaries

reference sends a guess to the server
ast-attendgru-gnn (LeClair et al, 2020) sends a guess to the socket
ast-attendgru-flat (LeClair et al, 2019) attempts to initiate a <UNK> guess

---

source code

```
public void sendGuess(String guess) {
  if( isConnected() ) {
    gui.statusBarInfo("Querying...", false);
    try {
      os.write( (guess + "\\r\\n").getBytes() );
      os.flush();
    } catch (IOException e) {
      gui.statusBarInfo("Failed to send guess.", true);
      System.err.println("IOException during send guess");
    }
  }
}
```

Example 23.1: The function `sendGuess()` and summary descriptions.

23.4.1.2 Model Architecture

The model architecture, as mentioned, is essentially a 2-hop graph2seq design based on a convolutional GNN. While we leave the details of the model to the relevant paper (LeClair et al, 2020), a bird's-eye view of the model is in Figure 23.2.

The model input is derived only from a single subroutine being described: the code as a sequence and the AST nodes and edges (Figure 23.2 area A). A word embedding projects tokens in the sequence and nodes in the AST into the same vector space, which is possible because the vocabulary is the same in both the sequence and the node input (area B). A 2-hop convolutional GNN is used to form a vectorized representation of the AST (area C). The output after the second hop is a matrix in which each column is a vector representing a node in the AST. A GRU is then applied to this matrix to capture information about the order in which the nodes appear. Meanwhile, a GRU is also applied to the sequence directly (area D). The decoder is a simple GRU representation of the summary (area H). Attention is applied between the decoder output and the sequence GRU output, as well as the GNN output (area E). The attended matrices are then concatenated into a context matrix (area F) and connected to an output dense layer (area G).

A key feature of the model is the attention between the decoder and the GNN output. The purpose of this attention is to highlight the nodes in the AST that are the most related to the words in the decoder sequence. We will describe below how this attention was made much more effective by the shared word embedding (area B).

23.4.1.3 Experiment

An experiment demonstrated improvement of the GNN model over various baselines, and explored the effects of various model design decisions. The experiment used a dataset of 2.1m Java methods and associated JavaDoc summaries (LeClair et al, 2020). Essentially the conditions were that 80% of the projects in the dataset were assigned for the training set, and 10% each for validation/testing. Duplicates and other defects were removed from the dataset in accordance with community standards (LeClair and McMillan, 2019). The model was trained with methods from the projects in the training set. The training ran for 10 epochs, and the model with the highest validation accuracy was selected for testing. The predictions from the tests were then compared with reference summaries.

Three findings stand out in findings reported by LeClair et al (2020). First, the GNN-based approaches outperform the most-similar baseline (ast-attendgru-flat) by about 1 BLEU point (about a 5% improvement). Since the only difference between the "flat" model and this GNN-based one is the AST encoder portion of the model, the improvement can be attributed to the use of the GNN (as opposed to an RNN) for the AST encoding. Improvement was also observed over two other baselines. The vanilla graph2seq model, which had only the AST and not the sequence encoder (Figure 23.2 area A), was roughly equivalent to the flat AST model in terms of aggregate BLEU score but this score obscures some details of the performance, which we will see in the next section.

The second key finding is that a hop distance of two results in the best over-all performance. While models with GNN iterations ranging between one and ten all achieve higher scores than the baselines, the model performs best with two iterations. One explanation is that nodes in the AST are only relevant to each other within a distance of about two. The AST is a tree, so information is propagated up and down levels of the tree. For two hops, this means information from a node will propagate to its parent in the first hop and then to its grandparent and siblings in the second hop. It is possible that nodes beyond this scope are not that relevant to the model for code summarization. However, another explanation is that the method of aggregating information in each hop is less efficient after two hops – this interpretation would be consistent with findings by Xu et al (2018c) that aggregation procedure is critical to GNN deployment. Either way, the practical advice for model designers is that the optimal number of GNN iterations for this task is not that high.

The third key finding is that the use of the GRU after the GNN layer (Figure 23.2 after area C) improves overall performance. The models labeled with the suffix +GRU use this GRU layer, as described in Section 23.4.1.2. The model labeled with the suffix +dense calculates attention between the decoder and the output matrix from the GNN. This model did not perform as well. A likely explanation is that source code has not only a tree structure via the AST – it also has an order from start to end. The GRU after the GNN captures this order and seems to result in a better representation of the code for summarization.

23.4.1.4 What benefit did the GNN bring?

A question remains regarding what benefit can be attributed to the use of a GNN. While we and others may observe an improvement in overall BLEU scores when using a GNN (LeClair et al, 2020; Zügner et al, 2021; Liu et al, 2021), a key point is that the GNN contributes *orthogonal* information to the model. This section explores how.

Concentration of Improvement:

The improvement is concentrated among a set of subroutines where the GNN adds significant improvement. It is not the case that the BLEU scores increase marginally for all subroutines – there is a set of subroutines that benefits the most. Consider Figure 23.3. The pie chart divides the test set into subroutines from the experiment describe above into five groups: one group where ast-attendgru-gnn performed the best, one group where ast-attendgru-flat performed the best, one group where they tied, one group for attendgru, and one group for other ties including when all models made the same prediction. For simplicity, we use BLEU-1 scores (BLEU-1 is unigram precision, single words predicted correctly).

What we observe is that each model achieves the highest BLEU-1 score for 20-25% of the subroutines. For about 12% of the subroutines, the AST-based models were tied, meaning that in total over 50% of the subroutines benefited from AST information (GNN plus flat AST models). But there still exists a large set of subroutines where attendgru outperformed all others. However, consider the bar chart in Figure 23.3. The "all" columns show the BLEU-1 score for that approach – note that ast-attendgru-gnn is only marginally higher than others. The "best" columns show the score for the set where that model achieved the highest BLEU-1 score (the set with that model's name indicated in the pie chart). We observe that the BLEU-1 scores for ast-attendgru-gnn are much higher for this set than others.

Demonstrating Improvement in Example 23.1:

A deeper dive into the subroutine `sendGuess()` from Example 23.1 demonstrates the improvement that a GNN provides. Recall that the ast-attendgru-gnn model calculates attention between each position in the decoder and each node in the output from the GNN (Section 23.4.1.2, Figure 23.2 area E). The result is an m x n matrix where m is the length of the decoder sequence and n is the number of nodes (in the implementation, $m=13$ and $n=100$). Thus each position in the attention matrix represents the relevance of an AST node to a word in the output summary. In fact, the attention matrix for ast-attendgru-flat has the same meaning: the models are identical except that ast-attendgru-gnn encodes the AST with a GNN then a GRU, while the flat model uses only the GRU. Comparing the values in these attention matrices provides a useful contrast of the two models because they show the contribution of the AST encoding to the prediction.

The benefit of a GNN becomes apparent in the attention networks in Figure 23.3. Both models have a very similar attention activation to the tokens in the source code sequence (Figures 23.3a and 23.3c). Both models show close attention to position 2 of the code sequence, which is the word "send". This is not surprising considering that "send" appears in the method's name. Yet, ast-attendgru-flat still incorrectly predicts the first word of the summary as "attempts", while ast-attendgru-gnn correctly predicts "sends." The explanation lies in the attention to AST nodes. The flat model focuses on node 37 (Figure 23.3d), which is an expr_stmt node immediately after the `try` block, just before the call to `os.write()`, indicated as area 1 in Figure 23.1. The reason for this focus suggested by the original paper on that model (LeClair et al, 2019) is that the flat AST model tends to learn broadly similar code structure such as "if-block, try-block, call to `os.write()`." Under this explanation, methods in the training set with this if-try-call-catch pattern are associated with the word "attempts."

In contrast, the GNN-based model focuses on position 8, which is the word "send" in the method name, just like in the attention to the code sequence (Figure 23.3b). The result is that the GNN-based AST encoding reinforces the attention paid to this word when predicting the first word of the output. Consider the method's AST in Figure 23.1. Position 8 is the node for "send" indicated at area 2. In a 2-hop GNN, this node will share information with its parent (name), grandparent (function), and sibling (guess). During training, the model learned that words associated with the AST nodes "function" and "name" are likely candidates for the first word of the summary, so the model knows to highlight this word.

In short, the GNN model outperformed because it conveys a lopsided benefit to a particular subset of the subroutines, and a likely reason it conveys this benefit is that it learns to associate AST tokens with particular locations in the code summary.

23.4.2 Directions for Improvement

The view of software as a graph described in Section 23.2 provides two directions for improvement: micro- and macro-level representations. Essentially the choice is whether to attempt to squeeze more information out of the source code being described (micro-level) or to draw upon more information from outside that source code (macro-level). If the aim is to generate summaries of a Java method, then one may learn more information about the details of that method, or one may use information from the classes, packages, dependencies, and etc., around the method. Micro- and macro-level improvements tend to be complementary rather than competitive. Learning more about the macro-level graph information benefits models of micro-level information and visa versa (Haque et al, 2020).

23.4.2.1 Example Micro-level Improvement

Liu et al (2021) present a notable example of an improvement to GNN-based code summarization using a richer micro-level graph representation of software. The essentials of the approach are similar to (LeClair et al, 2020) described above: the input to the model is the source code of a subroutine, and the output is a description of the subroutine. The encoder is based on a GNN, and the input to this GNN is the AST of the subroutine. The nodes in the graph are AST nodes, and the edges are the AST parent-child relationships. However, one novel aspect is that the model also considers other types of edges, namely control flow and data dependencies (these are unified as a Code Property Graph (Yamaguchi et al, 2014)). The benefit to this structure is that nodes in the AST will receive information directly from other relevant parts of the code, rather than only the nodes nearby in the AST.

Consider Figure 23.1 area 3, which is an AST node corresponding to the string variable "guess" in Example 23.1. The ast-attendgru-gnn approach would propagate information from that variable to the parents, grandparents, and siblings (in the two hops configuration). These would be the "name" and "decl" AST nodes. These nodes have locations in the word embedding associated with them, and these nodes also appear in practically every subroutine in the dataset. So, the model will learn how these nodes are used and associate them with what a human would call a variable declaration. The effect in this example is that the model will learn that the word "guess" is a variable name declaration.

The approach by Liu *et al.* improves over ast-attendgru-gnn because it can learn this relationship in addition to several others. The experiment with ast-attendgru-gnn showed evidence that AST structural information can lead to a better representation of code – it is useful to know that "guess" is a variable name declaration. But other relationships also exist. The variable "guess" is used in the call to `os.write()`. This relationship is a data dependency and is useful to human readers (Freeman, 2003). A human attempting to comprehend this code would likely note that whatever is passed into the subroutine as a parameter via the variable "guess" is subsequently written out via a method call. The benefit to Liu *et al.*'s approach is that it captures this relationship and uses it to form a more-complete GNN-based representation of the code.

A caveat is that as more edge types are added to the graph, more information will be propagated among nodes, which may have effects that are difficult to explain. Imagine in Figure 23.1 if an edge were to exist between "guess" at area 4 and "guess" at area 1, denoting a data dependency. A typical GNN design would propagate information across this edge. The result would be that the nodes around the location that uses "guess" would gain information from the nodes where "guess" is defined. But now imagine a control dependency from the `try` block start to the call to `os.write()`. The information would then also propagate from the `try` block to the use of "guess" over the control flow edge and then from the use of "guess" to the definition of "guess" over the data flow edge. This connection is difficult to explain – it is not clear what it means for a `try` block to be connected to the parameter list. A human may proffer an explanation for this particular subroutine, but a model such as ast-attendgru-gnn would always propagate information across these edges, even when it does not make sense to do so.

Liu *et al.* solve this problem by using an attentional GNN proposed by Zhu et al (2019b). Essentially, this GNN adds an attention layer as a gate prior to propagating information across an edge. The input to this gate includes the node embedding for the node at the origin of the edge, plus an edge embedding for that type of edge. The result is that the model learns during training when to propagate information from a node over a particular type of edge. That way, information from the, e.g., `try` block may or may not propagate to the parameter list, depending on whether that particular connection was useful during training. Liu *et al.* use the learned representation of code to help locate similar code comments in a database of those comments. However, the big picture idea is to use an attentional GNN to emphasize some edges in the code over others when the graph representation of code becomes large and complex, and this idea may serve as inspiration for a variety of software mining tasks. It is an example of how better micro-level representations of code can assist these software mining tasks.

23.4.2.2 Example Macro-level Improvement

One inspiration for macro-level improvement to neural code summarization is from (Aghamohammadi et al, 2020). Their approach focuses on generating summaries of code in Android projects. The approach is divided into two parts. The first part centers around an attentional encoder-decoder model similar to the attendgru baseline described by LeClair et al (2019). They use this model to generate an initial code summary based solely on the words inside the subroutine itself. The second part is to augment the initial summary with phrases from the summaries of other subroutines in the same project. The approach is to obtain a dynamic call graph of the Android program, which represents the actual runtime control flow from one subroutine to the next. Then a subset of the subroutines in this call graph is selected using PageRank – the idea is to emphasize the subroutines, which are called many times or hold other importance measurable from the structure of the call graph (McMillan et al, 2011). The summaries from these subroutines are then appended to the initial summary.

Aghamohammadi et al (2020)'s approach demonstrates an advantage to macro-level information. The macro-level information is the dynamic call graph of the entire program, and it is used to augment summaries created from the source code itself. The summaries tend to be longer and to provide more contextual information to readers. Recall `sendGuess()` in Example 23.1, for which ast-attendgru-gnn wrote "sends a guess to the socket." The approach by Aghamohammadi et al (2020) may (hypothetically) find that the subroutine that calls `sendGuess()` is a mouse click handler subroutine, and so would append, e.g., "called when the mouse is used to click the button." Human readers of documentation benefit from knowing how subroutines are used, so summaries that include this macro-level information tend to be considered more valuable by those readers (Holmes and Murphy, 2005; Ko et al, 2006; McBurney and McMillan, 2016).

Macro-level representations of code for software mining tasks are likely fertile ground for GNN-based technologies. The dynamic call graphs which Aghamohammadi et al (2020) extract contain information from actual runtime use, and a GNN may serve as a useful tool in generating a representation of this information. Yet, applications of GNNs to macro-level data for software mining tasks are still in their infancy.

23.5 Summary

In this chapter, we presented Software Mining Tasks as an application area for GNNs. A high-level view of any approach is to represent the software as a graph, then create a GNN model able to use this graph to learn to make predictions for a particular purpose. We present two views of software graphs: a micro- and macro-level representation. Micro-level representations predominate. For example, for the task of bug prediction in a subroutine, most approaches tend to look exclusively within those subroutines for patterns associated with that bug. Yet, evidence is emerging that macro-level representations may also benefit these tasks, as the context surrounding code is very likely to contain information necessary to comprehend that code. The future likely lies in combined GNN models of both micro- and macro-level graph representations of software.

We focus in this chapter on the task of source code summarization as an example of how GNN-based models help produce better predictions for software mining tasks. A straightforward approach is described in which the AST of subroutines is used to train a GNN, which leads to a better micro-level representation in many cases. An improvement based on an attentional GNN shows how much more complex graphs can also be exploited for better for this purpose. Yet, these improvements for code summarization likely herald improvements for many software mining tasks. Both code representational and code generational tasks depend heavily on understanding the nuances of the structure that code, and GNNs are a likely avenue for capturing this structure. This chapter has covered the history of this research, a specific target problem, and recommendations for future researchers.

Editor's Notes: AI for Code is a very fast-growing area in the recent years. Computer software or program is just like a second language compared to human language, which is not surprising that there are many shared attributes or aspects in both languages. Therefore, we have seen this trend that both NLP and Software communities start paying a large amount of attentions in applying GNNs for their domain applications and achieve the great successes in both domains. Just like GNNs for NLP, graph structure learning techniques in Chapter 14, GNN Methods in Chapter 4, GNN Scalability in Chapter 6, Heterogeneous GNNs in Chapter 16, GNN Robustness in Chapter 8 are all highly important building blocks for developing an effective and efficient approach with GNNs for code.

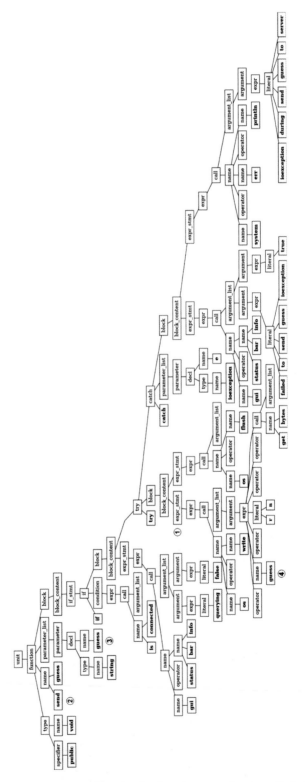

Figure 23.1: Abstract Syntax Tree for the function `sendGuess()` in Example 1.

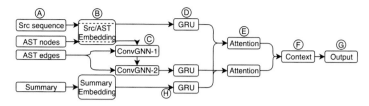

Figure 23.2: High-level diagram of the model architecture for 2-hop model.

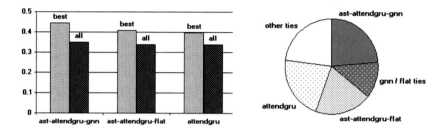

Figure 23.3: (left) Comparison of the BLEU-1 score for the subroutines where each method performed best, to BLEU-1 score for the whole test set. (right) Percent of test set for which each approach received the highest BLEU-1 score.

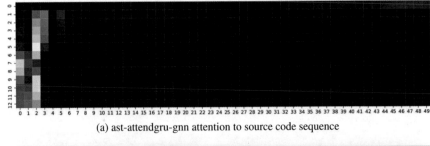

(a) ast-attendgru-gnn attention to source code sequence

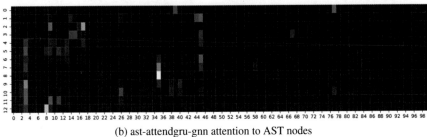

(b) ast-attendgru-gnn attention to AST nodes

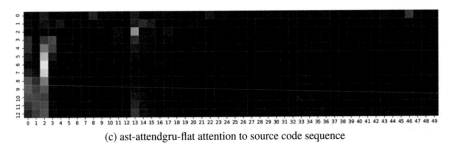

(c) ast-attendgru-flat attention to source code sequence

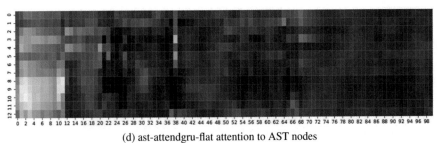

(d) ast-attendgru-flat attention to AST nodes

Figure 23.4: Visualization of attention network for ast-attendgru-gnn and ast-attendgru-flat for the subroutine `sendGuess()` in Example 23.1 and AST in Figure 23.1. Matrices are 13x100 because attention is applied between every position in the decoder output (length 13) and every position in the encoder (100 nodes or 100 code tokens). Bright areas indicate high attention. For example, position 2 in the code sequence is heavily emphasized for both models. Position 2 corresponds to the word "send" in the code sequence.

Chapter 24
GNN-based Biomedical Knowledge Graph Mining in Drug Development

Chang Su, Yu Hou, Fei Wang

Abstract Drug discovery and development (D^3) is an extremely expensive and time consuming process. It takes tens of years and billions of dollars to make a drug successfully on the market from scratch, which makes this process highly inefficient when facing emergencies such as COVID-19. At the same time, a huge amount of knowledge and experience has been accumulated during the D^3 process during the past decades. These knowledge are usually encoded in guidelines or biomedical literature, which provides an important resource containing insights that can be informative of the future D^3 process. Knowledge graph (KG) is an effective way of organizing the useful information in those literature so that they can be retrieved efficiently. It also bridges the heterogeneous biomedical concepts that are involved in the D^3 process. In this chapter we will review the existing biomedical KG and introduce how GNN techniques can facilitate the D^3 process on the KG. We will also introduce two case studies on Parkinson's disease and COVID-19, and point out future directions.

24.1 Introduction

Biomedicine is a discipline with lots of highly specialized knowledge accumulated from biological experiments and clinical practice. This knowledge is usually buried

Chang Su,
Department of Population Health Sciences, Weill Cornell Medicine, e-mail: `chs4001@med.cornell.edu`

Yu Hou,
Department of Population Health Sciences, Weill Cornell Medicine, e-mail: `yuh4001@med.cornell.edu`

Fei Wang,
Department of Population Health Sciences, Weill Cornell Medicine, e-mail: `few2001@med.cornell.edu`

© The Author(s), under exclusive license to Springer Nature Singapore Pte Ltd. 2022
L. Wu et al. (eds.), *Graph Neural Networks: Foundations, Frontiers, and Applications*,
https://doi.org/10.1007/978-981-16-6054-2_24

in massive biomedical literature and text books. This makes effective knowledge organization and efficient knowledge retrieval a challenging task. Knowledge graph is a recently emerged concept aiming at achieving this goal. A knowledge graph (KG) stores and represents knowledge by constructing a semantic network describing entities and the relationships between them. The basic elements comprising a knowledge graph are a set of ⟨head, relation, tail⟩ tuples, where the heads and tails are concept entities and relations link these entities with semantic relationships. In biomedicine, the typical entities could be diseases, drugs, genes, etc., and the relationships could be treats, binds, interactions, etc. Large scale biomedical KG makes efficient knowledge retrieval and inference possible.

Biomedical KG can effectively complement the biomedical data analytics processes. In particular, many different types of biomedical data are heterogeneous and noisy (Wang et al, 2019f; Wang and Preininger, 2019; Zhu et al, 2019e), which makes the data-driven models developed on these data not reliable for real practice. Biomedical KGs (BKGs) effectively encode the biomedical entities and their semantic relationships, which can serve as "prior knowledge" to guide the downstream data-driven analytics procedure and improve the quality of the model. On the other hand, we can also use BKGs to generate hypotheses (such as which drug can be used to treat which disease), and get them validated in real world health data (such as electronic health records).

In this chapter, we will review existing BKGs and present examples of how BKGs can be used for generating drug repurposing hypotheses, and point out future directions.

24.2 Existing Biomedical Knowledge Graphs

This section surveys the existing BKGs that are publicly available and the ways of BKG construction and curation (Table 24.3).

A common way for constructing a BKG is to extract and integrate data from data resources, usually, which are manually curated to summarize and organize the biomedical knowledge derived from biological experiments, clinical trials, genome wide association analyses, clinical practices, etc (Santos et al, 2020; Ioannidis et al, 2020; Himmelstein et al, 2017; Rizvi et al, 2019; Yu et al, 2019b; Zhu et al, 2020b; Zeng et al, 2020b,b; Domingo-Fernández et al, 2020; Wang et al, 2020e; Percha and Altman, 2018; Li et al, 2020d,b; Goodwin and Harabagiu, 2013; Rotmensch et al, 2017; Sun et al, 2020a). In table 24.2, we summarized some public data resources that have been commonly used in the construction of BKGs. For instance, Comparative Toxicogenomics Database (CTD) (Davis et al, 2019) is an open resource providing rich, manually curated chemical–gene, chemical–disease and gene–disease relational data, for the aim of advancing understanding the impacts of environmental exposures on human health. DrugBank (Wishart et al, 2018) is a database containing information of the approved drugs and drugs under trial, as well as the pharmacogenomic data (e.g., drug-target interactions). Ontology resources like Gene Ontology

(Ashburner et al, 2000) and Disease Ontology (Schriml et al, 2019) stored functional and semantic context of genes and diseases, respectively. By integrating data from these rich resources, a number of BKGs have been constructed (Santos et al, 2020; Ioannidis et al, 2020; Himmelstein et al, 2017; Rizvi et al, 2019; Yu et al, 2019b; Zhu et al, 2020b; Zeng et al, 2020b,b; Domingo-Fernández et al, 2020; Wang et al, 2020e). For example, Hetionet (Himmelstein et al, 2017), released in 2017, is a well-curated BKG that integrates 29 publicly available biomedical databases. It contains 11 types of 47031 biomedical entities and 24 types of over 2 million relations among thoses entities. Similar to Hetionet, Drug Repurposing Knowledge Graph (DRKG) (Ioannidis et al, 2020) was built by integrating data from six different existing biomedical databases, containing 13 types of about 100K entities and 107 types of over 5 million relationships. Zhu et al (2020b) constructed a drug-centric BKG by systematically integrating multiple drug databases such as Drug-Bank (Wishart et al, 2018) and PharmGKB (Whirl-Carrillo et al, 2012). Hetionet, DRKG, and BKGs have been used in accelerating computational drug repurposing. PreMedKB (Yu et al, 2019b) includes the information of disease, genes, variants, and drugs by integrating relational data among them from existing resources. By integrating multiple dietary related databases, Rizvi et al (2019) built a BKG, named Dietary Supplements Knowledge Base (iDISK), which covers knowledge of dietary supplements, including vitamins, herbs, minerals, etc. The Clinical Knowledge Graph (CKG)(Santos et al, 2020) was constructed by integrating relevant existing biomedical databases such as DrugBank (Wishart et al, 2018), Disease Ontology (Schriml et al, 2019), SIDER (Kuhn et al, 2016), etc. and knowledge extracted from scientific literature. It contains over 16 million nodes and over 220 million relationships. Compared to other BKGs, CKG has a finer granularity of knowledge as it involves more entity types such as metabolite, modified protein, molecule function, transcript, genetic variant, food, clinical variable, etc.

As the rapid development of biomedical research, a continuously increasing volume of biomedical articles have been published every day. Manually extracting knowledge from literature for BKG cuuration is no longer sufficient to meet current needs. To this end, efforts have been made in using text mining methods to extract biomedical knowledge from scientific literature to construct BKGs (Domingo-Fernández et al, 2020; Wang et al, 2020e; Percha and Altman, 2018; Li et al, 2020d). For example, Sun et al (2020a) constructed a knowledge graph by extracting biomedical entities and relationships from drug descriptions, medical dictionaries, and literature to identify suspected cases of Fraud, Waste, and Abuse from claim files. COVID-KG (Wang et al, 2020e) and COVID-19 Knowledge Graph (Domingo-Fernández et al, 2020) were built by extracting COVID-19 specific knowledge from biomedical literature. The resulting COVID-19 specific BKGs contain entities such as diseases, chemicals, genes, and pathways, along with their relationships. KGHC (Li et al, 2020d) is a BKG with the specific focus on hepatocellular carcinoma. It was built by extracting knowedge from literature and contents on the internet, as well as structured triples from SemMedDB (Kilicoglu et al, 2012). In addition, some studies (Goodwin and Harabagiu, 2013; Li et al, 2020b; Rotmensch et al, 2017; Sun et al, 2020a) tried to build BKGs from clinical data such as electronic health records (EHRs) and electronic medical records (EMRs). For instance, Rotmensch et al (2017) constructed a BKG by extracting disease-symptom associations from EHR data using the data-driven approach. Li et al (2020b) proposed a systematic pipeline for extracting BKG from large scale EMR data. Compared to other BKGs based on triplet structure, the resulting KG is based a quadruplet structure, i.e., $\langle head, relation, tail, property \rangle$. Here the property includes information such as co-occurrence number, co-occurrence probablity, specificity, and reliability of the corresponding $\langle head, relation, tail \rangle$ triplet.

Table 24.1: Summary of existing BKGs.

BKGs	Entities	Relations	Focus	Construction method	URL
Clinical Knowledge Graph (Santos et al, 2020)	16 million entities from 33 entity types	220 million relations from 51 relation types	General	Resources Integration	`https://github.com/MannLabs/CKG`
Drug Repurposing Knowledge Graph (Ioannidis et al, 2020)	97,238 entities from 13 entity types	5,874,261 relations from 107 relation types	General	Resources Integration	`https://github.com/gnn4dr/DRKG`
Hetionet (Himmelstein et al, 2017)	47,031 entities from 11 entity types	2,250,197 relations from 24 relation types	General	Resources Integration	`https://het.io/`
iDISK (Rizvi et al, 2019)	144,059 entities from 6 entity types	708,164 relations from 6 relation types	Dietary Supplements	Resources Integration	`https://conservancy.umn.edu/handle/11299/204783`
PreMedKB (Yu et al, 2019b)	404,904 entities from 4 entity types	496,689 relations from 52 relation types	General	Resources Integration	`http://www.fudan-pgx.org/premedkb/index.html#/home`
Zhu et al (2020b)	5 entity types	9 relation types	General	Resources Integration	-
Zeng et al (2020b)	145,179 entities from 4 entity types	15,018,067 relations from 39 relation types	General	Resources Integration	-
COVID-19 Knowledge Graph (Domingo-Fernández et al, 2020)	3,954 entities from 10 entity types	9,484 relations	COVID-19	Literature Mining	`https://github.com/covid19kg/covid19kg`
COVID-KG (Wang et al, 2020e)	67,217 entities from 3 entity types	85,126,762 relations from 3 relation types	COVID-19	Literature Mining	`http://blender.cs.illinois.edu/covid19/`
Global Network of Biomedical Relationships (Percha and Altman, 2018)	Three entity types (Chemical, Disease, Gene)	2,236,307 relations from 36 relation types	General	Literature Mining	`https://zenodo.org/record/1035500`
KGHC (Li et al, 2020d)	5,028 entities from 9 entity types	13,296 relations	Hepatocellular Carcinoma	Literature Mining	`http://202.118.75.18:18895/browser/`
Li et al (2020b)	22,508 entities from 9 entity types	579,094 relations	General	EHR Mining	-
QMKG (Goodwin and Harabagiu, 2013)	634,000 entities	1,390,000,000 relations	General	EHR Mining	-
Rotmensch et al (2017)	647 entities from 2 entity types	Disease-Symptom	General	EHR Mining	-
Sun et al (2020a)	1,616,549 entities from 62 entity types	5,963,444 relations from 202 relation types	General	EHR Mining	`https://web.archive.org/web/20191231152615if_/http://121.12.85.245:1347/kg_test/#/`

Table 24.2: Publicly available resources for BKG construction

Database	Entities	Relations	Short Description	URL
Bgee (Bastian et al, 2021)	60,072 Anatomy and Gene entities	11,731,369 relations in terms of presence/absence of expression	A database for Anatomy-Gene Expression	`https://bgee.org/`
Comparative Toxicoge-nomics Database (Davis et al, 2019)	73,922 Disease, Gene, Chemical, Pathway entities	38,344,568 Chemical-Gene, Chemical-Disease, Chemical-Pathway, Gene-Disease, Gene-Pathway, and Disease-Pathway relations	A database that is manually curated includes chemical-disease-gene-pathway relations	`http://ctdbase.org/`
Drug–Gene Interaction Database (Cotto et al, 2018)	160,054 Drug and Gene entities	96,924 Drug-Gene Interaction relations	A database for drug-gene interactions	`https://www.dgidb.org/`
DISEASES (Pletscher-Frankild et al, 2015)	22,216 Disease and Gene entities	543,405 relations	A database for Disease-Gene Association	`https://diseases.jensenlab.org/`
DisGeNET (Piñero et al, 2020)	159,052 Disease, Gene and Variant entities	839,138 Gene-Disease,Variant-Disease relations	A database that integrates data from expert-curated repositories for genes and variants associated with human diseases.	`https://www.disgenet.org/home/`
IntAct (Orchard et al, 2014)	119,281 Chemical and Gene entities	1,130,596 relations	A database for molecular interaction data	`https://www.ebi.ac.uk/intact/`
STRING (Szklarczyk et al, 2019)	24,584,628 Protein entities	3,123,056,667 Protein-Protein Interaction relations	A database for Protein-Protein Interaction netword	`https://string-db.org/`
SIDER (Kuhn et al, 2016)	7,298 Drug and Side-effect entities	139,756 Drug-Side effect relations	A database contains medicines and their recorded adverse drug reactions	`http://sideeffects.embl.de/`
SIGNOR (Licata et al, 2020)	7,095 entities from 10 entity types	26,523 relations	A database for signaling information published in the scientific literature	`https://signor.uniroma2.it/`
TISSUE (Palasca et al, 2018)	26,260 entities in Tissue and Gene	6,788,697 relations	A database for Tissue-Gene Expression by literature curated manually	`https://tissues.jensenlab.org/`
DrugBank (Wishart et al, 2018)	15,128 Drug entities	28,014 Drug-Target, Drug-Enzyme, Drug-Carrier, Drug-Transporter relations	A database for the information on drugs and drug targets	`https://go.drugbank.com/`
KEGG (Kanehisa and Goto, 2000)	33,756,186 entities in Drug, Pathway, Gene, etc.	-	A database for genomes, biological pathways, diseases, drugs, and chemical substances.	`https://www.kegg.jp/kegg/`
PharmGKB (Whirl-Carrillo et al, 2012)	43,112 entities in Genes, Variant, Drug/Chemical and Phenotype	61,616 relations	A database for drugs and drug-related relationships.	`https://www.pharmgkb.org/`
Reactome (Jassal et al, 2020)	21.087 Pathway entities	-	A manually curated database for peer-reviewed pathway	`https://reactome.org/`
Semantic MEDLINE Database (Kilicoglu et al, 2012)	-	109,966,978 relations	A database contains Semantic predictions from the literature	`https://skr3.nlm.nih.gov/index.html`
Gene Ontology (Ashburner et al, 2000)	44,085 Gene entities	-	An ontology the functions of genes	`http://geneontology.org/`

24.3 Inference on Knowledge Graphs

In KG inference, one usually needs to address two important attributes of KGs: 1) the KG's local and global structure properties, and 2) heterogeneity of entities and relations(Wang et al, 2017d; Cai et al, 2018b; Zhang et al, 2018c; Goyal and Ferrara, 2018; Su et al, 2020c; Zhao et al, 2019d). In this context, a standard pipeline for KG inference typically contains two major steps: 1) learning embeddings (i.e., representation vectors) for entities (and relations) while preserving their structural properties and entity and relation attributes in the KG; and 2) performing downstream tasks such as entity classification and link prediction using the learned embeddings. Of note, one can perform these two steps separately, but also build an end-to-end model that can jointly learn the embeddings and perform downstream tasks. In this section, we review the existing techniques for inference on KGs, including the conventional inference techniques and the GNN-based models.

24.3.1 Conventional KG inference techniques

This subsection surveys the conventional KG inference techniques.

Semantic matching models typically exploit the similarity-based energy functions by matching latent semantics of entities and relations in the embedding spaces. A well-known semantic matching model, RESCAL (Nickel et al, 2011; Jenatton et al, 2012), was proposed based on the idea that entities are similar if connected to similar entities via similar relations(Nickel and Tresp, 2013). By associating each relation r_k with a matrix M_k , it defines the energy function by a bilinear model $f(e_i, r_k, e_j) = \mathbf{h}_i^\top M_k \mathbf{h}_j$, where $\mathbf{h}_i, \mathbf{h}_j \in \mathbb{R}^d$ are d-dimensional embedding vectors for entities e_i and e_j, respectively. RESCAL jointly learns embedding results for entities by e_i and e_j and for relation by M_k. Another model, DistMult (Yang et al, 2015a) simplifies RESCAL by restricting matrix M_k for relation r_k as a diagonal matrix. Though DistMult is more efficient than RESCAL, it can only deal with the undirected graphs. To address this, HolE (Nickel et al, 2016b) composes e_i and e_j by their circular correlation. Consequently, power of RESCAL and efficiency of Dist-Mult are inherited by HolE. Other semantic matching models refer to the neural network architecture by considering embedding as the input layer and energy function as the output layer, such as the the semantic matching energy (SME) model (Bordes et al, 2014) and multi-layer perceptron (MLP) (Dong et al, 2014).

Translational distance models are based on the idea that, for each triplet (e_i, r_k, e_j) , the relation r_k can be considered as a translation from head entity e_i to tail entity e_j in the embedding space. Accordingly, they exploit distance-based energy functions to model the triplets in KG. In this context, TransE (Bordes et al, 2013) is the famous pioneer of the translational distance model family. It typically represents relation r_k as the translation vector $\mathbf{g}_k$, such that e_i and e_j are closely connected by r_k. Therefore, the energy function is defined as $f(e_i, r_k, e_j) = \|\mathbf{h}_i + \mathbf{g}_k + \mathbf{h}_j\|_2$. Since all parameters to learn are entity and relation embedding vectors lying in a same low-dimensional space, TransE is obviously easy to train. A drawback of TransE is that it cannot do well with N-to-1, N-to-1 and N-to- N structures in KGs. To address this issue, TransH (Wang et al, 2014) extends TransE by introducing a hyperplane for each relation r_k and projecting e_i and e_j into the hyperplane before constructing the translation scheme. In this way, TransH improves model capacity while preserving efficiency. Similarly, TransR (Lin et al, 2015) extends TransE by introducing the relation-specific space. Further, for more fine-grained embedding, TransD (Ji et al, 2015) extends TransE by constructing two matrices M_k^1 and M_k^2 for each r_k to project e_i and e_j, respectively. Hence it captures both entity diversity and relation diversity. Further, TranSparse (Ji et al, 2016) simplifies TransR by using adaptive sparse matrices to model different types of relations, and TransF (Feng et al, 2016) relaxes the translation restriction as $\mathbf{h}_i + \mathbf{g}_k \approx \alpha \mathbf{h}_j$.

Meta-path-based approaches. A potential issue for both semantic matching models and translational distance models is that they mainly focus on one-hop information (i.e., modeling neighboring entities within a triplet) and hence may ignore the global structure properties of KGs. To address this, the meta-path based models aim at capturing local and global structure properties, as well as entity and relation types for KG inference. Typically, a meta-path is defined as a sequence of node types separated by edge types (Sun et al, 2011). For example, a meta-path of length l is $a_1 \xrightarrow{b_1} a_2 \xrightarrow{b_2} ... \xrightarrow{b_{l-1}} a_l$, where $\{a_1, a_2, ..., a_l\}$ and $\{b_1, b_2, ..., b_{l-1}\}$ are the sets of node type and relation type, respectively. Following this idea, Heterogeneous Information Network Embedding (HINE) (Huang and Mamoulis, 2017) defines meta-path-based proximity. It preserves heterogeneous structure by minimizing the difference between meta-path-based proximity and expected proximity in the embedding space. Moreover, metapath2vec (Dong et al, 2017) formalizes meta-path-based random walks and extends the word embedding model SkipGram to learn entity embeddings, by considering each walk path as a sentence and entities as words.

Convolutional neural network (CNN) models have also been used to address the KG inference task. For example, ConvE (Dettmers et al, 2018) uses CNN architecture for link prediction in KGs. For each triplet (e_i, r_k, e_j), ConvE first reshapes embedding vectors of e_i and r_k as two matrices and concatenate them. The resulting matrix is then fed to the convolutional layers to produce feature maps, which are then transformed into the entity embedding space to match the embedding of e_j. In addition, ConvKB (Nguyen et al, 2017) directly concatenates embedding vectors of e_i, r_k, and e_j, for each triplet (e_i, r_k, e_j), into a 3-column matrix. Then the matrix is fed to the convolutional layers to learn the entity and relation embeddings.

24.3.2 GNN-based KG inference techniques

This subsection discusses KG inference techniques based on the novel GNN architectures.

Graph convolution network (GCN)-based architectures. A pioneer effort using and extending GCN in KG inference is the Relational GCN (R-GCN) (Schlichtkrull et al, 2018). In contrast to the original application scenario, the structure property of a KG is usually heterogeneous as having diverse entity types and relation types. To address this, R-GCN introduces two subtle modifications on the regular GCN architecture (Berg et al, 2017). Specifically, for each entity, instead of simply aggregating information from all of its neighbors, R-GCN uses a relation-specific transformation mechanism, which first gathers information from neighboring entities based on relation types and relation directions separately and then accumulates them together. Specifically,

$$\mathbf{h}_i^{(l+1)} = \sigma \left(\sum_{r_k \in \mathbb{R}} \sum_{j \in \mathcal{N}_i^k} \frac{1}{c_{i,k}} W_k^{(l)} \mathbf{h}_j^{(l)} + W_0^{(l)} \mathbf{h}_i^{(l)} \right) \tag{24.1}$$

Here $\mathbf{h}_i^{(l+1)}$ is the embedding vector of entity e_i at the $(l+1)$-th graph convolutional layer. $\mathcal{R}$ is the set of all relations and $\mathcal{N}_i^k$ is the neighbors of entity e_i under relation r_k. The problem-specific normalization coefficient $c_{i,k}$ can be either learned or pre-defined. Using softmax for each entity, R-GCN can be trained for entity classification. In link prediction, R-GCN is used as an encoder for learning embedding vectors of the entities while the factorization model, DistMult, is used as the decoder to predict missing links in the KG based on the learned entity embeddings. It resulted in a significantly improved performance compared to the baseline models like DistMult and TransE.

Cai et al (2019) proposed the TransGCN, which combines the GCN architecture with the translational distance models (e.g., TransE and RotatE) for link prediction in KGs. Compared to R-GCN, TransGCN aims to address the link prediction task without a task-specific decoder like R-GCN and learn both entity embeddings and relation embeddings simultaneously. For each triplet (e_i, r_k, e_j), TransGCN assumes that r_k is the transformation from the head e_i to the tail e_j in the embedding space. Then it extends the GCN layer to update e_i's embedding as

$$\mathbf{m}_i^{(l+1)} = \frac{1}{c_i} W_0^{(l)} \left(\sum_{(e_j, r_k, e_i) \in \mathcal{N}_i^{(in)}} \mathbf{h}_j^{(l)} \circ \mathbf{g}_k^{(l)} + \sum_{(e_i, r_k, e_j) \in \mathcal{N}_i^{(out)}} \mathbf{h}_j^{(l)} \star \mathbf{g}_k^{(l)} \right) \tag{24.2}$$

$$\mathbf{h}_i^{(l+1)} = \sigma \left(\mathbf{m}_i^{(l+1)} + \mathbf{h}_i^{(l)} \right) \tag{24.3}$$

where $\circ$ and $\star$ are transformation operators that can be defined based on specific translational mechanism used. $\mathcal{N}_i^{(in)}$ and $\mathcal{N}_i^{(out)}$ are incoming and outgoing triplet of e_i, respectively. The normalization constant c_i was defined by the total degree of entity e_i. Meanwhile, embedding of each relation r_k was updated by simply $\mathbf{g}_k^{(l+1)} = \sigma(W_1^{(l)} \mathbf{g}_k^{(l)})$. The authors engaged two translational mechanisms, TransE and RotatE, and defined $\circ$, $\star$, and scoring functions accordingly. Both resulting architectures, TransE-GCN and RotatE-GCN, showed higher performance than TransE, RotatE, and R-GCN in the experiments.

Structure-Aware Convolutional Network (SACN) (Shang et al, 2019) is another architecture for knowledge graph inference based on GCN. Similar to R-GCN, it engaged a weighted graph convolutional network (WGCN) as the encoder to capture the structure property of the KG. WGCN considers a KG with multiple relation types as a combination of multiple sub-graphs with single relation type. Then, the embedding vector of each entity e_i can be obtained by a weighted combination of information propagation based on each sub-graph,

$$\mathbf{h}_i^{(l+1)} = \sigma \left(\sum_{j \in \mathcal{N}_i} \alpha_k^{(l)} \mathbf{h}_j^{(l)} W^{(l)} + \mathbf{h}_i^{(l)} W^{(l)} \right) \tag{24.4}$$

where $\alpha_k^{(l)}$ is the weight of relation r_k at the l-th layer. The learned embedding from WGCN was then fed to a decoder, Conv-TransE, a CNN with TransE's translational mechanism, for link prediction.

Graph attention network (GAT)-based architectures. A potential drawback of the GCN architectures is that, for each entity, they treat the neighbors equally to gather information. However, different neighboring entities, relations or triplets may have different importances in indicating a specific entity, and the weights of neighboring entities under the same relation may be also distinct. To address this, GATs have been used to involved in the KG inference problems. One of the early efforts is the GATE-KG (i.e., graph attention-based embedding in KG) (Nathani et al, 2019). It introduces an extended and generalized attention mechanism as the encoder to produce the entity and relation embeddings while capturing the diverse relation type in KG. For each triplet (e_i, r_k, e_j), GATE-KG first produces a representation vector $\mathbf{c}_{ijk}^{(l)}$ of this triplet by

$$\mathbf{c}_{ijk}^{(l)} = W_1^{(l)} [\mathbf{h}_i^{(l)} || \mathbf{h}_j^{(l)} || \mathbf{g}_k^{(l)}] \tag{24.5}$$

Here $||$ is the concatenation operation. The attention coefficient α_{ijk} is obtained by

$$\beta_{ijk}^{(l)} = \text{LeakyReLU} \left(W_2^{(l)} \mathbf{c}_{ijk}^{(l)} \right) \tag{24.6}$$

$$\alpha_{ijk}^{(l)} = \frac{\exp(\beta_{ijk}^{(l)})}{\sum_{j' \in \mathcal{N}_i} \sum_{k' \in \mathcal{R}_{ij'}} \exp(\beta_{ij'k'}^{(l)})} \tag{24.7}$$

where $\mathcal{R}_{ij}$ is the set of all relations between e_i and e_j . By aggregating information from neighbors according to different relations, entity e_i's embedding vector $\mathbf{h}_i^{(l+1)}$ at the $(l+1)$-th layer can be calculated as

$$\mathbf{h}_i^{(l+1)} = \sigma \left(\sum_{j \in \mathcal{N}_i} \sum_{k \in \mathcal{R}_{ij}} \alpha_{ijk}^{(l)} \mathbf{c}_{ijk}^{(l)} \right) \tag{24.8}$$

In addition, by using the auxiliary relation between n-hop neighbors and iteratively accumulating information of n-hop neighbors at the n-th graph attention layer, GATE-KG gives high weights to the 1-hop neighbors while lower weights to the n-hop neighbors. Hence it captures the multi-hop structure information of KG.

Relational Graph neural network with Hierarchical ATtention (RGHAT) (Zhang et al, 2020i) is another GAT-based model to address link prediction in KGs. Specifically, it engages a two-level attention mechanism. First, a relational-level attention defines the weight of each relation r_k indicating a specific entity e_i as

$$\mathbf{a}_{ik} = W_1 \left[\mathbf{h}_i || \mathbf{g}_k \right] \tag{24.9}$$

$$\alpha_{ik} = \frac{\exp(\sigma(\mathbf{z}_1 \cdot \mathbf{a}_{ik}))}{\sum_{r_x \in \mathcal{N}_i} \exp(\sigma(\mathbf{z}_1 \cdot \mathbf{a}_{ix}))} \tag{24.10}$$

where $\mathbf{z}_1$ is a learnable parameter vector and σ is LeakyReLU. $\mathcal{N}_i$ is the neighboring relations of entity e_i. Second, it defines an entity-level attention as

$$\mathbf{b}_{ikj} = W_2 \left[\mathbf{a}_{ik} || \mathbf{h}_j \right] \tag{24.11}$$

$$\beta_{kj} = \frac{\exp(\sigma(\mathbf{z}_2 \cdot \mathbf{b}_{ikj}))}{\sum_{r_y \in \mathcal{N}_{i,k}} \exp(\sigma(\mathbf{z}_1 \cdot \mathbf{b}_{iyj}))} \tag{24.12}$$

where $\mathbf{z}_2$ is a learnable parameter vector and $\mathcal{N}_{i,k}$ denotes the set of tail entities of entity e_i under relation r_k. The final attention coefficient for gathering information via triplet (e_i, r_k, e_j) is calculated as $\mu_{ikj} = \alpha_{ik} \cdot \beta_{kj}$. Similar to GATE-KG, the RGHAT engages ConvE as the decoder for link prediction.

Wang et al (2019j) proposed the Knowledge Graph Attention Network (KGAT) for recommendation based on KG, which contains three types of layers. First, a embedding layer learns embeddings for entities and relations using TransR. Second, the attentive embedding propagation layers extend GAT to capture the high-order structure properties (i.e., multi-hop neighbor information) of KG. Specifically, they defined the attention coefficient for each triplet (e_i, r_k, e_j), depending on distance between e_i and e_j in the r_k's space, i.e.,

$$\beta_{ijk} = (W_k \mathbf{h}_i)^\top \tanh (W_k \mathbf{h}_j + \mathbf{g}_k) \tag{24.13}$$

$$\alpha_{ijk} = \frac{\exp(b_{ijk})}{\sum_{j' \in \mathcal{N}_i} \sum_{k' \in \mathcal{R}_{ij'}} \exp(\beta_{ij'k'})} \tag{24.14}$$

KGAT then stacks multiple attentive embedding propagation layers to capture information of multiple-hop neighbors of each entity, specifically, entity e_is embedding at the $(l+1)$-th layer, i.e., $\mathbf{h}_i^{(l+1)} = \sigma \left(\mathbf{h}_i^{(l)}, \mathbf{h}_{\mathcal{N}_i}^{(1)} \right)$, where $\mathbf{h}_{\mathcal{N}_i}^{(1)} = \sum_{(e_i, r_k, e_j) \in \mathcal{N}_i} \alpha_{ijk} \mathbf{h}_j^{(l)}$. Finally, a prediction layer concatenates embeddings at each graph attention layer for each entity to make prediction.

In addition, Heterogeneous graph Attention Network (HAN) (Wang et al, 2019m) uses GAT to address the node (i.e., entity) classification in the heterogeneous graphs (the KG can be considered as a specific type of heterogeneous graph). HAN couples graph attention mechanism with meta-paths to capture the heterogeneous structure properties. A hierarchical attention mechanism that contains a node-level attention and semantic-level attention was proposed. The node-level attention aims to learn the importance of the meta-path-based neighbors in indicating a node. Specifically, it first projects different types of entities into a same space by $\mathbf{h}_i = M_{\phi_i} \mathbf{h}'_i$, where ϕ_i is the type of entity e_i, and $\mathbf{h}_i$ and $\mathbf{h}'_i$ are the projected and original embeddings of e_i, respectively. It then calculates the attention weight α_{ij}^{Φ} of entity pair (e_i, e_j) under a specific meta-path Φ, as

$$\alpha_{ij}^{\Phi} = \frac{\exp(\mathbf{a}_{\Phi}^{\top} \cdot [\mathbf{h}_i || \mathbf{h}_j])}{\sum_{j' \in \mathcal{N}_i^{\Phi}} \exp(\mathbf{a}_{\Phi}^{\top} \cdot [\mathbf{h}_i || \mathbf{h}_{j'}])} \tag{24.15}$$

where $\mathcal{N}_i^{\Phi}$ is the neighbors of e_i under meta-path Φ and $\mathbf{a}_{\Phi}$ is the node-level attention vector. In addition, the semantic attention layer learns importance of each meta-path Φ in the task (i.e., classification) by

$$w_{\Phi} = \frac{1}{|\mathcal{V}|} \sum_{e_i \in \mathcal{V}} \mathbf{q}^{\top} \cdot \tanh \left(W \cdot \mathbf{z}_i^{\Phi} + \mathbf{b} \right) \tag{24.16}$$

where $\mathcal{V}$ is all entities, $\mathbf{q}$ is the learnable semantic-level attention vector, and $\mathbf{b}$ is the bias. Then the semantic-level attention weight is calculated as $\beta_{\Phi} = \frac{\exp(w_{\Phi})}{\sum_{\Phi'} \exp(w_{\Phi'})}$. The final embeddings of all entities, $Z = \sum_{\Phi} \beta_{\Phi} Z_{\Phi}$, are used for classification.

24.4 KG-based hypothesis generation in computational drug development

Generally, the drug repurposing procedure includes three major steps: hypothesis generation, assessment, and validation (Pushpakom et al, 2019). Among them, the first and foremost step is hypothesis generation. Typially, the hypothesis generation for drug repurposing aims at identifying candidate drugs that has a high confidence to be associated with the therapeutic indication of interest. Today's largly available BKGs, encoding huge volume of biomedical knowledge, have become a valuable resouce for drug repurposing. In KG, the hypothesis generation procedure can be formulated as a link prediction problem, i.e., computational identification of potential drug-target or drug-disease associations with a high confidence level based on existing knowledge (KG's structure properties). This section introduces some preliminary efforts of hypothesis generation for drug repurposing, using computational approaches in the BKGs.

24.4.1 A machine learning framework for KG-based drug repurposing

One of the previous efforts using computational inference in BKG for drug repurposing is Zhu et al.'s study (Zhu et al, 2020b). The main contributions of this study is two-fold: 1) KG construction via data integration, and 2) building the KG-based machine learning pipeline for drug repurposing.

First, by integrating six drug knowledge bases, including PharmGKB (Whirl-Carrillo et al, 2012), TTD (Yang et al, 2016a), KEGG DRUG (Kanehisa et al, 2007), DrugBank (Wishart et al, 2018), SIDER (Kuhn et al, 2016), and DID (Sharp, 2017), they curated a drug-centric KG consisting of five entity types including drugs, diseases, genes, pathways, and side-effects and nine relation types including drug-disease TREATS, drug-drug INTERACTS, and drug-gene REGULATES, BINDS, and ASSOCIATES, drug-side effect CAUSES relations, gene-gene ASSOCIATES, gene-disease ASSOCIATES, and gene-pathway PARTICIPATES relations.

Second, based on the drug-centric KG, a machine learning pipeline was built for drug repurposing. Specifically, the target of the proposed model was to predict the existence of relation between a pair of drug and disease entities. In this way, the task fell into the supervised classification setting where the input samples were the drug-disease pairs. To this end, representation for each sample (drug-disease pair) was calculated in two ways: 1) meta-path-based representation and 2) KG embedding-based representation. For meta-path-based representation, 99 possible meta-paths between drugs and diseases with length 2-4 were enumerated, such as Drug $\xrightarrow{\text{TREATS}}$ Gene $\xrightarrow{\text{ASSOCIATES}}$ Disease and Drug $\xrightarrow{\text{TREATS}}$ Gene $\xrightarrow{\text{ASSOCIATES}}$ Gene $\xrightarrow{\text{ASSOCIATES}}$ Disease. Then a 99-dimensional representation vector was calculated for a drug-disease pair, of which each element indicates the connectivity measure between this two entities based on a specific meta-path. In this study, four different connectivity measures were used, under a specific meta-path Φ, including

- Path count, $PC_\Phi(e_{dr}, e_{di})$, the number of paths between drug e_{dr} and disease e_{di};
- Head normalized path count $HNPC_\Phi = \frac{PC_\Phi(e_{dr},e_{di})}{PC_\Phi(e_{dr},*)}$;
- Tail normalized path count $TNPC_\Phi = \frac{PC_\Phi(e_{dr},e_{di})}{PC_\Phi(*,e_{di})}$;
- Normalized path count $NPC_\Phi = \frac{PC_\Phi(e_{dr},e_{di})}{PC_\Phi(e_{dr},*)+PC_\Phi(e_{dr},*)}$;

For KG embedding-based representation, three translational distance models, including TransE (Bordes et al, 2013), TransH (Wang et al, 2014), and TransR (Lin et al, 2015), were used. Specifically, for each pair of drug e_{dr} and disease e_{di}, using each of the three models, their embedding vectors $\mathbf{h}_{dr}$ and $\mathbf{h}_{di}$ were first learned. Then representation of the drug-disease pair (e_{dr}, e_{di}) was calculated by $\mathbf{h}_{di} - \mathbf{h}_{dr}$.

After that, a machine learning pipeline was built of which the input are representations of the drug-disease pairs. A drug-disease pair was labeled as positive if there is a relation between them. However, the drug-disease pair without a relation between them isn't really negative, instead, it was marked as unknown/unlabeled.

To address this, a positive and unlabeled (PU) learning framework (Elkan and Noto, 2008) was used. Decision Tree, Random Forest, and support vector machine (SVM) were used as basic classifiers of this PU learning framework, respectively. In this study, drug-disease relations related to eight diseases were used as the testing set, while the remaining drug-disease relations (positive) and 143,830 pairs associating the eight diseases with other drugs (unlabeled) were used as the training set. Experimental results showed that the KG-driven pipeline can produce high prediction results on known diabetes mellitus treatments with only using treatment information of other diseases.

24.4.2 Application of KG-based drug repurposing in COVID-19

The sudden outbreak of the human coronavirus disease 2019 (COVID-19) has led to a pandemic that heavily strikes the healthcare system and tremendously impacts people' life around the world. To date, many drugs have been under investigation to treat COVID-19, costing tremendous investment, however, very limited COVID-19 antiviral medications are approved. In this context, there is the urgent need for a more efficient and effective way for drug development against the pandemic, and computational drug repurposing can be a promising approach to address this.

Zeng et al.'s work (Zeng et al, 2020b) is a pioneer effort that computationally repurposes antiviral medications in COVID-19 based on KG inference. First of all, a comprehensive biomedical KG was constructed by integrating the two biomedical relational data resources, Global Network of Biomedical Relationships (GNBR) (Percha and Altman, 2018) and DrugBank (Wishart et al, 2018), and experimentally discovered COVID-gene relationships (Zhou et al, 2020f), resulting in a KG consisting of 145,179 entities of four types (drugs, disease, genes, and drug side information) and 15,018,067 relationships of 39 types. Secondly, a deep KG embedding model, RotatE, was performed to learn low-dimensional representations for the entities and relations. Using such learned embedding vectors, the top 100 drugs that are most close to the COVID-19 entity in the embedding space were prioritized as the candidate drugs. Using drugs in ongoing COVID-19 clinical trials (https://covid19-trials.com/) as a validation set, the results achieved a desirable performance with an area under the receiver operating characteristic curve (AUROC) of 0.85. Moreover, gene set enrichment analysis (GSEA), which involved transcriptome data from peripheral blood and Calu-3 cells, and proteome data from Caco-2 cells, was performed to validate the candidate drugs. Finally, 41 drugs were identified as potential repurposable candidates for COVID-19 therapy, especially 9 are under ongoing COVID-19 trials. Among the 41 candidates, three types of drugs were highlighted by the author: 1) the Anti-Inflammatory Agents such as dexamethasone, indomethacin, and melatonin; 2) the Selective Estrogen Receptor Modulators (SERMs) such as clomifene, bazedoxifene, and toremifene; and 3) the Antiparasitics including hydroxychloroquine and chloroquine phosphate.

Another work (Hsieh et al, 2020), has been focused on using GNN in KG to address the drug repurposing problem. By extracting and integrating drug-target interactions, pathways, gene/drug-phenotype interactions from CTD (Davis et al, 2019), a SARS-CoV-2 KG was built, which consists of 27 SARS-CoV-2 baits, 5,677 host genes, 3,635 drugs, and 1,285 phenotypes, as well as 330 virus-host protein-protein interactions, 13,423 gene-gene sharing pathway interactions, 16,972 drug-target interactions, 1,401 gene-phenotype associations, and 935 drug-phenotype associations. Nest, a variational graph autoencoder (Kipf and Welling, 2016), which engages R-GCN (Schlichtkrull et al, 2018) as encoder, was used to learn entity embeddings in the SARS-CoV-2 KG. Since the SARS-CoV-2 KG has a specific focus on COVID-19 related knowledge, some general yet meaningful biomedical knowledge may be missing. To address this, a transfer learning framework was introduced. Specifically, it first used entity embeddings of Zeng et al.'s work (Zeng et al, 2020b) that encode general biomedical knowledge to initialize entity embeddings in SARS-CoV-2 KG. Then the embeddings were fine-tuned in SARS-CoV-2 KG through the proposed GNN. Using a customized neural network ranking model, 300 drugs that are most relevant to the COVID-19 were selected as the candidate drugs. Similar to Zeng et al.'s work (Zeng et al, 2020b) , the authors engaged GSEA, retrospective in-vitro drug screening, and population-based treatment effect analysis in electronic health records (EHRs), to further validate the repurposable candidates. Through such a pipeline, 22 drugs were highlighted for potential COVID-19 treatment, including Azithromycin, Atorvastatin, Aspirin, Acetaminophen, and Albuterol.

In summary, these studies shed light on the importance of the KG-based computational approaches in drug repurposing to fight against the complex diseases like COVID-19. The reported good performance in terms of the high overlapping ratio between the repurposed candidate drug set and the drugs under ongoing COVID-19 trials, not only demonstrated the effectiveness of the KG-based techniques but also provided biological evidence of the ongoing clinical trials. Moreover, they proposed feasible ways using other publicly available data to validate or refine the hypothesis derived from KGs, which therefore enhances the usability of KG-based approaches.

24.5 Future directions

KGs have been playing a more and more important role in biomedicine. An increasing number of KG-based machine learning and deep learning approaches have been used in biomedical studies such as hypothesis generation in computational drug development. As one of the latest advances in artificial intelligence (AI), GNNs, which have led to tremendous progress in image and text data mining (Kipf and Welling, 2017b; Hamilton et al, 2017b; Veličković et al, 2018), have been introduced to address the KG inference problems. In this context, the use of GNN in biomedical KGs has a great potential in improving hypothesis generation in computational drug development. However, there remain significant gaps between the novel technique and the success of computational drug development. This section discusses the potential opportunities and future research possibilities in this field toward improvements of hypothesis generation for computational drug development.

24.5.1 KG quality control

The procedures of constructing and curating a biomedical KG typically include manually gathering, annotating, and extracting knowledge from text (e.g., literature or experimental reports), automatically or manually normalizing terminology to integrate multiple data resources, and automatically text mining for knowledge extraction, etc. However, none of them are perfect. Therefore, the quality issue has been challenging the KG inference approaches. In KG-based hypothesis generation for drug repurposing, a poor quality of KG will lead to uninformative or wrong representations and hence result in incorrect hypothesis generated (drug-disease associations) and even failure of the entire drug repurposing project. Therefore, there is an urgent need for accurate and appropriate KG quality control. In general, there are two categories of quality issues in KGs: the incorrectness and incompleteness.

Incorrectness refers to incorrect triplets in the KG, i.e., a triplet exists in KG but the corresponding relationship between the two entities is inconsistent with real-world evidence. To address this, a common strategy is manual annotation with sampled small subsets. Such a procedure is time- and cost-consuming, if one wants to evaluate sufficient triplets to reach the statistic criteria. To address this, for example, Gao et al (2019a) proposed an iterative evaluation framework for KG accuracy evaluation. Specifically, inspired by the properties of the annotation cost function observed in practice, the authors developed a cluster sampling strategy with unequal probability theory. Their framework resulted in a 60% shrunk annotation cost and can be easily extended to address evolving KG. In addition, the use of well-designed biomedical vocabularies such as the Unified Medical Language System (UMLS) (Bodenreider, 2004) will improve entity term normalization and hence reduce the risk of errors caused by the ambiguous biomedical entities. Moreover, learning based on KG structure to refine the KG is also a potential way to solve this issue. Early efforts, such as (Zhao et al, 2020d), have been focused on this field.

Incompleteness mainly refers to the missing of biologically or clinically meaningful triplets in the KG. To address the incompleteness in biomedical KG, a common way is to integrate multiple data resources, biomedical data bases, and biomedical KGs to construct and curate a more comprehensive one. CKG (Santos et al, 2020), Hetionet (Himmelstein et al, 2017), DRKG (Ioannidis et al, 2020), KG (Zhu et al, 2020b), etc. are good examples of this strategy. However, there is no guarantee that they are comprehensive enough to cover all biomedical knowledge. In addition, today's largely available biomedical literature and medical data (e.g., EHRs) are great treasure of biomedical knowledge. In this context, previous studies have been focused on deriving knowledge from biomedical literature (Zhao et al, 2020e; Xu et al, 2013; Zhang et al, 2018h; Sahu and Anand, 2018) and EHR data (Rotmensch et al, 2017; Chen et al, 2020e), and the derived knowledge could be a good complement for the biomedical KGs. Moreover, the computational methods such as the KG embedding models (e.g., TransE and TransH) and the GNNs (e.g., R-GCN) have been used in KG completion (Arora, 2020), which predict missing relations within a KG according to its structure properties.

24.5.2 Scalable inference

An ultimate goal of biomedical KGs is always to comprehensively incorporate the biomedical knowledge. For example, by integration of 26 publicly available biomedical databases, CKG (Santos et al, 2020) has included over 16 million biomedical entities connected by over 220 million relationships; another KG, DRKG (Ioannidis et al, 2020), integrating six databases and data collected from recent COVID-19 publications, has included 10K entities and 5.8 million relationships. Meanwhile, today's advanced high-throughput techniques as well as computer software and hardware have led to an inrush of a continuously increasing number of relational data interlinking biomedical entities like drugs, genes, proteins, chemical compounds, diseases and medical concepts extracted from clinical data. This largely enables us to extract knew knowledge to enrich the biomedical KGs and hence these KGs keep expanding constantly.

In this context, the huge and even continuously increasing volume of KGs may challenge the computational models like GNNs. To this end, there is an urgent need for scalable techniques to address the high memory- and time-cost in KGs. For example, Deep Graph Library (DGL, `https://www.dgl.ai`) (Wang et al, 2019f) is an open-source, free Python package designed by Amazon for facilitating the implementation of GNN family models, running on the top of several deep learning framework including PyTorch (Paszke et al, 2019), TensorFlow (Abadi et al, 2016), and MXNet (Chen et al, 2015). As of Mach 1, 2021, it has released the version 0.6. By distilling GNN's message passing procedure as the generalized sparse tensor operations, DGL provides the implementations of optimization techniques like kernel fusion, multi-thread and multi-process acceleration, and automatic sparse format tuning to speed up training process and reduce memory load. In addition to GNNs, DGL also released DGL-KE (`https://github.com/awslabs/dgl-ke`) (Zheng et al, 2020c), an easy-to-use framework for implementation of KG representation models such as TransE, DistMult, RotatE, etc., which has been used in existing KG-based drug-repurposing studies such as (Zeng et al, 2020b).

24.5.3 Coupling KGs with other biomedical data

Apart from the KGs, there is an enormous volume of other biomedical data available such as clinical data and omics data, which are also promising resources for computational drug repurposing. The clinical data is an important resource for healthcare

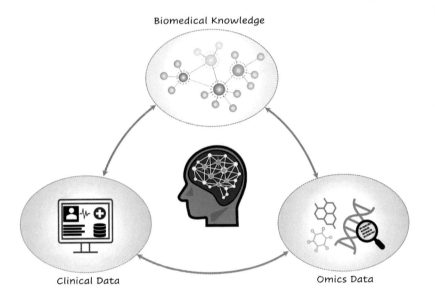

Figure 24.1: Coupling biomedical KGs with other biomedical data resources for improving computational drug development.

and medical research, mainly including EHR data, claim data, and clinical trial data, etc. The EHR data is routinely collected during the daily patient care, containing heterogeneous information of the patients, such as demographics, diagnoses, laboratory test results, medications, and clinical notes. Such rich information makes it possible for tracking patient's health condition changes, medication prescriptions, and clinical outcomes. In addition, a tremendous volume of EHR data has been collected and the volume is rapidly increasing, which largely strengthens the statistical power for EHR-based analysis. For this reason, beyond its common usage such as diagnostic and prognostic prediction (Xiao et al, 2018; Si et al, 2020; Su et al, 2020e,a), and phenotyping (Chiu and Hripcsak, 2017; Weng et al, 2020; Su et al, 2020d, 2021), EHR data has been used for computational drug repurposing (Hurle et al, 2013; Pushpakom et al, 2019). For example, Wu et al (2019d) identified some non-cancer drugs as the repurposable candidates to treat cancer using EHR; Gurwitz (Gurwitz, 2020) analyzed EHR data to repurpose drugs for treating COVID-19.

Advanced by the high throughput sequencing techniques, an enormous volume of omics data, including genomics, proteomics, transcriptomics, epigenomics, and metabolomics, have been collected and publicly available for analysis. Integrating and analyzing the omics data enable us to derive new biomedical insights and better understand human health and diseases at the molecular level (Subramanian et al, 2020; Nicora et al, 2020; Su et al, 2020b). Due to the wealth of the omics data, it has also been involved in computational drug development (Pantziarka and Meheus, 2018; Nicora et al, 2020; Issa et al, 2020). For example, via mining multiple omics data, Zhang et al (2016c) identified 18 proteins as the potential anti-Alzheimer's disease (AD) targets and prioritized 7 repurposable drugs inhibiting the targets. Mokou et al (2020) proposed a drug repurposing pipeline in bladder cancer based on patients' omics (proteomics and transcriptomics) signature data.

In this context, combining KGs, clinical data, and multi-omics data and jointly learning them is a promising route to advance computational drug development (Fig. 24.1). The benefits of combining of these data for inference can be two-way. First, computational models in clinical data and multi-omics data usually suffer from the data quality such as noise and limited cohort size especially for the population of a rare disease and model interpretability. The incorporation of KGs has been demonstrated to be able to address these issues effectively and accelerate the clinical data and omics data analysis. For example, Nelson et al (2019) linked EHR data with a biomedical KG and learned a barcode vector for each specific cohort (e.g., the obese cohort), which encodes both KG structure and EHR information and illustrates the importance of each biomedical entity (e.g., genes, symptoms, and medications) in indicating the cohort. Such cohort-specific barcode vectors further showed the effectiveness in link prediction (e.g., disease-gene associations prediction). Wang et al (2017c) bridged patient EHR data with the BKG and extended the KG embedding model for safe medicine recommendation, which comprehensively considered relevant knowledge such as drug-drug interactions. In addition, Santos et al (2020) developed an open platform that couples the CKG (i.e., Clinical Knowledge Graph) with the typical proteomics workflows. In this way, CKG facilitates analysis and interpretation of the protomics data. Second, the incorporation of clinical data and omics data can potentially improve KG inference. Current KG-based drug repurposing studies have involved the clinical data and omics data (Zeng et al, 2020b; Hsieh et al, 2020), which were typically used in an independent validation procedure to validate/refine the generated new hypotheses (i.e., novel disease-drug associations) . Moreover, previous studies have showcased that leveraging the clinical data (Rotmensch et al, 2017; Chen et al, 2020e; Pan et al, 2020c) and omics data (Ramos et al, 2019) can derive new knowledge. Therefore, we believe that incorporating clinical data and omics data in KG inference may largely reduce the impacts of KG quality issues especially the incompleteness. In total, when we design the next-generation GNN models for drug-repurposing, a considerable direction is the feasible and flexible architecture that can subtly harness KGs, clinical data, and multi-omics data to recursively improve each other.

Editor's Notes: Drug hypothesis generation aims to use biological and clinical knowledge to generate biomedical molecules. This knowledge is effectively stored in the form of knowledge graph (KG). The construction of KG is relevant to graph generation (Chapter 11) and some applications, such as text mining (Chapter 21). Based on KG, hypothesis generation process mainly contains graph representation learning (Chapter 2) and graph structure learning (Chapter 14). It can also be formulated as the link prediction (Chapter 10) problem and calculate the confidence level of candidate drugs. The future direction of drug developments focuses on scalability (Chapter 6) and interpretability (Chapter 7).

Table 24.3: Summary of existing BKGs.

Database	Number of Entities	Entity Types	Number of Relations	Relation Types	Focus	Available Formats	Source Type	URL
Clinical Knowledge Graph(Santos et al, 2020)	16 million	33 entity types, such as Drug, Gene, Disease, etc.	220 million	51 relation types, such as associate, has quantified protein, etc.	-	Neo4j	KG (Integration)	https://github.com/MannLabs/CKG
Drug Repurposing Knowledge Graph(Ioannidis et al, 2020)	97,238	13 entity types, such as Compound, Disease, etc.	5,874,261	107 relation types, such as interaction, etc.	-	TSV	KG (Integration)	https://github.com/gnn4dr/DRKG
Hetionet(Himmelstein et al, 2017)	47,031	11 entity types, such as Disease, Gene, Compound, etc.	2,250,197	24 relation types, such as treats, associates, etc.	-	Neo4j,TSV	KG (Integration)	https://het.io/
iDISK(Rizvi et al, 2019)	144,059	6 entity types, such as Semantic Dietary Supplement Ingredient, Dietary Supplement Product, Disease, etc.	708,164	6 relation types, such as has_adverse_reaction, is_effective_for, etc.	Dietary Supplements	Neo4j,RRF	KG (Integration)	https://conservancy.umn.edu/handle/11299/204783
PreMedKB(Yu et al, 2019b)	404,904	Drug, Variant, Gene, Disease	496,689	52 relation types, such as cause, associate, etc.	Variant	-	KG (Integration)	http://www.fudan-pgx.org/premedkb/index.html#/home
Zhu et al (2020b)	-	Drug, Side-effect, Disease, Gene, Pathway	-	9 relation types, such as Cause, Binds, Treats, etc.	Drug Repurposing	-	KG (Integration)	-
Zeng et al (2020b)	145,179	Drug, Gene, Disease, and Drug side	15,018,067	39 relation types, such as treatment, binding, etc.	Drug Repurposing	-	KG (Integration)	-

Database	Number of Entities	Entity Types	Number of Relations	Relation Types	Focus	Available Formats	Source Type	URL
COVID-19 Knowledge Graph(Domingo-Fernández et al, 2020)	3,954	10 entity types, such as proteins, genes, chemicals, etc.	9,484	Increases, Decreases, association, etc.	COVID-19	JSON	KG	https://github.com/covid19kg/covid19kg
COVID-KG(Wang et al, 2020e)	67,217	Diseases, Chemicals, Genes	85,126,762	Chemical-Gene, Chemical-Disease, Gene-Disease	-	CSV	KG	http://blender.cs.illinois.edu/covid19/
KGHC(Li et al, 2020d)	5,028	9 entity types, such as drug, protein, disease, etc.	13,296	Associate_with, Cause, etc.	Hepatocellular Carcinoma	Neo4j	KG	http://202.118.75.18:18895/browser/
Li et al (2020b)	22,508	9 entity types, such as disease, symptom, etc.	579,094	-	Disease-Symptom	-	KG	-
QMKG(Goodwin and Harabagiu, 2013)	634,000	-	1,390,000,000	-	-	-	KG	-
Rotmensch et al (2017)	647	Disease, Symptom	-	Disease-Symptom	The linkage between diseases and symptoms	-	KG	-
Sun et al (2020a)	1,616,549	62 entity types, such as Disease, Drug, etc	5,963,444	202 relation types	Clinical suspected claims detection	-	KG	https://web.archive.org/web/20191231152615if_/http://121.12.85.245:1347/kg_test/#/
Bgee(Bastian et al, 2021)	60,072	Anatomy, Gene	11,731,369	Expression_Present, Expression_Absent	Anatomy-Gene Expression	TSV	KB	https://bgee.org/
Comparative Toxicogenomics Database(Davis et al, 2019)	73,922	Disease, Gene, Chemical, Pathway	38,344,568	Chemical-Gene, Chemical-Disease, Chemical-Pathway, Gene-Disease, Gene-Pathway, Disease-Pathway	-	CSV,TSV	KB	http://ctdbase.org/

Database	Number of Entities	Entity Types	Number of Relations	Relation Types	Focus	Available Formats	Source Type	URL
Drug–Gene Interaction Database(Cotto et al, 2018)	160,054	Drug, Gene	96,924	-	Drug-Gene Interaction	TSV	KB	https://www.dgidb.org/
DISEASES(Pletscher-Frankild et al, 2015)	22,216	Disease, Gene	543,405	-	Disease-Gene Association	TSV	KB	https://diseases.jensenlab.org/
DisGeNET(Piñero et al, 2020)	159,052	Disease, Gene, Variant	839,138	Gene-Disease, Variant-Disease	Gene-Disease, Variant-Disease associations	TSV	KB	https://www.disgenet.org/home/
Global Network of Biomedical Relationships(Percha and Altman, 2018)	-	Chemical, Disease, Gene	2,236,307	36 relation types, such as causal mutations, treatment, etc.	-	TXT	KB	https://zenodo.org/record/1035500
IntAct(Orchard et al, 2014)	119,281	Chemical, Gene	1,130,596	-	Molecular Interaction	TXT	KB	https://www.ebi.ac.uk/intact/
STRING(Szklarczyk et al, 2019)	24,584,628	Protein	3,123,056,667	Protein-Protein Interaction	Protein-Protein Interaction	TXT	KB	https://string-db.org/
SIDER(Kuhn et al, 2016)	7,298	Drug, Side-effect	139,756	Drug-Side effect	Medicines and their recorded adverse drug reactions	TSV	KB	http://sideeffects.embl.de/
SIGNOR(Licata et al, 2020)	7,095	10 entity types, such as protein, chemical, etc.	26,523	-	Signaling information	TSV	KB	https://signor.uniroma2.it/
TISSUE(Palasca et al, 2018)	26,260	Tissue, Gene	6,788,697	Express	Tissue-Gene Expression	TSV	KB	https://tissues.jensenlab.org/
Catalogue of Somatic Mutations in Cancer(Tate et al, 2019)	12,339,359	Mutation	-	-	Somatic Mutations in Cancer	TSV	Database	https://cancer.sanger.ac.uk/cosmic

Database	Number of Entities	Entity Types	Number of Relations	Relation Types	Focus	Available Formats	Source Type	URL
ChEMBL(Mendez et al, 2019)	1,940,733	Molecule	-	-	Molecule	TXT	Database	https://www.ebi.ac.uk/chembl/
ChEBI(Hastings et al, 2016)	155,342	Molecule	-	-	Molecule	TXT	Database	https://www.ebi.ac.uk/chebi/init.do
DrugBank(Wishart et al, 2018)	15,128	Drug	28,014	Drug-Target, Drug-Enzyme, Drug-Carrier, Drug-Transporter	Drug	CSV	Database	https://go.drugbank.com/
Entrez Gene(Maglott et al, 2010)	30,896,060	Gene	-	-	Gene	TXT	Database	https://www.ncbi.nlm.nih.gov/gene/
HUGO Gene Nomenclature Committee(Braschi et al, 2017)	41,439	Gene	-	-	Gene	TXT	Database	https://www.genenames.org/
KEGG(Kanehisa and Goto, 2000)	33,756,186	Drug, Pathway, Gene, etc.	-	-	-	TXT	Database	https://www.kegg.jp/kegg/
PharmGKB(Whirl-Carrillo et al, 2012)	43,112	Genes, Variant, Drug/Chemical, Phenotype	61,616	-	-	TSV	Database	https://www.pharmgkb.org/
Reactome(Jassal et al, 2020)	21,087	Pathway	-	-	Pathway	TXT	Database	https://reactome.org/
Semantic MEDLINE Database(Kilicoglu et al, 2012)	-	-	109,966,978	Subject-Predicate-Object Triples	Semantic predictions from the literature	CSV	Database	https://skr3.nlm.nih.gov/index.html
UniProt(Bateman et al, 2020)	243,658	Protein	-	-	Protein	XML,TXT	Database	https://www.uniprot.org/
Brenda Tissue Ontology(Gremse et al, 2010)	6,478	Tissue	-	-	Tissue	OWL	Ontology	http://www.BTO.brenda-enzymes.org
Disease Ontology(Schriml et al, 2019)	10,648	Disease	-	-	Disease	OWL	Ontology	https://disease-ontology.org/

Database	Number of Entities	Entity Types	Number of Relations	Relation Types	Focus	Available Formats	Source Type	URL
Gene Ontology(Ashburner et al, 2000)	44,085	Gene	-	-	Gene	OWL	Ontology	http://geneontology.org/
Uberon(Mungall et al, 2012)	14,944	Anatomy	-	-	Anatomy	OWL	Ontology	http://uberon.github.io/publications.html

Chapter 25
Graph Neural Networks in Predicting Protein Function and Interactions

Anowarul Kabir and Amarda Shehu

Abstract Graph Neural Networks (GNNs) are becoming increasingly popular and powerful tools in molecular modeling research due to their ability to operate over non-Euclidean data, such as graphs. Because of their ability to embed both the inherent structure and preserve the semantic information in a graph, GNNs are advancing diverse molecular structure-function studies. In this chapter, we focus on GNN-aided studies that bring together one or more protein-centric sources of data with the goal of elucidating protein function. We provide a short survey on GNNs and their most successful, recent variants designed to tackle the related problems of predicting the biological function and molecular interactions of protein molecules. We review the latest methodological advances, discoveries, as well as open challenges promising to spur further research.

25.1 From Protein Interactions to Function: An Introduction

Molecular biology is now reaping the benefits of big data, as rapidly advancing high-throughput, automated wet-laboratory protocols have resulted in a vast amount of biological sequence, expression, interactions, and structure data (Stark, 2006; Zoete et al, 2011; Finn et al, 2013; Sterling and Irwin, 2015; Dana et al, 2018; Doncheva et al, 2018). Since functional characterization has lagged behind, we now have millions of protein products in databases for which no functional information is readily available; that is, we do not know what many of the proteins in our cells do (Gligorijevic et al, 2020).

Anowarul Kabir
Department of Computer Science, George Mason University, e-mail: akabir4@gmu.edu

Amarda Shehu
Department of Computer Science, George Mason University, e-mail: amarda@gmu.edu

Answering the question of what function a protein molecule performs is key not only to understanding our biology and protein-centric disorders, but also to advancing protein-targeted therapies. Hence, this question remains the driver of much wet- and dry-laboratory research in molecular biology (Radivojac et al, 2013; Jiang et al, 2016). Answering it can take many forms based on the detail sought or possible. The highest amount of detail provides an answer to the question by directly exposing the other molecules with which a target protein interacts in the cell, thus revealing what a protein does by elucidating the molecular partners to which it binds.

In this brief survey, we focus on how graph neural networks (GNNs) are advancing our ability to answer this question *in silico*. This chapter is organized as follows: First, a brief historical overview is provided, so that the reader understands the evolution of ideas and data that have made possible the application of machine learning to the problem of protein function prediction. Then, a brief overview of the (shallow) models prior to GNNs is provided. The rest of the survey is devoted to the GNN-based formulation of this question, a summary of state-of-the-art (SOTA) GNN-based methods, with a few selected methods highlighted where relevant, and an exposition of remaining challenges and potential ways forward via GNNs.

25.1.1 Enter Stage Left: Protein-Protein Interaction Networks

Historically, the earliest methods devised for protein function prediction related protein sequence similarity to protein function similarity. This led to important discoveries until remote homologs were identified, which are proteins with low sequence similarity but highly similar three-dimensional/tertiary structure and function. So methods evolved to utilize tertiary structure, but their applicability was limited, as determination of tertiary structure was and remains a laborious process. Other methods utilized patterns in gene expression data to infer interacting proteins, based on the insight that proteins interacting with one another need foremost to be expressed in the cell at the same time.

With the development of high-throughput technologies, such as two-hybrid analysis for the yeast protein interactome (Ito et al, 2001), tandem-affinity purification and mass spectrometry (TAP-MS) (Gavin et al, 2002) for characterizing multi-protein complexes and protein-protein associations (Huang et al, 2016a), high-throughput mass spectrometric protein complex identification (HMS-PCI) (Ho et al, 2002), co-immunoprecipitation coupled to mass spectrometry (Foltman and Sanchez-Diaz, 2016), protein-protein interaction (PPI) data suddenly became available, and in large amounts. PPI networks, with edges denoting interacting protein nodes, of many species, such as human, yeast, mouse, and others, suddenly became available to researchers. PPI networks, as small as a few nodes or as large as tens of thousands of nodes, gave a boost to machine learning methods and improved the performance of shallow models. Surveys such as Ref. (Shehu et al, 2016) provide a detailed history of the evolution of protein function prediction methods as different sources of wet-laboratory data became available to computational biologists.

25.1.2 Problem Formulation(s), Assumptions, and Noise: A Historical Perspective

A natural question arises. If we have access to PPI data, then what else remains to predict with regards to protein function? Despite significant progress, the reality remains that there are many unmapped PPIs. This is formally known as the *link prediction* problem. For various reasons, PPI networks are incomplete. They entirely miss information on a protein, or they may contain incomplete information on a protein. In particular, we now know that PPIs suffer from high type-I error, type-II error, and low inclusion (Luo et al, 2015; Byron and Vestergaard, 2015). The total number of PPI links that are experimentally determined is still moderate (Han et al, 2005). PPI data are inherently noisy as experimental methods often produce false-positive results (Hashemifar et al, 2018). Therefore, predicting protein function computationally remains an essential task.

The problem of protein function prediction is often formulated as that of link prediction, that is, *predicting whether or not there exists a connection between two nodes in a given PPI network*. While link prediction methods connect proteins on the basis of biological or network-based similarity, researchers report that interacting proteins are not necessarily similar and similar proteins do not necessarily interact (Kovács et al, 2019).

As indicated above, information on protein function can be provided at different levels of detail. There are several widely-used protein function annotation schemes, including the Gene Ontology (Lovell et al, 2003) (GO) Consortium, the Kyoto Encyclopedia of Genes and Genomes (Wang and Dunbrack, 2003) (KEGG), the Enzyme Commission (Rhodes, 2010) (EC) numbers, the Human Phenotype Ontology (Robinson et al, 2008), and others. It is beyond the scope of this paper to provide an explanation of these ontologies. However, we emphasize that the most popular one remains the GO annotation, which classifies proteins into hierarchically-related functional classes organized into 3 different ontologies: Molecular Function (MF), Biological Process (BP), and Cellular Component (CC), to describe different aspects of protein functions. Systematic benchmarking efforts via the Critical Assessment of Functional Annotation (CAFA) community-wide experiments (Radivojac et al, 2013; Jiang et al, 2016; Zhou et al, 2019b) and MouseFunc (Peña-Castillo et al, 2008) have been central to the automation of protein function annotation and rigorous assessment of devised methodologies.

25.1.3 Shallow Machine Learning Models over the Years

Many shallow machine learning approaches have been developed over the years. Xue-Wen and Mei propose a domain-based random forest of decision trees to infer protein interactions on the *Saccharomyces cerevisiae* dataset (Chen and Liu, 2005). Shinsuke *et al.* apply multiple support vector machines (SVMs) for predicting interactions between pairs of yeast proteins and pairs of human proteins by increasing more negative pairs than positives (Dohkan et al, 2006). Fiona *et al.* assess naïve bayes (NB), multi-layer perceptron (MLP) and k-nearest neighbour (KNN) methods on diverse, large-scale functional data to infer pairwise (PW) and module-based (MB) interaction networks (Browne et al, 2007). PRED_PPI provides a server developed on SVM for predicting PPIs in five organisms, such as humans, yeast, *Drosophila, Escherichia coli,* and *Caenorhabditis elegans* (Guo et al, 2010). Xiaotong and Xue-wen integrate features extracted from microarray expression measurements, GO labels and orthologous scores, and apply a tree-augmented NB classifier for human PPI predictions from model organisms (Lin and Chen, 2012). Zhu-Hong *et al.* propose a multi-scale local descriptor feature representation scheme to extract features from a protein sequence and use random forest (You et al, 2015a). Zhu-Hong *et al.* propose to apply SVM on a matrix-based representation of protein sequence, which fully considers the sequence order and dipeptide information of the protein primary sequence to detect PPIs (You et al, 2015b).

Although many advances were made by shallow models, as summarized in Table 25.1, the problem of protein function prediction is still a long way from being solved. Shallow machine learning methods depend greatly on feature extraction and feature computation, which hinder performance. The task of feature engineering, particularly when integrating different sources of data (sequence, expression, interactions) is complex, laborious, and ultimately limited by human creativity and domain-specific understanding of what may be determinants of protein function. In particular, feature-based shallow models cannot fully incorporate the rich, local and distal topological information present in one or more PPI networks. These reasons have prompted researchers to investigate GNNs for protein function prediction.

25.1.4 Enter Stage Right: Graph Neural Networks

This section first relates a general formulation of a GNN and forsakes detail in the interest of space, assuming readers are already somewhat familiar with GNNs. The rest of the section focuses on three task-specific formulations that allow leveraging GNNs for protein function prediction.

25.1.4.1 Preliminaries

Assume an undirected and unweighted molecular-interaction graph, i.e., a PPI network, is represented by $\mathscr{G} = (\mathscr{V}, \mathscr{E})$, where $\mathscr{V}$ and $\mathscr{E}$ denote the set of vertices representing proteins and the edges indicating interactions among proteins, respectively. Let the i-th protein be represented as an m-dimensional feature vector; that is, $p_i \in \mathbb{R}^m$. The objective of a GNN is to learn an embedding, h_i, using the message passing protocol which essentially aggregates and transforms neighboring information to update the current node's vector representation. Assuming f and g are two

Table 25.1: Summary of performance of shallow models as reported in (Chen and Liu, 2005; Guo et al, 2010; Lin and Chen, 2012; You et al, 2015a,b)

Literature	Model	Dataset	Sensitivity (%)	Specificity (%)	Accuracy (%)
Chen and Liu (2005)	RF	*Saccharomyces cerevisiae*	79.78	64.38	NA*
Guo et al (2010)	SVM	Human	89.17	92.17	90.67
		Yeast	88.17	89.81	88.99
		Drosophila	99.53	80.65	90.09
		Escherichia coli	95.11	90.35	92.73
		Caenorhabditis elegans	96.46	98.55	97.51
Lin and Chen (2012)	Tree-Augmented Naïve Bayes (TAN)	Human	88	70	NA*
You et al (2015a)	RF	*Saccharomyces cerevisiae*	94.34	NA*	94.72
You et al (2015b)	SVM	*Saccharomyces cerevisiae*	85.74	94.37	90.06

* Not available

parametric functions that compute the embedding and output considering a single protein, following (Scarselli et al, 2008), we formulate follows:

$$h_i = f(p_i, p_{e[i]}, p_{ne[i]}, h_{ne[i]}) \tag{25.1}$$

$$o_i = g(h_i, p_i) \tag{25.2}$$

where p_i, $p_{e[i]}$, $p_{ne[i]}$ and $h_{ne[i]}$ denote the feature representation of the i-th protein, features of all connected edges to the i-th protein, neighboring proteins' features and embeddings of neighborhood proteins of the i-th protein, respectively.

Let us now consider $|\mathcal{V}| = n$ proteins. All proteins are represented as a matrix, $P \in \mathbb{R}^{n \times m}$. The adjacency matrix $A \in \mathbb{R}^{n \times n}$ encodes the connectivity of the proteins; namely, $A_{i,j}$ indicates whether or not there exists a link between proteins i and j. Enforcing the self-loops with each protein, the updated adjacent matrix is $\tilde{A} = A + I$. The degree diagonal matrix, D, can then be defined, such that $D_{i,i} = \sum_{j=1}^{n} \tilde{A}_{i,j}$. From there, one can compute the symmetric Laplacian matrix $L = D - \tilde{A}$. Finally, one can then formulate the following iterative process:

$$H^{t+1} = F(H^t, P||A||L||X) \tag{25.3}$$

$$O = G(H, P||A||L||X) \tag{25.4}$$

where H^t denotes t-th iteration of H, $(\cdot||\cdot)$ indicates the aggregation operation based on the task at hand, and O is the final stacked output.

25.1.4.2 GNNs for Representation Learning

We now want to encode complex high-dimensional information, such as a protein, P, or a biological interaction, A, or an interaction network, $\mathcal{G}$, into low-dimensional embeddings, Z, by capturing linearity and non-linearity among nodes and edges. In principle, the representation should contain all the information for downstream machine learning tasks, such as link prediction, protein classification, protein cluster analysis, interaction prediction, etc.

Suppose we want to learn a graph embedding, Z, from the network $\mathcal{G}$. A graph auto-encoder neural network (Kipf and Welling, 2016) can be applied to learn Z:

$$Z = GNN(P, A; \theta_{gnn}) \tag{25.5}$$

where θ_{gnn} denotes GNN (encoder)-specific learnable parameters.

25.1.4.3 GNNs for the Link Prediction Problem

Given two proteins, we want to predict if there is a link between them, where probability $p(A_{i,j}) \approx 1$ indicates there exists an interaction with high confidence; conversely $p(A_{i,j}) \approx 0$ indicates a low interaction confidence. The prediction of a link between two given proteins can bet set up as a binary classification problem. The relations among nodes can be of several types; so, an edge of type r from node u to v can be defined as $u \xrightarrow{r} v \in \mathcal{E}$, which can be formulated as a multi-relational link prediction problem.

Using GNNs, one can map graph nodes into a low-dimensional vector space which may preserve both local graph structure and dissimilarities among node features. To address link prediction, one can employ a two layer encoder-decoder approach where the model learns Z from equation 25.5:

$$A' = DECODER(Z|P, A; \theta_{decoder}) \tag{25.6}$$

where $\theta_{decoder}$ denotes decoder (task)-specific learnable parameters, and $A'_{i,j}$ indicates the confidence score with the predicted link between protein i and j.

25.1.4.4 GNNs for Automated Function Prediction as a Multi-label Classification Problem

Given n-GO terms and m-proteins, $u = m - l$ proteins need to be annotated with term(s), whereas l proteins are already annotated. So for the i-th protein, the prediction will be $y_i = y_{i,1}, y_{i,2}, ..., y_{i,n}$ where $y_{i,j} \in \{0, 1\}$. This task can be considered as a binary multi-label classification problem, since a protein usually participates in multiple biological functions. This could be protein-centric, where GO-terms are annotated for each protein, or GO-term centric, where proteins are annotated for each GO-term, or protein-term pair centric, where a probability association score is predicted for each pair.

25.2 Highlighted Case Studies

In the following, we highlight three selected methods that exemplify SOTA techniques and performance.

25.2.1 Case Study 1: Prediction of Protein-Protein and Protein-Drug Interactions: The Link Prediction Problem

Liu et al (2019) apply a graph convolutional neural network (GCN) for PPI prediction as a supervised binary classification task. Learned representations of two proteins are fed to the model, and the model predicts the probability of interaction between the proteins. The model first captures position-specific information inside the PPI network and combines amino-acid sequence information to output final embeddings for each protein. The model encodes each amino acid as a one-hot vector and employs a graph convolutional layer to learn a hidden representation from the graph. To do that, Liu et al (2019) use the message passing protocol to update each protein embedding by aggregating the original features and first-hop neighbors' information, which is formulated as following:

$$X_1 = ReLU(D^{-1}\tilde{A}X_0W_0) \quad (25.7)$$

where $X_0 \in \mathbb{R}^{n \times n}$ is the original protein feature matrix which is an identity matrix; $X_1 \in \mathbb{R}^{n \times f}$ is the final output feature matrix, where f is the feature dimension of each protein after the graph convolution operation and W_0 is the trainable weight matrix. In the prediction phase, the authors utilize fully connected layers followed by batch normalization and dropout layers to extract high-level features; softmax is then used to predict the final interaction probability score. The experiments show that the method achieves mean AUPR (area under precision-recall curve) of 0.52 and 0.45 on yeast and human datasets, respectively, which outperforms sequence-based SOTA methods. Additionally, the authors report achieving 95% accuracy on yeast data under 93% sensitivity. Therefore, the extracted information from the PPI graph suggests that a single graph convolutional layer is capable of extracting useful information for the PPI prediction task.

Brockschmidt (Brockschmidt, 2020) proposes a novel GNN variant using feature-wise linear modulation (GNN-FiLM), originally introduced by Perez *et al.* (Perez et al, 2018) in the visual question-answering domain, and evaluates on three different tasks, including node-level classification of PPI networks. The targeted application in this work is the classification of proteins into known protein families or super-families, which is of great importance in numerous application domains, such as precision drug design. Typically, in GNN variants, the information is passed from the source to the target node considering the learned weights and the representation of the source node. However, the GNN-FiLM method proposes a hypernetwork, neural networks, that compute parameters for other networks (Ha et al, 2017), in graph settings, where the feature weights are learned dynamically based on the information that the target node holds. Therefore, considering function g as a learnable function to compute the parameters for the affine transformation, the update rule is defined for the l-th layer as follows:

$$\beta_{r,v}^{(l)} \gamma_{r,v}^{(l)} = g(h_v^{(l)}; \theta_{g,r})$$ (25.8)

$$h_v^{(l+1)} = \sigma \left(\sum_{u \xrightarrow{r} v \in \mathscr{E}} \gamma_{r,v}^{(l)} \odot W_r h_u^{(l)} + \beta_{r,v}^{(l)} \right)$$ (25.9)

where g is implemented as a single linear layer in practice considering $\beta_{e,v}^{(t)}$ and $\gamma_{e,v}^{(t)}$ as the hyperparameters of the message passing operation in GNN, and $u \xrightarrow{e} v$ indicates that message is passing from u to v through a type r edge. In experiments, GNN-FiLM achieves micro-averaged F1 score of 99% which outperforms other variants when evaluated on protein classification tasks.

Zitnik et al (2018) employ GCNs to predict polypharmacy side effects, which emerge from drug-drug interactions when using drug combinations on patients' treatments. The problem can be formulated as a multi-relational link prediction problem in multimodal graph structured data. Specifically, Zitnik et al (2018) consider two types of nodes, proteins and drugs, and construct the network using protein-protein, protein-drug, and drug-drug interactions as polypharmacy side effects, whereas each side effect can be of different types of edges, called Decagon. More precisely, a relation of type r between two nodes (proteins or drugs), u and v, is defined as $(u, r, v) \in \mathscr{E}$. Here, the relations can be a side effect between two proteins, binding affinity of two proteins, or relation between a protein and a drug. More formally, given a drug pair (u, v), the task is to predict the likelihood of an edge, $A_{u,v} = (u, r, v)$. For this purpose, they develop a non-linear and multi-layer graph convolutional encoder to compute the embeddings of each node using original node features, called Decagon. To update a node's representation, authors transform the information of neighboring nodes by aggregation and propagation operations over the edges. The update operator is defined using the following rule:

$$h_i^{(l+1)} = \phi \left(\sum_r \sum_{j \in \mathcal{N}_r^i} c_r^{i,j} W_r^{(l)} h_j^{(l)} + c_r^i h_i^{(l)} \right) \tag{25.10}$$

where ϕ denotes non-linear activation function, $h_i^{(l)}$ indicates hidden state of the i-th node at the l-th layer, $W_r^{(l)}$ means relation-type specific learnable parameter matrix, $j \in \mathcal{N}_r^i$ are the neighboring nodes of i, $c_r^{i,j} = \frac{1}{\sqrt{|\mathcal{N}_r^i||\mathcal{N}_r^i|}}$ and $c_r^i = \frac{1}{\sqrt{|\mathcal{N}_r^i|}}$ are the normalization constant. Finally, a tensor factorization model is used to predict the polypharmacy side effects using these embeddings. The probability of a link of type r between node u and v is defined as:

$$\mathbf{x}_r^{u,v} = \sigma(g(u,r,v)) \tag{25.11}$$

where σ is the sigmoid function and g is defined as follows:

$$g(u,r,v) = \begin{cases} z_u^T D_r R D_r z_v & \text{if } u \text{ and } v \text{ both denote drug nodes} \\ z_u^T M_r z_v & \text{if any of } u \text{ or } v \text{ is not drug node} \end{cases} \tag{25.12}$$

where D_r, R and M_r are parameter matrices, such that D_r defines side-effect-specific diagonal matrix, R is global drug-drug interaction matrix, and M_r is relation-type-specific parameter matrix. Decagon achieves an AUPR of 83% under 80% precision, outperforming other baselines by up to 69%. The authors attribute the large margin in improvement to two components, the graph-structured convolution encoder and the tensor factorization model.

25.2.2 Case Study 2: Prediction of Protein Function and Functionally-important Residues

Automated Function Prediction (AFP) problems are often formulated as a multi-label classification problems and are more nuanced than predicting interactions between two proteins. Many works report that proteins connected in the same molecular network share the same functions (Schwikowski et al, 2000), but recent developments show that interacting proteins are not necessarily similar, and similar proteins do not necessarily interact (Kovács et al, 2019). Moreover, more than 80% of proteins interact with other molecules while functioning (Berggård et al, 2007). Therefore, identifying or predicting the roles of proteins in organisms is vital, and community-wide challenges have been organized to advance research towards this goal. These include the Critical Assessment of Function Annotation (CAFA) (Radivojac et al, 2013; Jiang et al, 2016; Zhou et al, 2019b) and MouseFunc (Peña-Castillo et al, 2008).

Many computation methods have been developed to this end to analyze protein-function relationships. Traditional machine learning approaches, such as SVMs (Guan et al, 2008; Wass et al, 2012; Cozzetto et al, 2016), heuristic-based methods (Schug, 2002), high dimensional statistical methods (Koo and Bonneau, 2018), and hierarchical supervised clustering methods (Das et al, 2015) have been extensively studied in AFP tasks and found that integration of several features, such as gene and protein network or structure outperforms sequence-based features. However, these traditional approaches rely strongly on hand-engineered features.

Deep learning methods have become prevalent. For example, DeepSite (Jiménez et al, 2017), Torng and Altman (2018), and Enzynet (Amidi et al, 2018) apply 3D convolutonal neural networks (CNNs) for feature extraction and prediction from protein structure data. However, storing the high-resolution 3D representation of protein structure and applying 3D convolutions over the representation is inefficient (Gligorijevic et al, 2020). Very recently, GCNs (Kipf and Welling, 2017b) (Henaff et al, 2015; Bronstein et al, 2017) have been shown to generalize convolutional operations on graph-like molecular representations and overcome these limitations.

In particular, Ioannidis et al (2019) adapt the graph residual neural network (GRNN) approach for a semi-supervised learning task over multi-relational PPI graphs to address AFP. The authors formulate a multi-relational connectivity graph as an $n \times n \times I$ tensor S, where $S_{n,n',i}$ captures the edge between proteins v_n and $v_{n'}$ for the i-th relation. The n proteins are encoded in a feature matrix $X \in \mathbb{R}^{n \times f}$, where the i-th protein is represented as an $f \times 1$ feature vector. Furthermore, a label matrix $Y \in \mathbb{R}^{n \times k}$ encodes the k labels. Subsets of proteins are associated with true labels, and the task is to predict the labels of proteins with unavailable labels. The neighborhood aggregation for the n-th protein and the i-th relation at the l-th layer is defined by the following formula:

$$H_{n,i}^{(l)} = \sum_{n' \in \mathcal{N}_n^{(i)}} S_{n,n',i} \check{Z}_{n',i}^{(l-1)} \tag{25.13}$$

where n' denotes the neighboring nodes of the n-th protein, and $\check{Z}_{n',i}^{(l-1)}$ denotes the feature vector of the n-th protein in the i-th relation at the l-th to the first layer. Neighboring nodes are defined as one-hop only, which essentially incorporates one-hop diffusion. However, successive operations eventually spread the information across the network. To apply multi-relational graphs, the authors combine $H_{ni}^{(l)}$ across i as follows:

$$G_{n,i}^{(l)} = \sum_{i'=1}^{I} R_{i,i'}^{(l)} H_{n,i'}^{(l)} \tag{25.14}$$

where $R_{i,i'}^{(l)}$ is the learnable parameter. Then, a linear operation mixes the extracted features as follows:

$$Z^{(l)} = G_{n,i}^T W_{n,i}^{(l)} - 1 \tag{25.15}$$

where $W_{n,i}$ is the learnable parameter. In summary, the neighborhood convolution and propagation step can be shown as:

$$Z^{(l)} = f(Z^{(l-1)}; \theta_z^{(l)}) \tag{25.16}$$

where $\theta_z^{(l)}$ is comprised of two weight matrices, W and R, which linearly combine the information of neighboring nodes and the multi-relational information, respectively. Moreover, the authors incorporate residual connection to diffuse the input, X, across L-hop neighborhoods to capture multi-type diffusion; that is:

$$Z^{(l)} = f(Z^{(l-1)}; \theta_z^{(l)}) + f(X; \theta_x^{(l)}) \tag{25.17}$$

A softmax classification layer is used for the final prediction. The authors apply this model on three multi-relational networks, comprising generic, brain, and circulation cells. The model is shown to perform better than general graph convolutional neural networks.

Recently, Gligorijevic et al (2020) employ DeepFRI, based on GCNs, for functionally annotating protein sequences and structures. DeepFRI outputs probabilities for each function. A Long Short-Term Memory Language Model (LSTM-LM) (Graves, 2013) is pretrained on around 10 million protein sequences from protein family database (Pfam) (Finn et al, 2013) to extract residue-level position-context features. The following equation is used:

$$H^0 = H^{input} = ReLU(H^{LM}W^{LM} + XW^X + b) \tag{25.18}$$

where H^0 is the final residue-level feature representation and the first graph convolutional layer. W^{LM}, W^X and b are learnable parameters trained with the graph convolutional layers. Contact-map features, which encode tertiary protein structure, combined with LSTM-LM task-agnostic sequence-embeddings are fed to a GCN while keeping LSTM-LM frozen. The l-th layer of the convolution takes sequence-embeddings and the contact map A and outputs residue-level embeddings to the next, $(l+1)$-th, layer. Residue level features are extracted by propagating residue information to proximal residues. The rule for updating the node representation is:

$$H^{(l+1)} = ReLU(\tilde{D}^{-\frac{1}{2}}\tilde{A}\tilde{D}^{-\frac{1}{2}}H^{(l)}W^{(l)}) \tag{25.19}$$

The features are then concatenated into a single feature matrix as a protein embedding. Intuitively, embeddings from different layers can be thought as context-aware features. Additionally, the feature extraction strategy exploits linear or non-linear relationships from neighbouring residues, as well as residues distant in sequence but proximal in structure.

The learned protein representation is fed into two consecutive fully connected layers to obtain predictions as class probabilities for all the GO-terms. The authors evaluate their model on experimental and predicted structures and compare with existing baseline models, including CAFA-like BLAST (Wass et al, 2012) and CNN-based sequence-only DeepGOPlus (Kulmanov and Hoehndorf, 2019), on each sub-ontology of GO-terms and EC numbers and outperform in every category.

Zhou et al (2020b) apply a GCN model, DeepGOA, to predict maize protein functions. The authors exploit both GO structure information and protein sequence information for a multi-label classification task. Since GO organizes the functional annotation terms into a directed acyclic graph (DAG), the authors utilize the knowledge encoded in the GO hierarchy. First, amino acids of a protein are encoded into one-hot encodings, a 21-dimensional feature vector for each amino acid, as there are 20 amino acids and sometimes there are undetermined amino acids in a protein. Proteins might be different in length; therefore, the authors only extract the first 2000 amino acids for those proteins which are longer than that. Otherwise, the encodings are zero-padded. So the i-th protein is represented as

$$X_i = [x_{i1}, x_{i2}, x_{i3}, \ldots \ldots x_{i2000}] \qquad (25.20)$$

To learn the low-dimensional feature representation of each protein sequence, the authors apply CNNs of four different sizes of convolutional kernels, such as 8, 16, 24 and 32, to extract hypothetical non-linear secondary or tertiary structure information. The 1D convolution operation is formulated as follows:

$$c_{im} = f(w * x_{i(m:m+h)}), m \in [1, k-h] \qquad (25.21)$$

where h is the sliding window length, $w \in \mathbb{R}^{21 \times h}$ is a convolutional kernel, and $f(\cdot)$ is a non-linear activation function. Then, the authors incorporate the GO structure into the model. To do that, graph convolutional layers are deployed to generate the embeddings of the GO terms by propagating information among GO terms using neighboring terms in the GO hierarchy. For τ number of GO terms, initial one-hot feature description, $H^0 \in \mathbb{R}^{\tau \times \tau}$, and correlation matrix, $A \in \mathbb{R}^{\tau \times \tau}$ are computed as input. For the l-th layer's representation, H^l is updated using the following neighborhood information propagating equation:

$$H^l = f(\hat{A} H^{l-1} W^l) \qquad (25.22)$$

where $\hat{A} \in \mathbb{R}^{\tau \times \tau}$ is the symmetrically normalized correlation matrix derived from A, $f(\cdot)$ is a non-linear activation function, and $W^l \in \mathbb{R}^{d_{l-1} \times d_l}$ is the learnable transformation matrix. Then, such graph convolutional layers are stacked to capture high- and low-order information of the GO DAG. In this way, DeepGOA learns a semantic representation of GO-terms, $H \in \mathbb{R}^{\tau \times d}$, and protein sequence representation, $Z \in \mathbb{R}^{n \times d}$, in some d-dimensional semantic space. Dot product is used to then compute protein-term pair association probabilities as follows:

$$\hat{Y} = HZ^T \qquad (25.23)$$

Cross-entropy loss for the multi-label loss function is used to train the model end-to-end. The authors experiment on the Maize PH207 inbred line (Hirsch et al,

2016) and the human protein sequence dataset and show that DeepGOA outperforms SOTA methods.

25.2.3 Case Study 3: From Representation Learning to Multirelational Link Prediction in Biological Networks with Graph Autoencoders

Yang et al (2020a) employ signed variational graph auto-encoder (S-VGAE) to automatically learn graph representation, and incorporate protein sequence information as features for the PPI prediction task. The authors report SOTA performance compared to existing sequence-based models on several datasets.

The protein interaction network is encoded as an undirected graph, with different signs (i.e., positive, negative or neutral) added the edges in the adjacency matrix to extract fine-grained features, where the model is assumed to learn negative impact of highly negative interactions. Moreover, the authors consider only high-confidence interactions in the cost function, enabling the model to learn embeddings more accurately. First, protein sequences are encoded using the CT method (Shen et al, 2007). All amino acids are divided into seven categories considering their dipole and side-chain volumes. Each group represents analogous mutations due similar characteristics. Thus, a protein can be represented as a sequence of numbers representing a category. Then, a window of size 3 amino acids slides over the numeric sequence one step at a time and counts the number of occurrences of each triad. Thus, the size of a protein CT vector is 343(=m), which can be defined as follows:

$$V = [r_1, r_2, ...r_M] \tag{25.24}$$

where r_i is the number of occurrences of each triad type. For n proteins, the input features of each protein can be summarized in a matrix $X \in \mathbb{R}^{n \times m}$. Afterwards, S-VGAE is employed to extract protein embeddings by combining both graph structure and sequence information, following Kipf and Welling's (Kipf and Welling, 2016) variational graph auto-encoder. Considering the primary/sequence features, its neighborhood structures and positions in the graph, the encoder maps each protein x_i to a low-dimensional vector z_i. The idea is to map proteins' original features X into low dimensional embeddings Z using an augmented information adjacency matrix A. The encoding rule is formulated as follows:

$$q(Z|X,A) = \prod_{i=1}^{N} q(z_i|Z,A) \tag{25.25}$$

$$q(z_i|Z,A) = \mathcal{N}(z_i|\mu_i, diag(\sigma_i^2)) \tag{25.26}$$

Mean vector, μ_i, and standard deviation vector, σ_i, is defined as follows:

$$\mu = GCN_{\mu}(X,A) \tag{25.27}$$

$$\log \sigma = GCN_\sigma(X,A) \tag{25.28}$$

where GCN is a neighborhood aggregation propagation step formulated as below:

$$GCN(X,A) = AReLU(AXW_0) \tag{25.29}$$

$$GCN_\mu(X,A) = AReLU(AXW_1) \tag{25.30}$$

$$GCN_\sigma(X,A) = AReLU(AXW_2) \tag{25.31}$$

where W_0, W_1 and W_2 are trainable parameters and, GCN_μ and GCN_σ share W_0 to reduce parameters. The decoder predicts the classification label of protein i and j by taking the dot product of their lower-dimensional embeddings z_i and z_j; the interaction probability indicates whether there is a connection between two proteins. This is defined as follows:

$$p(A|Z) = \prod_{i=1}^{N}\prod_{j=1}^{N} p(A_{i,j}|z_i,z_j) \tag{25.32}$$

$$p(A_{i,j} = 1|z_i,z_j) = \sigma(z_i^T z_j) \tag{25.33}$$

where $\sigma(\cdot)$ is the logistic sigmoid function. Thus, the S-VGE learns to encode protein embeddings into low-dimensional features by solving the task of decoding the learned embeddings back to the original graph structure. Instead of using the decoder as the final classification layer, the authors utilize it as a generative model for learning latent features. Then, three fully connected layers perform the final classification task. Overall, the model achieves more than 98% accuracy on five different datasets.

Hasibi and Michoel (2020) propose a graph feature auto-encoder (GFAE) model, called FeatGraphConv, which is trained on a feature reconstruction task instead of graph reconstruction task. The model performs well on predicting unobserved node features on biological networks, such as transcriptional, protein-protein and genetic interaction networks. FeatGraphConv investigates how well GNNs might preserve node features. The authors aim to identify whether or not the graph structure and feature values encode similar information. The relationship between a graph G and latent embeddings Z can be formulated using graph convolutional layers as messaging passing protocol by aggregating neighborhood information as follows:

$$Z = GCN(G;\theta) = GCN(X,\tilde{A};\theta) \tag{25.34}$$

$$Z = \sigma(\tilde{A}ReLU(\tilde{A}XW_0)W_1) \tag{25.35}$$

where θ contains learnable weights, defined as $\theta = W_0;W_1;......W_i$, and σ is a nonlinear task-specific mapping function. The authors leverage four message passing and neighborhood information aggregation operations. The GCN update rule (Gilmer et al, 2017) is followed for the i-th protein's representation, h_i^k, at the l-th layer as follows:

$$h_i^l = \sum_{j \in \mathcal{N}(i) \cup i} \frac{1}{\sqrt{deg(i)} * \sqrt{deg(j)}} W h_j^{l-1} \tag{25.36}$$

The GraphSAGE (Hamilton et al, 2017b) update rule is then deployed:

$$h_i^l = W_1 h_i^{l-1} + W_2 Mean_{j \in \mathcal{N}(i) \cup i} h_j^{l-1} \tag{25.37}$$

Additionally, the authors employ the GraphConv (Morris et al, 2020b) operator:

$$h_i^l = W_1 h_i^{l-1} + \sum_{j \in \mathcal{N}(i)} W_2 h_j^{l-1} \tag{25.38}$$

A new update rule is also proposed:

$$h_i^l = W_2(W_1 h_i^{l-1} || Mean_{j \in \mathcal{N}(i) \cup i}(W_1 h_j^{l-1})) \tag{25.39}$$

where $(\cdot || \cdot)$ denotes a concatenation operation. The authors train the learnable parameters on the embeddings ability to reconstruct the adjacency matrix, which is formulated as follows:

$$\hat{A} = Sigmoid(ZZ^T) \tag{25.40}$$

Cross-entropy loss between A and $\hat{A}$ and gradient descent are used to update the weights. Finally, the embeddings Z are used to predict the class Y in predicting missing links in the adjacency matrix and thus in the graph.

25.3 Future Directions

As this survey indicates, many variants of GNNs have been applied to obtain information on protein function. Much work remains to be done. Future directions can be broadly divided into two categories, methodology-oriented and task-oriented.

Many existing GNN-based approaches are limited to proteins of the same size (number of amino acids). This essentially weakens model capacity for the particular task at hand. Therefore, future research needs to focus on size-agnostic, as well as task-agnostic models. Choosing the right model is always a difficult task. However, benchmark datasets and available packages are making it easier to develop models expediently.

Enhancing model explainability is also an important direction. Some community bias has been observed towards focusing model development on GCNs for learning semantic and topological information for the function prediction task. However, there are many other variants of GNNs. For instance, graph attention networks may prove useful. Existing literature also often ignores ablation studies, which are important to provide a strong rationale for choosing a particular component of the model over others.

Most of the PPI prediction tasks assume training a single model for an organism. Leveraging multi-organisms PPI networks provides more data and may result in better performance. In the same spirit, leveraging multi-omics data combined with sequence and structural data may advance the state of the art.

Finally, we draw attention to the site-specific function prediction task, which provides more information and highlights specific residues that are important for a particular function. This fine-grained function prediction task can be even more critical to support other tasks, such as drug design. Transfer learning across related tasks may additionally provide insights for learning important attributes.

This work is supported in part by National Science Foundation Grant No. 1907805 and Grant No. 1763233. This material is additionally based upon work by AS supported by (while serving at) the National Science Foundation. Any opinion, findings, and conclusions or recommendations expressed in this material are those of the author and do not necessarily reflect the views of the National Science Foundation.

Editor's Notes: In addition to small molecules introduced in Chapter 25, large molecules such as proteins and DNA represent another domain in bioinformatics that started to largely leverage the techniques from graph neural networks. The recent popularity of graph deep learning for small and large molecules seems to share similar reasons. The first reason is the well-formulated problem and the availability of benchmark datasets while the other is due to the high complexity of the problem and the insufficiency of existing techniques. On the other hand, there is also some subtle difference between them: The deep graph learning community seems dedicated to more extensive new models for small molecules than large ones previously. But in recent years, research frontiers tend to start to transfer the success in small molecules to benefit larger ones, with representative works such as AlphaFold.

Chapter 26
Graph Neural Networks in Anomaly Detection

Shen Wang, Philip S. Yu

Abstract Anomaly detection is an important task, which tackles the problem of discovering "different from normal" signals or patterns by analyzing a massive amount of data, thereby identifying and preventing major faults. Anomaly detection is applied to numerous high-impact applications in areas such as cyber-security, finance, e-commerce, social network, industrial monitoring, and many more mission-critical tasks. While multiple techniques have been developed in past decades in addressing unstructured collections of multi-dimensional data, graph-structure-aware techniques have recently attracted considerable attention. A number of novel techniques have been developed for anomaly detection by leveraging the graph structure. Recently, graph neural networks (GNNs), as a powerful deep-learning-based graph representation technique, has demonstrated superiority in leveraging the graph structure and been used in anomaly detection. In this chapter, we provide a general, comprehensive, and structured overview of the existing works that apply GNNs in anomaly detection.

26.1 Introduction

In the era of machine learning, sometimes, what stands out in the data is more important and interesting than the normal. This branch of task is called *anomaly detection*, which concentrates on discovering "different from normal" signals or patterns by analyzing a massive amount of data, thereby identifying and preventing major faults. This task plays a key on in several high-impact domains, such as cyber-security (network intrusion or network failure detection, malicious program

Shen Wang
Department of Computer Science, University of Illinois at Chicago, e-mail: `swang224@uic.edu`

Philip S. Yu
Department of Computer Science, University of Illinois at Chicago, e-mail: `psyu@uic.edu`

detection), finance (credit card fraud detection, malicious account detection, cashout user detection, loan fraud detection), e-commerce (reviews spam detection), social network (key player detection, anomaly user detection, real money trading detection), and industrial monitoring (fault detection).

In the past decades, many techniques have been developed for anomaly detection by leveraging the graph structure, a.k.a. graph-based anomaly detection. Unlike non-graph anomaly detection, they further take the inter-dependency among each data instance into consideration, where data instances in a wide range of disciplines, such as physics, biology, social sciences, and information systems, are inherently related to one another. Compare to the non-graph-based method, the performance of the graph-based method is greatly improved. Here, we provide an illustrative example of malicious program detection in the cyber-security domain in Figure 26.1. In a phishing email attack as shown in Figure 26.1, to steal sensitive data from the database of a computer/server, the attacker exploits a known venerability of Microsoft Office by sending a phishing email attached with a malicious .doc file to one of the IT staff of the enterprise. When the IT staff member opens the attached .doc file through the browser, a piece of a malicious macro is triggered. This malicious macro creates and executes a malware executable, which pretends to be an open-source Java runtime (Java.exe). This malware then opens a backdoor to the adversary, subsequently allowing the adversary to read and dump data from the target database via the affected computer. In this case, signature-based or behavior-based malware detection approaches generally do not work well in detecting the malicious program in our example. As the adversary can make the malicious program from scratch with binary obfuscation, signature-based approaches would fail due to the lack of known malicious signatures. Behavior-based approaches may not be effective unless the malware sample has previously been used to train the detection model. It might be possible to detect the malicious program using existing host-level anomaly detection techniques. These host-based anomaly detection methods can locally extract patterns from process events as the discriminators of abnormal behavior. However, such detection is based on observations of single operations, and it sacrifices the false positive rate to detect the malicious program. For example, the host-level anomaly detection can detect the fake "Java.exe" by capturing the database read. However, a Java-based SQL client may also exhibit the same operation. If we simply detect the database read, we may also classify normal Java-based SQL clients as abnormal program instances and generate false positives. In the enterprise environment, too many false positives can lead to the alert fatigue problem, causing cyber-analysts to fail to catch up with attacks. To accurately separate the database read of the malicious Java from the real Java instances, we need to consider the higher semantic-level context of the two Java instances. As shown in Figure ??, malicious Java is a very simple program and directly accesses the database. On the contrary, a real Java instance has to load a set of .DLL files in addition to the database read. By comparing the behavior graph of the fake Java instance with the normal ones, we can find that it is abnormal and precisely report it as a malicious program instance. Thus, leveraging the graph helps to identify the anomaly data instances.

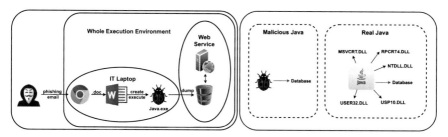

Figure 26.1: An illustrative example of malicious program detection in the cybersecurity domain. The left side shows an example of a phishing email attack: the hacker creates and executes a malware executable, which pretends to be an opensource Java runtime (Java.exe); this malware then opens a backdoor to the adversary, subsequently allowing the adversary to read and dump data from the target database via the affected computer. The right side demonstrates the behavior graph of the malicious Java.exe vs. normal Java runtime.

Specifically, the benefit of graph-based method is four-folded:

- *Inter-dependent Property* – Data instances in a wide range of disciplines, such as physics, biology, social sciences, and information systems, are inherently related to one another and can form a graph. These graph structures can provide additional side information to identify the anomalies in addition to the attributes of each data instance.
- *Relational Property* – The anomaly data instances sometimes can exhibit themselves as relational, e.g., in the fraud domain, the context of the anomaly data instance has a high probability of being abnormal; the anomaly data instance is closely related to a group of data instances. If we identify one anomaly data instance in the graph, some other anomaly data instances based on it can be detected.
- *Fruitful Data Structure* – The graph is a data structure encoding fruitful information. The graph consists of nodes and edges, enabling the incorporation of node and edge attributes/types for anomaly data instance identification. Besides, multiple paths exist between each pair of data instances, which allows the relation extraction in different ranges.
- *Robust Data Structure* – The graph is a more adversarially robust data structure, e.g., attacker or fraudster usually can only attack or fraud the specific data instance or its context and have a limited global view of the whole graph. In this case, the anomaly data instance is harder to fit into the graph as well as possible.

Recent years have witnessed a growing interest in developing deep-learning-based algorithms on the graph, including unsupervised methods (Grover and Leskovec, 2016; Liao et al, 2018; Perozzi et al, 2014) and supervised methods (Wang et al, 2016, 2017e; Hamilton et al, 2017b; Kipf and Welling, 2017b; Veličković et al, 2018). Among these deep-learning-based algorithms on the graph, the graph neural networks (GNNs) (Hamilton et al, 2017b; Kipf and Welling, 2017b; Veličković et al,

2018), as powerful deep graph representation learning techniques, have demonstrated superiority in leveraging the graph structure. The basic idea is to aggregate information from local neighborhoods in order to combine the content feature and graph structures to learn the new graph representation. In particular, GCN (Kipf and Welling, 2017b) leverages the "graph convolution" operation to aggregate the feature of one-hop neighbors and propagate multiple-hop information via the iterative "graph convolution". GraphSage (Hamilton et al, 2017b) develops the graph neural network in an inductive setting, which performs neighborhood sampling and aggregation to generate new node representation efficiently. GAT (Veličković et al, 2018) further incorporates attention mechanism into GCN to perform the attentional aggregation of the neighborhoods. Given the importance of graph-based anomaly detection and the success of graph neural networks, both academia and industry are interested in applying GNNs to tackle the problem of anomaly detection. In recent years, some researchers have successfully applied GNNs in several important anomaly detection tasks. In this book chapter, we summarize different GNN-based anomaly detection approaches and provide taxonomies for them according to various criteria. Despite the more than 10+ papers published in the last three years, several challenges remain unsolved until now, which we summarize and introduce in this chapter as below.

- *Issues* Unlike GNNs applications in other domains, the GNNs applications in anomaly detection have several unique issues, which comes from data, task, and model. We briefly discuss and summarize them to provide a comprehensive understanding of the difficulties of the problems.
- *Pipeline* There are various GNN-based anomaly detection works. It is challenging and time-consuming to understand the big pictures of all these works. To facilitate an easy understanding of existing research on this line, we summarize the general pipeline of GNN-based anomaly detection approaches.
- *Taxonomy* There are already several works in the domain of GNN-based anomaly detection. Compared with other GNN applications, GNN-based anomaly detection is more complicated due to unique challenges and problem definitions. To provide a quick understanding of the similarity and differences between existing works, we list some representative works and summarize novel elaborated taxonomies according to various criteria.
- *Case Studies* We provide the case studies of some representative GNN-based anomaly detection approaches.

The rest of this chapter is organized as follows. Section 26.2 discusses and summarizes the issues of the GNN-based anomaly detection. Section 26.3 provides the unified pipeline of the GNN-based anomaly detection. Section 26.4 provides the taxonomies of existing GNN-based anomaly detection approaches. Section 26.5 provides the case studies of some representative GNN-based anomaly detection approaches. In the last section, we provide the discussion and future directions.

26.2 Issues

In this section, we provide a brief discussion and summary of the issues in GNN-based anomaly detection. In particular, we group them into three: (i) data-specific issues, (ii) task-specific issues, and (iii) model-specific issues.

26.2.1 Data-specific issues

As the anomaly detection systems usually work with real-world data, they demonstrate high volume, high dimensionality, high heterogeneity, high complexity, and dynamic property.

High Volume – With the advance of information storage, it is much easier to collect large amounts of data. For example, in an e-commerce platform like Xianyu, there are over 1 billion second-hand goods published by over ten millions users; in an enterprise network monitoring system, the system event data collected from a single computer system in one day can easily reach 20 GB, and the number of events related to one specific program can easily reach thousands. It is prohibitively expensive to perform the analytic task on such massive data in terms of both time and space.

High Dimensionality – Also, benefit from the advance of the information storage, rich amount of information is collected. It results in high dimensionality of the attributes for each data instance. For example, in an e-commerce platform like Xianyu, different types of attributes are collected for each data instance, such as user demographics, interests, roles, as well as different types of relations; in an enterprise network monitoring system, each collected system event is associated with hundreds of attributes, including information of involved system entities and their relationships, which causes the curse of dimensionality.

High Heterogeneity – As rich types of information are collected, it results in high heterogeneity of the attributes for each data instance: the feature of each data instance can be multi-view or multi-sourced. For example, in an e-commerce platform like Xianyu, multiple types of data are collected from the user, such as personal profile, purchase history, explore history, and so on. Nevertheless, multi-view data like social relations and user attributes have different statistical properties. Such heterogeneity poses a great challenge to integrate multi-view data.

High Complexity – As we can collect more and more information, the collected data is complex in content: it can be categorical or numerical, which increases the difficulty of leveraging all the contents jointly.

Dynamic Property – The data collection is usually conducted every day or continuously. For example, billions of credit card transactions are performed every day; billions of click-through traces of web users are generated each day. This kind of data can be thought of as streaming data, and it demonstrates dynamic property.

The above data-specific issues are general and apply to all kinds of data. So we also need to discuss the graph-data-specific issues, including relational property, graph heterogeneity, graph dynamic, variety of definitions, lack of intrinsic distance/similarity metrics, and search space size.

Relational Property – The relational property of the data makes it challenging to quantify the anomalousness of graph objects. While in traditional outlier detection, the objects or data instances are treated as independent and identically distributed (i.i.d.) from each other, the data instances in graph data are pair-wise correlated. Thus, the "spreading activation" of anomalousness or "guilt by associations" needs to be carefully accounted for. For example, the cash-out users not only have abnormal features, but also behavior abnormally in interaction relations. They may simultaneously have many transactions and fund transfer interactions with particular merchants, which is hard to be exploited by traditional feature extraction.

Graph Heterogeneity – Similar to the general data-specific issues of high heterogeneity, the graph instance type, and relation type are usually heterogeneous. For example, in a computer system graph, there are three types of entities: process (P), file (F), and INETSocket (I) and multiple types of relations: a process forking another process (P→P), a process accessing a file (P→F), a process connecting to an Internet socket (P→I), and so on. Due to the heterogeneity of entities (nodes) and dependencies (edges) in a heterogeneous graph, the diversities between different dependencies vary dramatically, significantly increasing the difficulty of jointly leveraging these nodes and edges.

Graph Dynamic – As the data are collected periodically or continuously, the constructed graph also demonstrates the dynamic property. It is challenging to detect the anomaly due to its dynamic nature. Some anomalous operations show some explicit patterns but try to hide them in a large graph, while others are with implicit patterns. Take an explicit anomaly pattern in a recommender system as an example. As anomalous users usually control multiple accounts to promote the target items, the edges between these accounts and items may compose a dense subgraph, which emerges in a short time period. In addition, although the accounts which involve the anomaly perform anomalous operations sometimes, these accounts perform normally most of the time, which hides their long-term anomalous behavior and increases the difficulty of detection.

Variety of Definitions – The definitions of anomalies in graphs are much more diverse than in traditional outlier detection, given the rich representation of graphs. For example, novel types of anomalies related to graph substructures are of interest for many applications, e.g., money-laundering rings in trading networks.

Lack of Intrinsic Distance/Similarity Metrics – The intrinsic distance/similarity metrics are not clear. For example, in real computer systems, given two programs with thousands of system events related to them, it is a difficult task to measure their distance/similarity.

Search Space Size – The main issue associated with more complex anomalies such as graph substructures is that the search space is huge, as in many graph-theoretical problems associated with graph search. The enumeration of possible substructures is combinatorial, making the problem of finding out the anomalies a much harder task. This search space is enlarged even more when the graphs are attributed as the possibilities span both the graph structure and the attribute space.

As a result, the graph-based anomaly detection algorithms need to be designed not only for effectiveness but also for efficiency and scalability.

26.2.2 Task-specific Issues

Due to the unique characteristics of the anomaly detection task, the issues also come from the problems, including labels quantity and quality, class imbalance and asymmetric error, and novel anomalies.

Labels Quantity and Quality – The major issue of anomaly detection is that the data often has no or very few class labels. It is unknown which data is abnormal or normal. Usually, it is costly and time-consuming to obtain ground-truth labels from the domain expert. Moreover, due to the complexity of the data, the produced label may be noisy and biased. Therefore, this issue limits the performance of the supervised machine learning algorithm. What is more, the lack of true clean labels, i.e., ground truth data, also makes the evaluation of anomaly detection techniques challenging.

Class Imbalance and Asymmetric Error – Since the anomalies are rare and only a small fraction of the data is excepted to be abnormal, the data is extremely imbalanced. Moreover, the cost of mislabeling a good data instance versus a bad instance may change depending on the application and further could be hard to estimate beforehand. For example, mis-predicting a cash-out fraudster as a normal user is essentially harmful to the whole financial system or even the national security, while mis-predicting a normal user as a fraudster could cause customer loss fidelity. Therefore, the class imbalance and asymmetric error affect the machine-learning-based method seriously.

Novel Anomalies – In some domain, such as fraud detection or malware detection, the anomalies are created by the human. They are created by analyzing the detection system and designed to be disguised as a normal instance to bypass the detection. As a result, not only should the algorithms be adaptive to changing and growing data over time, they should also be adaptive to and be able to detect novel anomalies in the face of adversaries.

26.2.3 Model-specific Issues

Apart from data-specific and task-specific issues, it is also challenging to apply the graph neural network directly to anomaly detection task sdue to its unique model properties, such as homogeneous focus and vulnerability.

Homogeneous Focus – Most graph neural network models are designed for homogeneous graph, which considers a single type of nodes and edges. In many real-world applications, data can be naturally represented as heterogeneous graphs. However, traditional GNNs treat different features equally. All the features are mapped and propagated together to get the representations of nodes. Considering that the role of each node is just a one-dimensional feature in the high dimensional feature space, there exist more features that are not related to the role, e.g., age, gender, and education. Thus the representation of applicants with neighbors of different roles has no distinction in representation space after neighbor aggregation, which causes the traditional GNNs to fail.

Vulnerability – Recently theoretical studies prove the limitations and vulnerabilities of GNNs, when graphs have noisy nodes and edges. Therefore, a small change to the node features may cause a dramatic performance drop and failing to tackle the camouflage, where fraudsters would sabotage the GNN-based fraud detectors.

26.3 Pipeline

In this section, we introduce the standard pipeline of the GNN-based anomaly detection. Typically, GNN-based anomaly detection methods consist of three important components, including graph construction and transformation, graph representation learning, and prediction.

26.3.1 Graph Construction and Transformation

As discussed in the previous section, a real-world anomaly detection system has some data-specific issues. Therefore, it requires data analysis on the raw data to address them. Then the graph can be constructed to capture the complex interactions and eliminate the data redundancies. Based on the type of the data instance and relations, the graph can be constructed as a homogeneous graph or heterogeneous graph, where a homogeneous graph only has a single-typed data instance and relation, and a heterogeneous graph has multi-typed data instances and relations. Based on the availability of the timestamp, the graph can be constructed as a static graph or a dynamic graph, where a static graph refers to the graph that has fixed nodes and edges, and a dynamic graph refers to the graph that has nodes and/or edges change over time. Based on the availability of the node and/or edge attributes, the constructed graph can be a plain graph or an attributed graph, where the plain graph only contains the structure information and the attributed graph has attributes on nodes and/or edges.

When the constructed graph is heterogeneous, simply aggregating neighbors cannot capture the semantic and structural correlations among different types of entities. To address the graph heterogeneity issue, a graph transformation is performed to transform the heterogeneous graph to a multi-channel graph guided by the meta-paths, where a meta-path (Sun et al, 2011) is a path that connects entity types via a sequence of relations over a heterogeneous network. For example, in a computer system, a meta-path can be the system events (P→P, P→F, and P→I), with each one defining a unique relationship between two entities. The multi-channel graph is a graph with each channel constructed via a certain type of meta-path. Formally, given a heterogeneous graph $\mathscr{G}$ with a set of meta-paths $\mathscr{M} = \{M_1, ..., M_{|\mathscr{M}|}\}$, the transformed multi-channel network $\hat{\mathscr{G}}$ is defined as follows:

$$\hat{\mathscr{G}} = \{\mathscr{G}_i | \mathscr{G}_i = (\mathscr{V}_i, \mathscr{E}_i, A_i), i = 1, 2, ..., |\mathscr{M}|)\} \tag{26.1}$$

where $\mathscr{E}_i$ denotes the homogeneous links between the entities in $\mathscr{V}_i$, which are connected through the meta-path M_i. Each channel graph $\mathscr{G}_i$ is associated with an adjacency matrix A_i. $|\mathscr{M}|$ indicates the number of meta-paths. Notice that the potential meta-paths induced from the heterogeneous network can be infinite, but not everyone is relevant and useful for the specific task of interest. Fortunately, there are some algorithms (Chen and Sun, 2017) proposed recently for automatically selecting the meta-paths for particular tasks.

26.3.2 Graph Representation Learning

After the graph is constructed and transformed, graph representation learning is conducted to get the proper new representation of the graph. Generally GNNs are built by stacking seven types of basic operations, including neural aggregator function $AGG()$, linear mapping function $MAP_{linear}()$, nonlinear mapping function $MAP_{nonlinear}()$, multilayer perceptron function $MLP()$, feature concatenation $CONCAT()$, attentional feature fusion $COMB_{att}$, and readout function $Readout()$. Among these operations, linear mapping function, nonlinear mapping function, multilayer perceptron function, feature concatenation, and attentional feature fusion are typical operations used in traditional deep learning algorithms. Their formal descriptions are described as follows:

Linear Mapping Function $MAP_{linear}()$:

$$MAP_{linear}(\mathbf{x}) = \mathbf{Wx} \tag{26.2}$$

where $\mathbf{x}$ is the input feature vector, and $\mathbf{W}$ is the trainable weight matrix.

Nonlinear Mapping Function $MAP_{nonlinear}()$:

$$MAP_{nonlinear}(\mathbf{x}) = \sigma(\mathbf{Wx}) \tag{26.3}$$

where $\mathbf{x}$ is the input feature vector, $\mathbf{W}$ is the trainable weight matrix, and $\sigma()$ represents the non-linear activation function.

Multilayer Perceptron Function $MLP()$:

$$MLP(\mathbf{x}) = \sigma(\mathbf{W}^k \cdots \sigma(\mathbf{W}^1 \mathbf{x})) \tag{26.4}$$

where $\mathbf{x}$ is the input feature vector, $\mathbf{W^i}$ with $i = 1, ..., k$ is the trainable weight matrix, k indicates the number of layers, and $\sigma()$ represents the non-linear activation function.

Feature Concatenation CONCAT():

$$CONCAT(\mathbf{x}_1, \cdots \mathbf{x}_n) = [\mathbf{x}_1, \cdots \mathbf{x}_n] \tag{26.5}$$

where n indicates the number of the features.

Attentional Feature Fusion $COMB_{att}()$:

$$COMB_{att}(\mathbf{x}_1, \cdots \mathbf{x}_n) = \sum_{i=1}^{n} softmax(\mathbf{x}_i)\mathbf{x}_i \tag{26.6}$$

$$softmax(\mathbf{x}_i) = \frac{\exp(MAP(\mathbf{x}_i))}{\sum_{j=1}^{n} \exp(MAP(\mathbf{x}_j))} \tag{26.7}$$

where $MAP()$ can be linear or nonlinear.

Different from traditional deep learning algorithm, the GNNs have its unique operation–neural aggregation function $AGG()$. Based on the level of object to aggregate, it can be categorized into three specific types: node-wise neural aggregator $AGG_{node}()$, layer-wise neural aggregator $AGG_{layer}()$, and path-wise neural aggregator $AGG_{path}()$.

Node-wise Neural Aggregator $AGG_{node}()$ is the GNN module that aims to aggregate the node neighborhoods, which can be described as follows,

$$\mathbf{h}_v^{(i)(k)} = AGG_{node}(\mathbf{h}_v^{(i)(k-1)}, \{\mathbf{h}_u^{(i)(k-1)}\}_{u \in \mathcal{N}_v^i}) \tag{26.8}$$

where i is meta-path (relation) indicator, $k \in \{1, 2, ...K\}$ is the layer indicator, $\mathbf{h}_v^{(i)(k)}$ is the feature vector of node v for relation M_i at the k-th layer, $\mathcal{N}_v^i$ indicates the neighbourhoods of node v under the relation M_i. Based on the way the the node neighborhoods are aggregated, typically, the node-level neural aggregator can be GCN $AGG^{GCN}()$ (Kipf and Welling, 2017b), GAT $AGG^{GAT}()$ (Veličković et al, 2018) or Message-Passing $AGG^{MPNN}()$ (Gilmer et al, 2017). For the GCN and GAT, the formulations can be described by Equation 8. While for the Message-Passing, the edges are also used during the node-level aggregation. Formally, it can be described as follows,

$$\mathbf{h}_v^{(i)(k)} = AGG_{node}(\mathbf{h}_v^{(i)(k-1)}, \{\mathbf{h}_v^{(i)(k-1)}, \mathbf{h}_u^{(i)(k-1)}, \mathbf{h}_{vu}^{(i)(k-1)}\}_{u \in \mathcal{N}_v^i}) \tag{26.9}$$

where $\mathbf{h}_{vu}^{(i)(k-1)}$ denotes the edge embedding between the target node v and its neighbor node u, and $\{\}$ indicates a fusion function to combine the target node, its neighbor node and the corresponding edge between them.

Layer-wise Neural Aggregator $AGG_{layer}()$ is the GNN module that aims to aggregate the context information from different hops. For example, if layer number $k = 2$, the GNN gets 1-hop neighborhood information, and if layer number

$k = K + 1$, the GNN gets K-hop neighborhood information. The larger the k is, the more global information the GNN obtains. Formally, this function can be described as follows,

$$\mathbf{l}_v^{(i)(k)} = AGG_{layer}(\mathbf{l}_v^{(i)(k-1)}, \mathbf{h}_v^{(i)(k)}) \tag{26.10}$$

where $\mathbf{l}_v^{(i)(k)}$ is the aggregated representation of $(k-1)-$hop neighborhood node v for relation M_i at the k-th layer.

   Path-wise Neural Aggregator $AGG_{layer}()$ is the GNN module that aims to aggregate the context information from different relations. Generally, the relation can be described by meta-path (Sun et al, 2011) based contextual search. Formally, this function can be described as follows,

$$\mathbf{p}_v^{(i)} = \mathbf{l}_v^{(i)(K)} \tag{26.11}$$

$$\mathbf{p}_v = AGG_{path}(\mathbf{p}_v^{(1)}, \ldots \mathbf{p}_v^{(|\mathcal{M}|)}) \tag{26.12}$$

where $\mathbf{p}_v^{(i)}$ is the aggregated final layer representation of node v for relation M_i.

   Then the final node representation is described by the fusion representation from different meta-paths (relations) as follows,

$$\mathbf{h}_v^{(final)} = \mathbf{p}_v \tag{26.13}$$

Based on the task, we can also compute the graph representation by performing readout function *Readout*() to aggregate all the nodes' final representations, which can be described as follows,

$$\mathbf{g} = Readout(\mathbf{h}_{v_1}^{(final)}, \ldots \mathbf{h}_{v_V}^{(final)}) \tag{26.14}$$

Typically, we can obtain different levels of graph representations, including node-level, edge-level, and graph-level. The node-level and edge-level representation are the most preliminary representations, which can be learned via graph neural network. The graph-level representation is a higher-level representation, which can be obtained by performing the readout function to the node-level and edge-level representations. Based on the target of the task, the specific level of graph representations is fed to the next stage.

26.3.3 Prediction

After the graph representation is learned, they are fed to the prediction stage. Depends on the task and the target label, there are two types of prediction: classification and matching. In the classification-based prediction, it assumes that enough labeled anomaly data instances are provided. A good classifier can be trained to identify if the given graph target is abnormal or not. As mentioned in the issues section, there might be no or few anomaly data instances. In this case, the matching-based prediction is usually used. If there are very few anomaly samples, we learn the representation of them, and when the candidate sample is similar to one of the anomaly samples, an alarm is triggered. If there is no anomaly sample, we learn the representation of the normal data instance. When the candidate sample is not similar to any of the normal samples, an alarm is triggered.

26.4 Taxonomy

In this section, we provide the taxonomies of existing GNN-based anomaly detection approaches. Due to the variety of graph data and anomalies, the GNN-based anomaly detection can have multiple taxonomies. Here we provided four types of taxonomy in order to give a quickly understand of the similarity and difference between existing works, including static/dynamic graph taxonomy, homogeneous/heterogeneous graph taxonomy, plain/attributed graph taxonomy, object taxonomy, and task taxonomy.

In **task taxonomy**, the exiting works can be categorized into GNN-based anomaly detection in financial networks, GNN-based anomaly detection in computer networks, GNN-based anomaly detection in telecom networks, GNN-based anomaly detection in social networks, GNN-based anomaly detection in opinion networks, and GNN-based anomaly detection in sensor networks.

In **anomaly taxonomy**, the existing works can be categorized into node-level anomaly detection, edge-level anomaly detection, and graph-level anomaly detection.

In **static/dynamic graph taxonomy**, the existing works can be categorized into static GNN-based anomaly detection and dynamic GNN-based anomaly detection.

In **homogeneous/heterogeneous graph taxonomy**, the exiting works can be categorized into homogeneous GNN-based anomaly detection and heterogeneous GNN-based anomaly detection.

In **plain/attributed graph taxonomy**, the exiting works can be categorized into plain GNN-based anomaly detection and attributed GNN-based anomaly detection.

In **object taxonomy**, the exiting works can be categorized into: classification-based approach and matching-based approach.

We present our taxonomy with more details in Table 1.

26.5 Case Studies

In this section, we provide the case studies to give the details of some representative GNN-based anomaly detection approaches.

Table 26.1: Summary of GNN-based anomaly detection approaches.

Approach	Year	Venue	Task	Anomaly	Static Dynamic	Homogeneous Heterogeneous	Plain Attributed	Model	Object
GEM (Liu et al, 2018f)	2018	CIKM	Malicious Account Detection	Node	Static	Heterogeneous	Attributed	GCN, Attention$_{(path)}$	Classification
HACUD (Hu et al, 2019b)	2019	AAAI	Cashout User Detection	Node	Static	Heterogeneous	Attributed	GCN, Attention$_{(feature,path)}$	Classification
DeepHGNN (Wang et al, 2019h)	2019	SDM	Malicious Program Detection	Node	Static	Heterogeneous	Attributed	GCN, Attention$_{(path)}$	Classification
MatchGNet (Wang et al, 2019i).	2019	IJCAI	Malicious Program Detection	Graph	Static	Heterogeneous	Attributed	GCN, Attention$_{(node,layer,path)}$	Matching
AddGraph (Zheng et al, 2019)	2019	IJCAI	Malicious Connection Detection	Edge	Dynamic	Homogeneous	Plain	GCN, GRU$_{att}$	Matching
SemiGNN (Wang et al, 2019b)	2019	ICDM	Malicious Account Detection	Node	Static	Heterogeneous	Attributed	GCN, Attention$_{(node,path)}$	Classification, Matching
MVAN (Tao et al, 2019)	2019	KDD	Real Money Trading Detection	Node	Static	Heterogeneous	Attributed	GAT, Attention$_{(path,view)}$	Classification
GAS (Li et al, 2019a)	2019	CIKM	Spam Detection	Edge	Static	Heterogeneous	Attributed	MPNN, Attention$_{(message)}$	Classification
iDetective (Zhang et al, 2019a)	2019	CIKM	Key Player Detection	Node	Static	Heterogeneous	Attributed	GCN, Attention$_{(path)}$	Classification
GAL (Zhao et al, 2020f)	2020	CIKM	Anomaly User Detection	Node	Static	Homogeneous	Attributed	GCN/GAT	Matching
CARE-GNN (Dou et al, 2020)	2020	CIKM	Fraud Detection	Node	Static	Heterogeneous	Attributed	GCN, Attention$_{(node)}$	Classification

26.5.1 Case Study 1: Graph Embeddings for Malicious Accounts Detection

Graph embeddings for malicious accounts detection (GEM) (Liu et al, 2018f) is the first attempt to apply the GNN to anomaly detection. The aim of GEM is to detect the malicious account at Alipay pay, a mobile cashless payment platform.

The graph constructed from the raw data is static and heterogeneous. The construed graph $\mathscr{G} = (\mathscr{V}, \mathscr{E})$ consists of 7 types of nodes, including account typed nodes (U) and 6 types of device typed nodes (phone number (PN), User Machine ID (UMID), MAC address (MACA), International Mobile Subscriber Identity (IMSI), Alipay Device ID (APDID) and a random number generated via IMSI and IMEI (TID), such that $\mathscr{V} = U \cup PN \cup UMID \cup MACA \cup IMSI \cup APDID \cup TID$. To overcome the heterogeneous graph challenge and make GNN applicable to the graph, through graph transformation, GEM constructs a 6-channel graph $\hat{\mathscr{G}} = \{\mathscr{G}_i | \mathscr{G}_i = (\mathscr{V}_i, \mathscr{E}_i, A_i), i = 1, 2, ..., |\mathscr{M}|\}$ with $|\mathscr{M}| = 6$. In particular, 6 types of edges are specifically modeled to capture the edge heterogeneity, e.g., account connects phone number ($U \rightarrow PN$), account connects UMID ($U \rightarrow UMID$), account connects MAC address ($U \rightarrow MACA$), account connects IMSI ($U \rightarrow IMSI$), account connects Alipay Device ID ($U \rightarrow APDID$) and account connects TID ($U \rightarrow TID$). As the activity attributes are constructed, the constructed graph is an attributed graph. After the graphs are constructed and transformed, GEM performs a graph convolutional network to aggregate the neighborhood on each channel graph. As each channel graph is treated as a homogeneous graph corresponding to a specific relation, GNN can be directly applied to each channel graph.

During the graph representation learning stage, the node aggregated representation $\mathbf{h}_v^{(i)(k)}$ is computed by performing a GCN aggregator $AGG^{GCN}()$. To get the path aggregated representation, it adopts the attentionally feature fusion to fuse the node aggregated representation obtained in each channel graph $\mathscr{G}^i$. Besides, an activity feature for each node is constructed, and it adds the linear mapping of this activity feature to the attentional feature fusion of the path aggregated representations. Formally, the GNN operations can be described as follow.

Node-wise aggregation:

$$
\begin{aligned}
\mathbf{h}_v^{(i)(k)} &= AGG_{node}(\mathbf{h}_v^{(i)(k-1)}, \{\mathbf{h}_u^{(i)(k-1)}\}_{u \in \mathcal{N}_v^i}) \\
&= AGG^{GCN}(\mathbf{h}_v^{(i)(k-1)}, \{\mathbf{h}_u^{(i)(k-1)}\}_{u \in \mathcal{N}_v^i})
\end{aligned}
\tag{26.15}
$$

Path-wise aggregation:

$$
\mathbf{p}_v^{(k)} = MAP_{linear}(\mathbf{x}_v) + COMB_{att}(\mathbf{h}_v^{(1)(k)}, ..., \mathbf{h}_v^{(|\mathcal{M}|)(k)})
\tag{26.16}
$$

Layer-wise aggregation:

$$
\mathbf{l}_v^{(K)} = \mathbf{p}_v^{(K)}
\tag{26.17}
$$

Final node representation:

$$
\mathbf{h}_v^{(final)} = \mathbf{l}_v^{(K)}
\tag{26.18}
$$

where K indicates the number of the layers.

The object of GEM is classification. It feeds the learned account node embedding to a standard logistic loss function.

26.5.2 Case Study 2: Hierarchical Attention Mechanism based Cash-out User Detection

Hierarchical attention mechanism based cash-out user detection (HACUD) (Hu et al, 2019b) applied the GNN to the fraud user detection at Credit Payment Services platform, where the fraud user performs the cash-out fraud, that pursues cash gains with illegal or insincere intent.

HACUD also constructs a static heterogeneous graph from the raw data. Specifically, it consists of multiple types of nodes (i.e., User (U), Merchant (M), Device (D)) with rich attributes and relations (i.e., fund transfer relation between users and transaction relation between users and merchants). Different from the way GEM deal with the graph heterogeneity issues, during the graph transformation stage, HACUD only models the user nodes and considers two specific types of meta-paths (relations) between pairwise of users, including User-(fund transfer)-User (UU) and User-(transaction)-Merchant-(transaction)-User (UMU) and constructs a 2-channel graph, such that $\mathscr{G} = \{\mathscr{G}_i | \mathscr{G}_i = (\mathcal{V}_i, \mathscr{E}_i, A_i), i = 1, ..., |\mathcal{M}|\}$ with $|\mathcal{M}| = 2$ and $\mathcal{V}_i \in U$.

The two selected meta-paths capture different semantics. For example, the UU path connects users having fund transfers from one to another, while the UMU connects users having transactions with the same merchants. Then each channel graph is homogeneous and can work with GNN directly. As the user attributes are available, the constructed graph is attributed.

In the graph representation stage, the node-wise aggregation is performed to each channel graph via a convolutional graph network. Different from GEM (Liu et al, 2018f), it adds and joins the user feature $\mathbf{x}_v$ to the aggregated node representation in an attentional way. Then the node-wise aggregation extends to a 3-step procedure, including (a) initial node-wise aggregation, (b) feature fusion, and (c) feature attention. After the initial aggregated node representation $\tilde{\mathbf{h}}_v^{(i)}$ is computed vis GCN $AGG^{GCN}()$, it is fused with user feature $\mathbf{x}_v$ through a feature fusion. Next, it performs the feature attention. Since only 1-hop neighborhoods are considered, there is no layer-wise aggregation, and the final node-wise aggregated representations $\mathbf{h}_v^{(i)}$ are fed to the path-wise aggregation directly. Formally, it can be described as follows,

Node-wise aggregation:

(a)*Initial node-wise aggregation:*

$$\tilde{\mathbf{h}}_v^{(i)} = AGG_{node}(\mathbf{h}_v^{(i)}, \{\mathbf{h}_u^{(i)}\}_{u \in \mathcal{N}_v^i})$$
$$= AGG^{GNN}(\mathbf{h}_v^{(i)}, \{\mathbf{h}_u^{(i)}\}_{u \in \mathcal{N}_v^i}) \tag{26.19}$$

(b)*Feature fusion:*

$$\mathbf{f}_v^{(i)} = MAP_{nonlinear}(CONCAT(MAP_{linear}(\tilde{\mathbf{h}}_v^{(i)}), MAP_{linear}(\mathbf{x}_v))) \tag{26.20}$$

(c)*Feature attention:*

$$\alpha_v^{(i)} = MAP_{nonlinear}(MAP_{nonlinear}(CONCAT(MAP_{linear}(\mathbf{x}_v), \mathbf{f}_v^{(i)}))) \tag{26.21}$$

$$\mathbf{h}_v^{(i)} = softmax(\alpha_v^{(i)}) \odot \mathbf{f}_v^{(i)} \tag{26.22}$$

Path-wise aggregation:

$$\mathbf{p}_v = AGG_{path}(\mathbf{h}_v^{(0)}, \mathbf{h}_v^{(1)})$$
$$= COMB_{att}(\mathbf{h}_v^{(0)}, \mathbf{h}_v^{(1)}) \tag{26.23}$$

Final node representation:

$$\mathbf{h}_v^{(final)} = MLP(\mathbf{p}_v) \tag{26.24}$$

where $\odot$ denotes the element-wise product. As only one-hop information is used, there is no layer indicator k.

As same as GEM, the object of HACUD is classification. It feeds the learned user node embedding to a standard logistic loss function.

26.5.3 Case Study 3: Attentional Heterogeneous Graph Neural Networks for Malicious Program Detection

Attentional heterogeneous graph neural network for malicious program detection (DeepHGNN) (Wang et al, 2019h) applied the GNN to the malicious program detection in a computer system of an enterprise network.

The raw data is a large volume of system behavioral data with rich information on program/process level events. A static heterogeneous graph is constructed to model the program behaviors. Formally, given the program event data across many machines within a time window (*e.g.*, 1 day), a heterogeneous graph $\mathscr{G} = (\mathscr{V}, \mathscr{E})$ is constructed for the target program. $\mathscr{V}$ denotes a set of nodes, with each one representing an entity of three types: process (P), file (F), and INETSocket (I). Namely, $\mathscr{V} = P \cup F \cup I$. $\mathscr{E}$ denotes a set of edges (v_s, v_d, r) between the source entity v_s and destination entity v_d with relation r. To address the heterogeneous graph challenges, it takes three types of relations, including: (1) a process forking another process (P→P), (2) a process accessing a file (P→F), and (3) a process connecting to an Internet socket (P→I). Similar to GEM, DeepHGNN designs a graph transformation module to transform the heterogeneous graph to a 3-channel graph guided by above three meta-paths (relations), such that $\hat{\mathscr{G}} = \{\mathscr{G}_i | \mathscr{G}_i = (\mathscr{V}_i, \mathscr{E}_i, A_i), i = 1, 2, ..., |\mathscr{M}|\}$ with $|\mathscr{M}| = 3$ and $\mathscr{V}_i = \mathscr{V}$. The attributes are constructed for each node. Since the process node, file node, and INETSocket node has quite different attributes, the graph statistic features $\mathbf{x}_v^{(i)(gstat)}$ are constructed and act as the node attributes.

Similar to the GEM and HACUD, DeepHGNN also adopts the graph convolutional network $AGG^{GCN}()$ for node-wise aggregation. Three layers are used in order to capture program behavior within 3-hop contexts. Different from GEM and HACUD, DeepHGNN uses the graph statistic node attributes as the initialization of the node representation for each channel graph. After the three node-wise aggregation and layer-wise aggregation, the node representations from different channel graphs are fused via the attentional feature fusion as GEM and HACUD. Formally, it can be described as follows,

Node-wise aggregation:

$$\mathbf{h}_v^{(i)(0)} = \mathbf{x}_v^{(i)(gstat)} \tag{26.25}$$

$$\begin{aligned}
\mathbf{h}_v^{(i)(k)} &= AGG_{node}(\mathbf{h}_v^{(i)(k-1)}, \{\mathbf{h}_u^{(i)(k-1)}\}_{u \in \mathscr{N}_v^i}) \\
&= AGG^{GNN}(\mathbf{h}_v^{(i)(k-1)}, \{\mathbf{h}_u^{(i)(k-1)}\}_{u \in \mathscr{N}_v^i})
\end{aligned} \tag{26.26}$$

Layer-wise aggregation:

$$\mathbf{l}_v^{(i)(k)} = \mathbf{h}_v^{(i)(k)} \tag{26.27}$$

Path-wise aggregation:

$$\mathbf{p}_v = COMB_{att}(\mathbf{l}_v^{(1)(K)}, ..., \mathbf{l}_v^{(|\mathcal{M}|)(K)}) \tag{26.28}$$

Final node representation:

$$\mathbf{h}_v^{(final)} = \mathbf{p}_v \tag{26.29}$$

The object of DeepHGNN is classification. However, it is different from GEM and HACUD, which simply build single classifiers for all the samples. DeepHGNN formulates the problem of program reidentification in malicious program detection. The graph representation learning aims to learn the representation of the normal target program, and each target program learns a unique classifier. Given a target program with corresponding event data during a time window $U = \{e_1, e_2, ...\}$ and a claimed name/ID, the system checks whether it belongs to the claimed name/ID. If it matches the behavior pattern of the claimed name/ID, the predicted label should be $+1$; otherwise, it should be -1.

26.5.4 Case Study 4: Graph Matching Framework to Learn the Program Representation and Similarity Metric via Graph Neural Networks for Unknown Malicious Program Detection

Graph matching framework to learn the program representation and similarity metric via graph neural network (MatchGNet) (Wang et al, 2019i) is another GNN-based anomaly detection approach for malicious program detection in a computer system of an enterprise network. MatchGNet is different from DeepHGNN in five aspects: (1) after the graph transformation, the resulted channel graph only keep the target type node – process node, which is similar to HACUD, (2) the raw program attributes are used as the program node representation initialization, (3) the GNN aggregation is conducted hierarchically in node-wise, layer-wise, and path-wise, (4) the anomaly target is the subgraph of the target program (5) the final graph representation is fed to a similarity learning framework with contrastive loss to deal with the unknown anomaly.

It follows a similar style to construct the static heterogeneous graph from system behavioral data. In the graph transformation, it adopts three meta-paths (relations): a process forking another process $(P \rightarrow P)$, two processes accessing the same file $(P \leftarrow F \rightarrow P)$, and two processes opening the same internet socket $(P \leftarrow I \rightarrow P)$ with each one defining a unique relationship between two processes. Based on them, a 3-channel graph is constructed from the the heterogeneous graph, such that $\hat{\mathcal{G}} = \{\mathcal{G}_i | \mathcal{G}_i = (\mathcal{V}_i, \mathcal{E}_i, A_i), i = 1, ..., |\mathcal{M}|\}$ with $|\mathcal{M}| = 3$ and $\mathcal{V}_i \in P$. Then the GNN can be directly applied to each channel graph. As only process typed nodes are available, we use the raw attributes of these process $\mathbf{x}_v$ as the node representation initialization.

During the graph representation stage, a hierarchical attentional graph neural network is designed, including node-wise attentional neural aggregator, layer-wise dense-connected neural aggregator, and path-wise attentional neural aggregator. In particular, the node-wise attentional neural aggregator aims to generate node embeddings by selectively aggregating the entities in each channel graph based on random walk scores $\alpha_{(u)}^i$. Layer-wise dense-connected neural aggregator aggregates the node embeddings generated from different layers towards a dense-connected node embedding. Path-wise attentional neural aggregator performs attentional feature fusion of the layer-wise dense-connected representations. In the end, the final node representation is used as the graph representation. Formally, it can be described as follows,

Node-wise aggregation:

$$\mathbf{h}_v^{(i)(0)} = \mathbf{x}_v \tag{26.30}$$

$$
\begin{aligned}
\mathbf{h}_v^{(i)(k)} &= AGG_{node}(\mathbf{h}_v^{(i)(k-1)}, \{\mathbf{h}_u^{(i)(k-1)}\}_{u \in \mathcal{N}_v^i}) \\
&= MLP((1+\varepsilon^{(k)})\mathbf{h}_v^{(i)(k-1)} + \sum_{u \in \mathcal{N}_v^i} \alpha_{(u)(:)}^i \mathbf{h}_u^{(i)(k-1)})
\end{aligned}
\tag{26.31}
$$

Layer-wise aggregation:

$$
\begin{aligned}
\mathbf{l}_v^{(i)(k)} &= AGG_{layer}(\mathbf{h}_v^{(i)(0)}, \mathbf{l}_v^{(i)(1)}, \dots \mathbf{l}_v^{(i)(k)}) \\
&= MLP(CONCAT(\mathbf{h}_v^{(i)(0)}; \mathbf{l}_v^{(i)(1)}; \dots \mathbf{l}_v^{(i)(k)}))
\end{aligned}
\tag{26.32}
$$

Path-wise aggregation:

$$\mathbf{p}_v = COMB_{att}(\mathbf{l}_v^{(i)(K)}, \dots, \mathbf{l}_v^{(|\mathcal{M}|)(K)}) \tag{26.33}$$

Final node representation:

$$\mathbf{h}_v^{(final)} = \mathbf{p}_v \tag{26.34}$$

Final graph representation:

$$\mathbf{h}_{G_v} = \mathbf{h}_v^{(final)} \tag{26.35}$$

where k indicates the number of layers, and ε is a small number. Different from GEM, HACUD, and DeepHGNN, the object of MatchGNet is matching. The final graph representation is fed to a similarity learning framework with contrastive loss to deal with the unknown anomaly. During the training, P pairs of program graph snapshots $(\mathcal{G}_{i(1)}, \mathcal{G}_{i(2)}), i \in \{1, 2, \dots P\}$ are collected with corresponding ground truth pairing information $y_i \in \{+1, -1\}$. If the pair of graph snapshots belong to the same program, the ground truth label is $y_i = +1$; otherwise, its ground truth label is $y_i = -1$. For each pair of program snapshots, a cosine score function is used to measure the similarity of the two program embeddings, and the output is defined as follows:

$$Sim(\mathscr{G}_{i(1)}, \mathscr{G}_{i(2)}) = cos((\mathbf{h}_{\mathscr{G}_{i(1)}}, \mathbf{h}_{\mathscr{G}_{i(2)}}))$$
$$= \frac{\mathbf{h}_{\mathscr{G}_{i(1)}} \cdot \mathbf{h}_{\mathscr{G}_{i(2)}}}{||\mathbf{h}_{\mathscr{G}_{i(1)}}|| \cdot ||\mathbf{h}_{\mathscr{G}_{i(2)}}||} \quad (26.36)$$

Correspondingly, our objective function can be formulated as:

$$\ell = \sum_{i=1}^{P} (Sim(\mathscr{G}_{i(1)}, \mathscr{G}_{i(2)}) - y_i)^2 \quad (26.37)$$

26.5.5 Case Study 5: Anomaly Detection in Dynamic Graph Using Attention-based Temporal GCN

Anomaly detection in dynamic graph using attention-based temporal GCN (Add-Graph) (Zheng et al, 2019) is the first work that applies the GNN to solve the problem of anomaly edge detection in the dynamic graph. It focuses on the modeling of the dynamic graph via GNN and performs anomaly connection detection in telecom networks and social networks. The graphs are constructed from the edge stream data, and the constructed graphs are dynamic, homogeneous, and plain.

The basic idea is to build a framework to describe the normal edges by using all possible features in the graph snapshots in the training phase, including structural, content, and temporal features. Then at the prediction stage, the matching objective is used similar to MatchGNet. In particular, AddGraph applies GCN $AGG^{GCN}()$ to compute the new current state of a node $\mathbf{c}_v^t$ by aggregating its neighborhoods in the current snapshot graph, which can be described as follows,

$$\mathbf{c}_v^t = AGG^{GCN}(\mathbf{h}_v^{t-1}) \quad (26.38)$$

As the state of a node $\mathbf{c}_v^t$ can be computed by aggregating the neighboring hidden states in the previous timestamp $t-1$, the node hidden states in a short window w can be obtained and combined to get the short-term embedding $\mathbf{s}_v^t$. In particular, an attentional feature fusion is used to combine these node hidden states in a short window, as follows,

$$\mathbf{s}_v^t = COMB_{att}(\mathbf{h}_v^{t-w}, ..., \mathbf{h}_v^{t-1}) \quad (26.39)$$

Then short-term embedding $\mathbf{s}_v^t$ and current state $\mathbf{c}_v^t$ are fed to GRU, a classic recurrent neural network, to compute the current hidden state that encoding the dynamics within the graph. This stage can be described as follows:

$$\mathbf{h}_v^t = GRU(\mathbf{c}_v^t, \mathbf{s}_v^t) \quad (26.40)$$

The object of AddGraph is matching. The hidden state of the nodes at each times-tamp are used to calculate the anomalous probabilities of an existing edge and a negative sampled edge, and then feed them to a margin loss.

26.5.6 Case Study 6: GCN-based Anti-Spam for Spam Review Detection

GCN-based anti-spam (GAS) (Li et al, 2019a) applies the GNN in the spam re-view detection at the e-commerce platform Xianyu. Similar to previous works, the constructed graph is static, heterogeneous and attributed, such that $\mathscr{G} = (\mathscr{U}, \mathscr{I}, \mathscr{E})$. There are two types of nodes: user nodes $\mathscr{U}$ and item nodes $\mathscr{I}$. The edges $\mathscr{E}$ are a set of comments. Different from previous works, the edges $\mathscr{E}$ are the anomalies tar-gets. Moreover, as each edge represents a sentence, edge modeling is complicated, and the number of edge types increases dramatically. To better capture the edge representation, the message-passing-like GNN is used. The edge-wise aggregation is proposed by concatenation of previous representation of the edge itself $\mathbf{h}_{iu}^{k-1}$ and corresponding user node representation $\mathbf{h}_u^{k-1}$, item node representation $\mathbf{h}_i^{k-1}$ To get the initial attributes of edge, the word2vec word embedding for each word in the comments of the edges is extracted via the embedding function pre-training on a million-scale comment dataset. Then the word embedding of each words in an edge of comments $\mathbf{w}_0, \mathbf{w}_1, ...\mathbf{w}_n$ is fed to $TextCNN()$ function to get the comments em-bedding $\mathbf{h}_{iu}^0$, which is used as the initial attributes of edge. Then the edge-wise ag-gregation is defined as:

Edge-wise aggregation:

$$\mathbf{h}_{iu}^0 = TextCNN(w_0, w_1, ...w_n) \tag{26.41}$$

$$\mathbf{h}_{iu}^k = MAP_{nonlinear}(CONCAT(\mathbf{h}_{iu}^{k-1}, \mathbf{h}_i^{k-1}, \mathbf{h}_u^{k-1})) \tag{26.42}$$

On the other hand, the node-wise aggregation also needs to take the edges into con-sideration. The node-wise aggregation is performed by attention feature fusion of the target node and its connected edge followed by a non-linear mapping, which can be described with (a) user node-wise aggregation, and (b) item node-wise aggrega-tion as follows:

Node-wise aggregation:

(a)*User node-wise aggregation*:

$$\mathbf{h}_u^k = CONCAT(MAP_{linear}(\mathbf{h}_u^{k-1}), MAP_{nonlinear}(COMB_{att}(\mathbf{h}_u^{k-1}, CONCAT(\mathbf{h}_{iu}^{k-1}, \mathbf{h}_i^{k-1} \tag{26.43}$$

(b)*Item node-wise aggregation*:

$$\mathbf{h}_i^k = CONCAT(MAP_{linear}(\mathbf{h}_i^{k-1}), MAP_{nonlinear}(COMB_{att}(\mathbf{h}_i^{k-1}, CONCAT(\mathbf{h}_{iu}^{k-1}, \mathbf{h}_u^{k-1}))))$$

$$(26.44)$$

where k is the layer indicator. The final edge representation is computed by concatenation of the raw edge embedding $\mathbf{h}_{iu}^0$, new edge embedding $\mathbf{h}_{iu}^K$, corresponding new user node embedding $\mathbf{h}_u^K$, and corresponding new item node embedding $\mathbf{h}_i^K$ as follows:

Final edge representation:

$$\mathbf{h}_{iu}^{final} = CONCAT(\mathbf{h}_{vu}^0, \mathbf{h}_{vu}^K, \mathbf{h}_u^K, \mathbf{h}_i^K) \qquad (26.45)$$

The object of GAS is classification, and the final edge representation is fed to a standard logistic loss function.

26.6 Future Directions

GNNs on anomaly detection is an important research direction, which leverages multi-source, multi-view features extracted from both content and structure for anomaly sample analysis and detection. It plays a key role in numerous high-impact applications in areas such as cyber-security, finance, e-commerce, social network, industrial monitoring, and many more mission-critical tasks. Due to the multiple issues from data, model and task, it still needs a lot of effort in the field. The future works are mainly lying in two perspectives: anomaly analysis and machine learning.

From an anomaly analysis perspective, there are still a lot of research questions. How to define and identify the anomalies in the graph in the different tasks? How to effectively convert the large-scale raw data to the graph? How to effectively leverage the attributes? How to model the dynamic during the graph construction? How to keep the heterogeneity during the graph construction? Recently, due to the data-specific and task-specific issues, the applications of GNN-based anomaly detection are still limited. There is still a lot of potential scenarios that can be applied.

From a machine learning perspective, lots of issues need to be addressed. How to model the graph? How to represent the graph? How to leverage the context? How to fuse the content and structure features? Which part of the structure to capture, local or global? How to provide the model explainability? How to protect the model from adversarial attacks? How to overcome the time-space scalability bottleneck. Recently, lots of contributions have been made from the machine learning perspective. However, due to the unique characteristics of the anomaly detection problem, which GNNs to use and how to apply GNNs are still critical questions. Further work will also benefit from the new findings and new models in the graph machine learning community.

Editor's Notes: Graph neural networks for anomaly detection can be considered as a downstream task of graph representation learning, where the long-term challenges in anomaly detection are coupled with the vulnerability of graph neural networks such as scalability discussed in Chapter 6 and robustness discussed in Chapter 8. Graph neural networks for anomaly detection also further benefits a wide range of downstream tasks in various interesting, important, yet usually challenging areas such as anomaly detection in dynamic networks, spam review detection for recommender system, and malware program detection, which are highly relevant to the topics introduced in Chapters 15, 19, and 22.

Chapter 27
Graph Neural Networks in Urban Intelligence

Yanhua Li, Xun Zhou, and Menghai Pan

Abstract In recent years, smart and connected urban infrastructures have undergone a fast expansion, which increasingly generates huge amounts of urban big data, such as human mobility data, location-based transaction data, regional weather and air quality data, social connection data. These heterogeneous data sources convey rich information about the city and can be naturally linked with or modeled by graphs, e.g., urban social graph, transportation graph. These urban graph data can enable intelligent solutions to solve various urban challenges, such as urban facility planning, air pollution, etc. However, it is also very challenging to manage, analyze, and make sense of such big urban graph data. Recently, there have been many studies on advancing and expanding Graph Neural Networks (GNNs) approaches for various urban intelligence applications. In this chapter, we provide a comprehensive overview of the graph neural network (GNN) techniques that have been used to empower urban intelligence, in four application categories, namely, (i) urban anomaly and event detection, (ii) urban configuration and transportation planning, (iii) urban traffic prediction, and (iv) urban human behavior inference. The chapter also discusses future directions of this line of research. The chapter is (tentatively) organized as follows.

Yanhua Li
Computer Science Department, Worcester Polytechnic Institute, e-mail: yli15@wpi.edu

Xun Zhou
Tippie College of Business, University of Iowa e-mail: un-zhou@uiowa.edu

Menghai Pan
Computer Science Department, Worcester Polytechnic Institute, e-mail: mpan@wpi.edu

27.1 Graph Neural Networks for Urban Intelligence

27.1.1 Introduction

According to the report (Desa, 2018) published by the United Nations in 2018, the urban population in the world reached 55 percent in 2018, which is growing rapidly over time. By 2050, the world will be one-third rural (34 percent) and two-thirds urban (66 percent). Moreover, thanks to the fast development of sensing technologies in recent years, various sensors are widely deployed in the urban areas, e.g., the GPS sets on vehicles, personal devices, air quality monitoring stations, gas pressure regulators, etc. Stimulated by the large urban population and the wide use of the sensors, there are massive data generated in the urban environment, for example, the trajectory data of the vehicles in ride-sharing services, the air quality monitoring data. Given a large amount of heterogeneous urban data, the question to answer is what and how can we benefit from these data. For instance, can we use the GPS data of the vehicles to help urban planners better design the road network? Can we infer the air quality index across the city based on a limited number of existing monitoring stations? To answer these practical questions, the interdisciplinary research area, *Urban Intelligence*, has been extensively studied in recent years. In general, *Urban Intelligence*, which is also referred as *urban computing*, is a process of acquisition, integration, and analysis of big and heterogeneous data generated by a diversity of sources in urban spaces, such as sensors, devices, vehicles, buildings, and humans, to tackle the major issues in cities (Zheng et al, 2014).

Data analytics (e.g., data mining, machine learning, optimization) techniques are usually employed to analyze numerous types of data generated in the urban scenarios for prediction, pattern discovery, and decision-making purposes. How to represent urban data is an essential question for the design and implementation of these techniques. Given the heterogeneity of urban big data, various data structures can be used to represent them. For example, spatial data in an urban area can be represented as raster data (like images), where the area is partitioned into grid cells (pixels) with attribute functions imposed on them (Pan et al, 2020b; Zhang et al, 2019, 2020b,a; Pan et al, 2019, 2020a). Spatial data can also be represented as a collection of objects (e.g., vehicles, point-of-interests, and trajectory GPS points) with their locations and topological relationships defined (Ding et al, 2020b).

Moreover, the intrinsic structures of many urban big data enable people to represent them with graphs. For instance, the structure of urban road network helps people model the traffic data with graphs (Xie et al, 2019b; Dai et al, 2020; Cui et al, 2019; Chen et al, 2019b; Song et al, 2020a; Zhang et al, 2020e; Zheng et al, 2020a; Diao et al, 2019; Guo et al, 2019b; Li et al, 2018e; Yu et al, 2018a; Zhang et al, 2018e); the pipeline of gas supply network enable people to model the gas pressure monitoring data with graph (Yi and Park, 2020); people can also represent the data on the map with a graph by dividing the city into functional regions (Wang et al, 2019o; Yi and Park, 2020; Geng et al, 2019; Bai et al, 2019a; Xie et al, 2016). Representing urban data with graphs can capture the intrinsic topological information and knowledge in the data, and plenty of techniques are developed to analyze the urban graph data.

Graph Neural Networks (GNNs) are naturally employed to solve various real-world problems with urban graph data. For example, Convolutional Graph Neural Networks (ConvGNN) (Kipf and Welling, 2017b) are used to capture the spatial dependencies of the urban graph data, and Recurrent Graph Neural Networks (RecGNN) (Li et al, 2016b) are for the temporal dependencies. Spatial-temporal Graph Neural Networks (STGNN) (Yu et al, 2018a) can capture both spatial and temporal dependencies in the data, which are widely used in dealing with many urban intelligence problems, e.g., predicting traffic status based on urban traffic data (Zhang et al, 2018e; Li et al, 2018e; Yu et al, 2018a). The traffic data are modeled as spatial-temporal graphs where the nodes are sensors on road segments, and each node has the average traffic speed within a window as dynamic input features.

In the following sections, we first summarize the general application scenarios in urban intelligence, followed by the graph representations in urban scenarios. Then, we provide more details on GNN for urban configuration and transportation planning, urban anomaly and event detection, and urban human behavior inference, respectively.

27.1.2 Application scenarios in urban intelligence

The diverse application domains in urban intelligence include urban planning, transportation, environment, energy, human behavior analysis, economy, and event detection, etc. In the following paragraphs, we will introduce the practical problems and the common datasets in these domains. The problems and examples highlighted below are not exhaustive, here we just introduce some critical problems and typical examples from literature, which are summarized in Table 27.1.

1) Urban configuration. Urban configuration is essential for enabling smart cities. It deals with the design problem of the entire urban area, such as, the land use, the layout of human settlements, design of road networks, etc. The problems in this domain includes estimating the impact of a construction (Zhang et al, 2019c), discovering the functional regions of the city (Yuan et al, 2012), detecting city boundaries (Ratti et al, 2010), etc. In (Zhang et al, 2019c), the authors employ and analyze the historical taxi GPS data and the road network data, where they define the off-deployment traffic estimation problem as a traffic generation problem, and develop a novel deep generative model TrafficGAN that captures the shared patterns across spatial regions of how traffic conditions evolve according to travel demand changes and underlying road network structures. This problem is important to city planners to evaluate and develop urban deployment plans. In (Yuan et al, 2012), the authors propose a DRoF framework that Discovers Regions of different Functions in a city using human mobility between regions with data collected from the GPS set in Taxis in Beijing and points of interest (POIs) located in the city. The understanding of functional regions in a city can calibrate urban planning and facilitate

Table 27.1: Application domain and examples in urban intelligence.

Application domain	Example task	Example data source
Urban configuration	Estimate impact of construction (Zhang et al, 2019c)	Taxi GPS, road network.
	Discover functional regions (Yuan et al, 2012)	Taxi GPS, POIs.
Transportation	Improve efficiency of taxi drivers (Pan et al, 2019)	Taxi GPS, road network.
Environment	Infer air quality(Zheng et al, 2013)	Air quality data from monitor stations, road network, POIs.
Energy consumption	Estimate gas consumption (Shang et al, 2014)	Taxi GPS.
Human behavior	Estimate user similarity(Li et al, 2008)	GPS data from phones.
Economy	Place retail store (Karamshuk et al, 2013)	POIs, human mobility data.
Public Safety	Detect anomalous traffic pattern (Pang et al, 2011)	Taxi GPS, road network.

other applications, such as choosing a location for a business. In (Ratti et al, 2010), the authors propose a model to detect the city's boundary by analyzing the human network inferred from a large telecommunications database in Great Britain. Answering this question can help the city planner get a sense on what the exact range the urban area is within as the urban area changes fast over time.

2) Transportation. Transportation plays an important role in the urban area. Urban intelligence deals with several problems regarding the transportation in the city, e.g., routing for the drivers, estimating the travel time, improving the efficiency of taxi system and the public transit system, etc. In (Yuan et al, 2010), the authors propose a T-Drive system, that provides personalized driving directions that adapt to weather, traffic conditions, and a person's own driving habits. The system is built based on historical trajectory data of taxicabs. In (Pan et al, 2019), the authors propose a solution framework to analyze the learning curve of taxi drivers. The proposed method first learns the driver's preference to different profiles and habit features in each time period, then analyzes the preference dynamics of different groups of drivers. The results illustrate that taxi drivers tend to change their preference to some habit features to improve their operation efficiency. This finding can help the new drivers improve their operation efficiency faster. The authors in (Watkins et al, 2011) conducted a study on the impact of providing real-time bus arrival information directly on riders' mobile phones and found it to reduce not only the perceived wait time of those already at a bus stop, but also the actual wait time experienced by customers who plan their journey using such information.

3) Urban Environment. Urban intelligence can deal with the potential threat to the environment caused by the fast pace of urbanization. The environment is essential for people's health, for example, air quality, noise, etc. In (Zheng et al, 2013), the authors infer the real-time and fine-grained air quality information throughout a city based on the (historical and real-time) air quality data reported by existing monitor stations and a variety of data sources observed in the city, such as meteorology, traffic flow, human mobility, structure of road networks, and POIs. The results can be used to suggest people when and where to conduct outdoor activities, e.g., jogging. Also, the result can infer suitable locations for deploying new air quality monitoring stations. Noise pollution is usually serious in the urban area. It has impacts to both the mental and physical health of human beings. Santini et al (2008) assess environmental noise pollution in urban areas by using the monitoring data from wireless sensor networks.

4) Energy supply and consumption. Another application domain of urban intelligence is energy consumption in the urban area, which usually deals with the problem of sensing city-scale energy cost, improving energy infrastructures, and finally reducing energy consumption. The common energy include gas and electricity. Shang et al (2014) inferred the gas consumption and pollution emission of vehicles traveling on a city's road network in the current time slot using GPS trajectories from a sample of vehicles (e.g., taxicabs). The knowledge can be used not only to suggest cost-efficient driving routes but also to identify the road segments where gas has been wasted significantly. Momtazpour et al (2012) proposes a framework to predict electronic vehicle (EV) charging needs based on owners' activities, EV charging demands at different locations in the city and available charge of EV batteries, and design distributed mechanisms that manage the movements of EVs to different charging stations.

5) Urban human behavior analysis. With the popularization of smart devices, people can generate massive location-embedded information every day, such as, location-tagged text, image, video, check-ins, GPS trajectories. The first question in this domain is estimating user similarity, and similar users can be recommended as friends. Li et al (2008) connects users with similar interests even when they may not have known each other previously, and community discovery, which employs the GPS trajectories collected from GPS equipped devices like phones.

6) Economy. Urban intelligence can benefit the urban economy. The human mobility and the statistics of POIs can reflect the economy of the city. For example, the average price of a dinner in the restaurants can indicate the income level and the power of consumption. In (Karamshuk et al, 2013), the authors study the problem of optimal retail store placement in the context of location-based social networks. They collected human mobility data from Foursquare and analyzed it to understand how the popularity of three retail store chains in New York is shaped in terms of number of check-ins. The result indicates that some POIs, like train station and airport, can imply the popularity of the location, also, the number of competitive stores is an indicator for the popularity.

7) Public safety. Public safety and security in the urban area is always attracting people's concerns. The availability of different data enable us to learn from history how to deal with public safety problems, e.g., traffic accident (Yuan et al, 2018), large event (Vahedian et al, 2019; Khezerlou et al, 2021, 2017; Vahedian et al, 2017), pandemic (Bao et al, 2020), etc., and we can use the data to detect and predict abnormal events. Pang et al (2011) detects the anomalous traffic pattern from the spatial-temporal data of vehicles. The authors partition a city into uniform grids and counted the number of vehicles arriving in a grid over a time period. The objective was to identify contiguous sets of cells and time intervals that have the largest statistically significant departure from expected behavior (i.e., the number of vehicles).

27.1.3 Representing urban systems as graphs

Various data structures and models can be employed to define the spatial settings of urban systems. For example, a simple model is a grid structure, where the urban area is partitioned into grid cells, with a set of attribute values of interest (e.g., average traffic speed, number of taxis, population, rainfall) associated with each cell. While such a model is simple to implement, it ignores many intrinsic and important relationships existing in urban data. For example, a grid structure may lose the information of road connectivity in the underlying traffic system of the city. In many scenarios, instead, graph is an elegant choice to capture the intrinsic topological information and knowledge in the data. Many urban system components can be represented as graphs. Additional attributes may be associated with nodes and/or edges. In this section, we introduce graph representations of various urban system scenarios, which are summarized in Table 27.2. The application domains covered include **1)** Urban transportation and configuration planning, **2)** Urban environment monitoring, **3)** Urban energy supply and consumption, **4)** Urban event and anomaly detection, and **5)** Urban human behavior analysis.

1) Urban transportation and configuration planing. Modeling urban transportation system as a graph is widely used in solving real-world urban intelligence problems, e.g., traffic flow prediction (Xie et al, 2019b; Dai et al, 2020; Cui et al, 2019; Chen et al, 2019b; Song et al, 2020a; Zhang et al, 2020e; Zheng et al, 2020a; Diao et al, 2019; Guo et al, 2019b; Li et al, 2018e; Yu et al, 2018a; Zhang et al, 2018e), parking availability problem (Zhang et al, 2020h), etc. The graphs are usually built based on the real-world road network. To solve the problem of *traffic flow prediction*, in (Cui et al, 2019), the authors employ an undirected graph to predict the traffic state, the nodes are the traffic sensing locations, e.g., sensor stations, road segments, and the edges are the intersections or road segments connecting those traffic sensing locations. Xie et al (2019b); Dai et al (2020) model the urban traffic network as a directed graph with attributes to predict the traffic speed, the nodes are the road segments, and the edges are the intersections. Road segment width, length, and direction are the attributes of the nodes, and the type of intersection, and whether there are traffic lights, toll gates are the attributes of the edges. For *urban configuration,* Wu et al (2020c) incorporates a hierarchical GNN framework to learn Road Network Representation in different levels. The nodes in the hierarchical graph include road segments, structural regions, and functional zones, and the edges are intersections and hyperedges. There are some works about *predicting parking availability*. Zhang et al (2020h) models the parking lots and the surrounding POIs and population features as a graph to predict the parking availability for

Table 27.2: Graph representations in urban systems

Application domain	Nodes	Edges	Examples
Transportation & configuration planning	Road segments	Intersections	Traffic flow prediction (Xie et al, 2019b) (Dai et al, 2020) (Cui et al, 2019) (Chen et al, 2019b) (Song et al, 2020a) (Zhang et al, 2020e) (Zheng et al, 2020a) (Diao et al, 2019) (Guo et al, 2019b) (Li et al, 2018e) (Yu et al, 2018a) (Zhang et al, 2018e)
	Functional zones	Road connections	Learning road network representation (Wu et al, 2020c)
	POIs	Road connections	Parking availability prediction, POI recommendation (Zhang et al, 2020h) (Chang et al, 2020a)
Environment monitoring	Monitoring sensors	Proximity	Air quality inference (Wang et al, 2020h) (Li et al, 2017f)
Energy supply & consumption	Regulators	Pipelines	Gas pressure monitoring (Yi and Park, 2020)
Event & anomaly detection	Urban regions	Proximity	Traffic accident prediction (Zhou et al, 2020g) (Zhou et al, 2020h) (Yu et al, 2021b)
Human behavior analysis	Sessions, locations, objects	Event stream	User behavior modeling (Wang et al, 2020a)
	Urban regions	Proximity	Passenger demand prediction (Wang et al, 2019o) (Yi and Park, 2020) (Geng et al, 2019) (Bai et al, 2019a) (Xie et al, 2016)

the parking lots. The nodes are the parking lots, and the edges are determined by the connectivity between each two parking lots whose on-road distance is smaller than a threshold. Context features, e.g., POI distribution, population, etc., are the attributes of the nodes.

2) Urban environment monitoring system. People model the air quality monitoring system as a graph to forecast the air quality in the urban area(Wang et al, 2020h; Li et al, 2017f). For example, Wang et al (2020h) proposed the PM2.5-GNN to forecast the PM2.5 index in different locations. The nodes are locations determined by latitude, longitude, altitude, and there exists an edge between two nodes if the distance and difference of altitudes between them are less than threshholds respectively (e.g., distance $< 300\ km$ and difference of altitudes $< 1200\ m$). The node attributes include Planetary Boundary Layer (PBL) height, K index, wind speed, 2m temperature, relative humidity, precipitation, and surface pressure. Edge attributes include wind speed of source node, distance between source and sink, wind direction of source node, and direction from source to sink.

3) Urban energy supply and consumption. GNN is also employed in analyzing urban energy supply and consuming systems. For example, Yi and Park (2020) proposed a framework to predict the gas pressure in the gas supply network. The gas regulators are considered as the nodes, and the pipelines that connect every two regulators are the edges.

4) Urban event and anomaly detection. Urban event and anomaly detection is a hot topic in urban intelligence. People employ machine learning models to detect or predict the events occurring in the urban area, e.g., traffic accident prediction(Zhou et al, 2020g,h; Yu et al, 2021b). In (Zhou et al, 2020g), the authors proposed a framework to predict traffic accident in different regions of the city. The urban area is divided into subregions, i.e., grids, and if the traffic elements within two subregions have strong correlations, there is a connection.

5) Urban human behavior analysis. Studying human behavior in urban region can benefit people in many aspects, for example, demographic attribute prediction, personalized recommendation, passenger demand prediction, etc. Some works proposed GNN to study ***Human behavior modeling.*** Human behavior modeling is essential for many real-world applications such as demographic attribute prediction, content recommendation, and target advertising. In (Wang et al, 2020a), the authors model human behavior via a tripartite graph. The nodes include user's sessions, locations and items. There exists an edge between a session node and a location node if the user started the session at this location. Similarly, there exists an edge between a session node and an item node if the user interacted with this item within the session. Each edge possesses a time attribute indicating the temporal signal of the interaction between two nodes. Another application of analysing human behavior is ***passenger demand prediction.*** Understanding human behavior in daily transits can help improve the efficiency of urban transportation system. For example, predicting the passenger demand in the ride-sharing system can help the ride-sharing company and the drivers improve their operation efficiency. And in recent publications, many researchers employ graph neural networks to solve the problem of predicting human mobility (Wang et al, 2019o; Yi and Park, 2020; Geng et al, 2019; Bai et al, 2019a; Xie et al, 2016), and usually the nodes of the graph are subregions of the city, and the edges are usually defined based on spatial proximity.

27.1.4 Case Study 1: Graph Neural Networksin urban configuration and transportation

Urban intelligence can help urban planners design urban configuration, and benefit the urban transportation system from different perspectives, e.g., operation effi-

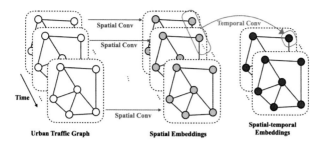

Figure 27.1: CNN-based STGNN

ciency, safety, environmental protection, etc. To enable urban intelligence in urban configuration and transportation planning, researchers developed practical machine learning approaches, including graph neural networks (GNN), to deal with real-world problems. In this section, we introduce some state-of-the-art (SOTA) designs of GNN targeting on solving the real-world urban configuration and transportation problems.

Urban traffic prediction. Predicting traffic status, e.g., speed, volume, is important in enabling urban intelligence. The traffic prediction problem is a typical time-series prediction problem:

Definition 27.1. Urban traffic prediction problem. *Given* historical traffic observations and context features of the road network, *predicting* the traffic status (e.g., speed, flow, etc.) in future time slots over the road network.

To address the traffic prediction problem, Spatial-temporal Graph Neural Networks (STGNN) are usually employed. The road segments are the nodes, and the traffic status is the attributes of the nodes. The traffic status in different time slots are corresponding to the temporal dynamics of the graph. Usually, graph convolution operation is used to capture the spatial dependencies among the nodes, and a 1D-convolution operation is then employed to capture the temporal dependencies among different time slots. The framework of CNN-based STGNN is illustrated in Fig.27.1. The spatial-temporal embeddings can be used to predict the traffic status.

Another design of STGNN is based on Recurrent Neural Networks (RNN), which can also predict traffic status in Spatial-temporal graphs. Most RNN-based approaches capture spatial-temporal dependencies by filtering inputs and hidden states passed to a recurrent unit using graph convolution operations. The basic RNN can be formulated in Eq. (27.1).

$$H^{(t)} = \sigma(WX^{(t)} + UH^{(t-1)} + \mathbf{b}), \qquad (27.1)$$

where $X^{(t)}$ is the node feature matrix at time step t. H is the hidden state. W, U, and $\mathbf{b}$ are the network parameters. Then, the STGNN based on RNN can be formulated as Eq. (27.2):

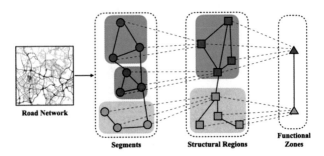

Figure 27.2: Hierarchical road network graph

$$H^{(t)} = \sigma(Gconv(X^{(t)}, A; W) + Gconv(H^{(t-1)}, A; U) + \mathbf{b}), \qquad (27.2)$$

where $Gconv(\cdot)$ is the graph convolution operation, and A is the graph adjacency matrix. Both designs of STGNN can be employed to predict the node attributes, i.e., traffic status, given the spatial-temporal graph of traffic.

Urban configuration. An urban road network is a vital component in urban configuration. How to represent it is essential for many analyses and researches related to real-world applications. As a real-world road network is a complex system with hierarchical structures, long-range dependency among units, and functional roles, it is challenging to design effective representation learning methods. The road network representation learning problem can be defined like this:

Definition 27.2. Road network representation learning problem. *Given* a road network, the *target* is to construct the corresponding graphs that can represent the structure and topological information of the road network.

Benefit from the topology of graph, we can represent road network with hierarchical graphs. In (Wu et al, 2020c), the authors propose to represent urban road networks with a hierarchical graph with three levels, and the node in each level corresponds to road segments, structural region, and functional zone, respectively, as illustrated in Fig.27.2. The structural region is the aggregation of some connected road segments, which serves as some specific traffic roles, e.g., intersection, overpass. And functional zone is the aggregation of structural regions, which can represent some functional facilities in the city, e.g., transportation hub, shopping area. To learn the hierarchical graph representation, the road segments are first represented by contextual embedding, e.g., road type, lane number, segment length, etc. Then, graph clustering and network reconstruction techniques are employed to form the structural region graph. And vehicle trajectory data is employed to capture the functional zones over structural regions.

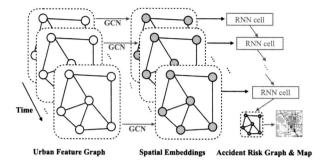

Figure 27.3: Example GNN framework for traffic accident prediction.

27.1.5 Case Study 2: Graph Neural Networks in urban anomaly and event detection

Public safety and security in the urban area always attracts people's concerns. The availability of different data enables us to learn from history how to deal with public safety problems, e.g., traffic accidents, crime, large events, pandemic, etc., and we can use the data to detect and predict abnormal events.

Traffic accident prediction. Traffic accident prediction is of great significance to improve the safety of the road network. Although "accident" is a word related to "randomness", there exist a significant correlation between the occurrence of traffic accidents and the surrounding environmental features, e.g., traffic flow, road network structure, weather, etc. Thus, machine learning approaches, like GNN, can be employed to predict or forecast traffic accidents over the city, which can help enable urban intelligence.

The problem of traffic accident prediction is as follows:

Definition 27.3. Traffic accident prediction problem. *Given* the road network data and the historical environmental features, the *target* is to predict the traffic accident risk over the city in the future.

The environmental features include the traffic conditions, surrounding POIs, etc. In recent publications (Zhou et al, 2020g,h; Yu et al, 2021b), GNN is employed to solve this problem.

The graphs in solving traffic accident problem are usually constructed based on dividing the urban area into grids, and each grid is considered as a node. If the traffic conditions between two nodes have a strong correlation, there is an edge between them. The context environmental features are the attributes with each grid. After the graphs are constructed in different historical time slots, graph convolutional neural networks (GCNs) are usually used to extract the hidden embedding in each time slot. Then, methods dealing with time-series inputs can be employed to capture the temporal dependencies, e.g., RNN-based neural networks. Finally, the spatial-temporal information is used to predict traffic accident risk over the city. Overall,

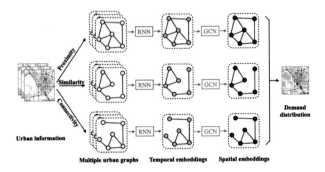

Figure 27.4: Example STGNN framework for passenger demand prediction.

the solution framework can be considered as an STGNN as illustrate in Fig. 27.3. For more details, please refer to (Zhou et al, 2020g,h; Yu et al, 2021b).

27.1.6 Case Study 3: Graph Neural Networks in urban human behavior inference

Human behavior analysis plays an important role in enabling urban intelligence, for example, studying the behavior of drivers can help improve the efficiency of urban transportation system, analysing passenger behaviors can help improve the operation efficiency of the drivers in taxi or ride-hailing services, and understanding user behavior pattern can help improve personal recommendation of commercial items, which will benefit the urban economy. In this section, we demonstrate how GNN works in analyzing urban human behaviors via two real-world applications, i.e., passenger demand prediction and user behavior modeling.

Passenger demand prediction. Passenger demand prediction is mostly conducted at the region-level, i.e., the urban area is divided into small grids. The problem can be defined as follows:

Definition 27.4. Passenger demand prediction problem. *Given* the historical demands and context features distributions, the task is to *predict* the passenger demand in each region.

Different from most traffic graphs which construct the graphs with road segments as nodes, here in passenger demand prediction problem, people usually construct the graph with grids as the nodes. The edges, i.e., the correlations between each pair of nodes, are determined by spatial proximity, similarity of contextual environment, or road network connectivity for distant grids.

Spatial-temporal Graph Neural Networks (STGNN) are the most popular GNN models employed in predicting passenger demand. In (Geng et al, 2019), the authors propose the spatiotemporal multi-graph convolution network (ST-MGCN) to predict the passenger demand in the ride-hailing service. The overall framework can be illustrated as in Fig.27.4. First, multiple graphs are constructed based on different aspects of relationships between each two grids, i.e., proximity, functional similarity, and transportation connectivity. Then, a RNN is used to aggregate observations in different times considering the global contextual information. After that, GCN is used to model the non-Euclidean correlations among regions. Finally, the aggregated embeddings are used to predict the passenger demand over the city.

User behavior modeling. Modeling human behavior is important for many real-world applications, e.g., demographic attribute prediction, content recommendation, and target advertising, etc. Studying human behavior in the urban scenario can benefit urban intelligence in many aspects, e.g., economy, transportation, etc. Here, we introduce an example of modeling spatial-temporal user behavior with tripartite graphs (Wang et al, 2020a).

Take the urban user online browsing behavior as an example, the spatial-temporal user behavior can be defined on a set of users U, a set of sessions S, a set of items V, and a set of locations L. Each user's behavior log can be represented by a set of session-location tuples, and each session contains multiple item-timestamp tuples. Then a user's spatial-temporal behavior can be captured via a tripartite graph as illustrated in Fig.27.5. The nodes of this tripartite graph include user's sessions S, locations L, and items V. The edges include session-item edges and session-location edges.

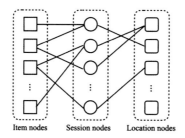

Item nodes   Session nodes   Location nodes

Figure 27.5: Spatial-temporal user behavior graph

To extract the user representation from each user's spatial-temporal behavior graph, GNN can be employed. The idea is to extract session embeddings from the items within each session, and RNN can be employed to aggregate the information of items. Then session embeddings are further aggregated into temporal embeddings of different time span, e.g., day, week. Also, the session embeddings and locations are composed to produce the spatial embeddings. Last, the spatial and temporal embeddings are fused into one embedding which can represent the user's behavior. For more details, we would like to refer you to (Wang et al, 2020a).

27.1.7 Future Directions

It is inspiring that GNNs have obtained significant achievements on urban intelligence. For future research, we envision that there exist several potential directions as following.

Interpretability of the GNNs model on urban intelligence. The applications of GNNs on urban intelligence are closely related to real-world problems. Besides improving the performance of the GNNs model, it is necessary to enhance the interpritability of the GNNs model. For example, in the application predicting traffic flow, it is important to identify hidden factors (e.g., structure of road network) that can affect the traffic flow. These hidden factors may also help urban planners better design road network to balance the traffic flow.

Recent advances in interpretable AI and machine learning research have led to the development of numerous intrinsic or post-hoc interpretable graph neural network models (Huang et al, 2020c). However, few of them are designed for GNNs on urban problems. Designing interpretable urban GNNs is non-trivial due to the unique properties of urban big data. For example, urban data are usually heterogeneous, i.e., the interpretation of learned relationships between the input features and target variables vary over space. For example, the risk factors for traffic accidents may shift when moving from a densely populated area to a non-residential area. Also, the interpretation model of GNN at nearby locations (e.g., neighboring nodes) share similarities due to the auto-correlation of spatial data (Pan et al, 2020b). These factors should be considered when designing interpretable urban GNNs.

New applications for GNNs on urban intelligence. As introduced above, GNNs have demonstrated their effectiveness and efficiency in many applications domains in urban intelligence, e.g., transportation, environment, energy, safety, human behavior. There exist potential applications of GNNs on urban scenario, such as, improving urban power (electricity) supply, contact tracing of patients of infectious diseases (e.g., COVID-19), and modeling responses to complex environmental and climate events (e.g., flood, Hurricane, etc).

Editor's Notes: Urban intelligence covers a wide range of macro-scale physical networks such as transportation networks and power grids. They are typical cases of spatial networks, which are networks whose nodes and edges are embedded in space probably under spatial constraints (e.g., planarity). So it is not a surprise that urban intelligence could largely benefit from deep learning techniques for spatial data and network data. Different from most of the application domains introduced in Chapters 19-27, there are usually well-designed computational models for many subareas in urban intelligence, so it is important to explore how deep graph learning techniques can contribute and compensate for the weakness of the existing strategies.

References

Abadi M, Barham P, Chen J, Chen Z, Davis A, Dean J, Devin M, Ghemawat S, Irving G, Isard M, et al (2016) Tensorflow: A system for large-scale machine learning. In: 12th {USENIX} Symposium on Operating Systems Design and Implementation ({OSDI} 16), pp 265–283

Abbe E (2017) Community detection and stochastic block models: recent developments. Journal of Machine Learning Research 18(1):6446–6531

Abbe E, Sandon C (2015) Community detection in general stochastic block models: Fundamental limits and efficient algorithms for recovery. In: IEEE 56th Annual Symposium on Foundations of Computer Science, pp 670–688

Abboud R, Ceylan ii, Grohe M, Lukasiewicz T (2020) The surprising power of graph neural networks with random node initialization. CoRR abs/2010.01179

Abdelaziz I, Dolby J, McCusker JP, Srinivas K (2020) Graph4Code: A machine interpretable knowledge graph for code. arXiv preprint arXiv:200209440

Abdollahpouri H, Adomavicius G, Burke R, Guy I, Jannach D, Kamishima T, Krasnodebski J, Pizzato L (2020) Multistakeholder recommendation: Survey and research directions. User Modeling and User-Adapted Interaction 30(1):127–158

Abid NJ, Dragan N, Collard ML, Maletic JI (2015) Using stereotypes in the automatic generation of natural language summaries for c++ methods. In: 2015 IEEE International Conference on Software Maintenance and Evolution (ICSME), IEEE, pp 561–565

Abney S (2007) Semisupervised learning for computational linguistics. CRC Press

Adamic LA, Adar E (2003) Friends and neighbors on the web. Social networks 25(3):211–230

Adams RP, Zemel RS (2011) Ranking via sinkhorn propagation. arXiv preprint arXiv:11061925

Aghamohammadi A, Izadi M, Heydarnoori A (2020) Generating summaries for methods of event-driven programs: An android case study. Journal of Systems and Software 170:110,800

Ahmad MA, Eckert C, Teredesai A (2018) Interpretable machine learning in health-care. In: Proceedings of the 2018 ACM international conference on bioinformatics, computational biology, and health informatics, pp 559–560

Ahmad WU, Chakraborty S, Ray B, Chang KW (2020) A transformer-based approach for source code summarization. arXiv preprint arXiv:200500653

Ahmed A, Shervashidze N, Narayanamurthy S, Josifovski V, Smola AJ (2013) Distributed large-scale natural graph factorization. In: Proceedings of the 22nd international conference on World Wide Web, pp 37–48

Aho AV, Lam MS, Sethi R, Ulman JD (2006) Compilers: principles, techniques and tools. Pearson Education

Ain QU, Butt WH, Anwar MW, Azam F, Maqbool B (2019) A systematic review on code clone detection. IEEE Access 7:86,121–86,144

Airoldi EM, Blei DM, Fienberg SE, Xing EP (2008) Mixed membership stochastic blockmodels. Journal of Machine Learning Research 9(Sep):1981–2014

Akoglu L, Tong H, Koutra D (2015) Graph based anomaly detection and description: a survey. Data mining and knowledge discovery 29(3):626–688

Al Hasan M, Zaki MJ (2011) A survey of link prediction in social networks. In: Social network data analytics, Springer, pp 243–275

Albert R, Barabási AL (2002) Statistical mechanics of complex networks. Reviews of modern physics 74(1):47

Albooyeh M, Goel R, Kazemi SM (2020) Out-of-sample representation learning for knowledge graphs. In: Empirical Methods in Natural Language Processing: Findings, pp 2657–2666

Ali H, Tran SN, Benetos E, Garcez ASd (2018) Speaker recognition with hybrid features from a deep belief network. Neural Computing and Applications 29(6):13–19

Allamanis M (2019) The adverse effects of code duplication in machine learning models of code. In: Proceedings of the 2019 ACM SIGPLAN International Symposium on New Ideas, New Paradigms, and Reflections on Programming and Software, pp 143–153

Allamanis M, Barr ET, Devanbu P, Sutton C (2018a) A survey of machine learning for big code and naturalness. ACM Computing Surveys (CSUR) 51(4):1–37

Allamanis M, Brockschmidt M, Khademi M (2018b) Learning to represent programs with graphs. In: International Conference on Learning Representations (ICLR)

Allamanis M, Barr ET, Ducousso S, Gao Z (2020) Typilus: neural type hints. In: Proceedings of the 41st ACM SIGPLAN Conference on Programming Language Design and Implementation, pp 91–105

Alon U, Brody S, Levy O, Yahav E (2019a) code2seq: Generating sequences from structured representations of code. International Conference on Learning Representations

Alon U, Zilberstein M, Levy O, Yahav E (2019b) code2vec: Learning distributed representations of code. Proceedings of the ACM on Programming Languages 3(POPL):1–29

Amidi A, Amidi S, Vlachakis D, et al (2018) EnzyNet: enzyme classification using 3d convolutional neural networks on spatial representation. PeerJ 6:e4750

Amizadeh S, Matusevych S, Weimer M (2018) Learning to solve circuit-sat: An unsupervised differentiable approach. In: International Conference on Learning Representations

Anand N, Huang PS (2018) Generative modeling for protein structures. In: Proceedings of the 32nd International Conference on Neural Information Processing Systems, pp 7505–7516

Arjovsky M, Chintala S, Bottou L (2017) Wasserstein generative adversarial networks. In: International Conference on Machine Learning, pp 214–223

Arora S (2020) A survey on graph neural networks for knowledge graph completion. arXiv preprint arXiv:200712374

Arvind V, Köbler J, Rattan G, Verbitsky O (2015) On the power of color refinement. In: International Symposium on Fundamentals of Computation Theory, pp 339–350

Arvind V, Fuhlbrück F, Köbler J, Verbitsky O (2019) On weisfeiler-leman invariance: subgraph counts and related graph properties. In: International Symposium on Fundamentals of Computation Theory, Springer, pp 111–125

Ashburner M, Ball CA, Blake JA, Botstein D, Butler H, Cherry JM, Davis AP, Dolinski K, Dwight SS, Eppig JT, et al (2000) Gene ontology: tool for the unification of biology. Nature genetics 25(1):25–29

Aynaz Taheri TBW Kevin Gimpel (2018) Learning graph representations with recurrent neural network autoencoders. In: KDD'18 Deep Learning Day

Azizian W, Lelarge M (2020) Characterizing the expressive power of invariant and equivariant graph neural networks. arXiv preprint arXiv:200615646

Babai L (2016) Graph isomorphism in quasipolynomial time. In: Proceedings of the Forty-Eighth Annual ACM Symposium on Theory of Computing, pp 684–697

Babai L, Kucera L (1979) Canonical labelling of graphs in linear average time. In: Foundations of Computer Science, 1979., 20th Annual Symposium on, IEEE, pp 39–46

Bach S, Binder A, Montavon G, Klauschen F, Müller KR, Samek W (2015) On pixel-wise explanations for non-linear classifier decisions by layer-wise relevance propagation. PloS one 10(7):e0130,140

Badihi S, Heydarnoori A (2017) Crowdsummarizer: Automated generation of code summaries for java programs through crowdsourcing. IEEE Software 34(2):71–80

Bahdanau D, Cho K, Bengio Y (2015) Neural machine translation by jointly learning to align and translate. In: 3rd International Conference on Learning Representations

Bai L, Yao L, Kanhere SS, Wang X, Liu W, Yang Z (2019a) Spatio-temporal graph convolutional and recurrent networks for citywide passenger demand prediction. In: Proceedings of the 28th ACM International Conference on Information and Knowledge Management, Association for Computing Machinery, CIKM '19, p 2293–2296, DOI 10.1145/3357384.3358097

Bai X, Zhu L, Liang C, Li J, Nie X, Chang X (2020a) Multi-view feature selection via nonnegative structured graph learning. Neurocomputing 387:110–122

Bai Y, Ding H, Sun Y, Wang W (2018) Convolutional set matching for graph similarity. In: NeurIPS 2018 Relational Representation Learning Workshop

Bai Y, Ding H, Bian S, Chen T, Sun Y, Wang W (2019b) Simgnn: A neural network approach to fast graph similarity computation. In: Proceedings of the Twelfth ACM International Conference on Web Search and Data Mining, pp 384–392

Bai Y, Ding H, Qiao Y, Marinovic A, Gu K, Chen T, Sun Y, Wang W (2019c) Unsupervised inductive graph-level representation learning via graph-graph proximity. arXiv preprint arXiv:190401098

Bai Y, Ding H, Gu K, Sun Y, Wang W (2020b) Learning-based efficient graph similarity computation via multi-scale convolutional set matching. In: Proceedings of the AAAI Conference on Artificial Intelligence, pp 3219–3226

Bai Y, Xu D, Wang A, Gu K, Wu X, Marinovic A, Ro C, Sun Y, Wang W (2020c) Fast detection of maximum common subgraph via deep q-learning. arXiv preprint arXiv:200203129

Bajaj M, Wang L, Sigal L (2019) G3raphground: Graph-based language grounding. In: Proceedings of the IEEE/CVF International Conference on Computer Vision, pp 4281–4290

Baker B, Gupta O, Naik N, Raskar R (2016) Designing neural network architectures using reinforcement learning. arXiv preprint arXiv:161102167

Baker CF, Ellsworth M (2017) Graph methods for multilingual framenets. In: Proceedings of TextGraphs-11: the Workshop on Graph-based Methods for Natural Language Processing, pp 45–50

Balcilar M, Renton G, Héroux P, Gaüzère B, Adam S, Honeine P (2021) Analyzing the expressive power of graph neural networks in a spectral perspective. In: International Conference on Learning Representations

Baldassarre F, Azizpour H (2019) Explainability techniques for graph convolutional networks. arXiv preprint arXiv:190513686

Balinsky H, Balinsky A, Simske S (2011) Document sentences as a small world. In: 2011 IEEE International Conference on Systems, Man, and Cybernetics, IEEE, pp 2583–2588

Banarescu L, Bonial C, Cai S, Georgescu M, Griffitt K, Hermjakob U, Knight K, Koehn P, Palmer M, Schneider N (2013) Abstract meaning representation for sembanking. In: Proceedings of the 7th linguistic annotation workshop and interoperability with discourse, pp 178–186

Bao H, Zhou X, Zhang Y, Li Y, Xie Y (2020) Covid-gan: Estimating human mobility responses to covid-19 pandemic through spatio-temporal conditional generative adversarial networks. In: Proceedings of the 28th International Conference on Advances in Geographic Information Systems, pp 273–282

Barabási AL (2013) Network science. Philosophical Transactions of the Royal Society A: Mathematical, Physical and Engineering Sciences 371(1987):20120,375

Barabási AL, Albert R (1999) Emergence of scaling in random networks. science 286(5439):509–512

Barabasi AL, Oltvai ZN (2004) Network biology: Understanding the cell's functional organization. Nature Reviews Genetics 5(2):101–113

Barber D (2004) Probabilistic modelling and reasoning: The junction tree algorithm. Course Notes

Barceló P, Kostylev EV, Monet M, Pérez J, Reutter J, Silva JP (2019) The logical expressiveness of graph neural networks. In: International Conference on Learning Representations

Bastian FB, Roux J, Niknejad A, Comte A, Fonseca Costa SS, De Farias TM, Moretti S, Parmentier G, De Laval VR, Rosikiewicz M, et al (2021) The bgee suite: integrated curated expression atlas and comparative transcriptomics in animals. Nucleic Acids Research 49(D1):D831–D847

Bastings J, Titov I, Aziz W, Marcheggiani D, Sima'an K (2017) Graph convolutional encoders for syntax-aware neural machine translation. arXiv preprint arXiv:170404675

Batagelj V, Zaversnik M (2003) An o(m) algorithm for cores decomposition of networks. arXiv preprint cs/0310049

Bateman A, Martin MJ, Orchard S, Magrane M, Agivetova R, Ahmad S, Alpi E, Bowler-Barnett EH, Britto R, Bursteinas B, et al (2020) Uniprot: the universal protein knowledgebase in 2021. Nucleic Acids Research

Battaglia P, Pascanu R, Lai M, Rezende DJ, kavukcuoglu K (2016) Interaction networks for learning about objects, relations and physics. In: Proceedings of the 30th International Conference on Neural Information Processing Systems, pp 4509–4517

Battaglia PW, Hamrick JB, Bapst V, Sanchez-Gonzalez A, Zambaldi V, Malinowski M, Tacchetti A, Raposo D, Santoro A, Faulkner R, et al (2018) Relational inductive biases, deep learning, and graph networks. arXiv preprint arXiv:180601261

Beaini D, Passaro S, Létourneau V, Hamilton WL, Corso G, Liò P (2020) Directional graph networks. CoRR abs/2010.02863

Beck D, Haffari G, Cohn T (2018) Graph-to-sequence learning using gated graph neural networks. arXiv preprint arXiv:180609835

Belghazi MI, Baratin A, Rajeswar S, Ozair S, Bengio Y, Hjelm RD, Courville AC (2018) Mutual information neural estimation. In: Dy JG, Krause A (eds) Proceedings of the 35th International Conference on Machine Learning, ICML 2018, Stockholmsmässan, Stockholm, Sweden, July 10-15, 2018, PMLR, Proceedings of Machine Learning Research, vol 80, pp 530–539

Belkin M, Niyogi P (2002) Laplacian eigenmaps and spectral techniques for embedding and clustering. In: Advances in neural information processing systems, pp 585–591

Bengio Y (2008) Neural net language models. Scholarpedia 3(1):3881

Bengio Y, Senécal JS (2008) Adaptive importance sampling to accelerate training of a neural probabilistic language model. IEEE Transactions on Neural Networks 19(4):713–722

Bennett J, Lanning S, et al (2007) The netflix prize. In: Proceedings of KDD cup and workshop, New York, vol 2007, p 35 van den Berg R, Kipf TN, Welling M (2018) Graph convolutional matrix completion. KDD18 Deep Learning Day

Berg Rvd, Kipf TN, Welling M (2017) Graph convolutional matrix completion. arXiv preprint arXiv:170602263

Berger P, Hannak G, Matz G (2020) Efficient graph learning from noisy and incomplete data. IEEE Trans Signal Inf Process over Networks 6:105–119

Berggård T, Linse S, James P (2007) Methods for the detection and analysis of protein–protein interactions. PROTEOMICS 7(16):2833–2842

Berline N, Getzler E, Vergne M (2003) Heat kernels and Dirac operators. Springer Science & Business Media

Bian R, Koh YS, Dobbie G, Divoli A (2019) Network embedding and change modeling in dynamic heterogeneous networks. In: Proceedings of the 42nd International ACM SIGIR Conference on Research and Development in Information Retrieval, pp 861–864

Bianchi FM, Grattarola D, Alippi C (2020) Spectral clustering with graph neural networks for graph pooling. In: International Conference on Machine Learning, ACM, pp 2729–2738

Bielik P, Raychev V, Vechev M (2017) Learning a static analyzer from data. In: International Conference on Computer Aided Verification, Springer, pp 233–253

Biggs N, Lloyd EK, Wilson RJ (1986) Graph Theory, 1736-1936. Oxford University Press

Bingel J, Søgaard A (2017) Identifying beneficial task relations for multi-task learning in deep neural networks. In: Proceedings of the 15th Conference of the European Chapter of the Association for Computational Linguistics: Volume 2, Short Papers, pp 164–169

Bishop CM (2006) Pattern recognition and machine learning. springer

Bizer C, Lehmann J, Kobilarov G, Auer S, Becker C, Cyganiak R, Hellmann S (2009) Dbpedia-a crystallization point for the web of data. Journal of web semantics 7(3):154–165

Blitzer J, McDonald R, Pereira F (2006) Domain adaptation with structural correspondence learning. In: Proceedings of the 2006 conference on empirical methods in natural language processing, pp 120–128

Bodenreider O (2004) The unified medical language system (umls): integrating biomedical terminology. Nucleic acids research 32(suppl_1):D267–D270

Bojchevski A, Günnemann S (2019) Adversarial attacks on node embeddings via graph poisoning. In: International Conference on Machine Learning, PMLR, pp 695–704

Bojchevski A, Günnemann S (2019) Certifiable robustness to graph perturbations. In: Wallach H, Larochelle H, Beygelzimer A, d'Alché-Buc F, Fox E, Garnett R (eds) Advances in Neural Information Processing Systems, Curran Associates, Inc., vol 32

Bojchevski A, Matkovic Y, Günnemann S (2017) Robust spectral clustering for noisy data: Modeling sparse corruptions improves latent embeddings. In: Proceedings of the 23rd ACM SIGKDD International Conference on Knowledge Discovery and Data Mining, pp 737–746

Bojchevski A, Shchur O, Zügner D, Günnemann S (2018) Netgan: Generating graphs via random walks. arXiv preprint arXiv:180300816

Bojchevski A, Klicpera J, Günnemann S (2020a) Efficient robustness certificates for discrete data: Sparsity-aware randomized smoothing for graphs, images and more. In: International Conference on Machine Learning, PMLR, pp 1003–1013

Bojchevski A, Klicpera J, Perozzi B, Kapoor A, Blais M, Rózemberczki B, Lukasik M, Günnemann S (2020b) Scaling graph neural networks with approximate pagerank. In: Proceedings of the 26th ACM SIGKDD International Conference on Knowledge Discovery & Data Mining, pp 2464–2473

Bollacker K, Tufts P, Pierce T, Cook R (2007) A platform for scalable, collaborative, structured information integration. In: Intl. Workshop on Information Integration on the Web (IIWeb'07), pp 22–27

Bollobás B (2013) Modern graph theory, vol 184. Springer Science & Business Media

Bollobás B, Béla B (2001) Random graphs. 73, Cambridge university press

Bollobás B, Janson S, Riordan O (2007) The phase transition in inhomogeneous random graphs. Random Structures & Algorithms 31(1):3–122

Bordes A, Usunier N, Garcia-Duran A, Weston J, Yakhnenko O (2013) Translating embeddings for modeling multi-relational data. In: Neural Information Processing Systems, pp 1–9

Bordes A, Glorot X, Weston J, Bengio Y (2014) A semantic matching energy function for learning with multi-relational data. Machine Learning 94(2):233–259

Borgwardt KM, Ong CS, Schönauer S, Vishwanathan SVN, Smola AJ, Kriegel HP (2005) Protein function prediction via graph kernels. Bioinformatics 21(Supplement 1):i47–i56

Borgwardt KM, Ghisu ME, Llinares-López F, O'Bray L, Rieck B (2020) Graph kernels: State-of-the-art and future challenges. Found Trends Mach Learn 13(5-6)

Bose A, Hamilton W (2019) Compositional fairness constraints for graph embeddings. In: International Conference on Machine Learning, PMLR, pp 715–724

Bottou L (1998) Online learning and stochastic approximations. On-line learning in neural networks 17(9):142

Bourgain J (1985) On lipschitz embedding of finite metric spaces in hilbert space. Israel Journal of Mathematics 52(1-2):46–52

Bourigault S, Lagnier C, Lamprier S, Denoyer L, Gallinari P (2014) Learning social network embeddings for predicting information diffusion. In: Proceedings of the 7th ACM international conference on Web search and data mining, pp 393–402

Bouritsas G, Frasca F, Zafeiriou S, Bronstein MM (2020) Improving graph neural network expressivity via subgraph isomorphism counting. CoRR abs/2006.09252, 2006.09252

Boyd S, Boyd SP, Vandenberghe L (2004) Convex optimization. Cambridge university press

Braschi B, Denny P, Gray K, Jones T, Seal R, Tweedie S, Yates B, Bruford E (2017) Genenames. org: the hgnc and vgnc resources in

Brauckmann A, Goens A, Ertel S, Castrillon J (2020) Compiler-based graph representations for deep learning models of code. In: Proceedings of the 29th International Conference on Compiler Construction, pp 201–211

Braude EJ, Bernstein ME (2016) Software engineering: modern approaches. Waveland Press

Brin S, Page L (1998) The anatomy of a large-scale hypertextual web search engine. Computer networks and ISDN systems 30(1-7):107–117

Brin S, Page L (2012) Reprint of: The anatomy of a large-scale hypertextual web search engine. Computer networks 56(18):3825–3833

Brockschmidt M (2020) GNN-FiLM: Graph neural networks with feature-wise linear modulation. In: III HD, Singh A (eds) Proceedings of the 37th International Conference on Machine Learning, PMLR, Virtual, Proceedings of Machine Learning Research, vol 119, pp 1144–1152

Bronstein MM, Bruna J, LeCun Y, Szlam A, Vandergheynst P (2017) Geometric deep learning: going beyond euclidean data. IEEE Signal Processing Magazine 34(4):18–42

Browne F, Wang H, Zheng H, et al (2007) Supervised statistical and machine learning approaches to inferring pairwise and module-based protein interaction networks. In: 2007 IEEE 7th International Symposium on BioInformatics and BioEngineering, pp 1365–1369, DOI 10.1109/BIBE.2007.4375748

Bruna J, Zaremba W, Szlam A, LeCun Y (2014) Spectral networks and deep locally connected networks on graphs. In: 2nd International Conference on Learning Representations, ICLR 2014

Bui TN, Chaudhuri S, Leighton FT, Sipser M (1987) Graph bisection algorithms with good average case behavior. Combinatorica 7(2):171–191

Bunke H (1997) On a relation between graph edit distance and maximum common subgraph. Pattern Recognition Letters 18(8):689–694

Burt RS (2004) Structural holes and good ideas. American journal of sociology 110(2):349–399

Byron O, Vestergaard B (2015) Protein–protein interactions: a supra-structural phenomenon demanding trans-disciplinary biophysical approaches. Current Opinion in Structural Biology 35:76 – 86, catalysis and regulation • Protein-protein interactions

Cai D, Lam W (2020) Graph transformer for graph-to-sequence learning. In: Proceedings of the AAAI Conference on Artificial Intelligence, vol 34, pp 7464–7471

Cai H, Chen T, Zhang W, Yu Y, Wang J (2018a) Efficient architecture search by network transformation. In: Proceedings of the AAAI Conference on Artificial Intelligence, vol 32

Cai H, Zheng VW, Chang KCC (2018b) A comprehensive survey of graph embedding: Problems, techniques, and applications. IEEE Transactions on Knowledge and Data Engineering 30(9):1616–1637

Cai H, Gan C, Wang T, Zhang Z, Han S (2020a) Once for all: Train one network and specialize it for efficient deployment. In: ICLR

Cai H, Gan C, Zhu L, Han S (2020b) Tinytl: Reduce memory, not parameters for efficient on-device learning. Advances in Neural Information Processing Systems 33

Cai JY, Fürer M, Immerman N (1992) An optimal lower bound on the number of variables for graph identification. Combinatorica 12(4):389–410

Cai L, Ji S (2020) A multi-scale approach for graph link prediction. In: Proceedings of the AAAI Conference on Artificial Intelligence, vol 34, pp 3308–3315

Cai L, Yan B, Mai G, Janowicz K, Zhu R (2019) Transgcn: Coupling transformation assumptions with graph convolutional networks for link prediction. In: Proceedings of the 10th International Conference on Knowledge Capture, pp 131–138

Cai L, Li J, Wang J, Ji S (2020c) Line graph neural networks for link prediction. arXiv preprint arXiv:201010046

Cai T, Luo S, Xu K, He D, Liu Ty, Wang L (2020d) Graphnorm: A principled approach to accelerating graph neural network training. arXiv preprint arXiv:200903294

Cai X, Han J, Yang L (2018c) Generative adversarial network based heterogeneous bibliographic network representation for personalized citation recommendation. In: Proceedings of the AAAI Conference on Artificial Intelligence, vol 32

Cai Z, Wen L, Lei Z, Vasconcelos N, Li SZ (2014) Robust deformable and occluded object tracking with dynamic graph. IEEE Transactions on Image Processing 23(12):5497–5509

Cairong Z, Xinran Z, Cheng Z, Li Z (2016) A novel dbn feature fusion model for cross-corpus speech emotion recognition. Journal of Electrical and Computer Engineering 2016

Cangea C, Velickovic P, Jovanovic N, Kipf T, Liò P (2018) Towards sparse hierarchical graph classifiers. CoRR abs/1811.01287

Cao S, Lu W, Xu Q (2015) Grarep: Learning graph representations with global structural information. In: Proceedings of the 24th ACM international on conference on information and knowledge management, pp 891–900

Cao Y, Peng H, Philip SY (2020) Multi-information source hin for medical concept embedding. In: Pacific-Asia Conference on Knowledge Discovery and Data Mining, Springer, pp 396–408

Cao Z, Simon T, Wei SE, Sheikh Y (2017) Realtime multi-person 2d pose estimation using part affinity fields. In: Proceedings of the IEEE conference on computer vision and pattern recognition, pp 7291–7299

Cao Z, Hidalgo G, Simon T, Wei SE, Sheikh Y (2019) Openpose: realtime multi-person 2d pose estimation using part affinity fields. IEEE transactions on pattern analysis and machine intelligence 43(1):172–186

Cappart Q, Chételat D, Khalil E, Lodi A, Morris C, Veličković P (2021) Combinatorial optimization and reasoning with graph neural networks. CoRR abs/2102.09544

Carlini N, Wagner D (2017) Towards Evaluating the Robustness of Neural Networks. IEEE Symposium on Security and Privacy pp 39–57, DOI 10.1109/SP. 2017.49

Caron M, Bojanowski P, Joulin A, Douze M (2018) Deep clustering for unsupervised learning of visual features. In: Proceedings of the European Conference on Computer Vision (ECCV), pp 132–149

Carreira J, Zisserman A (2017) Quo vadis, action recognition? a new model and the kinetics dataset. In: proceedings of the IEEE Conference on Computer Vision and Pattern Recognition, pp 6299–6308

Cartwright D, Harary F (1956) Structural balance: a generalization of heider's theory. Psychological review 63(5):277

Cen Y, Zou X, Zhang J, Yang H, Zhou J, Tang J (2019) Representation learning for attributed multiplex heterogeneous network. In: Proceedings of the 25th ACM SIGKDD International Conference on Knowledge Discovery & Data Mining, pp 1358–1368

Cetoli A, Bragaglia S, O'Harney A, Sloan M (2017) Graph convolutional networks for named entity recognition. In: Proceedings of the 16th International Workshop on Treebanks and Linguistic Theories, pp 37–45

Chakrabarti D, Faloutsos C (2006) Graph mining: Laws, generators, and algorithms. ACM computing surveys (CSUR) 38(1)

Chami I, Ying Z, Ré C, Leskovec J (2019) Hyperbolic graph convolutional neural networks. In: Advances in neural information processing systems, pp 4868–4879

Chami I, Abu-El-Haija S, Perozzi B, Ré C, Murphy K (2020) Machine learning on graphs: A model and comprehensive taxonomy. CoRR abs/2005.03675

Chang B, Jang G, Kim S, Kang J (2020a) Learning graph-based geographical latent representation for point-of-interest recommendation. In: Proceedings of the 29th ACM International Conference on Information & Knowledge Management, pp 135–144

Chang H, Rong Y, Xu T, Huang W, Zhang H, Cui P, Zhu W, Huang J (2020b) A Restricted Black-Box Adversarial Framework Towards Attacking Graph Embedding Models. In: AAAI Conference on Artificial Intelligence, vol 34, pp 3389–3396, DOI 10.1609/aaai.v34i04.5741

Chang J, Scherer S (2017) Learning representations of emotional speech with deep convolutional generative adversarial networks. In: 2017 IEEE International Conference on Acoustics, Speech and Signal Processing (ICASSP), IEEE, pp 2746–2750

Chang S (2018) Scaling knowledge access and retrieval at airbnb. Airbnb Engineering and Data Science

Chang S, Han W, Tang J, Qi GJ, Aggarwal CC, Huang TS (2015) Heterogeneous network embedding via deep architectures. In: Proceedings of the 21th ACM SIGKDD international conference on knowledge discovery and data mining, pp 119–128

Chao YW, Vijayanarasimhan S, Seybold B, Ross DA, Deng J, Sukthankar R (2018) Rethinking the faster r-cnn architecture for temporal action localization. In: Proceedings of the IEEE Conference on Computer Vision and Pattern Recognition, pp 1130–1139

Chen B, Sun L, Han X (2018a) Sequence-to-action: End-to-end semantic graph generation for semantic parsing. arXiv preprint arXiv:180900773

Chen B, Barzilay R, Jaakkola T (2019a) Path-augmented graph transformer network. ICML 2019 Workshop on Learning and Reasoning with Graph-Structured Data

Chen B, Zhang J, Zhang X, Tang X, Cai L, Chen H, Li C, Zhang P, Tang J (2020a) Coad: Contrastive pre-training with adversarial fine-tuning for zero-shot expert linking. arXiv preprint arXiv:201211336

Chen C, Li K, Teo SG, Zou X, Wang K, Wang J, Zeng Z (2019b) Gated residual recurrent graph neural networks for traffic prediction. In: Proceedings of the AAAI Conference on Artificial Intelligence, vol 33, pp 485–492

Chen D, Lin Y, Li L, Li XR, Zhou J, Sun X, et al (2020b) Distance-wise graph contrastive learning. arXiv preprint arXiv:201207437

Chen D, Lin Y, Li W, Li P, Zhou J, Sun X (2020c) Measuring and relieving the over-smoothing problem for graph neural networks from the topological view. In: The Thirty-Fourth AAAI Conference on Artificial Intelligence, AAAI 2020, The Thirty-Second Innovative Applications of Artificial Intelligence Conference, IAAI 2020, The Tenth AAAI Symposium on Educational Advances in Artificial Intelligence, EAAI 2020, New York, NY, USA, February 7-12, 2020, AAAI Press, pp 3438–3445

Chen H, Yin H, Wang W, Wang H, Nguyen QVH, Li X (2018b) Pme: projected metric embedding on heterogeneous networks for link prediction. In: Proceedings of the 24th ACM SIGKDD International Conference on Knowledge Discovery & Data Mining, pp 1177–1186

Chen H, Xu Y, Huang F, Deng Z, Huang W, Wang S, He P, Li Z (2020d) Label-aware graph convolutional networks. In: The 29th ACM International Conference on Information and Knowledge Management, pp 1977–1980

Chen IY, Agrawal M, Horng S, Sontag D (2020e) Robustly extracting medical knowledge from ehrs: A case study of learning a health knowledge graph. In: Pac Symp Biocomput, World Scientific, pp 19–30

Chen J, Ma T, Xiao C (2018c) Fastgcn: Fast learning with graph convolutional networks via importance sampling. In: International Conference on Learning Representations

Chen J, Zhu J, Song L (2018d) Stochastic training of graph convolutional networks with variance reduction. In: International Conference on Machine Learning, PMLR, pp 942–950

Chen J, Chen Y, Zheng H, Shen S, Yu S, Zhang D, Xuan Q (2020f) MGA: Momentum Gradient Attack on Network. IEEE Transactions on Computational Social Systems pp 1–10, DOI 10.1109/TCSS.2020.3031058

Chen J, Lei B, Song Q, Ying H, Chen DZ, Wu J (2020g) A hierarchical graph network for 3d object detection on point clouds. In: Proceedings of the IEEE/CVF Conference on Computer Vision and Pattern Recognition, pp 392–401

Chen J, Lin X, Shi Z, Liu Y (2020h) Link Prediction Adversarial Attack Via Iterative Gradient Attack. IEEE Transactions on Computational Social Systems 7(4):1081–1094, DOI 10.1109/TCSS.2020.3004059

Chen J, Lin X, Xiong H, Wu Y, Zheng H, Xuan Q (2020i) Smoothing Adversarial Training for GNN. IEEE Transactions on Computational Social Systems pp 1–12, DOI 10.1109/TCSS.2020.3042628

Chen J, Xu H, Wang J, Xuan Q, Zhang X (2020j) Adversarial Detection on Graph Structured Data. In: Workshop on Privacy-Preserving Machine Learning in Practice

Chen L, Tan B, Long S, Yu K (2018e) Structured dialogue policy with graph neural networks. In: Proceedings of the 27th International Conference on Computational Linguistics, pp 1257–1268

Chen L, Chen Z, Bruna J (2020k) On graph neural networks versus graph-augmented mlps. arXiv preprint arXiv:201015116

Chen M, Wei Z, Huang Z, Ding B, Li Y (2020l) Simple and deep graph convolutional networks. In: International Conference on Machine Learning, PMLR, pp 1725–1735

Chen Q, Zhou M (2018) A neural framework for retrieval and summarization of source code. In: 2018 33rd IEEE/ACM International Conference on Automated Software Engineering (ASE), IEEE, pp 826–831

Chen T, Sun Y (2017) Task-guided and path-augmented heterogeneous network embedding for author identification. In: Proceedings of the Tenth ACM International Conference on Web Search and Data Mining, pp 295–304

Chen T, Li M, Li Y, Lin M, Wang N, Wang M, Xiao T, Xu B, Zhang C, Zhang Z (2015) Mxnet: A flexible and efficient machine learning library for heterogeneous distributed systems. arXiv preprint arXiv:151201274

Chen T, Bian S, Sun Y (2019c) Are powerful graph neural nets necessary? a dissection on graph classification. arXiv preprint arXiv:190504579

Chen X, Ma H, Wan J, Li B, Xia T (2017) Multi-view 3d object detection network for autonomous driving. In: Proceedings of the IEEE conference on Computer Vision and Pattern Recognition, pp 1907–1915

Chen XW, Liu M (2005) Prediction of protein–protein interactions using random decision forest framework. Bioinformatics 21(24):4394–4400

Chen Y, Rohrbach M, Yan Z, Shuicheng Y, Feng J, Kalantidis Y (2019d) Graph-based global reasoning networks. In: Proceedings of the IEEE/CVF Conference on Computer Vision and Pattern Recognition, pp 433–442

Chen Y, Wu L, Zaki M (2020m) Iterative deep graph learning for graph neural networks: Better and robust node embeddings. Advances in Neural Information Processing Systems 33

Chen Y, Wu L, Zaki MJ (2020n) Graphflow: Exploiting conversation flow with graph neural networks for conversational machine comprehension. In: Proceedings of the Twenty-Ninth International Joint Conference on Artificial Intelligence, pp 1230–1236

Chen Y, Wu L, Zaki MJ (2020o) Reinforcement learning based graph-to-sequence model for natural question generation. In: 8th International Conference on Learning Representations

Chen Y, Wu L, Zaki MJ (2020p) Toward subgraph guided knowledge graph question generation with graph neural networks. arXiv preprint arXiv:200406015

Chen YC, Bansal M (2018) Fast abstractive summarization with reinforce-selected sentence rewriting. arXiv preprint arXiv:180511080

Chen YW, Song Q, Hu X (2021) Techniques for automated machine learning. ACM SIGKDD Explorations Newsletter 22(2):35–50

Chen Z, Kommrusch SJ, Tufano M, Pouchet LN, Poshyvanyk D, Monperrus M (2019e) Sequencer: Sequence-to-sequence learning for end-to-end program repair. IEEE Transactions on Software Engineering pp 1–1, DOI 10.1109/TSE. 2019.2940179

Chen Z, Villar S, Chen L, Bruna J (2019f) On the equivalence between graph isomorphism testing and function approximation with gnns. In: Advances in Neural Information Processing Systems, pp 15,868–15,876

Chen Z, Chen L, Villar S, Bruna J (2020q) Can graph neural networks count substructures? vol 33

Chenxi Liu FSHAWHAYLFF Liang-Chieh Chen (2019) Auto-deeplab: Hierarchical neural architecture search for semantic image segmentation. arXiv preprint arXiv:190102985

Chiang WL, Liu X, Si S, Li Y, Bengio S, Hsieh CJ (2019) Cluster-gcn: An efficient algorithm for training deep and large graph convolutional networks. In: ACM SIGKDD International Conference on Knowledge Discovery and Data Mining (KDD), pp 257–266

Chibotaru V, Bichsel B, Raychev V, Vechev M (2019) Scalable taint specification inference with big code. In: Proceedings of the 40th ACM SIGPLAN Conference on Programming Language Design and Implementation, pp 760–774

Chidambaram M, Yang Y, Cer D, Yuan S, Sung YH, Strope B, Kurzweil R (2019) Learning cross-lingual sentence representations via a multi-task dual-encoder model. ACL 2019 p 250

Chien E, Peng J, Li P, Milenkovic O (2021) Adaptive universal generalized pagerank graph neural network. In: International Conference on Learning Representations

Chiu PH, Hripcsak G (2017) Ehr-based phenotyping: bulk learning and evaluation. Journal of biomedical informatics 70:35–51

Cho K, van Merriënboer B, Gulcehre C, Bahdanau D, Bougares F, Schwenk H, Bengio Y (2014a) Learning phrase representations using RNN encoder–decoder for statistical machine translation. In: Proceedings of the 2014 Conference on Empirical Methods in Natural Language Processing (EMNLP), Association for Computational Linguistics, Doha, Qatar, pp 1724–1734, DOI 10.3115/v1/D14-1179

Cho M, Lee J, Lee KM (2010) Reweighted random walks for graph matching. In: European conference on Computer vision, Springer, pp 492–505

Cho M, Sun J, Duchenne O, Ponce J (2014b) Finding matches in a haystack: A max-pooling strategy for graph matching in the presence of outliers. In: IEEE Conference on Computer Vision and Pattern Recognition, pp 2083–2090

Choi E, Xu Z, Li Y, Dusenberry M, Flores G, Xue E, Dai AM (2020) Learning the graphical structure of electronic health records with graph convolutional transformer. In: The Thirty-Fourth AAAI Conference on Artificial Intelligence, pp 606–613

Choromanski K, Likhosherstov V, Dohan D, Song X, Gane A, Sarlós T, Hawkins P, Davis J, Mohiuddin A, Kaiser L, Belanger D, Colwell L, Weller A (2021) Rethinking attention with performers. In: International Conference on Learning Representations

Chorowski J, Weiss RJ, Bengio S, van den Oord A (2019) Unsupervised speech representation learning using wavenet autoencoders. IEEE/ACM transactions on audio, speech, and language processing 27(12):2041–2053

Chung F (2007) The heat kernel as the pagerank of a graph. Proceedings of the National Academy of Sciences 104(50):19,735–19,740

Chung J, Gulcehre C, Cho K, Bengio Y (2014) Empirical evaluation of gated recurrent neural networks on sequence modeling. arXiv preprint arXiv:14123555

Cohen J, Rosenfeld E, Kolter Z (2019) Certified adversarial robustness via randomized smoothing. In: International Conference on Machine Learning, PMLR, pp 1310–1320

Cohen N, Shashua A (2016) Convolutional rectifier networks as generalized tensor decompositions. In: International Conference on Machine Learning, PMLR, pp 955–963

Collard ML, Decker MJ, Maletic JI (2011) Lightweight transformation and fact extraction with the srcml toolkit. In: Source Code Analysis and Manipulation (SCAM), 2011 11th IEEE International Working Conference on, IEEE, pp 173–184

Collobert R, Weston J, Bottou L, Karlen M, Kavukcuoglu K, Kuksa P (2011) Natural language processing (almost) from scratch. Journal of machine learning research 12(ARTICLE):2493–2537

Colson B, Marcotte P, Savard G (2007) An overview of bilevel optimization. Annals of operations research 153(1):235–256

mypy Contributors (2021) mypy - optional static typing for Python. http://mypy-lang.org/, accessed: 2021-01-30

Corso G, Cavalleri L, ini D, Liò P, Velickovic P (2020) Principal neighbourhood aggregation for graph nets. CoRR abs/2004.05718

Cortés-Coy LF, Linares-Vásquez M, Aponte J, Poshyvanyk D (2014) On automatically generating commit messages via summarization of source code changes. In: 2014 IEEE 14th International Working Conference on Source Code Analysis and Manipulation, IEEE, pp 275–284

Cosmo L, Kazi A, Ahmadi SA, Navab N, Bronstein M (2020) Latent patient network learning for automatic diagnosis. arXiv preprint arXiv:200313620

Costa F, De Grave K (2010) Fast neighborhood subgraph pairwise distance kernel. In: International Conference on Machine Learning, Omnipress, pp 255–262

Cotto KC, Wagner AH, Feng YY, Kiwala S, Coffman AC, Spies G, Wollam A, Spies NC, Griffith OL, Griffith M (2018) Dgidb 3.0: a redesign and expansion of the drug–gene interaction database. Nucleic acids research 46(D1):D1068–D1073

Cozzetto D, Minneci F, Currant H, et al (2016) FFPred 3: feature-based function prediction for all gene ontology domains. Scientific Reports 6(1)

Cucurull G, Taslakian P, Vazquez D (2019) Context-aware visual compatibility prediction. In: Proceedings of the IEEE/CVF Conference on Computer Vision and Pattern Recognition, pp 12,617–12,626

Cui J, Kingsbury B, Ramabhadran B, Sethy A, Audhkhasi K, Cui X, Kislal E, Mangu L, Nussbaum-Thom M, Picheny M, et al (2015) Multilingual representations for low resource speech recognition and keyword search. In: 2015 IEEE Workshop on Automatic Speech Recognition and Understanding (ASRU), IEEE, pp 259–266

Cui P, Wang X, Pei J, Zhu W (2018) A survey on network embedding. IEEE Transactions on Knowledge and Data Engineering 31(5):833–852

Cui Z, Henrickson K, Ke R, Wang Y (2019) Traffic graph convolutional recurrent neural network: A deep learning framework for network-scale traffic learning and forecasting. IEEE Transactions on Intelligent Transportation Systems 21(11):4883–4894

Cummins C, Fisches ZV, Ben-Nun T, Hoefler T, Leather H (2020) Programl: Graph-based deep learning for program optimization and analysis. arXiv preprint arXiv:200310536

Cussens J (2011) Bayesian network learning with cutting planes. In: Proceedings of the Twenty-Seventh Conference on Uncertainty in Artificial Intelligence, pp 153–160

Cvitkovic M, Singh B, Anandkumar A (2018) Deep learning on code with an unbounded vocabulary. In: Machine Learning for Programming

Cybenko G (1989) Approximation by superpositions of a sigmoidal function. Mathematics of control, signals and systems 2(4):303–314

Cygan M, Pilipczuk M, Pilipczuk M, Wojtaszczyk JO (2012) Sitting closer to friends than enemies, revisited. In: International Symposium on Mathematical Foundations of Computer Science, Springer, pp 296–307

Dabkowski P, Gal Y (2017) Real time image saliency for black box classifiers. arXiv preprint arXiv:170507857

Dahl G, Ranzato M, Mohamed Ar, Hinton GE (2010) Phone recognition with the mean-covariance restricted boltzmann machine. Advances in neural information processing systems 23:469–477

Dai B, Zhang Y, Lin D (2017) Detecting visual relationships with deep relational networks. In: Proceedings of the IEEE conference on computer vision and Pattern recognition, pp 3076–3086

Dai H, Dai B, Song L (2016) Discriminative embeddings of latent variable models for structured data. In: International conference on machine learning, PMLR, pp 2702–2711

Dai H, Li H, Tian T, Huang X, Wang L, Zhu J, Song L (2018a) Adversarial attack on graph structured data. In: International conference on machine learning, PMLR, pp 1115–1124

Dai H, Tian Y, Dai B, Skiena S, Song L (2018b) Syntax-directed variational autoencoder for structured data. arXiv preprint arXiv:180208786

Dai Q, Li Q, Tang J, Wang D (2018c) Adversarial network embedding. In: Proceedings of the AAAI Conference on Artificial Intelligence, vol 32

Dai R, Xu S, Gu Q, Ji C, Liu K (2020) Hybrid spatio-temporal graph convolutional network: Improving traffic prediction with navigation data. In: Proceedings of the 26th ACM SIGKDD International Conference on Knowledge Discovery & Data Mining, pp 3074–3082

Daitch SI, Kelner JA, Spielman DA (2009) Fitting a graph to vector data. In: Proceedings of the 26th Annual International Conference on Machine Learning, pp 201–208

Damonte M, Cohen SB (2019) Structural neural encoders for amr-to-text generation. arXiv preprint arXiv:190311410

Dana JM, Gutmanas A, Tyagi N, et al (2018) SIFTS: updated structure integration with function, taxonomy and sequences resource allows 40-fold increase in coverage of structure-based annotations for proteins. Nucleic Acids Research 47(D1):D482–D489

Das S, Lee D, Sillitoe I, et al (2015) Functional classification of CATH superfamilies: a domain-based approach for protein function annotation. Bioinformatics 31(21):3460–3467

Dasgupta SS, Ray SN, Talukdar P (2018) Hyte: Hyperplane-based temporally aware knowledge graph embedding. In: Empirical Methods in Natural Language Processing, pp 2001–2011

Davidson TR, Falorsi L, De Cao N, Kipf T, Tomczak JM (2018) Hyperspherical variational auto-encoders. In: 34th Conference on Uncertainty in Artificial Intelligence 2018, UAI 2018, Association For Uncertainty in Artificial Intelligence (AUAI), pp 856–865

Davis AP, Grondin CJ, Johnson RJ, Sciaky D, McMorran R, Wiegers J, Wiegers TC, Mattingly CJ (2019) The comparative toxicogenomics database: update 2019. Nucleic acids research 47(D1):D948–D954

De Cao N, Kipf T (2018) Molgan: An implicit generative model for small molecular graphs. arXiv preprint arXiv:180511973

De Lucia A, Di Penta M, Oliveto R, Panichella A, Panichella S (2012) Using ir methods for labeling source code artifacts: Is it worthwhile? In: 2012 20th IEEE International Conference on Program Comprehension (ICPC), IEEE, pp 193–202

Dearman D, Cox A, Fisher M (2005) Adding control-flow to a visual data-flow representation. In: 13th International Workshop on Program Comprehension (IWPC'05), IEEE, pp 297–306

Debnath AK, Lopez de Compadre RL, Debnath G, Shusterman AJ, Hansch C (1991) Structure-activity relationship of mutagenic aromatic and heteroaromatic nitro compounds. correlation with molecular orbital energies and hydrophobicity. Journal of medicinal chemistry 34(2):786–797

Defferrard M, X B, Vandergheynst P (2016) Convolutional neural networks on graphs with fast localized spectral filtering. In: Advances in Neural Information Processing Systems, pp 3844–3852

Delaney JS (2004) Esol: estimating aqueous solubility directly from molecular structure. Journal of chemical information and computer sciences 44(3):1000–1005

Deng C, Zhao Z, Wang Y, Zhang Z, Feng Z (2020) Graphzoom: A multi-level spectral approach for accurate and scalable graph embedding. In: International Conference on Learning Representations

Deng Z, Dong Y, Zhu J (2019) Batch Virtual Adversarial Training for Graph Convolutional Networks. In: ICML 2019 Workshop: Learning and Reasoning with Graph-Structured Representations

Desa U (2018) Revision of world urbanization prospects. UN Department of Economic and Social Affairs 16

Dettmers T, Minervini P, Stenetorp P, Riedel S (2018) Convolutional 2d knowledge graph embeddings. In: Proceedings of the AAAI Conference on Artificial Intelligence, vol 32

Devlin J, Chang MW, Lee K, Toutanova K (2019) BERT: Pre-training of deep bidirectional transformers for language understanding. In: Proceedings of the 2019 Conference of the North American Chapter of the Association for Computational Linguistics: Human Language Technologies, Volume 1 (Long and Short Papers), Association for Computational Linguistics, Minneapolis, Minnesota, pp 4171–4186, DOI 10.18653/v1/N19-1423

Dhillon IS, Guan Y, Kulis B (2007) Weighted graph cuts without eigenvectors a multilevel approach. IEEE Transactions on Pattern Analysis and Machine Intelligence 29(11):1944–1957

Diao Z, Wang X, Zhang D, Liu Y, Xie K, He S (2019) Dynamic spatial-temporal graph convolutional neural networks for traffic forecasting. In: Proceedings of the AAAI Conference on Artificial Intelligence, vol 33, pp 890–897

Dinella E, Dai H, Li Z, Naik M, Song L, Wang K (2020) Hoppity: Learning graph transformations to detect and fix bugs in programs. In: International Conference on Learning Representations (ICLR)

Ding M, Zhou C, Chen Q, Yang H, Tang J (2019a) Cognitive graph for multi-hop reading comprehension at scale. In: Proceedings of the 57th Annual Meeting of the Association for Computational Linguistics, pp 2694–2703

Ding S, Qu S, Xi Y, Sangaiah AK, Wan S (2019b) Image caption generation with high-level image features. Pattern Recognition Letters 123:89–95

Ding Y, Yao Q, Zhang T (2020a) Propagation model search for graph neural networks. arXiv preprint arXiv:201003250

Ding Y, Zhou X, Bao H, Li Y, Hamann C, Spears S, Yuan Z (2020b) Cycling-net: A deep learning approach to predicting cyclist behaviors from geo-referenced egocentric video data. Association for Computing Machinery, SIGSPATIAL '20, p 337–346, DOI 10.1145/3397536.3422258

Do K, Tran T, Venkatesh S (2019) Graph transformation policy network for chemical reaction prediction. In: Proceedings of the 25th ACM SIGKDD International Conference on Knowledge Discovery & Data Mining, pp 750–760

Doersch C, Gupta A, Efros AA (2015) Unsupervised visual representation learning by context prediction. In: 2015 IEEE International Conference on Computer Vision, ICCV 2015, Santiago, Chile, December 7-13, 2015, IEEE Computer Society, pp 1422–1430, DOI 10.1109/ICCV.2015.167

Dohkan S, Koike A, Takagi T (2006) Improving the performance of an svm-based method for predicting protein-protein interactions. In Silico Biology 6:515–529, 6

Domingo-Fernández D, Baksi S, Schultz B, Gadiya Y, Karki R, Raschka T, Ebeling C, Hofmann-Apitius M, et al (2020) Covid-19 knowledge graph: a computable, multi-modal, cause-and-effect knowledge model of covid-19 pathophysiology. BioRxiv

Donahue C, McAuley J, Puckette M (2018) Synthesizing audio with generative adversarial networks. arXiv preprint arXiv:180204208 1

Donahue J, Anne Hendricks L, Guadarrama S, Rohrbach M, Venugopalan S, Saenko K, Darrell T (2015) Long-term recurrent convolutional networks for visual recognition and description. In: Proceedings of the IEEE conference on computer vision and pattern recognition, pp 2625–2634

Doncheva NT, Morris JH, Gorodkin J, Jensen LJ (2018) Cytoscape StringApp: Network analysis and visualization of proteomics data. Journal of Proteome Research 18(2):623–632

Dong X, Gabrilovich E, Heitz G, Horn W, Lao N, Murphy K, Strohmann T, Sun S, Zhang W (2014) Knowledge vault: A web-scale approach to probabilistic knowledge fusion. In: Proceedings of the 20th ACM SIGKDD international conference on Knowledge discovery and data mining, pp 601–610

Dong X, Thanou D, Frossard P, Vandergheynst P (2016) Learning laplacian matrix in smooth graph signal representations. IEEE Transactions on Signal Processing 64(23):6160–6173

Dong X, Thanou D, Rabbat M, Frossard P (2019) Learning graphs from data: A signal representation perspective. IEEE Signal Processing Magazine 36(3):44–63

Dong Y, Chawla NV, Swami A (2017) metapath2vec: Scalable representation learning for heterogeneous networks. In: Proceedings of the 23rd ACM SIGKDD international conference on knowledge discovery and data mining, pp 135–144

Donsker M, Varadhan S (1976) Asymptotic evaluation of certain markov process expectations for large time—iii. Communications on Pure and Applied Mathematics 29(4):389–461, copyright: Copyright 2016 Elsevier B.V., All rights reserved.

Dos Santos C, Gatti M (2014) Deep convolutional neural networks for sentiment analysis of short texts. In: Proceedings of COLING 2014, the 25th International Conference on Computational Linguistics: Technical Papers, pp 69–78

Dosovitskiy A, Springenberg JT, Riedmiller MA, Brox T (2014) Discriminative unsupervised feature learning with convolutional neural networks. In: Ghahramani Z, Welling M, Cortes C, Lawrence ND, Weinberger KQ (eds) Advances in Neural Information Processing Systems 27: Annual Conference on Neural Information Processing Systems 2014, December 8-13 2014, Montreal, Quebec, Canada, pp 766–774

Dosovitskiy A, et al (2021) An image is worth 16x16 words: Transformers for image recognition at scale. ICLR

Dou Y, Liu Z, Sun L, Deng Y, Peng H, Yu PS (2020) Enhancing graph neural network-based fraud detectors against camouflaged fraudsters. In: Proceedings of the 29th ACM International Conference on Information & Knowledge Management, pp 315–324

Du M, Liu N, Yang F, Hu X (2019) Learning credible deep neural networks with rationale regularization. In: 2019 IEEE International Conference on Data Mining (ICDM), IEEE, pp 150–159

Du M, Yang F, Zou N, Hu X (2020) Fairness in deep learning: A computational perspective. IEEE Intelligent Systems

Duvenaud DK, Maclaurin D, Iparraguirre J, Bombarell R, Hirzel T, Aspuru-Guzik A, Adams RP (2015a) Convolutional networks on graphs for learning molecular fingerprints. In: Advances in neural information processing systems, pp 2224–2232

Duvenaud DK, Maclaurin D, Iparraguirre J, Bombarell R, Hirzel T, Aspuru-Guzik A, Adams RP (2015b) Convolutional networks on graphs for learning molecular fingerprints. In: Advances in Neural Information Processing Systems, pp 2224–2232

Dvijotham KD, Hayes J, Balle B, Kolter Z, Qin C, Gyorgy A, Xiao K, Gowal S, Kohli P (2020) A framework for robustness certification of smoothed classifiers using f-divergences. In: International Conference on Learning Representations, ICLR

Dwivedi VP, Joshi CK, Laurent T, Bengio Y, Bresson X (2020) Benchmarking graph neural networks. arXiv preprint arXiv:200300982

Dyer C, Ballesteros M, Ling W, Matthews A, Smith NA (2015) Transition-based dependency parsing with stack long short-term memory. arXiv preprint arXiv:150508075

Easley D, Kleinberg J, et al (2012) Networks, crowds, and markets: Reasoning about a highly connected world. Significance 9(1):43–44

Eksombatchai C, Jindal P, Liu JZ, Liu Y, Sharma R, Sugnet C, Ulrich M, Leskovec J (2018) Pixie: A system for recommending 3+ billion items to 200+ million users in real-time. In: Proceedings of the 2018 world wide web conference, pp 1775–1784

Elinas P, Bonilla EV, Tiao L (2020) Variational inference for graph convolutional networks in the absence of graph data and adversarial settings. In: Advances in Neural Information Processing Systems, vol 33, pp 18,648–18,660

Elkan C, Noto K (2008) Learning classifiers from only positive and unlabeled data. In: Proceedings of the 14th ACM SIGKDD international conference on Knowledge discovery and data mining, pp 213–220

Elman JL (1990) Finding structure in time. Cognitive Science 14(2):179–211

Elmsallati A, Clark C, Kalita J (2016) Global alignment of protein-protein interaction networks: A survey. IEEE/ACM Trans Comput Biol Bioinformatics 13(4):689–705

Entezari N, Al-Sayouri SA, Darvishzadeh A, Papalexakis EE (2020) All you need is low (rank) defending against adversarial attacks on graphs. In: Proceedings of the 13th International Conference on Web Search and Data Mining, pp 169–177

Erdős P, Rényi A (1959) On random graphs i. Publ Math Debrecen 6:290–297

Erdős P, Rényi A (1960) On the evolution of random graphs. Publ Math Inst Hung
    Acad Sci 5(1):17–60

Erkan G, Radev DR (2004) Lexrank: Graph-based lexical centrality as salience in
    text summarization. Journal of artificial intelligence research 22:457–479

Ernst MD, Perkins JH, Guo PJ, McCamant S, Pacheco C, Tschantz MS, Xiao C
    (2007) The Daikon system for dynamic detection of likely invariants. Science of
    computer programming 69(1-3):35–45

Eykholt K, Evtimov I, Fernandes E, Li B, Rahmati A, Xiao C, Prakash A, Kohno T,
    Song D (2018) Robust physical-world attacks on deep learning visual classifica-
    tion. In: IEEE Conference on Computer Vision and Pattern Recognition, CVPR,
    pp 1625–1634

Faghri F, Fleet DJ, Kiros JR, Fidler S (2017) Vse++: Improving visual-semantic
    embeddings with hard negatives. arXiv preprint arXiv:170705612

Fan Y, Hou S, Zhang Y, Ye Y, Abdulhayoglu M (2018) Gotcha-sly malware! scor-
    pion a metagraph2vec based malware detection system. In: Proceedings of the
    24th ACM SIGKDD International Conference on Knowledge Discovery & Data
    Mining, pp 253–262

Fang Y, Sun S, Gan Z, Pillai R, Wang S, Liu J (2020) Hierarchical graph network
    for multi-hop question answering. In: Proceedings of the 2020 Conference on
    Empirical Methods in Natural Language Processing (EMNLP), pp 8823–8838

Fatemi B, Asri LE, Kazemi SM (2021) Slaps: Self-supervision improves structure
    learning for graph neural networks. arXiv preprint arXiv:210205034

Feng B, Wang Y, Wang Z, Ding Y (2021) Uncertainty-aware Attention Graph Neu-
    ral Network for Defending Adversarial Attacks. In: AAAI Conference on Artifi-
    cial Intelligence

Feng F, He X, Tang J, Chua T (2019a) Graph adversarial training: Dynamically
    regularizing based on graph structure. TKDE pp 1–1

Feng J, Huang M, Wang M, Zhou M, Hao Y, Zhu X (2016) Knowledge graph
    embedding by flexible translation. In: Proceedings of the Fifteenth International
    Conference on Principles of Knowledge Representation and Reasoning, pp 557–
    560

Feng W, Zhang J, Dong Y, Han Y, Luan H, Xu Q, Yang Q, Kharlamov E, Tang J
    (2020) Graph random neural networks for semi-supervised learning on graphs. In:
    Advances in Neural Information Processing Systems, vol 33, pp 22,092–22,103

Feng X, Zhang Y, Glass J (2014) Speech feature denoising and dereverberation via
    deep autoencoders for noisy reverberant speech recognition. In: 2014 IEEE inter-
    national conference on acoustics, speech and signal processing (ICASSP), IEEE,
    pp 1759–1763

Feng Y, Lv F, Shen W, Wang M, Sun F, Zhu Y, Yang K (2019b) Deep session interest
    network for click-through rate prediction. arXiv preprint arXiv:190506482

Feng Y, You H, Zhang Z, Ji R, Gao Y (2019c) Hypergraph neural networks. In:
    Proceedings of the AAAI Conference on Artificial Intelligence, vol 33, pp 3558–
    3565

Feurer M, Hutter F (2019) Hyperparameter optimization. In: Automated Machine
    Learning, Springer, Cham, pp 3–33

Févotte C, Idier J (2011) Algorithms for nonnegative matrix factorization with the β-divergence. Neural computation 23(9):2421–2456

Fey M, Lenssen JE (2019) Fast graph representation learning with PyTorch Geometric. CoRR abs/1903.02428

Fey M, Lenssen JE, Weichert F, Müller H (2018) Splinecnn: Fast geometric deep learning with continuous b-spline kernels. In: Proceedings of the IEEE Conference on Computer Vision and Pattern Recognition, pp 869–877

Fey M, Lenssen JE, Morris C, Masci J, Kriege NM (2020) Deep graph matching consensus. In: International Conference on Learning Representations

Finn RD, Bateman A, Clements J, et al (2013) Pfam: the protein families database. Nucleic Acids Research 42(D1):D222–D230

Foggia P, Percannella G, Vento M (2014) Graph matching and learning in pattern recognition in the last 10 years. International Journal of Pattern Recognition and Artificial Intelligence 28(01):1450,001

Foltman M, Sanchez-Diaz A (2016) Studying protein–protein interactions in budding yeast using co-immunoprecipitation. In: Yeast Cytokinesis, Springer, pp 239–256, DOI 10.1007/978-1-4939-3145-3_17

Fong RC, Vedaldi A (2017) Interpretable explanations of black boxes by meaningful perturbation. In: Proceedings of the IEEE International Conference on Computer Vision, pp 3429–3437

Fortin S (1996) The graph isomorphism problem

Fortunato S (2010) Community detection in graphs. Physics reports 486(3-5):75–174

Fouss F, Pirotte A, Renders JM, Saerens M (2007) Random-walk computation of similarities between nodes of a graph with application to collaborative recommendation. IEEE Transactions on knowledge and data engineering 19(3):355–369

Fowkes J, Chanthirasegaran P, Ranca R, Allamanis M, Lapata M, Sutton C (2017) Autofolding for source code summarization. IEEE Transactions on Software Engineering 43(12):1095–1109

Franceschi L, Niepert M, Pontil M, He X (2019) Learning discrete structures for graph neural networks. In: Proceedings of the 36th International Conference on Machine Learning, vol 97, pp 1972–1982

Freeman LA (2003) A refresher in data flow diagramming: an effective aid for analysts. Commun ACM 46(9):147–151, DOI 10.1145/903893.903930

Freeman LC (2000) Visualizing social networks. Journal of social structure 1(1):4

Fröhlich H, Wegner JK, Sieker F, Zell A (2005) Optimal assignment kernels for attributed molecular graphs. In: International Conference on Machine Learning, pp 225–232

Fu R, Zhang Z, Li L (2016) Using lstm and gru neural network methods for traffic flow prediction. In: 2016 31st Youth Academic Annual Conference of Chinese Association of Automation (YAC), IEEE, pp 324–328

Fu Ty, Lee WC, Lei Z (2017) Hin2vec: Explore meta-paths in heterogeneous information networks for representation learning. In: Proceedings of the 2017 ACM on Conference on Information and Knowledge Management, pp 1797–1806

Fu X, Zhang J, Meng Z, King I (2020) Magnn: metapath aggregated graph neural network for heterogeneous graph embedding. In: Proceedings of The Web Conference 2020, pp 2331–2341

Fu Y, Ma Y (2012) Graph embedding for pattern analysis. Springer Science & Business Media

Gabrié M, Manoel A, Luneau C, Barbier J, Macris N, Krzakala F, Zdeborová L (2019) Entropy and mutual information in models of deep neural networks. Journal of Statistical Mechanics: Theory and Experiment 2019(12):124,014

Gao D, Li K, Wang R, Shan S, Chen X (2020a) Multi-modal graph neural network for joint reasoning on vision and scene text. In: Proceedings of the IEEE/CVF Conference on Computer Vision and Pattern Recognition, pp 12,746–12,756

Gao H, Ji S (2019) Graph u-nets. In: International Conference on Machine Learning, PMLR, pp 2083–2092

Gao H, Wang Z, Ji S (2018a) Large-scale learnable graph convolutional networks. In: Proceedings of the 24th ACM SIGKDD International Conference on Knowledge Discovery & Data Mining, ACM, pp 1416–1424

Gao J, Yang Z, Nevatia R (2017) Cascaded boundary regression for temporal action detection. arXiv preprint arXiv:170501180

Gao J, Li X, Xu YE, Sisman B, Dong XL, Yang J (2019a) Efficient knowledge graph accuracy evaluation. arXiv preprint arXiv:190709657

Gao S, Chen C, Xing Z, Ma Y, Song W, Lin SW (2019b) A neural model for method name generation from functional description. In: 2019 IEEE 26th International Conference on Software Analysis, Evolution and Reengineering (SANER), IEEE, pp 414–421

Gao X, Hu W, Qi GJ (2021) Unsupervised learning of topology transformation equivariant representations

Gao Y, Guo X, Zhao L (2018b) Local event forecasting and synthesis using unpaired deep graph translations. In: Proceedings of the 2nd ACM SIGSPATIAL Workshop on Analytics for Local Events and News, pp 1–8

Gao Y, Wu L, Homayoun H, Zhao L (2019c) Dyngraph2seq: Dynamic-graph-to-sequence interpretable learning for health stage prediction in online health forums. In: 2019 IEEE International Conference on Data Mining (ICDM), IEEE, pp 1042–1047

Gao Y, Yang H, Zhang P, Zhou C, Hu Y (2020b) Graph neural architecture search. In: International Joint Conference on Artificial Intelligence, pp 1403–1409

Garcia V, Bruna J (2017) Few-shot learning with graph neural networks. arXiv preprint arXiv:171104043

García-Durán A, Dumančić S, Niepert M (2018) Learning sequence encoders for temporal knowledge graph completion. In: Proceedings of the 2018 Conference on Empirical Methods in Natural Language Processing, pp 4816–4821, DOI 10.18653/v1/D18-1516

Garey MR (1979) A guide to the theory of np-completeness. Computers and intractability

Garey MR, Johnson DS (2002) Computers and intractability, vol 29. wh freeman New York

Garg V, Jegelka S, Jaakkola T (2020) Generalization and representational limits of graph neural networks. In: International Conference on Machine Learning, PMLR, pp 3419–3430

Gaudelet T, Day B, Jamasb AR, Soman J, Regep C, Liu G, Hayter JBR, Vickers R, Roberts C, Tang J, Roblin D, Blundell TL, Bronstein MM, Taylor-King JP (2020) Utilising graph machine learning within drug discovery and development. CoRR abs/2012.05716

Gavin AC, Bösche M, Krause R, et al (2002) Functional organization of the yeast proteome by systematic analysis of protein complexes. Nature 415(6868):141–147

Geisler S, Zügner D, Günnemann S (2020) Reliable graph neural networks via robust aggregation. Advances in Neural Information Processing Systems 33

Geisler S, Zügner D, Bojchevski A, Günnemann S (2021) Attacking Graph Neural Networks at Scale. In: Deep Learning for Graphs at AAAI Conference on Artificial Intelligence

Gema RP, Robles G, Alexander S, Zaidman A, Germán DM, Gonzalez-Barahona JM (2020) How bugs are born: a model to identify how bugs are introduced in software components. Empirical Software Engineering 25(2):1294–1340

Geng X, Li Y, Wang L, Zhang L, Yang Q, Ye J, Liu Y (2019) Spatiotemporal multigraph convolution network for ride-hailing demand forecasting. In: Proceedings of the AAAI conference on artificial intelligence, vol 33, pp 3656–3663

Ghosal D, Hazarika D, Majumder N, Roy A, Poria S, Mihalcea R (2020) Kingdom: Knowledge-guided domain adaptation for sentiment analysis. arXiv preprint arXiv:200500791

Gidaris S, Singh P, Komodakis N (2018) Unsupervised representation learning by predicting image rotations. In: 6th International Conference on Learning Representations, ICLR 2018, Vancouver, BC, Canada, April 30 - May 3, 2018, Conference Track Proceedings, OpenReview.net

Gilbert EN (1959) Random graphs. The Annals of Mathematical Statistics 30(4):1141–1144

Gilmer J, Schoenholz SS, Riley PF, Vinyals O, Dahl GE (2017) Neural message passing for quantum chemistry. In: Precup D, Teh YW (eds) Proceedings of the 34th International Conference on Machine Learning, ICML 2017, Sydney, NSW, Australia, 6-11 August 2017, PMLR, Proceedings of Machine Learning Research, vol 70, pp 1263–1272

Girvan M, Newman ME (2002) Community structure in social and biological networks. Proceedings of the national academy of sciences 99(12):7821–7826

Gligorijevic V, Renfrew PD, Kosciolek T, Leman JK, Berenberg D, Vatanen T, Chandler C, Taylor BC, Fisk IM, Vlamakis H, et al (2020) Structure-based function prediction using graph convolutional networks. bioRxiv p 786236

Goel R, Kazemi SM, Brubaker M, Poupart P (2020) Diachronic embedding for temporal knowledge graph completion. In: Proceedings of the AAAI Conference on Artificial Intelligence, vol 34, pp 3988–3995

Gold S, Rangarajan A (1996) A graduated assignment algorithm for graph matching. IEEE Transactions on pattern analysis and machine intelligence 18(4):377–388

Goldberg D, Nichols D, Oki BM, Terry D (1992) Using collaborative filtering to weave an information tapestry. Communications of the ACM 35(12):61–70

Gong X, Chang S, Jiang Y, Wang Z (2019) Autogan: Neural architecture search for generative adversarial networks. In: Proceedings of the IEEE/CVF International Conference on Computer Vision, pp 3224–3234

Gong Y, Jiang Z, Feng Y, Hu B, Zhao K, Liu Q, Ou W (2020) Edgerec: Recommender system on edge in mobile taobao. In: Proceedings of the 29th ACM International Conference on Information & Knowledge Management, pp 2477–2484

Goodfellow I, Shlens J, Szegedy C (2015) Explaining and harnessing adversarial examples. In: International Conference on Learning Representations

Goodfellow IJ, Pouget-Abadie J, Mirza M, Bing X, Bengio Y (2014a) Generative adversarial nets. MIT Press

Goodfellow IJ, Pouget-Abadie J, Mirza M, Xu B, Warde-Farley D, Ozair S, Courville A, Bengio Y (2014b) Generative adversarial networks. arXiv preprint arXiv:14062661

Goodwin T, Harabagiu SM (2013) Automatic generation of a qualified medical knowledge graph and its usage for retrieving patient cohorts from electronic medical records. In: 2013 IEEE Seventh International Conference on Semantic Computing, IEEE, pp 363–370

Gori M, Monfardini G, Scarselli F (2005) A new model for learning in graph domains. In: IEEE International Joint Conference on Neural Networks, vol 2, pp 729–734, DOI 10.1109/IJCNN.2005.1555942

Goyal P, Ferrara E (2018) Graph embedding techniques, applications, and performance: A survey. Knowledge-Based Systems 151:78–94

Grattarola D, Alippi C (2020) Graph neural networks in TensorFlow and Keras with Spektral. CoRR abs/2006.12138, 2006.12138

Graves A (2013) Generating sequences with recurrent neural networks. CoRR abs/1308.0850

Graves A, Fernández S, Schmidhuber J (2005) Bidirectional lstm networks for improved phoneme classification and recognition. In: International Conference on Artificial Neural Networks, Springer, pp 799–804

Grbovic M, Cheng H (2018) Real-time personalization using embeddings for search ranking at airbnb. In: Proceedings of the 24th ACM SIGKDD International Conference on Knowledge Discovery & Data Mining, pp 311–320

Greff K, Srivastava RK, Koutník J, Steunebrink BR, Schmidhuber J (2016) Lstm: A search space odyssey. IEEE transactions on neural networks and learning systems 28(10):2222–2232

Gremse M, Chang A, Schomburg I, Grote A, Scheer M, Ebeling C, Schomburg D (2010) The brenda tissue ontology (bto): the first all-integrating ontology of all organisms for enzyme sources. Nucleic acids research 39(suppl_1):D507–D513

Grohe M (2017) Descriptive complexity, canonisation, and definable graph structure theory, vol 47. Cambridge University Press

Grohe M, Otto M (2015) Pebble games and linear equations. The Journal of Symbolic Logic pp 797–844

Grover A, Leskovec J (2016) node2vec: Scalable feature learning for networks. In: Proceedings of the 22nd ACM SIGKDD international conference on Knowledge discovery and data mining, pp 855–864

Grover A, Zweig A, Ermon S (2019) Graphite: Iterative generative modeling of graphs. In: International Conference on Machine Learning, pp 2434–2444

Gu J, Cai J, Joty SR, Niu L, Wang G (2018) Look, imagine and match: Improving textual-visual cross-modal retrieval with generative models. In: Proceedings of the IEEE Conference on Computer Vision and Pattern Recognition, pp 7181–7189

Gu S, Lillicrap T, Ghahramani Z, Turner RE, Levine S (2016) Q-prop: Sample-efficient policy gradient with an off-policy critic. arXiv preprint arXiv:161102247

Guan Y, Myers CL, Hess DC, et al (2008) Predicting gene function in a hierarchical context with an ensemble of classifiers. Genome Biology 9(Suppl 1):S3

Gui H, Liu J, Tao F, Jiang M, Norick B, Han J (2016) Large-scale embedding learning in heterogeneous event data. In: 2016 IEEE 16th International Conference on Data Mining (ICDM), IEEE, pp 907–912

Gui T, Zou Y, Zhang Q, Peng M, Fu J, Wei Z, Huang XJ (2019) A lexicon-based graph neural network for chinese ner. In: Proceedings of the 2019 Conference on Empirical Methods in Natural Language Processing and the 9th International Joint Conference on Natural Language Processing (EMNLP-IJCNLP), pp 1039–1049

Guille A, Hacid H, Favre C, Zighed DA (2013) Information diffusion in online social networks: A survey. ACM Sigmod Record 42(2):17–28

Gulrajani I, Ahmed F, Arjovsky M, Dumoulin V, Courville A (2017) Improved training of wasserstein gans. arXiv preprint arXiv:170400028

Guo G, Ouyang S, He X, Yuan F, Liu X (2019a) Dynamic item block and prediction enhancing block for sequential recommendation. In: Proceedings of the International Joint Conference on Artificial Intelligence, pp 1373–1379

Guo H, Tang R, Ye Y, Li Z, He X (2017) Deepfm: a factorization-machine based neural network for ctr prediction. In: Proceedings of the International Joint Conference on Artificial Intelligence, pp 1725–1731

Guo M, Chou E, Huang DA, Song S, Yeung S, Fei-Fei L (2018a) Neural graph matching networks for fewshot 3d action recognition. In: Proceedings of the European Conference on Computer Vision (ECCV), pp 653–669

Guo S, Lin Y, Feng N, Song C, Wan H (2019b) Attention based spatial-temporal graph convolutional networks for traffic flow forecasting. In: Proceedings of the AAAI Conference on Artificial Intelligence, vol 33, pp 922–929

Guo X, Wu L, Zhao L (2018b) Deep graph translation. arXiv preprint arXiv:180509980

Guo X, Zhao L, Nowzari C, Rafatirad S, Homayoun H, Dinakarrao SMP (2019c) Deep multi-attributed graph translation with node-edge co-evolution. In: 2019 IEEE International Conference on Data Mining (ICDM), IEEE, pp 250–259

Guo Y, Li M, Pu X, et al (2010) Pred_ppi: a server for predicting protein-protein interactions based on sequence data with probability assignment. BMC Research Notes 3(1):145

Guo Z, Zhang Y, Lu W (2019d) Attention guided graph convolutional networks for relation extraction. In: Proceedings of the 57th Annual Meeting of the Association for Computational Linguistics, pp 241–251

Guo Z, Zhang Y, Teng Z, Lu W (2019e) Densely connected graph convolutional networks for graph-to-sequence learning. Transactions of the Association for Computational Linguistics 7:297–312

Gurwitz D (2020) Repurposing current therapeutics for treating covid-19: A vital role of prescription records data mining. Drug development research 81(7):777–781

Gutmann M, Hyvärinen A (2010) Noise-contrastive estimation: A new estimation principle for unnormalized statistical models. In: Proceedings of the International Conference on Artificial Intelligence and Statistics

Ha D, Dai A, Le QV (2017) Hypernetworks. In: Proceedings of the International Conference on Learning Representations (ICLR)

Haghighi A, Ng AY, Manning CD (2005) Robust textual inference via graph matching. In: Proceedings of Human Language Technology Conference and Conference on Empirical Methods in Natural Language Processing, pp 387–394

Haiduc S, Aponte J, Moreno L, Marcus A (2010) On the use of automated text summarization techniques for summarizing source code. In: 2010 17th Working Conference on Reverse Engineering, IEEE, pp 35–44

Haldar R, Wu L, Xiong J, Hockenmaier J (2020) A multi-perspective architecture for semantic code search. arXiv preprint arXiv:200506980

Hamaguchi T, Oiwa H, Shimbo M, Matsumoto Y (2017) Knowledge transfer for out-of-knowledge-base entities: a graph neural network approach. In: Proceedings of the 26th International Joint Conference on Artificial Intelligence, pp 1802–1808

Hamilton W, Ying Z, Leskovec J (2017a) Inductive representation learning on large graphs. In: Advances in Neural Information Processing Systems, vol 30

Hamilton WL (2020) Graph representation learning. Synthesis Lectures on Artificial Intelligence and Machine Learning 14(3):1–159

Hamilton WL, Ying R, Leskovec J (2017b) Inductive representation learning on large graphs. In: Advances in Neural Information Processing Systems, pp 1025–1035

Hamilton WL, Ying R, Leskovec J (2017c) Representation learning on graphs: Methods and applications. IEEE Data Engineering Bulletin 40(3):52–74

Hammond DK, Vandergheynst P, Gribonval R (2011) Wavelets on graphs via spectral graph theory. Applied and Computational Harmonic Analysis 30(2):129–150

Han J, Luo P, Wang X (2019) Deep self-learning from noisy labels. In: 2019 IEEE/CVF International Conference on Computer Vision, ICCV 2019, Seoul, Korea (South), October 27 - November 2, 2019, IEEE, pp 5137–5146, DOI 10.1109/ICCV.2019.00524

Han JDJ, Dupuy D, Bertin N, et al (2005) Effect of sampling on topology predictions of protein-protein interaction networks. Nature Biotechnology 23(7):839–844

Han K, Wang Y, Chen H, Chen X, Guo J, Liu Z, Tang Y, Xiao A, Xu C, Xu Y, et al (2020) A survey on visual transformer. arXiv preprint arXiv:201212556

Han X, Zhu H, Yu P, Wang Z, Yao Y, Liu Z, Sun M (2018) Fewrel: A large-scale supervised few-shot relation classification dataset with state-of-the-art evaluation. In: Proceedings of the 2018 Conference on Empirical Methods in Natural Language Processing, pp 4803–4809

Haque S, LeClair A, Wu L, McMillan C (2020) Improved automatic summarization of subroutines via attention to file context. International Conference on Mining Software Repositories p 300–310

Hart PE, Nilsson NJ, Raphael B (1968) A formal basis for the heuristic determination of minimum cost paths. IEEE transactions on Systems Science and Cybernetics 4(2):100–107

Hashemifar S, Neyshabur B, Khan AA, et al (2018) Predicting protein–protein interactions through sequence-based deep learning. Bioinformatics 34(17):i802–i810

Hasibi R, Michoel T (2020) Predicting gene expression from network topology using graph neural networks. arXiv preprint arXiv:200503961

Hassan AE, Xie T (2010) Software intelligence: the future of mining software engineering data. In: Proceedings of the FSE/SDP workshop on Future of software engineering research, pp 161–166

Hassani K, Khasahmadi AH (2020) Contrastive multi-view representation learning on graphs. In: International Conference on Machine Learning, PMLR, pp 4116–4126

Hastings J, Owen G, Dekker A, Ennis M, Kale N, Muthukrishnan V, Turner S, Swainston N, Mendes P, Steinbeck C (2016) Chebi in 2016: Improved services and an expanding collection of metabolites. Nucleic acids research 44(D1):D1214–D1219

Haveliwala TH (2002) Topic-sensitive pagerank. In: Proceedings of the 11th international conference on World Wide Web, ACM, pp 517–526

He K, Zhang X, Ren S, Sun J (2016a) Deep residual learning for image recognition. In: Proceedings of the IEEE conference on computer vision and pattern recognition, pp 770–778

He K, Gkioxari G, Dollár P, Girshick R (2017a) Mask r-cnn. In: Proceedings of the IEEE international conference on computer vision, pp 2961–2969

He Q, Chen B, Agarwal D (2016b) Building the linkedin knowledge graph. Engineering linkedin com

He X, Niyogi P (2004) Locality preserving projections. Advances in neural information processing systems 16(16):153–160

He X, Liao L, Zhang H, Nie L, Hu X, Chua TS (2017b) Neural collaborative filtering. In: Proceedings of the 26th international conference on world wide web, pp 173–182

He X, Deng K, Wang X, Li Y, Zhang Y, Wang M (2020) Lightgcn: Simplifying and powering graph convolution network for recommendation. In: Proceedings of the 43rd International ACM SIGIR Conference on Research and Development in Information Retrieval, pp 639–648

He Y, Song Y, Li J, Ji C, Peng J, Peng H (2019) Hetespaceywalk: A heterogeneous spacey random walk for heterogeneous information network embedding. In: Proceedings of the 28th ACM International Conference on Information and Knowledge Management, pp 639–648

Hearst MA, Dumais ST, Osuna E, Platt J, Scholkopf B (1998) Support vector machines. IEEE Intelligent Systems and their applications 13(4):18–28

Heimer RZ, Myrseth KOR, Schoenle RS (2019) Yolo: Mortality beliefs and household finance puzzles. The Journal of Finance 74(6):2957–2996

Helfgott HA, Bajpai J, Dona D (2017) Graph isomorphisms in quasi-polynomial time. arXiv preprint arXiv:171004574

Helgason S (1979) Differential geometry, Lie groups, and symmetric spaces. Academic press

Hellendoorn VJ, Bird C, Barr ET, Allamanis M (2018) Deep learning type inference. In: Proceedings of the 2018 26th ACM joint meeting on european software engineering conference and symposium on the foundations of software engineering, pp 152–162

Hellendoorn VJ, Devanbu PT, Polozov O, Marron M (2019a) Are my invariants valid? a learning approach. arXiv preprint arXiv:190306089

Hellendoorn VJ, Sutton C, Singh R, Maniatis P, Bieber D (2019b) Global relational models of source code. In: International Conference on Learning Representations

Henaff M, Bruna J, LeCun Y (2015) Deep convolutional networks on graph-structured data. arXiv preprint arXiv:150605163

Henderson K, Gallagher B, Eliassi-Rad T, Tong H, Basu S, Akoglu L, Koutra D, Faloutsos C, Li L (2012) Rolx: structural role extraction & mining in large graphs. In: the ACM SIGKDD international conference on Knowledge discovery and data mining, pp 1231–1239

Hensman S (2004) Construction of conceptual graph representation of texts. In: Proceedings of the Student Research Workshop at HLT-NAACL 2004, pp 49–54

Hermann KM, Hill F, Green S, Wang F, Faulkner R, Soyer H, Szepesvari D, Czarnecki WM, Jaderberg M, Teplyashin D, et al (2017) Grounded language learning in a simulated 3d world. arXiv preprint arXiv:170606551

Herzig R, Levi E, Xu H, Gao H, Brosh E, Wang X, Globerson A, Darrell T (2019) Spatio-temporal action graph networks. In: 2019 IEEE/CVF International Conference on Computer Vision Workshop (ICCVW), pp 2347–2356, DOI 10.1109/ICCVW.2019.00288

Hidasi B, Karatzoglou A, Baltrunas L, Tikk D (2015) Session-based recommendations with recurrent neural networks. arXiv preprint arXiv:151106939

Higgins I, Matthey L, Pal A, Burgess C, Glorot X, Botvinick M, Mohamed S, Lerchner A (2017) beta-vae: Learning basic visual concepts with a constrained variational framework. ICLR

Himmelstein DS, Lizee A, Hessler C, Brueggeman L, Chen SL, Hadley D, Green A, Khankhanian P, Baranzini SE (2017) Systematic integration of biomedical knowledge prioritizes drugs for repurposing. Elife 6:e26,726

Hinton GE, Osindero S, Teh YW (2006) A fast learning algorithm for deep belief nets. Neural computation 18(7):1527–1554

Hirsch CN, Hirsch CD, Brohammer AB, et al (2016) Draft assembly of elite inbred line PH207 provides insights into genomic and transcriptome diversity in maize. The Plant Cell 28(11):2700–2714

Hjelm RD, Fedorov A, Lavoie-Marchildon S, Grewal K, Bachman P, Trischler A, Bengio Y (2018) Learning deep representations by mutual information estimation and maximization. arXiv preprint arXiv:180806670

Ho Y, Gruhler A, Heilbut A, et al (2002) Systematic identification of protein complexes in saccharomyces cerevisiae by mass spectrometry. Nature 415(6868):180–183

Hochreiter S, Schmidhuber J (1997) Long short-term memory. Neural computation 9(8):1735–1780

Hoff PD, Raftery AE, Handcock MS (2002) Latent space approaches to social network analysis. Journal of the american Statistical association 97(460):1090–1098

Hoffart J, Suchanek FM, Berberich K, Lewis-Kelham E, De Melo G, Weikum G (2011) Yago2: exploring and querying world knowledge in time, space, context, and many languages. In: Proceedings of the 20th international conference companion on World wide web, pp 229–232

Hoffman MD, Blei DM, Wang C, Paisley J (2013) Stochastic variational inference. The Journal of Machine Learning Research 14(1):1303–1347

Hogan A, Blomqvist E, Cochez M, d'Amato C, de Melo G, Gutierrez C, Gayo JEL, Kirrane S, Neumaier S, Polleres A, et al (2020) Knowledge graphs. arXiv preprint arXiv:200302320

Holland PW, Laskey KB, Leinhardt S (1983) Stochastic blockmodels: First steps. Social networks 5(2):109–137

Holmes R, Murphy GC (2005) Using structural context to recommend source code examples. In: Proceedings. 27th International Conference on Software Engineering, 2005. ICSE 2005., IEEE, pp 117–125

Hong D, Gao L, Yao J, Zhang B, Plaza A, Chanussot J (2020a) Graph convolutional networks for hyperspectral image classification. IEEE Transactions on Geoscience and Remote Sensing pp 1–13, DOI 10.1109/TGRS.2020.3015157

Hong H, Guo H, Lin Y, Yang X, Li Z, Ye J (2020b) An attention-based graph neural network for heterogeneous structural learning. In: Proceedings of the AAAI Conference on Artificial Intelligence, vol 34, pp 4132–4139

Hornik K, Stinchcombe M, White H, et al (1989) Multilayer feedforward networks are universal approximators. Neural Networks 2(5):359–366

Horton T (1992) Object-oriented analysis & design. Englewood Cliffs (New Jersey): Prentice-Hall

Hosseini A, Chen T, Wu W, Sun Y, Sarrafzadeh M (2018) Heteromed: Heterogeneous information network for medical diagnosis. In: Proceedings of the 27th ACM International Conference on Information and Knowledge Management, pp 763–772

Hou S, Ye Y, Song Y, Abdulhayoglu M (2017) Hindroid: An intelligent android malware detection system based on structured heterogeneous information network. In: Proceedings of the 23rd ACM SIGKDD international conference on knowledge discovery and data mining, pp 1507–1515

Houlsby N, Giurgiu A, Jastrzebski S, Morrone B, De Laroussilhe Q, Gesmundo A, Attariyan M, Gelly S (2019) Parameter-efficient transfer learning for nlp. In: International Conference on Machine Learning, PMLR, pp 2790–2799

Hsieh K, Wang Y, Chen L, Zhao Z, Savitz S, Jiang X, Tang J, Kim Y (2020) Drug repurposing for covid-19 using graph neural network with genetic, mechanistic, and epidemiological validation. arXiv preprint arXiv:200910931

Hsu WN, Zhang Y, Glass J (2017) Unsupervised learning of disentangled and interpretable representations from sequential data. In: Proceedings of the 31st International Conference on Neural Information Processing Systems, pp 1876–1887

Hsu WN, Zhang Y, Weiss RJ, Chung YA, Wang Y, Wu Y, Glass J (2019) Disentangling correlated speaker and noise for speech synthesis via data augmentation and adversarial factorization. In: ICASSP 2019-2019 IEEE International Conference on Acoustics, Speech and Signal Processing (ICASSP), IEEE, pp 5901–5905

Hu B, Shi C, Zhao WX, Yu PS (2018a) Leveraging meta-path based context for top-n recommendation with a neural co-attention model. In: Proceedings of the 24th ACM SIGKDD International Conference on Knowledge Discovery & Data Mining, pp 1531–1540

Hu B, Fang Y, Shi C (2019a) Adversarial learning on heterogeneous information networks. In: Proceedings of the 25th ACM SIGKDD International Conference on Knowledge Discovery & Data Mining, pp 120–129

Hu B, Zhang Z, Shi C, Zhou J, Li X, Qi Y (2019b) Cash-out user detection based on attributed heterogeneous information network with a hierarchical attention mechanism. In: Proceedings of the AAAI Conference on Artificial Intelligence, vol 33, pp 946–953

Hu L, Xu S, Li C, Yang C, Shi C, Duan N, Xie X, Zhou M (2020a) Graph neural news recommendation with unsupervised preference disentanglement. In: Proceedings of the 58th Annual Meeting of the Association for Computational Linguistics, pp 4255–4264

Hu R, Aggarwal CC, Ma S, Huai J (2016) An embedding approach to anomaly detection. In: 2016 IEEE 32nd International Conference on Data Engineering (ICDE), IEEE, pp 385–396

Hu W, Fey M, Zitnik M, Dong Y, Ren H, Liu B, Catasta M, Leskovec J (2020b) Open graph benchmark: Datasets for machine learning on graphs. arXiv preprint arXiv:200500687

Hu W, Liu B, Gomes J, Zitnik M, Liang P, Pande VS, Leskovec J (2020c) Strategies for pre-training graph neural networks. In: 8th International Conference on Learning Representations, ICLR 2020, Addis Ababa, Ethiopia, April 26-30, 2020, OpenReview.net

Hu X, Chiueh Tc, Shin KG (2009) Large-scale malware indexing using function-call graphs. In: Proceedings of the 16th ACM Conference on Computer and Communications Security (CCS), Association for Computing Machinery, New York, NY, USA, p 611–620

Hu X, Li G, Xia X, Lo D, Jin Z (2018b) Deep code comment generation. In: Proceedings of the 26th Conference on Program Comprehension, ACM, pp 200–210

Hu X, Li G, Xia X, Lo D, Lu S, Jin Z (2018c) Summarizing source code with transferred api knowledge. In: Proceedings of the 27th International Joint Conference on Artificial Intelligence, AAAI Press, pp 2269–2275

Hu Z, Fan C, Chen T, Chang KW, Sun Y (2019c) Pre-training graph neural networks for generic structural feature extraction. arXiv preprint arXiv:190513728

Hu Z, Dong Y, Wang K, Chang KW, Sun Y (2020d) Gpt-gnn: Generative pre-training of graph neural networks. In: Proceedings of the 26th ACM SIGKDD International Conference on Knowledge Discovery & Data Mining, pp 1857–1867

Hu Z, Dong Y, Wang K, Sun Y (2020e) Heterogeneous graph transformer. In: Proceedings of The Web Conference 2020, pp 2704–2710

Huang D, Chen P, Zeng R, Du Q, Tan M, Gan C (2020a) Location-aware graph convolutional networks for video question answering. In: The Thirty-Fourth AAAI Conference on Artificial Intelligence, AAAI Press, pp 11,021–11,028

Huang G, Liu Z, Van Der Maaten L, Weinberger KQ (2017a) Densely connected convolutional networks. In: Proceedings of the IEEE conference on computer vision and pattern recognition, pp 4700–4708

Huang H, Wang X, Yi Z, Ma X (2000) A character recognition based on feature extraction. Journal of Chongqing University (Natural Science Edition) 23:66–69

Huang H, Alvarez S, Nusinow DA (2016a) Data on the identification of protein interactors with the evening complex and PCH1 in arabidopsis using tandem affinity purification and mass spectrometry (TAP–MS). Data in Brief 8:56–60

Huang J, Li Z, Li N, Liu S, Li G (2019) Attpool: Towards hierarchical feature representation in graph convolutional networks via attention mechanism. In: IEEE/CVF International Conference on Computer Vision, pp 6479–6488

Huang JT, Sharma A, Sun S, Xia L, Zhang D, Pronin P, Padmanabhan J, Ottaviano G, Yang L (2020b) Embedding-based retrieval in facebook search. In: Proceedings of the 26th ACM SIGKDD International Conference on Knowledge Discovery & Data Mining, pp 2553–2561

Huang L, Ma D, Li S, Zhang X, Houfeng W (2019a) Text level graph neural network for text classification. In: Proceedings of the 2019 Conference on Empirical Methods in Natural Language Processing and the 9th International Joint Conference on Natural Language Processing (EMNLP-IJCNLP), pp 3435–3441

Huang Q, Yamada M, Tian Y, Singh D, Yin D, Chang Y (2020c) Graphlime: Local interpretable model explanations for graph neural networks. arXiv preprint arXiv:200106216

Huang S, Kang Z, Tsang IW, Xu Z (2019b) Auto-weighted multi-view clustering via kernelized graph learning. Pattern Recognition 88:174–184

Huang W, Zhang T, Rong Y, Huang J (2018) Adaptive sampling towards fast graph representation learning. Advances in Neural Information Processing Systems 31:4558–4567

Huang X, Alzantot M, Srivastava M (2019c) Neuroninspect: Detecting backdoors in neural networks via output explanations. arXiv preprint arXiv:191107399

Huang X, Song Q, Li Y, Hu X (2019d) Graph recurrent networks with attributed random walks. In: Proceedings of the 25th ACM SIGKDD International Conference on Knowledge Discovery & Data Mining, pp 732–740

Huang Y, Wang W, Wang L (2017b) Instance-aware image and sentence matching with selective multimodal lstm. In: Proceedings of the IEEE Conference on Computer Vision and Pattern Recognition, pp 2310–2318

Huang Z, Mamoulis N (2017) Heterogeneous information network embedding for meta path based proximity. arXiv preprint arXiv:170105291

Huang Z, Xu W, Yu K (2015) Bidirectional lstm-crf models for sequence tagging. arXiv preprint arXiv:150801991

Huang Z, Zheng Y, Cheng R, Sun Y, Mamoulis N, Li X (2016b) Meta structure: Computing relevance in large heterogeneous information networks. In: Proceedings of the 22nd ACM SIGKDD International Conference on Knowledge Discovery and Data Mining, pp 1595–1604

Hurle M, Yang L, Xie Q, Rajpal D, Sanseau P, Agarwal P (2013) Computational drug repositioning: from data to therapeutics. Clinical Pharmacology & Therapeutics 93(4):335–341

Hussein R, Yang D, Cudré-Mauroux P (2018) Are meta-paths necessary? revisiting heterogeneous graph embeddings. In: Proceedings of the 27th ACM International Conference on Information and Knowledge Management, pp 437–446

Hutchins WJ (1995) Machine translation: A brief history. In: Concise history of the language sciences, Elsevier, pp 431–445

Ioannidis VN, Marques AG, Giannakis GB (2019) Graph neural networks for predicting protein functions. In: 2019 IEEE 8th International Workshop on Computational Advances in Multi-Sensor Adaptive Processing (CAMSAP), pp 221–225, DOI 10.1109/CAMSAP45676.2019.9022646

Ioannidis VN, Song X, Manchanda S, Li M, Pan X, Zheng D, Ning X, Zeng X, Karypis G (2020) Drkg - drug repurposing knowledge graph for covid-19. https://github.com/gnn4dr/DRKG/

Ioffe S, Szegedy C (2015) Batch normalization: Accelerating deep network training by reducing internal covariate shift. In: International Conference on Machine Learning, pp 448–456

Irving G, Szegedy C, Alemi AA, Eén N, Chollet F, Urban J (2016) DeepMath - deep sequence models for premise selection. Advances in neural information processing systems 29:2235–2243

Irwin JJ, Sterling T, Mysinger MM, Bolstad ES, Coleman RG (2012) Zinc: a free tool to discover chemistry for biology. Journal of Chemical Information and Modeling 52(7):1757–1768

Issa NT, Stathias V, Schürer S, Dakshanamurthy S (2020) Machine and deep learning approaches for cancer drug repurposing. In: Seminars in cancer biology, Elsevier

Ito T, Chiba T, Ozawa R, et al (2001) A comprehensive two-hybrid analysis to explore the yeast protein interactome. Proceedings of the National Academy of Sciences of the United States of America 98(8):4569–4574

Iyer S, Konstas I, Cheung A, Zettlemoyer L (2016) Summarizing source code using a neural attention model. In: Proceedings of the 54th Annual Meeting of the Association for Computational Linguistics (Volume 1: Long Papers), pp 2073–2083

Jaakkola T, Sontag D, Globerson A, Meila M (2010) Learning bayesian network structure using lp relaxations. In: Proceedings of the Thirteenth International Conference on Artificial Intelligence and Statistics, JMLR Workshop and Conference Proceedings, pp 358–365

Jabri A, Owens A, Efros AA (2020) Space-time correspondence as a contrastive random walk. arXiv preprint arXiv:200614613

Jacob Y, Denoyer L, Gallinari P (2014) Learning latent representations of nodes for classifying in heterogeneous social networks. In: Proceedings of the 7th ACM international conference on Web search and data mining, pp 373–382

Jain A, Zamir AR, Savarese S, Saxena A (2016a) Structural-RNN: Deep learning on spatio-temporal graphs. In: IEEE Conference on Computer Vision and Pattern Recognition, pp 5308–5317

Jain A, Zamir AR, Savarese S, Saxena A (2016b) Structural-rnn: Deep learning on spatio-temporal graphs. In: Proceedings of the ieee conference on computer vision and pattern recognition, pp 5308–5317

Jaitly N, Hinton G (2011) Learning a better representation of speech soundwaves using restricted boltzmann machines. In: 2011 IEEE International Conference on Acoustics, Speech and Signal Processing (ICASSP), IEEE, pp 5884–5887

Jang E, Gu S, Poole B (2017) Categorical reparameterization with gumbel-softmax. In: 5th International Conference on Learning Representations

Jang S, Moon SE, Lee JS (2019) Brain signal classification via learning connectivity structure. arXiv preprint arXiv:190511678

Jassal B, Matthews L, Viteri G, Gong C, Lorente P, Fabregat A, Sidiropoulos K, Cook J, Gillespie M, Haw R, et al (2020) The reactome pathway knowledgebase. Nucleic acids research 48(D1):D498–D503

Jean S, Cho K, Memisevic R, Bengio Y (2014) On using very large target vocabulary for neural machine translation. arXiv preprint arXiv:14122007

Jebara T, Wang J, Chang SF (2009) Graph construction and b-matching for semi-supervised learning. In: Proceedings of the 26th annual international conference on machine learning, pp 441–448

Jeh G, Widom J (2002) Simrank: a measure of structural-context similarity. In: Proceedings of the eighth ACM SIGKDD international conference on Knowledge discovery and data mining, ACM, pp 538–543

Jeh G, Widom J (2003) Scaling personalized web search. In: the International Conference on World Wide Web, pp 271–279

Jenatton R, Le Roux N, Bordes A, Obozinski G (2012) A latent factor model for highly multi-relational data. In: Advances in Neural Information Processing Systems 25 (NIPS 2012), pp 3176–3184

Ji G, He S, Xu L, Liu K, Zhao J (2015) Knowledge graph embedding via dynamic mapping matrix. In: Proceedings of the 53rd annual meeting of the association for computational linguistics and the 7th international joint conference on natural language processing, pp 687–696

Ji G, Liu K, He S, Zhao J (2016) Knowledge graph completion with adaptive sparse transfer matrix. In: Proceedings of the AAAI Conference on Artificial Intelligence, vol 30

Jia J, Wang B, Cao X, Gong NZ (2020) Certified robustness of community detection against adversarial structural perturbation via randomized smoothing. In: The Web Conference, pp 2718–2724

Jia X, De Brabandere B, Tuytelaars T, Gool LV (2016) Dynamic filter networks. Advances in neural information processing systems 29:667–675

Jiang B, Sun P, Tang J, Luo B (2019a) GLMNet: Graph learning-matching networks for feature matching. arXiv preprint arXiv:191107681

Jiang B, Zhang Z, Lin D, Tang J, Luo B (2019b) Semi-supervised learning with graph learning-convolutional networks. In: Proceedings of the IEEE Conference on Computer Vision and Pattern Recognition, pp 11,313–11,320

Jiang C, Coenen F, Sanderson R, Zito M (2010) Text classification using graph mining-based feature extraction. In: Research and Development in Intelligent Systems XXVI, Springer, pp 21–34

Jiang S, Balaprakash P (2020) Graph neural network architecture search for molecular property prediction. arXiv preprint arXiv:200812187

Jiang S, McMillan C, Santelices R (2016) Do programmers do change impact analysis in debugging? Empirical Software Engineering pp 1–39

Jiang S, Armaly A, McMillan C (2017) Automatically generating commit messages from diffs using neural machine translation. In: Proceedings of the 32nd IEEE/ACM International Conference on Automated Software Engineering, IEEE Press, pp 135–146

Jiménez J, Doerr S, Martínez-Rosell G, et al (2017) DeepSite: protein-binding site predictor using 3d-convolutional neural networks. Bioinformatics 33(19):3036–3042

Jin H, Zhang X (2019) Latent Adversarial Training of Graph Convolution Networks. In: ICML 2019 Workshop: Learning and Reasoning with Graph-Structured Representations

Jin H, Song Q, Hu X (2019a) Auto-keras: An efficient neural architecture search system. In: Proceedings of the 25th ACM SIGKDD International Conference on Knowledge Discovery & Data Mining, pp 1946–1956

Jin H, Shi Z, Peruri VJSA, Zhang X (2020a) Certified robustness of graph convolution networks for graph classification under topological attacks. Advances in Neural Information Processing Systems 33

Jin J, Qin J, Fang Y, Du K, Zhang W, Yu Y, Zhang Z, Smola AJ (2020b) An efficient neighborhood-based interaction model for recommendation on heterogeneous graph. In: Proceedings of the 26th ACM SIGKDD International Conference on Knowledge Discovery & Data Mining, pp 75–84

Jin L, Gildea D (2020) Generalized shortest-paths encoders for amr-to-text generation. In: Proceedings of the 28th International Conference on Computational Linguistics, pp 2004–2013

Jin M, Chang H, Zhu W, Sojoudi S (2019b) Power up! robust graph convolutional network against evasion attacks based on graph powering. CoRR abs/1905.10029, 1905.10029

Jin W, Barzilay R, Jaakkola T (2018a) Junction tree variational autoencoder for molecular graph generation. In: Proceedings of the 35th International Conference on Machine Learning, pp 2323–2332

Jin W, Barzilay R, Jaakkola TS (2018b) Junction tree variational autoencoder for molecular graph generation. In: International Conference on Machine Learning, pp 2328–2337

Jin W, Yang K, Barzilay R, Jaakkola T (2018c) Learning multimodal graph-to-graph translation for molecular optimization. arXiv preprint arXiv:181201070

Jin W, Barzilay R, Jaakkola T (2020c) Composing molecules with multiple property constraints. arXiv preprint arXiv:200203244

Jin W, Derr T, Liu H, Wang Y, Wang S, Liu Z, Tang J (2020d) Self-supervised learning on graphs: Deep insights and new direction. arXiv preprint arXiv:200610141

Jin W, Ma Y, Liu X, Tang X, Wang S, Tang J (2020e) Graph structure learning for robust graph neural networks. In: The 26th ACM SIGKDD Conference on Knowledge Discovery and Data Mining, pp 66–74

Jin W, Derr T, Wang Y, Ma Y, Liu Z, Tang J (2021) Node similarity preserving graph convolutional networks. In: Proceedings of the 14th ACM International Conference on Web Search and Data Mining, pp 148–156

Johansson FD, Dubhashi D (2015) Learning with similarity functions on graphs using matchings of geometric embeddings. In: ACM SIGKDD International Conference on Knowledge Discovery and Data Mining, pp 467–476

Johnson D, Larochelle H, Tarlow D (2020) Learning graph structure with a finite-state automaton layer. In: Larochelle H, Ranzato M, Hadsell R, Balcan MF, Lin H (eds) Advances in Neural Information Processing Systems, Curran Associates, Inc., vol 33, pp 3082–3093

Jonas E (2019) Deep imitation learning for molecular inverse problems. Advances in Neural Information Processing Systems 32:4990–5000

Jurafsky D (2000) Speech & language processing. Pearson Education India

Kagdi H, Collard ML, Maletic JI (2007) A survey and taxonomy of approaches for mining software repositories in the context of software evolution. Journal of software maintenance and evolution: Research and practice 19(2):77–131

Kahneman D (2011) Thinking, fast and slow. Macmillan

Kalchbrenner N, Grefenstette E, Blunsom P (2014) A convolutional neural network for modelling sentences. In: Proceedings of the 52nd Annual Meeting of the Association for Computational Linguistics, Association for Computational Linguistics, pp 655–665, DOI 10.3115/v1/P14-1062

Kalliamvakou E, Gousios G, Blincoe K, Singer L, German DM, Damian D (2014) The promises and perils of mining github. In: Proceedings of the 11th working conference on mining software repositories, pp 92–101

Kalofolias V (2016) How to learn a graph from smooth signals. In: Artificial Intelligence and Statistics, PMLR, pp 920–929

Kalofolias V, Perraudin N (2019) Large scale graph learning from smooth signals. In: 7th International Conference on Learning Representations

Kaluza MCDP, Amizadeh S, Yu R (2018) A neural framework for learning dag to dag translation. In: NeurIPS'2018 Workshop

Kampffmeyer M, Chen Y, Liang X, Wang H, Zhang Y, Xing EP (2019) Rethinking knowledge graph propagation for zero-shot learning. In: Proceedings of the IEEE/CVF Conference on Computer Vision and Pattern Recognition, pp 11,487–11,496

Kandasamy K, Neiswanger W, Schneider J, Poczos B, Xing E (2018) Neural architecture search with bayesian optimisation and optimal transport. In: Advances in Neural Information Processing Systems

Kanehisa M, Goto S (2000) Kegg: kyoto encyclopedia of genes and genomes. Nucleic acids research 28(1):27–30

Kanehisa M, Araki M, Goto S, Hattori M, Hirakawa M, Itoh M, Katayama T, Kawashima S, Okuda S, Tokimatsu T, et al (2007) Kegg for linking genomes to life and the environment. Nucleic acids research 36(suppl_1):D480–D484

Kang U, Tong H, Sun J (2012) Fast random walk graph kernel. In: SIAM International Conference on Data Mining, pp 828–838

Kang WC, McAuley J (2018) Self-attentive sequential recommendation. In: 2018 IEEE International Conference on Data Mining (ICDM), IEEE, pp 197–206

Kang Z, Pan H, Hoi SC, Xu Z (2019) Robust graph learning from noisy data. IEEE transactions on cybernetics 50(5):1833–1843

Karampatsis RM, Sutton C (2020) How often do single-statement bugs occur? the ManySStuBs4J dataset. In: Proceedings of the 17th International Conference on Mining Software Repositories, pp 573–577

Karamshuk D, Noulas A, Scellato S, Nicosia V, Mascolo C (2013) Geo-spotting: mining online location-based services for optimal retail store placement. In: Proceedings of the 19th ACM SIGKDD international conference on Knowledge discovery and data mining, pp 793–801

Karita S, Watanabe S, Iwata T, Ogawa A, Delcroix M (2018) Semi-supervised end-to-end speech recognition. In: Interspeech, pp 2–6

Karpathy A, Fei-Fei L (2015) Deep visual-semantic alignments for generating image descriptions. In: Proceedings of the IEEE conference on computer vision and pattern recognition, pp 3128–3137

Karypis G, Kumar V (1995) Multilevel graph partitioning schemes. In: ICPP (3), pp 113–122

Karypis G, Kumar V (1998) A fast and high quality multilevel scheme for partitioning irregular graphs. SIAM Journal on scientific Computing 20(1):359–392

Katharopoulos A, Vyas A, Pappas N, Fleuret F (2020) Transformers are rnns: Fast autoregressive transformers with linear attention. In: International Conference on Machine Learning, PMLR, pp 5156–5165

Katz L (1953) A new status index derived from sociometric analysis. Psychometrika 18(1):39–43

Kawahara J, Brown CJ, Miller SP, Booth BG, Chau V, Grunau RE, Zwicker JG, Hamarneh G (2017) Brainnetcnn: Convolutional neural networks for brain networks; towards predicting neurodevelopment. NeuroImage 146:1038–1049

Kazemi E, Hassani SH, Grossglauser M (2015) Growing a graph matching from a handful of seeds. Proc VLDB Endow 8(10):1010–1021

Kazemi SM, Poole D (2018) Simple embedding for link prediction in knowledge graphs. In: Neural Information Processing Systems, p 4289–4300

Kazemi SM, Goel R, Eghbali S, Ramanan J, Sahota J, Thakur S, Wu S, Smyth C, Poupart P, Brubaker M (2019) Time2vec: Learning a vector representation of time. arXiv preprint arXiv:190705321

Kazemi SM, Goel R, Jain K, Kobyzev I, Sethi A, Forsyth P, Poupart P (2020) Representation learning for dynamic graphs: A survey. Journal of Machine Learning Research 21(70):1–73

Kazi A, Cosmo L, Navab N, Bronstein M (2020) Differentiable graph module (dgm) graph convolutional networks. arXiv preprint arXiv:200204999

Kearnes S, McCloskey K, Berndl M, Pande V, Riley P (2016) Molecular graph convolutions: moving beyond fingerprints. Journal of computer-aided molecular design 30(8):595–608

Keriven N, Peyré G (2019) Universal invariant and equivariant graph neural networks. In: Advances in Neural Information Processing Systems, pp 7090–7099

Kersting K, Kriege NM, Morris C, Mutzel P, Neumann M (2016) Benchmark data sets for graph kernels

Khezerlou AV, Zhou X, Li L, Shafiq Z, Liu AX, Zhang F (2017) A traffic flow approach to early detection of gathering events: Comprehensive results. ACM Transactions on Intelligent Systems and Technology (TIST) 8(6):1–24

Khezerlou AV, Zhou X, Tong L, Li Y, Luo J (2021) Forecasting gathering events through trajectory destination prediction: A dynamic hybrid model. IEEE Transactions on Knowledge and Data Engineering 33(3):991–1004, DOI 10.1109/TKDE.2019.2937082

Khrulkov V, Novikov A, Oseledets I (2018) Expressive power of recurrent neural networks. In: International Conference on Learning Representations

Kiefer S, Schweitzer P, Selman E (2015) Graphs identified by logics with counting. In: International Symposium on Mathematical Foundations of Computer Science, pp 319–330

Kilicoglu H, Shin D, Fiszman M, Rosemblat G, Rindflesch TC (2012) Semmeddb: a pubmed-scale repository of biomedical semantic predications. Bioinformatics 28(23):3158–3160

Kim B, Koyejo O, Khanna R, et al (2016) Examples are not enough, learn to criticize! criticism for interpretability. In: NIPS, pp 2280–2288

Kim D, Oh A (2021) How to find your friendly neighborhood: Graph attention design with self-supervision. In: International Conference on Learning Representations

Kim J, Kim T, Kim S, Yoo CD (2019) Edge-labeling graph neural network for few-shot learning. In: Proceedings of the IEEE/CVF Conference on Computer Vision and Pattern Recognition, pp 11–20

Kingma DP, Welling M (2013) Auto-encoding variational bayes. arXiv preprint arXiv:13126114

Kingma DP, Welling M (2014) Auto-encoding variational bayes. In: 2nd International Conference on Learning Representations

Kingma DP, Rezende DJ, Mohamed S, Welling M (2014) Semi-supervised learning with deep generative models. In: Proceedings of the 27th International Conference on Neural Information Processing Systems-Volume 2, pp 3581–3589

Kingsbury PR, Palmer M (2002) From treebank to propbank. In: LREC, Citeseer, pp 1989–1993

Kipf T, Fetaya E, Wang KC, Welling M, Zemel R (2018) Neural relational inference for interacting systems. In: International Conference on Machine Learning, pp 2688–2697

Kipf TN, Welling M (2016) Variational graph auto-encoders. arXiv preprint arXiv:161107308

Kipf TN, Welling M (2017a) Semi-supervised classification with graph convolutional networks. In: International Conference on Learning Representations

Kipf TN, Welling M (2017b) Semi-supervised classification with graph convolutional networks. In: 5th International Conference on Learning Representations, ICLR 2017, Toulon, France, April 24-26, 2017, Conference Track Proceedings, OpenReview.net

Kireev DB (1995) ChemNet: A novel neural network based method for graph/property mapping. Journal of Chemical Information and Computer Sciences 35(2):175–180

Klicpera J, Bojchevski A, Günnemann S (2019a) Predict then propagate: Graph neural networks meet personalized pagerank. In: International Conference on Learning Representations

Klicpera J, Weißenberger S, Günnemann S (2019b) Diffusion improves graph learning. In: Advances in Neural Information Processing Systems, pp 13,333–13,345

Klicpera J, Groß J, Günnemann S (2020) Directional message passing for molecular graphs. In: International Conference on Learning Representations

Ko AJ, Myers BA, Coblenz MJ, Aung HH (2006) An exploratory study of how developers seek, relate, and collect relevant information during software maintenance tasks. IEEE Transactions on software engineering 32(12):971–987

Koch O, Kriege NM, Humbeck L (2019) Chemical similarity and substructure searches. In: Encyclopedia of Bioinformatics and Computational Biology, Academic Press, Oxford, pp 640–649

Kohavi R, John GH (1995) Automatic parameter selection by minimizing estimated error. In: Machine Learning Proceedings 1995, Elsevier, pp 304–312

Koivisto M, Sood K (2004) Exact bayesian structure discovery in bayesian networks. The Journal of Machine Learning Research 5:549–573

Koncel-Kedziorski R, Bekal D, Luan Y, Lapata M, Hajishirzi H (2019) Text generation from knowledge graphs with graph transformers. In: Proceedings of the 2019 Conference of the North American Chapter of the Association for Computational Linguistics: Human Language Technologies, Volume 1 (Long and Short Papers), pp 2284–2293

Koo DCE, Bonneau R (2018) Towards region-specific propagation of protein functions. Bioinformatics 35(10):1737–1744

Kool W, Van Hoof H, Welling M (2019) Stochastic beams and where to find them: The gumbel-top-k trick for sampling sequences without replacement. In: International Conference on Machine Learning, PMLR, pp 3499–3508

Koren Y (2008) Factorization meets the neighborhood: a multifaceted collaborative filtering model. In: Proceedings of the 14th ACM SIGKDD international conference on Knowledge discovery and data mining, ACM, pp 426–434

Koren Y (2009) Collaborative filtering with temporal dynamics. In: Proceedings of the 15th ACM SIGKDD international conference on Knowledge discovery and data mining, pp 447–456

Koren Y, Bell R, Volinsky C (2009) Matrix factorization techniques for recommender systems. Computer 42(8):30–37

Korte BH, Vygen J, Korte B, Vygen J (2011) Combinatorial optimization, vol 1. Springer

Kosugi S, Yamasaki T (2020) Unpaired image enhancement featuring reinforcement-learning-controlled image editing software. In: Proceedings of the AAAI Conference on Artificial Intelligence, vol 34, pp 11,296–11,303

Kovács IA, Luck K, Spirohn K, et al (2019) Network-based prediction of protein interactions. Nature Communications 10(1)

Kremenek T, Ng AY, Engler DR (2007) A factor graph model for software bug finding. In: IJCAI, pp 2510–2516

Kriege N, Mutzel P (2012) Subgraph matching kernels for attributed graphs. In: Proceedings of the 29th International Coference on International Conference on Machine Learning, Omnipress, Madison, WI, USA, ICML'12, p 291–298

Kriege NM, P-L G, Wilson RC (2016) On valid optimal assignment kernels and applications to graph classification. In: Advances in Neural Information Processing Systems, pp 1615–1623

Kriege NM, Johansson FD, Morris C (2020) A survey on graph kernels. Applied Network Science 5(1):6

Krishnan A (2018) Making search easier: How amazon's product graph is helping customers find products more easily. ed Amazon Blog

Krishnapuram R, Medasani S, Jung SH, Choi YS, Balasubramaniam R (2004) Content-based image retrieval based on a fuzzy approach. IEEE transactions on knowledge and data engineering 16(10):1185–1199

Krizhevsky A, Sutskever I, Hinton GE (2012) Imagenet classification with deep convolutional neural networks. Advances in neural information processing systems 25:1097–1105

Kuhn M, Letunic I, Jensen LJ, Bork P (2016) The sider database of drugs and side effects. Nucleic acids research 44(D1):D1075–D1079

Kulmanov M, Hoehndorf R (2019) DeepGOPlus: improved protein function prediction from sequence. Bioinformatics

Kumar S, Spezzano F, Subrahmanian V, Faloutsos C (2016) Edge weight prediction in weighted signed networks. In: 2016 IEEE 16th International Conference on Data Mining (ICDM), IEEE, pp 221–230

Kumar S, Ying J, de Miranda Cardoso JV, Palomar D (2019a) Structured graph learning via laplacian spectral constraints. In: Advances in Neural Information Processing Systems, pp 11,651–11,663

Kumar S, Zhang X, Leskovec J (2019b) Predicting dynamic embedding trajectory in temporal interaction networks. In: ACM SIGKDD International Conference on Knowledge Discovery & Data Mining, pp 1269–1278

Kumar S, Ying J, de Miranda Cardoso JV, Palomar DP (2020) A unified framework for structured graph learning via spectral constraints. Journal of Machine Learning Research 21(22):1–60

Kusner MJ, Paige B, Hernández-Lobato JM (2017) Grammar variational autoencoder. In: International Conference on Machine Learning, pp 1945–1954

Lacroix T, Obozinski G, Usunier N (2020) Tensor decompositions for temporal knowledge base completion. In: International Conference on Learning Representations

Lake B, Tenenbaum J (2010) Discovering structure by learning sparse graphs. In: Proceedings of the Annual Meeting of the Cognitive Science Society, vol 32

Lamb LC, Garcez A, Gori M, Prates M, Avelar P, Vardi M (2020) Graph neural networks meet neural-symbolic computing: A survey and perspective. In: Proceedings of IJCAI-PRICAI 2020

Lan Z, Chen M, Goodman S, Gimpel K, Sharma P, Soricut R (2020) ALBERT: A lite BERT for self-supervised learning of language representations. In: 8th International Conference on Learning Representations, ICLR 2020, Addis Ababa, Ethiopia, April 26-30, 2020, OpenReview.net

Lanczos C (1950) An iteration method for the solution of the eigenvalue problem of linear differential and integral operators. United States Governm. Press Office Los Angeles, CA

Landrieu L, Simonovsky M (2018) Large-scale point cloud semantic segmentation with superpoint graphs. In: Proceedings of the IEEE Conference on Computer Vision and Pattern Recognition, pp 4558–4567

Latif S, Rana R, Khalifa S, Jurdak R, Epps J (2019) Direct modelling of speech emotion from raw speech. In: Proceedings of the 20th Annual Conference of the International Speech Communication Association (INTERSPEECH 2019), International Speech Communication Association (ISCA), pp 3920–3924

Lawler EL (1963) The quadratic assignment problem. Management science 9(4):586–599

Le Cun Y, Boser B, Denker JS, Henderson D, Howard RE, Hubbard W, Jackel LD (1989) Handwritten digit recognition with a back-propagation network. In: Neural Information Processing Systems, pp 396–404

Le-Khac PH, Healy G, Smeaton AF (2020) Contrastive representation learning: a framework and review. IEEE Access 8:1–28

Leblay J, Chekol MW (2018) Deriving validity time in knowledge graph. In: Companion Proceedings of the The Web Conference 2018, pp 1771–1776

LeClair A, McMillan C (2019) Recommendations for datasets for source code summarization. In: Proceedings of the 2019 Conference of the North American Chapter of the Association for Computational Linguistics: Human Language Technologies, Volume 1 (Long and Short Papers), pp 3931–3937

LeClair A, Jiang S, McMillan C (2019) A neural model for generating natural language summaries of program subroutines. In: Proceedings of the 41st International Conference on Software Engineering, IEEE Press, pp 795–806

LeClair A, Haque S, Wu L, McMillan C (2020) Improved code summarization via a graph neural network. In: 28th ACM/IEEE International Conference on Program Comprehension (ICPC'20)

LeCun Y, Boser B, Denker JS, Henderson D, Howard RE, Hubbard W, Jackel LD (1989) Backpropagation applied to handwritten zip code recognition. Neural computation 1(4):541–551

Lecuyer M, Atlidakis V, Geambasu R, Hsu D, Jana S (2019) Certified robustness to adversarial examples with differential privacy. In: IEEE Symposium on Security and Privacy, DOI 10.1109/SP.2019.00044

Lee G, Yuan Y, Chang S, Jaakkola TS (2019a) Tight certificates of adversarial robustness for randomly smoothed classifiers. In: Wallach HM, Larochelle H, Beygelzimer A, d'Alché-Buc F, Fox EB, Garnett R (eds) Advances in Neural Information Processing Systems 32: Annual Conference on Neural Information Processing Systems 2019, NeurIPS 2019, December 8-14, 2019, Vancouver, BC, Canada, pp 4911–4922

Lee J, Lee I, Kang J (2019b) Self-attention graph pooling. In: International Conference on Machine Learning, PMLR, pp 3734–3743

Lee JB, Rossi RA, Kim S, Ahmed NK, Koh E (2019c) Attention models in graphs: A survey. ACM Transactions on Knowledge Discovery from Data (TKDD) 13(6):1–25

Lee JB, Rossi RA, Kong X, Kim S, Koh E, Rao A (2019d) Graph convolutional networks with motif-based attention. In: 28th ACM International Conference on Information, pp 499–508

Lee S, Park C, Yu H (2019e) Bhin2vec: Balancing the type of relation in heterogeneous information network. In: Proceedings of the 28th ACM International Conference on Information and Knowledge Management, pp 619–628

Lei T, Jin W, Barzilay R, Jaakkola T (2017a) Deriving neural architectures from sequence and graph kernels. In: Proceedings of the 34th International Conference on Machine Learning-Volume 70, pp 2024–2033

Lei T, Zhang Y, Wang SI, Dai H, Artzi Y (2017b) Simple recurrent units for highly parallelizable recurrence. arXiv preprint arXiv:170902755

Leordeanu M, Hebert M (2005) A spectral technique for correspondence problems using pairwise constraints. In: IEEE International Conference on Computer Vision, pp 1482–1489

Leskovec J, Grobelnik M, Milic-Frayling N (2004) Learning sub-structures of document semantic graphs for document summarization. In: LinkKDD Workshop, pp 133–138

Leskovec J, Chakrabarti D, Kleinberg J, Faloutsos C, Ghahramani Z (2010) Kronecker graphs: an approach to modeling networks. Journal of Machine Learning Research 11(2)

Letovsky S (1987) Cognitive processes in program comprehension. Journal of Systems and software 7(4):325–339

Levi FW (1942) Finite geometrical systems: six public lectues delivered in February, 1940, at the University of Calcutta. University of Calcutta

Levie R, Monti F, Bresson X, Bronstein MM (2019) Cayleynets: Graph convolutional neural networks with complex rational spectral filters. IEEE Trans Signal Process 67(1):97–109

Levin E, Pieraccini R, Eckert W (2000) A stochastic model of human-machine interaction for learning dialog strategies. IEEE Transactions on speech and audio processing 8(1):11–23

Levy O, Goldberg Y (2014) Neural word embedding as implicit matrix factorization. In: Advances in neural information processing systems, pp 2177–2185

Lewis HR, et al (1983) Michael r. garey, david s. johnson, computers and intractability. a guide to the theory of np-completeness. Journal of Symbolic Logic 48(2):498–500

Lewis M, Liu Y, Goyal N, Ghazvininejad M, Mohamed A, Levy O, Stoyanov V, Zettlemoyer L (2020) BART: Denoising sequence-to-sequence pre-training for natural language generation, translation, and comprehension. In: Proceedings of the 58th Annual Meeting of the Association for Computational Linguistics, p 7871, DOI 10.18653/v1/2020.acl-main.703

Li A, Qin Z, Liu R, Yang Y, Li D (2019a) Spam review detection with graph convolutional networks. In: Proceedings of the 28th ACM International Conference on Information and Knowledge Management, pp 2703–2711

Li C, Ma J, Guo X, Mei Q (2017a) Deepcas: An end-to-end predictor of information cascades. In: Proceedings of the 26th international conference on World Wide Web, pp 577–586

Li C, Liu Z, Wu M, Xu Y, Zhao H, Huang P, Kang G, Chen Q, Li W, Lee DL (2019b) Multi-interest network with dynamic routing for recommendation at tmall. In: Proceedings of the 28th ACM International Conference on Information and Knowledge Management, pp 2615–2623

Li F, Gan C, Liu X, Bian Y, Long X, Li Y, Li Z, Zhou J, Wen S (2017b) Temporal modeling approaches for large-scale youtube-8m video understanding. arXiv preprint arXiv:170704555

Li G, Muller M, Thabet A, Ghanem B (2019c) Deepgcns: Can gcns go as deep as cnns? In: Proceedings of the IEEE/CVF International Conference on Computer Vision, pp 9267–9276

Li J, Wang Y, Lyu MR, King I (2018a) Code completion with neural attention and pointer networks. In: Proceedings of the 27th International Joint Conference on Artificial Intelligence, pp 4159–25

Li J, Yang F, Tomizuka M, Choi C (2020a) Evolvegraph: Multi-agent trajectory prediction with dynamic relational reasoning. Advances in Neural Information Processing Systems 33

Li L, Feng H, Zhuang W, Meng N, Ryder B (2017c) Cclearner: A deep learning-based clone detection approach. In: 2017 IEEE International Conference on Software Maintenance and Evolution (ICSME), IEEE, pp 249–260

Li L, Tang S, Deng L, Zhang Y, Tian Q (2017d) Image caption with global-local attention. In: Proceedings of the AAAI Conference on Artificial Intelligence, vol 31

Li L, Gan Z, Cheng Y, Liu J (2019d) Relation-aware graph attention network for visual question answering. In: Proceedings of the IEEE/CVF International Conference on Computer Vision, pp 10,313–10,322

Li L, Wang P, Yan J, Wang Y, Li S, Jiang J, Sun Z, Tang B, Chang TH, Wang S, et al (2020b) Real-world data medical knowledge graph: construction and applications. Artificial intelligence in medicine 103:101,817

Li L, Zhang Y, Chen L (2020c) Generate neural template explanations for recommendation. In: Proceedings of the 29th ACM International Conference on Information & Knowledge Management, pp 755–764

Li M, Chen S, Chen X, Zhang Y, Wang Y, Tian Q (2019e) Actional-structural graph convolutional networks for skeleton-based action recognition. In: IEEE/CVF Conference on Computer Vision and Pattern Recognition, pp 3595–3603

Li N, Yang Z, Luo L, Wang L, Zhang Y, Lin H, Wang J (2020d) Kghc: a knowledge graph for hepatocellular carcinoma. BMC Medical Informatics and Decision Making 20(3):1–11

Li P, Chien I, Milenkovic O (2019f) Optimizing generalized pagerank methods for seed-expansion community detection. In: Advances in Neural Information Processing Systems, pp 11,705–11,716

Li P, Wang Y, Wang H, Leskovec J (2020e) Distance encoding: Design provably more powerful neural networks for graph representation learning. Advances in Neural Information Processing Systems 33

Li Q, Zheng Y, Xie X, Chen Y, Liu W, Ma WY (2008) Mining user similarity based on location history. In: Proceedings of the 16th ACM SIGSPATIAL international conference on Advances in geographic information systems, pp 1–10

Li Q, Han Z, Wu XM (2018b) Deeper insights into graph convolutional networks for semi-supervised learning. In: Proceedings of the AAAI Conference on Artificial Intelligence, vol 32

Li R, Tapaswi M, Liao R, Jia J, Urtasun R, Fidler S (2017e) Situation recognition with graph neural networks. In: Proceedings of the IEEE International Conference on Computer Vision, pp 4173–4182

Li R, Wang S, Zhu F, Huang J (2018c) Adaptive graph convolutional neural networks. In: Proceedings of the AAAI Conference on Artificial Intelligence, vol 32

Li S, Wu L, Feng S, Xu F, Xu F, Zhong S (2020f) Graph-to-tree neural networks for learning structured input-output translation with applications to semantic parsing and math word problem. In: Findings of the Association for Computational Linguistics: EMNLP 2020, Association for Computational Linguistics, Online, pp 2841–2852, DOI 10.18653/v1/2020.findings-emnlp.255, URL https://www.aclweb.org/anthology/2020.findings-emnlp.255

Li X, Cheng Y, Cong G, Chen L (2017f) Discovering pollution sources and propagation patterns in urban area. In: Proceedings of the 23rd ACM SIGKDD International Conference on Knowledge Discovery and Data Mining, pp 1863–1872

Li X, Kao B, Ren Z, Yin D (2019g) Spectral clustering in heterogeneous information networks. In: Proceedings of the AAAI Conference on Artificial Intelligence, vol 33, pp 4221–4228

Li X, Wang C, Tong B, Tan J, Zeng X, Zhuang T (2020g) Deep time-aware item evolution network for click-through rate prediction. In: Proceedings of the 29th ACM International CIKM, pp 785–794

Li Y, Gupta A (2018) Beyond grids: Learning graph representations for visual recognition. In: Proceedings of the 32nd International Conference on Neural Information Processing Systems, pp 9245–9255

Li Y, King I (2020) Autograph: Automated graph neural network. In: International Conference on Neural Information Processing, Springer, pp 189–201

Li Y, Tarlow D, Brockschmidt M, Zemel R (2016a) Gated graph seqrlence neural networks. In: International Conference on Learning Representations

Li Y, Tarlow D, Brockschmidt M, Zemel R (2016b) Gated graph sequence neural networks. In: International Conference on Learning Representations (ICLR)

Li Y, Vinyals O, Dyer C, Pascanu R, Battaglia P (2018d) Learning deep generative models of graphs. arXiv preprint arXiv:180303324

Li Y, Yu R, Shahabi C, Liu Y (2018e) Diffusion convolutional recurrent neural network: Data-driven traffic forecasting. In: International Conference on Learning Representations

Li Y, Zhang L, Liu Z (2018f) Multi-objective de novo drug design with conditional graph generative model. Journal of cheminformatics 10(1):1–24

Li Y, Gu C, Dullien T, Vinyals O, Kohli P (2019h) Graph matching networks for learning the similarity of graph structured objects. In: International Conference on Machine Learning, PMLR, pp 3835–3845

Li Y, Liu M, Yin J, Cui C, Xu XS, Nie L (2019i) Routing micro-videos via a temporal graph-guided recommendation system. In: Proceedings of the 27th ACM International Conference on Multimedia, pp 1464–1472

Li Y, Lin Y, Madhusudan M, Sharma A, Xu W, Sapatnekar SS, Harjani R, Hu J (2020h) A customized graph neural network model for guiding analog ic placement. In: International Conference On Computer Aided Design, IEEE, pp 1–9

Liang S, Srikant R (2017) Why deep neural networks for function approximation? In: 5th International Conference on Learning Representations, ICLR 2017

Liang Y, Zhu KQ (2018) Automatic generation of text descriptive comments for code blocks. In: McIlraith SA, Weinberger KQ (eds) Proceedings of the Thirty-Second AAAI Conference on Artificial Intelligence (AAAI-18), AAAI Press, pp 5229–5236

Liao L, He X, Zhang H, Chua TS (2018) Attributed social network embedding. IEEE Transactions on Knowledge and Data Engineering 30(12):2257–2270

Liao R, Li Y, Song Y, Wang S, Nash C, Hamilton WL, Duvenaud D, Urtasun R, Zemel RS (2019a) Efficient graph generation with graph recurrent attention networks. arXiv preprint arXiv:191000760

Liao R, Zhao Z, Urtasun R, Zemel RS (2019b) Lanczosnet: Multi-scale deep graph convolutional networks. arXiv preprint arXiv:190101484

Liao R, Urtasun R, Zemel R (2021) A pac-bayesian approach to generalization bounds for graph neural networks. In: International Conference on Learning Representations

Liben-Nowell D, Kleinberg J (2007) The link-prediction problem for social networks. Journal of the American society for information science and technology 58(7):1019–1031

Licata L, Lo Surdo P, Iannuccelli M, Palma A, Micarelli E, Perfetto L, Peluso D, Calderone A, Castagnoli L, Cesareni G (2020) Signor 2.0, the signaling network open resource 2.0: 2019 update. Nucleic acids research 48(D1):D504–D510

Lillicrap TP, Hunt JJ, Pritzel A, Heess N, Erez T, Tassa Y, Silver D, Wierstra D (2015) Continuous control with deep reinforcement learning. arXiv preprint arXiv:150902971

Lin C, Sun GJ, Bulusu KC, Dry JR, Hernandez M (2020a) Graph neural networks including sparse interpretability. arXiv preprint arXiv:200700119

Lin G, Wen S, Han QL, Zhang J, Xiang Y (2020b) Software vulnerability detection using deep neural networks: a survey. Proceedings of the IEEE 108(10):1825–1848

Lin P, Sun P, Cheng G, Xie S, Li X, Shi J (2020c) Graph-guided architecture search for real-time semantic segmentation. In: Proceedings of the IEEE/CVF Conference on Computer Vision and Pattern Recognition, pp 4203–4212

Lin T, Zhao X, Shou Z (2017) Single shot temporal action detection. In: Proceedings of the 25th ACM international conference on Multimedia, pp 988–996

Lin W, Ji S, Li B (2020d) Adversarial Attacks on Link Prediction Algorithms Based on Graph Neural Networks. In: ACM Asia Conference on Computer and Communications Security

Lin X, Chen X (2012) Heterogeneous data integration by tree-augmented naïve bayes for protein-protein interactions prediction. PROTEOMICS 13(2):261–268

Lin Y, Liu Z, Sun M, Liu Y, Zhu X (2015) Learning entity and relation embeddings for knowledge graph completion. In: Proceedings of the AAAI Conference on Artificial Intelligence, vol 29

Lin Y, Ren P, Chen Z, Ren Z, Yu D, Ma J, Rijke Md, Cheng X (2020e) Meta matrix factorization for federated rating predictions. In: Proceedings of the 43rd International ACM SIGIR Conference on Research and Development in Information Retrieval, pp 981–990

Lin ZH, Huang SY, Wang YCF (2020f) Convolution in the cloud: Learning deformable kernels in 3d graph convolution networks for point cloud analysis. In: Proceedings of the IEEE/CVF Conference on Computer Vision and Pattern Recognition, pp 1800–1809

Ling X, Ji S, Zou J, Wang J, Wu C, Li B, Wang T (2019) DEEPSEC: A uniform platform for security analysis of deep learning model. In: 2019 IEEE Symposium on Security and Privacy (S&P), IEEE, pp 673–690

Ling X, Wu L, Wang S, Ma T, Xu F, Liu AX, Wu C, Ji S (2020) Multi-level graph matching networks for deep graph similarity learning. arXiv preprint arXiv:200704395

Ling X, Wu L, Wang S, Pan G, Ma T, Xu F, Liu AX, Wu C, Ji S (2021) Deep graph matching and searching for semantic code retrieval. ACM Transactions on Knowledge Discovery from Data (TKDD)

Linial N, London E, Rabinovich Y (1995) The geometry of graphs and some of its algorithmic applications. Combinatorica 15(2):215–245

Linmei H, Yang T, Shi C, Ji H, Li X (2019) Heterogeneous graph attention networks for semi-supervised short text classification. In: Proceedings of the 2019 Conference on Empirical Methods in Natural Language Processing and the 9th International Joint Conference on Natural Language Processing (EMNLP-IJCNLP), pp 4823–4832

Liu A, Xu N, Zhang H, Nie W, Su Y, Zhang Y (2018a) Multi-level policy and reward reinforcement learning for image captioning. In: IJCAI, pp 821–827

Liu B, Niu D, Lai K, Kong L, Xu Y (2017a) Growing story forest online from massive breaking news. In: Proceedings of the 2017 ACM on Conference on Information and Knowledge Management, pp 777–785

Liu B, Niu D, Wei H, Lin J, He Y, Lai K, Xu Y (2019a) Matching article pairs with graphical decomposition and convolutions. In: Proceedings of the 57th Annual Meeting of the Association for Computational Linguistics, pp 6284–6294

Liu B, Han FX, Niu D, Kong L, Lai K, Xu Y (2020a) Story forest: Extracting events and telling stories from breaking news. ACM Transactions on Knowledge Discovery from Data (TKDD) 14(3):1–28

Liu C, Zoph B, Neumann M, Shlens J, Hua W, Li LJ, Fei-Fei L, Yuille A, Huang J, Murphy K (2018b) Progressive neural architecture search. In: Proceedings of the European conference on computer vision, pp 19–34

Liu H, Simonyan K, Vinyals O, Fernando C, Kavukcuoglu K (2017b) Hierarchical representations for efficient architecture search. arXiv preprint arXiv:171100436

Liu H, Simonyan K, Yang Y (2018c) Darts: Differentiable architecture search. arXiv preprint arXiv:180609055

Liu J, Chi Y, Zhu C (2015) A dynamic multiagent genetic algorithm for gene regulatory network reconstruction based on fuzzy cognitive maps. IEEE Transactions on Fuzzy Systems 24(2):419–431

Liu J, Kumar A, Ba J, Kiros J, Swersky K (2019b) Graph normalizing flows. arXiv preprint arXiv:190513177

Liu L, Ma Y, Zhu X, et al (2019) Integrating sequence and network information to enhance protein-protein interaction prediction using graph convolutional networks. In: 2019 IEEE International Conference on Bioinformatics and Biomedicine (BIBM), pp 1762–1768, DOI 10.1109/BIBM47256.2019.8983330

Liu L, Ouyang W, Wang X, Fieguth P, Chen J, Liu X, Pietikäinen M (2020b) Deep learning for generic object detection: A survey. International journal of computer vision 128(2):261–318

Liu M, Gao H, Ji S (2020c) Towards deeper graph neural networks. In: Proceedings of the 26th ACM SIGKDD International Conference on Knowledge Discovery & Data Mining, pp 338–348

Liu N, Tan Q, Li Y, Yang H, Zhou J, Hu X (2019a) Is a single vector enough? exploring node polysemy for network embedding. In: Proceedings of the 25th ACM

SIGKDD International Conference on Knowledge Discovery & Data Mining, pp 932–940

Liu N, Du M, Hu X (2020d) Adversarial machine learning: An interpretation perspective. arXiv preprint arXiv:200411488

Liu P, Chang S, Huang X, Tang J, Cheung JCK (2019b) Contextualized non-local neural networks for sequence learning. In: Proceedings of the AAAI Conference on Artificial Intelligence, vol 33, pp 6762–6769

Liu Q, Allamanis M, Brockschmidt M, Gaunt AL (2018d) Constrained graph variational autoencoders for molecule design. arXiv preprint arXiv:180509076

Liu S, Yang N, Li M, Zhou M (2014) A recursive recurrent neural network for statistical machine translation. In: Proceedings of the 52nd Annual Meeting of the Association for Computational Linguistics, ACL 2014, June 22-27, 2014, Baltimore, MD, USA, Volume 1: Long Papers, The Association for Computer Linguistics, pp 1491–1500

Liu S, Chen Y, Xie X, Siow JK, Liu Y (2021) Retrieval-augmented generation for code summarization via hybrid gnn. In: 9th International Conference on Learning Representations

Liu X, Si S, Zhu X, Li Y, Hsieh CJ (2019c) A Unified Framework for Data Poisoning Attack to Graph-based Semi-supervised Learning. In: Neural Information Processing Systems, NeurIPS

Liu X, Pan H, He M, Song Y, Jiang X, Shang L (2020e) Neural subgraph isomorphism counting. In: Proceedings of the 26th ACM SIGKDD International Conference on Knowledge Discovery & Data Mining, pp 1959–1969

Liu X, Zhang F, Hou Z, Wang Z, Mian L, Zhang J, Tang J (2020f) Self-supervised learning: Generative or contrastive. arXiv preprint arXiv:200608218 1(2)

Liu Y, Lee J, Park M, Kim S, Yang E, Hwang SJ, Yang Y (2018e) Learning to propagate labels: Transductive propagation network for few-shot learning. arXiv preprint arXiv:180510002

Liu Y, Wan B, Zhu X, He X (2020g) Learning cross-modal context graph for visual grounding. In: Proceedings of the AAAI Conference on Artificial Intelligence, vol 34, pp 11,645–11,652

Liu Y, Zhang F, Zhang Q, Wang S, Wang Y, Yu Y (2020h) Cross-view correspondence reasoning based on bipartite graph convolutional network for mammogram mass detection. In: Proceedings of the IEEE/CVF Conference on Computer Vision and Pattern Recognition, pp 3812–3822

Liu Z, Chen C, Yang X, Zhou J, Li X, Song L (2018f) Heterogeneous graph neural networks for malicious account detection. In: Proceedings of the 27th ACM International Conference on Information and Knowledge Management, pp 2077–2085

Livshits B, Nori AV, Rajamani SK, Banerjee A (2009) Merlin: specification inference for explicit information flow problems. ACM Sigplan Notices 44(6):75–86

Locatelli A, Sieniutycz S (2002) Optimal control: An introduction. Appl Mech Rev 55(3):B48–B49

Loiola EM, de Abreu NMM, Boaventura-Netto PO, Hahn P, Querido T (2007) A survey for the quadratic assignment problem. European journal of operational research 176(2):657–690

Lops P, De Gemmis M, Semeraro G (2011) Content-based recommender systems: State of the art and trends. In: Recommender systems handbook, Springer, pp 73–105

Loukas A (2020) What graph neural networks cannot learn: depth vs width. In: International Conference on Learning Representations

Lovász L, et al (1993) Random walks on graphs: A survey. Combinatorics, Paul erdos is eighty 2(1):1–46

Lovell SC, Davis IW, Arendall WB, et al (2003) Structure validation by c geometry: , and c deviation. Proteins: Structure, Function, and Bioinformatics 50(3):437–450

Loyola P, Marrese-Taylor E, Matsuo Y (2017) A neural architecture for generating natural language descriptions from source code changes. In: Proceedings of the 55th Annual Meeting of the Association for Computational Linguistics (Volume 2: Short Papers), pp 287–292

Lü L, Zhou T (2011) Link prediction in complex networks: A survey. Physica A: statistical mechanics and its applications 390(6):1150–1170

Lu X, Wang B, Zheng X, Li X (2017a) Exploring models and data for remote sensing image caption generation. IEEE Transactions on Geoscience and Remote Sensing 56(4):2183–2195

Lu Y, Zhao Z, Li G, Jin Z (2017b) Learning to generate comments for api-based code snippets. In: Software Engineering and Methodology for Emerging Domains, Springer, pp 3–14

Lucic A, ter Hoeve M, Tolomei G, de Rijke M, Silvestri F (2021) Cf-gnnexplainer: Counterfactual explanations for graph neural networks. arXiv preprint arXiv:210203322

Luo D, Cheng W, Xu D, Yu W, Zong B, Chen H, Zhang X (2020) Parameterized explainer for graph neural network. arXiv preprint arXiv:201104573

Luo D, Cheng W, Yu W, Zong B, Ni J, Chen H, Zhang X (2021) Learning to Drop: Robust Graph Neural Network via Topological Denoising. In: International Conference on Web Search and Data Mining, WSDM

Luo R, Liao W, Huang X, Pi Y, Philips W (2016) Feature extraction of hyperspectral images with semisupervised graph learning. IEEE Journal of Selected Topics in Applied Earth Observations and Remote Sensing 9(9):4389–4399

Luo R, Tian F, Qin T, Chen EH, Liu TY (2018) Neural architecture optimization. In: Advances in neural information processing systems

Luo X, You Z, Zhou M, et al (2015) A highly efficient approach to protein inter-actome mapping based on collaborative filtering framework. Scientific Reports 5(1):7702

Luong T, Pham H, Manning CD (2015) Effective approaches to attention-based neural machine translation. In: Proceedings of the 2015 Conference on Empirical Methods in Natural Language Processing, Association for Computational Linguistics, Lisbon, Portugal, pp 1412–1421, DOI 10.18653/v1/D15-1166

Ma G, Ahmed NK, Willke TL, Yu PS (2019a) Deep graph similarity learning: A survey. arXiv preprint arXiv:191211615

Ma H, Bian Y, Rong Y, Huang W, Xu T, Xie W, Ye G, Huang J (2020a) Multi-view graph neural networks for molecular property prediction. arXiv e-prints pp arXiv–2005

Ma J, Tang W, Zhu J, Mei Q (2019b) A flexible generative framework for graph-based semi-supervised learning. In: Advances in Neural Information Processing Systems, pp 3281–3290

Ma J, Zhou C, Cui P, Yang H, Zhu W (2019c) Learning disentangled representations for recommendation. In: Wallach HM, Larochelle H, Beygelzimer A, d'Alché-Buc F, Fox EB, Garnett R (eds) Advances in Neural Information Processing Systems 32: Annual Conference on Neural Information Processing Systems 2019, NeurIPS 2019, December 8-14, 2019, Vancouver, BC, Canada, pp 5712–5723

Ma J, Ding S, Mei Q (2020b) Towards more practical adversarial attacks on graph neural networks. In: Larochelle H, Ranzato M, Hadsell R, Balcan M, Lin H (eds) Advances in Neural Information Processing Systems 33: Annual Conference on Neural Information Processing Systems 2020, NeurIPS 2020, December 6-12, 2020, virtual

Ma T, Chen J, Xiao C (2018) Constrained generation of semantically valid graphs via regularizing variational autoencoders. In: Advances in Neural Information Processing Systems, pp 7113–7124

Ma Y, Wang S, Aggarwal CC, Tang J (2019d) Graph convolutional networks with eigenpooling. In: ACM SIGKDD International Conference on Knowledge Discovery & Data Mining, ACM, pp 723–731

Maalej W, Tiarks R, Roehm T, Koschke R (2014) On the comprehension of program comprehension. ACM Transactions on Software Engineering and Methodology (TOSEM) 23(4):1–37

Maddison C, Mnih A, Teh Y (2017) The concrete distribution: A continuous relaxation of discrete random variables. International Conference on Learning Representations

Madry A, Makelov A, Schmidt L, Tsipras D, Vladu A (2017) Towards deep learning models resistant to adversarial attacks. arXiv preprint arXiv:170606083

Maglott D, Ostell J, Pruitt KD, Tatusova T (2010) Entrez gene: gene-centered information at ncbi. Nucleic acids research 39(suppl_1):D52–D57

Malewicz G, Austern MH, Bik AJ, Dehnert JC, Horn I, Leiser N, Czajkowski G (2010) Pregel: a system for large-scale graph processing. In: Proceedings of the 2010 ACM SIGMOD International Conference on Management of data, pp 135–146

Malliaros FD, Vazirgiannis M (2013) Clustering and community detection in directed networks: A survey. Physics reports 533(4):95–142

Man T, Shen H, Liu S, Jin X, Cheng X (2016) Predict anchor links across social networks via an embedding approach. In: Ijcai, vol 16, pp 1823–1829

Manessi F, Rozza A (2020) Graph-based neural network models with multiple self-supervised auxiliary tasks. arXiv preprint arXiv: 201107267

Manessi F, Rozza A, Manzo M (2020) Dynamic graph convolutional networks. Pattern Recognition 97:107,000

Mangal R, Zhang X, Nori AV, Naik M (2015) A user-guided approach to program analysis. In: Proceedings of the 2015 10th Joint Meeting on Foundations of Software Engineering, pp 462–473

Manning C, Schutze H (1999) Foundations of statistical natural language processing. MIT press

Marcheggiani D, Titov I (2017) Encoding sentences with graph convolutional networks for semantic role labeling. In: EMNLP 2017-Conference on Empirical Methods in Natural Language Processing, Proceedings, pp 1506–1515

Marcheggiani D, Bastings J, Titov I (2018) Exploiting semantics in neural machine translation with graph convolutional networks. arXiv preprint arXiv:180408313

Maretic HP, Thanou D, Frossard P (2017) Graph learning under sparsity priors. In: 2017 IEEE International Conference on Acoustics, Speech and Signal Processing (ICASSP), Ieee, pp 6523–6527

Markovitz A, Sharir G, Friedman I, Zelnik-Manor L, Avidan S (2020) Graph embedded pose clustering for anomaly detection. In: Proceedings of the IEEE/CVF Conference on Computer Vision and Pattern Recognition, pp 10,539–10,547

Maron H, Ben-Hamu H, Shamir N, Lipman Y (2018) Invariant and equivariant graph networks. In: International Conference on Learning Representations

Maron H, Ben-Hamu H, Serviansky H, Lipman Y (2019a) Provably powerful graph networks. In: Advances in Neural Information Processing Systems, pp 2153–2164

Maron H, Fetaya E, Segol N, Lipman Y (2019b) On the universality of invariant networks. In: International Conference on Machine Learning, pp 4363–4371

Mathew B, Sikdar S, Lemmerich F, Strohmaier M (2020) The polar framework: Polar opposites enable interpretability of pre-trained word embeddings. In: Proceedings of The Web Conference 2020, pp 1548–1558

Matsuno R, Murata T (2018) Mell: effective embedding method for multiplex networks. In: Companion Proceedings of the The Web Conference 2018, pp 1261–1268

Matuszek C (2018) Grounded language learning: Where robotics and nlp meet (invited talk). In: Proceedings of the 27th International Joint Conference on Artificial Intelligence, pp 5687–5691

Maziarka Ł, Danel T, Mucha S, Rataj K, Tabor J, Jastrzebski S (2020a) Molecule attention transformer. arXiv preprint arXiv:200208264

Maziarka Ł, Pocha A, Kaczmarczyk J, Rataj K, Danel T, Warchoł M (2020b) Molcyclegan: a generative model for molecular optimization. Journal of Cheminformatics 12(1):1–18

McBurney PW, McMillan C (2014) Automatic documentation generation via source code summarization of method context. In: Proceedings of the 22nd International Conference on Program Comprehension, ACM, pp 279–290

McBurney PW, McMillan C (2016) Automatic source code summarization of context for java methods. IEEE Transactions on Software Engineering 42(2):103–119

McBurney PW, Liu C, McMillan C (2016) Automated feature discovery via sentence selection and source code summarization. Journal of Software: Evolution and Process 28(2):120–145

McMillan C, Grechanik M, Poshyvanyk D, Xie Q, Fu C (2011) Portfolio: finding relevant functions and their usage. In: Proceedings of the 33rd International Conference on Software Engineering, pp 111–120

Mcmillan C, Poshyvanyk D, Grechanik M, Xie Q, Fu C (2013) Portfolio: Searching for relevant functions and their usages in millions of lines of code. ACM Transactions on Software Engineering and Methodology (TOSEM) 22(4):1–30

McNee SM, Riedl J, Konstan JA (2006) Being accurate is not enough: how accuracy metrics have hurt recommender systems. In: CHI'06 extended abstracts on Human factors in computing systems, pp 1097–1101

Mendez D, Gaulton A, Bento AP, Chambers J, De Veij M, Félix E, Magariños MP, Mosquera JF, Mutowo P, Nowotka M, et al (2019) Chembl: towards direct deposition of bioassay data. Nucleic acids research 47(D1):D930–D940

Merkwirth C, Lengauer T (2005) Automatic generation of complementary descriptors with molecular graph networks. Journal of Chemical Information and Modeling 45(5):1159–1168

Mesquita DPP, Jr AHS, Kaski S (2020) Rethinking pooling in graph neural networks. In: Advances in Neural Information Processing Systems

Mihalcea R, Tarau P (2004) Textrank: Bringing order into text. In: Proceedings of the 2004 conference on empirical methods in natural language processing, pp 404–411

Miikkulainen R, Liang J, Meyerson E, Rawal A, Fink D, Francon O, Raju B, Shahrzad H, Navruzyan A, Duffy N, et al (2019) Evolving deep neural networks. In: Artificial Intelligence in the Age of Neural Networks and Brain Computing, Elsevier, pp 293–312

Mikolov T, Karafiát M, Burget L, Cernocký J, Khudanpur S (2010) Recurrent neural network based language model. In: Kobayashi T, Hirose K, Nakamura S (eds) INTERSPEECH 2010, 11th Annual Conference of the International Speech Communication Association, Makuhari, Chiba, Japan, September 26-30, 2010, ISCA, pp 1045–1048

Mikolov T, Deoras A, Kombrink S, Burget L, Cernocký J (2011a) Empirical evaluation and combination of advanced language modeling techniques. In: INTERSPEECH 2011, 12th Annual Conference of the International Speech Communication Association, Florence, Italy, August 27-31, 2011, ISCA, pp 605–608

Mikolov T, Kombrink S, Burget L, Černocký J, Khudanpur S (2011b) Extensions of recurrent neural network language model. In: 2011 IEEE international conference on acoustics, speech and signal processing (ICASSP), IEEE, pp 5528–5531

Mikolov T, Chen K, Corrado G, Dean J (2013a) Efficient estimation of word representations in vector space. arXiv preprint arXiv:13013781

Mikolov T, Sutskever I, Chen K, Corrado GS, Dean J (2013b) Distributed representations of words and phrases and their compositionality. In: Advances in neural information processing systems, pp 3111–3119

Mikolov T CGDJ Chen K (2013) Efficient estimation of word representations in vector space. In: International Conference on Learning Representations

Miller BA, Çamurcu M, Gomez AJ, Chan K, Eliassi-Rad T (2019) Improving Robustness to Attacks Against Vertex Classification. In: Deep Learning for Graphs at AAAI Conference on Artificial Intelligence

Miller GA (1995) Wordnet: a lexical database for english. Communications of the ACM 38(11):39–41

Miller T (2019) Explanation in artificial intelligence: Insights from the social sciences. Artificial intelligence 267:1–38

Milo R, Shen-Orr S, Itzkovitz S, Kashtan N, Chklovskii D, Alon U (2002) Network motifs: simple building blocks of complex networks. Science 298(5594):824–827

Min S, Gao Z, Peng J, Wang L, Qin K, Fang B (2021) Stgsn—a spatial–temporal graph neural network framework for time-evolving social networks. Knowledge-Based Systems 214:106,746

Mir AM, Latoskinas E, Proksch S, Gousios G (2021) Type4Py: Deep similarity learning-based type inference for Python. arXiv preprint arXiv:210104470

Mirza M, Osindero S (2014) Conditional generative adversarial nets. arXiv preprint arXiv:14111784

Mnih A, Salakhutdinov RR (2008) Probabilistic matrix factorization. In: Advances in neural information processing systems, pp 1257–1264

Mnih V, Kavukcuoglu K, Silver D, Rusu AA, Veness J, Bellemare MG, Graves A, Riedmiller M, Fidjeland AK, Ostrovski G, et al (2015) Human-level control through deep reinforcement learning. Nature 518(7540):529–533

Mokou M, Lygirou V, Angelioudaki I, Paschalidis N, Stroggilos R, Frantzi M, Latosinska A, Bamias A, Hoffmann MJ, Mischak H, et al (2020) A novel pipeline for drug repurposing for bladder cancer based on patients' omics signatures. Cancers 12(12):3519

Momtazpour M, Butler P, Hossain MS, Bozchalui MC, Ramakrishnan N, Sharma R (2012) Coordinated clustering algorithms to support charging infrastructure design for electric vehicles. In: Proceedings of the ACM SIGKDD International Workshop on Urban Computing, pp 126–133

Montavon G, Samek W, Müller KR (2018) Methods for interpreting and understanding deep neural networks. Digital Signal Processing 73:1–15

Monti F, Bronstein M, Bresson X (2017) Geometric matrix completion with recurrent multi-graph neural networks. In: Advances in Neural Information Processing Systems, pp 3700–3710

Monti F, Frasca F, Eynard D, Mannion D, Bronstein MM (2019) Fake news detection on social media using geometric deep learning. In: Workshop on Representation Learning on Graphs and Manifolds

Moreno L, Aponte J, Sridhara G, Marcus A, Pollock L, Vijay-Shanker K (2013) Automatic generation of natural language summaries for java classes. In: 2013 21st International Conference on Program Comprehension (ICPC), IEEE, pp 23–32

Moreno L, Bavota G, Di Penta M, Oliveto R, Marcus A, Canfora G (2014) Automatic generation of release notes. In: Proceedings of the 22nd ACM SIGSOFT International Symposium on Foundations of Software Engineering, ACM, pp 484–495

Morris C, Kersting K, Mutzel P (2017) Glocalized Weisfeiler-Lehman kernels: Global-local feature maps of graphs. In: IEEE International Conference on Data Mining, IEEE, pp 327–336

Morris C, Ritzert M, Fey M, Hamilton WL, Lenssen JE, Rattan G, Grohe M (2019) Weisfeiler and leman go neural: Higher-order graph neural networks. In: the AAAI Conference on Artificial Intelligence, vol 33, pp 4602–4609

Morris C, Kriege NM, Bause F, Kersting K, Mutzel P, Neumann M (2020a) TU-Dataset: A collection of benchmark datasets for learning with graphs. CoRR abs/2007.08663

Morris C, Rattan G, Mutzel P (2020b) Weisfeiler and leman go sparse: Towards scalable higher-order graph embeddings. Advances in Neural Information Processing Systems 33

Mueller J, Thyagarajan A (2016) Siamese recurrent architectures for learning sentence similarity. In: Proceedings of the AAAI Conference on Artificial Intelligence, vol 30

Mungall CJ, Torniai C, Gkoutos GV, Lewis SE, Haendel MA (2012) Uberon, an integrative multi-species anatomy ontology. Genome biology 13(1):1–20

Murphy R, Srinivasan B, Rao V, Ribeiro B (2019a) Relational pooling for graph representations. In: International Conference on Machine Learning, pp 4663–4673

Murphy RL, Srinivasan B, Rao VA, Ribeiro B (2019b) Janossy pooling: Learning deep permutation-invariant functions for variable-size inputs. In: International Conference on Learning Representations

Murphy RL, Srinivasan B, Rao VA, Ribeiro B (2019c) Relational pooling for graph representations. In: International Conference on Machine Learning, pp 4663–4673

Nair V, Hinton GE (2010) Rectified linear units improve restricted boltzmann machines. In: Fürnkranz J, Joachims T (eds) Proceedings of the 27th International Conference on Machine Learning (ICML-10), June 21-24, 2010, Haifa, Israel, Omnipress, pp 807–814

Nathani D, Chauhan J, Sharma C, Kaul M (2019) Learning attention-based embeddings for relation prediction in knowledge graphs. arXiv preprint arXiv:190601195

Nelson CA, Butte AJ, Baranzini SE (2019) Integrating biomedical research and electronic health records to create knowledge-based biologically meaningful machine-readable embeddings. Nature communications 10(1):1–10

Neville J, Jensen D (2000) Iterative classification in relational data. In: Proc. AAAI-2000 workshop on learning statistical models from relational data, pp 13–20

Newman M (2010) Networks: an introduction. Oxford university press

Newman M (2018) Networks. Oxford university press

Newman ME (2006a) Finding community structure in networks using the eigenvectors of matrices. Physical review E 74(3):036,104

Newman ME (2006b) Modularity and community structure in networks. Proceedings of the national academy of sciences 103(23):8577–8582

Ng A (2011) Machine learning

Nguyen DQ, Nguyen TD, Nguyen DQ, Phung D (2017) A novel embedding model for knowledge base completion based on convolutional neural network. arXiv preprint arXiv:171202121

Nguyen HV, Bai L (2010) Cosine similarity metric learning for face verification. In: Asian conference on computer vision, Springer, pp 709–720

Nickel M, Tresp V (2013) Tensor factorization for multi-relational learning. In: Joint European Conference on Machine Learning and Knowledge Discovery in Databases, Springer, pp 617–621

Nickel M, Tresp V, Kriegel HP (2011) A three-way model for collective learning on multi-relational data. In: Proceedings of the 28th International Conference on International Conference on Machine Learning, Omnipress, Madison, WI, USA, ICML'11, p 809–816

Nickel M, Jiang X, Tresp V (2014) Reducing the rank in relational factorization models by including observable patterns. In: Advances in Neural Information Processing Systems, pp 1179–1187

Nickel M, Murphy K, Tresp V, Gabrilovich E (2016a) A review of relational machine learning for knowledge graphs. Proceedings of the IEEE 104(1):11–33

Nickel M, Rosasco L, Poggio T (2016b) Holographic embeddings of knowledge graphs. In: Proceedings of the AAAI Conference on Artificial Intelligence, vol 30

Nicora G, Vitali F, Dagliati A, Geifman N, Bellazzi R (2020) Integrated multi-omics analyses in oncology: a review of machine learning methods and tools. Frontiers in oncology 10:1030

Nie P, Rai R, Li JJ, Khurshid S, Mooney RJ, Gligoric M (2019) A framework for writing trigger-action todo comments in executable format. In: Proceedings of the 2019 27th ACM Joint Meeting on European Software Engineering Conference and Symposium on the Foundations of Software Engineering, ACM, pp 385–396

Nielson F, Nielson HR, Hankin C (2015) Principles of program analysis. Springer

Niepert M, Ahmed M, Kutzkov K (2016) Learning convolutional neural networks for graphs. In: International Conference on Machine Learning, pp 2014–2023

Nikolentzos G, Meladianos P, Vazirgiannis M (2017) Matching node embeddings for graph similarity. In: AAAI Conference on Artificial Intelligence, pp 2429–2435

Ning X, Karypis G (2011) Slim: Sparse linear methods for top-n recommender systems. In: 2011 IEEE 11th International Conference on Data Mining, IEEE, pp 497–506

Niu C, Wu F, Tang S, Hua L, Jia R, Lv C, Wu Z, Chen G (2020) Billion-scale federated learning on mobile clients: A submodel design with tunable privacy. In: Proceedings of the 26th Annual International Conference on Mobile Computing and Networking, pp 1–14

Norcliffe-Brown W, Vafeias S, Parisot S (2018) Learning conditioned graph structures for interpretable visual question answering. In: Advances in neural information processing systems, pp 8334–8343

Noroozi M, Favaro P (2016) Unsupervised learning of visual representations by solving jigsaw puzzles. In: European conference on computer vision, Springer, pp 69–84

Nowozin S, Cseke B, Tomioka R (2016) f-gan: Training generative neural samplers using variational divergence minimization. In: Advances in Neural Information Processing Systems, vol 29

Noy N, Gao Y, Jain A, Narayanan A, Patterson A, Taylor J (2019) Industry-scale knowledge graphs: lessons and challenges. Communications of the ACM 62(8):36–43

NT H, Maehara T (2019) Revisiting graph neural networks: All we have is low-pass filters. arXiv preprint arXiv:190509550

Nunes M, Pappa GL (2020) Neural architecture search in graph neural networks. In: Brazilian Conference on Intelligent Systems, Springer, pp 302–317

Oda Y, Fudaba H, Neubig G, Hata H, Sakti S, Toda T, Nakamura S (2015) Learning to generate pseudo-code from source code using statistical machine translation (t). In: 2015 30th IEEE/ACM International Conference on Automated Software Engineering (ASE), IEEE, pp 574–584

Ok S (2020) A graph similarity for deep learning. In: Larochelle H, Ranzato M, Hadsell R, Balcan MF, Lin H (eds) Advances in Neural Information Processing Systems, Curran Associates, Inc., vol 33, pp 1–12

Olah C, Satyanarayan A, Johnson I, Carter S, Schubert L, Ye K, Mordvintsev A (2018) The building blocks of interpretability. Distill DOI 10.23915/distill.00010, https://distill.pub/2018/building-blocks

On K, Kim E, Heo Y, Zhang B (2020) Cut-based graph learning networks to discover compositional structure of sequential video data. In: The Thirty-Fourth AAAI Conference on Artificial Intelligence, pp 5315–5322

Oono K, Suzuki T (2020) Graph neural networks exponentially lose expressive power for node classification. In: International Conference on Learning Representations

Oord Avd, Kalchbrenner N, Vinyals O, Espeholt L, Graves A, Kavukcuoglu K (2016) Conditional image generation with pixelcnn decoders. In: Proceedings of the 30th International Conference on Neural Information Processing Systems, pp 4797–4805

Oord Avd, Li Y, Vinyals O (2018) Representation learning with contrastive predictive coding. arXiv preprint arXiv:180703748

Orchard S, Ammari M, Aranda B, Breuza L, Briganti L, Broackes-Carter F, Campbell NH, Chavali G, Chen C, Del-Toro N, et al (2014) The mintact project—intact as a common curation platform for 11 molecular interaction databases. Nucleic acids research 42(D1):D358–D363

Ottenstein KJ, Ottenstein LM (1984) The program dependence graph in a software development environment. ACM Sigplan Notices 19(5):177–184

Ou M, Cui P, Wang F, Wang J, Zhu W (2015) Non-transitive hashing with latent similarity components. In: Proceedings of the 21th ACM SIGKDD International Conference on Knowledge Discovery and Data Mining, pp 895–904

Ou M, Cui P, Pei J, Zhang Z, Zhu W (2016) Asymmetric transitivity preserving graph embedding. In: Proceedings of the 22nd ACM SIGKDD international conference on Knowledge discovery and data mining, pp 1105–1114

Oyetunde T, Zhang M, Chen Y, Tang YJ, Lo C (2017) Boostgapfill: improving the fidelity of metabolic network reconstructions through integrated constraint and pattern-based methods. Bioinformatics 33(4):608–611

Page L, Brin S, Motwani R, Winograd T (1999) The pagerank citation ranking: Bringing order to the web. Tech. rep., Stanford InfoLab

Paige CC, Saunders MA (1981) Towards a generalized singular value decomposition. SIAM Journal on Numerical Analysis 18(3):398–405

Pal S, Malekmohammadi S, Regol F, Zhang Y, Xu Y, Coates M (2020) Nonparametric graph learning for bayesian graph neural networks. In: Conference on Uncertainty in Artificial Intelligence, PMLR, pp 1318–1327

Palasca O, Santos A, Stolte C, Gorodkin J, Jensen LJ (2018) Tissues 2.0: an integrative web resource on mammalian tissue expression. Database 2018

Palaz D, Collobert R, et al (2015a) Analysis of cnn-based speech recognition system using raw speech as input. Tech. rep., Idiap

Palaz D, Doss MM, Collobert R (2015b) Convolutional neural networks-based continuous speech recognition using raw speech signal. In: 2015 IEEE International Conference on Acoustics, Speech and Signal Processing (ICASSP), IEEE, pp 4295–4299

Paliwal A, Gimeno F, Nair V, Li Y, Lubin M, Kohli P, Vinyals O (2020) Reinforced genetic algorithm learning for optimizing computation graphs. In: International Conference on Learning Representations

Pan M, Li Y, Zhou X, Liu Z, Song R, Lu H, Luo J (2019) Dissecting the learning curve of taxi drivers: A data-driven approach. In: Proceedings of the 2019 SIAM International Conference on Data Mining, SIAM, pp 783–791

Pan M, Huang W, Li Y, Zhou X, Liu Z, Song R, Lu H, Tian Z, Luo J (2020a) Dhpa: Dynamic human preference analytics framework: A case study on taxi drivers' learning curve analysis. ACM Trans Intell Syst Technol 11(1), DOI 10.1145/3360312

Pan M, Huang W, Li Y, Zhou X, Luo J (2020b) Xgail: Explainable generative adversarial imitation learning for explainable human decision analysis. In: Proceedings of the 26th ACM SIGKDD International Conference on Knowledge Discovery amp; Data Mining, Association for Computing Machinery, KDD '20, p 1334–1343, DOI 10.1145/3394486.3403186

Pan S, Wu J, Zhu X, Zhang C, Wang Y (2016) Tri-party deep network representation. In: Proceedings of the Twenty-Fifth International Joint Conference on Artificial Intelligence, pp 1895–1901

Pan S, Hu R, Long G, Jiang J, Yao L, Zhang C (2018) Adversarially regularized graph autoencoder for graph embedding. In: Proceedings of the 27th International Joint Conference on Artificial Intelligence, pp 2609–2615

Pan W, Su C, Chen K, Henchcliffe C, Wang F (2020c) Learning phenotypic associations for parkinson's disease with longitudinal clinical records. medRxiv

Pandi IV, Barr ET, Gordon AD, Sutton C (2020) OptTyper: Probabilistic type inference by optimising logical and natural constraints. arXiv preprint arXiv:200400348

Pang L, Lan Y, Guo J, Xu J, Wan S, Cheng X (2016) Text matching as image recognition. In: Proceedings of the AAAI Conference on Artificial Intelligence, vol 30

Pang LX, Chawla S, Liu W, Zheng Y (2011) On mining anomalous patterns in road traffic streams. In: International conference on advanced data mining and applications, Springer, pp 237–251

Panichella S, Aponte J, Di Penta M, Marcus A, Canfora G (2012) Mining source code descriptions from developer communications. In: 2012 20th IEEE International Conference on Program Comprehension (ICPC), IEEE, pp 63–72

Paninski L (2003) Estimation of entropy and mutual information. Neural computation 15(6):1191–1253

Pantziarka P, Meheus L (2018) Omics-driven drug repurposing as a source of innovative therapies in rare cancers. Expert Opinion on Orphan Drugs 6(9):513–517

Park C, Kim D, Zhu Q, Han J, Yu H (2019) Task-guided pair embedding in heterogeneous network. In: Proceedings of the 28th ACM International Conference on Information and Knowledge Management, pp 489–498

Parthasarathy S, Busso C (2017) Jointly predicting arousal, valence and dominance with multi-task learning. In: Interspeech, vol 2017, pp 1103–1107

Pascanu R, Mikolov T, Bengio Y (2013) On the difficulty of training recurrent neural networks. In: International conference on machine learning, PMLR, pp 1310–1318

Paszke A, Gross S, Massa F, Lerer A, Bradbury J, Chanan G, Killeen T, Lin Z, Gimelshein N, Antiga L, Desmaison A, Kopf A, Yang E, DeVito Z, Raison M, Tejani A, Chilamkurthy S, Steiner B, Fang L, Bai J, Chintala S (2019) Pytorch: An imperative style, high-performance deep learning library. In: Advances in Neural Information Processing Systems, vol 32

Pathak D, Krahenbuhl P, Donahue J, Darrell T, Efros AA (2016) Context encoders: Feature learning by inpainting. In: Proceedings of the IEEE conference on computer vision and pattern recognition, pp 2536–2544

Peña-Castillo L, Tasan M, Myers CL, et al (2008) A critical assessment of mus musculus gene function prediction using integrated genomic evidence. Genome Biology 9(Suppl 1):S2, DOI 10.1186/gb-2008-9-s1-s2

Peng H, Li J, He Y, Liu Y, Bao M, Wang L, Song Y, Yang Q (2018) Large-scale hierarchical text classification with recursively regularized deep graph-cnn. In: Proceedings of the 2018 world wide web conference, pp 1063–1072

Peng H, Pappas N, Yogatama D, Schwartz R, Smith N, Kong L (2021) Random feature attention. In: International Conference on Learning Representations

Peng Z, Dong Y, Luo M, Wu XM, Zheng Q (2020) Self-supervised graph representation learning via global context prediction. arXiv preprint arXiv:200301604

Pennington J, Socher R, Manning CD (2014) Glove: Global vectors for word representation. In: Proceedings of the 2014 conference on empirical methods in natural language processing (EMNLP), pp 1532–1543

Percha B, Altman RB (2018) A global network of biomedical relationships derived from text. Bioinformatics 34(15):2614–2624

Perez E, Strub F, De Vries H, Dumoulin V, Courville A (2018) Film: Visual reasoning with a general conditioning layer. In: Proceedings of the AAAI Conference on Artificial Intelligence, vol 32

Perozzi B, Al-Rfou R, Skiena S (2014) Deepwalk: Online learning of social representations. In: Proceedings of the 20th ACM SIGKDD international conference on Knowledge discovery and data mining, pp 701–710

Petar V, Guillem C, Arantxa C, Adriana R, Pietro L, Yoshua B (2018) Graph attention networks. In: International Conference on Learning Representations

Pham H, Guan M, Zoph B, Le Q, Dean J (2018) Efficient neural architecture search via parameter sharing. In: International Conference on Machine Learning, pp 4092–4101

Pham T, Tran T, Phung D, Venkatesh S (2017) Column networks for collective classification. In: Proceedings of the Thirty-First AAAI Conference on Artificial Intelligence, AAAI Press, AAAI'17, p 2485–2491

Piñero J, Ramírez-Anguita JM, Saüch-Pitarch J, Ronzano F, Centeno E, Sanz F, Furlong LI (2020) The disgenet knowledge platform for disease genomics: 2019 update. Nucleic acids research 48(D1):D845–D855

Pires DE, Blundell TL, Ascher DB (2015) pkcsm: predicting small-molecule pharmacokinetic and toxicity properties using graph-based signatures. Journal of medicinal chemistry 58(9):4066–4072

Pletscher-Frankild S, Pallejà A, Tsafou K, Binder JX, Jensen LJ (2015) Diseases: Text mining and data integration of disease–gene associations. Methods 74:83–89

Pogancic MV, Paulus A, Musil V, Martius G, Rolinek M (2020) Differentiation of blackbox combinatorial solvers. In: International Conference on Learning Representations, OpenReview.net

Pope PE, Kolouri S, Rostami M, Martin CE, Hoffmann H (2019) Explainability methods for graph convolutional neural networks. In: Proceedings of the IEEE/CVF Conference on Computer Vision and Pattern Recognition, pp 10,772–10,781

Pradel M, Sen K (2018) Deepbugs: A learning approach to name-based bug detection. Proceedings of the ACM on Programming Languages 2(OOPSLA):1–25

Pradel M, Gousios G, Liu J, Chandra S (2020) TypeWriter: Neural type prediction with search-based validation. In: Proceedings of the 28th ACM Joint Meeting on European Software Engineering Conference and Symposium on the Foundations of Software Engineering, pp 209–220

Pushpakom S, Iorio F, Eyers PA, Escott KJ, Hopper S, Wells A, Doig A, Guilliams T, Latimer J, McNamee C, et al (2019) Drug repurposing: progress, challenges and recommendations. Nature reviews Drug discovery 18(1):41–58

Putra JWG, Tokunaga T (2017) Evaluating text coherence based on semantic similarity graph. In: Proceedings of TextGraphs-11: the Workshop on Graph-based Methods for Natural Language Processing, pp 76–85

Qi Y, Bar-Joseph Z, Klein-Seetharaman J (2006) Evaluation of different biological data and computational classification methods for use in protein interaction prediction. Proteins: Structure, Function, and Bioinformatics 63(3):490–500

Qiu J, Dong Y, Ma H, Li J, Wang K, Tang J (2018) Network embedding as matrix factorization: Unifying deepwalk, line, pte, and node2vec. In: Proceedings of the eleventh ACM international conference on web search and data mining, pp 459–467

Qiu J, Chen Q, Dong Y, Zhang J, Yang H, Ding M, Wang K, Tang J (2020a) Gcc: Graph contrastive coding for graph neural network pre-training. In: Proceedings of the 26th ACM SIGKDD International Conference on Knowledge Discovery & Data Mining, pp 1150–1160

Qiu J, Cen Y, Chen Q, Zhou C, Zhou J, Yang H, Tang J (2021) Local clustering graph neural networks. OpenReview

Qiu X, Sun T, Xu Y, Shao Y, Dai N, Huang X (2020b) Pre-trained models for natural language processing: A survey. Science China Technological Sciences pp 1–26

Qu Y, Cai H, Ren K, Zhang W, Yu Y, Wen Y, Wang J (2016) Product-based neural networks for user response prediction. In: 2016 IEEE 16th International Conference on Data Mining (ICDM), IEEE, pp 1149–1154

Radford A, Narasimhan K, Salimans T, Sutskever I (2018) Improving language understanding with unsupervised learning. Tech. rep., OpenAI

Radivojac P, Clark WT, Oron TR, et al (2013) A large-scale evaluation of computational protein function prediction. Nature Methods 10(3):221–227

Raghothaman M, Kulkarni S, Heo K, Naik M (2018) User-guided program reasoning using Bayesian inference. In: Proceedings of the 39th ACM SIGPLAN Conference on Programming Language Design and Implementation, pp 722–735

Rahman TA, Surma B, Backes M, Zhang Y (2019) Fairwalk: Towards fair graph embedding. In: IJCAI, pp 3289–3295

Ramakrishnan R, Dral PO, Rupp M, Von Lilienfeld OA (2014) Quantum chemistry structures and properties of 134 kilo molecules. Scientific data 1(1):1–7

Ramos PIP, Arge LWP, Lima NCB, Fukutani KF, de Queiroz ATL (2019) Leveraging user-friendly network approaches to extract knowledge from high-throughput omics datasets. Frontiers in genetics 10:1120

Rastkar S, Murphy GC (2013) Why did this code change? In: Proceedings of the 2013 International Conference on Software Engineering, IEEE Press, pp 1193–1196

Rastkar S, Murphy GC, Bradley AW (2011) Generating natural language summaries for crosscutting source code concerns. In: 2011 27th IEEE International Conference on Software Maintenance (ICSM), IEEE, pp 103–112

Rastkar S, Murphy GC, Murray G (2014) Automatic summarization of bug reports. IEEE Transactions on Software Engineering 40(4):366–380

Ratti C, Sobolevsky S, Calabrese F, Andris C, Reades J, Martino M, Claxton R, Strogatz SH (2010) Redrawing the map of great britain from a network of human interactions. PloS one 5(12)

Raychev V, Vechev M, Yahav E (2014) Code completion with statistical language models. In: Proceedings of the 35th ACM SIGPLAN Conference on Programming Language Design and Implementation, pp 419–428

Raychev V, Vechev M, Krause A (2015) Predicting program properties from Big Code. In: Principles of Programming Languages (POPL)

Real E, Moore S, Selle A, Saxena S, Suematsu YL, Tan J, Le Q, Kurakin A (2017) Large-scale evolution of image classifiers. arXiv preprint arXiv:170301041

Real E, Aggarwal A, Huang Y, Le QV (2019) Regularized evolution for image classifier architecture search. In: Proceedings of the AAAI Conference on Artificial Intelligence, vol 33, pp 4780–4789

Ren H, Hu W, Leskovec J (2020) Query2box: Reasoning over knowledge graphs in vector space using box embeddings. In: International Conference on Learning Representations

Ren S, He K, Girshick R, Sun J (2015) Faster r-cnn: towards real-time object detection with region proposal networks. In: Proceedings of the 28th International Conference on Neural Information Processing Systems-Volume 1, pp 91–99

Ren Z, Wang X, Zhang N, Lv X, Li LJ (2017) Deep reinforcement learning-based image captioning with embedding reward. In: Proceedings of the IEEE conference on computer vision and pattern recognition, pp 290–298

Rendle S (2010) Factorization machines. In: 10th IEEE International Conference on Data Mining (ICDM), IEEE, pp 995–1000

Rezende DJ, Mohamed S, Wierstra D (2014) Stochastic backpropagation and approximate inference in deep generative models. In: International conference on machine learning, PMLR, pp 1278–1286

Rhodes G (2010) Crystallography made crystal clear: a guide for users of macromolecular models. Elsevier

Ribeiro LF, Saverese PH, Figueiredo DR (2017) struc2vec: Learning node representations from structural identity. In: the ACM SIGKDD International Conference on Knowledge Discovery and Data Mining, pp 385–394

Ribeiro MT, Ziviani N, Moura ESD, Hata I, Lacerda A, Veloso A (2014) Multiobjective pareto-efficient approaches for recommender systems. ACM Transactions on Intelligent Systems and Technology (TIST) 5(4):1–20

Ribeiro MT, Singh S, Guestrin C (2016) " why should i trust you?" explaining the predictions of any classifier. In: Proceedings of the 22nd ACM SIGKDD international conference on knowledge discovery and data mining, pp 1135–1144

Richiardi J, Achard S, Bunke H, Van De Ville D (2013) Machine learning with brain graphs: predictive modeling approaches for functional imaging in systems neuroscience. IEEE Signal Processing Magazine 30(3):58–70

Riesen K (2015) Structural Pattern Recognition with Graph Edit Distance Approximation Algorithms and Applications. Springer

Riesen K, Fankhauser S, Bunke H (2007) Speeding up graph edit distance computation with a bipartite heuristic. In: MLG, Citeseer, pp 21–24

Riloff E (1996) Automatically generating extraction patterns from untagged text. In: Proceedings of the national conference on artificial intelligence, pp 1044–1049

Rink B, Bejan CA, Harabagiu SM (2010) Learning textual graph patterns to detect causal event relations. In: FLAIRS Conference

Rizvi RF, Vasilakes JA, Adam TJ, Melton GB, Bishop JR, Bian J, Tao C, Zhang R (2019) Integrated dietary supplement knowledge base (idisk)

Robinson PN, Köhler S, Bauer S, et al (2008) The human phenotype ontology: A tool for annotating and analyzing human hereditary disease. The American Journal of Human Genetics 83(5):610–615

Rocco I, Cimpoi M, Arandjelović R, Torii A, Pajdla T, Sivic J (2018) Neighbourhood consensus networks. In: Advances in Neural Information Processing Systems, vol 31

Rodeghero P, McMillan C, McBurney PW, Bosch N, D'Mello S (2014) Improving automated source code summarization via an eye-tracking study of programmers. In: Proceedings of the 36th international conference on Software engineering, ACM, pp 390–401

Rodeghero P, Jiang S, Armaly A, McMillan C (2017) Detecting user story information in developer-client conversations to generate extractive summaries. In: 2017 IEEE/ACM 39th International Conference on Software Engineering (ICSE), IEEE, pp 49–59

Roehm T, Tiarks R, Koschke R, Maalej W (2012) How do professional developers comprehend software? In: 2012 34th International Conference on Software Engineering (ICSE), IEEE, pp 255–265

Rogers D, Hahn M (2010) Extended-connectivity fingerprints. Journal of Chemical Information and Modeling 50(5):742–754

Rolínek M, Swoboda P, Zietlow D, Paulus A, Musil V, Martius G (2020) Deep graph matching via blackbox differentiation of combinatorial solvers. In: European Conference on Computer Vision, Springer, pp 407–424

Rong Y, Bian Y, Xu T, Xie W, Wei Y, Huang W, Huang J (2020a) Self-supervised graph transformer on large-scale molecular data. Advances in Neural Information Processing Systems 33

Rong Y, Huang W, Xu T, Huang J (2020b) Dropedge: Towards deep graph convolutional networks on node classification. In: International Conference on Learning Representations

Rong Y, Xu T, Huang J, Huang W, Cheng H, Ma Y, Wang Y, Derr T, Wu L, Ma T (2020c) Deep graph learning: Foundations, advances and applications. In: Proceedings of the 26th ACM SIGKDD International Conference on Knowledge Discovery & Data Mining, ACM, Virtual Event, pp 3555–3556

Rossi A, Barbosa D, Firmani D, Matinata A, Merialdo P (2021) Knowledge graph embedding for link prediction: A comparative analysis. ACM Transactions on Knowledge Discovery from Data (TKDD) 15(2):1–49

Rossi E, Chamberlain B, Frasca F, Eynard D, Monti F, Bronstein M (2020) Temporal graph networks for deep learning on dynamic graphs. arXiv preprint arXiv:200610637

Rotmensch M, Halpern Y, Tlimat A, Horng S, Sontag D (2017) Learning a health knowledge graph from electronic medical records. Scientific reports 7(1):5994

Rousseau F, Vazirgiannis M (2013) Graph-of-word and tw-idf: new approach to ad hoc ir. In: Proceedings of the 22nd ACM international conference on Information & Knowledge Management, pp 59–68

Rousseau F, Kiagias E, Vazirgiannis M (2015) Text categorization as a graph classification problem. In: Proceedings of the 53rd Annual Meeting of the Association for Computational Linguistics and the 7th International Joint Conference on Natural Language Processing (Volume 1: Long Papers), pp 1702–1712

Roweis ST, Saul LK (2000) Nonlinear dimensionality reduction by locally linear embedding. science 290(5500):2323–2326

Rubner Y, Tomasi C, Guibas LJ (1998) A metric for distributions with applications to image databases. In: Sixth International Conference on Computer Vision (IEEE Cat. No. 98CH36271), IEEE, pp 59–66

Rue H, Held L (2005) Gaussian Markov random fields: theory and applications. CRC press

Rui SCLDJZJL T (2005) A character recognition based on feature extraction. Journal of Chinese Computer Systems, 26(2), 289-292 26(2):289–292

Sabour S, Frosst N, Hinton GE (2017) Dynamic routing between capsules. In: Proceedings of the 31st International Conference on Neural Information Processing Systems, pp 3859–3869

Sachdev S, Li H, Luan S, Kim S, Sen K, Chandra S (2018) Retrieval on source code: a neural code search. In: Proceedings of the 2nd ACM SIGPLAN International Workshop on Machine Learning and Programming Languages, pp 31–41

Sahu S, Gupta R, Sivaraman G, AbdAlmageed W, Espy-Wilson C (2017) Adversarial auto-encoders for speech based emotion recognition. Proc Interspeech 2017 pp 1243–1247

Sahu SK, Anand A (2018) Drug-drug interaction extraction from biomedical texts using long short-term memory network. Journal of biomedical informatics 86:15–24

Saire D, Ramírez Rivera A (2019) Graph learning network: A structure learning algorithm. In: Workshop on Learning and Reasoning with Graph-Structured Data (ICMLW 2019)

Samanta B, Abir D, Jana G, Chattaraj PK, Ganguly N, Rodriguez MG (2019) Nevae: A deep generative model for molecular graphs. In: Proceedings of the AAAI Conference on Artificial Intelligence, vol 33, pp 1110–1117

Sanchez-Lengeling B, Wei J, Lee B, Reif E, Wang P, Qian W, McCloskey K, Colwell L, Wiltschko A (2020) Evaluating attribution for graph neural networks. In: Larochelle H, Ranzato M, Hadsell R, Balcan MF, Lin H (eds) Advances in Neural Information Processing Systems, Curran Associates, Inc., vol 33, pp 5898–5910

Sandryhaila A, Moura JF (2013) Discrete signal processing on graphs. IEEE Trans Signal Process 61(7):1644–1656

Sangeetha J, Jayasankar T (2019) Emotion speech recognition based on adaptive fractional deep belief network and reinforcement learning. In: Cognitive Informatics and Soft Computing, Springer, pp 165–174

Santini S, Ostermaier B, Vitaletti A (2008) First experiences using wireless sensor networks for noise pollution monitoring. In: Proceedings of the workshop on Real-world wireless sensor networks, pp 61–65

Santos A, Colaço AR, Nielsen AB, Niu L, Geyer PE, Coscia F, Albrechtsen NJW, Mundt F, Jensen LJ, Mann M (2020) Clinical knowledge graph integrates proteomics data into clinical decision-making. bioRxiv

Sato R (2020) A survey on the expressive power of graph neural networks. arXiv preprint arXiv:200304078

Sato R, Yamada M, Kashima H (2021) Random features strengthen graph neural networks. In: Proceedings of the 2021 SIAM International Conference on Data Mining (SDM), SIAM, pp 333–341

Satorras VG, Estrach JB (2018) Few-shot learning with graph neural networks. In: International Conference on Learning Representations

Scarselli F, Gori M, Tsoi AC, Hagenbuchner M, Monfardini G (2008) The graph neural network model. IEEE transactions on neural networks 20(1):61–80

Schenker A, Last M, Bunke H, Kandel A (2003) Clustering of web documents using a graph model. In: Web Document Analysis: Challenges and Opportunities, World Scientific, pp 3–18

Schlichtkrull M, Kipf TN, Bloem P, Van Den Berg R, Titov I, Welling M (2018) Modeling relational data with graph convolutional networks. In: European semantic web conference, Springer, pp 593–607

Schlichtkrull MS, De Cao N, Titov I (2021) Interpreting graph neural networks for nlp with differentiable edge masking. In: International Conference on Learning Representations

Schnake T, Eberle O, Lederer J, Nakajima S, Schütt KT, Müller KR, Montavon G (2020) Xai for graphs: Explaining graph neural network predictions by identifying relevant walks. arXiv preprint arXiv:200603589

Schneider N, Flanigan J, O'Gorman T (2015) The logic of amr: Practical, unified, graph-based sentence semantics for nlp. In: Proceedings of the 2015 Conference of the North American Chapter of the Association for Computational Linguistics: Tutorial Abstracts, pp 4–5

Schriml LM, Mitraka E, Munro J, Tauber B, Schor M, Nickle L, Felix V, Jeng L, Bearer C, Lichenstein R, et al (2019) Human disease ontology 2018 update: classification, content and workflow expansion. Nucleic acids research 47(D1):D955–D962

Schuchardt J, Bojchevski A, Klicpera J, Günnemann S (2021) Collective robustness certificates. In: International Conference on Learning Representations, ICLR

Schug J (2002) Predicting gene ontology functions from ProDom and CDD protein domains. Genome Research 12(4):648–655

Schulman J, Wolski F, Dhariwal P, Radford A, Klimov O (2017) Proximal policy optimization algorithms. arXiv preprint arXiv:170706347

Schuster M, Paliwal KK (1997) Bidirectional recurrent neural networks. IEEE Transactions on Signal Processing 45(11):2673–2681

Schwarzenberg R, Hübner M, Harbecke D, Alt C, Hennig L (2019) Layerwise relevance visualization in convolutional text graph classifiers. In: Proceedings of the EMNLP 2019 Workshop on Graph-Based Natural Language Processing

Schweidtmann AM, Rittig JG, König A, Grohe M, Mitsos A, Dahmen M (2020) Graph neural networks for prediction of fuel ignition quality. Energy & Fuels 34(9):11,395–11,407

Schwikowski B, Uetz P, Fields S (2000) A network of protein–protein interactions in yeast. Nature Biotechnology 18(12):1257–1261

Seide F, Li G, Yu D (2011) Conversational speech transcription using context-dependent deep neural networks. In: Twelfth annual conference of the international speech communication association

Seidman SB (1983) Network structure and minimum degree. Social Networks 5(3):269–287

Selsam D, Bjørner N (2019) Guiding high-performance SAT solvers with unsat-core predictions. In: International Conference on Theory and Applications of Satisfiability Testing, Springer, pp 336–353

Semasaba AOA, Zheng W, Wu X, Agyemang SA (2020) Literature survey of deep learning-based vulnerability analysis on source code. IET Software

Seo Y, Defferrard M, Vandergheynst P, Bresson X (2018) Structured sequence modeling with graph convolutional recurrent networks. In: Neural Information Processing, Springer, pp 362–373

Shah M, Chen X, Rohrbach M, Parikh D (2019) Cycle-consistency for robust visual question answering. In: Proceedings of the IEEE/CVF Conference on Computer Vision and Pattern Recognition, pp 6649–6658

Shang C, Tang Y, Huang J, Bi J, He X, Zhou B (2019) End-to-end structure-aware convolutional networks for knowledge base completion. In: Proceedings of the AAAI Conference on Artificial Intelligence, vol 33, pp 3060–3067

Shang J, Zheng Y, Tong W, Chang E, Yu Y (2014) Inferring gas consumption and pollution emission of vehicles throughout a city. In: Proceedings of the 20th ACM SIGKDD international conference on Knowledge discovery and data mining, pp 1027–1036

Shanthamallu US, Thiagarajan JJ, Spanias A (2021) Uncertainty-Matching Graph Neural Networks to Defend Against Poisoning Attacks. In: AAAI Conference on Artificial Intelligence

Sharp ME (2017) Toward a comprehensive drug ontology: extraction of drug-indication relations from diverse information sources. Journal of biomedical semantics 8(1):1–10

Shehu A, Barbará D, Molloy K (2016) A survey of computational methods for protein function prediction. In: Wong KC (ed) Big Data Analytics in Genomics, Springer Verlag, pp 225–298

Shen J, Zhang J, Luo X, et al (2007) Predicting protein-protein interactions based only on sequences information. Proceedings of the National Academy of Sciences 104(11):4337–4341

Shen K, Wu L, Xu F, Tang S, Xiao J, Zhuang Y (2020) Hierarchical attention based spatial-temporal graph-to-sequence learning for grounded video description. In:

Bessiere C (ed) Proceedings of the Twenty-Ninth International Joint Conference on Artificial Intelligence, IJCAI-20, International Joint Conferences on Artificial Intelligence Organization, pp 941–947, main track

Shen YL, Huang CY, Wang SS, Tsao Y, Wang HM, Chi TS (2019) Reinforcement learning based speech enhancement for robust speech recognition. In: ICASSP 2019-2019 IEEE International Conference on Acoustics, Speech and Signal Processing (ICASSP), IEEE, pp 6750–6754

Shen Z, Zhang M, Zhao H, Yi S, Li H (2021) Efficient attention: Attention with linear complexities. In: Proceedings of the IEEE/CVF Winter Conference on Applications of Computer Vision, pp 3531–3539

Shervashidze N, Schweitzer P, van Leeuwen EJ, Mehlhorn K, Borgwardt KM (2011a) Weisfeiler-Lehman graph kernels. Journal of Machine Learning Research 12:2539–2561

Shervashidze N, Schweitzer P, Leeuwen EJv, Mehlhorn K, Borgwardt KM (2011b) Weisfeiler-lehman graph kernels. Journal of Machine Learning Research 12(Sep):2539–2561

Shi C, Li Y, Zhang J, Sun Y, Philip SY (2016) A survey of heterogeneous information network analysis. IEEE Transactions on Knowledge and Data Engineering 29(1):17–37

Shi C, Hu B, Zhao WX, Philip SY (2018a) Heterogeneous information network embedding for recommendation. IEEE Transactions on Knowledge and Data Engineering 31(2):357–370

Shi C, Xu M, Zhu Z, Zhang W, Zhang M, Tang J (2019a) Graphaf: a flow-based autoregressive model for molecular graph generation. In: International Conference on Learning Representations

Shi J, Malik J (2000) Normalized cuts and image segmentation. IEEE Transactions on Pattern Analysis and Machine Intelligence 22(8):888–905, DOI 10.1109/34. 868688

Shi L, Zhang Y, Cheng J, Lu H (2019b) Skeleton-based action recognition with directed graph neural networks. In: IEEE/CVF Conference on Computer Vision and Pattern Recognition, pp 7912–7921

Shi M, Wilson DA, Zhu X, Huang Y, Zhuang Y, Liu J, Tang Y (2020) Evolutionary architecture search for graph neural networks. arXiv preprint arXiv:200910199

Shi W, Rajkumar R (2020) Point-gnn: Graph neural network for 3d object detection in a point cloud. In: Proceedings of the IEEE/CVF conference on computer vision and pattern recognition, pp 1711–1719

Shi Y, Gui H, Zhu Q, Kaplan L, Han J (2018b) Aspem: Embedding learning by aspects in heterogeneous information networks. In: Proceedings of the 2018 SIAM International Conference on Data Mining, SIAM, pp 144–152

Shi Y, Zhu Q, Guo F, Zhang C, Han J (2018c) Easing embedding learning by comprehensive transcription of heterogeneous information networks. In: Proceedings of the 24th ACM SIGKDD International Conference on Knowledge Discovery & Data Mining, pp 2190–2199

Shibata N, Kajikawa Y, Sakata I (2012) Link prediction in citation networks. Journal of the American society for information science and technology 63(1):78–85

Shorten C, Khoshgoftaar TM (2019) A survey on image data augmentation for deep learning. Journal of Big Data 6(1):1–48

Shou Z, Wang D, Chang SF (2016) Temporal action localization in untrimmed videos via multi-stage cnns. In: Proceedings of the IEEE conference on computer vision and pattern recognition, pp 1049–1058

Shou Z, Chan J, Zareian A, Miyazawa K, Chang SF (2017) Cdc: Convolutional-de-convolutional networks for precise temporal action localization in untrimmed videos. In: Proceedings of the IEEE conference on computer vision and pattern recognition, pp 5734–5743

Shrivastava S (2017) Bring rich knowledge of people places things and local businesses to your apps. Bing Blogs

Shu DW, Park SW, Kwon J (2019) 3d point cloud generative adversarial network based on tree structured graph convolutions. In: Proceedings of the IEEE/CVF International Conference on Computer Vision, pp 3859–3868

Shu K, Mahudeswaran D, Wang S, Liu H (2020) Hierarchical propagation networks for fake news detection: Investigation and exploitation. In: International AAAI Conference on Web and Social Media

Shuman DI, Narang SK, Frossard P, Ortega A, Vandergheynst P (2013) The emerging field of signal processing on graphs: Extending high-dimensional data analysis to networks and other irregular domains. IEEE Signal Process Mag 30(3):83–98

Si X, Dai H, Raghothaman M, Naik M, Song L (2018) Learning loop invariants for program verification. Advances in Neural Information Processing Systems 31:7751–7762

Si Y, Du J, Li Z, Jiang X, Miller T, Wang F, Zheng J, Roberts K (2020) Deep representation learning of patient data from electronic health records (ehr): A systematic review. Journal of Biomedical Informatics pp 103,671–103,671

Siddharth N, Paige B, van de Meent JW, Desmaison A, Goodman ND, Kohli P, Wood F, Torr PH (2017) Learning disentangled representations with semi-supervised deep generative models. In: Proceedings of the 31st International Conference on Neural Information Processing Systems, pp 5927–5937

Siegelmann HT, Sontag ED (1995) On the computational power of neural nets. Journal of computer and system sciences 50(1):132–150

Silander T, Myllymäki P (2006) A simple approach for finding the globally optimal bayesian network structure. In: Proceedings of the Twenty-Second Conference on Uncertainty in Artificial Intelligence, pp 445–452

Sillito J, Murphy GC, De Volder K (2008) Asking and answering questions during a programming change task. IEEE Transactions on Software Engineering 34(4):434–451

Silva J (2012) A vocabulary of program slicing-based techniques. ACM computing surveys (CSUR) 44(3):1–41

Silver D, Lever G, Heess N, Degris T, Wierstra D, Riedmiller M (2014) Deterministic policy gradient algorithms. In: International conference on machine learning, PMLR, pp 387–395

Simonovsky M, Komodakis N (2017) Dynamic edge-conditioned filters in convolutional neural networks on graphs. In: IEEE Conference on Computer Vision and Pattern Recognition, pp 29–38

Simonovsky M, Komodakis N (2018) Graphvae: Towards generation of small graphs using variational autoencoders. arXiv preprint arXiv:180203480

Simonyan K, Zisserman A (2014a) Two-stream convolutional networks for action recognition in videos. In: Proceedings of the 27th International Conference on Neural Information Processing Systems, pp 568–576

Simonyan K, Zisserman A (2014b) Very deep convolutional networks for large-scale image recognition. arXiv preprint arXiv:14091556

Simonyan K, Vedaldi A, Zisserman A (2013) Deep inside convolutional networks: Visualising image classification models and saliency maps. arXiv preprint arXiv:13126034

Singhal A (2012) Introducing the knowledge graph: things, not strings. Official google blog 5:16

Skarding J, Gabrys B, Musial K (2020) Foundations and modelling of dynamic networks using dynamic graph neural networks: A survey. arXiv preprint arXiv:200507496

Smilkov D, Thorat N, Kim B, Viégas F, Wattenberg M (2017) Smoothgrad: removing noise by adding noise. Workshop on Visualization for Deep Learning, ICML

Socher R, Huang EH, Pennington J, Ng AY, Manning CD (2011) Dynamic pooling and unfolding recursive autoencoders for paraphrase detection. In: NIPS, vol 24, pp 801–809

Socher R, Chen D, Manning CD, Ng A (2013) Reasoning with neural tensor networks for knowledge base completion. In: Advances in neural information processing systems, Citeseer, pp 926–934

Sohn K, Lee H, Yan X (2015) Learning structured output representation using deep conditional generative models. Advances in neural information processing systems 28:3483–3491

Song C, Lin Y, Guo S, Wan H (2020a) Spatial-temporal synchronous graph convolutional networks: A new framework for spatial-temporal network data forecasting. In: Proceedings of the AAAI Conference on Artificial Intelligence, vol 34, pp 914–921

Song L, Zhang Y, Wang Z, Gildea D (2018) A graph-to-sequence model for amr-to-text generation. In: Proceedings of the 56th Annual Meeting of the Association for Computational Linguistics (Volume 1: Long Papers), pp 1616–1626

Song L, Wang A, Su J, Zhang Y, Xu K, Ge Y, Yu D (2020b) Structural information preserving for graph-to-text generation. In: Proceedings of the 58th Annual Meeting of the Association for Computational Linguistics, pp 7987–7998

Song W, Xiao Z, Wang Y, Charlin L, Zhang M, Tang J (2019a) Session-based social recommendation via dynamic graph attention networks. In: ACM International Conference on Web Search and Data Mining, pp 555–563

Song X, Sun H, Wang X, Yan J (2019b) A survey of automatic generation of source code comments: Algorithms and techniques. IEEE Access 7:111,411–111,428

Sridhara G, Pollock L, Vijay-Shanker K (2011) Automatically detecting and describing high level actions within methods. In: Proceedings of the 33rd International Conference on Software Engineering, ACM, pp 101–110

Srinivasan B, Ribeiro B (2020a) On the equivalence between node embeddings and structural graph representations. In: International Conference on Learning Representations

Srinivasan B, Ribeiro B (2020b) On the equivalence between positional node embeddings and structural graph representations. In: 8th International Conference on Learning Representations, ICLR 2020, Addis Ababa, Ethiopia, April 26-30, 2020, OpenReview.net

Srivastava N, Hinton G, Krizhevsky A, Sutskever I, Salakhutdinov R (2014) Dropout: a simple way to prevent neural networks from overfitting. The journal of machine learning research 15(1):1929–1958

Stanfield Z, Coşkun M, Koyutürk M (2017) Drug response prediction as a link prediction problem. Scientific reports 7(1):1–13

Stanic A, van Steenkiste S, Schmidhuber J (2021) Hierarchical relational inference. In: Proceedings of the AAAI Conference on Artificial Intelligence

Stark C (2006) BioGRID: a general repository for interaction datasets. Nucleic Acids Research 34(90001):D535–D539

van Steenkiste S, Chang M, Greff K, Schmidhuber J (2018) Relational neural expectation maximization: Unsupervised discovery of objects and their interactions. In: International Conference on Learning Representations

Sterling T, Irwin JJ (2015) ZINC 15 – ligand discovery for everyone. Journal of Chemical Information and Modeling 55(11):2324–2337

Stokes J, Yang K, Swanson K, Jin W, Cubillos-Ruiz A, Donghia N, MacNair C, French S, Carfrae L, Bloom-Ackerman Z, Tran V, Chiappino-Pepe A, Badran A, Andrews I, Chory E, Church G, Brown E, Jaakkola T, Barzilay R, Collins J (2020) A deep learning approach to antibiotic discovery. Cell 180:688–702.e13

Su C, Aseltine R, Doshi R, Chen K, Rogers SC, Wang F (2020a) Machine learning for suicide risk prediction in children and adolescents with electronic health records. Translational psychiatry 10(1):1–10

Su C, Tong J, Wang F (2020b) Mining genetic and transcriptomic data using machine learning approaches in parkinson's disease. npj Parkinson's Disease 6(1):1–10

Su C, Tong J, Zhu Y, Cui P, Wang F (2020c) Network embedding in biomedical data science. Briefings in bioinformatics 21(1):182–197

Su C, Xu Z, Hoffman K, Goyal P, Safford MM, Lee J, Alvarez-Mulett S, Gomez-Escobar L, Price DR, Harrington JS, et al (2020d) Identifying organ dysfunction trajectory-based subphenotypes in critically ill patients with covid-19. medRxiv

Su C, Xu Z, Pathak J, Wang F (2020e) Deep learning in mental health outcome research: a scoping review. Translational Psychiatry 10(1):1–26

Su C, Zhang Y, Flory JH, Weiner MG, Kaushal R, Schenck EJ, Wang F (2021) Novel clinical subphenotypes in covid-19: derivation, validation, prediction, temporal patterns, and interaction with social determinants of health. medRxiv

Subramanian I, Verma S, Kumar S, Jere A, Anamika K (2020) Multi-omics data integration, interpretation, and its application. Bioinformatics and biology insights 14:1177932219899,051

Sugiyama M, Borgwardt KM (2015) Halting in random walk kernels. In: Advances in Neural Information Processing Systems, pp 1639–1647

Sukhbaatar S, Fergus R, et al (2016) Learning multiagent communication with backpropagation. Advances in neural information processing systems 29:2244–2252

Sun C, Gong Y, Wu Y, Gong M, Jiang D, Lan M, Sun S, Duan N (2019) Joint type inference on entities and relations via graph convolutional networks. In: Proceedings of the 57th Annual Meeting of the Association for Computational Linguistics, pp 1361–1370

Sun H, Xiao J, Zhu W, He Y, Zhang S, Xu X, Hou L, Li J, Ni Y, Xie G (2020a) Medical knowledge graph to enhance fraud, waste, and abuse detection on claim data: Model development and performance evaluation. JMIR Medical Informatics 8(7):e17,653

Sun J, Jiang Q, Lu C (2020b) Recursive social behavior graph for trajectory prediction. In: Proceedings of the IEEE/CVF Conference on Computer Vision and Pattern Recognition, pp 660–669

Sun K, Lin Z, Zhu Z (2020c) Multi-stage self-supervised learning for graph convolutional networks on graphs with few labeled nodes. In: Proceedings of the AAAI Conference on Artificial Intelligence, vol 34, pp 5892–5899

Sun M, Li P (2019) Graph to graph: a topology aware approach for graph structures learning and generation. In: The 22nd International Conference on Artificial Intelligence and Statistics, PMLR, pp 2946–2955

Sun S, Zhang B, Xie L, Zhang Y (2017) An unsupervised deep domain adaptation approach for robust speech recognition. Neurocomputing 257:79–87

Sun Y, Han J (2013) Mining heterogeneous information networks: a structural analysis approach. Acm Sigkdd Explorations Newsletter 14(2):20–28

Sun Y, Han J, Yan X, Yu PS, Wu T (2011) Pathsim: Meta path-based top-k similarity search in heterogeneous information networks. Proceedings of the VLDB Endowment 4(11):992–1003

Sun Y, Wang S, Tang X, Hsieh TY, Honavar V (2020d) Adversarial attacks on graph neural networks via node injections: A hierarchical reinforcement learning approach. In: Proceedings of The Web Conference 2020, Association for Computing Machinery, WWW '20, p 673–683, DOI 10.1145/3366423.3380149

Sun Y, Yuan F, Yang M, Wei G, Zhao Z, Liu D (2020e) A generic network compression framework for sequential recommender systems. In: Proceedings of the 43rd International ACM SIGIR Conference on Research and Development in Information Retrieval, pp 1299–1308

Sundararajan M, Taly A, Yan Q (2017) Axiomatic attribution for deep networks. In: International Conference on Machine Learning, PMLR, pp 3319–3328

Sutskever I, Vinyals O, Le QV (2014) Sequence to sequence learning with neural networks. Advances in Neural Information Processing Systems 27:3104–3112

Sutton RS, Barto AG (2018) Reinforcement learning: An introduction. MIT press

Sutton RS, McAllester DA, Singh SP, Mansour Y (2000) Policy gradient methods for reinforcement learning with function approximation. In: Advances in Neural Information Processing Systems, pp 1057–1063

Swietojanski P, Li J, Renals S (2016) Learning hidden unit contributions for unsupervised acoustic model adaptation. IEEE/ACM Transactions on Audio, Speech, and Language Processing 24(8):1450–1463

Szegedy C, Liu W, Jia Y, Sermanet P, Reed S, Anguelov D, Erhan D, Vanhoucke V, Rabinovich A (2015) Going deeper with convolutions. In: Proceedings of the IEEE conference on computer vision and pattern recognition, pp 1–9

Szklarczyk D, Gable AL, Lyon D, Junge A, Wyder S, Huerta-Cepas J, Simonovic M, Doncheva NT, Morris JH, Bork P, et al (2019) String v11: protein–protein association networks with increased coverage, supporting functional discovery in genome-wide experimental datasets. Nucleic acids research 47(D1):D607–D613

Takahashi T (2019) Indirect adversarial attacks via poisoning neighbors for graph convolutional networks. In: 2019 IEEE International Conference on Big Data (Big Data), IEEE, pp 1395–1400

Tang J, Wang K (2018) Personalized top-n sequential recommendation via convolutional sequence embedding. In: Proceedings of the Eleventh ACM International Conference on Web Search and Data Mining, pp 565–573

Tang J, Qu M, Mei Q (2015a) Pte: Predictive text embedding through large-scale heterogeneous text networks. In: Proceedings of the 21th ACM SIGKDD international conference on knowledge discovery and data mining, pp 1165–1174

Tang J, Qu M, Wang M, Zhang M, Yan J, Mei Q (2015b) Line: Large-scale information network embedding. In: Proceedings of the 24th international conference on world wide web, pp 1067–1077

Tang R, Du M, Liu N, Yang F, Hu X (2020a) An embarrassingly simple approach for trojan attack in deep neural networks. In: Proceedings of the 26th ACM SIGKDD International Conference on Knowledge Discovery & Data Mining, pp 218–228

Tang X, Li Y, Sun Y, Yao H, Mitra P, Wang S (2020b) Transferring robustness for graph neural network against poisoning attacks. In: Proceedings of the 13th International Conference on Web Search and Data Mining, pp 600–608

Tao J, Lin J, Zhang S, Zhao S, Wu R, Fan C, Cui P (2019) Mvan: Multi-view attention networks for real money trading detection in online games. In: Proceedings of the 25th ACM SIGKDD International Conference on Knowledge Discovery & Data Mining, pp 2536–2546

Tarlow D, Moitra S, Rice A, Chen Z, Manzagol PA, Sutton C, Aftandilian E (2020) Learning to fix build errors with Graph2Diff neural networks. In: Proceedings of the IEEE/ACM 42nd International Conference on Software Engineering Workshops, pp 19–20

Tate JG, Bamford S, Jubb HC, Sondka Z, Beare DM, Bindal N, Boutselakis H, Cole CG, Creatore C, Dawson E, et al (2019) Cosmic: the catalogue of somatic mutations in cancer. Nucleic acids research 47(D1):D941–D947

Te G, Hu W, Zheng A, Guo Z (2018) Rgcnn: Regularized graph cnn for point cloud segmentation. In: Proceedings of the 26th ACM international conference on Multimedia, pp 746–754

Tenenbaum JB, De Silva V, Langford JC (2000) A global geometric framework for nonlinear dimensionality reduction. science 290(5500):2319–2323

Teru K, Denis E, Hamilton W (2020) Inductive relation prediction by subgraph reasoning. In: International Conference on Machine Learning, PMLR, pp 9448–9457

Thomas S, Seltzer ML, Church K, Hermansky H (2013) Deep neural network features and semi-supervised training for low resource speech recognition. In: 2013 IEEE international conference on acoustics, speech and signal processing, IEEE, pp 6704–6708

Tian Z, Guo M, Wang C, Liu X, Wang S (2017) Refine gene functional similarity network based on interaction networks. BMC bioinformatics (16)

Torng W, Altman RB (2018) High precision protein functional site detection using 3d convolutional neural networks. Bioinformatics 35(9):1503–1512

Train K (1986) Qualitative choice analysis: Theory, econometrics, and an application to automobile demand, vol 10. MIT press

Tramer F, Carlini N, Brendel W, Madry A (2020) On adaptive attacks to adversarial example defenses. In: Larochelle H, Ranzato M, Hadsell R, Balcan MF, Lin H (eds) Advances in Neural Information Processing Systems, Curran Associates, Inc., vol 33, pp 1633–1645

Tran D, Bourdev L, Fergus R, Torresani L, Paluri M (2015) Learning spatiotemporal features with 3d convolutional networks. In: Proceedings of the IEEE international conference on computer vision, pp 4489–4497

Trivedi R, Dai H, Wang Y, Song L (2017) Know-evolve: Deep temporal reasoning for dynamic knowledge graphs. In: International Conference on Machine Learning, PMLR, pp 3462–3471

Trivedi R, Farajtabar M, Biswal P, Zha H (2019) Dyrep: Learning representations over dynamic graphs. In: International Conference on Learning Representations

Trouillon T, Welbl J, Riedel S, Gaussier É, Bouchard G (2016) Complex embeddings for simple link prediction. In: International Conference on Machine Learning, pp 2071–2080

Tsai YHH, Bai S, Yamada M, Morency LP, Salakhutdinov R (2019) Transformer dissection: An unified understanding for transformer's attention via the lens of kernel. In: Proceedings of the 2019 Conference on Empirical Methods in Natural Language Processing and the 9th International Joint Conference on Natural Language Processing (EMNLP-IJCNLP), pp 4335–4344

Tsuyuzaki K, Nikaido I (2017) Biological systems as heterogeneous information networks: a mini-review and perspectives. WSDM HeteroNAM 18 - International Workshop on Heterogeneous Networks Analysis and Mining

Tu C, Zhang W, Liu Z, Sun M, et al (2016) Max-margin deepwalk: Discriminative learning of network representation. In: IJCAI, vol 2016, pp 3889–3895

Tu K, Cui P, Wang X, Wang F, Zhu W (2018) Structural deep embedding for hypernetworks. In: Proceedings of the AAAI Conference on Artificial Intelligence, vol 32

Tu M, Wang G, Huang J, Tang Y, He X, Zhou B (2019) Multi-hop reading comprehension across multiple documents by reasoning over heterogeneous graphs. In:

Proceedings of the 57th Annual Meeting of the Association for Computational Linguistics, pp 2704–2713

Tufano M, Drain D, Svyatkovskiy A, Sundaresan N (2020) Generating accurate assert statements for unit test cases using pretrained transformers. arXiv preprint arXiv:200905634

Tzirakis P, Zhang J, Schuller BW (2018) End-to-end speech emotion recognition using deep neural networks. In: 2018 IEEE international conference on acoustics, speech and signal processing (ICASSP), IEEE, pp 5089–5093

Ulutan O, Iftekhar A, Manjunath BS (2020) Vsgnet: Spatial attention network for detecting human object interactions using graph convolutions. In: Proceedings of the IEEE/CVF Conference on Computer Vision and Pattern Recognition, pp 13,617–13,626

Vahedian A, Zhou X, Tong L, Li Y, Luo J (2017) Forecasting gathering events through continuous destination prediction on big trajectory data. In: Proceedings of the 25th ACM SIGSPATIAL International Conference on Advances in Geographic Information Systems, pp 1–10

Vahedian A, Zhou X, Tong L, Street WN, Li Y (2019) Predicting urban dispersal events: A two-stage framework through deep survival analysis on mobility data. In: Proceedings of the AAAI Conference on Artificial Intelligence, vol 33, pp 5199–5206

Van Hasselt H, Guez A, Silver D (2016) Deep reinforcement learning with double q-learning. In: Proceedings of the AAAI Conference on Artificial Intelligence, vol 30

Van Oord A, Kalchbrenner N, Kavukcuoglu K (2016) Pixel recurrent neural networks. In: International Conference on Machine Learning, pp 1747–1756

Vashishth S, Yadati N, Talukdar P (2019) Graph-based deep learning in natural language processing. In: Proceedings of the 2019 Conference on Empirical Methods in Natural Language Processing and the 9th International Joint Conference on Natural Language Processing (EMNLP-IJCNLP): Tutorial Abstracts

Vashishth S, Sanyal S, Nitin V, Talukdar P (2020) Composition-based multi-relational graph convolutional networks. In: International Conference on Learning Representations

Vasic M, Kanade A, Maniatis P, Bieber D, Singh R (2018) Neural program repair by jointly learning to localize and repair. In: International Conference on Learning Representations

Vaswani A, Shazeer N, Parmar N, Uszkoreit J, Jones L, Gomez AN, Kaiser u, Polosukhin I (2017) Attention is all you need. In: Proceedings of the 31st International Conference on Neural Information Processing Systems, Curran Associates Inc., Red Hook, NY, USA, NIPS'17, p 6000–6010

Veličković P, Cucurull G, Casanova A, Romero A, Lio P, Bengio Y (2018) Graph attention networks. In: International Conference on Learning Representations

Veličković P, Cucurull G, Casanova A, Romero A, Liò P, Bengio Y (2018) Graph Attention Networks. In: International Conference on Learning Representations (ICLR)

Velickovic P, Fedus W, Hamilton WL, Liò P, Bengio Y, Hjelm RD (2019) Deep graph infomax. In: ICLR (Poster)

Veličković P, Ying R, Padovano M, Hadsell R, Blundell C (2019) Neural execution of graph algorithms. In: International Conference on Learning Representations

Velickovic P, Buesing L, Overlan M, Pascanu R, Vinyals O, Blundell C (2020) Pointer graph networks. In: Larochelle H, Ranzato M, Hadsell R, Balcan MF, Lin H (eds) Advances in Neural Information Processing Systems, Curran Associates, Inc., vol 33, pp 2232–2244

Veličković P, Fedus W, Hamilton WL, Liò P, Bengio Y, Hjelm RD (2019) Deep graph infomax. In: International Conference on Learning Representations

Vento M, Foggia P (2013) Graph matching techniques for computer vision. In: Image Processing: Concepts, Methodologies, Tools, and Applications, IGI Global, chap 21, pp 381–421

Vignac C, Loukas A, Frossard P (2020a) Building powerful and equivariant graph neural networks with structural message-passing. arXiv e-prints pp arXiv–2006

Vignac C, Loukas A, Frossard P (2020b) Building powerful and equivariant graph neural networks with structural message-passing. In: Larochelle H, Ranzato M, Hadsell R, Balcan MF, Lin H (eds) Advances in Neural Information Processing Systems, Curran Associates, Inc., vol 33, pp 14,143–14,155

Vincent P, Larochelle H, Bengio Y, Manzagol P (2008) Extracting and composing robust features with denoising autoencoders. In: Cohen WW, McCallum A, Roweis ST (eds) Machine Learning, Proceedings of the Twenty-Fifth International Conference (ICML 2008), Helsinki, Finland, June 5-9, 2008, ACM, ACM International Conference Proceeding Series, vol 307, pp 1096–1103, DOI 10.1145/1390156.1390294

Vinyals O, Fortunato M, Jaitly N (2015) Pointer networks. In: Neural Information Processing Systems (NeurIPS), pp 2692–2700

Vinyals O, Bengio S, Kudlur M (2016) Order matters: Sequence to sequence for sets. In: International Conference on Learning Representations

Vishwanathan SVN, Schraudolph NN, Kondor R, Borgwardt KM (2010) Graph kernels. Journal of Machine Learning Research 11(Apr):1201–1242

VONLUXBURG U (2007) A tutorial on spectral clustering. Statistics and Computing 17:395–416

Vrandečić D, Krötzsch M (2014) Wikidata: a free collaborative knowledgebase. Communications of the ACM 57(10):78–85

Vu MN, Thai MT (2020) Pgm-explainer: Probabilistic graphical model explanations for graph neural networks. arXiv preprint arXiv:201005788

Wald J, Dhamo H, Navab N, Tombari F (2020) Learning 3d semantic scene graphs from 3d indoor reconstructions. In: Proceedings of the IEEE/CVF Conference on Computer Vision and Pattern Recognition, pp 3961–3970

Wan S, Lan Y, Guo J, Xu J, Pang L, Cheng X (2016) A deep architecture for semantic matching with multiple positional sentence representations. In: Proceedings of the AAAI Conference on Artificial Intelligence, vol 30

Wan Y, Zhao Z, Yang M, Xu G, Ying H, Wu J, Yu PS (2018) Improving automatic source code summarization via deep reinforcement learning. In: Proceedings of the 33rd ACM/IEEE International Conference on Automated Software Engineering, ACM, pp 397–407

Wang B, Gong NZ (2019) Attacking graph-based classification via manipulating the graph structure. In: Proceedings of the 2019 ACM SIGSAC Conference on Computer and Communications Security, pp 2023–2040

Wang C, Pan S, Long G, Zhu X, Jiang J (2017a) Mgae: Marginalized graph autoencoder for graph clustering. In: Proceedings of the 2017 ACM on Conference on Information and Knowledge Management, pp 889–898

Wang D, Cui P, Zhu W (2016) Structural deep network embedding. In: Proceedings of the 22nd ACM SIGKDD international conference on Knowledge discovery and data mining, pp 1225–1234

Wang D, Jamnik M, Lio P (2019a) Abstract diagrammatic reasoning with multiplex graph networks. In: International Conference on Learning Representations

Wang D, Lin J, Cui P, Jia Q, Wang Z, Fang Y, Yu Q, Zhou J, Yang S, Qi Y (2019b) A semi-supervised graph attentive network for financial fraud detection. In: 2019 IEEE International Conference on Data Mining (ICDM), IEEE, pp 598–607

Wang D, Jiang M, Syed M, Conway O, Juneja V, Subramanian S, Chawla NV (2020a) Calendar graph neural networks for modeling time structures in spatiotemporal user behaviors. In: Proceedings of the 26th ACM SIGKDD International Conference on Knowledge Discovery & Data Mining, pp 2581–2589

Wang F, Preininger A (2019) Ai in health: State of the art, challenges, and future directions. Yearbook of medical informatics 28(1):16–26

Wang F, Zhang C (2007) Label propagation through linear neighborhoods. IEEE Transactions on Knowledge and Data Engineering 20(1):55–67

Wang G, Dunbrack RL (2003) PISCES: a protein sequence culling server. Bioinformatics 19(12):1589–1591, DOI 10.1093/bioinformatics/btg224

Wang H, Huan J (2019) Agan: Towards automated design of generative adversarial networks. arXiv preprint arXiv:190611080

Wang H, Schmid C (2013) Action recognition with improved trajectories. In: Proceedings of the IEEE international conference on computer vision, pp 3551–3558

Wang H, Wang J, Wang J, Zhao M, Zhang W, Zhang F, Xie X, Guo M (2018a) Graphgan: Graph representation learning with generative adversarial nets. In: Proceedings of the AAAI conference on artificial intelligence, vol 32

Wang H, Zhang F, Zhang M, Leskovec J, Zhao M, Li W, Wang Z (2019c) Knowledge-aware graph neural networks with label smoothness regularization for recommender systems. In: KDD'19, pp 968–977

Wang H, Zhao M, Xie X, Li W, Guo M (2019d) Knowledge graph convolutional networks for recommender systems. In: The world wide web conference, pp 3307–3313

Wang H, Zhao M, Xie X, Li W, Guo M (2019e) Knowledge graph convolutional networks for recommender systems. In: WWW'19, pp 3307–3313

Wang H, Wang K, Yang J, Shen L, Sun N, Lee HS, Han S (2020b) Gcn-rl circuit designer: Transferable transistor sizing with graph neural networks and reinforcement learning. In: Design Automation Conference, IEEE, pp 1–6

Wang J, Zheng VW, Liu Z, Chang KCC (2017b) Topological recurrent neural network for diffusion prediction. In: 2017 IEEE International Conference on Data Mining (ICDM), IEEE, pp 475–484

Wang J, Huang P, Zhao H, Zhang Z, Zhao B, Lee DL (2018b) Billion-scale commodity embedding for e-commerce recommendation in alibaba. In: Proceedings of the 24th ACM SIGKDD International Conference on Knowledge Discovery & Data Mining, pp 839–848

Wang J, Oh J, Wang H, Wiens J (2018c) Learning credible models. In: Proceedings of the 24th ACM SIGKDD International Conference on Knowledge Discovery & Data Mining, pp 2417–2426

Wang J, Luo M, Suya F, Li J, Yang Z, Zheng Q (2020c) Scalable attack on graph data by injecting vicious nodes. Data Mining and Knowledge Discovery 34(5):1363–1389

Wang K, Singh R, Su Z (2018d) Dynamic neural program embeddings for program repair. In: International Conference on Learning Representations

Wang M, Liu M, Liu J, Wang S, Long G, Qian B (2017c) Safe medicine recommendation via medical knowledge graph embedding. arXiv preprint arXiv:171005980

Wang M, Yu L, Zheng D, Gan Q, Gai Y, Ye Z, Li M, Zhou J, Huang Q, Ma C, Huang Z, Guo Q, Zhang H, Lin H, Zhao J, Li J, Smola AJ, Zhang Z (2019f) Deep graph library: Towards efficient and scalable deep learning on graphs. International Conference on Learning Representations Workshop on Representation Learning on Graphs and Manifolds

Wang M, Lin Y, Lin G, Yang K, Wu Xm (2020d) M2grl: A multi-task multi-view graph representation learning framework for web-scale recommender systems. In: Proceedings of the 26th ACM SIGKDD International Conference on Knowledge Discovery & Data Mining, pp 2349–2358

Wang Q, Mao Z, Wang B, Guo L (2017d) Knowledge graph embedding: A survey of approaches and applications. IEEE Transactions on Knowledge and Data Engineering 29(12):2724–2743

Wang Q, Li M, Wang X, Parulian N, Han G, Ma J, Tu J, Lin Y, Zhang H, Liu W, et al (2020e) Covid-19 literature knowledge graph construction and drug repurposing report generation. arXiv preprint arXiv:200700576

Wang R, Yan J, Yang X (2019g) Learning combinatorial embedding networks for deep graph matching. In: Proceedings of the IEEE/CVF International Conference on Computer Vision, pp 3056–3065

Wang R, Zhang T, Yu T, Yan J, Yang X (2020f) Combinatorial learning of graph edit distance via dynamic embedding. arXiv preprint arXiv:201115039

Wang S, He L, Cao B, Lu CT, Yu PS, Ragin AB (2017e) Structural deep brain network mining. In: Proceedings of the 23rd ACM SIGKDD International Conference on Knowledge Discovery and Data Mining, pp 475–484

Wang S, Tang J, Aggarwal C, Chang Y, Liu H (2017f) Signed network embedding in social media. In: Proceedings of the 2017 SIAM international conference on data mining, SIAM, pp 327–335

Wang S, Chen Z, Li D, Li Z, Tang LA, Ni J, Rhee J, Chen H, Yu PS (2019h) Attentional heterogeneous graph neural network: Application to program reidentification. In: Proceedings of the 2019 SIAM International Conference on Data Mining, SIAM, pp 693–701

Wang S, Chen Z, Yu X, Li D, Ni J, Tang L, Gui J, Li Z, Chen H, Yu PS (2019i) Heterogeneous graph matching networks for unknown malware detection. In: Proceedings of the Twenty-Eighth International Joint Conference on Artificial Intelligence, IJCAI, pp 3762–3770

Wang S, Li BZ, Khabsa M, Fang H, Ma H (2020g) Linformer: Self-attention with linear complexity. CoRR abs/2006.04768

Wang S, Li Y, Zhang J, Meng Q, Meng L, Gao F (2020h) Pm2. 5-gnn: A domain knowledge enhanced graph neural network for pm2. 5 forecasting. In: Proceedings of the 28th International Conference on Advances in Geographic Information Systems, pp 163–166

Wang S, Wang R, Yao Z, Shan S, Chen X (2020i) Cross-modal scene graph matching for relationship-aware image-text retrieval. In: Proceedings of the IEEE/CVF Winter Conference on Applications of Computer Vision, pp 1508–1517

Wang T, Ling H (2017) Gracker: A graph-based planar object tracker. IEEE transactions on pattern analysis and machine intelligence 40(6):1494–1501

Wang T, Liu H, Li Y, Jin Y, Hou X, Ling H (2020j) Learning combinatorial solver for graph matching. In: Proceedings of the IEEE/CVF conference on computer vision and pattern recognition, pp 7568–7577

Wang T, Wan X, Jin H (2020k) Amr-to-text generation with graph transformer. Transactions of the Association for Computational Linguistics 8:19–33

Wang X, Gupta A (2018) Videos as space-time region graphs. In: Proceedings of the European conference on computer vision (ECCV), pp 399–417

Wang X, Cui P, Wang J, Pei J, Zhu W, Yang S (2017g) Community preserving network embedding. In: Proceedings of the AAAI Conference on Artificial Intelligence, vol 31

Wang X, Girshick R, Gupta A, He K (2018e) Non-local neural networks. In: Proceedings of the IEEE conference on computer vision and pattern recognition, pp 7794–7803

Wang X, Ye Y, Gupta A (2018f) Zero-shot recognition via semantic embeddings and knowledge graphs. In: Proceedings of the IEEE conference on computer vision and pattern recognition, pp 6857–6866

Wang X, He X, Cao Y, Liu M, Chua TS (2019j) Kgat: Knowledge graph attention network for recommendation. In: KDD'19, pp 950–958

Wang X, He X, Wang M, Feng F, Chua TS (2019k) Neural graph collaborative filtering. In: Proceedings of the 42nd international ACM SIGIR conference on Research and development in Information Retrieval, pp 165–174

Wang X, Ji H, Shi C, Wang B, Ye Y, Cui P, Yu PS (2019l) Heterogeneous graph attention network. In: The World Wide Web Conference, pp 2022–2032

Wang X, Ji H, Shi C, Wang B, Ye Y, Cui P, Yu PS (2019m) Heterogeneous graph attention network. In: The World Wide Web Conference, pp 2022–2032

Wang X, Zhang Y, Shi C (2019n) Hyperbolic heterogeneous information network embedding. In: Proceedings of the AAAI conference on artificial intelligence, vol 33, pp 5337–5344

Wang X, Bo D, Shi C, Fan S, Ye Y, Yu PS (2020l) A survey on heterogeneous graph embedding: Methods, techniques, applications and sources. arXiv preprint arXiv:201114867

Wang X, Lu Y, Shi C, Wang R, Cui P, Mou S (2020m) Dynamic heterogeneous information network embedding with meta-path based proximity. IEEE Transactions on Knowledge and Data Engineering pp 1–1, DOI 10.1109/TKDE.2020.2993870

Wang X, Wang R, Shi C, Song G, Li Q (2020n) Multi-component graph convolutional collaborative filtering. In: Proceedings of the AAAI Conference on Artificial Intelligence, vol 34, pp 6267–6274

Wang X, Wu Y, Zhang A, He X, seng Chua T (2021) Causal screening to interpret graph neural networks

Wang Y, Ni X, Sun JT, Tong Y, Chen Z (2011) Representing document as dependency graph for document clustering. In: Proceedings of the 20th ACM international conference on Information and knowledge management, pp 2177–2180

Wang Y, Shen H, Liu S, Gao J, Cheng X (2017h) Cascade dynamics modeling with attention-based recurrent neural network. In: Proceedings of the 26th International Joint Conference on Artificial Intelligence, pp 2985–2991

Wang Y, Che W, Guo J, Liu T (2018g) A neural transition-based approach for semantic dependency graph parsing. In: Proceedings of the AAAI Conference on Artificial Intelligence, vol 32

Wang Y, Yin H, Chen H, Wo T, Xu J, Zheng K (2019o) Origin-destination matrix prediction via graph convolution: a new perspective of passenger demand modeling. In: Proceedings of the 25th ACM SIGKDD International Conference on Knowledge Discovery & Data Mining, pp 1227–1235

Wang Y, Liu S, Yoon M, Lamba H, Wang W, Faloutsos C, Hooi B (2020o) Provably robust node classification via low-pass message passing. In: 2020 IEEE International Conference on Data Mining (ICDM), pp 621–630, DOI 10.1109/ICDM50108.2020.00071

Wang Z, Zhang J, Feng J, Chen Z (2014) Knowledge graph embedding by translating on hyperplanes. In: Proceedings of the AAAI Conference on Artificial Intelligence, vol 28

Wang Z, Zheng L, Li Y, Wang S (2019p) Linkage based face clustering via graph convolution network. In: Proceedings of the IEEE/CVF Conference on Computer Vision and Pattern Recognition, pp 1117–1125

Wass MN, Barton G, Sternberg MJE (2012) CombFunc: predicting protein function using heterogeneous data sources. Nucleic Acids Research 40(W1):W466–W470

Watkins KE, Ferris B, Borning A, Rutherford GS, Layton D (2011) Where is my bus? impact of mobile real-time information on the perceived and actual wait time of transit riders. Transportation Research Part A: Policy and Practice 45(8):839–848

Watts DJ, Strogatz SH (1998) Collective dynamics of 'small-world'networks. nature 393(6684):440–442

Wei J, Goyal M, Durrett G, Dillig I (2019) LambdaNet: Probabilistic type inference using graph neural networks. In: International Conference on Learning Representations

Wei X, Yu R, Sun J (2020) View-gcn: View-based graph convolutional network for 3d shape analysis. In: Proceedings of the IEEE/CVF Conference on Computer Vision and Pattern Recognition, pp 1850–1859

Weihua Hu MZYDHRBLMCJL Matthias Fey (2020) Open graph benchmark: Datasets for machine learning on graphs. arXiv preprint arXiv:200500687

Weininger D (1988) Smiles, a chemical language and information system. 1. introduction to methodology and encoding rules. Journal of chemical information and computer sciences 28(1):31–36

Weisfeiler B (1976) On Construction and Identification of Graphs. Lecture Notes in Mathematics, Vol. 558, Springer

Weisfeiler B, Leman A (1968) The reduction of a graph to canonical form and the algebra which appears therein. Nauchno-Technicheskaya Informatsia 2(9):12–16

Weisfeiler B, Leman A (1968) The reduction of a graph to canonical form and the algebra which appears therein. NTI, Series 2(9):12–16

Weng C, Shah NH, Hripcsak G (2020) Deep phenotyping: embracing complexity and temporality—towards scalability, portability, and interoperability. Journal of biomedical informatics 105:103,433

Weston J, Bengio S, Usunier N (2010) Large scale image annotation: learning to rank with joint word-image embeddings. Machine learning 81(1):21–35

Whirl-Carrillo M, McDonagh EM, Hebert J, Gong L, Sangkuhl K, Thorn C, Altman RB, Klein TE (2012) Pharmacogenomics knowledge for personalized medicine. Clinical Pharmacology & Therapeutics 92(4):414–417

White M, Tufano M, Vendome C, Poshyvanyk D (2016) Deep learning code fragments for code clone detection. In: 2016 31st IEEE/ACM International Conference on Automated Software Engineering (ASE), IEEE, pp 87–98

Williams RJ (1992) Simple statistical gradient-following algorithms for connectionist reinforcement learning. Machine learning 8(3-4):229–256

Wishart DS, Feunang YD, Guo AC, Lo EJ, Marcu A, Grant JR, Sajed T, Johnson D, Li C, Sayeeda Z, et al (2018) Drugbank 5.0: a major update to the drugbank database for 2018. Nucleic acids research 46(D1):D1074–D1082

Wold S, Esbensen K, Geladi P (1987) Principal component analysis. Chemometrics and intelligent laboratory systems 2(1-3):37–52

Woźnica A, Kalousis A, Hilario M (2010) Adaptive matching based kernels for labelled graphs. In: Advances in Knowledge Discovery and Data Mining, Springer, Lecture Notes in Computer Science, vol 6119, pp 374–385

Wu B, Xu C, Dai X, Wan A, Zhang P, Tomizuka M, Keutzer K, Vajda P (2020a) Visual transformers: Token-based image representation and processing for computer vision. arXiv preprint arXiv:200603677

Wu F, Souza A, Zhang T, Fifty C, Yu T, Weinberger K (2019a) Simplifying graph convolutional networks. In: International conference on machine learning, PMLR, pp 6861–6871

Wu H, Wang C, Tyshetskiy Y, Docherty A, Lu K, Zhu L (2019b) Adversarial examples for graph data: Deep insights into attack and defense. In: Proceedings of the Twenty-Eighth International Joint Conference on Artificial Intelligence,

IJCAI-19, International Joint Conferences on Artificial Intelligence Organization, pp 4816–4823

Wu H, Ma Y, Xiang Z, Yang C, He K (2021a) A spatial-temporal graph neural network framework for automated software bug triaging. arXiv preprint arXiv:210111846

Wu J, Cao M, Cheung JCK, Hamilton WL (2020b) Temp: Temporal message passing for temporal knowledge graph completion. In: Proceedings of the 2020 Conference on Empirical Methods in Natural Language Processing (EMNLP), pp 5730–5746

Wu L, Chen Y, Ji H, Li Y (2021b) Deep learning on graphs for natural language processing. In: Proceedings of the 2021 Conference of the North American Chapter of the Association for Computational Linguistics: Human Language Technologies: Tutorials, pp 11–14

Wu L, Chen Y, Shen K, Guo X, Gao H, Li S, Pei J, Long B (2021c) Graph neural networks for natural language processing: A survey. arXiv preprint arXiv:210606090

Wu N, Zhao XW, Wang J, Pan D (2020c) Learning effective road network representation with hierarchical graph neural networks. In: Proceedings of the 26th ACM SIGKDD International Conference on Knowledge Discovery & Data Mining, pp 6–14

Wu S, Tang Y, Zhu Y, Wang L, Xie X, Tan T (2019c) Session-based recommendation with graph neural networks. In: Proceedings of the AAAI Conference on Artificial Intelligence, vol 33, pp 346–353

Wu T, Ren H, Li P, Leskovec J (2020d) Graph information bottleneck. In: Larochelle H, Ranzato M, Hadsell R, Balcan MF, Lin H (eds) Advances in Neural Information Processing Systems, Curran Associates, Inc., vol 33, pp 20,437–20,448

Wu Y, Warner JL, Wang L, Jiang M, Xu J, Chen Q, Nian H, Dai Q, Du X, Yang P, et al (2019d) Discovery of noncancer drug effects on survival in electronic health records of patients with cancer: a new paradigm for drug repurposing. JCO clinical cancer informatics 3:1–9

Wu Z, Ramsundar B, Feinberg EN, Gomes J, Geniesse C, Pappu AS, Leswing K, Pande V (2018) MoleculeNet: A benchmark for molecular machine learning. Chemical Science 9:513–530

Wu Z, Pan S, Chen F, Long G, Zhang C, Yu PS (2019e) A comprehensive survey on graph neural networks. CoRR abs/1901.00596

Wu Z, Pan S, Chen F, Long G, Zhang C, Philip SY (2021d) A comprehensive survey on graph neural networks. IEEE Transactions on Neural Networks and Learning Systems 32(1):4–24

Xhonneux LP, Qu M, Tang J (2020) Continuous graph neural networks. In: Proceedings of the International Conference on Machine Learning

Xia R, Liu Y (2015) A multi-task learning framework for emotion recognition using 2d continuous space. IEEE Transactions on Affective Computing 8(1):3–14

Xiao C, Choi E, Sun J (2018) Opportunities and challenges in developing deep learning models using electronic health records data: a systematic review. Journal of the American Medical Informatics Association 25(10):1419–1428

Xie L, Yuille A (2017) Genetic cnn. In: Proceedings of the IEEE International Conference on Computer Vision, pp 1379–1388

Xie M, Yin H, Wang H, Xu F, Chen W, Wang S (2016) Learning graph-based poi embedding for location-based recommendation. In: Proceedings of the 25th ACM International on Conference on Information and Knowledge Management, Association for Computing Machinery, CIKM '16, p 15–24, DOI 10.1145/2983323.2983711

Xie S, Kirillov A, Girshick R, He K (2019a) Exploring randomly wired neural networks for image recognition. In: Proceedings of the IEEE/CVF International Conference on Computer Vision, pp 1284–1293

Xie T, Grossman JC (201f8) Crystal graph convolutional neural networks for an accurate and interpretable prediction of material properties. Physical Review Letters 120:145,301

Xie Y, Xu Z, Wang Z, Ji S (2021) Self-supervised learning of graph neural networks: A unified review. arXiv preprint arXiv:210210757

Xie Z, Lv W, Huang S, Lu Z, Du B, Huang R (2019b) Sequential graph neural network for urban road traffic speed prediction. IEEE Access 8:63,349–63,358

Xiu H, Yan X, Wang X, Cheng J, Cao L (2020) Hierarchical graph matching network for graph similarity computation. arXiv preprint arXiv:200616551

Xu D, Zhu Y, Choy CB, Fei-Fei L (2017a) Scene graph generation by iterative message passing. In: Proceedings of the IEEE conference on computer vision and pattern recognition, pp 5410–5419

Xu D, Cheng W, Luo D, Liu X, Zhang X (2019a) Spatio-temporal attentive rnn for node classification in temporal attributed graphs. In: International Joint Conference on Artificial Intelligence, pp 3947–3953

Xu D, Ruan C, Korpeoglu E, Kumar S, Achan K (2020a) Inductive representation learning on temporal graphs. In: International Conference on Learning Representations

Xu H, Jiang C, Liang X, Li Z (2019b) Spatial-aware graph relation network for large-scale object detection. In: Proceedings of the IEEE/CVF Conference on Computer Vision and Pattern Recognition, pp 9298–9307

Xu J, Gan Z, Cheng Y, Liu J (2020b) Discourse-aware neural extractive text summarization. In: Proceedings of the 58th Annual Meeting of the Association for Computational Linguistics, pp 5021–5031

Xu K, Ba J, Kiros R, Cho K, Courville A, Salakhudinov R, Zemel R, Bengio Y (2015) Show, attend and tell: Neural image caption generation with visual attention. In: International conference on machine learning, PMLR, pp 2048–2057

Xu K, Li C, Tian Y, Sonobe T, Kawarabayashi K, Jegelka S (2018a) Representation learning on graphs with jumping knowledge networks. In: International Conference on Machine Learning, pp 5453–5462

Xu K, Wu L, Wang Z, Feng Y, Sheinin V (2018b) Sql-to-text generation with graph-to-sequence model. arXiv preprint arXiv:180905255

Xu K, Wu L, Wang Z, Feng Y, Witbrock M, Sheinin V (2018c) Graph2seq: Graph to sequence learning with attention-based neural networks. arXiv preprint arXiv:180400823

Xu K, Wu L, Wang Z, Yu M, Chen L, Sheinin V (2018d) Exploiting rich syntactic information for semantic parsing with graph-to-sequence model. In: Proceedings of the 2018 Conference on Empirical Methods in Natural Language Processing, Association for Computational Linguistics, Brussels, Belgium, pp 918–924

Xu K, Chen H, Liu S, Chen PY, Weng TW, Hong M, Lin X (2019c) Topology attack and defense for graph neural networks: An optimization perspective. In: Proceedings of the Twenty-Eighth International Joint Conference on Artificial Intelligence, IJCAI-19, International Joint Conferences on Artificial Intelligence Organization, pp 3961–3967, DOI 10.24963/ijcai.2019/550

Xu K, Hu W, Leskovec J, Jegelka S (2019d) How powerful are graph neural networks? In: International Conference on Learning Representations

Xu K, Wang L, Yu M, Feng Y, Song Y, Wang Z, Yu D (2019e) Cross-lingual knowledge graph alignment via graph matching neural network. In: Proceedings of the 57th Annual Meeting of the Association for Computational Linguistics, pp 3156–3161

Xu K, Li J, Zhang M, Du SS, Kawarabayashi Ki, Jegelka S (2020c) What can neural networks reason about? In: International Conference on Learning Representations

Xu L, Wei X, Cao J, Yu PS (2017b) Embedding of embedding (eoe) joint embedding for coupled heterogeneous networks. In: Proceedings of the Tenth ACM International Conference on Web Search and Data Mining, pp 741–749

Xu M, Li L, Wai D, Liu Q, Chao LS, et al (2020d) Document graph for neural machine translation. arXiv preprint arXiv:201203477

Xu Q, Sun X, Wu CY, Wang P, Neumann U (2020e) Grid-gcn for fast and scalable point cloud learning. In: Proceedings of the IEEE/CVF Conference on Computer Vision and Pattern Recognition, pp 5661–5670

Xu R, Li L, Wang Q (2013) Towards building a disease-phenotype knowledge base: extracting disease-manifestation relationship from literature. Bioinformatics 29(17):2186–2194

Yamaguchi F, Golde N, Arp D, Rieck K (2014) Modeling and discovering vulnerabilities with code property graphs. In: 2014 IEEE Symposium on Security and Privacy, IEEE, pp 590–604

Yan J, Yin XC, Lin W, Deng C, Zha H, Yang X (2016) A short survey of recent advances in graph matching. In: Proceedings of the 2016 ACM on International Conference on Multimedia Retrieval, pp 167–174

Yan J, Yang S, Hancock E (2020a) Learning for graph matching and related combinatorial optimization problems. In: Bessiere C (ed) Proceedings of the Twenty-Ninth International Joint Conference on Artificial Intelligence, IJCAI-20, International Joint Conferences on Artificial Intelligence Organization, pp 4988–4996

Yan S, Xiong Y, Lin D (2018a) Spatial temporal graph convolutional networks for skeleton-based action recognition. In: AAAI Conference on Artificial Intelligence, vol 32

Yan X, Han J (2002) gspan: Graph-based substructure pattern mining. In: Proceedings of IEEE International Conference on Data Mining, IEEE, pp 721–724

Yan Y, Mao Y, Li B (2018b) Second: Sparsely embedded convolutional detection. Sensors 18(10):3337

Yan Y, Zhang Q, Ni B, Zhang W, Xu M, Yang X (2019) Learning context graph for person search. In: Proceedings of the IEEE/CVF Conference on Computer Vision and Pattern Recognition, pp 2158–2167

Yan Y, Qin J, Chen J, Liu L, Zhu F, Tai Y, Shao L (2020b) Learning multi-granular hypergraphs for video-based person re-identification. In: Proceedings of the IEEE/CVF Conference on Computer Vision and Pattern Recognition, pp 2899–2908

Yanardag P, Vishwanathan S (2015) Deep graph kernels. In: Proceedings of the 21th ACM SIGKDD International Conference on Knowledge Discovery and Data Mining, ACM, pp 1365–1374

Yang B, Yih W, He X, Gao J, Deng L (2015a) Embedding entities and relations for learning and inference in knowledge bases. In: Bengio Y, LeCun Y (eds) 3rd International Conference on Learning Representations, ICLR 2015, San Diego, CA, USA, May 7-9, 2015, Conference Track Proceedings

Yang B, Luo W, Urtasun R (2018a) Pixor: Real-time 3d object detection from point clouds. In: Proceedings of the IEEE conference on Computer Vision and Pattern Recognition, pp 7652–7660

Yang C, Liu Z, Zhao D, Sun M, Chang EY (2015b) Network representation learning with rich text information. In: IJCAI, vol 2015, pp 2111–2117

Yang C, Zhuang P, Shi W, Luu A, Li P (2019a) Conditional structure generation through graph variational generative adversarial nets. In: NeurIPS, pp 1338–1349

Yang F, Fan K, Song D, et al (2020a) Graph-based prediction of protein-protein interactions with attributed signed graph embedding. BMC Bioinformatics 21(1):323

Yang H, Qin C, Li YH, Tao L, Zhou J, Yu CY, Xu F, Chen Z, Zhu F, Chen YZ (2016a) Therapeutic target database update 2016: enriched resource for bench to clinical drug target and targeted pathway information. Nucleic acids research 44(D1):D1069–D1074

Yang J, Zheng WS, Yang Q, Chen YC, Tian Q (2020b) Spatial-temporal graph convolutional network for video-based person re-identification. In: Proceedings of the IEEE/CVF Conference on Computer Vision and Pattern Recognition, pp 3289–3299

Yang K, Swanson K, Jin W, Coley C, Eiden P, Gao H, Guzman-Perez A, Hopper T, Kelley B, Mathea M, et al (2019b) Analyzing learned molecular representations for property prediction. Journal of chemical information and modeling 59(8):3370–3388

Yang L, Kang Z, Cao X, Jin D, Yang B, Guo Y (2019c) Topology optimization based graph convolutional network. In: Proceedings of the Twenty-Eighth International Joint Conference on Artificial Intelligence, pp 4054–4061

Yang L, Zhan X, Chen D, Yan J, Loy CC, Lin D (2019d) Learning to cluster faces on an affinity graph. In: Proceedings of the IEEE/CVF Conference on Computer Vision and Pattern Recognition, pp 2298–2306

Yang Q, Liu Y, Chen T, Tong Y (2019e) Federated machine learning: Concept and applications. ACM Transactions on Intelligent Systems and Technology (TIST) 10(2):1–19

Yang S, Li G, Yu Y (2019f) Dynamic graph attention for referring expression comprehension. In: Proceedings of the IEEE/CVF International Conference on Computer Vision, pp 4644–4653

Yang S, Liu J, Wu K, Li M (2020c) Learn to generate time series conditioned graphs with generative adversarial nets. arXiv preprint arXiv:200301436

Yang X, Tang K, Zhang H, Cai J (2019g) Auto-encoding scene graphs for image captioning. In: Proceedings of the IEEE/CVF Conference on Computer Vision and Pattern Recognition, pp 10,685–10,694

Yang Y, Abrego GH, Yuan S, Guo M, Shen Q, Cer D, Sung YH, Strope B, Kurzweil R (2019h) Improving multilingual sentence embedding using bi-directional dual encoder with additive margin softmax. In: Proceedings of the 28th International Joint Conference on Artificial Intelligence, AAAI Press, pp 5370–5378

Yang Z, Cohen W, Salakhudinov R (2016b) Revisiting semi-supervised learning with graph embeddings. In: International conference on machine learning, PMLR, pp 40–48

Yang Z, Qi P, Zhang S, Bengio Y, Cohen W, Salakhutdinov R, Manning CD (2018b) Hotpotqa: A dataset for diverse, explainable multi-hop question answering. In: Proceedings of the 2018 Conference on Empirical Methods in Natural Language Processing, pp 2369–2380

Yang Z, Zhao J, Dhingra B, He K, Cohen WW, Salakhutdinov RR, LeCun Y (2018c) Glomo: Unsupervised learning of transferable relational graphs. In: Advances in Neural Information Processing Systems, pp 8950–8961

Yang Z, Ding M, Zhou C, Yang H, Zhou J, Tang J (2020d) Understanding negative sampling in graph representation learning. In: Proceedings of the 26th ACM SIGKDD International Conference on Knowledge Discovery & Data Mining, pp 1666–1676

Yao L, Wang L, Pan L, Yao K (2016) Link prediction based on common-neighbors for dynamic social network. Procedia Computer Science 83:82–89

Yao L, Mao C, Luo Y (2019) Graph convolutional networks for text classification. In: Proceedings of the AAAI Conference on Artificial Intelligence, vol 33, pp 7370–7377

Yao S, Wang T, Wan X (2020) Heterogeneous graph transformer for graph-to-sequence learning. In: Proceedings of the 58th Annual Meeting of the Association for Computational Linguistics, pp 7145–7154

Yao T, Pan Y, Li Y, Mei T (2018) Exploring visual relationship for image captioning. In: Proceedings of the European conference on computer vision (ECCV), pp 684–699

Yarotsky D (2017) Error bounds for approximations with deep relu networks. Neural Networks 94:103–114

Yasunaga M, Liang P (2020) Graph-based, self-supervised program repair from diagnostic feedback. In: International Conference on Machine Learning, PMLR, pp 10,799–10,808

Ye Y, Hou S, Chen L, Lei J, Wan W, Wang J, Xiong Q, Shao F (2019a) Out-of-sample node representation learning for heterogeneous graph in real-time android malware detection. In: Proceedings of the Twenty-Eighth International Joint Conference on Artificial Intelligence, IJCAI-19, International Joint Conferences on Artificial Intelligence Organization, pp 4150–4156

Ye Y, Wang X, Yao J, Jia K, Zhou J, Xiao Y, Yang H (2019b) Bayes embedding (bem): Refining representation by integrating knowledge graphs and behavior-specific networks. In: Proceedings of the 28th ACM International Conference on Information and Knowledge Management, Association for Computing Machinery, CIKM '19, p 679–688, DOI 10.1145/3357384.3358014

Yefet N, Alon U, Yahav E (2020) Adversarial examples for models of code. Proceedings of the ACM on Programming Languages 4(OOPSLA):1–30

Yeung DY, Chang H (2007) A kernel approach for semisupervised metric learning. IEEE Transactions on Neural Networks 18(1):141–149

Yi J, Park J (2020) Hypergraph convolutional recurrent neural network. In: Proceedings of the 26th ACM SIGKDD International Conference on Knowledge Discovery & Data Mining, pp 3366–3376

YILMAZ B, Genc H, Agriman M, Demirdover BK, Erdemir M, Simsek G, Karagoz P (2020) Recent trends in the use of graph neural network models for natural language processing. In: Deep Learning Techniques and Optimization Strategies in Big Data Analytics, IGI Global, pp 274–289

Ying J, de Miranda Cardoso JV, Palomar D (2020a) Nonconvex sparse graph learning under laplacian constrained graphical model. Advances in Neural Information Processing Systems 33

Ying R, He R, Chen K, Eksombatchai P, Hamilton WL, Leskovec J (2018a) Graph convolutional neural networks for web-scale recommender systems. In: Proceedings of the 24th ACM SIGKDD International Conference on Knowledge Discovery & Data Mining, pp 974–983

Ying R, He R, Chen K, Eksombatchai P, Hamilton WL, Leskovec J (2018b) Graph convolutional neural networks for web-scale recommender systems. In: Proceedings of the 24th ACM SIGKDD International Conference on Knowledge Discovery & Data Mining, pp 974–983

Ying R, Bourgeois D, You J, Zitnik M, Leskovec J (2019) Gnnexplainer: Generating explanations for graph neural networks. Advances in neural information processing systems 32:9240

Ying R, Lou Z, You J, Wen C, Canedo A, Leskovec J, et al (2020b) Neural subgraph matching. arXiv preprint arXiv:200703092

Ying Z, You J, Morris C, Ren X, Hamilton W, Leskovec J (2018c) Hierarchical graph representation learning with differentiable pooling. In: Advances in Neural Information Processing Systems, pp 4800–4810

YLow, JGonzalez, AKyrola, DBickson, CGuestrin, JHellerstein (2012) Distributed graphlab: A framework for machine learning in the cloud. PVLDB 5(8):716–727

You J, Liu B, Ying Z, Pande V, Leskovec J (2018a) Graph convolutional policy network for goal-directed molecular graph generation. In: Advances in Neural Information Processing Systems, pp 6412–6422

You J, Ying R, Ren X, Hamilton W, Leskovec J (2018b) Graphrnn: Generating realistic graphs with deep auto-regressive models. In: International Conference on Machine Learning, PMLR, pp 5708–5717

You J, Ying R, Leskovec J (2019) Position-aware graph neural networks. In: International Conference on Machine Learning, PMLR, pp 7134–7143

You J, Ying Z, Leskovec J (2020a) Design space for graph neural networks. Advances in Neural Information Processing Systems 33

You J, Gomes-Selman J, Ying R, Leskovec J (2021) Identity-aware graph neural networks. CoRR abs/2101.10320, 2101.10320

You Y, Chen T, Sui Y, Chen T, Wang Z, Shen Y (2020b) Graph contrastive learning with augmentations. In: Larochelle H, Ranzato M, Hadsell R, Balcan MF, Lin H (eds) Advances in Neural Information Processing Systems, Curran Associates, Inc., vol 33, pp 5812–5823

You Y, Chen T, Wang Z, Shen Y (2020c) When does self-supervision help graph convolutional networks? In: International Conference on Machine Learning, PMLR, pp 10,871–10,880

You ZH, Chan KCC, Hu P (2015a) Predicting protein-protein interactions from primary protein sequences using a novel multi-scale local feature representation scheme and the random forest. PLOS ONE 10:1–19

You ZH, Li J, Gao X, et al (2015b) Detecting protein-protein interactions with a novel matrix-based protein sequence representation and support vector machines. BioMed Research International 2015:1–9

Yu B, Yin H, Zhu Z (2018a) Spatio-temporal graph convolutional networks: a deep learning framework for traffic forecasting. In: Proceedings of the 27th International Joint Conference on Artificial Intelligence, pp 3634–3640

Yu D, Fu J, Mei T, Rui Y (2017a) Multi-level attention networks for visual question answering. In: Proceedings of the IEEE Conference on Computer Vision and Pattern Recognition, pp 4709–4717

Yu D, Zhang R, Jiang Z, Wu Y, Yang Y (2021a) Graph-revised convolutional network. In: Hutter F, Kersting K, Lijffijt J, Valera I (eds) Machine Learning and Knowledge Discovery in Databases, Springer International Publishing, Cham, pp 378–393

Yu H, Wu Z, Wang S, Wang Y, Ma X (2017b) Spatiotemporal recurrent convolutional networks for traffic prediction in transportation networks. Sensors 17(7):1501

Yu J, Lu Y, Qin Z, Zhang W, Liu Y, Tan J, Guo L (2018b) Modeling text with graph convolutional network for cross-modal information retrieval. In: Pacific Rim Conference on Multimedia, Springer, pp 223–234

Yu L, Du B, Hu X, Sun L, Han L, Lv W (2021b) Deep spatio-temporal graph convolutional network for traffic accident prediction. Neurocomputing 423:135–147

Yu T, Wang R, Yan J, Li B (2020) Learning deep graph matching with channel-independent embedding and hungarian attention. In: International conference on learning representations

Yu Y, Chen J, Gao T, Yu M (2019a) Dag-gnn: Dag structure learning with graph neural networks. In: International Conference on Machine Learning, pp 7154–7163

Yu Y, Wang Y, Xia Z, Zhang X, Jin K, Yang J, Ren L, Zhou Z, Yu D, Qing T, et al (2019b) Premedkb: an integrated precision medicine knowledgebase for interpreting relationships between diseases, genes, variants and drugs. Nucleic acids research 47(D1):D1090–D1101

Yuan F, He X, Karatzoglou A, Zhang L (2020a) Parameter-efficient transfer from sequential behaviors for user modeling and recommendation. In: Proceedings of the 43rd International ACM SIGIR Conference on Research and Development in Information Retrieval, pp 1469–1478

Yuan H, Tang J, Hu X, Ji S (2020b) Xgnn: Towards model-level explanations of graph neural networks. In: Proceedings of the 26th ACM SIGKDD International Conference on Knowledge Discovery & Data Mining, pp 430–438

Yuan J, Zheng Y, Zhang C, Xie W, Xie X, Sun G, Huang Y (2010) T-drive: driving directions based on taxi trajectories. In: Proceedings of the 18th SIGSPATIAL International conference on advances in geographic information systems, pp 99–108

Yuan J, Zheng Y, Xie X (2012) Discovering regions of different functions in a city using human mobility and pois. In: Proceedings of the 18th ACM SIGKDD international conference on Knowledge discovery and data mining, pp 186–194

Yuan Y, Liang X, Wang X, Yeung DY, Gupta A (2017) Temporal dynamic graph lstm for action-driven video object detection. In: Proceedings of the IEEE international conference on computer vision, pp 1801–1810

Yuan Z, Zhou X, Yang T (2018) Hetero-convlstm: A deep learning approach to traffic accident prediction on heterogeneous spatio-temporal data. In: Proceedings of the 24th ACM SIGKDD International Conference on Knowledge Discovery & Data Mining, pp 984–992

Yue-Hei Ng J, Hausknecht M, Vijayanarasimhan S, Vinyals O, Monga R, Toderici G (2015) Beyond short snippets: Deep networks for video classification. In: Proceedings of the IEEE conference on computer vision and pattern recognition, pp 4694–4702

Yun S, Jeong M, Kim R, Kang J, Kim HJ (2019) Graph transformer networks. Advances in Neural Information Processing Systems 32:11,983–11,993

Zaheer M, Kottur S, Ravanbakhsh S, Poczos B, Salakhutdinov RR, Smola AJ (2017) Deep sets. In: Advances in Neural Information Processing Systems, pp 3391–3401

Zanfir A, Sminchisescu C (2018) Deep learning of graph matching. In: Proceedings of the IEEE conference on computer vision and pattern recognition, pp 2684–2693

Zelnik-Manor L, Perona P (2004) Self-tuning spectral clustering. Advances in neural information processing systems 17:1601–1608

Zeng H, Zhou H, Srivastava A, Kannan R, Prasanna V (2020a) Graphsaint: Graph sampling based inductive learning method. In: International Conference on Learning Representations

Zeng R, Huang W, Tan M, Rong Y, Zhao P, Huang J, Gan C (2019) Graph convolutional networks for temporal action localization. In: Proceedings of the IEEE/CVF International Conference on Computer Vision, pp 7094–7103

Zeng X, Song X, Ma T, Pan X, Zhou Y, Hou Y, Zhang Z, Li K, Karypis G, Cheng F (2020b) Repurpose open data to discover therapeutics for covid-19 using deep learning. Journal of proteome research 19(11):4624–4636

Zeng Z, Tung AK, Wang J, Feng J, Zhou L (2009) Comparing stars: On approximating graph edit distance. Proceedings of the VLDB Endowment 2(1):25–36

Zhang B, Hill E, Clause J (2016a) Towards automatically generating descriptive names for unit tests. In: Proceedings of the 31st IEEE/ACM International Conference on Automated Software Engineering, ACM, pp 625–636

Zhang C, Huang C, Yu L, Zhang X, Chawla NV (2018a) Camel: Content-aware and meta-path augmented metric learning for author identification. In: Proceedings of the 2018 World Wide Web Conference, pp 709–718

Zhang C, Chao WL, Xuan D (2019a) An empirical study on leveraging scene graphs for visual question answering. arXiv preprint arXiv:190712133

Zhang C, Song D, Huang C, Swami A, Chawla NV (2019b) Heterogeneous graph neural network. In: Proceedings of the 25th ACM SIGKDD International Conference on Knowledge Discovery & Data Mining, pp 793–803

Zhang C, Swami A, Chawla NV (2019c) Shne: Representation learning for semantic-associated heterogeneous networks. In: Proceedings of the Twelfth ACM International Conference on Web Search and Data Mining, pp 690–698

Zhang D, Yin J, Zhu X, Zhang C (2016b) Collective classification via discriminative matrix factorization on sparsely labeled networks. In: Proceedings of the 25th ACM International on Conference on Information and Knowledge Management, pp 1563–1572

Zhang D, Yin J, Zhu X, Zhang C (2018b) Metagraph2vec: Complex semantic path augmented heterogeneous network embedding. In: Pacific-Asia conference on knowledge discovery and data mining, Springer, pp 196–208

Zhang D, Yin J, Zhu X, Zhang C (2018c) Network representation learning: A survey. IEEE transactions on Big Data 6(1):3–28

Zhang G, He H, Katabi D (2019d) Circuit-GNN: Graph neural networks for distributed circuit design. In: International Conference on Machine Learning, pp 7364–7373

Zhang H, Zheng T, Gao J, Miao C, Su L, Li Y, Ren K (2019e) Data poisoning attack against knowledge graph embedding. In: Proceedings of the Twenty-Eighth International Joint Conference on Artificial Intelligence, IJCAI-19, International Joint Conferences on Artificial Intelligence Organization, pp 4853–4859

Zhang J (2020) Graph neural distance metric learning with graph-bert. arXiv preprint arXiv:200203427

Zhang J, Bargal SA, Lin Z, Brandt J, Shen X, Sclaroff S (2018d) Top-down neural attention by excitation backprop. International Journal of Computer Vision 126(10):1084–1102

Zhang J, Shi X, Xie J, Ma H, King I, Yeung DY (2018e) Gaan: Gated attention networks for learning on large and spatiotemporal graphs. arXiv preprint arXiv:180307294

Zhang J, Wang X, Zhang H, Sun H, Wang K, Liu X (2019f) A novel neural source code representation based on abstract syntax tree. In: 2019 IEEE/ACM 41st International Conference on Software Engineering (ICSE), IEEE, pp 783–794

Zhang J, Zhang H, Xia C, Sun L (2020a) Graph-bert: Only attention is needed for learning graph representations. arXiv preprint arXiv:200105140

Zhang L, Lu H (2020) A Feature-Importance-Aware and Robust Aggregator for GCN. In: ACM International Conference on Information & Knowledge Management, DOI 10.1145/3340531.3411983

Zhang M, Chen Y (2018a) Link prediction based on graph neural networks. In: Advances in Neural Information Processing Systems, pp 5165–5175

Zhang M, Chen Y (2018b) Link prediction based on graph neural networks. In: Proceedings of the 32nd International Conference on Neural Information Processing Systems, pp 5171–5181

Zhang M, Chen Y (2019) Inductive matrix completion based on graph neural networks. In: International Conference on Learning Representations

Zhang M, Chen Y (2020) Inductive matrix completion based on graph neural networks. In: International Conference on Learning Representations

Zhang M, Schmitt-Ulms G, Sato C, Xi Z, Zhang Y, Zhou Y, St George-Hyslop P, Rogaeva E (2016c) Drug repositioning for alzheimer's disease based on systematic 'omics' data mining. PloS one 11(12):e0168,812

Zhang M, Cui Z, Neumann M, Chen Y (2018f) An end-to-end deep learning architecture for graph classification. In: Association for the Advancement of Artificial Intelligence

Zhang M, Cui Z, Neumann M, Chen Y (2018g) An end-to-end deep learning architecture for graph classification. In: the AAAI Conference on Artificial Intelligence, pp 4438–4445

Zhang M, Hu L, Shi C, Wang X (2020b) Adversarial label-flipping attack and defense for graph neural networks. In: 2020 IEEE International Conference on Data Mining (ICDM), IEEE, pp 791–800

Zhang M, Li P, Xia Y, Wang K, Jin L (2020c) Revisiting graph neural networks for link prediction. arXiv preprint arXiv:201016103

Zhang N, Deng S, Li J, Chen X, Zhang W, Chen H (2020d) Summarizing chinese medical answer with graph convolution networks and question-focused dual attention. In: Proceedings of the 2020 Conference on Empirical Methods in Natural Language Processing: Findings, pp 15–24

Zhang Q, Chang J, Meng G, Xiang S, Pan C (2020e) Spatio-temporal graph structure learning for traffic forecasting. In: Proceedings of the AAAI Conference on Artificial Intelligence, vol 34, pp 1177–1185

Zhang R, Isola P, Efros AA (2016d) Colorful image colorization. In: European conference on computer vision, Springer, pp 649–666

Zhang S, Hu Z, Subramonian A, Sun Y (2020f) Motif-driven contrastive learning of graph representations. arXiv preprint arXiv:201212533

Zhang W, Tang S, Cao Y, Pu S, Wu F, Zhuang Y (2019g) Frame augmented alternating attention network for video question answering. IEEE Transactions on Multimedia 22(4):1032–1041

Zhang W, Fang Y, Liu Z, Wu M, Zhang X (2020g) mg2vec: Learning relationship-preserving heterogeneous graph representations via metagraph embedding. IEEE Transactions on Knowledge and Data Engineering 14(8):1

Zhang W, Liu H, Liu Y, Zhou J, Xiong H (2020h) Semi-supervised hierarchical recurrent graph neural network for city-wide parking availability prediction. In: Proceedings of the AAAI Conference on Artificial Intelligence, vol 34, pp 1186–1193

Zhang W, Wang XE, Tang S, Shi H, Shi H, Xiao J, Zhuang Y, Wang WY (2020i) Relational graph learning for grounded video description generation. In: Proceedings of the 28th ACM International Conference on Multimedia, pp 3807–3828

Zhang X, Zitnik M (2020) Gnnguard: Defending graph neural networks against adversarial attacks. Advances in Neural Information Processing Systems 33

Zhang X, Li Y, Zhou X, Luo J (2019) Unveiling taxi drivers' strategies via cgail: Conditional generative adversarial imitation learning. In: 2019 IEEE International Conference on Data Mining (ICDM), pp 1480–1485, DOI 10.1109/ICDM.2019.00194

Zhang X, Li Y, Zhou X, Luo J (2020a) cgail: Conditional generative adversarial imitation learning—an application in taxi drivers' strategy learning. IEEE Transactions on Big Data pp 1–1, DOI 10.1109/TBDATA.2020.3039810

Zhang X, Li Y, Zhou X, Zhang Z, Luo J (2020b) Trajgail: Trajectory generative adversarial imitation learning for long-term decision analysis. In: 2020 IEEE International Conference on Data Mining (ICDM), pp 801–810, DOI 10.1109/ICDM50108.2020.00089

Zhang Y, Zheng W, Lin H, Wang J, Yang Z, Dumontier M (2018h) Drug–drug interaction extraction via hierarchical rnns on sequence and shortest dependency paths. Bioinformatics 34(5):828–835

Zhang Y, Fan Y, Ye Y, Zhao L, Shi C (2019a) Key player identification in underground forums over attributed heterogeneous information network embedding framework. In: Proceedings of the 28th ACM International Conference on Information and Knowledge Management, pp 549–558

Zhang Y, Khan S, Coates M (2019b) Comparing and detecting adversarial attacks for graph deep learning. In: Representation Learning on Graphs and Manifolds Workshop at ICLR

Zhang Y, Li Y, Zhou X, Kong X, Luo J (2019c) Trafficgan: Off-deployment traffic estimation with traffic generative adversarial networks. 2019 IEEE International Conference on Data Mining (ICDM) pp 1474–1479

Zhang Y, Pal S, Coates M, Ustebay D (2019d) Bayesian graph convolutional neural networks for semi-supervised classification. In: Proceedings of the AAAI Conference on Artificial Intelligence, vol 33, pp 5829–5836

Zhang Y, Defazio D, Ramesh A (2020a) Relex: A model-agnostic relational model explainer. arXiv preprint arXiv:200600305

Zhang Y, Deng W, Wang M, Hu J, Li X, Zhao D, Wen D (2020b) Global-local gcn: Large-scale label noise cleansing for face recognition. In: Proceedings of the IEEE/CVF Conference on Computer Vision and Pattern Recognition, pp 7731–7740

Zhang Y, Guo Z, Teng Z, Lu W, Cohen SB, Liu Z, Bing L (2020c) Lightweight, dynamic graph convolutional networks for amr-to-text generation. In: Proceedings of the 2020 Conference on Empirical Methods in Natural Language Processing (EMNLP), pp 2162–2172

Zhang Y, Yu X, Cui Z, Wu S, Wen Z, Wang L (2020d) Every document owns its structure: Inductive text classification via graph neural networks. In: Proceedings of the 58th Annual Meeting of the Association for Computational Linguistics, pp 334–339

Zhang Z, Wang M, Xiang Y, Huang Y, Nehorai A (2018i) Retgk: Graph kernels based on return probabilities of random walks. In: Advances in Neural Information Processing Systems, pp 3964–3974

Zhang Z, Cui P, Zhu W (2020e) Deep learning on graphs: A survey. IEEE Transactions on Knowledge and Data Engineering pp 1–1, DOI 10.1109/TKDE.2020.2981333

Zhang Z, Zhang Z, Zhou Y, Shen Y, Jin R, Dou D (2020f) Adversarial attacks on deep graph matching. Advances in Neural Information Processing Systems 33

Zhang Z, Zhao Z, Lin Z, Huai B, Yuan NJ (2020g) Object-aware multi-branch relation networks for spatio-temporal video grounding. arXiv preprint arXiv:200806941

Zhang Z, Zhao Z, Zhao Y, Wang Q, Liu H, Gao L (2020h) Where does it exist: Spatio-temporal video grounding for multi-form sentences. In: Proceedings of the IEEE/CVF Conference on Computer Vision and Pattern Recognition, pp 10,668–10,677

Zhang Z, Zhuang F, Zhu H, Shi Z, Xiong H, He Q (2020i) Relational graph neural network with hierarchical attention for knowledge graph completion. In: Proceedings of the AAAI Conference on Artificial Intelligence, vol 34, pp 9612–9619

Zhao H, Du L, Buntine W (2017) Leveraging node attributes for incomplete relational data. In: International Conference on Machine Learning, pp 4072–4081

Zhao H, Zhou Y, Song Y, Lee DL (2019a) Motif enhanced recommendation over heterogeneous information network. In: Proceedings of the 28th ACM international conference on information and knowledge management, pp 2189–2192

Zhao H, Wei L, Yao Q (2020a) Simplifying architecture search for graph neural network. In: Conrad S, Tiddi I (eds) Proceedings of the CIKM 2020 Workshops co-located with 29th ACM International Conference on Information and Knowledge Management (CIKM 2020), Galway, Ireland, October 19-23, 2020, CEUR-WS.org, CEUR Workshop Proceedings, vol 2699

Zhao J, Zhou Z, Guan Z, Zhao W, Ning W, Qiu G, He X (2019b) Intentgc: a scalable graph convolution framework fusing heterogeneous information for recommendation. In: Proceedings of the 25th ACM SIGKDD International Conference on Knowledge Discovery & Data Mining, pp 2347–2357

Zhao J, Wang X, Shi C, Liu Z, Ye Y (2020b) Network schema preserving heterogeneous information network embedding. In: Bessiere C (ed) Proceedings of the Twenty-Ninth International Joint Conference on Artificial Intelligence, IJCAI-20, International Joint Conferences on Artificial Intelligence Organization, pp 1366–1372

Zhao J, Wang X, Shi C, Hu B, Song G, Ye Y (2021) Heterogeneous graph structure learning for graph neural networks. In: Proceedings of the AAAI Conference on Artificial Intelligence

Zhao K, Bai T, Wu B, Wang B, Zhang Y, Yang Y, Nie JY (2020c) Deep adversarial completion for sparse heterogeneous information network embedding. In: Proceedings of The Web Conference 2020, pp 508–518

Zhao L, Akoglu L (2019) Pairnorm: Tackling oversmoothing in gnns. In: International Conference on Learning Representations

Zhao L, Song Y, Zhang C, Liu Y, Wang P, Lin T, Deng M, Li H (2019c) T-GCN: A temporal graph convolutional network for traffic prediction. IEEE Transactions on Intelligent Transportation Systems 21(9):3848–3858

Zhao M, Wang D, Zhang Z, Zhang X (2015) Music removal by convolutional denoising autoencoder in speech recognition. In: 2015 Asia-Pacific Signal and Information Processing Association Annual Summit and Conference (APSIPA), IEEE, pp 338–341

Zhao S, Su C, Sboner A, Wang F (2019d) Graphene: A precise biomedical literature retrieval engine with graph augmented deep learning and external knowledge empowerment. In: Proceedings of the 28th ACM International Conference on Information and Knowledge Management, pp 149–158

Zhao S, Qin B, Liu T, Wang F (2020d) Biomedical knowledge graph refinement with embedding and logic rules. arXiv preprint arXiv:201201031

Zhao S, Su C, Lu Z, Wang F (2020e) Recent advances in biomedical literature mining. Briefings in Bioinformatics

Zhao T, Deng C, Yu K, Jiang T, Wang D, Jiang M (2020f) Error-bounded graph anomaly loss for gnns. In: Proceedings of the 29th ACM International Conference on Information & Knowledge Management, pp 1873–1882

Zhao Y, Wang D, Gao X, Mullins R, Lio P, Jamnik M (2020g) Probabilistic dual network architecture search on graphs. arXiv preprint arXiv:200309676

Zheng C, Fan X, Wang C, Qi J (2020a) Gman: A graph multi-attention network for traffic prediction. In: Proceedings of the AAAI Conference on Artificial Intelligence, vol 34, pp 1234–1241

Zheng C, Zong B, Cheng W, Song D, Ni J, Yu W, Chen H, Wang W (2020b) Robust graph representation learning via neural sparsification. In: International Conference on Machine Learning, pp 11,458–11,468

Zheng D, Song X, Ma C, Tan Z, Ye Z, Dong J, Xiong H, Zhang Z, Karypis G (2020c) Dgl-ke: Training knowledge graph embeddings at scale. In: Proceedings of the 43rd International ACM SIGIR Conference on Research and Development in Information Retrieval, pp 739–748

Zheng L, Lu CT, Jiang F, Zhang J, Yu PS (2018a) Spectral collaborative filtering. In: Proceedings of the 12th ACM Conference on Recommender Systems, ACM, pp 311–319

Zheng L, Li Z, Li J, Li Z, Gao J (2019) Addgraph: Anomaly detection in dynamic graph using attention-based temporal gcn. In: Proceedings of the Twenty-Eighth International Joint Conference on Artificial Intelligence, IJCAI-19, pp 4419–4425

Zheng X, Aragam B, Ravikumar PK, Xing EP (2018b) Dags with no tears: Continuous optimization for structure learning. Advances in Neural Information Processing Systems 31:9472–9483

Zheng Y, Liu F, Hsieh HP (2013) U-air: When urban air quality inference meets big data. In: Proceedings of the 19th ACM SIGKDD international conference on Knowledge discovery and data mining, pp 1436–1444

Zheng Y, Capra L, Wolfson O, Yang H (2014) Urban computing: Concepts, methodologies, and applications 5(3), DOI 10.1145/2629592

Zhou C, Liu Y, Liu X, Liu Z, Gao J (2017) Scalable graph embedding for asymmetric proximity. In: Proceedings of the AAAI Conference on Artificial Intelligence, vol 31

Zhou C, Bai J, Song J, Liu X, Zhao Z, Chen X, Gao J (2018a) Atrank: An attention-based user behavior modeling framework for recommendation. In: Proceedings of the AAAI Conference on Artificial Intelligence, vol 32

Zhou C, Ma J, Zhang J, Zhou J, Yang H (2020a) Contrastive learning for debiased candidate generation in large-scale recommender systems. arXiv preprint csIR/200512964

Zhou D, Bousquet O, Lal TN, Weston J, Schölkopf B (2004) Learning with local and global consistency. Advances in neural information processing systems 16(16):321–328

Zhou F, De la Torre F (2012) Factorized graph matching. In: 2012 IEEE Conference on Computer Vision and Pattern Recognition, IEEE, pp 127–134

Zhou G, Zhu X, Song C, Fan Y, Zhu H, Ma X, Yan Y, Jin J, Li H, Gai K (2018b) Deep interest network for click-through rate prediction. In: Proceedings of the 24th ACM SIGKDD, pp 1059–1068

Zhou G, Wang J, Zhang X, Guo M, Yu G (2020b) Predicting functions of maize proteins using graph convolutional network. BMC Bioinformatics 21(16):420

Zhou J, Cui G, Zhang Z, Yang C, Liu Z, Sun M (2018c) Graph neural networks: A review of methods and applications. arXiv preprint arXiv:181208434

Zhou K, Song Q, Huang X, Hu X (2019a) Auto-gnn: Neural architecture search of graph neural networks. arXiv preprint arXiv:190903184

Zhou K, Dong Y, Wang K, Lee WS, Hooi B, Xu H, Feng J (2020c) Understanding and resolving performance degradation in graph convolutional networks. arXiv preprint arXiv:200607107

Zhou K, Huang X, Li Y, Zha D, Chen R, Hu X (2020d) Towards deeper graph neural networks with differentiable group normalization. In: Advances in Neural Information Processing Systems, vol 33

Zhou K, Song Q, Huang X, Zha D, Zou N, Hu X (2020e) Multi-channel graph neural networks. In: International Joint Conference on Artificial Intelligence, pp 1352–1358

Zhou N, Jiang Y, Bergquist TR, et al (2019b) The CAFA challenge reports improved protein function prediction and new functional annotations for hundreds of genes through experimental screens. Genome Biology 20(1), DOI 10.1186/s13059-019-1835-8

Zhou T, Lü L, Zhang YC (2009) Predicting missing links via local information. The European Physical Journal B 71(4):623–630

Zhou Y, Tuzel O (2018) Voxelnet: End-to-end learning for point cloud based 3d object detection. In: Proceedings of the IEEE Conference on Computer Vision and Pattern Recognition, pp 4490–4499

Zhou Y, Hou Y, Shen J, Huang Y, Martin W, Cheng F (2020f) Network-based drug repurposing for novel coronavirus 2019-ncov/sars-cov-2. Cell discovery 6(1):1–18

Zhou Z, Kearnes S, Li L, Zare RN, Riley P (2019c) Optimization of molecules via deep reinforcement learning. Scientific reports 9(1):1–10

Zhou Z, Wang Y, Xie X, Chen L, Liu H (2020g) Riskoracle: A minute-level citywide traffic accident forecasting framework. In: Proceedings of the AAAI Conference on Artificial Intelligence, vol 34, pp 1258–1265

Zhou Z, Wang Y, Xie X, Chen L, Zhu C (2020h) Foresee urban sparse traffic accidents: A spatiotemporal multi-granularity perspective. IEEE Transactions on Knowledge and Data Engineering pp 1–1, DOI 10.1109/TKDE.2020.3034312

Zhu D, Cui P, Wang D, Zhu W (2018) Deep variational network embedding in wasserstein space. In: Proceedings of the 24th ACM SIGKDD International Conference on Knowledge Discovery & Data Mining, pp 2827–2836

Zhu D, Zhang Z, Cui P, Zhu W (2019a) Robust graph convolutional networks against adversarial attacks. In: Proceedings of the 25th ACM SIGKDD International Conference on Knowledge Discovery amp; Data Mining, Association for Computing Machinery, KDD '19, p 1399–1407, DOI 10.1145/3292500.3330851

Zhu J, Li J, Zhu M, Qian L, Zhang M, Zhou G (2019b) Modeling graph structure in transformer for better AMR-to-text generation. In: Proceedings of the 2019 Conference on Empirical Methods in Natural Language Processing and the 9th International Joint Conference on Natural Language Processing (EMNLP-IJCNLP), Association for Computational Linguistics, Hong Kong, China, pp 5459–5468

Zhu JY, Park T, Isola P, Efros AA (2017) Unpaired image-to-image translation using cycle-consistent adversarial networks. In: Proceedings of the IEEE international conference on computer vision, pp 2223–2232

Zhu Q, Du B, Yan P (2020a) Self-supervised training of graph convolutional networks. arXiv preprint arXiv:200602380

Zhu R, Zhao K, Yang H, Lin W, Zhou C, Ai B, Li Y, Zhou J (2019c) Aligraph: a comprehensive graph neural network platform. Proceedings of the VLDB Endowment 12(12):2094–2105

Zhu S, Yu K, Chi Y, Gong Y (2007) Combining content and link for classification using matrix factorization. In: Proceedings of the 30th annual international ACM SIGIR conference on Research and development in information retrieval, pp 487–494

Zhu S, Zhou C, Pan S, Zhu X, Wang B (2019d) Relation structure-aware heterogeneous graph neural network. In: 2019 IEEE International Conference on Data Mining (ICDM), IEEE, pp 1534–1539

ZHU X (2002) Learning from labeled and unlabeled data with label propagation. Tech Report

Zhu Y, Elemento O, Pathak J, Wang F (2019e) Drug knowledge bases and their applications in biomedical informatics research. Briefings in bioinformatics 20(4):1308–1321

Zhu Y, Che C, Jin B, Zhang N, Su C, Wang F (2020b) Knowledge-driven drug repurposing using a comprehensive drug knowledge graph. Health Informatics Journal 26(4):2737–2750

Zhu Y, Xu Y, Yu F, Liu Q, Wu S, Wang L (2020c) Deep graph contrastive representation learning. arXiv preprint arXiv:200604131

Zhu Y, Xu Y, Yu F, Liu Q, Wu S, Wang L (2021) Graph Contrastive Learning with Adaptive Augmentation. In: Proceedings of The Web Conference 2021, ACM, WWW '21

Zhuang Y, Jain R, Gao W, Ren L, Aizawa K (2017) Panel: cross-media intelligence. In: Proceedings of the 25th ACM international conference on Multimedia, pp 1173–1173

Zimmermann T, Zeller A, Weissgerber P, Diehl S (2005) Mining version histories to guide software changes. IEEE Transactions on Software Engineering 31(6):429–445

Zitnik M, Leskovec J (2017) Predicting multicellular function through multi-layer tissue networks. Bioinformatics 33(14):i190–i198

Zitnik M, Agrawal M, Leskovec J (2018) Modeling polypharmacy side effects with graph convolutional networks. Bioinformatics 34(13):i457–i466

Zoete V, Cuendet MA, Grosdidier A, Michielin O (2011) SwissParam: A fast force field generation tool for small organic molecules. Journal of Computational Chemistry 32(11):2359–2368

Zoph B, Le QV (2016) Neural architecture search with reinforcement learning. arXiv preprint arXiv:161101578

Zoph B, Yuret D, May J, Knight K (2016) Transfer learning for low-resource neural machine translation. In: Proceedings of the 2016 Conference on Empirical Methods in Natural Language Processing, pp 1568–1575

Zoph B, Vasudevan V, Shlens J, Le QV (2018) Learning transferable architectures for scalable image recognition. In: Proceedings of the IEEE Conference on Computer Vision and Pattern Recognition, pp 8697–8710

Zügner D, Günnemann S (2019) Adversarial attacks on graph neural networks via meta learning. In: International Conference on Learning Representations, ICLR

Zügner D, Günnemann S (2019) Certifiable robustness and robust training for graph convolutional networks. In: Proceedings of the 25th ACM SIGKDD International Conference on Knowledge Discovery & Data Mining, pp 246–256

Zügner D, Günnemann S (2020) Certifiable robustness of graph convolutional networks under structure perturbations. In: Proceedings of the 26th ACM SIGKDD International Conference on Knowledge Discovery amp; Data Mining, Association for Computing Machinery, KDD '20, p 1656–1665, DOI 10.1145/3394486.3403217

Zügner D, Akbarnejad A, Günnemann S (2018) Adversarial attacks on neural networks for graph data. In: Proceedings of the 24th ACM SIGKDD International Conference on Knowledge Discovery & Data Mining, pp 2847–2856

Zügner D, Borchert O, Akbarnejad A, Günnemann S (2020) Adversarial attacks on graph neural networks: Perturbations and their patterns. ACM Trans Knowl Discov Data 14(5):57:1–57:31

Zügner D, Kirschstein T, Catasta M, Leskovec J, Günnemann S (2021) Language-agnostic representation learning of source code from structure and context. In: International Conference on Learning Representations

Printed in the United States
by Baker & Taylor Publisher Services